Agriculture 4.0

Franco da Silveira • Jayme Garcia Arnal Barbedo • Chrysanthi Charatsari • Evagelos D. Lioutas
Editors

Agriculture 4.0

Towards a New Agri-Food Ecosystem

Springer

Editors

Franco da Silveira
EMBRAPA Digital Agriculture
Campinas, São Paulo, Brazil

Jayme Garcia Arnal Barbedo
EMBRAPA Digital Agriculture
Campinas, São Paulo, Brazil

Chrysanthi Charatsari
School of Agriculture
Aristotle University of Thessaloniki
Thessaloniki, Greece

Evagelos D. Lioutas
Department of Supply Chain Management
International Hellenic University
Katerini, Greece

ISBN 978-3-032-24551-9 ISBN 978-3-032-24552-6 (eBook)
https://doi.org/10.1007/978-3-032-24552-6

This Springer imprint is published by the registered company Springer Nature Switzerland AG
The registered company address is: Gewerbestrasse 11, 6330 Cham, Switzerland

Foreword

Less than a decade ago, the concept of "digital agriculture" was largely aspirational: a vision to integrate advanced technologies into a sector that has sustained humanity for millennia. Today, that vision is becoming a reality in some regions and sub-sectors. The digitalisation of agri-food systems is now well underway, driven by significant public and private sector investment as technologies evolve. These developments are reshaping the socio-technical systems that underpin agriculture and food production across the world at an unprecedented pace.

At the Commonwealth Scientific and Industrial Research Organisation (CSIRO), we have been actively exploring what 'Agriculture 4.0' means for Australia and sharing these insights internationally. This collection continues that important work. It brings together leading scholars from across the globe to examine both the opportunities and challenges inherent in this dynamic transformation. We commend the editors for not only capturing the current state of digital agri-food systems across different contexts but also illuminating potential pathways towards the future. These agri-food ecosystems will influence the production and consumption systems inherited by generations to come, so understanding this transition gives us the opportunity to shape these systems in responsible and inclusive ways.

Digitalisation is not occurring uniformly across regions, sectors, or scales. The value of technology adoption varies according to the context in which these tools are developed and deployed. What excites us about this collection, however, is its recognition of the complexities of this large-scale transition, together with its commitment to addressing sustainability challenges alongside technological innovation. We applaud the developments made on critical themes such as 'agri-work', 'servitization', mental health challenges, start-ups, agro-technological districts, and the governance of innovation ecosystems.

It is rare to see a collection of work that covers so many regions, sectors, scales, and themes in such a compelling manner. It moves beyond conventional narratives of agtech adoption in developed nations to consider how and why digital agricultural technologies are increasingly relevant for emerging economies. Topics often underrepresented in the literature are given attention, for example gender, the viability of technologies, or portfolio-level assessments. The recurring emphasis on responsible and inclusive innovation reflects the principles that underpin our own research and collaboration efforts.

The implications of 'Agriculture 4.0' extend far beyond technological adoption and efficiency. They touch on questions of equity, resilience, and governance. How do we ensure that digital tools enhance rather than erode the livelihoods of farmers? How do we design systems that balance automation with human well-being? These are not abstract concerns as they are central to shaping agri-food ecosystems that are sustainable, inclusive, and adaptive in the face of climate change and global uncertainty.

We anticipate that this book will serve as a valuable resource for researchers, technology developers, advisors, and all those operating at the intersection of digital technology and agri-food systems. The insights presented here will inform co-innovation processes as we strive to realise the socio-economic benefits of digitalisation. For example, through automation, improved safety, and cost efficiencies, whilst simultaneously seeking to improve the environmental sustainability of agri-food systems.

The journey towards a new agri-food ecosystem is underway. It is complex, uneven, and yet full of promise. This collection invites you to engage as active participants with that journey across the world.

With optimism,

Commonwealth Scientific and Industrial Research Organisation (CSIRO)
Canberra, Australia

Simon Fielke
Emma Jakku

Contents

Part I
Structural Transformation of the Agri-Food Ecosystem

What We Do Not Talk About When We Talk About AgTech

Seeing the Social, Asking Better Questions, and Designing Fairer Technologies

Jenny Melo-Velasco

> *"If we want to change power relations, e.g., to make them more inclusive, we must understand where and how to engage" Gaventa. (2006: 23). Finding the Spaces for Change: A Power Analysis.*

1 Introduction: Beyond Adoption—Toward a Relational View of Digital Agriculture

In every conference on Digital Agriculture[1] and rural development I have attended, two questions inevitably surface, whether in keynote speeches, panels, or coffee breaks: How do we achieve adoption? And how do we develop technologies that are actually used and meet the needs of all kinds of farmers, including the smallest-scale producers? These concerns implicitly acknowledge what we already know: that solutions rarely scale as hoped, technologies often bypass the most marginalized, and agricultural inefficiencies persist. The sense of unease is palpable: the current wave of innovation has yet to fulfill its promises. Despite early enthusiasm and the recognition of substantial structural barriers, there is little clarity about why the expected benefits have proven elusive and what path forward we should take. Perhaps part of the problem lies in how we understand technology itself.

Innovation in agriculture is often imagined as a one-way street: a need is perceived, a technology is developed, and then the challenge becomes getting people

[1] Throughout this chapter, I use the term Digital Agriculture and Agtech to refer to all kind of digital technologies in agriculture. I adopted a broad definition of digital agriculture which encompasses "any form of Information and Communication Technologies (ICTs) used in agriculture, either for automated or non-automated decision-making and practices" (Duncan et al. 2021).

J. Melo-Velasco (✉)
Huella Delta Impacto Colectivo SAS BIC, Cali, Colombia
e-mail: jenny@huelladelta.com

F. da Silveira et al. (eds.), *Agriculture 4.0*,
https://doi.org/10.1007/978-3-032-24552-6_1

to accept it. This mindset assumes that technology is neutral, that more of it is inherently better, and that social issues—when acknowledged—are seen not as structural concerns but as frictions to be overcome or realities to be reshaped to fit predesigned business models. This is a profoundly instrumental vision: technology is a tool, farmers are its users, and success is measured by uptake.

This chapter offers a different lens. It begins with the recognition that technology is never neutral. It is created in specific contexts, by specific actors, with particular ideas of what agriculture should become. Digital technologies do not just deliver services; they reorganize relationships, redistribute agency, and shape who gets to decide what farming looks like. In short, they express and enact power. From this starting point, I propose a shift in how researchers, developers, engineers, and funders approach AgTech, not only as a technical endeavor but as a deeply social and political one.

To build this lens and support this shift, we need to make three moves to deepen our conversation on innovation: first, shift from an instrumental to a social understanding; second, see power more clearly and how it manifests; and third, reflect critically on the entry points we are designing.

Moving Beyond the Instrumental View

The first step is to challenge the instrumental view of technology. When we treat AgTech as a neutral input, we overlook how it is embedded in systems of knowledge, economics, and control. This limits our ability to understand why technologies fail, who they serve, and what kinds of futures they make possible. As scholars have argued in other domains (Umoja Noble 2018; Zuboff 2019), technologies often reflect and reinforce existing inequalities. In agriculture, this can mean consolidating power among larger actors (Bronson 2022), marginalizing alternative knowledges, or designing tools based on the assumptions of farmers' uniformity.

Seeing the Social Is Seeing Power

Recognizing the social dimensions of technology means seeing power at work. Power is not an add-on—it is constitutive of how digital agriculture operates. To help see this more clearly, I bring in the Socioethical Awareness Framework for Digital Agriculture (Melo-Velasco et al. 2025), which examines three interrelated domains: who shapes dominant narratives, who is included or excluded in access, and how technologies are governed. These questions prove far from secondary, as they determine whose realities count, whose needs shape design, and who gains or loses influence. Power is relational and asymmetrical (Roscigno 2011): it emerges in interaction, is expressed through inclusion and exclusion, and is embedded in infrastructure, platforms, and contracts. By naming these dynamics explicitly, the framework helps shift the conversation from adoption toward justice and accountability.

Toward an Integrative Perspective

To build better AgTech, we need better questions: not just about what works, but for whom, in what context, and to what consequences. This final section proposes an integrative approach that complements critical reflection with concrete entry points. It focuses on two practical dimensions often overlooked: first, the type of service being offered to farmers, specifically what kind of relationship it affects, and how it fits into existing practices; and second, the business models of producers, which shape what kinds of technologies are useful, viable, or even relevant. Services are not just technical artifacts; they reorganize flows of knowledge, trust, and resources. Meanwhile, business models differ across contexts and markets, influencing how technologies land on the ground. Together, these lenses help align innovation with real-world dynamics and support more context-sensitive, inclusive, and effective solutions.

This chapter draws on both academic research and insights from applied work in rural Latin America. My academic research spans the United States and Latin American countries, combining a rural sociology perspective with attention to how economic structures, institutional arrangements, and power relations shape agricultural systems. Developed in previous publications (Melo-Velasco et al. 2022, 2025), this work adopts a pragmatic and critical stance, aimed at exposing tensions and supporting actionable change. On the applied side, my work has been carried out primarily as an independent consultant and in later years through Huella Delta, a Latin American social enterprise I cofounded, which supports community-led rural businesses in developing strategic clarity about their operations to identify and pursue fairer and more sustainable market relationships. This chapter brings together these strands as a reflection that bridges theory and practice, to advance a vision of digital agriculture that is smarter and more inclusive and equitable.

I aim to speak directly to those designing, implementing, and funding AgTech solutions, to enrich the conversation and broaden the realm of what is possible. This integrative lens deepens our understanding of the social structures in which technology operates, enabling us to build solutions that are not only technically robust but also socially grounded and inclusive.

2 Moving Beyond the Instrumental View

> *"Agricultural digitalization is inevitable; what is unknown is how dynamic and inclusive it will be".*
>
> ECLAC, FAO, IICA (2021:89). The Outlook for Agriculture and Rural Development in the Americas: A Perspective on Latin America and the Caribbean 2021–2022.

A dominant narrative in the AgTech ecosystems presents technologies as neutral instruments, designed to solve problems of productivity, traceability, and resource efficiency. This view treats these technologies as existing independently of the social, political, and economic contexts into which they are introduced.

This thinking aligns with the "thesis of neutrality" (Miller 2021): the belief that technologies, in themselves, are value-free and only take on ethical or political significance once they are used. Under this assumption, technology development is framed as a strictly technical process, while questions of governance, legitimacy, or equity are postponed to implementation stages, or overlooked entirely.

This instrumental view is supported by a mainstream narrative that circulates through policy reports, media coverage, and development discourse. A review of high-level documents from the World Bank, the Food and Agriculture Organization of the United Nations (FAO), and the Organisation for Economic Co-operation and Development (OECD) (Lajoie-O'Malley et al. 2020) found a consistent message: digital technologies are framed as technical solutions to global food scarcity, promising higher yields, better incomes, and reduced harmful inputs. The argument is compelling: in a world expecting two billion more people by 2050, and with agriculture increasingly constrained by climate change, the solution to hunger lies in sensors, apps, and platforms. In interviews with American farmers, a rural sociologist (Carolan 2017) found that this narrative is echoed by farmers and entrepreneurs who justify digitalization based on the need to feed a global population of nine billion. This optimistic outlook is not limited to policy documents; it also dominates public discourse across traditional and social media. For instance, analyses of media coverage in Germany and English-language tweets reveal that arguments in favor of digitalization vastly outnumber critical or neutral perspectives, often emphasizing productivity, environmental benefits, and positive emotions (Ancín et al. 2022; Mohr and Höhler 2023). Such framings, while compelling, treat technology as an apolitical force for good, leaving unexamined the deeper questions of whose knowledge counts, who benefits, and under what conditions these technologies operate.

This instrumental view is both an oversimplification and a serious limitation. It conceals how technologies are developed by specific actors with particular interests, under conditions shaped by concentrated power in agri-food systems (Mooney 2018; Rotz et al. 2019; Bronson 2022; Fairbairn et al. 2025). Digital tools are never neutral: they embed assumptions about what kind of agriculture is desirable, what forms of knowledge count, and who gets to define problems and solutions, thereby institutionalizing certain pathways while foreclosing others.

This limitation becomes particularly acute when these digital technologies are deployed within a food system already shaped by high levels of corporate concentration. Consolidated power constrains both what is produced and how it is produced, narrowing the range of choices available to farmers and consumers alike. In essence, a handful of large global firms control food decisions (Hendrickson et al. 2019). This corporate concentration in food systems undermines agency —the ability to make choices and act on them to advance desired outcomes— through three key pathways: restricting available choices, reducing visibility and transparency, and weakening countervailing power (Clapp et al. 2025). This limits how technologies are adopted and shapes who gets to define food system futures.

Recognizing this dual entanglement between the politics of technology and the politics of the food system is essential for understanding the real implications of digital agriculture. Moving beyond the instrumental view is not just a theoretical concern but essential to understanding how digital technologies actually shape and are shaped by the food systems they enter. Presenting technology as neutral erases the political, economic, and knowledge-based choices embedded in its design, development, and deployment. It also blinds us to the realities of a food system marked by deep structural inequalities and concentrated power. As Fairclough (Fairclough 2017) reminds us, narratives do more than describe the world: they actively construct it. How we frame technology influences what is imagined as possible, fundable, and desirable. Recognizing this opens the door to a more situated, relational, and political perspective that invites us to ask not just what technologies do, but who they empower, who they exclude, and under what terms.

3 Seeing the Social Is Seeing Power

To understand the social implications of digital agriculture, we must begin with a clear grasp of power. Power encompasses far more than coercion or domination; it is multidimensional and woven into social structures and everyday interactions. As sociologist Vincent Roscigno defines it, power is best understood as "an unequal relation, or inequality, based on personal attributes, institutional positioning, and statuses that are defined, codified and acted upon within historical and cultural contexts" (p. 353) (Roscigno 2011). This view underscores two key aspects: first, power operates through relationships and is always asymmetrical; second, it is deeply contextual, shaped by the cultural, political, and economic conditions in which actors are situated.

In the context of digital agriculture, power can be analyzed through multiple and overlapping lenses. Drawing from critical social science, we can distinguish three intertwined expressions of power: discursive, structural, and relational.

Discursive power refers to the capacity to shape narratives, produce dominant framings, and define what counts as legitimate knowledge (IPES-Food 2016). In digital agriculture, this is evident in the persistent framing of technologies as neutral, inevitable, and universally beneficial. Large agrifood corporations and development agencies often promote narratives of "feeding the world" and "climate-smart innovation," which obscure the contested nature of both problems and solutions. For instance, the way Artificial Intelligence (AI) is described as enhancing precision farming tends to erase questions about who owns the algorithms, how decisions are made, and what kind of farming it enables or disables.

Structural power manifests in the configuration of the agri-food system and the digital ecosystem that overlays it. This includes who gets to design technologies, who owns the infrastructure and platforms, who can afford to adopt and repair tools, and who controls access to and ownership of data. For example, the fact that most farm management platforms are owned by a handful of multinational companies like Bayer or John Deere creates dependencies that reduce farmers' autonomy and limit alternative pathways. Structural power thus operates through infrastructure, capital, and rules; often invisible, but highly consequential.

Relational power emerges in the dynamics between actors, particularly in the design, deployment, and appropriation of digital tools. It manifests through every-

Table 1 Socioethical awareness framework for digital agriculture

Power facets: shaping the narratives and maintaining current imbalances	Factors shaping access: who is included and who is not?	Factors on technologies governance: who governs technologies and how?
Aspects of inquiry:		
How and by whom is the narrative about these technologies shaped? What interests are at play, and how do they influence the distribution of benefits and risks in the development and use of digital agricultural technologies? How does the process of designing, testing, and deploying technologies in the market unfold? Are there shifts in relationships a consequence of using digital technology?	How does infrastructure support access to digital technology, and to what extent? What are intersectional factors limiting or extending access to digital products? What is the relationship between individual's capabilities, motivations, and opportunities to use a digital technology? What worldviews, livelihoods, productions system and crops are being represented in available data? What are the relationships between those possessing data—collecting, storing, and mining—and those whose data is collected? How are gender and cultural barriers addressed in relation to access to digital technologies, and who is working to overcome these challenges?	What types of data are being gathered, and what challenges do they face regarding ownership, access, and privacy? What data is being collected, and what challenges exist regarding availability, accessibility, and usability? What relationships and exchanges are encouraged in the business model supporting the expansion of this technology? What ethical frameworks are being used (or should be used) to ensure that the development, deployment, and regulation of agricultural technologies respect farmers' rights, particularly in vulnerable communities?

Source: Melo-Velasco et al. (2025)

day asymmetries: the ways small farmers engage with technology providers, how extension agents mediate data for local contexts, and how Indigenous knowledge is systematically excluded from digital mapping systems. These power relations shape access, voice, and recognition.

To center power in both analysis and practice, colleagues at the University of Missouri (Melo-Velasco et al. 2025) and I introduced the *Socioethical Awareness Framework for Digital Agriculture*. As shown in Table 1, this framework offers a structured set of questions to examine how power is expressed, distributed, and materialized within digital agricultural systems. It focuses on three critical dimensions: the discursive and structural forces that shape dominant narratives; the dynamics of access and exclusion—who is included, who is not, and under what conditions; and the governance of technologies—who owns, controls, and defines their development and use.

Who Shapes the Narratives?

This dimension examines how narratives around digital agriculture are constructed, identifying which actors can define problems, propose solutions, and influence the public agenda. It looks at whose voices are amplified or silenced, which visions of the future are promoted, and how these narratives affect policy priorities and research investments.

Who Is Included and Who Is Not?
This dimension examines the multiple layers of inclusion and exclusion in digital agriculture, beyond physical access. It incorporates the digital divide (who has connectivity and infrastructure), the gender gap (who is recognized as a user or knowledge holder), and the data gap (whose realities are made visible or rendered invisible by data collection). These factors intersect with social identities and structural inequalities to determine who can meaningfully engage with digital tools and who remains peripheral.

Who Governs the Technology?
Governance-related power concerns who designs, owns, controls, and profits from agricultural technologies. This includes decisions about platform architectures, data ownership, pricing models, repairability, and regulatory oversight. When control concentrates in private hands, transparency, accountability, and democratic participation are constrained, making it difficult for farmers, communities, or public institutions to influence technological evolution or ensure alignment with broader public goals such as equity, sustainability, or food sovereignty.

By naming these dimensions concretely, the framework expands the conversation about power in digital agriculture and offers a sharper lens for examining how social dynamics are embedded in technology. It invites practitioners, engineers, and decision-makers to move beyond abstract ideals and engage with specific questions: who shapes the system, who is included or excluded, and how technologies are governed. This approach opens space for more grounded, intentional, and socially responsive design and implementation.

4 Toward an Integrative Perspective

To move beyond narrow tech-centric narratives and toward more socially grounded innovation, the following section offers a practical lens to help researchers, developers, and decision-makers ask sharper questions and design more inclusive innovations. The Socioethical Awareness Framework for Digital Agriculture, outlined earlier, is guided by broad questions about power, narrative, and governance. Yet these questions risk remaining abstract unless they are linked to the concrete ways in which technology takes shape in everyday agricultural life. To anchor this framework in practice, I introduce two analytical entry points: the technological services through which digital technologies materialize, and the business models that mediate their access and use by producers. Together, these dimensions render the framework more tangible, revealing how dynamics of power, meaning, and decision-making are embedded in specific technologies that respond to the situated challenges farmers face.

4.1 Seeing Relationships Through Services

Not all technologies do the same thing, they provide specific services to farmers and rural communities. These are not isolated tools but interventions within existing webs of social and economic relations. Digital tools are often described in terms of their technical features—AI, blockchain, Internet of Things (IoT)—but what matters most is the functions they serve and the relationships they transform. Such services never emerge in a vacuum; they intervene in pre-existing networks of support, knowledge, and exchange, reshaping roles and relationships already in place.

Drawing on a typology synthesized from recent literature and AgTech mappings, Porciello et al. (2021); Digital Agri Hub (2022). I identify the following five key types of digital services:

Digital Advisory and Extension These services provide farmers with information to support decision-making. While they vary in sophistication, their core function is to guide agricultural practices through agronomic, climatic, or market-based recommendations. Tools range from SMS and WhatsApp groups to web-based platforms. By mediating the relationship between expert knowledge and everyday practice, they reshape who advises, how, and under what terms.

Digitized Farm Tools This category includes technologies that enable the collection, analysis, and use of farm-level data to improve production. Tools may involve sensors, drones, digital platforms, or robotic equipment. These services alter the relationship between farmers and their land, making production more legible, measurable, and subject to new forms of control and intervention.

Digital Financial Services These tools aim to expand access to financial services—payments, credit, insurance, subsidies—often through data-driven systems. More than just enabling transactions, they transform relationships between farmers and financial institutions, redefining inclusion, risk assessment, and trust.

Digital Market Linkages These platforms connect farmers with buyers, input providers, and service actors. They include e-commerce sites, digital marketplaces, and logistics services. By intermediating transactions, they reshape commercial relationships, influence negotiation power, and determine access to differentiated markets.

Supply Chain Management These solutions focus on facilitating information flows across the agri-food system. They include tools for traceability, transparency, and coordination, such as QR codes, blockchain, and supply chain software. In doing so, they transform how value is tracked, trust is distributed, and accountability is enforced along the chain.

Framing technology in terms of services makes visible what is being transformed, who is affected, and how trust, dependency, or autonomy are redistributed. It also broadens participation in the conversation, enabling farmers and other actors to engage not through technical jargon, but through the concrete shifts they experience in their daily interactions.

4.2 Seeing Farmers Through Their Business Models and Aspirations

Farmers are not a homogeneous group; they build diverse business models and market relationships shaped by their histories, resources, and visions of what is viable and desirable. In dominant market-oriented and global value chain frameworks, producers are often positioned as raw material suppliers within linear and extractive commodity chains. This framing obscures producer agency and overlooks the possibility that farmers may aspire to different forms of participation in markets, ones grounded in autonomy, reciprocity, or territorial stewardship.

Field experience and empirical research reveal a more complex reality, where producers navigate multiple economic logics simultaneously (Gibson-Graham 2008; van der Ploeg 2018). These different orientations translate into distinct technological needs, not only in terms of tools, but in how technology mediates relationships, redistributes control, and enables specific strategies. The following examples illustrate this diversity.

A coffee producers' organization supplying green coffee to fair-trade cooperatives or large exporters operates within a relatively standardized commodity relationship. Their strategy prioritizes volume, consistency, and meeting certification requirements. Technological needs center on operational efficiency: coordinating harvest schedules across member farms, tracking volumes delivered, managing payments, and documenting compliance with certification standards. Here, technology supports coordination within an existing market structure rather than altering the terms of exchange.

In contrast, a coffee cooperative selling directly to multiple specialty roasters has pursued a differentiated path, one centered on quality, identity, and control. Initially, the cooperative adopted a traceability platform provided by one of its buyers, which simplified transactions but gave the buyer full visibility into the cooperative's production data. Over time, members realized that this arrangement undermined their negotiating position: the buyer knew their costs and capacities better than they knew the buyer's. In response, they developed their own traceability system, deciding what data to share, with whom, and when. Their technological needs are thus not limited to quality control but extend to autonomy and bargaining power within direct market relationships.

Finally, other producer organizations operate within short and territorially grounded market circuits, where the goal is not scale but connection. Networks of agroecological producers that sell directly to local restaurants, community stores, or consumer groups depend on digital tools that enable coordination and visibility at a small but relational scale. Their technological needs focus on logistics, collective marketing, and communication—tools for managing orders, coordinating deliveries, and conveying the story of origin and practice that gives their products meaning and value. These systems often emerge from within the organizations themselves, combining adaptable, low-cost tools that reflect commitments to transparency, trust, and mutual accountability. In such contexts, digital technologies sustain proximity rather than expansion; they strengthen local economies and reinforce producer control over how their territories and products are represented.

These examples highlight that there is no single "digital agriculture" but multiple configurations of technology aligned with distinct economic orientations and social aspirations. Recognizing this diversity shifts the question from how to achieve adoption to how to design technologies that respect and enhance farmers' chosen paths of development—whether oriented toward efficiency, differentiation, or territorial stewardship.

4.3 Making Socioethical Awareness Work in Practice

By bringing together the insights developed throughout this chapter—the limits of the instrumental view, the need to make power visible, the importance of framing digital tools as services, and the diversity of farmer business models—we open the door to a richer and more grounded conversation. One that moves beyond adoption metrics to ask: what kind of innovation do we need, and for whom? This is what we do not talk about when we talk about AgTech. And it is what we must begin to discuss if we hope to build more inclusive, transformative, and sustainable food systems.

Making socioethical awareness operational requires creating spaces and mechanisms where technology is not only designed or evaluated for performance, but discussed as a social and political process. Building on ideas previously developed in research workshops (Melo-Velasco 2022), the following practices show how socioethical awareness can be enacted across different scales of innovation.

4.3.1 Micro Scale: Producers, Researchers, and Entrepreneurs

At the microlevel, the focus is on integrating reflection, inclusion, and responsibility into the everyday practices of those designing, using, or studying digital tools.

Responsible innovation panels: establish small review panels—similar to ethics committees—within innovation projects. These groups (farmers, developers, researchers) periodically assess how a technology is evolving in practice, examining its implications for power, inclusion, and governance, and generating concrete recommendations for adaptation.

Situated technology documentation: support the creation of "technology diaries" or microcase studies documenting how one digital tool interacts with specific production systems or communities. These records reveal patterns of exclusion, creativity, and adaptation often invisible in conventional evaluations.

Equity-focused design cycles: during early development, integrate reflexive checkpoints—structured workshops where teams revisit who benefits, who is burdened, and what dependencies are emerging. The outcome should be design adjustments, not just discussions.

Recognition and incentives: create prizes or research calls that explicitly reward responsible, inclusive design processes rather than purely technological novelty.

4.3.2 Meso Scale: Networks, Sectoral Platforms, and Territorial Initiatives

At the meso level, socioethical awareness translates into collaborative governance, coordination, and monitoring practices across organizations, value chains, and territories.

Technology review panels: convene periodic multiactor panels within agri-food sectors or regions to review how digital tools are reshaping market relations, data control, and access to services. These panels generate public reports and joint recommendations for collective action within specific value chains (e.g., coffee, cocoa, dairy).

Inclusive data on the innovation ecosystem: develop shared observatories that track who is developing and offering digital services, what business models they follow, and which producers they target. Mapping concentration, diversity, and alignment with farmer needs provides evidence for more equitable innovation policies.

Participatory territorial agendas: facilitate territorial innovation agendas where cooperatives, local governments, and private actors identify shared priorities for investment and coordination, balancing efficiency goals with inclusion and data sovereignty.

Networked learning platforms: support spaces within regional networks (such as Latin American AgTech network) to document cases, exchange tools for participatory evaluation, and train local facilitators to apply the socioethical lens.

4.3.3 Macro Scale: Policy, Funding, and International Cooperation

At the macrolevel, the goal is to align regulatory frameworks, investment priorities, and cooperation programs with technological justice and producer autonomy.

Regulatory frameworks for data and digital rights: advance clear rules on data ownership, privacy, and platform accountability, ensuring that smallholders and cooperatives retain agency over their digital assets.

Public investment priorities: redefine digital agriculture funding criteria to privilege projects that demonstrate inclusive governance, open data practices, or collective ownership models.

Policy review panels: create standing committees—integrated by producer organizations, researchers, and civil society—to periodically review national and regional digital agriculture policies, identifying asymmetries, and proposing corrective actions.

International cooperation platforms: encourage donors and multilateral institutions to support cross-country learning on inclusive digital governance and to channel resources toward public-interest digital infrastructure.

Competing Interests The author(s) has no competing interests to declare that are relevant to the content of this manuscript.

References

Ancín M, Pindado E, Sánchez M (2022) New trends in the global digital transformation process of the agri-food sector: an exploratory study based on twitter. Agric Syst 203:103520

Bronson K (2022) The immaculate conception of data: agribusiness, activists, and their shared politics of the future. McGill-Queen's University Press, Montreal [Quebec]; Kingston [Ontario]; London; Chicago [Illinois]. 206 p

Carolan M (2017) Publicising food: big data, precision agriculture, and co-experimental techniques of addition. Sociol Rural 57(2):135–154

Clapp J, Vriezen R, Laila A, Conti C, Gordon L, Hicks C et al (2025) Corporate concentration and power matter for agency in food systems. Food Policy 134:102897

Digital Agri Hub (2022) Dashboard. Cultivating tomorrow's agriculture: your ultimate digital solutions reference hub. https://digitalagrihub.org/

Duncan E, Abdulai AR, Fraser EDG (2021) Modernizing agriculture through digital technologies: prospects and challenges. In: James HS Jr (ed) Handbook on the human impact of agriculture [Internet]. Edward Elgar Publishing. [cited 2024 Apr 12]. Available from: https://china.elgaronline.com/view/edcoll/9781839101731/9781839101731.00018.xml

Fairbairn M, Faxon HO, Montenegro De Wit M, Bronson K, Kish Z, Ruder SL et al (2025) Digital agriculture will perpetuate injustice unless led from the grassroots. Nat Food 6(4):312–315

Fairclough N (2017) Critical discourse analysis: the critical study of language. Routledge, London

Gibson-Graham JK (2008) Diverse economies: performative practices for 'other worlds'. Prog Hum Geogr 32(5):613–632

Hendrickson M, Howard P, Constance D (2019) Power, food and agriculture: implications for farmers, consumers and communities. In: In defense of farmers: the future of agriculture in the shadow of corporate power [Internet]. University of Nebraska Press. Available from: https://doi.org/10.2307/j.ctvgs0crb.7

IPES-Food (2016) From uniformity to diversity: a paradigm shift from industrial agriculture to diversified agroecological systems. International Panel of Experts on Sustainable Food systems

Lajoie-O'Malley A, Bronson K, van der Burg S, Klerkx L (2020) The future(s) of digital agriculture and sustainable food systems: an analysis of high-level policy documents. Ecosyst Serv 45:101183

Melo Velasco J, Palacios N, Villota E, Palacios EA (2022) How difficult is it to build a sustainable and inclusive company in the global south? Reflections on a case in Chocó (Colombia). In: ICSD 2021 [Internet]. MDPI. [cited 2024 Apr 11]. p. 41. Available from: https://www.mdpi.com/2673-4931/15/1/41

Melo-Velasco (2022) ¿Qué debemos saber sobre las implicaciones sociales de las tecnologías digitales en la Agricultura? Presentation offered virtually at the Red Aurora Network

Melo-Velasco J, Wilson K, Hendrickson M, Valdivia C (2025) Understanding the social implications of digital agricultural technologies. J Agric Food Syst Community Dev:1–16

Miller B (2021) Is technology value-neutral? Sci Technol Hum Values 46(1):53–80

Mohr S, Höhler J (2023) Media coverage of digitalization in agriculture - an analysis of media content. Technol Forecast Soc Change 187:122238

Mooney P. Blocking the chain: industrial food chain concentration, Big Data platforms and food sovereignty solutions. 2018.;43 Seiten

Porciello J, Coggins S, Otunda-Payne G, Mubaya E (2021) Digital agriculture services in low- and middle-income countries: a systematic scoping review. Glob Food Secur 34(1):100640. https://doi.org/10.1016/j.gfs.2022.100640

Roscigno VJ (2011) Power, revisited. Soc Forces 90(2):349–374

Rotz S, Duncan E, Small M, Botschner J, Dara R, Mosby I et al (2019) The politics of digital agricultural technologies: a preliminary review. Sociol Rural 59(2):203–229

Umoja Noble S (2018) Algorithms of oppression. How search engines reinforce racism, New York, NYU Press

van der Ploeg JD (2018) The New Peasantries: Rural Development in Times of Globalization. Routledge: London

Zuboff S (2019) The age of surveillance capitalism the fight for a human future at the new frontier of power. PublicAffairs, New York

The Fourth Agricultural Revolution: How AI Transforms Strategic Decision Support Systems

Maria Kamariotou and Fotis Kitsios

1 Introduction

The Fourth Agricultural Revolution refers to the continuous transformation of agriculture, which is predominantly influenced by advancements in Artificial Intelligence (AI), data analytics, and digital technologies. Previous agricultural revolutions primarily focused on mechanisation, whereas this new revolution centres on the integration of chemical innovations and enhancements. AI and intelligent technologies are being utilised to enhance agricultural practices, improve decision-making, and ensure sustainability (Ennaji et al. 2023; Pandey and Mishra 2024; Kpodo and Nejadhashemi 2025).

Strategic Decision Support Systems (SDSS) are gaining traction as potent instruments in this context. These systems facilitate the process of making informed decisions in the agricultural sector. These systems, which have historically depended on human expertise and historical data, are currently being enhanced by AI algorithms that are capable of analysing enormous amounts of data. Predict outcomes, provide actionable insights, and analyse vast quantities of real-time data. The outcome is a radical change in the manner in which decisions are made at all levels of the agricultural value chain, from farm management to policy-making and beyond (Ennaji et al. 2023; Pandey and Mishra 2024; Kpodo and Nejadhashemi 2025).

In light of the increasing complexity and competitiveness of the business environment, it is imperative to make decisions in a timely manner. The implementation of decision support technology is becoming increasingly significant and necessitating enhanced efficiency and reduced complexity. The significance of Information Technology (IT) in facilitating decision-making was the primary focus of research

M. Kamariotou · F. Kitsios (✉)
Department of Applied Informatics, University of Macedonia, Thessaloniki, Greece
e-mail: mkamariotou@uom.edu.gr; kitsios@uom.gr

F. da Silveira et al. (eds.), *Agriculture 4.0*,
https://doi.org/10.1007/978-3-032-24552-6_2

in this field. This technology aims to help decision-makers make faster decisions. Accorsi et al. (2014) highlight the benefits of employing computer-based systems to streamline logistics management, especially in transportation and storage.

The agrifood sector has undergone numerous societal, economic, and technological changes recently. These transformations significantly impact the entire food processing chain, from agricultural production to food refining and consumer distribution. Subsequent firms in this sector are dependent on innovation, which is essential for their sustainability. This innovation is crucial for improving the competitiveness of firms in the agrifood sector (Baregheh et al. 2012).

Although Small and Medium Enterprises (SMEs) are the primary pillar of many rural economies, researchers have focused on examining the innovation practices of SMEs in the agrifood sector (Baregheh et al. 2012). This sector is the largest one within the EU and is one of the key drivers of the EU economy, contributing to both economic output and employment (Baregheh et al. 2012; Gold et al. 2017). This sector is one of the primary drivers of the EU economy and is the greatest within the EU to both employment and economic output. Whereas Supply Chain Management underscores the importance of viewing the entire supply chain as a single system. The primary focus of current research is to enhance the position of agrifood actors in the development countries (Gold et al. 2017). Additionally, the literature focusses on the specific obstacles that market constraints and inadequate infrastructure create for smallholder producers in gaining access to global supply chains. Additionally, insufficient resources hinder smallholder producers from gaining access to global chains (Gold et al. 2017).

The purpose of this paper is to discuss how AI can transform SDSS in the agriculture industry.

2 Logistics Strategy

Recent research has focused on the strategic significance of logistics. Organisations that prioritise customer service as their primary objective develop significant strategic implications from logistics. The administration of information and material is a comprehensive aspect of logistics. Meeting consumers' needs is crucial. The logistics strategy must align with the business strategy, as it is essential for identifying the selection of products, services, and markets while determining the organisation's logistics objectives. Sustainable Supply Chain Management is significant because it enables businesses to incorporate sustainable objectives into their supply chains (Shi et al. 2015).

As managers engage in strategic planning, they establish the organisation's primary objectives to remain competitive. The logistics strategic planning process begins with a situational analysis for businesses that operate in a volatile environment. This analysis encompasses the development of the vision, strategic objectives, strategies, and action plans. The objective of this analysis is to determine the strengths, vulnerabilities, threats, and opportunities by examining both the business environment and the logistic system. Furthermore, strengths and weaknesses influence the examination of inventory management, logistical structure, and logistical costs. This includes transportation, information systems (IS), organisational structure, collaboration with other corporations, materials management, and various logistical functions. The availability of resources determines the long-term direction of both the environment and the logistics function (Shi et al. 2015).

Specifically, the vision outlines the logistics organisation's desired future state. The strategic objectives and vision guide the logistics implementation. Both the logistic vision and the activities must be consistent with the business vision. The objectives and strategies facilitate the conversion of logistics vision into precise operational metrics and performance metrics. These models include inventory management, order processing, transportation, customer service, warehousing, information systems, and organisation. We synchronise the establishment of objectives and the development of strategies with other business functions, such as marketing and production, to enhance business benefits. Additionally, action plans provide a comprehensive account of the operational activities and short-term objectives that are necessary for the strategy's implementation to examine the environment, which has a substantial impact on the logistics function (Shi et al. 2015).

Three phases comprise the logistics strategic management process. The initial phase involves the identification of trends and the assessment of their urgency and impact. The subsequent phase involves the identification of trends, followed by the assessment of priorities and the final phase, the development of a strategy that addresses the challenges (Shi et al. 2015).

The strategic management of logistics can generate strategic decision outcomes or environmental forces. To address the issues, it may be necessary to modify the vision, objectives, strategies, or action plans. The strategic issues management process supports periodic planning. Strategic management formats design their procedure to balance the advantages of strategic planning with the adaptability of ongoing strategic administration. Additionally, it integrates the logistics processes of the organisation with a strategic orientation and the capacity to confront both internal and external developments (Shi et al. 2015).

Innovation appears to significantly influence the agrifood sector in terms of business success. Agrifood enterprises have historically prioritised the following: reducing production costs rather than providing benefits to their ultimate customers. Recently, there have been numerous pressures; the necessity to guarantee the safety of food, the content of its nutrients, and the preferences of consumers is a result of globalisation for the sake of convenience, quality, and variety, in addition to the novel opportunities presented by biotechnology. All of these revolutions have resulted in a shift in perspective. Consequently, the agrifood sector is becoming more focused on the creation of products that contemplate the needs of customers (Fortuin and Omta 2009).

3 Strategic DSS Models in Logistics

The complexity of the design of a distribution/logistics network is implied by strategic decisions, as it entails substantial investments in resources over the course of several years. Strategic logistics planning aims to reduce inventory-related costs while maintaining necessary customer service levels. This involves managing the production and storage of products as they move from manufacturers to consumers. As a consequence, in the logistics industry, strategy is particularly important for establishing long-term competitive advantages. The distribution network is crucial for the cost of inventory and transportation. Additionally, it is critical to customer satisfaction in terms of logistics response (Kengpol 2008; Kitsios et al. 2020).

In this field, prior research has concentrated on the significance of IT in assisting decision-makers to improve the efficiency of decisions and to increase their effectiveness. Particularly, surveys conducted in the past concentrated on the advantages of employing computer-based systems to facilitate logistics administration, especially in warehousing and transportation (Kengpol 2008; Songbai et al. 2010; Salam and Khan 2016). The domains of inventory and product forecasting have been the subject of limited surveys (Accorsi et al. 2014).

The functionalities of various DSSs in logistics are summarised in Table 1. Subsequently, the similarities among the following paragraphs are discussed.

Table 1 DSS in logistics

DSS	DSS functionalities	References
DSS for operational and tactical decisions in logistics	Data used is related with products and services prices, resource and budget allocation, payroll cost, cost per product Simulation events such as demands, departures and arrivals of means of transportation at terminals and acquisitions and releases of resources by vehicles Identification of the performances of the systems Evaluate the selected parameters which can improve the performance indices	Fanti et al. (2015)
DSS model for vehicle routing	Demand analysis Analysis of data (number of drivers, strength of vehicle, mileage per vehicle) Decision analysis for the transportation personnel requirements, vehicle demands, path choosing optimisation and resource transportation information	Songbai et al. (2010)
Logistics distribution network	Preliminary analysis (information such as GMS locations, transportation costs of listed distribution centres and customers) Evaluation of the alternatives for the logistics distribution network Estimation of the delivery time, quality, unexpected demand Calculation of the transportation cost The implementation and feedback	Kengpol (2008)

The studies under review show how flexible DSSs are in meeting various logistics decision-making requirements. While each model concentrates on different operational situations, they all aim to maximise performance and resource allocation.

The integration of cost, product pricing, and resource allocation data is emphasised in the DSS for operational and tactical decisions in logistics (Fanti et al. 2015). Its ability to simulate real-world occurrences, like resource acquisitions and vehicle arrivals, gives decision-makers the ability to evaluate system performance in dynamic scenarios. This method emphasises how crucial scenario analysis is for pinpointing crucial elements that boost productivity.

On the other hand, Songbai et al. (2010) DSS model for vehicle routing tackles a more focused issue: transportation activity optimisation. This system offers straightforward solutions for reducing inefficiencies in vehicle use by concentrating on demand analysis, staffing needs, and path optimisation. Its pragmatic approach to operational decision-making is demonstrated by the thorough examination of drivers, vehicle capacity, and mileage.

By examining preliminary data, like distribution centre locations and transportation costs, to assess network alternatives, the logistics distribution network model (Kengpol 2008) broadens the scope. Estimating delivery dates, service quality, and unforeseen demand are all part of its all-inclusive approach. The focus on implementation and feedback sets this model apart from the other two since it evaluates the ideas' practicality in addition to suggesting solutions.

Overall, the goal of increasing logistical efficiency is shared by all three DSS applications; however, their approaches and scopes vary. Kengpol (2008) discusses the construction of strategic distribution networks, Songbai et al. (2010) focus on vehicle-level optimisation, and Fanti et al. (2015) offer a comprehensive tactical approach. When taken as a whole, they show how DSS may help with decision-making at all levels of logistics management, from strategic to operational.

The initial phase of the strategic management process involves scanning of both internal and external environment. The second stage entails the assessment of the urgency and impact of the issues. This stage includes evaluating the priorities of the previous issues and determining the appropriate responses to these concerns. The final stage involves the preparation of the necessary responses for the strategic concerns. DSSs are employed to determine the extent to which strategic issues affect logistics and to pinpoint the most significant factors. The development of the systems necessitates the consideration of certain fundamental features. There are numerous components to DSS.

The basic components are as follows. The data component typically includes a database management system from the outset Database Management System (DBMS). The DBMS includes general programming languages and modelling tools. Cross-sectional data or time series data may be implemented, whether internally or externally. Data generated internally from the organisation's functions includes information on product and service prices, financial data, and resource allocation details such as payroll costs and cost per unit. The external issue at hand pertains to government arrangements, competition, and market share. The external sources include government agencies, market research firms, and information from the internet. The data that is maintained in the DSS—the optimisation processes associated with models—is informed by the database. The data files in the DSS, whether they are internal or external to the enterprise, serve as sources of information. The subsequent module is the model component, which comprises a mathematical model and a simulation model. Additionally, the system employs optimisation algorithms to facilitate the examination of its response to the selected options (Fanti et al. 2015). Specifically, DSSs implement a variety of algorithms, theories, models, and methods to develop and analyse alternative decisions. Examples of these methodologies include the use of genetic algorithms, simulated and fuzzy modelling, intelligent data analysis, fuzzy theory, neural networks, and decision-making theory (Kondratenko et al. 2014).

The evaluation of the supply chain is a significant application of DSS in logistics. The current changes in global production have exacerbated the complexity of the supply chain and increased the argument that logistics strategies are substantial components of business strategy. Recent business dealings highlight the importance of building relationships with suppliers to improve sustainable administration. The objective of the supplier selection and assessment process is to mitigate risk and enhance the overall value to the consumer. People view supplier selection as a complex process involving numerous criteria. This process presents an even more complex challenge within the decision-making framework. To implement a more effective strategy, the maker must analyse a substantial volume of data that takes into account a variety of factors. To decrease expenses and enhance profitability, businesses must consider each variable. Profit is affected by the heightened globalisation of commerce and the expansion of competition. Businesses develop strategies for the supplier selection process that prioritise sustainability to address the increased level of competition, as well as environmental responsibility requirements. Numerous managers must take sustainability into account in the selection of suppliers and the assessment of their performance.

Therefore, managers consider the following criteria when selecting and evaluating suppliers: the initial category pertains to quality, the second category pertains to price, and the subsequent category pertains to the supply/delivery capability, while the fourth type encompasses factors regarding the service. An additional category that decision makers consider when selecting suppliers is the assessment of environmental protection (Karthik et al. 2015).

4 The Role of AI in Agricultural Decision-Making

The impact of AI on strategic decision-making stems from its ability to manage intricate, multifaceted situations. We can resolve dimensional problems by processing large datasets with precision and speed. These systems have the capacity to incorporate data from a diverse range of sources, including satellites, drones, remote sensing technologies, and IoT sensors, to monitor soil health, weather patterns, crop conditions, and insect outbreaks. The system also monitors water consumption. This plethora of information, once overwhelming, now undergoes synthesis and analysis by AI, allowing farmers and agribusinesses to make data-driven, evidence-based decisions (Ennaji et al. 2023; Pandey and Mishra 2024; Kpodo and Nejadhashemi 2025).

For instance, SDSS that are AI driven can assist producers in determining the optimal time to plant, irrigate, and harvest crops by predicting optimal conditions using real-time variables and historical data. In the same way, AI algorithms can suggest the appropriate timing and quantity of fertiliser or pesticide to apply. This approach minimises waste and reduces environmental impact (Ennaji et al. 2023; Pandey and Mishra 2024; Kpodo and Nejadhashemi 2025).

The impact of AI on precision is a significant aspect of its contribution to agricultural decision-making. Agriculture is a method that prioritises optimising field variations to enhance productivity and reduce resource usage. Through AI, SDSS can predict crop yields, anticipate potential diseases or insect infestations, and recommend specific interventions for different components of a farm. This approach utilises precise data from a variety of sources (Ennaji et al. 2023; Pandey and Mishra 2024; Kpodo and Nejadhashemi 2025).

For instance, AI facilitates predictive analytics, which can predict market trends and supply chain dynamics. This allows producers to modify their production strategies in advance during disruptions. Models trained to identify patterns in input costs, market demands, and weather data form the basis of these predictions. In the end, these predictions aid farmers and agribusinesses in preserving a competitive advantage amidst dynamic market circumstances (Ennaji et al. 2023; Pandey and Mishra 2024; Kpodo and Nejadhashemi 2025).

AI's ability to automate decision-making forms the foundation of the Fourth Agricultural Revolution. Machine learning algorithms have the capacity to continuously analyse data and refine recommendations over time. AI learns from past decisions to enhance future outcomes. This continuous learning can lead to better efficiency in operations and less need for human involvement (Ennaji et al. 2023; Pandey and Mishra 2024; Kpodo and Nejadhashemi 2025).

For instance, AI-powered systems have the ability to autonomously adapt in large-scale farming operations. This includes managing irrigation schedules, controlling mechanisation, and detecting crop diseases early to implement targeted remedies before more significant issues arise. This level of automation frees up human decision-making capacity. This automation allows decision-makers to focus on higher-level strategic decisions while simultaneously managing operational duties effectively. The Fourth Agricultural Revolution not only focusses on increasing profitability, but also strives to enhance the sustainability of agriculture. AI-driven decision assistance systems are crucial in ensuring the optimisation of agricultural practices for both environmental impact and productivity. Through the utilisation of real-time data, AI can assist producers in minimising the use of resources (e.g., water, fertilisers, and pesticides) while maximising crop yields, which will result in more sustainable agricultural practices (Ennaji et al. 2023; Kpodo and Nejadhashemi 2025; Pandey and Mishra 2024).

Additionally, AI-powered SDSS can facilitate more efficient land management by assisting producers in adapting. Producers can adjust to changing environmental circumstances, like climate change, by improving soil health, cutting down on waste, and lowering carbon emissions. Policymakers and environmentalists may also benefit from AI. Organisations make decisions regarding land use policies and conservation initiatives that are supported by data (Ennaji et al. 2023; Kpodo and Nejadhashemi 2025; Pandey and Mishra 2024).

5 The Future of SDSS in Agriculture

The future of SDSS in agriculture is promising, as the potential applications for these systems are expanding with the ongoing development of AI technologies. Systems in agriculture are virtually limitless. AI is changing how agriculture is managed and organised, helping small farmers grow and large agribusinesses improve their operations, while also encouraging countries to adopt better practices and boost crop production (Ennaji et al. 2023; Kpodo and Nejadhashemi 2025; Pandey and Mishra 2024).

Additionally, the integration of AI with emerging technologies, including blockchain, 5G, and the cloud, will further enhance the capability of SDSS through computation, resulting in increased accessibility, reliability, and scalability. However, the forthcoming iteration of AI-powered decision systems will not only facilitate tactical and operational decisions but will also be essential for the formulation of long-term strategic plans and policies. This includes global food security initiatives (Ennaji et al. 2023; Kpodo and Nejadhashemi 2025; Pandey and Mishra 2024).

Figure 1 presents the interrelation between AI and DSS in supply chains.

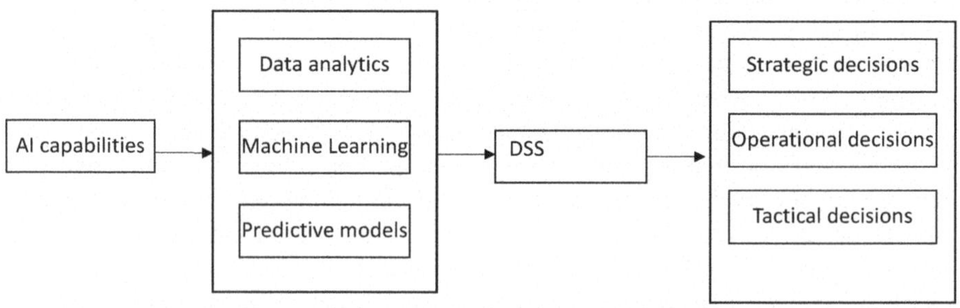

Fig. 1 The interrelation between AI and DSS in supply chains

6 Conclusions

6.1 Theoretical and Practical Contribution

The research contribution of this paper is dual. Initially, it endeavours to address the lacuna in the literature with respect to the relationship between DSS and Strategic Information Systems Planning processes. Additionally, the phases that provide support are defined. The paper presents the implications for future research on managers' decision-making. Managers and academics can assess the efficacy of Strategic Information Systems Planning steps because it is necessary to develop, visualise, and test a model that will enhance the effectiveness of the decision-making process.

By emphasising how artificial intelligence can improve the efficacy of DSS in strategic decision-making, this research advances the field of supply chain management in the agrifood industry. In particular, it increases knowledge of how managers might be assisted by AI-driven technologies in handling the increasing complexity of agrifood logistics, ranging from food delivery to agricultural production. The study offers fresh perspectives on process optimisation, enhancing responsiveness to outside changes, and bolstering the general resilience and effectiveness of agrifood supply chains by fusing AI with strategic information systems planning.

6.2 Suggestions for Future Research

The manner in which decisions are made is being fundamentally altered by the Fourth Agricultural Revolution, which is being driven by AI. Through the use of AI-enhanced SDSS, producers, agribusinesses and policymakers can make more informed, accurate, and timely decisions that enhance profitability, sustainability, and productivity. AI will assume a significant role as it continues to develop a more significant role in the future of agriculture, assisting in the resolution of some of the most urgent challenges confronting the sector, including environmental sustainability and nutritional security.

In the end, AI is not only changing our farming practices but also transforming our understanding of agriculture in the twenty-first century. It is offering a vision of a more sustainable, productive, and efficient future. The agricultural landscape will be essential in satisfying the needs of an expanding global population.

The strategic decision-making process can be substantially enhanced by the integration of DSS and strategic planning to achieve efficiency. A meticulous design is essential for achieving the benefits of SDSS. The DSS research area and IS will provide new motivations for effective SDSS development.

DSSs are based on the needs for information of the existing organisational functions. In the future, DSS will strive to incorporate tools that are responsive to environmental changes and information requirements.

Assist decision-makers in adjusting their work practices to meet future needs. Additionally, the DSS can provide customers with information regarding the procurement process if agrifood producers implement it. By implementing production methods, producers can closely monitor their customer base and make necessary supply-chain adjustments. A sustainable competitive advantage is achieved through chain visibility and transparency.

Competing Interests The author(s) has no competing interests to declare that are relevant to the content of this manuscript.

References

Accorsi R, Manzini R, Maranesi F (2014) A decision-support system for the design and management of warehousing systems. Comput Ind 65:175–186

Baregheh A, Rowley J, Sambrook S, Davies D (2012) Innovation in food sector SMEs. J Small Bus Enterp Dev 19:300–321

Ennaji O, Vergütz L, El Allali A (2023) Machine learning in nutrient management: a review. Artif Intell Agric 9:1–11

Fanti MP, Iacobellis G, Ukovich W, Boschian V, Georgoulas G, Stylios C (2015) A simulation based Decision Support System for logistics management. J Comput Sci 10:86–96

Fortuin FT, Omta SWF (2009) Innovation drivers and barriers in food processing. Br Food J 111:839–851

Gold S, Kunz N, Reiner G (2017) Sustainable global agrifood supply chains: exploring the barriers. J Ind Ecol 21:249–260

Karthik B, Raut RD, Kamble SS, Kharat MG, Kamble SJ (2015) Decision support system framework for performance based evaluation and ranking system of carry and forward agents. Strateg Outsourcing Int J 8:23–52

Kengpol A (2008) Design of a decision support system to evaluate logistics distribution network in Greater Mekong Subregion Countries. Int J Prod Econ 115:388–399

Kitsios F, Kamariotou M, Madas MA, Fouskas K, Manthou V (2020) Information systems strategy in SMEs: critical factors of strategic planning in logistics. Kybernetes 49:1197–1212

Kondratenko YP, Klymenko LP, Sidenko IV (2014) Comparative analysis of evaluation algorithms for decision-making in transport logistics. In: Advance trends in soft computing, Springer, Switzerland, pp 203–217

Kpodo J, Nejadhashemi AP (2025) Navigating challenges/opportunities in developing smart agricultural extension platforms: multi-media data mining techniques. Artif Intell Agric. (in press)

Pandey DK, Mishra R (2024) Towards sustainable agriculture: harnessing AI for global food security. Artif Intell Agric. (in press)

Salam MA, Khan SA (2016) Simulation based decision support system for optimization: a case of Thai logistics service provider. Ind Manag Data Syst 116:236–254

Shi P, Yan B, Shi S, Ke C (2015) A decision support system to select suppliers for a sustainable supply chain based on a systematic DEA approach. Inf Technol Manag 16:39–49

Songbai H, Yajun W, Dianxiang Y, Yaqing A, Ke Z (2010) The design and realization of vehicle transportation support DSS under contingency logistics. In: Proceedings of the 2nd international conference on advanced computer control, Shenyang, Liaoning, March 2010, pp 463–466

Agriculture 5.0: Building Resilience by Integrating Social, Technological and Economic Innovations

Maleeka Singh, Olivia Willoughby, Maria G. Corradini, Lenore Newman, Rene Van Acker, Rickey Y. Yada, and Evan Fraser

1 Introduction

Overpopulation, climate change, urbanization, and an aging population are some of the challenges agriculture is facing today (Bigonah et al. 2025). For example, by 2050, the global demand for food will increase by 70% to accommodate the needs of a projected global population of more than 10 billion (Bissadu et al. 2024). Additionally, future forecasting suggests that extreme weather conditions will have adverse effects on global agricultural production (Fraser and Campbell 2019; Verma

M. Singh (✉)
Department of Food Science, University of Guelph, Guelph, ON, Canada
e-mail: maleeka@uoguelph.ca

O. Willoughby
Department of Population Medicine, University of Guelph, Guelph, ON, Canada

M. G. Corradini
Department of Food Science, University of Guelph, Guelph, ON, Canada

Arrell Food Institute, University of Guelph, Guelph, ON, Canada

L. Newman
Food and Agriculture Institute, University of the Fraser Valley, Abbotsford, BC, Canada

R. Van Acker
Department of Plant Agriculture, University of Guelph, Guelph, ON, Canada

R. Y. Yada
Faculty of Agricultural, Life and Environmental Sciences, University of Alberta, Edmonton, AB, Canada

E. Fraser
Arrell Food Institute, University of Guelph, Guelph, ON, Canada

Department of Geography, Environment and Geomatics, University of Guelph, Guelph, ON, Canada

F. da Silveira et al. (eds.), *Agriculture 4.0*,
https://doi.org/10.1007/978-3-032-24552-6_3

et al. 2025). For example, in 2023, agriculture faced disruptive weather events across Europe due to winds, hail, flooding and landslides, all of which severely impacted crop yields, infrastructure and resources worldwide (Holzinger et al. 2024). Indirect crises, including overall rising inflation levels, global conflicts, and world health events (e.g., the COVID-19 pandemic, H5N1 outbreaks, etc.), also affect global agricultural performance. The prolonged nature of these overlapping crises and the seemingly never-ending onslaught of new challenges in food production has led to what has been termed the "permacrisis," i.e. permanently being in a crisis state without the possibility of escape or reprise. Lastly, agriculture is constrained by limited farmable land, access to labour, crop susceptibility to climate and pests, and inefficiencies/insufficiencies of the currently available technologies. As such, improvements in methods to decrease the vulnerabilities of modern agriculture to present (and ongoing) challenges necessitate the development of a more resilient and adaptable system.

Modernizing agricultural practices, technologies, and policies is critical in the existing global climate to improve food production resiliency. The current system must be transformed to meet the growing demand for food and to overcome barriers and limitations. Enhancements in technology, such as Artificial Intelligence (AI) and applications of new human-centric methodologies, promote sustainable production that can overcome the extant limitations of current agricultural practices. As such, the discourse should not stop at adapting to change, but should expand to predicting, anticipating and innovating to offset future agricultural disruptions (Holzinger et al. 2024).

Agriculture 5.0 is the next step in innovation to ensure future human resiliency and survival. However, the conceptualization of Agriculture 5.0 requires understanding the driving forces behind major agricultural historical movements as well as the emergence of, and intersections with, Industry 5.0 and Society 5.0. This chapter provides a brief history of agriculture through its major technological and societal revolutions, puts forth a definition of Agriculture 5.0 and its technological and behavioural driving forces (Industry 5.0 and Society 5.0), while examining enabling tools for this paradigm shift and opportunities and barriers for the future of food. Although agriculture can be discussed from farm to fork, including at preharvest, postharvest, processing, retail, and consumer stages, this chapter primarily focuses on the farm level, particularly as it relates to field crops such as oilseeds, grains and pulses. Figure 1 provides an overview of the key enabling factors of agriculture over time, including Agriculture 5.0 and its potential interactions with Society 5.0, and Industry 5.0 to ensure resilient, effective and sustainable agricultural practices.

2 Agriculture 1.0–3.0: History of Agriculture

Modern-day agriculture is the result of continuous improvements and refinements of technologies and their applications over approximately 10,000 years. The different revolutions throughout agricultural history can be broken down into four main

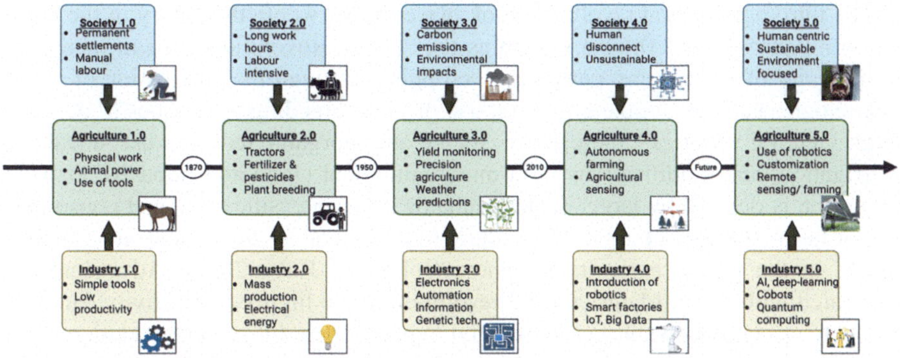

Fig. 1 The revolution and evolution of agriculture over time and its industrial and societal contributions (figure created with Biorender). (Adapted from Bissadu et al. 2024 and Haloui et al. 2024)

phases, spanning from the Neolithic revolution (about 10,000 years ago) to the current state of agriculture in the context of the digital age. The original agricultural revolution, which saw the beginning of "Agriculture 1.0", is defined by the gradual replacement of the hunter-gatherer, nomadic lifestyle common to Neolithic humans with more permanent settlements, as well as the domestication of the first plants and animals (Quealy et al. 2022). This shift can be viewed as an early instance of humans moving to create controlled and productive food systems (Scanes 2018). This era is defined by the emergence of stationary farming, manual labour, simple tools, and overall low productivity (Quealy et al. 2022). This period of agricultural technology had many phases, including an expansion of large-scale, trade-oriented farming across much of western Europe during the Roman period, followed by a contraction in trade and the sophistication of farming after the collapse of Roman authority (Fraser 2011). Nevertheless, much remained constant about farming systems worldwide during this period, notably the dependence on human and draught labour, relatively low yields, and simple tools.

The second agricultural revolution involved the application of basic science to the management of farms. Spanning from the seventeenth to the nineteenth century, Agriculture 2.0 was influenced by the technological advances of the Industrial Revolution, which allowed for the introduction of basic mechanized farming tools and the expansion of larger-scale farming operations. The implementation of rotational grazing and cropping, as well as selective plant and animal breeding, characterized Agriculture 2.0 (Quealy et al. 2022). In addition to technological breakthroughs, the second agricultural revolution was driven by changes in land access and the way that rural people interacted with food; agricultural workers were not just farming for their own use but, with greater productivity, they sought to market their products and participate in their respective economies (Clegg and Lucas 2020).

The third agricultural revolution took place in the twentieth century in the post-war era. Agriculture 3.0, also known as the "Green Revolution", led to increased productivity of agriculture, particularly cropping, through the application of novel technologies such as double cropping, use of plant breeding techniques to develop "high-yielding varieties", increased pesticide and inorganic fertilizer use, advances in irrigation, and modifications in farming equipment (John and Babu 2021). This revolution is defined by larger-scale mechanization, intensification, and consolidation of farms (Quealy et al. 2022), leading to the arrival of what some call "factory farms" in the global agricultural sphere (John and Babu 2021). Advancements associated with the third agricultural revolution led to higher productivity, but also started to ensue perceivable environmental impacts, including soil degradation. This transition saw a marked decrease in the diversity of farmed crops and livestock species, resulting in the current predominance (in North America, for example) of approximately five major crops (i.e., wheat, corn, soybeans, sugar beets and dry beans) and three livestock species (i.e., cattle, chickens, pigs). Although the Green Revolution began as a response to the desire from producers, governments, and consumers for more sustainable farming, the technologies of this era were limited by their reliance on fossil fuels and the prolonged dependence on extractive farming methods (Quealy et al. 2022). Additionally, Agriculture 3.0 increased the separation of consumers from farmers (Quealy et al. 2022).

3 Agriculture 4.0: Current Agriculture Within the Realm of the Digital Revolution

Agriculture 4.0, a product of the fourth and current technological revolution, represents a shift in the agricultural sector towards the use of digital and data technologies to increase productivity and efficiency. This shift is a result of the need for substantial and ongoing increases in agricultural production to meet ongoing growing demands for food (Bissadu et al. 2024). The shift to digital agriculture was driven by farmers seeking to adopt technologies to increase productivity in the current global climate, while lowering production costs. Said to have started in the 2010's, Agriculture 4.0 refers to the implementation of technologies including remote sensing, the Internet of Things (IoT), Artificial Intelligence (AI), and cloud computing, for the generation and processing of Big Data to be used in decision-making (Quealy et al. 2022; Haloui et al. 2024). This agricultural revolution has been driven by an aspiration for precision, rather than supporting greater scale per se, and the technologies used in Agriculture 4.0 are often referred to as "precision agriculture technologies" (Cisternas et al. 2020). Precision agriculture is able to produce immediate benefits from its application while also reducing environmental impacts of, for example, the overapplication of farming inputs such as pesticides and fertilizer (Cisternas et al. 2020). Agriculture 5.0 builds on the Agriculture 4.0 concept. In this chapter, we discuss Agriculture 4.0 and 5.0 simultaneously to explore their similarities and differences, as outlined in Table 1.

Table 1 Agriculture 4.0 vs 5.0: focus, practices, goals, technology and drivers

	Agriculture 4.0	Agriculture 5.0
Focus	Technology, green solutions, precision farming, efficiency of input use	Greener solutions, hyperintensive food production, human/societal well-being, consumer demands
Practices	Those replacing/complementing technologies associated with Ag 3.0	More radically different from Ag 4.0. Human-focused, requiring collaboration among humans, machines, and AI
Goals	Building on Ag 3.0's mass production of food, increased productivity, and economic growth. Improve precision agriculture	Increased productivity, growth, sustainability and efficiency, maintaining social and environmental license, and a high degree of specificity/customization in agriculture
Supportive technologies	AI, Big Data, Wireless Sensor Networks (WSN), IoT	6G technologies, cobots, robots, remote sensing, IoT, Blockchain, quantum computing
Drivers	Information, the need for precision agriculture, and increased production	Unsustainable production, significant social issues, increased consumer demands and the need for resilience

Adapted from Bissadu et al. (2024)

4 The Future of Agriculture: Agriculture 5.0 and the Fifth Industrial Revolution

There is no consensual agreement on the definition of Agriculture 5.0 in the literature, particularly when it comes to differentiating it from its predecessor, i.e., Agriculture 4.0 (Bissadu et al. 2024). That being said, there are two overarching trends in defining Agriculture 5.0. The first applies a technological focus and, in this definition, Agriculture 5.0 integrates smart technologies, such as AI and robotics, through interconnected devices, using wireless technologies and cloud computing for data analysis and automation with the aim of improving agricultural practices that benefit the environment and human development (Bissadu et al. 2024). Thus, Agriculture 5.0 aims to increase agricultural production in sustainable ways, using advanced interconnected technologies with everywhere, everything and every time capabilities (Fountas et al. 2024). In the second definition, Agriculture 5.0 prioritizes humans over machines to attain resilient and sustainable agricultural practices while maintaining human, cultural and societal values (Bissadu et al. 2024). Therefore, under this conceptualization, Agriculture 5.0 aims to produce healthy, sustainable and affordable foods, while resolving social and political issues in food production (Baryshnikova et al. 2022).

Notwithstanding this discourse in opinion, a consensus is emerging that Agriculture 5.0 lies at the intersection of Industry 5.0 and Society 5.0. Unlike Agriculture 4.0, which examines the relationships between agriculture and industry and agriculture and society separately, Agriculture 5.0 aims to move toward more human-centred and environmentally focused agricultural approaches using smart and interconnected technologies. Agriculture 5.0, therefore, seeks to address issues that Agriculture 4.0 fails to resolve by having more harmonious human-machine interactions (Bissadu et al. 2024). In this way, Agriculture 5.0 aims to overcome societal and environmental issues that have arisen in recent years, in a more holistic approach by integrating cyber and physical spaces.

5 Drivers of Agriculture 5.0 and Key Enabling Industry 5.0 Technologies

Agriculture 4.0 has been criticized for being very technologically oriented, while simultaneously removing human elements from production, i.e. machines replacing human roles. Additionally, it has been criticized for failing to address land degradation, climate change mitigation, environmental impacts and overall sustainability (Bissadu et al. 2024). Societal demands, in addition to the increased demand for food, have necessitated the shift of Agriculture 4.0 toward Agriculture 5.0, which is essentially adopting Industry 5.0 technologies within a Society 5.0 framework.

Agriculture 5.0 builds on essential technologies that defined Agriculture 4.0. One of the main attributes of Agriculture 4.0 is digital farming. This incorporates the use of several technologies, such as IoT, Big Data, cloud computing, robotics and AI, to automate many agricultural processes and practices (Araújo et al. 2021). However, Agriculture 5.0 goes beyond the use and applications of new technology and equipment, as its innovation lies in the ability to gather, use and exchange data in real-time and remotely (Javaid et al. 2022). Figure 2 outlines some of the key technologies ingrained in Industry 5.0 that support the Agriculture 5.0 concept.

5.1 Sixth Generation of Internet of Things (6G-IoT) and Cloud Computing

Interconnected, wireless and smart technologies have promoted the development of IoT, which allows for the sensing and communication among millions of physical devices or objects through the internet (Nguyen et al. 2022; Polymeni et al. 2023), everywhere, at any time, in real time. IoT is essential in advancing smart agriculture and is capable of gathering, using and transmitting a large amount of data through wireless technologies for continuous monitoring (Polymeni et al. 2023). Some of these networks include WiFi, Bluetooth, Radio Frequency Identification (RFID), Sigfox and Zigbee, which improve end-to-end latency with increased capacity and

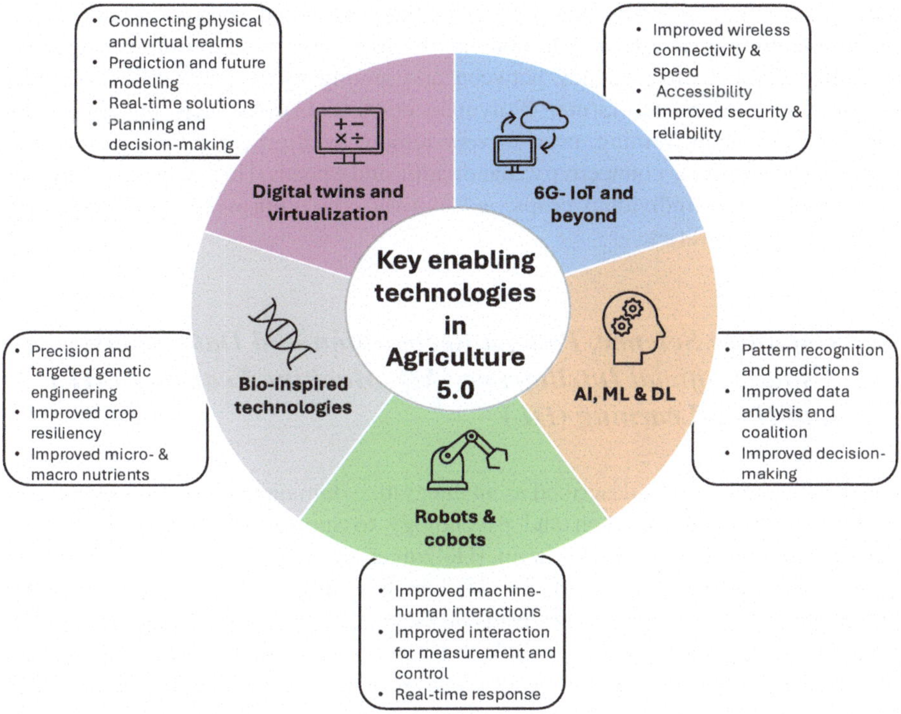

Fig. 2 Key enabling Industry 5.0 technologies involved in Agriculture 5.0 and their contributions

connection through 5G networks (Haloui et al. 2024). The 5G network represents the fifth generation of wireless communication technology that constitutes the basis of Industry 4.0's smart factories. However, there are several limitations to this technology (Haloui et al. 2024). In particular, 5G infrastructure requires a larger number of base stations to provide adequate and even coverage (Haloui et al. 2024), which may increase costs and reduce accessibility in remote areas. Additionally, there are concerns about data security, breaches, and lack of privacy. As such, the vision and advent of the 6G network aims to provide ubiquitous wireless connectivity to larger populations, narrowing the socioeconomic gap, improving accessibility and useability (Haloui et al. 2024) with higher efficiency and security.

With advancements in 6G networks through AI, including but not limited to Machine Learning (ML), an increase in the interconnectivity of devices and data traffic is projected to optimize networks, improve data reliability, provide more automated control, optimize service delivery, and overall boost user experience (Polymeni et al. 2023; Chataut et al. 2024). Such technologies are supposed to improve the link between physical and digital objects through very rapid and responsive communication where monitoring is continuous and in real time (Haloui et al. 2024). For example, 6G-IoT is projected to improve the connections among a large number of sensors to a single cell tower, allowing farmers to gather information item-by-item (e.g. plant-by-plant) by installing individual sensors (Zhang et al. 2020; Mahmood et al. 2021). Additionally, 6G-IoT is expected to improve security, reliability and ultra-low-latency between and among robots, as well as between the robot system and the farmer (Polymeni et al. 2023). As such, the ability for smart and precision farming, even in very remote locations, may be possible due to improved network connectivity, bandwidth and speed. This may give farmers more control over individual crops and cultivars, by improving monitoring and surveillance capabilities.

5.2 Cognitive Science, Pattern Recognition and Data Analysis Using Artificial Intelligence (AI), Machine Learning (ML) and Deep Learning (DL)

Cognitive science can be described as an integrative domain that combines language processing, visual recognition and psychology to support thinking and decision making, similar to the human brain (Holzinger et al. 2024). In fact, cognitive science provided the framework for AI, which is inspired by the ways that the human brain makes decisions, performs tasks, and continuously learns. AI-enabled programs, software or devices, are able to trigger actions, without requiring human interference, as they can automatically analyse and contextualize data (CE 2025) based on prior experience and knowledge. As such, accelerated decision-making and automation of tasks are possible using AI-driven technologies. Often, AI-enabled methods apply ML and Deep Learning (DL) for pattern recognition and predictions without human input.

As it relates to Agriculture 5.0, the use of AI may lead to smart and autonomous farms and agricultural systems (Abdul et al. 2024). For example, predicting crop yield, weather patterns and remote video monitoring of crops, farms and fields are some of the basic necessities of Agriculture 5.0 (Polymeni et al. 2023), which are enhanced with the advancements of AI. Additionally, with the introduction of 6G-IoT, decisions, communications, and devices are able to be more integrated, allowing for real-time or instantaneous suggestions, course corrections and outcomes. Akkem et al. (2023) reviewed the use of AI, particularly ML and DL, in solving current agricultural challenges such as water management, crop selection and yield, soil status, as well as commodity price prediction, improving the discussion of AI in smart farming. Convolutional Neural Networks (CNNs) and recurrent neural networks are some of the most applied DL architectures to address agricultural challenges such as sorting, classification and detection (Martos et al. 2021; Taha et al. 2025). For example, a CNN model was used to classify and identify multiple diseases on plant leaves with over 75% accuracy, achieving a 12% improvement compared to previously tested methods (Arnal 2019). CNN has also been used to predict the maturity of citrus fruits (Itakura et al. 2019), detect water contamination and pollution (Chen et al. 2020), classify hyperspectral images of agricultural lands (Hsieh and Kiang 2020), multispectral satellite imagery for detecting crop anomalies (Mendoza-Bernal et al. 2024), and even for managing postharvest process monitoring such as product safety, quality and identity (Singh et al. 2025a).

Recent advancements in DL models have given rise to Natural Language Processing (NLP), which allows computers and devices to understand, condense and find useful information from language and text more rapidly and accurately (Fountas et al. 2024). Large Language Models (LLMs), which are a type of NLP, go one step further and allow for the interaction between the human and machine with prompts and task-specific responses, as seen with the emergence and use of chatbots, such as ChatGPT (Fountas et al. 2024; Shahriar et al. 2025). As such, LLMs, integrated with IoT, have been used to enhance data-driven decision-making to improve agricultural yield, quality and efficiency (Feng et al. 2024). NLP and LLMs are able to aggregate and interpret data from various sources, including social media, scientific literature, sensors, market trends, environmental conditions, weather forecasts and historical crop performance to provide farmers with insights into best agricultural practices (Hrushikesava et al. 2018).

The ability of remote and wireless sensing is one of the most discussed applications of AI. In agriculture, remote sensors, connected over Wireless Sensor Networks (WSNs), can monitor individual crops, weather, and predict the potential of diseases when analysed with AI tools. Remote and WSNs have been used to monitor field status (98%) and weather conditions (94%) to improve crop production, yield and quality, using AI for prediction (Akilan and Baalamurugan 2024). This technology has also been used to analyse satellite images for land and greenhouse monitoring and detection (Li et al. 2020). The use of AI to analyse agricultural status, using WSNs and remote sensors, is expected to improve real-life decision-making systems, ultimately leading to smarter farming, which is core to the Agriculture 5.0 aspiration. Lastly, in the past, AI has been performed as a "black box", meaning that its workings are difficult to disentangle and understand. However, explainable AI (Ex-AI) is expected to ensure that algorithms are clear, understandable (Spoeth 2025) and more conducive to human interaction in the future. Therefore, with AI, ML and DL methods, farmers may have access to tools that can condense a wide range of information, in more conducive ways, so that real-time decisions and solutions, such as the amount of water, fertilizer, etc., can be made to improve productivity and quality. This not only provides support for farmers, but also potentially boost "brand" safety and quality, and subsequently, its reputation.

5.3 Advancements in Robots and Collaborative Robots (Cobots) Towards Agriculture 5.0

In recent years, robotics has become increasingly utilized in various sectors, such as manufacturing, to remove the need for humans to perform strenuous and repetitive tasks. This is extending into agriculture, leading to improved aerial spraying, autonomous farming and field monitoring (Haloui et al. 2024). The long tradition of adopting mechanization and robots in agriculture has been successful in improving production, while also cutting costs. However, there has been criticism on robots removing humans from industry. Industry 5.0 and Agriculture 5.0 aim to reintroduce humans into manufacturing and production, such that robotic systems "partner" with humans, also referred to as collaborative robots or cobots. The Agriculture 5.0 framework anticipates an increased number of these AI-based cobots performing tasks such as harvesting, detecting defects, deficiencies and anomalies in plants, soil, seeds and crops, executing aerial spraying and monitoring the status of large fields or land (Polymeni et al. 2023; Haloui et al. 2024). The aim of cobots is to place humans at the core of manufacturing and production, such that their skills and capabilities are complemented (by the cobots) and streamlined, to improve processes, rather than for laborious tasks (Siddiqui et al. 2025).

In Industry 4.0 and Agriculture 4.0, machines took precedence over activities, processes and even outcomes. Agriculture 5.0 aims to maximize human-machine interactions to empower farmers and recognize their value, instead of being pushed out by technology (Bissadu et al. 2024) and the narrowing margins resulting from the technology-facilitated commodification of farming. Agriculture 5.0 adopts the Industry 5.0 framework, where human intelligence will be used for tasks requiring critical thinking and complex logic processes, while cobots will be primarily used for physically demanding, hazardous, repetitive and labour-intensive tasks (Maddikunta et al. 2022). However, there are still concerns about the adoption of these technologies as it may limit the number of operators able to make financial profits from cobot implementation, within agriculture, and the global food system. Cobots often contain sensors and are highly responsive, which allows them to be controlled, monitored or stopped instantaneously if a human detects an error, improving their applicability and safety compared to traditional robots (Maddikunta et al. 2022; Siddiqui et al. 2025).

The increased demand for more precise and labour-efficient agricultural production has propelled the use of autonomous mobile robots and agricultural robots to improve precision crop monitoring and perform farming tasks (Ganesh and Kanakaprabha 2025; Taha et al. 2025). These cobots often allow for autonomous decision-making abilities, advanced perception, control and precise execution capabilities, which can often be performed in dangerous environments (Cheng et al. 2023). Agricultural robots and cobots are often classified into field robots, fruit and vegetable robots and animal husbandry robots (Cheng et al. 2023). For example, tilling soil can be quite arduous. Thus, better connectivity and communication between the user and the tilling machine, facilitated by software as proposed by Panarin and Khvorova (2022), can improve this task. Cobots are also being developed to detect pests and diseases on crops, monitor phenotypes, and harvest crops, e.g. picking tomatoes and weeds (Cubero et al. 2020; Cheng et al. 2023). Agricultural Unmanned Aerial Vehicles (UAVs) such as drones are increasingly being employed for crop protection measures like spraying, field and herd monitoring and plant pollination due to their ease of use, cost and low environmental impact (Ghobadpour et al. 2022). These UAVs can also ensure the protection of farmers' health. For example, spraying drones reduce the need for farmers to be in the field while the crop is being sprayed, reducing their exposure to chemicals such as pesticides. While these cobots have specific applications conducive to their implementation goals, their core technologies, i.e. multisensor collaboration, use of AI, advanced visual image processing technologies, locomotion control and interconnectivity over IoT technologies, are similar across the board (Cheng et al. 2023).

Lastly, current agricultural machinery, which includes tractors, agricultural robots and a broad array of equipment, is often powered by fossil fuels. With Agriculture 5.0, electric or hybrid machinery, such as hybrid electric vehicles, is expected to become more prevalent, reducing emissions by up to 17% and improving sustainability (Ghobadpour et al. 2022). Despite their benefits, there are concerns about the actual implementation of these green technologies on farms and fields as they require long charging times, demand a large amount of energy in short periods, can be costly, and contain low-durability batteries (Ghobadpour et al. 2022). However, further research into other energy sources, such as using hydrogen fuel cells, is currently underway, with a significant reduction of costs expected by 2030 (Gonzalez-de-Soto et al. 2016; Taloba and Rayan 2025). The advances made in this regard in the automotive sector can translate to serve other sectors, including agriculture.

6 Virtual or Digital Agriculture

Virtual or digital agriculture has been increasingly applied to digitally or virtually represent a product (crop), field or farm (sometimes this is called creating a "digital model" "digital ghost" or "digital twin" depending on the level of feedback applied between the digital and physical versions). The digital agriculture framework is organized into four layers: (1) the foundation layer, which comprises the sensing devices and cameras, (2) the application layer, which consists of automation devices, machines and robots, and (3) the decision layer, which combines physics-based modelling, AI, and data integration strategies (Naikwade et al. 2021). These layers allow for the estimation of outcomes, evaluation of scenarios and effective connection and feedback between the virtual and physical realms.

6.1 Big Data and Digital Twins in Agriculture 5.0

Farmers often collect data on a range of factors including the status of crops and livestock, sometimes on a daily basis. However, with the improvement of technologies such as computer-aided design, IoT, sensors, AI and robotics, there are significantly more tools and data that facilitate collection and integration, not only daily but hourly, or even by the minute. Agricultural Big Data sources such as sensors for pH, temperature, humidity and light, and satellite data may be used to develop and model virtual objects or systems, known as digital models, ghosts or twins. Digital Twins (DT) are virtual representations of physical entities that can range from a plant or piece of machinery to a field or a process (e.g. harvesting) that benefits from real-time monitoring with the physical object (Corradini et al. 2024; Udugama et al. 2023; Zhang et al. 2025). They rely on the integration of information, physical principles, and communication technologies such as IoT, Big Data, AI, remote sensors and Geographic Information Systems (GIS) (Kim and Heo 2024). DTs and virtualization are expected to be major contributors to the Agriculture 5.0 framework by improving, optimizing and allowing for better decision-making based on the predictions of the virtual model or system. As such, DTs not only virtually represent the current state, but they can model historical states as well as assess hypothetical scenarios or conditions and facilitate research. They have been proposed to improve food safety risks, quality assessments, traceability, logistics and reduce food waste (Singh et al. 2025b).

As it relates to farming and agriculture, DTs have been applied in crop protection to model pest management systems. For example, Dai et al. (2024) integrated an ML model, i.e. random forest coupled with a genetic algorithm, into a DT of controlled growing chambers to predict aphid populations and design pest control interventions on peppers. The model had a prediction accuracy of approximately 86%, with a 20% improvement in economic efficiency (Dai et al. 2024). DTs also show promise in monitoring and predicting the output of very large fields for fruit and vegetable quality and quantity. Kim and Heo (2024) developed a DT of approximately 185,000 hectares of mandarins, by collecting data from multiple different sources (e.g. spatial information and soil components, frequency and time of fertilization, pruning, mulching and spraying, location of crop, fruit quality and weather). The developed DT was able to predict fruit quality, diseases, and pest outbreaks, as well as suggest preventative measures. Lastly, DTs are expected to improve sustainability and environmental impact. For example, it is estimated that 70% of the world's fresh water is used for agriculture, of which approximately 40% is lost to overirrigation (Alves et al. 2019). Alves et al. (2019) developed a DT using data collected from soil sensing probes and the Smart Water Management Platform to aid farmers in streamlining their resources and equipment more efficiently. As such, this system allows farmers to not only improve the quality and yield of crops, but to manage resources, decreasing overall environmental impact, a key target in Agriculture 5.0. However, with most DT, scalability is often a major issue as the accuracy is highly dependent on the quality, amount and availability of the data. Carlo et al. (2025) developed an ML-enhanced DT for improving the accuracy and scalability of an irrigation management system for olive cultivation.

6.2 eXtended Reality (XR) for Linking the Virtual and Physical Space in Agriculture 5.0

XR or eXtended Reality is a combination of Virtual Reality (VR), augmented reality (AR) and mixed reality (MR) technologies to link the virtual and physical spaces (Maddikunta et al. 2022; Fountas et al. 2024). These XR technologies are essential to Industry 5.0, particularly as they relate to remote assistance, navigation, education, training, and drone/UAV control (Maddikunta et al. 2022). XR aims to connect humans to remote, physical spaces virtually, a core concept in Industry 5.0 and, subsequently, Agriculture 5.0. As it relates to agriculture, VR and AR have been used to educate farmers and to provide instructions on maintaining agricultural machinery (Figueredo et al. 2019; de Oliveira and Corrêa 2020). XR technology has been used to improve productivity, safety and quality, reduce cost and ensure environmental sustainability. For example, AR has been used to detect strawberry ripeness by combining AI and IoT technologies for improved classification, to monitor plant growth, and to visualize agrometeorological data of greenhouses in real-time and remotely (Chai et al. 2023; Bigonah et al. 2025). With the use of robots and cobots, physical machines on a farm or field can be accessed remotely to perform a range of tasks, subsequently reducing labour, as explained earlier in this chapter. However, some challenges of XR may include the initial cost of the interacting objects, such as the drones, UAVs or cobots, the time investments for training and understanding how to use the technology, as well as mental and neurological effects concerns regarding long-term use (Bigonah et al. 2025).

7 Smart and Precision Agriculture/Farming

Precision agriculture uses different information technologies to collect data from various agricultural sources, e.g. environmental conditions, soil, weather and/or yields, to inform decisions related to agricultural production. Since precision agriculture's first appearance in the late twentieth century, with the inception of remote sensing, Geographic Information Systems (GIS) and global navigation satellite systems, the integration of IoT, WSN, robotics and Big Data, constitutive of Agriculture 4.0, allowed precision crop monitoring to become more fully realized (Holzinger et al. 2024). These tools are used to improve decision-making, particularly as it relates to optimizing the workforce, water and nutrient use, as well as the application of fertilizers and pesticides (Taha et al. 2025), allowing for more precise crop management and reduced waste. Real-world precision agriculture models have been used for pest detection using machine vision systems and crop farming (Taha et al. 2025).

Each crop has different needs, conditions and technologies required for optimal growth and production (Yost et al. 2017; Xia et al. 2018). As such, in Agriculture 5.0, more individualized crop monitoring strategies will ensure a sustainable future of food while enhancing profitability, productivity and minimizing labour and environmental impacts. Precision farming in Agriculture 5.0 relies on (1) remote and smart sensors, (2) information technologies such as IoT, 6G and distributed ledgers (e.g., Blockchain), (3) AI and ML tools and (4) robots, cobots and automation (Fountas et al. 2024). When these tools are integrated, a more efficient agricultural supply chain can be realized. For example, Holzinger et al. (2024) proposed an

Agriculture 5.0 framework that also incorporates the use of digital twins and human-centred AI to drive decision-making, safety, training and eco-efficiency. Table 2 overviews the research and surveys related to Agriculture 4.0 and 5.0 from 2020–2025 in crops, grains and pulses.

7.1 Smart Pest Management

Pests pose a major threat to crop yields and agricultural productivity, often leading to significant economic losses and a decline in agricultural productivity. Pesticides are commonly required to manage pests. However, the use of pesticides raises concerns about their adverse effects on the environment and human health (Rashid et al. 2010; Raffa and Chiampo 2021). Smart pest monitoring aids in the more precise use of pesticides; thus, it is important in the Agriculture 5.0 concept.

7.2 Smart Crop Management and Harvesting

DL technologies, such as CNN and computer vision systems, are some of the most utilized methods in ensuring crop yield and productivity. Crop management systems should include every part of farming (Naikwade et al. 2021). For example, Chukkapalli et al. (2020) proposed an architecture for a smart farming co-operative ecosystem that connects a multilayered physical system with individual member farms through a cloud-based system. The framework uses AI for training and crop classification, as well as a DT for estimations, predictions and self-corrections. Lastly, smart farming aided by DT can deliver real-time predictions of crop yield and provide solutions to deal with adverse aftermaths such as climate events, improving supply chain monitoring preharvest and postharvest (Guidani et al. 2024).

7.3 Smart Soil Mapping and Analysis

Soil, microbiota and ecological diversity analyses (e.g., moisture, types of microorganisms, beneficial insects) have been shown to correlate with crop yield. In recent years, agroecological food webs have been widely discussed for their potential to ensure agricultural resiliency and offset climate impacts. The intricate interactions between soil microorganisms (bacteria, fungi), beneficial insects and plants can play essential roles in promoting sustainable agriculture and ecological restoration (Malard et al. 2020; Jagadesh et al. 2024). Soil microbiota play an important role in plant health as they regulate soil fertility, cycling of nutrients and biodiversity (Jagadesh et al. 2024). The future of soil analysis in precision farming will incorporate AI and ML to map and predict soil properties, including microorganisms

Table 2 Research and surveys related to Agriculture 4.0 and 5.0 from 2020 to 2025 in crops, grains and pulses

Target industry revolution		Target agriculture revolution		Focused technologies							Application				Food quality and nutrition		References
4.0	5.0	4.0	5.0	Big data	Artificial intelli-gence	Robotics	Bio-inspired	Digital twin	6G and beyond	Others[a]	Energy	Land	Water	Food waste	Quality	Nutrients	
✓		✓		✓		✓				5, 6		✓	✓				Cisternas et al. (2020)
✓	✓	✓	✓		✓	✓				1, 2, 5, 6		✓					Pathan et al. (2020)
		✓	✓	✓	✓	✓				1, 2, 4, 5, 6		✓					Saiz-Rubio and Rovira-Más (2020)
		✓		✓	✓					1, 2, 4, 5, 6		✓					Sishodia et al. (2020)
✓		✓		✓	✓	✓				1, 5, 6	✓	✓		✓	✓		Zhai, et al. (2020)
		✓		✓	✓	✓				1, 2, 4, 5, 6, 7, 8		✓					Araújo et al. (2021)
✓		✓		✓	✓	✓				1, 2, 3, 4, 5, 4, 6, 7	✓	✓					Dayioğlu and Turker (2021)
✓	✓	✓	✓		✓	✓				5, 6					✓		Herzig et al. (2021)
✓		✓		✓	✓	✓				1, 2, 3, 5, 6, 7		✓					Javaid et al. (2022)
✓		✓			✓			✓		1, 2				✓	✓		Melesse et al. (2022)

(continued)

Table 2 (continued)

Target industry revolution		Target agriculture revolution		Focused technologies							Application				Food quality and nutrition		References
4.0	5.0	4.0	5.0	Big data	Artificial intelli-gence	Robotics	Bio-inspired	Digital twin	6G and beyond	Others[a]	Energy	Land	Water	Food waste	Quality	Nutrients	
	✓							✓		–				✓	✓		Onwude et al. (2022)
✓		✓			✓					1, 2, 6		✓			✓		Akkem et al. (2023)
✓	✓	✓	✓	✓	✓			✓		1, 2, 4, 5			✓				Alves et al. (2023)
✓		✓		✓	✓	✓		✓		2		✓			✓		Gallego-García et al. (2023)
		✓	✓		✓	✓				1, 2, 5, 6	✓	✓					Juwono et al. (2023)
		✓		✓	✓					1, 5, 6					✓		Yoosefzadeh et al. (2023)
			✓						✓	1, 2, 4, 5, 6, 7, 9					✓		Polymeni et al. (2023)
			✓	✓	✓	✓		✓		2, 4, 7, 9							Bissadu et al. (2024)
		✓				✓				1, 2, 5, 4, 6, 7		✓	✓		✓		Bongomin et al. (2024)
			✓	✓	✓					1, 8					✓	✓	Fang (2024)
		✓	✓	✓	✓	✓			✓	1, 2, 7, 8							Fountas et al. (2024)

✓	✓	✓	✓	✓	✓			✓		2	✓	✓	✓	✓	✓	✓	Guidani et al. (2024)
✓	✓	✓	✓	✓	✓	✓	✓	✓	✓	1, 2, 4, 5, 6	✓	✓					Haloui et al. (2024)
✓	✓	✓	✓	✓	✓	✓		✓		1, 2, 3, 4, 5, 6, 7	✓	✓					Holzinger et al. (2024)
✓		✓	✓	✓	✓	✓				1, 2, 4, 6, 7, 9							Maraveas et al. (2024)
✓	✓	✓	✓	✓	✓	✓				1, 2, 4, 5, 6, 7		✓					Victor et al. (2024)
			✓	✓	✓		✓			1, 5					✓	✓	Darlan et al. (2025)
✓	✓	✓	✓	✓	✓	✓			✓	1, 2, 4, 5, 6		✓	✓		✓	✓	Taha et al. (2025)

Adapted from Haloui et al. 2024, Guidani et al. 2024, and information reviewed in this paper

[a]Includes technologies such as (1) Deep Learning (DL), (2) Internet of Things (IoT), (3) 3D printing, (4) cloud computing, (5) remote sensing, (6) Unmanned Aerial Vehicles (7) blockchain, (8) Large Language Models (LLM) or Natural Language Processing (NLP), (9) quantum computing

and water management. For example, Barel et al. (2023) tested six ML models to understand the microbial trophic interactions constituting the microbiomes of soils. The study found that understanding the microbial interactions of the soil was vital for ecosystem sustainability. As AI, ML and DL methods improve, more precise management of soil properties is expected. CNNs can be used to process soil or satellite images to identify phenotypic patterns and adsorption behaviour, while self-learning soil sensors can predict moisture content, pH and temperature (Minasny and McBratney 2025). ML models, used to map soil micronutrient health, such as the available minerals and organic matter, which have direct impacts on crop yields, human and animal health, are also being researched (Dasgupta et al. 2023; Chaudhry et al. 2024). When these data sources are integrated into DTs, improved land stewardship, crop resiliency, sustainability and crop yield are expected to increase, contributing to the Society 5.0 intentions. Therefore, the use of AI, ML and DL will be necessary for soil analysis and mapping to ensure future agricultural sustainability within Agriculture 5.0.

8 Breeding 5.0 Within the AI-Driven Revolution to Improve Crop Production Efficiency and Nutrition

Plant breeding has tremendously increased crop yields over decades and created much more disease-resistant crops. More recently, advanced genetic technologies have been used in conjunction with traditional plant breeding methods and techniques to further improve yields, disease resistance, environmental resilience, and sustainability of crop cultivars (Khan et al. 2024). With an ongoing need to increase production and with accumulating global challenges, such as climate change, improvements to plant breeding effectiveness remain important. Plant breeding has evolved from phenotypic trait selection to genotype selection for performance improvements (Breeding 3.0) to recombinant DNA technologies and targeted gene editing (e.g., Clustered Regularly Interspaced Short Palindromic Repeats (CRISPR)-Cas9 system), enabling more precise and more rapid plant breeding advances (Breeding 4.0) (Fang 2024). However, contemporary crop cultivars often fall short of their theoretical maximum potential (Khan et al. 2024). As such, Breeding 5.0 aims to integrate technologies such as genetic information and editing technologies, high-throughput phenotypic analysis, AI, ML, Big Data and other data analysis tools to optimize breeding, precision and efficiency along with precision and data-driven agronomy to realize new potentials (Fang 2024). This has the potential to propel agricultural sustainability and ongoing production gains.

The use of AI, ML and data analysis is crucial in the Breeding 5.0 paradigm as these technologies will improve the analysis and interpretation of increasingly large amounts of genetic and phenotypic data. They will allow for the extraction of important information from complex datasets to drive decision-making on which genotypic and phenotypic traits should be selected for yield improvement (Singh and Prasad 2021; Khan et al. 2022). For example, crop production is limited by the availability of the resources needed for photosynthesis and plant metabolism and suffers when these resources are not available. As such, improving these processes through genetic engineering will allow for crop production to be more resilient even when resources are less abundant (e.g., droughts), allowing plants to grow and thrive with fewer inputs. Modifying, or "hacking", photosynthesis is a novel strategy to improve the efficiency of cropped plants through the modification of photosynthetic processes. By using genetic modification techniques to manipulate the size of the antennae complexes in photosynthetic organs, or broaden the spectrum of light that can be absorbed, photosynthesis and plant efficiency may be improved (Croce et al. 2024; Nazari et al. 2024). DL and AI have been used to integrate environmental and genotypic data to estimate, predict and design crop ideotypes to improve chlorophyll content or respond to climate conditions (Farooq et al. 2024). The use of LLMs for designing proteins through the creation of the first de novo gene editor, OpenCRISPR-1, which was trained on >1 million CRISPR operons, was studied by Ruffolo et al. (2024). As such, LLMs show promise in condensing and analysing large amounts of genomic data, which can also be applied for gene editing and crop enhancement, allowing for the production of more resilient plants through genetic modification. For example, an LLM gene function prediction model can be used to understand the pathways of differently expressed genes in plants under salt stress (Huang et al. 2023). Breeding 5.0 will allow for more precise and customized breeding approaches, crucial for precision management based on the individual needs of the farms, while also considering environmental conditions.

8.1 *Genomic Selection to Improve Nutrition, Nutrition Management and Pest Resistance*

Advances in high-throughput remote sensing and plant "omics" have allowed scientists to address problems using multidimensional datasets and ML for prebreeding strategies, boosting genetic diversity and for marker-assisted selection (Farooq et al. 2024). With over seven million germplasm accessions, from >1750 global gene banks, the use of AI-driven predictive genomics shows significant promise in allowing for testing, selection and utilization of these genomic big data sources (McCouch et al. 2020; Farooq et al. 2024). Globally, it is estimated that 2 billion people suffer from nutritional deficiencies, with insufficient macronutrients and micronutrients in their diets (Darlan et al. 2025). The future of breeding programs may lie in the ability to utilize AI for targeted metabolomics to boost plant nutrition outputs, so that they are more nutritious than prior cultivars (Ferrão et al. 2023). For example, vitamins, minerals, starch, and proteins can be selected for improved nutrition. Additionally, similar methods can be used to choose cultivars and genomic regions with enhanced pest resistance, reducing the need for the overuse of pesticides. Lastly, precision nutrient management can improve the future of plant breeding, where AI can be used to select and predict the crop, genetic regions and nutrient mixture that will optimize yield. For example, Darlan et al. (2025) proposed a SmartBerry AI-based classification to improve the nutritional management of strawberries during cultivation. The model aimed to assess the gaps in the current datasets, literature and methodologies for sustainable nutrient management, reducing the need for fertilizer, while simultaneously improving crop yield (Darlan et al. 2025). If genomic methods are also used to select for crop or genetic regions to enhance crop phenotypes, crops may be able to be grown with fewer nutritional inputs, while maintaining or improving nutritional outputs. As such, this will not only improve production and yield, but ensure that foods are more nutrient dense, for a healthier population.

9 Opportunities in and Barriers to Agriculture 5.0

The transition from Agriculture 4.0 to Agriculture 5.0 is expected to generate opportunities for improving food security, environmental sustainability and human well-being, while alleviating productivity, unemployment and social crises (Taha et al. 2025). The evolution from the Agriculture 4.0 era to Agriculture 5.0 is necessary because of challenges in Agriculture 4.0, including increased front-end costs, the aging agricultural workforce, lack of human-machine interactions and issues with environmental sustainability. Agriculture 5.0 aims to usher in change in terms of how agriculture is performed by reintroducing humans into the food supply chain, integrating human workers with intelligent machines, while also improving profitability, efficiency and sustainability. This has the potential to provide job opportunities and improved job satisfaction as agriculture could finally be more generally perceived as a highly technical sector, requiring a trained and skilled workforce, as opposed to a low-status and unskilled sector which is labour-focused (Humayun 2021). As such, there may be a net increase in people working in the agri-food sector. Additionally, the use of Industry 5.0 technologies allows for the customization of every detail in the agriculture supply chain by enabling the tailoring of crop cultivars to given environments, and improving the crops to meet more specific and varied end-use, to meet customers' needs. With increased climate events projected in the future, crop production needs to be resilient to excessive heat, water scarcity and irregular seasons (Fraser and Campbell 2019; Mesías-Ruiz et al. 2023). Increased nutrition is also anticipated in the Agriculture 5.0 framework, contributing to Society 5.0 by allowing consumers to understand and participate more closely in the food supply chain. The customization potential in Agriculture 5.0 can revolutionize the way that food is produced and the way that customers interact with their food supply chain. Figure 3 proposes a framework for the application of Agriculture 5.0 on farms.

However, many challenges may arise during the transition to Agriculture 5.0 that may prevent these ideals from being fully realized. As will be discussed below, policy and regulatory frameworks are needed to channel these technologies' potential into human and environmental outcomes. Another challenge is data

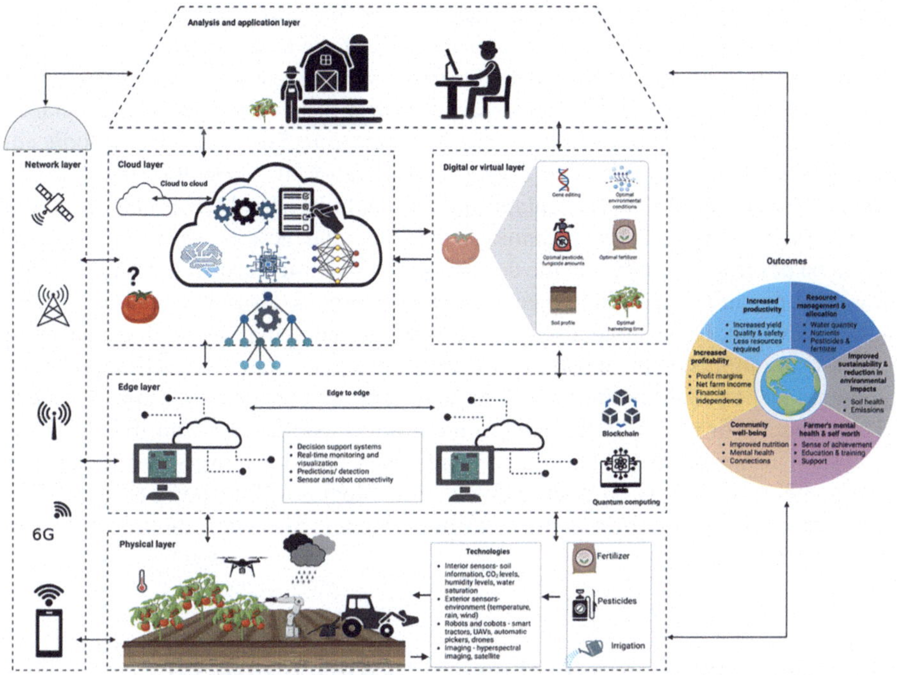

Fig. 3 An Agriculture 5.0 framework on a farm/field, including inputs, layers, connecting technologies and outputs (figure created with Biorender). (Adapted from Gupta et al. 2020, Singh et al. 2025b, and information reviewed in this paper)

management, particularly as it relates to data security, privacy and access control. As we transition towards more digitized computation, there are increased risks of privacy leaks, unsafe and nonsecure human-machine interactions, data breaches and hacking (Maddikunta et al. 2022). Data authentication and integrity are essential for the secure connection among billions of devices, particularly when transferring data over third-party networks (Maddikunta et al. 2022). The use of AI and ML for automation requires that the data being used is trusted and protected, as proper decision-making and algorithm execution are essential for operation. However, with increased interactions between machines and humans, significant ethical issues such as ensuring human privacy and human data rights protection are at the forefront of discussions (Vesnic-Alujevic et al. 2020). Blockchain technology may allow for secure IoT transactions by encrypting peer-to-peer communications, providing a higher level of protection for data by decentralizing agricultural operations (Taha et al. 2025). Quantum technologies also show promise in Agriculture 5.0 due to their potential for enhanced data security and ultra-fast data processing and transmission (Taha et al. 2025). However, the adoption of new technologies requires the development of new skills and competencies (Mourtzis 2024). For example, the use of Blockchain and quantum computing technologies may require extensive upskilling for those already working in agriculture, including farmers. As such, not only are investments needed for the upfront costs of the technologies and tools, but the costs associated with education and training should be considered. Some in the agriculture sector, including farmers, may also show instances of anxiety, reluctance and confusion towards Agriculture 5.0. How these technologies are introduced, adopted and supported will need careful consideration if Agriculture 5.0 is to be realized (Mishra et al. 2024). Scalability and flexibility should also be considered when designing agricultural technology systems, and new (or evolved) systems should be developed to adapt to the continuously changing needs required for future growth. For example, a digital object may not perform as desired once implemented in the real world. LLMs may also be susceptible to hallucinations or the creation of false data or outputs that fit a pattern rather than being accurate. As such, appropriate validation steps should be implemented, and care and consideration must be taken when selecting or choosing the data to input into models to prevent data bias. As the system also expands to incorporate more nodes, additional users or even a larger farm or field, the ability to integrate new technologies and respond to increasing data volumes is essential for continuous improvement and innovation (Tombe and Smuts 2023).

10 Opportunities in and Barriers to Agriculture 5.0

Future research should consider the regulatory framework and ethical considerations for the implementation of Agriculture 5.0 (Fig. 4). For example, policies and guidelines should be discussed to ensure data security, privacy, inclusiveness and fair market competition, while creating a system that benefits all stakeholders, integrating technology without compromising ethical values (Fountas et al. 2024). As Agriculture 5.0 aims to use Industry 5.0 technologies to reintroduce humans into agriculture (Society 5.0) and integrate cobots and smart machines, regulations are essential (Lima et al. 2019). For example, policies and regulations should foster the creation of environments that support the deployment of Industry 5.0 technologies for agricultural purposes, such as the use of drones or cobots in farming. Additionally, quality standards and laws should be in place for the use of AI and ML technologies, data curation and data handling to ensure standardization, promoting an increased adoption, adaptation, management and application of these technologies in the fields. A robust and secure framework for Agriculture 5.0 should also include the implementation of distributed ledgers (e.g., Blockchain), and quantum computing technologies for data preservation, faster processing, and improved end-to-end encryption to ensure security (and sovereignty) of data.

If a new (and technologically advanced) agricultural system only rewards the agricultural sector (including farmers) for producing more food, then the system will only be geared toward production gains, rather than the more holistic range of goals of Society 5.0, such as greater biodiversity, carbon sequestration, the maintenance of rural communities, etc. In the reality of narrow-margin agriculture, farmers

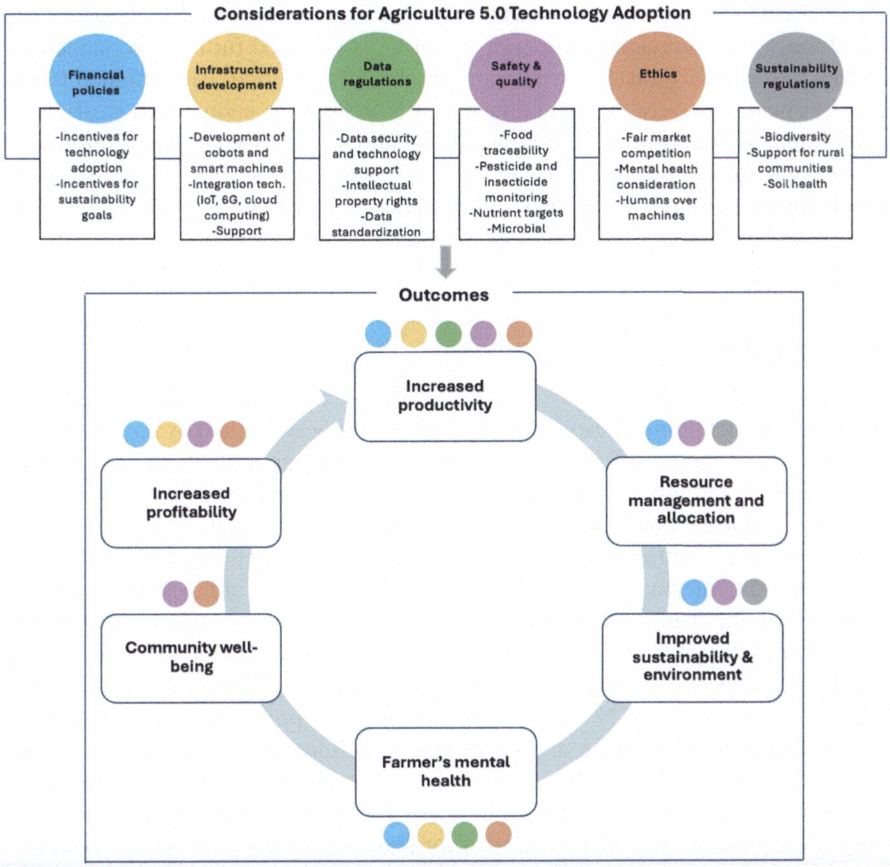

Fig. 4 Considerations for Agriculture 5.0 technology adoption and the social, economic and environmental outcomes

commonly find themselves renting land, often at high prices, in unstable markets, to expand their crop acreage to be profitable and competitive (Mosby et al. 2020). These narrow margins can hinder the adoption of technologies and can reduce the focus on broader goals (beyond production) (Mosby et al. 2020). Additionally, there are also concerns about data ownership of data generated on farms. Often, companies or developers own the data that farmers are generating on their own farms, and farmers may have to pay to access their own data. Policies to protect and incentivize farmers may be an important part of achieving Agriculture 5.0.

Regional adaptation of Agriculture 5.0 technologies, such as remote sensors, IoT and AI, is pivotal for the successful implementation in the field. Agricultural solutions should be tailored depending on the local environment and the needs of each individual farm. This may depend on the region (from an agro-ecological perspective) and the jurisdiction (from a policy and economic perspective). Agriculture

4.0 saw the advancement and adoption of production and precision technologies in agriculture with a production imperative, but it also created further disconnections between farmers and consumers, and consumers and their food. Agriculture 5.0 may usher in a new era where the mental health of both farmers, people who work in the agriculture and food sector, and consumers can be considered. For example, the use and implementation of these technologies may improve land stewardship for farmers (as an example) while also incentivizing, via the high-technology nature of Agriculture 5.0, the next generation to consider a career in agriculture.

11 Conclusion

Throughout agricultural history, food production has been limited by environmental conditions (e.g., land, soil, water, and weather), the amount of labour (human, draught animal and mechanical), and the technologies used to produce food. However, while environmental and labour conditions are increasingly constrained, new technologies are creating new opportunities for food production and sustainability while addressing the burgeoning issue of food security. Agriculture 5.0 aims to generate opportunities for farming and agriculture by reducing constraints. This chapter defines Agriculture 5.0 as evolving from the Agriculture 4.0 concept, by integrating Industry 5.0 and Society 5.0, using an array of new and emerging technologies such as sensors, IoT, WSN and AI to improve crop production and sustainably while taking a human-centric approach. In Agriculture 5.0, advancements in 6G-IoT technologies, AI and robotics (cobots) are expected to be at the forefront. Virtualization of agriculture (e.g., DTs) will link virtual and physical spaces to improve crop production and sustainability and expand possibilities for research. However, while Agriculture 5.0 shows substantial promise to improve agriculture, its barriers and challenges, including data privacy and security, upfront costs and adaptation and adoption of new high-technology systems, must be considered in order for Agriculture 5.0 to happen. Policies that support and incentivize farmers and the agricultural industry will also be important to create favourable environments for the deployment of Industry 5.0 technologies. Agriculture 5.0, if implemented, has the capacity to create a frame shift from scarcity to abundance within our food systems.

Acknowledgments The authors gratefully acknowledge support from the Natural Science and Engineering Research Council of Canada (NSERC) Discovery program (RY - RGPIN-2018-04598). Funding for this chapter, including the research from which these insights were derived, was provided by the Arrell Family Foundation, the SMART Training Program, Food from Thought (University of Guelph's Canada First Research Excellence Fund), and the Social Science and Humanities Research Council of Canada (SSHRC).

Competing Interests The authors declare no conflicts of interest.

References

Abdul Razak SF, Yogarayan S, Sayeed MS, Mohd Derafi MIF (2024) Agriculture 5.0 and explainable AI for smart agriculture: a scoping review. Emerg Sci J 8(2):744–760. https://doi.org/10.28991/ESJ-2024-08-02-024

Akilan T, Baalamurugan KM (2024) Automated weather forecasting and field monitoring using GRU-CNN model along with IoT to support precision agriculture. Expert Syst Appl 249(PA):123468. https://doi.org/10.1016/j.eswa.2024.123468

Akkem Y, Biswas SK, Varanasi A (2023) Smart farming using artificial intelligence: a review. Eng Appl Artif Intell 120(January):105899. https://doi.org/10.1016/j.engappai.2023.105899

Alves RG, Souza G, Maia RF, Tran ALH, Kamienski C, Soininen JP et al (2019) Digital twin for smart farming. In: 2019 IEEE global humanitarian technology conference (GHTC). IEEE, pp 1–4. https://doi.org/10.2174/9789815196177124010004

Alves RG, Maia RF, Lima F (2023) Development of a Digital Twin for smart farming: irrigation management system for water saving. J Clean Prod 388:135920. https://doi.org/10.1016/j.jclepro.2023.135920

Araújo SO, Peres RS, Barata J, Lidon F, Ramalho JC (2021) Characterising the agriculture 4.0 landscape—emerging trends, challenges and opportunities. Agronomy 11(4):1–37. https://doi.org/10.3390/agronomy11040667

Arnal Barbedo JG (2019) Plant disease identification from individual lesions and spots using deep learning. Biosyst Eng 180(2016):96–107. https://doi.org/10.1016/j.biosystemseng.2019.02.002

Barel JM, Petchey OL, Ghaffouli A, Jassey VEJ (2023) Uncovering microbial food webs using machine learning. Soil Biol Biochem 186:109174. https://doi.org/10.1016/j.soilbio.2023.109174

Baryshnikova N, Altukhov P, Naidenova N, Shkryabina A (2022) Ensuring global food security: transforming approaches in the context of Agriculture 5.0. IOP Conf Ser: Earth Environ Sci 988(3):0–6. https://doi.org/10.1088/1755-1315/988/3/032024

Bigonah M, Jamshidi F, Pant A, Poudel S, Nallapareddy SR, Langroudi AC, Marghitu D (2025) A systematic review of extended reality (XR) Technologies in Agriculture and Related Sectors (2022-2024). IEEE Access 13(February):49721–49734. https://doi.org/10.1109/ACCESS.2025.3550891

Bissadu KD, Sonko S, Hossain G (2024) Society 5.0 enabled agriculture: drivers, enabling technologies, architectures, opportunities, and challenges. Inf Process Agric 12(April):112–124. https://doi.org/10.1016/j.inpa.2024.04.003

Bongomin O, Lamo J, Guina JM, Okello C, Ocen GG, Obura M et al (2024) UAV image acquisition and processing for high-throughput phenotyping in agricultural research and breeding programs. Plant Phenom J 7(1). https://doi.org/10.1002/ppj2.20096

Carlo Tancredi GP, Preite L, Vignali G (2025) Digital twin enhanced with machine learning algorithms for irrigation management using sensor data. Procedia Comput Sci 253:2419–2427. https://doi.org/10.1016/j.procs.2025.01.302

CE (2025) Artificial Intelligence (AI) vs. Machine Learning. Retrieved April 25, 2025, from https://ai.engineering.columbia.edu/ai-vs-machine-learning/

Chai JJK, Xu J-L, O'Sullivan C (2023) Real-time detection of strawberry ripeness using augmented reality and deep learning. Sensors 23(17):7639. https://doi.org/10.3390/s23177639

Chataut R, Nankya M, Akl R (2024) 6G networks and the AI revolution—exploring technologies, applications, and emerging challenges. Sensors 24(6):1–29. https://doi.org/10.3390/s24061888

Chaudhry H, Vasava HB, Chen S, Saurette D, Beri A, Gillespie A, Biswas A (2024) Evaluating the soil quality index using three methods to assess soil fertility. Sensors 24(3):864. https://doi.org/10.3390/s24030864

Chen H, Chen A, Xu L, Xie H, Qiao H, Lin Q, Cai K (2020) A deep learning CNN architecture applied in smart near-infrared analysis of water pollution for agricultural irrigation resources. Agric Water Manag 240(April):106303. https://doi.org/10.1016/j.agwat.2020.106303

Cheng C, Fu J, Su H, Ren L (2023) Recent advancements in agriculture robots: benefits and challenges. Machines 11(1):1–24. https://doi.org/10.3390/machines11010048
Chukkapalli SSL, Mittal S, Gupta M, Abdelsalam M, Joshi A, Sandhu R, Joshi K (2020) Ontologies and artificial intelligence systems for the cooperative smart farming ecosystem. IEEE Access 8:164045–164064. https://doi.org/10.1109/ACCESS.2020.3022763
Cisternas I, Velásquez I, Caro A, Rodríguez A (2020) Systematic literature review of implementations of precision agriculture. Comput Electron Agric 176:105626. https://doi.org/10.1016/j.compag.2020.105626
Clegg J, Lucas R (2020) Three agricultural revolutions. South Atl Q 119(1):95–111. https://doi.org/10.1215/00382876-8007677
Corradini G, Homez-jara AK, Chen C (2024) Virtualization and digital twins of the food supply chain for enhanced food safety. In: Advances in food and nutrition research, vol 111. Elsevier. https://doi.org/10.1016/bs.afnr.2024.06.001
Croce R, Carmo-Silva E, Cho YB, Ermakova M, Harbinson J, Lawson T et al (2024) Perspectives on improving photosynthesis to increase crop yield. Plant Cell 36(10):3944–3973. https://doi.org/10.1093/plcell/koae132
Cubero S, Marco-noales E, Aleixos N, Barbé S, Blasco J (2020) Robhortic: a field robot to detect pests and diseases in horticultural crops by proximal sensing. Agriculture (Switzerland) 10(7):1–13. https://doi.org/10.3390/agriculture10070276
Dai M, Shen Y, Li X, Liu J, Zhang S, Miao H (2024) Digital twin system of pest management driven by data and model fusion. Agriculture (Switzerland) 14(7):1099. https://doi.org/10.3390/agriculture14071099
Darlan D, Ajani OS, An JW, Bae NY, Lee B, Park T, Mallipeddi R (2025) SmartBerry for AI-based growth stage classification and precision nutrition management in strawberry cultivation. Sci Rep 15(1):1–10. https://doi.org/10.1038/s41598-025-97168-z
Dasgupta S, Debnath S, Das A, Biswas A, Weindorf DC, Li B et al (2023) Developing regional soil micronutrient management strategies through ensemble learning based digital soil mapping. Geoderma 433:116457. https://doi.org/10.1016/j.geoderma.2023.116457
Dayioğlu MA, Turker U (2021) Digital transformation for sustainable future - agriculture 4.0: a review. J Agric Sci 27(4):373–399. https://doi.org/10.15832/ankutbd.986431
de Oliveira ME, Corrêa CG (2020) Virtual reality and augmented reality applications in agriculture: a literature review. In 22nd symposium on virtual and augmented reality (SVR), Porto de Galinhas, Brazil, pp 1–9. https://doi.org/10.1109/SVR51698.2020.00017
Fang J (2024) Breeding 5.0 AI-driven revolution in designed plant breeding. Mol Plant Breed 15(1):27–33. https://doi.org/10.5376/mpb.2024.15.0004
Farooq MA, Gao S, Hassan MA, Huang Z, Rasheed A, Hearne S et al (2024) Artificial intelligence in plant breeding. Trends Genet 40(10):891–908. https://doi.org/10.1016/j.tig.2024.07.001
Feng T, Shen H, Yang X, Nianga J, Wang Z (2024) Integration of large language models with IoT in smart agriculture to improve efficiency, yield, and quality. Ind Sci Eng 1(4):15–35. http://www.aeph.press/uploadfile/202408/29eca53f55b1.pdf
Ferrão LFV, Dhakal R, Dias R, Tieman D, Whitaker V, Gore MA et al (2023) Machine learning applications to improve flavor and nutritional content of horticultural crops through breeding and genetics. Curr Opin Biotechnol 83:102968. https://doi.org/10.1016/j.copbio.2023.102968
Figueredo V, dos Reis AV, Garcia F, Araújo FC (2019) Virtual reality for agribusiness in the development of a maintenance simulator for agricultural machinery for senar goias. In: 2019 21st symposium on virtual and augmented reality (SVR). IEEE, pp 17–19. https://doi.org/10.1109/SVR.2019.00019
Fountas S, Espejo-García B, Kasimati A, Gemtou M, Panoutsopoulos H, Anastasiou E (2024) Agriculture 5.0: cutting-edge technologies, trends, and challenges. IT Professional 26(1):40–47. https://doi.org/10.1109/MITP.2024.3358972
Fraser EDG (2011) Can economic, land use and climatic stresses lead to famine, disease, warfare and death? Using Europe's calamitous 14th century as a parable for the modern age. Ecol Econ 70(7):1269–1279. https://doi.org/10.1016/j.ecolecon.2010.02.010

Fraser EDG, Campbell M (2019) Agriculture 5.0: reconciling production with planetary health. One Earth 1(3):278–280. https://doi.org/10.1016/j.oneear.2019.10.022

Gallego-García S, Gallego-García D, García-García M (2023) Sustainability in the agri-food supply chain: a combined digital twin and simulation approach for farmers. Procedia Comput Sci 217:1280–1295. https://doi.org/10.1016/j.procs.2022.12.326

Ganesh Kumar G, Kanakaprabha S (2025) Intelligent robots and Cobots. In: Ramasamy V, Balamurugan S, Peng S-L (eds) Intelligent robots and cobots. Wiley, pp 131–157. https://doi.org/10.1002/9781394198252.ch7

Ghobadpour A, Monsalve G, Cardenas A, Mousazadeh H (2022) Off-road electric vehicles and autonomous robots in agricultural sector: trends, challenges, and opportunities. Vehicles 4(3):843–864. https://doi.org/10.3390/vehicles4030047

Gonzalez-de-Soto M, Emmi L, Benavides C, Garcia I, Gonzalez-de-Santos P (2016) Reducing air pollution with hybrid-powered robotic tractors for precision agriculture. Biosyst Eng 143:79–94. https://doi.org/10.1016/j.biosystemseng.2016.01.008

Guidani B, Ronzoni M, Accorsi R (2024) Virtual agri-food supply chains: a holistic digital twin for sustainable food ecosystem design, control and transparency. Sustain Prod Consum 46:161–179. https://doi.org/10.1016/j.spc.2024.01.016

Gupta M, Abdelsalam M, Khorsandroo S, Mittal S (2020) Security and privacy in smart farming: challenges and opportunities. IEEE Access 8:34564–34584. https://doi.org/10.1109/ACCESS.2020.2975142

Haloui D, Oufaska K, Oudani M, El Yassini K (2024) Bridging industry 5.0 and agriculture 5.0: historical perspectives, opportunities, and future perspectives. Sustainability (Switzerland) 16(9):1–27. https://doi.org/10.3390/su16093507

Herzig P, Borrmann P, Knauer U, Klück H-C, Kilias D, Seiffert U et al (2021) Evaluation of RGB and multispectral unmanned aerial vehicle (UAV) imagery for high-throughput phenotyping and yield prediction in barley breeding. Remote Sens 13(14):2670. https://doi.org/10.3390/rs13142670

Holzinger A, Fister I, Fister I, Kaul HP, Asseng S (2024) Human-centered AI in smart farming: toward Agriculture 5.0. IEEE Access 12(April):62199–62214. https://doi.org/10.1109/ACCESS.2024.3395532

Hrushikesava Raju S, Nagabhushana Rao M, Sudheer N, Kavitharani P (2018) Agri-Iot: a sustainable environment for improvement of crops revenue in the field of agriculture using Iot. Int J Eng Technol 7(2.32):439. https://doi.org/10.14419/ijet.v7i2.32.15735

Hsieh TH, Kiang JF (2020) Comparison of CNN algorithms on hyperspectral image classification in agricultural lands. Sensors (Switzerland) 20(6). https://doi.org/10.3390/s20061734

Huang J, Lin Q, Fei H, He Z, Xu H, Li Y et al (2023) Discovery of deaminase functions by structure-based protein clustering. Cell 186(15):3182–3195.e14. https://doi.org/10.1016/j.cell.2023.05.041

Humayun M (2021) Industrial Revolution 5.0 and the role of cutting edge technologies. Int J Adv Comput Sci Appl 12(12):10.14569/IJACSA.2021.0121276

Itakura K, Saito Y, Suzuki T, Kondo N, Hosoi F (2019) Estimation of citrus maturity with fluorescence spectroscopy using deep learning. Horticulturae 5(1):1–9. https://doi.org/10.3390/horticulturae5010002

Jagadesh M, Dash M, Kumari A, Singh SK, Verma KK, Kumar P et al (2024) Revealing the hidden world of soil microbes: metagenomic insights into plant, bacteria, and fungi interactions for sustainable agriculture and ecosystem restoration. Microbiol Res 285:127764. https://doi.org/10.1016/j.micres.2024.127764

Javaid M, Haleem A, Singh RP, Suman R (2022) Enhancing smart farming through the applications of agriculture 4.0 technologies. Int J Intell Netw 3(September):150–164. https://doi.org/10.1016/j.ijin.2022.09.004

John DA, Babu GR (2021) Lessons from the aftermaths of green revolution on food system and health. Front Sustain Food Syst 5:644559. https://doi.org/10.3389/fsufs.2021.644559

Juwono FH, Wong WK, Verma S, Shekhawat N, Lease BA, Apriono C (2023) Machine learning for weed–plant discrimination in agriculture 5.0: An in-depth review. Artif Intell Agric 10:13–25. https://doi.org/10.1016/j.aiia.2023.09.002

Khan MHU, Wang S, Wang J, Ahmar S, Saeed S, Khan SU et al (2022) Applications of artificial intelligence in climate-resilient smart-crop breeding. Int J Mol Sci 23(19):11156. https://doi.org/10.3390/ijms231911156

Khan N, Choi S-H, Lee C-H, Qu M, Jeon J-S (2024) Photosynthesis: genetic strategies adopted to gain higher efficiency. Int J Mol Sci 25(16):8933. https://doi.org/10.3390/ijms25168933

Kim S, Heo S (2024) An agricultural digital twin for mandarins demonstrates the potential for individualized agriculture. Nat Commun 15(1):1–15. https://doi.org/10.1038/s41467-024-45725-x

Li M, Zhang Z, Lei L, Wang X, Guo X (2020) Agricultural greenhouses detection in high-resolution satellite images based on convolutional neural networks: comparison of faster R-CNN, YOLO v3 and SSD. Sensors (Switzerland) 20(17):1–14. https://doi.org/10.3390/s20174938

Lima F, De Carvalho CN, Acardi MBS, Dos Santos EG, De Miranda GB, Maia RF, Massote AA (2019) Digital manufacturing tools in the simulation of collaborative robots: towards Industry 4.0. Braz J Oper Prod Manag 16(2):261–280. https://doi.org/10.14488/BJOPM.2019.v16.n2.a8

Maddikunta PKR, Pham QV, Prabadevi B, Deepa N, Dev K, Gadekallu TR et al (2022) Industry 5.0: a survey on enabling technologies and potential applications. J Ind Inf Integr 26(August 2021):100257. https://doi.org/10.1016/j.jii.2021.100257

Mahmood NH, Böcker S, Moerman I, López OA, Munari A, Mikhaylov K et al (2021) Machine type communications: key drivers and enablers towards the 6G era. EURASIP J Wirel Commun Netw 2021(1):134. https://doi.org/10.1186/s13638-021-02010-5

Malard JJ, Adamowski JF, Rojas Díaz M, Nassar JB, Anandaraja N, Tuy H et al (2020) Agroecological food web modelling to evaluate and design organic and conventional agricultural systems. Ecol Model 421:108961. https://doi.org/10.1016/j.ecolmodel.2020.108961

Maraveas C, Rajarajan M, Arvanitis KG, Vatsanidou A (2024) Cybersecurity threats and mitigation measures in agriculture 4.0 and 5.0. Smart Agric Technol 9:100616. https://doi.org/10.1016/j.atech.2024.100616

Martos V, Ahmad A, Cartujo P, Ordoñez J (2021) Ensuring agricultural sustainability through remote sensing in the era of agriculture 5.0. Appl Sci (Switzerland) 11(13):5911. https://doi.org/10.3390/app11135911

McCouch S, Navabi ZK, Abberton M, Anglin NL, Barbieri RL, Baum M et al (2020) Mobilizing crop biodiversity. Mol Plant 13(10):1341–1344. https://doi.org/10.1016/j.molp.2020.08.011

Melesse TY, Bollo M, Di Pasquale V, Centro F, Riemma S (2022) Machine learning-based digital twin for monitoring fruit quality evolution. Procedia Comput Sci 200(3rd International Conference on Industry 4.0 and Smart Manufacturing):13–20. https://doi.org/10.1016/j.procs.2022.01.200

Mendoza-Bernal J, González-Vidal A, Skarmeta AF (2024) A convolutional neural network approach for image-based anomaly detection in smart agriculture. Expert Syst Appl 247:123210. https://doi.org/10.1016/j.eswa.2024.123210

Mesías-Ruiz GA, Pérez-Ortiz M, Dorado J, de Castro AI, Peña JM (2023) Boosting precision crop protection towards agriculture 5.0 via machine learning and emerging technologies: a contextual review. Front Plant Sci 14:1143326. https://doi.org/10.3389/fpls.2023.1143326

Minasny B, McBratney AB (2025) Machine learning and artificial intelligence applications in soil science. Eur J Soil Sci 76(2):e70093. https://doi.org/10.1111/ejss.70093

Mishra N, Bhandari N, Maraseni T, Devkota N, Khanal G, Bhusal B et al (2024) Technology in farming: unleashing farmers' behavioral intention for the adoption of agriculture 5.0. PLoS One 19(8):e0308883. https://doi.org/10.1371/journal.pone.0308883

Mosby I, Rotz S, Fraser ED (2020) Chapter 6: Milk. In: Uncertain harvest: the future of food on a warming planet. University of Regina Press, pp 139–162. https://doi.org/10.1525/abt.2023.85.7.409

Mourtzis D (2024) Challenges and opportunities of the transition from Industry 4.0 to Industry 5.0. In: Manufacturing from Industry 4.0 to Industry 5.0. Elsevier, pp 97–131. https://doi.org/10.1016/B978-0-443-13924-6.00004-1

Naikwade RR, Patle BK, Joshi VS, Pagar ND, Hirwe SB (2021) Agriculture 5.0: future of smart farming. In: National conference on innovative global technology trends in art, design, technology, management, vedic science, education and architecture, film & media, pp 1–6. Retrieved from https://www.researchgate.net/publication/362125676

Nazari M, Kordrostami M, Ghasemi-Soloklui AA, Eaton-Rye JJ, Pashkovskiy P, Kuznetsov V, Allakhverdiev SI (2024) Enhancing photosynthesis and plant productivity through genetic modification. Cells 13(16):1319. https://doi.org/10.3390/cells13161319

Nguyen DC, Ding M, Pathirana PN, Seneviratne A, Li J, Niyato D et al (2022) 6G Internet of Things: a comprehensive survey. IEEE Internet Things J 9(1):359–383. https://doi.org/10.1109/JIOT.2021.3103320

Onwude D, Bahrami F, Shrivastava C, Berry T, Cronje P, North J et al (2022) Physics-driven digital twins to quantify the impact of pre- and postharvest variability on the end quality evolution of orange fruit. Resour Conserv Recycl 186:106585. https://doi.org/10.1016/j.resconrec.2022.106585

Panarin RN, Khvorova LA (2022) Software development for agricultural tillage robot based on technologies of machine intelligence. In: Jordan V, Tarasov I, Faerman V (eds) High-performance computing systems and technologies in scientific research, automation of control and production. HPCST 2021, Communications in Computer and Information Science. Springer International Publishing, pp 354–367. https://doi.org/10.1007/978-3-030-94141-3_28

Pathan M, Patel N, Yagnik H, Shah M (2020) Artificial cognition for applications in smart agriculture: a comprehensive review. Artif Intell Agric 4:81–95. https://doi.org/10.1016/j.aiia.2020.06.001

Polymeni S, Plastras S, Skoutas DN, Kormentzas G, Skianis C (2023) The impact of 6G-IoT technologies on the development of Agriculture 5.0: a review. Electronics (Switzerland) 12:2651. https://doi.org/10.3390/electronics12122651

Quealy S, Lynch PJ, Hasan N (2022) Agriculture 4.0. In: Mottram T (ed) Digital agritechnology. Academic Press, pp 219–241. https://doi.org/10.1016/B978-0-12-817634-4.00011-2

Raffa CM, Chiampo F (2021) Bioremediation of agricultural soils polluted with pesticides: a review. Bioengineering 8(7):92. https://doi.org/10.3390/bioengineering8070092

Rashid B, Husnain T, Riazuddin S (2010) Herbicides and pesticides as potential pollutants: a global problem. In: Ashraf M, Ozturk M, Ahmad M (eds) Plant adaptation and phytoremediation. Springer Netherlands, Dordrecht, pp 427–447. https://doi.org/10.1007/978-90-481-9370-7_19

Ruffolo JA, Nayfach S, Gallagher J, Bhatnagar A, Beazer J, Hussain R et al (2024) Design of highly functional genome editors by modeling the universe of CRISPR-Cas sequences. BioRxiv 2024–04. https://doi.org/10.1101/2024.04.22.590591

Saiz-Rubio V, Rovira-Más F (2020) From smart farming towards agriculture 5.0: a review on crop data management. Agronomy 10(2):207. https://doi.org/10.3390/agronomy10020207

Scanes CG (2018) The neolithic revolution, animal domestication, and early forms of animal agriculture. In: Scanes CG, Toukhsati SR (eds) Animals and human society. Elsevier, pp 103–131. https://doi.org/10.1016/B978-0-12-805247-1.00006-X

Shahriar S, Corradini MG, Sharif S, Moussa M, Dara R (2025) The role of generative artificial intelligence in digital agri-food. J Agric Food Res 20:101787. https://doi.org/10.1016/j.jafr.2025.101787

Siddiqui AF, Paul AJ, Mishra S, Balamurugan S (2025) Cobots for Industry 5.0 transformation. In: Ramasamy V, Balamurugan S, Peng S-L (eds) Intelligent robots and cobots. Wiley, pp 1–17. https://doi.org/10.1002/9781394198252.ch1

Singh RK, Prasad M (2021) Big genomic data analysis leads to more accurate trait prediction in hybrid breeding for yield enhancement in crop plants. Plant Cell Rep 40(10):2009–2011. https://doi.org/10.1007/s00299-021-02761-x

Singh M, Zhang M, Espinal-Ruiz M, Rathnayake S, Xue J, Shi J et al (2025a) Maple syrup adulteration: fluorescence fingerprints as a source of information for enhanced detection. J AOAC Int:qsaf029. https://doi.org/10.1093/jaoacint/qsaf029

Singh M, Wang K, Chen C, Corradini MG (2025b) Luminescence-based techniques to monitor pre-, post-harvest and processing changes in fresh produce and derivatives: principles, applications and trends. Trends Food Sci Technol, 162(May), 105077. https://doi.org/10.1016/j.tifs.2025.105077

Sishodia RP, Ray RL, Singh SK (2020) Applications of remote sensing in precision agriculture: a review. Remote Sens 12(19):1–31. https://doi.org/10.3390/rs12193136

Spoeth A (2025) What is Ex-AI (Explainable AI)? Retrieved April 30, 2025, from https://www.theainavigator.com/blog/what-is-ex-ai-explainable-ai

Taha MF, Mao H, Zhang Z, Elmasry G, Awad MA, Abdalla A et al (2025) Emerging technologies for precision crop management towards Agriculture 5.0: a comprehensive overview. Agriculture (Switzerland) 15(6):1–30. https://doi.org/10.3390/agriculture15060582

Taloba AI, Rayan A (2025) Machine learning based on reliable and sustainable electricity supply from renewable energy sources in the agriculture sector. J Radiat Res Appl Sci 18(1):101282. https://doi.org/10.1016/j.jrras.2024.101282

Tombe R, Smuts H (2023) Society 5.0-inspired digitalization framework for resilient and sustainable agriculture. EPiC Ser Comput 93:216–227. https://doi.org/10.29007/xc5q

Udugama IA, Kelton W, Bayer C (2023) Digital twins in food processing: a conceptual approach to developing multi-layer digital models. Digit Chem Eng 7:100087. https://doi.org/10.1016/j.dche.2023.100087

Verma KK, Song X, Kumari A, Jagadesh M, Singh SK, Bhatt R et al (2025) Climate change adaptation: challenges for agricultural sustainability. Plant Cell Environ 48(4):2522–2533. https://doi.org/10.1111/pce.15078

Vesnic-Alujevic L, Nascimento S, Pólvora A (2020) Societal and ethical impacts of artificial intelligence: critical notes on European policy frameworks. Telecommun Policy 44(6):101961. https://doi.org/10.1016/j.telpol.2020.101961

Victor N, Maddikunta PKR, Mary DRK, Murugan R, Chengoden R, Gadekallu TR et al (2024) Remote sensing for agriculture in the era of industry 5.0—a survey. IEEE J Sel Top Appl Earth Obs Remote Sens 17:5920–5945. https://doi.org/10.1109/JSTARS.2024.3370508

Xia JA, Yang YW, Cao HX, Han C, Ge DK, Zhang WY (2018) Visible-near infrared spectrum-based classification of apple chilling injury on cloud computing platform. Comput Electron Agric 145(December 2017):27–34. https://doi.org/10.1016/j.compag.2017.12.012

Yoosefzadeh Najafabadi M, Hesami M, Eskandari M (2023) Machine learning-assisted approaches in modernized plant breeding programs. Genes 14(4):777. https://doi.org/10.3390/genes14040777

Yost MA, Kitchen NR, Sudduth KA, Sadler EJ, Drummond ST, Volkmann MR (2017) Long-term impact of a precision agriculture system on grain crop production. Precis Agric 18(5):823–842. https://doi.org/10.1007/s11119-016-9490-5

Zhai Z, Martínez JF, Beltran V, Martínez NL (2020) Decision support systems for agriculture 4.0: survey and challenges. Comput Electron Agric 170:105256. https://doi.org/10.1016/j.compag.2020.105256

Zhang S, Liu J, Guo H, Qi M, Kato N (2020) Envisioning device-to-device communications in 6G. IEEE Netw 34(3):86–91. https://doi.org/10.1109/MNET.001.1900652

Zhang R, Zhu H, Chang Q, Mao Q (2025) A comprehensive review of digital twins technology in agriculture. Agriculture 15(9):903. https://doi.org/10.3390/agriculture15090903

Agriculture 4.0 and Transformation in Smallholder Farming Systems

Thomas Daum

1 Introduction

Across the world, 510 million farms—representing 85% of all farms—operate on less than two hectares of land (Lowder et al. 2021). While smallholder farming can be associated with decent livelihoods, it is often characterized by poverty and food insecurity (e.g., Rapsomanikis 2015; Castaneda et al. 2016) and two-thirds of the world's extreme poor are smallholder farmers (Fuglie et al. 2019). These farms must be at the heart of efforts aimed at achieving the sustainable development goals (e.g., Lowder et al. 2025).

There are large efforts to promote agricultural transformation in smallholder farming systems to improve the livelihoods of farming households as well as generate positive spillover effects to the broader (rural) economy (e.g., de Janvry and Sadoulet 2019; Diao et al. 2024). Agricultural transformation aims to "get agriculture moving" with the help of technologies and practices that raise land and labor productivity and initiate a "process during which the agricultural sector develops from a low-productivity, subsistence-oriented sector to a modern, commercially oriented one" (Diao et al. 2024, p. iii).

While agricultural transformation is already underway—or high on the policy agenda—in parts of the Global South, transformation efforts are confronted by a range of economic, institutional, political, and biophysical barriers (e.g., de Janvry and Sadoulet 2019; Fuglie et al. 2019). In addition, technology-heavy, Green-Revolution-style transformation visions are increasingly questioned for their ability to contribute to sustainable and inclusive development (e.g., Holt-Giménez and Altieri 2013). There are high hopes that Agriculture 4.0 can contribute to addressing

T. Daum (✉)
School of Global Studies, University of Gothenburg, Gothenburg, Sweden
e-mail: thomas.daum@gu.se

F. da Silveira et al. (eds.), *Agriculture 4.0*,
https://doi.org/10.1007/978-3-032-24552-6_4

barriers to transformation and make it more sustainable and inclusive (e.g., CTA 2019; Kim et al. 2020; Schroeder et al. 2021).

There are two alternative conceptions of Agriculture 4.0. The first equates Agriculture 4.0 with the rise of digital agriculture, that is, the use of digital technologies along the agricultural value chain (Birner et al. 2021). At the farm level, this conception is linked to "smart" and "precision" farming and informed by concepts such as Industry 4.0., cyber-physical systems and Internet of Things (Birner et al. 2021). Digital agriculture can combine data sensing, data processing and automation, drawing on rapid developments in big data analytics, machine learning, predictive modeling, and artificial intelligence, as well as agricultural robots and drones (e.g., da Silveira et al. 2021). However, digital agriculture can take different forms in less mechanized farming systems, as discussed below, as well as also target the upstream and downstream agricultural value chain (Birner et al. 2021; Abate et al. 2023). The second conception views Agriculture 4.0 as encompassing a wider range of "game-changing" innovations including biotechnologies such as gene editing and synthetic food production—whose spread is framed within the narrative of a "Fourth Agricultural Revolution" (Klerkx and Rose 2020; da Silveira et al. 2021; Rose et al. 2021; Barrett and Rose 2022). This chapter focuses on the first conception.

Proponents argue that (digital) Agriculture 4.0 can help to address barriers to agricultural transformation and make it more sustainable and inclusive (e.g., Kim et al. 2020; Schroeder et al. 2021). CTA (2019) argues that digital agriculture "can be a game-changer in supporting and accelerating agricultural transformation" and does so "sustainably and inclusively" (p. 17). However, there are also concerns that Agriculture 4.0 can lead to a digital divide where smallholder farmers are left behind, can shift power relations where their data ends up in the hands of big companies, and can undermine more agroecological approaches on their farms (e.g., Abdulai 2022; Hidalgo et al. 2024; Milian Gómez and Byttebier 2025). Beyond these optimistic and pessimistic views, skeptics argue that the promises of Agriculture 4.0 for smallholder farmers might be overstated, arguing that it offers limited practical value, has so far failed to scale, and cannot address more structural barriers to (inclusive) transformation (e.g., Parlasca et al. 2022; Abate et al. 2023; Abdulai et al. 2023).

In this chapter, I explore the potential of Agriculture 4.0 for addressing barriers to agricultural transformation in smallholder farming systems as well as its potential and risks for sustainability and inclusiveness. In Sect. 2, I unpack the concept of Agriculture 4.0, showing that it comes in different forms and with different capabilities. In Sect. 3, I examine the entry points through which Agriculture 4.0 might address barriers to transformation. In Sect. 4, I discuss potential risks and limitations and identify governance needs. Section 5 concludes.

2 Disentangling Agriculture 4.0

As noted above, this chapter focuses on Agriculture 4.0 as digital agriculture. Following Daum et al. (2022) and Njuguna et al. (2025), one could classify Agriculture 4.0 in smallholder farming systems into three categories: (1) tools for farmers to optimize production; (2) tools connecting farmers with other actors along the value chain, in the agricultural innovation system, and from the broader institutional environment; (3) tools for other actors in the value chain, the agricultural innovation systems, and from the broader institutional environment, which may indirectly affect farmers. Table 1 provides an overview of the different categories of Agriculture 4.0 in smallholder farming systems, which are discussed in more detail in the subsequent sections.

2.1 Tools for Farmers to Optimize Production

One can distinguish four types of digital tools related to improving farm activities: first, "simple" tools, which provide generic information; second, "smart" tools, which offer farm-tailored information based on data entered by farmers or workers; third, "smart and connected" tools, which use data from sensors; and fourth, "automated" tools, which integrate data sensing, analytics, and automation (Table 1). "Simple," "smart," and "smart and connected" tools are mainly disembodied, relying only on software and apps—and computers and (smart)phones. Since these tools are not embodied in physical farm hardware, they are highly applicable even in less mechanized smallholder farming systems. In contrast, "automated" tools rely on sensors and actuators that are physically integrated into farm machinery or equipment.

So far, most tools are "simple" tools, which require no or limited farm-level data such as apps that provide farmers with market price updates, weather forecasts, or generic extension advice. Such tools are "low hanging fruits" (Aker et al. 2016) as they require no technologies other than mobile- or smartphones—which are spreading rapidly—and limited digital literacy. The costs of scaling such tools are low since all users receive the same information; however, this can make it difficult to find financially sustainable business models for providers (Birner et al. 2021).

"Smart" tools use "humans as sensors" to provide more farm-tailored advice without the need for sensors such as those embodied in farm technologies (Rajkhowa and Qaim 2021; Daum et al. 2022). Smart digital tools could be described as a second generation of digital tools for farm optimization in the context of smallholder farming. Such tools are also "low hanging fruits" as they require no tools other than mobile- or smartphones. However, their potential to provide farm-specific information hinges on farmers' capacity and willingness to enter sufficiently high-quality data (Bateki et al. 2021; Daum et al. 2022).

Table 1 Classification of Agriculture 4.0 in smallholder farming systems

Category		Type	Main source of farm data	Main capability	Main users	Examples
Tools for farmers to optimize production	*Simple*	Disembodied	None	Generic information	Farmers	Digital market updates, weather forecasts, generic extension advice
	Smart	Disembodied	People (i.e., farmers, workers)	Farm-specific information	Farmers	Decision support tools using "humans as sensors" (e.g., tools to optimize feeding based on data entered by livestock keepers)
	Smart and connected	Disembodied/embodied	Sensors	Farm-specific information	Farmers	Decision support tools using real sensors (e.g., tools advising on soil fertility management based on soil scans, tools suggesting grazing routes based on satellite maps)
	Automated	Embodied	Sensors	Farm-specific information, automation	Farmers	Automated irrigation system
Tools connecting farmers		Disembodied	None, people, sensors	Actor-integrating	Farmers, other actors	Uber for tractors, digital marketplaces, mobile payments, peer-to-peer extension
Tools for other actors		Disembodied/embodied	None, people, sensors	Multiple	Other actors	Digital credit management, digital supply chain management, e-governance

Adopted from Daum et al. (2022) and Njuguna et al. (2025)

"Smart and connected" tools use data from remote sensors (e.g., satellite imaging) and/or proximal sensors (e.g., sensors embedded in machinery or hand-held devices). While tools relying on remote sensors come with little or no requirements for technology and technological capacity at the farm level, tools based on proximal sensors can require more advanced equipment and skills. These challenges can be addressed through institutional solutions such as service provision or cooperative models (Njuguna et al. 2025).

"Automated" tools are physically integrated into farm machinery or equipment such as tractors, irrigation systems, and milking machines. Sensors are used to collect "big" data, which is then processed to control actuators that, for example, adjust application rates in input sprayers and spreaders or regulate valves in irrigation systems. These tools are closely related to concepts such as "smart" and "precision" farming and linked to approaches such as the Internet of Things. In most cases, "automated" tools have certain processes automated (e.g., input application) but still require human operators (e.g., tractor drivers). Full automation such in the form of robotic milking systems and automated irrigation that operate without direct human intervention are still rare. While some "automated" tools, such as irrigation systems (Nigussie et al. 2020), are becoming more common in certain smallholder farming systems, the use of embodied forms of Agriculture 4.0 is more limited in smallholder contexts due to low levels of mechanization, which constrains the use of sensors and actuators. The recent "drone revolution" in agriculture, which has advanced rapidly in parts of the Global South, is mainly associated with the broadcast application of inputs and drones are mostly operated manually - although there are also some examples of variable-rate input application (Belton et al. 2025).

2.2 Tools Connecting Farmers

Some digital tools aim to reduce the time, costs, and risks for farmers when interacting with upstream value chain segments related to the provision of inputs and services (e.g., improved seeds, finance) and downstream segments related to trading, processing, retailing, and marketing (Daum 2019; Abate et al. 2023). On the upstream side, Uber-type tools, for example, seek to lower transaction costs in mechanization service markets to better connect smallholder farmers with service providers (Daum et al. 2021). On the downstream side, digital marketing platforms aim to link farmers with traders or consumers (Kim et al. 2020; Abate et al. 2023). There are also tools designed to improve connections between farmers and extension officers (e.g., Fu and Akter 2016; van Campenhout et al. 2017) and between farmers and peers (Khan et al. 2025).

2.3 Tools for Other Actors

The focus so far has been on tools used directly by farmers. However, there is a wide range of digital tools not designed for farmers but for other actors along the value chain and within agricultural innovation systems. These tools can have an equal or even greater impact on agricultural transformation. A wide range of tools exists to help improve the work by private value chain actors such as input manufactures, agro-dealers, banks, insurance companies, mechanization service providers, traders, processors, and retailers. For example, input manufactures can use Quick Response (QR) codes or Radio-Frequency Identification (RFID) tags using electromagnetic fields to track input packages and containers for quality assurance (Birner et al. 2021), banks can use digital credit scoring systems (Kim et al. 2020), and mechanization service providers can deploy fleet management systems (Daum et al. 2021). On the downstream side, traders, processors, and retailers may adopt blockchain technology and other digital platforms for supply chain management, traceability, and certification (e.g., Kos and Kloppenburg 2019; Steinke et al. 2024). Agriculture 4.0 tools can also be used in the context of contract farming and cooperatives. Daum et al. (2022) studied "Herdman," which aims to improve the performance of veterinarians and artificial insemination agents hired by private dairy companies and cooperatives to work with their farmers. Another area of use is agricultural research, education, and extension. For example, Agriculture 4.0 creates new opportunities for data collection and analysis in crop and animal breeding, while in extension, a wide range of tools exists to help public extension agents provide better advice to farmers (e.g., Oyinbo et al. 2022). Lastly, e-governance tools are a highly relevant area, for example, e-vouchers to manage input subsidy programs, e-diaries to monitor field staff, and satellites for land titling and environmental monitoring (e.g., Daum 2019; Namyenya et al. 2022).

3 Entry Points for Agriculture 4.0 to Address Barriers to Transformation

Agricultural transformation is confronted with a wide range of barriers (e.g., de Janvry and Sadoulet 2019; Fuglie et al. 2019). Table 2 provides an overview of these barriers and lists examples of Agriculture 4.0 that aim to address them, as discussed in more detail in the following subsections. First, farm management can be limited by challenges in accessing timely, relevant information and extension services, as well as missing opportunities for developing new knowledge and skills. Second, the nature of rural and agricultural markets (e.g., high transaction costs) can make it challenging for smallholder farmers to connect to upstream value chain segments related to the provision of inputs and services (e.g., improved seeds, mechanization services). Third, related constraints arise in accessing finance and insurance. Fourth, farmers may face challenges in connecting to downstream value

Table 2 Barriers to transformation and entry points for Agriculture 4.0

Barriers to transformation, related to ...	Examples of entry points for Agriculture 4.0
... farm management	*Tools for farmers to optimize farm production* Digital information (markets, weather) Decision support tools Precision/smart farming *Tools connecting farmers* Peer-to-peer platforms
... production factors (machinery, inputs, labor)	*Digital tools connecting farmers* Uber for tractors Input market platforms Virtual labor markets *Tools for other actors* Digital fleet management Digital quality assurance Digital contract farming
... finance and insurance	*Digital tools connecting farmers* Mobile payments/savings/credits Crowd funding Digital insurance *Tools for other actors* Credit risk management Collateral tracking Loss verification
... output markets	*Digital tools connecting farmers* Marketing platforms Harvest pooling *Tools for other actors* Traceability and certification Digital supply chain management
... institutional framework	*Tools for other actors* Digital extension E-governance Digital tools for AIS

chain segments such as trading, processing, retailing, and marketing. Fifth, there is also a range of more deep-rooted, systemic challenges linked to institutional framework conditions, including limited investment and governance challenges in agricultural research, education, and extension, inadequate rural infrastructure (roads, electrification, irrigation), and unfavorable political conditions - such as market, trade, and land policies. Finally, it is important to note that transformation efforts take place within agroecological and biophysical constraints, such as soil degradation and climate stress. While some of these challenges can be mitigated through improved technologies and practices, others are inherent and difficult to overcome.

3.1 Farm Management

Successful farm management requires that farmers possess adequate knowledge and skills, have access to timely information, and can obtain extension advice when needed (e.g., Fuglie et al. 2019). Ensuring these aspects is often challenging due to farmers' highly localized needs, their geographic dispersion in hard-to-reach areas, and persistent underinvestment and governance issues in public service provision (e.g., Nakasone and Torero 2016). Agriculture 4.0 offers potential solutions by providing tools that directly support farmers' farm management and by enabling other actors to deliver more timely, targeted, and effective services to farmers (see also Sect. 3.5).

There is a wide range of tools for farmers to support their decision-making—ranging from "simple" tools, which provide generic information to "smart" and "smart and connected" tools, which offer more farm-tailored advice (see Sect. 2). "Simple" tools help farmers to make better-informed choices by providing timely information such as market price updates or weather forecasts (e.g., Aker and Ksoll 2016; Etwire et al. 2017; Camacho and Conover 2019). There are also digital tools that aim to enable farmers to share and discuss information with peers using dedicated apps such as WeFarm and social media platforms such as WhatsApp (Khan et al. 2025).

"Smart" tools provide more farm-specific information by using "humans as sensors." Examples include Farmtree in India, which advises farmers on herd management based on regularly entered livestock data (e.g., yields, feed, health); Fodjan in Kenya, which offers recommendations for optimizing livestock feeding based on farmer-entered data; and iCow in Kenya, which provides general animal husbandry advice tailored to the lactation cycle of registered animals (Marwa et al. 2020; Bateki et al. 2021; Daum et al. 2022).

"Smart and connected" tools draw on the use of sensors collecting farm-specific data. Examples are AfriScout in Kenya, which provides pastoralists with vegetation and water maps to support grazing decisions; SoilScanner in Kenya, which provides recommendations on soil fertility management using soil scanners; and Plantix in India, which helps farmers identify pests and diseases through image recognition (Machado et al. 2020; Daum 2025; Njuguna et al. 2025). There is some evidence that such "personalized digital extension service" can help optimize production and increase productivity, thereby contributing to transformation (e.g., Arouna et al. 2020; Marwa et al. 2020; Rajkhowa and Qaim 2021).

"Automated" tools can also play a role in some setting. For example, automated irrigation systems are increasingly popular in some smallholder farming systems (Nigussie et al. 2020). In general, embodied forms of Agriculture 4.0 remain limited in smallholder contexts, however, due to low levels of mechanization.

3.2 Production Factors (Machinery, Inputs, Labor)

Limited access to technologies such as improved seeds, fertilizers, mechanization, and irrigation can undermine agricultural land and labor productivity and have been identified as a key barrier to agricultural transformation (Fuglie et al. 2019). One reason is the high transaction costs in rural markets (Deichmann et al. 2016). Digital tools that aim to connect farmers with other value chain actors by reducing transaction costs have therefore received considerable attention (Deichmann et al. 2016; Sekabira and Qaim 2017). Barrett et al. (2017) argue that such tools "unlock further potential, in particular by making markets more efficient" (p. 21).

In the context of mechanization, Uber-type solutions such as Hello Tractor in Nigeria and EM3 in India promise to better link (smallholder) farmers who cannot afford to buy machinery with mechanization service providers. Daum et al. (2021) find that such tools can indeed help to reduce transaction costs in some setting, making it easier for service providers to work with smallholders. However, they also argue that such tools face a range of thorny challenges related to the nature of rural markets such as the seasonality and synchronicity of agriculture, lacking rural infrastructure, and limited digital ownership, infrastructure, literacy, and trust.

There are also digital tools that aim to improve farmers' access to inputs such as fertilizers and seeds such as Kobiri in Guinea and Lersha in Ethiopia (Abate et al. 2023). Digital tools can also play a role in addressing challenges related to information asymmetries and moral hazard problems in input and service markets that can lead to substandard products and services (e.g., Abate et al. 2023).

3.3 Finance and Insurance

Limited access to financial services is a major barrier to transformation, as it restricts farmers' ability to invest in technologies such as tractors, irrigation, improved storage and inputs like seeds and fertilizers as well as undermines financial transactions between farmers and other value chain actors. Financial institutions often perceive smallholder farmers as high-risk and low-return clients, due to lacking collaterals (e.g., land titles), missing credit histories, and the risky and seasonal nature of agriculture. Various Agriculture 4.0 tools seek to address these constraints. Mobile financial services—most notably M-PESA in Kenya—have attracted significant attention. While they have been successful in reducing the time, costs, and risks associated with payments, with notable benefits for farmers (e.g., Kikulwe et al. 2014; Sekabira and Qaim 2017), their role in improving access to credit remains limited (Abdul-Rahaman and Abdulai 2022; Parlasca et al., 2022).

Some tools do target credit access more directly, such as Enda Tamweel in Tunisia and Agri-Wallet in Kenya (Kim et al. 2020; Abate et al. 2023), however, providing credit in rural areas remains constrained by classic market failures (Daum 2019). Some tools such as FarmDrive in Kenya aim to assist lenders by reducing transaction costs through digital credit scoring systems and risk monitoring (Kim et al. 2020). Other platforms like DrumNet in Kenya or Farmable in Ghana aim to improve financial inclusion by facilitating value chain financing and crowdfunding solutions linked to urban consumers (Bateki et al. 2021).

Agricultural insurance can enhance the resilience of smallholder farmers but delivering insurance to this group remains challenging due to several factors, including covariate risks, adverse selection, moral hazard, and the difficulty of validating claims from geographically dispersed farmers. This is problematic as shocks such as extreme weather events—which will become more frequent and severe with climate change—can significantly affect smallholder farmers, thereby hindering transformation. Agriculture 4.0 offers new opportunities to address some of these constraints. For instance, digital tools can help reduce costs associated with information asymmetries by utilizing satellite data, weather station records, or geo-tagged images for loss verification (e.g., Ceballos et al. 2019), and mobile payment solutions can help to disburse insurance payouts (Baumüller 2018).

3.4 Output Markets

Market barriers related to limited market information and high transaction costs, and poor rural infrastructure have been identified as key barriers to agricultural transformation (e.g., de Janvry and Sadoulet 2019; Fuglie et al. 2019). Digital tools can help to improve farmers' market knowledge such as market prices, thereby allowing them to make better production decision and reducing information asymmetries vis-a-vis buyers (e.g., Muto and Yamano 2009; Jensen 2010; Rajkhowa and Baumüller 2024). In a study in Ghana, Hildebrandt et al. (2020) found that digital market updates helped smallholder farmers to receive 9% higher prices. On a broader level, this can reduce spatial-temporal price differences and increase market efficiency (e.g., Rajkhowa and Baumüller 2024).

There are also digital platforms to better connect farmers with traders or directly with consumers such as Freshfarm in Egypt, Hmizate in Morocco, Farmshine in Kenya, Zowasel in Nigeria, and Farmster in Tanzania (Kim et al. 2020; Abate et al. 2023). Baumüller (2018) highlights the potential of Agriculture 4.0 for pooling farm produce, thereby allowing farmers to improve their bargaining positions. Daum et al. (2022) discuss the potential of Agriculture 4.0 tools for addressing challenges in the management of dairy cooperatives. There is also a wide range of tools for supply chain management, certification, and traceability—mainly in the context of high-value export crops—such as QualiTrace in Ghana, which could help address barriers to integrating smallholder farmers while also creating some new hurdles (Kim et al. 2020; Baumüller et al. 2023; Steinke et al. 2024).

3.5 *Institutional Framework*

Institutional framework conditions such as effective agricultural research, education, and extension services, supportive rural infrastructure (roads, electrification, irrigation), and favorable market, trade, and land policies have been shown to play a key role in facilitating agricultural transformation (e.g., de Janvry and Sadoulet 2019; Fuglie et al. 2019). Agriculture 4.0 cannot contribute to addressing all these aspects; however, it can potentially play an essential role in some of them. For instance, Agriculture 4.0 can play a key role in agricultural extension, which is critical for helping farmers acquire knowledge, develop skills, and learn about new technologies (Fuglie et al. 2019). Digital tools can enable farmers and extension officers to communicate across time and space, including using digital data such as maps, images, and videos (e.g., Fu and Akter 2016; van Campenhout et al. 2017; Voss et al. 2021; Oyinbo et al. 2022). Likewise, Agriculture 4.0 can support agricultural education and vocational training (Yameogo et al. 2024) as well as play a role in agricultural research and development (Birner et al. 2021).

Many of the institutional framework conditions are closely linked to the functioning of public agencies, including those involved in research, education, and extension (see above), as well as other agencies that play a role in agricultural transformation, such as those responsible for land administration, infrastructure development, and environmental monitoring, among others. These organizations can be affected by governance challenges, which Agriculture 4.0 could help address. For example, Agriculture 4.0 tools such as blockchain, e-vouchers, and satellites can improve the governance of state-led input subsidy programs (e.g., Tende et al. 2022) and support public land administration activities (e.g., Huggins and Frosina 2017). Digital tools can also help tackle long-standing governance issues affecting agricultural extension (e.g., Steinke et al. 2021). For example, in Uganda, a pilot project by the Ministry of Agriculture, Animal Industry, and Fisheries is exploring the use of an e-diary incorporating field records, GPS coordinates, and verification photos to monitor mobile extension agents (Namyenya et al. 2022). Likewise, farmers could be empowered to verify and rate extension visits, increasing upward accountability (Aker et al. 2016).

4 Risks, Limits, and Governance Needs

Agriculture 4.0 is the subject of critical debate. The following subsections discuss risks and limitations to better understand the governance needs and measures needed to harness its potential while minimizing its risks. Key concerns include digital divides (Sect. 4.1), data-related issues (Sect. 4.2), inadequate digital infrastructure (Sect. 4.3), a possible lock-in into unsustainable agricultural development pathways (Sect. 4.4), and the limited ability to address deeper structural barriers (Sect. 4.5).

4.1 Digital Divides

Proponents of Agriculture 4.0 argue that it can make agricultural transformation more inclusive. Section 3 has shown that there is indeed a wide range of entry points for Agriculture 4.0 to provide smallholder farmers with better access to information, markets, and services, among other benefits. However, there are also risks that smallholder farmers may be less able to afford digital tools such as smartphones and apps, may lack the digital literacy needed to use them effectively, or may be unable to harness their potential due to persistent structural barriers—as compared to larger, wealthier, better-educated, and more connected farmers (Aker et al. 2016; Camacho and Conover 2019; Coggins et al. 2022). If this is the case, Agriculture 4.0 could lead to a digital divide, benefiting large farms to the disadvantage of small ones. This pattern is not unprecedented: historically, many new technologies—for example, Green Revolution technologies—often benefited large-scale farmers more than smallholder farmers (e.g., Pingali 2012).

Accompanying policies and investments can play a key role in ensuring that Agriculture 4.0 reduces rather than increases divides. In the case of the above-mentioned Green Revolution, many negative side effects could have been avoided with better institutional framework conditions (Pingali 2012). To make Agriculture 4.0 more inclusive, the three C's—connectivity, content, and capacity—are essential (Aker et al. 2016; Nakasone and Torero 2016). Connectivity refers to the availability, accessibility, and affordability of digital tools and infrastructure (Daum 2019; Coggins et al. 2022). Governments play a key role in the provision of digital infrastructure and making sure it is set up in ways to ensure inclusiveness. For example, using mechanisms such as Universal Service and Access Funds can correct the tendency of private network providers to concentrate on more densely populated areas (Birner et al. 2021). A digital divide can also be addressed by adapting the content—for example, by using audio-visual elements (Coggins et al. 2022). Lastly, smallholder farmers need the capacity to use such tools, including in critical ways, which requires digital literacy (Coggins et al. 2022; Daum 2025).

While the three C are essential, Sect. 3 also suggests that smallholder farmers can benefit from Agriculture 4.0 even if they do not own modern digital devices themselves (Fu and Akter 2016; Oyinbo et al. 2022). For example, Hello Tractor works with booking agents to reach farmers who do not own mobile- or smartphones and who may lack digital literacy (Daum et al. 2021). Operating semi-digital business models allows to reach farmers who do not own modern Information and Communication Technologies (ICTs) but also means that the full potential of using digitalization is not fulfilled yet—an aspect that may change in the future with growing digital ownership and digital literacy.

4.2 Data Sovereignty

Agriculture 4.0 raises many concerns related to data security and control (e.g., Fraser 2019). In extreme cases, these issues can increase farmers' dependency due to path dependencies and "digital lock-ins," affecting power relations within value chains. While such concerns are most pronounced with data-heavy forms of (automated) Agriculture 4.0 in the Global North, there is also evidence of similar risks in smallholder farming systems (Abdulai 2022; Daum et al. 2022). Ensuring data sovereignty helps guarantee that farmers retain ownership of their data and can decide whether to keep it private or share it, which can enable private companies, public organizations, or data cooperatives to use the obtained big data to improve inputs and services (Daum et al. 2022). Governments can help ensure data sovereignty by establishing clear data ownership regulations, promoting digital literacy, and requiring interoperability and data portability so that farmers can easily transfer their data between different tools to avoid lock-ins (e.g., Birner et al. 2021). Chichaibelu et al. (2023) suggest that more efforts are needed to ensure data sovereignty in smallholder farming systems.

Another concern related to Agriculture 4.0 is that it can lead to market concentration in upstream and downstream agricultural value chains due to self-reinforcing processes such as competitive advantages by big data and network effects (Birner et al. 2021). In the Global North, Agriculture 4.0 is driven primarily by companies that manufacture agricultural inputs (e.g., BASF, Bayer, Syngenta) and machinery (e.g., AGCO, John Deere), which increasingly form joint ventures to strengthen their ability to collect and leverage big data (Birner et al. 2021). In the Global South, these multinational corporations are active as well, but there is also a vibrant start-up culture suggesting substantial competition (Birner et al. 2021). Nevertheless, tendencies toward market concentration must be carefully monitored and antitrust regulation can be an essential tool to address excessive market concentration (Birner et al. 2021). Moreover, alternative ways of implementing Agriculture 4.0 by farmers themselves can provide an important counterbalance—either individually or collectively, for example, through data cooperatives (e.g., Fraser 2022).

4.3 Techno-Fixes

There are two different visions for the development of smallholder farming: sustainable intensification and agroecology (e.g., Bernard and Lux 2017; Mockshell and Kamanda 2018). Sustainable intensification is generally open to the use of modern technologies and market integration to increase yield on existing lands. This approach is linked to Green Revolution-style agricultural transformation, with a strong focus on genetic improvements of crops and livestock, albeit with a much stronger emphasis on environmental sustainability, particularly through practices that promote input efficiency and substitution and the adoption of "good agricultural practices" (Pretty 2008; Pretty and Bharucha 2014). Sustainable intensification is compatible with the mainstream idea of agricultural transformation as developing agriculture to become "modern, commercially oriented" (Diao et al., p. iii). In this vision, many farmers eventually leave for more productive sectors, while the remaining farms expand and rely more on capital (Barrett et al. 2017).

In contrast, agroecology advocates for a paradigm shift in agriculture, grounded in ecological principles, traditional knowledge, and the strengthening of local value chains (Altieri 2002; Wezel et al. 2009). Agroecology can be understood primarily as a science or a set of practices—focused on how to best manage farms. This more apolitical understanding of agroecology partly overlaps with the idea of sustainable intensification, which also encourages agricultural practices such as crop diversity, intercropping, and agroforestry (e.g., Wezel et al. 2015; Mockshell and Kamanda 2018). In turn, political agroecology is highly skeptical of modern technologies and market integration, which is seen as conflicting with food sovereignty and locally driven, inclusive transformation—an understanding that is at odds with sustainable intensification and agricultural transformation (Holt-Giménez and Altieri 2013; Wezel et al. 2009).

While Agriculture 4.0 can take many different forms and is not inherently aligned with any specific pathway, many of its manifestations are more closely aligned with the paradigm of sustainable intensification, as it is linked to a techno-optimistic perspective that focuses on enhancing or optimizing—but not eliminating—external inputs and assumes continued market engagement (e.g., Abdulai 2022). As such, Agriculture 4.0 is also confronted with critical views warning that it reinforces industrial farming models that are perceived by some as harmful for smallholders and sustainability (e.g., Holt-Giménez and Altieri 2013; Abdulai 2022) as well as undermine traditional knowledge and skills (e.g., Hidalgo et al. 2024; Daum 2025). However, Agriculture 4.0 can take a variety of different forms as shown by the emerging literature on "digital agroecology" (e.g., Wittman et al. 2020). In smallholder farming systems, tools such as soil scanners for monitoring soil health, mobile apps for plant health or digital platforms for farmer knowledge sharing can support agroecological practices and knowledge and skills development if well designed (e.g., Daum 2025; Njuguna et al. 2025). Moreover, many concerns related to Agriculture 4.0 are addressable with governance mechanisms (see Sects. 4.1 and 4.2)

4.4 Limitation to Address Structural Barriers

A key concern is also that the attention on Agriculture 4.0 distracts from the need to tackle deeper, structural ("analog") barriers to agricultural transformation. Many challenges such as limited investment in rural infrastructure (e.g., roads, electrification, irrigation) and agricultural research, education, and extension cannot be addressed by digital tools, although there might be some scope to use Agriculture 4.0 to improve the governance of public sector organizations (see Sect. 3.5.). Some of the above-mentioned structural barriers can also undermine both the uptake and the effectiveness of Agriculture 4.0. For example, a tool like iCow, which provides livestock farmers with recommendations on how to optimize feeding, may have little practical value if farmers cannot access the recommended feeds (Bateki et al. 2021). Similarly, while the Uber-for-tractors tool Hello Tractor can connect farmers with tractor owners, adequate roads remain essential so that tractors can actually reach farms (Daum et al. 2021). Hence, it is not surprising that Lio and Liu (2006) find that the impacts of digital agriculture are approximately twice as large in high-income countries relative to low-income countries, where these barriers are more significant (see also Camacho and Conover 2019). Bundling different services within digital tools can help to simultaneously address multiple binding constraints—for example, tools that combine agronomic advice, credit provision, input marketplaces, and market platforms to sell outputs (e.g., Bateki et al. 2021).

4.5 Digital Infrastructure

Digital infrastructure, such as affordable and reliable mobile networks and satellite coverage, is essential for enabling Agriculture 4.0 (CTA 2019; Birner et al. 2021; Schroeder et al. 2021). Governments also play a key role in ensuring an enabling business environment for the development and scaling of Agriculture 4.0. For this, investments into knowledge and skills development not only for farmers but for all agricultural value chain actors as well as those working in agricultural innovation systems, such as researchers, software developers, and entrepreneurs are essential (CTA 2019; Birner et al. 2021; Schroeder et al. 2021). Likewise, public investments in the creation of publicly accessible datasets such as soil maps, feed composition databases, and weather records can help to support developers of Agriculture 4.0 and ensure that not only large companies are able to develop solutions (CTA 2019; Birner et al. 2021; Schroeder et al. 2021). In addition, governments can help to ensure streamlined registration processes, good access to finance, protection of intellectual property rights, and support for cross-border business operations (CTA 2019; Birner et al. 2021; Schroeder et al. 2021).

5 Future Research

Significant knowledge gaps persist on Agriculture 4.0 and smallholder transformation. Table 3 groups possible research avenues into four themes: (i) adoption and diffusion; (ii) supply-side trends and business models; (iii) sustainability impacts, structural barriers, and complementarity; and (iv) governance and policy implications. For each theme, example research questions are presented. The list of research questions is illustrative and meant to spark ideas, rather than being exhaustive.

Table 3 Possible future research themes and questions for research on Agriculture 4.0 and smallholder transformation

Themes	Examples of key research questions
Adoption and diffusion	How is the adoption and diffusion of different types of Agriculture 4.0 ("simple," "smart," "smart and connected," "automated") by (i) geography/region, (ii) agro-ecological zones, (iii) farming systems, (iv) value chains, and (v) smallholder typologies (e.g., farm size, location, age, gender)? What factors (economic, social, agroecological, behavioral, institutional) drive or hinder the adoption of Agriculture 4.0? Who adopts which types of tools and who is left behind? How do different factors intersect to create potential digital divides? Which structural bottlenecks constrain the scaling of Agriculture 4.0? How do digital literacy, trust, and social networks influence uptake? Which models (e.g., service provision, cooperative solutions, social enterprises, public-private partnerships, public programs) are best suited to make Agriculture 4.0 inclusive? How is the adoption and diffusion of Agriculture 4.0 in upstream and downstream segments of agricultural value chains? How is the adoption and diffusion of e-governance tools that focus on transparency, accountability, and service delivery in agricultural research, education, extension, subsidy programs, land management, and regulation/monitoring?
Supply-side trends and business models	Which providers—input companies, machinery manufacturers, processors and supermarkets, startups—deliver Agriculture 4.0 to smallholder farmers? What incentives do they face, and what are the implications for business models, design, and impacts? Which models (e.g., service provision, cooperative solutions, social enterprises, public-private partnerships, public programs) work best in which context (e.g., Uber-type platforms in seasonal, low-density rural markets)? Which business models—e.g., value-chain-funded "free" provision, subscriptions, pay-per-use/transaction fees, or freemium—are being used? Which models work best for which tool type - "simple," "smart," "smart and connected," and "automated"? In what ways does the nature of smallholder farming and rural areas influence business models and product/service design? To what extent can hybrid (digital-analog) solutions address smallholders' needs and the thorny challenges of rural markets? How should business models and tool design look like to include smallholders and address challenges such as gender inequality?

(continued)

Table 3 (continued)

Themes	Examples of key research questions
Sustainability impacts, structural barriers, and complementarity	How do different categories of tools ("simple," "smart," "smart and connected," "automated") affect farm productivity, profitability, food security, rural livelihoods, labor, gender relations, environmental sustainability? Does Agriculture 4.0 make agricultural transformation more or less inclusive for smallholder farmers? How does this differ depending on tool types, models, business models, design, complementary policy actions, and contextual conditions? How do smallholder systems evolve under alternative digitalization pathways (inclusive vs. concentrated)? How can digital tools contribute to agroecological transitions (soil health, biodiversity, water efficiency)? Do they lead to input use efficiency or substitution—or do they lock farmers into input-intensive agriculture and create rebound effects? How do smallholders engage with digital tools, data, and advice (e.g., decision support advice)? What unintended behaviors or dependencies do digital tools create (e.g., overreliance on algorithmic advice)? How do "analog" enablers—roads, electrification, connectivity, mechanization—as well as knowledge and market access affect uptake and effectiveness of digital tools? Can bundled tools (e.g., advice, input markets, finance access, market linkage) overcome multiple simultaneous constraints to transformation? Which structural barriers (e.g., infrastructure, extension capacity) can Agriculture 4.0 realistically help to address?
Governance and policy implications	What combination of public investments and market incentives best supports inclusive digital infrastructure and ensures the "three C's"—connectivity, content, and capacity? What are the implications of Agriculture 4.0 for agricultural innovation systems (AIS)? What changes are needed to support Agriculture 4.0 such as through capacity building and the creation and stewardship of digital public goods (e.g., public databases on soils and weather records)? How does Agriculture 4.0 affect power dynamics in value chains—e.g., by dis/empowering smallholders or consolidating control by agribusinesses? How can competition and antitrust policies prevent monopolization? What data is collected, who owns it, who can access it, and can it be transferred? What are the risks related to data security, privacy, and sovereignty? How can ownership and portability of data and interoperability between tools be safeguarded? What role can farmer cooperatives or "data commons" play in countering digital lock-ins? How can regulation balance innovation with protection of data sovereignty, competition, and other goals such as inclusivity (smallholders, gender)? What mechanisms ensure responsible innovation in designing and deploying Agriculture 4.0? What cybersecurity risks exist, and which low-cost governance and technical measures best reduce these risks? Which complementary policies (e.g., environment, health, labor) are needed to make sure that Agriculture 4.0 contributes to sustainability gains? To what extent can Agriculture 4.0 e-governance tools address governance challenges in agricultural research, education, extension, subsidy programs, land administration, and regulation/monitoring- and what new challenges might they create?

6 Conclusion

Agriculture 4.0 can play a key role in helping to address some of the most persistent barriers to agricultural transformation in smallholder farming systems, although it is no silver bullet for solving all challenges, and many structural barriers require different solutions. While embodied forms of Agriculture 4.0—such as automated smart and precision farming approaches—are for now relevant mainly in specific niches given the limited spread of (modern) mechanization in smallholder farming systems, more disembodied forms of Agriculture 4.0 already play an important role. These include tools that support farmers' decision-making and help connect them to services and markets by reducing the high transaction costs typical of rural economies, thereby addressing a long-standing barrier to agricultural transformation. An often-overlooked dimension is that Agriculture 4.0 can improve the activities not only of farmers but also by other actors in the agricultural value chain and agricultural innovation systems, which can play a role in driving agricultural transformation. While Agriculture 4.0 is still less widespread than often assumed, rapid technological developments are likely to accelerate its adoption. Public investments and effective regulation will be essential to support Agriculture 4.0 and ensure it contributes to sustainable and inclusive transformation.

Competing Interests The author(s) has no competing interests to declare that are relevant to the content of this manuscript.

References

Abate GT, Abay KA, Chamberlin J, Kassim Y, Spielman DJ, Tabe-Ojong MPJ (2023) Digital tools and agricultural market transformation in Africa: why are they not at scale yet, and what will it take to get there? Food Policy 116(102439):1–12. https://doi.org/10.1016/j.foodpol.2023.102439

Abdulai AR (2022) A new green revolution (GR) or neoliberal entrenchment in agri-food systems? Exploring narratives around digital agriculture (DA), food systems, and development in sub-sahara Africa. J Dev Stud 58(8):1588–1604. https://doi.org/10.1080/00220388.2022.2032673

Abdulai AR, Tetteh Quarshie P, Duncan E, Fraser E (2023) Is agricultural digitization a reality among smallholder farmers in Africa? Unpacking farmers' lived realities of engagement with digital tools and services in rural Northern Ghana. Agric Food Secur 12(11):1–14. https://doi.org/10.1186/s40066-023-00416-6

Abdul-Rahaman A, Abdulai A (2022) Mobile money adoption, input use, and farm output among smallholder rice farmers in Ghana. Agribusiness 38(1):236–255. https://doi.org/10.1002/agr.21721

Aker JC, Ksoll C (2016) Can mobile phones improve agricultural outcomes? Evidence from a randomized experiment in Niger. Food Policy 60:44–51. https://doi.org/10.1016/j.foodpol.2015.03.006

Aker JC, Ghosh I, Burrell J (2016) The promise (and pitfalls) of ICT for agriculture initiatives. Agric Econ 47(S1):35–48. https://doi.org/10.1111/agec.12301

Altieri MA (2002) Agroecology: the science of natural resource management for poor farmers in marginal environments. Agric Ecosyst Environ 93(1–3):1–24. https://doi.org/10.1016/S0167-8809(02)00085-3

Arouna A, Michler JD, Yergo WG, Saito K (2020) One size fits all? Experimental evidence on the digital delivery of personalized extension advice in Nigeria. Am J Agric Econ 103:596–619. https://doi.org/10.1111/ajae.12151

Barrett H, Rose DC (2022) Perceptions of the fourth agricultural revolution: what's in, what's out, and what consequences are anticipated? Sociol Rural 62(2):162–189. https://doi.org/10.1111/soru.12324

Barrett CB, Christiaensen L, Sheahan M, Shimeles A (2017) On the structural transformation of rural Africa. J Afr Econ 26(S1):11–35. https://doi.org/10.1093/jae/ejx009

Bateki CA, Daum T, Salvatierra-Rojas A, Müller J, Birner R, Dickhoefer U (2021) Of milk and mobiles: assessing the potential of cellphone applications to reduce cattle milk yield gaps in Africa using a case study. Comput Electron Agric 191(106516):1–10. https://doi.org/10.1016/j.compag.2021.106516

Baumüller H (2018) The little we know: an exploratory literature review on the utility of mobile phone-enabled services for smallholder farmers. J Int Dev 30(1):134–154. https://doi.org/10.1002/jid.3314

Baumüller H, Ikpi U, Jumpah ET, Kamau G, Kergna AO, Mose L et al (2023) Building digital bridges in African value chains: exploring linkages between ICT use and social capital in agricultural marketing. J Rural Stud 100(103002). https://doi.org/10.1016/j.jrurstud.2023.03.010

Belton B, Baldiga L, Justice S, Minten B, Narayanan S, Reardon T (2025) Can the global drone revolution make agriculture more sustainable? Science 389(6764):972–976. https://doi.org/10.1126/science.ady1791

Bernard B, Lux A (2017) How to feed the world sustainably: an overview of the discourse on agroecology and sustainable intensification. Reg Environ Chang 17(5):1279–1290. https://doi.org/10.1007/s10113-016-1027-y

Birner R, Daum T, Pray C (2021) Who drives the digital revolution in agriculture? A review of supply-side trends, players and challenges. Appl Econ Perspect Policy 43(4):1260–1285. https://doi.org/10.1002/aepp.13145

Camacho A, Conover E (2019) The impact of receiving SMS price and weather information on small scale farmers in Colombia. World Dev 123(104596):1–15. https://doi.org/10.1016/j.worlddev.2019.06.020

Castaneda R, Doan D, Newhouse DL, Nguyen M, Uematsu H, Azevedo JP (2016) Who are the poor in the developing world? Policy research working paper 7844. World Bank, Washington, DC

Ceballos F, Kramer B, Robles M (2019) The feasibility of picture-based insurance (PBI): smartphone pictures for affordable crop insurance. Dev Eng 4(100042). https://doi.org/10.1016/j.deveng.2019.100042

Chichaibelu BB, Baumüller H, Matschuck MA (2023) Protecting the data of African agricultural producers: a review of national laws, compliance and perceptions. Law Innov Technol 15(2):617–661. https://doi.org/10.1080/17579961.2023.2245673

Coggins S, McCampbell M, Sharma A, Sharma R, Haefele SM, Karki E et al (2022) How have smallholder farmers used digital extension tools? Developer and user voices from Sub-Saharan Africa, South Asia and Southeast Asia. Glob Food Sec 32(100577). https://doi.org/10.1016/j.gfs.2021.100577

CTA (2019) The digitalization of African agriculture report. Technical Center for Agriculture and Rural Cooperation (CTA)

da Silveira F, Lermen FH, Amaral FG (2021) An overview of agriculture 4.0 development: systematic review of descriptions, technologies, barriers, advantages, and disadvantages. Comput Electron Agric 189(106405). https://doi.org/10.1016/j.compag.2021.106405

Daum T (2019) ICT applications in agriculture. In: Ferranti P, Berry E, Jock A (eds) Encyclopedia of food security and sustainability, vol 1, pp 255–260. Amsterdam, Netherlands.

Daum T (2025) Digitalization and skills in agriculture. Outlook Agric 54(2):171–181. https://doi.org/10.1177/00307270251336474
Daum T, Villalba R, Anidi O, Mayienga SM, Gupta S, Birner R (2021) Uber for tractors? Opportunities and challenges of digital tools for tractor hire in India and Nigeria. World Dev 144(105480):1–15. https://doi.org/10.1016/j.worlddev.2021.105480
Daum T, Ravichandran T, Kariuki J, Chagunda M, Birner R (2022) Connected cows and cyber chickens? Stocktaking and case studies of digital livestock tools in Kenya and India. Agric Syst 196(103353):1–13. https://doi.org/10.1016/j.agsy.2021.103353
de Janvry A, Sadoulet E (2019) Transforming developing country agriculture: removing adoption constraints and promoting inclusive value chain development. FERDI Working Paper 253. Fondation pour les études et recherches sur le développement international
Deichmann U, Goyal A, Mishra D (2016) Will digital technologies transform agriculture in developing countries? Agric Econ 47(S1):21–33. https://doi.org/10.1111/agec.12300
Diao X, Jones E, Pauw K, Thurlow J, Xu W (2024) The agricultural transformation index. IFPRI, Washington, DC
Etwire PM, Buah S, Ouédraogo M, Zougmoré R, Partey ST, Martey E et al (2017) An assessment of mobile phone-based dissemination of weather and market information in the Upper West Region of Ghana. Agric Food Secur 6:1–9. https://doi.org/10.1186/s40066-016-0088-y
Fraser A (2019) Land grab/data grab: precision agriculture and its new horizons. J Peasant Stud 46(5):893–912. https://doi.org/10.1080/03066150.2017.1415887
Fraser A (2022) 'You can't eat data'?: moving beyond the misconfigured innovations of smart farming. J Rural Stud 91:200–207. https://doi.org/10.1016/j.jrurstud.2021.06.010
Fu X, Akter S (2016) The impact of mobile phone technology on agricultural extension services delivery: evidence from India. J Dev Stud 52(11):1561–1576. https://doi.org/10.1080/00220388.2016.1146700
Fuglie K, Gautam M, Goyal A, Maloney WF (2019) Harvesting prosperity: technology and productivity growth in agriculture. World Bank, Washington, DC
Hidalgo F, Birkenberg A, Daum T, Bosch C, Quiñones-Ruiz XF (2024) How do coffee farmers engage with digital technologies? A capabilities perspective. Agric Hum Values 41(4):1707–1723. https://doi.org/10.1007/s10460-024-10574-3
Hildebrandt N, Nyarko Y, Romagnoli G, Soldani E (2020) Price information, inter-village networks, and 'bargaining spillovers': experimental evidence from Ghana. NYU Stern School of Business
Holt-Giménez E, Altieri MA (2013) Agroecology, food sovereignty, and the new green revolution. Agroecol Sustain Food Syst 37(1):90–102. https://doi.org/10.1080/10440046.2012.716388
Huggins C, Frosina N (2017) ICT-driven projects for land governance in Kenya: disruption and e-government frameworks. GeoJournal 82:643–663. https://doi.org/10.1007/s10708-016-9710-6
Jensen R (2010) Information, efficiency, and welfare in agricultural markets. Agric Econ 41(S1):203–216. https://doi.org/10.1111/j.1574-0862.2010.00501.x
Khan RP, Gupta S, Daum T, Birner R, Ringler C (2025) Levelling the field: a review of the ICT revolution and agricultural extension in the global south. J Int Dev 37(1):1–21. https://doi.org/10.1002/jid.3949
Kikulwe EM, Fischer E, Qaim M (2014) Mobile money, smallholder farmers, and household welfare in Kenya. PLoS One 9(10):e109804. https://doi.org/10.1371/journal.pone.0109804
Kim J, Shah P, Gaskell JC, Prasann A, Luthra A (2020) Scaling up disruptive agricultural technologies in Africa. World Bank, Washington, DC
Klerkx L, Rose D (2020) Dealing with the game-changing technologies of Agriculture 4.0: how do we manage diversity and responsibility in food system transition pathways? Glob Food Sec 24(100347):1–7. https://doi.org/10.1016/j.gfs.2019.100347
Kos D, Kloppenburg S (2019) Digital technologies, hyper-transparency and smallholder farmer inclusion in global value chains. Curr Opin Environ Sustain 41:56–63. https://doi.org/10.1016/j.cosust.2019.10.011
Lio M, Liu MC (2006) ICT and agricultural productivity: evidence from cross-country data. Agric Econ 34(3):221–228. https://doi.org/10.1111/j.1574-0864.2006.00120.x

Lowder SK, Sánchez MV, Bertini R (2021) Which farms feed the world and has farmland become more concentrated? World Dev 142(105455):1–15. https://doi.org/10.1016/j.worlddev.2021.105455

Lowder SK, Bhalla G, Davis B (2025) Decreasing farm sizes and the viability of smallholder farmers: implications for resilient and inclusive rural transformation. Glob Food Sec 45(100854):1–10. https://doi.org/10.1016/j.gfs.2025.100854

Machado EA, Purcell H, Simons AM et al (2020) The quest for greener pastures: evaluating the livelihoods impacts of providing vegetation condition maps to pastoralists in Eastern Africa. Ecol Econ 175(106708). https://doi.org/10.1016/j.ecolecon.2020.106708

Marwa ME, Mburu J, Oburu REJ, Mwai O, Kahumbu S (2020) Impact of ICT based extension services on dairy production and household welfare: the case of iCow service in Kenya. J Agric Sci 12(3):1–12. https://doi.org/10.5539/jas.v12n3p141

Milian Gómez JF, Byttebier K (2025) Agroecological sustainability: exploring the intersection of digital agriculture, ethics and the right to food. Discov Agric 3(1):1–16. https://doi.org/10.1007/s44279-025-00228-1

Mockshell J, Kamanda J (2018) Beyond the agroecological and sustainable agricultural intensification debate: is blended sustainability the way forward? Int J Agric Sustain 16(2):127–149. https://doi.org/10.1080/14735903.2018.1448047

Muto M, Yamano T (2009) The impact of mobile phone coverage expansion on market participation: panel data evidence from Uganda. World Dev 37(12):1887–1896. https://doi.org/10.1016/j.worlddev.2009.05.004

Nakasone E, Torero M (2016) A text message away: ICTs as a tool to improve food security. Agric Econ 47(S1):49–59. https://doi.org/10.1111/agec.12314

Namyenya A, Daum T, Rwamigisa PB, Birner R (2022) E-diary: a digital tool for strengthening accountability in agricultural extension. Inf Technol Dev 28(2):319–345. https://doi.org/10.1080/02681102.2021.1875186

Nigussie E, Olwal T, Musumba G, Tegegne T, Lemma A, Mekuria F (2020) IoT-based irrigation management for smallholder farmers in rural Sub-Saharan Africa. Procedia Comput Sci 177:86–93. https://doi.org/10.1016/j.procs.2020.10.015

Njuguna E, Daum T, Birner R, Mburu J (2025) Silicon Savannah and smallholder farming: how can digitalization contribute to sustainable agricultural transformation in Africa? Agric Syst 222(104180):1–15. https://doi.org/10.1016/j.agsy.2024.104180

Oyinbo O, Chamberlin J, Abdoulaye T, Maertens M (2022) Digital extension, price risk, and farm performance: experimental evidence from Nigeria. Am J Agric Econ 104(2):831–852. https://doi.org/10.1111/ajae.12242

Parlasca MC, Johnen C, Qaim M (2022) Use of mobile financial services among farmers in Africa: insights from Kenya. Glob Food Sec 32(100590):1–10. https://doi.org/10.1016/j.gfs.2021.100590

Pingali PL (2012) Green revolution: impacts, limits, and the path ahead. Proc Natl Acad Sci 109(31):12302–12308. https://doi.org/10.1073/pnas.0912953109

Pretty J (2008) Agricultural sustainability: concepts, principles and evidence. Philos Trans R Soc B 363(1491):447–465. https://doi.org/10.1098/rstb.2007.2163

Pretty J, Bharucha ZP (2014) Sustainable intensification in agricultural systems. Ann Bot 114(8):1571–1596. https://doi.org/10.1093/aob/mcu205

Rajkhowa P, Baumüller H (2024) Assessing the potential of ICT to increase land and labour productivity in agriculture: global and regional perspectives. J Agric Econ 75(2):477–503. https://doi.org/10.1111/1477-9552.12566

Rajkhowa P, Qaim M (2021) Personalized digital extension services and agricultural performance: evidence from smallholder farmers in India. PLoS One 16(10):e0259319. https://doi.org/10.1371/journal.pone.0259319

Rapsomanikis G (2015) The economic lives of smallholder farmers. FAO, Rome

Rose DC, Wheeler R, Winter M, Lobley M, Chivers CA (2021) Agriculture 4.0: making it work for people, production, and the planet. Land Use Policy 100(104933). https://doi.org/10.1016/j.landusepol.2020.104933

Schroeder K, Lampietti J, Elabed G (2021) What's cooking: digital transformation of the agrifood system. World Bank, Washington, DC

Sekabira H, Qaim M (2017) Mobile money, agricultural marketing, and off-farm income in Uganda. Agric Econ 48(5):597–611. https://doi.org/10.1111/agec.12360

Steinke J, van Etten J, Müller A, Ortiz-Crespo B, van de Gevel J, Silvestri S, Priebe J (2021) Tapping the full potential of the digital revolution for agricultural extension: an emerging innovation agenda. Int J Agric Sustain 19(5–6):549–565. https://doi.org/10.1080/14735903.2020.1738754

Steinke J, Ivanova Y, Jones SK, Minh T, Sánchez A, Sánchez-Choy J, Mockshell J (2024) Digital sustainability tracing in smallholder context: ex-ante insights from the Peruvian cocoa supply chain. World Dev Sustain 5(100185):1–15. https://doi.org/10.1016/j.wds.2024.100185

Tende IG, Aburada K, Yamaba H, Katayama T, Okazaki N (2022) Proposal for a blockchain based agricultural input voucher system. Artif Life Robot 27(3):461–471. https://doi.org/10.1007/s10015-021-00724-9

van Campenhout B, Vandevelde S, Walukano W, van Asten P (2017) Agricultural extension messages using video on portable devices increased knowledge about seed selection, storage and handling among smallholder potato farmers in southwestern Uganda. PLoS One 12(1):e0169557. https://doi.org/10.1371/journal.pone.0169557

Voss RC, Jansen T, Mané B, Shennan C (2021) Encouraging technology adoption using ICTs and farm trials in Senegal: lessons for gender equity and scaled impact. World Dev 146:105620. https://doi.org/10.1016/j.worlddev.2021.105620

Wezel A, Bellon S, Doré T, Francis C, Vallod D, David C (2009) Agroecology as a science, a movement and a practice. A review. Agron Sustain Dev 29:503–515. https://doi.org/10.1051/agro/2009004

Wezel A, Soboksa G, McClelland S, Delespesse F, Boissau A (2015) The blurred boundaries of ecological, sustainable, and agroecological intensification: a review. Agron Sustain Dev 35(4):1283–1295

Wittman H, James D, Mehrabi Z (2020) Advancing food sovereignty through farmer-driven digital agroecology. Ciencia e investigación agraria: revista latinoamericana de ciencias de la agricultura 47(3):235–248. https://doi.org/10.7764/ijanr.v47i3.2299

Yameogo V, Zossou RC, Scheiterle L, Daum T, Adegbola YC, Daudu C, Kergna AO, Mulinge W, Nientao A, Angara US, Fatunbi O, Birner R (2024) Sustainability in agricultural education in Sub-Saharan Africa: perspectives from students and teachers. Hohenheim Working Papers on Social and Institutional Change in Agricultural Development 023-2024. University of Hohenheim

Transformations in Working Conditions in the Era of Agriculture 4.0

Franco da Silveira, Daniel Hideki Shibuya, Stanley Robson de Medeiros Oliveira, Fernando Gonçalves Amaral, and Jayme Garcia Arnal Barbedo

1 An Overview of the Modern Agri-Food System

The modern agri-food system faces growing, complex, and interconnected challenges that undermine its ability to ensure food security, environmental sustainability, and broad-based, equitable socioeconomic inclusion. Rapid population growth (Lidicker Jr 2020), geopolitical conflicts (Hossain et al. 2024), and the increasingly severe effects of climate change (Bouteska et al. 2024) not only strain the global food supply but also exacerbate social and economic inequalities, increasing the vulnerability of rural communities and historically marginalized populations. These factors hinder equitable access to food, productive resources, and development opportunities, directly undermining efforts toward inclusion in the modern agri-food system (Ndege et al. 2024). In this context, it becomes urgent to promote structural transformations in conventional agricultural production models, adopting solutions that integrate productivity, environmental sustainability, climate resilience, and socioeconomic justice.

Agriculture 4.0—also known as the Fourth Agricultural Revolution—has increasingly been recognized in scientific debates as a promising alternative to address current and future challenges faced by the modern agri-food system (da Silveira et al. 2021; da Silveira et al. 2025). Based on the integration of emerging technologies such as the Internet of Things (IoT), Artificial Intelligence (AI), Big Data, automation, robotics, drones, and smart sensors (Hassoun et al.

F. da Silveira (✉) · S. R. de Medeiros Oliveira · J. G. A. Barbedo
EMBRAPA Digital Agriculture, Campinas, São Paulo, Brazil

D. H. Shibuya
University of Campinas, Campinas, Brazil

F. G. Amaral
Federal University of Rio Grande do Sul, Porto Alegre, Brazil

F. da Silveira et al. (eds.), *Agriculture 4.0*,
https://doi.org/10.1007/978-3-032-24552-6_5

2025), this new technological configuration aims to increase productivity, reduce consumption and waste of natural resources, and make processes throughout the sector more sustainable (Klerkx and Rose 2020). However, the effectiveness of this technological transition depends on its ability to inclusively and equitably integrate all stakeholders involved in the modern agri-food system—especially regarding the reorganization of agricultural labor. Without this capacity, technologies may not be used correctly or even adopted, compromising expected benefits. By profoundly transforming agricultural operations, Agriculture 4.0 also reshapes occupational structures and working conditions of stakeholders within this system, imposing new demands for qualification, organization, and adaptation to emerging technological dynamics (Prause 2021). Therefore, it is essential to advance discussions on how Agriculture 4.0 technologies impact working conditions, productive dynamics, and training requirements in rural areas—giving rise to what this chapter calls Agri-Work 4.0.

The concept of Agri-Work 4.0 refers to the reconfiguration of labor in the context of Agriculture 4.0, characterized by intensive automation, digitization of processes, increasing reliance on data, and use of intelligent systems, among others. Crucially, Agri-Work 4.0 applies not only to farmers but to all actors involved in the Agriculture 4.0 ecosystem—including technical staff, consultants, operators, and other stakeholders—reflecting the transformation of work, skills, and employment across the entire agri-food system. This transformation directly impacts occupational profiles, forms of employment, skill requirements, and labor relations in the agri-food system. While it may create opportunities for new jobs and technical specializations, this transition also imposes significant barriers to the inclusion of groups that already face historical limitations in access to formal education, digital infrastructure, and social protection. In contexts marked by structural inequalities, Agri-Work 4.0 tends to reinforce a division in the labor market: on one side, a limited group of highly skilled workers—mainly large producers and operators with high levels of capitalization—engaged in technological activities; on the other, smallholder farmers and low-scale rural workers, often relegated to precarious roles, mediated by digital platforms, outsourcing, or even replaced by machines. In developing countries, this scenario is even more complex due to the persistent lack of digital and technological literacy among farmers, compounded by universities whose curricula remain outdated and disconnected from the demands of Agriculture 4.0 (da Silveira et al. 2023). Thus, the transition to Agri-Work 4.0 risks deepening existing inequalities, hindering the inclusion and capacity-building of the majority of the rural workforce, and ultimately compromising the socioeconomic benefits that the Fourth Agricultural Revolution promises to deliver (Abdulai et al. 2023; Miine et al. 2023).

Without deliberate and coordinated strategies for digital inclusion, professional training, and social protection, Agriculture 4.0 runs the serious risk of reproducing—and even intensifying—the structural inequalities that have historically shaped the modern agri-food system. Rather than serving as a driver of democratized innovation and promotion of sustainable rural development, this new technological configuration may consolidate itself as an additional mechanism of exclusion, concentrating benefits in the hands of a few highly capitalized stakeholders while relegating the majority of the rural workforce to increasingly precarious conditions. From this perspective, Agri-Work 4.0 ceases to represent a just transition for the future of work in the modern agri-food system and becomes a critical turning point—capable of deepening socioeconomic fragmentation and undermining the pillars of sustainability, equity, and social justice that the Fourth Agricultural Revolution claims to uphold.

The realization of an inclusive and equitable Agri-Work 4.0 critically depends on the articulation of institutional arrangements capable of promoting digital inclusion, professional training, and social protection for workers—not only in rural areas but across all segments of the modern agri-food system. These arrangements require coordinated efforts among various stakeholders—governments, educational institutions, the private sector, civil society organizations, and labor unions—to establish aligned policies, programs, and initiatives that address historical barriers to accessing technology, education, and basic services, ensuring that changes in working conditions are effective across all levels of the modern agri-food system.

Government institutions play a key role in formulating and implementing public policies that guarantee adequate digital infrastructure, access to up-to-date technical training, and social protection mechanisms to support workers through technological transitions. Universities and research and development centers must update their curricula and foster technological innovation tailored to local realities, training professionals capable of meeting the demands of Agriculture 4.0. The private sector can contribute through investments, the development of accessible technological solutions, and training and qualification programs. Civil society organizations and labor unions are essential for representing the interests of rural workers, ensuring their voices are heard and their rights are upheld.

Therefore, the absence of effective institutional arrangements not only limits the capacity to foster inclusion and skills development but also compromises the creation of an environment conducive to responsible and ethical innovation in Agriculture 4.0. Without strong and coordinated institutions, emerging technologies risk being adopted in a fragmented and unequal manner, weakening the governance of the modern agri-food system and hindering the development of social protection mechanisms adapted to new labor dynamics. Moreover, the lack of coordination among institutional stakeholders prevents the development of integrated public policies that balance productive efficiency, environmental sustainability, and social justice, making the transition to Agri-Work 4.0 vulnerable to significant socioeconomic and environmental risks.

In this context, this chapter aims to analyze the impacts of Agriculture 4.0 on working conditions within the modern agri-food system, highlighting how the adoption of emerging technologies is reshaping labor organization, professional profiles, and employment relations. It seeks to understand both the opportunities and the risks associated with this transformation, particularly in terms of inclusion, skills development, and social protection for workers. The chapter then emphasizes the fundamental role of institutional arrangements in guiding and consolidating this transformative journey, ensuring that it is inclusive, sustainable, and socially just.

Finally, it proposes short and medium-term strategic actions to promote professional training, digital inclusion, and multisectoral cooperation in the development of Agri-Work 4.0. The contribution of this chapter to the Agriculture 4.0 literature can be emphasized by highlighting its focus on institutional arrangements and by providing a clear definition of Agri-Work 4.0, which may serve as a conceptual reference for future research.

2 Agri-Work 4.0: Opportunities and Risks

The adoption of Agriculture 4.0 technologies goes beyond the transformation of production processes and redefines working conditions, which become essential pillars for the integrated functioning of the modern agri-food system (Trendov et al. 2019; Prause 2021; Howard and Lincoln 2023). Within the multiple dimensions of this system—production, logistics, distribution, and consumption—new forms of work play a central role in articulating the various stakeholders, information flows, and resources (Carolan 2024). Furthermore, these transformations generate new occupations, skills, and modes of work organization, shaping Agri-Work 4.0, which directly impacts the sector's efficiency, innovation, and resilience, ensuring its sustainability and capacity to respond to contemporary challenges. However, these new labor arrangements also raise important issues regarding access, qualification, and working conditions, highlighting the need for policies and institutional arrangements that ensure an inclusive and fair technological transition (Birner et al. 2021; Bellon-Maurel et al. 2023).

Among the positive impacts of Agri-Work 4.0, the significant reduction of physical strain in tasks traditionally characterized by high levels of repetitiveness, risk, and physical demand within the agri-food system is particularly noteworthy. Advanced technologies—such as autonomous tractors, self-driving harvesters, and drones for aerial spraying and mapping—are increasingly replacing labor-intensive manual activities, thereby minimizing workers' exposure to harsh environmental conditions, including extreme heat, direct contact with agrochemicals, and prolonged working hours in open-field settings (Marinoudi et al. 2019; Rejeb et al. 2022; Júnior et al. 2024). Moreover, automated milking systems reduce the need for physically demanding manual labor in dairy farming, while simultaneously enhancing hygiene standards and sanitary control (Gargiulo et al. 2022). In horticultural production, innovations such as automated vertical farming systems and smart greenhouses—equipped with temperature, humidity, and light sensors—diminish the requirement for continuous human intervention and enable more precise control over cultivation conditions (Saad et al. 2021). In parallel, digital tools such as farm management software, mobile applications (e.g., for machinery tracking), climate-precision platforms, and AI-based agronomic decision-support systems foster greater operational efficiency and facilitate strategic planning (Mendes et al. 2020; Hachimi et al. 2022; Mandal et al. 2024). These technologies enable both workers and managers to monitor agricultural operations in real time and from remote locations, thereby improving operational comfort, productivity, and data-driven decision-making capabilities. Collectively, these emerging technological advancements contribute not only to improved working conditions in the agricultural sector but also to the professionalization of the workforce, elevating the value of skilled labor and requiring the development of new digital and analytical competencies. Consequently, labor in the modern agri-food system is becoming less burdensome, safer, more technologically sophisticated, and increasingly attractive to younger generations.

However, the impacts of Agri-Work 4.0 are not distributed evenly and reveal significant tensions across the various chains that comprise the agri-food system (Benegiamo 2024). The replacement of manual labor by machines, sensors, robots, and algorithms may result in the elimination of traditional job roles, particularly among low-skilled workers. In this sense, the technological transition carries inherent structural contradictions (Soma and Nuckchady 2021). The increasing demand for technical qualifications, in a context marked by asymmetries in access to education, digital infrastructure, and connectivity, tends to exacerbate the exclusion of already vulnerable groups such as small-scale farmers, rural youth, and informal workers (da Silveira et al. 2023). Furthermore, within the framework of Agri-Work 4.0, the fragmentation and intensification of labor are increasingly evident. New roles associated with digitalized operations often require the simultaneous use of multiple platforms, systems, and real-time performance metrics. This can lead to heightened pressure for productivity, increased task surveillance, and cognitive overload—even in environments that were previously considered automated.

The adoption of digital platforms for task and productivity management may also result in flexible working hours, variable compensation, and the erosion of labor rights, giving rise to dynamics akin to the phenomenon of "agricultural uberization" (Daum et al. 2021). In certain cases, algorithmic surveillance systems monitor real-time performance, introducing new forms of control and psychological pressure, while simultaneously reducing workers' autonomy. These dynamics expose the ambiguity inherent in the technological discourse—while narratives of modernization and improved living conditions are promoted, in practice, many workers may experience increased precariousness, instability, and a lack of social protection.

Thus, this paradox cannot be overlooked. Rather than treating Agriculture 4.0 technologies as a neutral force, it is essential to recognize them as the outcome of political, economic, and institutional choices that shape who benefits and who is excluded. Therefore, beyond simply adopting emerging technologies, it is crucial to reorient the pathways of transformation within the modern agri-food system to ensure that such changes are grounded in principles of social justice, equity, and dignity in labor. The challenge, therefore, is not solely technological but also deeply social and political in nature. Table 1 presents a synthesized overview of opportunities and risks associated with Agri-Work 4.0 in the agri-food system. The organization of elements was developed by the authors based on recurring evidence identified in the specialized literature and public discussions on the topic (e.g., Spykman et al. 2022; Bellon-Maurel et al. 2023; da Silveira et al. 2023; Howard and Lincoln 2023; Benegiamo 2024; Klerkx and Villalobos 2024; Nettle et al. 2025).

Table 1 Overview of key opportunities and risks of Agri-Work 4.0 in the modern agri-food system

Professional aspects in Agriculture 4.0	Opportunities	Risks
New skills and training	Creation of new professions such as drone operators, rural data analysts, and agricultural automation technicians. Farmers can improve their management skills with courses in precision and digital agriculture.	Workers and farmers with low education levels face difficulties adapting. Lack of accessible public training policies.
Transformation of agricultural work	Reduction of manual labor and improvement of physical conditions through automation. Farmers begin to act as technological managers of their production systems.	Small farmers may be excluded due to a lack of resources to invest in technology. Replacement of workers without professional repositioning.
Entrepreneurship and professional autonomy	Farmers can become digital entrepreneurs (e.g., direct online sales, agro-influencing, remote consulting). Expansion of AgriFoodTech startups generates collaborative work opportunities between rural and urban areas.	High financial risks for farmers investing in technologies without support or guaranteed return. Overload from having to accumulate technical and management functions.
Digitalization of farm management	Improved decision-making using farm management software and applications. Detailed production control, traceability, and access to credit via digital platforms.	Information overload can create dependency and insecurity for farmers with little digital familiarity. Risks of cybersecurity breaches and data loss.
Inclusion of youth and generation of new leadership	Agriculture 4.0 attracts digitally skilled youth, renewing rural areas, and encouraging family succession. Opportunities for more diverse leadership in rural areas (e.g., women, youth, trained technicians).	Resistance to change from older farmers. Difficulty in retaining youth in rural areas without adequate digital infrastructure.
Working conditions and quality of life	Less exposure to sun, agrochemicals, and heavy loads. Remote monitoring allows more time for planning and leisure.	New psychological pressures, such as anxiety over productivity or difficulty dealing with technological failures. Intensification of work pace and surveillance.
Access to markets and digital platforms	Farmers can sell products directly to consumers, improving profit margins. Integration with logistics and e-commerce systems.	Need for digital marketing and platform management skills. Risk of exclusion for farmers who cannot integrate into these digital networks.

Interdisciplinarity in universities	Development of integrated academic programs connecting agronomy, engineering, Information Technology, and management, forming professionals capable of working in the modern food system with a multidisciplinary perspective. Encouragement of applied research and collaborative innovation to solve real challenges of the food system in the Agriculture 4.0 era. Strengthening universities' capacity to train talents aligned with the technological, environmental, and social demands of the food system.	Institutional and bureaucratic resistance is hindering the implementation of interdisciplinary curricula within the food system context. Lack of preparation and training of faculty to work in multidisciplinary approaches and emerging technologies related to the food system. Misalignment between academic theory and the practical needs of the agri-food market, reducing graduates' employability. Risk of curricular overload or fragmentation of knowledge, compromising the depth of multidisciplinary training.
Transformations in agricultural consultancy work	Expansion of consultancy roles, with growing demand for specialized services in digital technologies, data analysis, and systems integration within the Agriculture 4.0 context. Strengthening of the consultant's role as a strategic intermediary between producers, tech companies, public policies, and applied science. Opportunity to develop more flexible business models, including remote consultancy, digital platforms, and tailored services.	Increased workload and constant pressure to stay technologically up to date, often without adequate institutional support. Risk of exclusion for professionals who fail to keep up with digital innovations or adapt to new service delivery formats. Weakening of traditional, trust-based relationships with farmers due to the excessive digitalization of consultancy processes.

3 The Role of Institutional Arrangements

Emerging technologies are transforming the agri-food system, generating a wide range of new job opportunities across the entire sector. The implementation of Agriculture 4.0 technologies supports data-driven decision-making, increases productivity, reduces physical labor, and promotes more sustainable production systems. However, many farmers still lack confidence in replacing traditional methods with new technologies (Jellason et al. 2021). This lack of confidence often arises from limited knowledge of emerging technologies, insufficient technical assistance, financial barriers, and restricted access to digital tools. Furthermore, the next generation of agricultural workers will need to acquire knowledge and skills related to these emerging technologies in order to apply them effectively in their daily tasks (Bampasidou et al. 2024; Kondoyanni et al. 2024).

Stakeholders across the agri-food system, including educational institutions, private and public service providers, government agencies, and farming communities, are progressively contributing to the development of these innovations. Nonetheless, coordinated efforts involving all stakeholders in the agri-food system act as key drivers towards the implementation of Agri-Work 4.0. In this context, institutional arrangements within the agri-food system refer to the interaction among these stakeholders to facilitate the adoption of emerging technologies. Such an arrangement ensures equitable access to technology, fosters continuous innovation, strengthens knowledge networks, and creates new job opportunities, key elements for building a sustainable and inclusive Agriculture 4.0 ecosystem. Figure 1 presents a conceptual representation of these dynamics. It is not intended as a definitive model, but rather as an evolving framework that reflects the direction in which the sector is moving to meet the demands of Agri-Work 4.0—namely, integration, coordination, and innovation across all levels of the agri-food system.

3.1 Regulatory Frameworks and Institutional Coordination

Regulatory frameworks refer to a structured set of laws, regulations, guidelines, and institutions that define how specific activities should be conducted within a given sector. In the context of Agriculture 4.0, the emergence of novel job roles and technology-driven services necessitates the development of appropriate regulatory mechanisms to address growing demand and to support the responsible and secure integration of technological innovations. For instance, emerging professions such as drone pilots and agricultural robotics operators currently lack formal licensing procedures and certified training programs, unlike established regulatory systems for conventional technologies—such as the requirement of driver's licenses for automobile drivers and operators of agricultural machinery. This regulatory gap between conventional and emerging technologies contributes to uncertainty and a lack of confidence among farmers in adopting innovations (Eastwood and Renwick 2020).

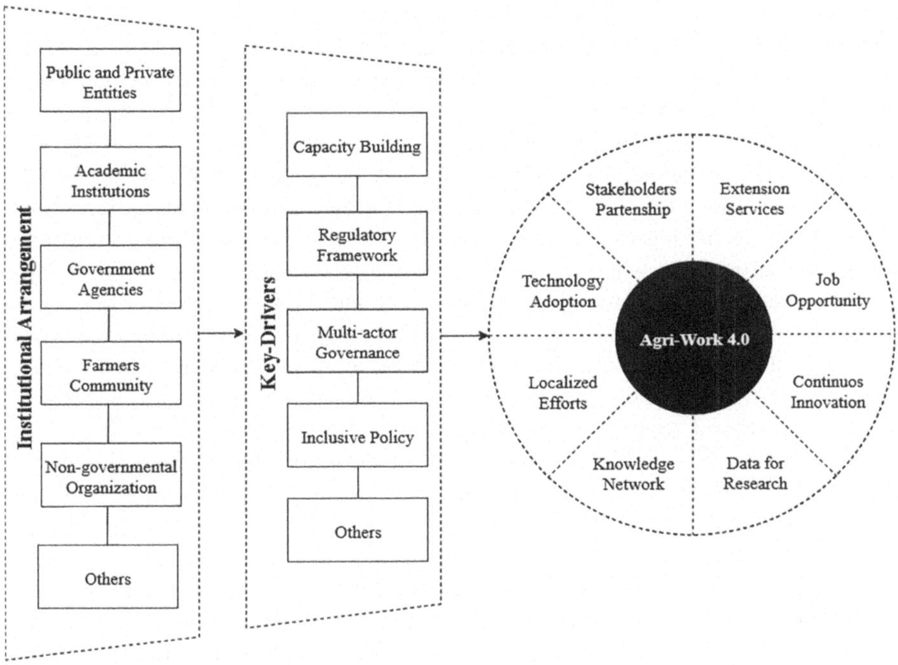

Fig. 1 Framework of institutional arrangement and key drivers towards the Agri-Work 4.0

While private entities, agricultural extension services, and education institutions can play a complementary role by providing technical training capacity-building and qualification initiatives, government agencies must assume the central role in formulating comprehensive regulatory frameworks, integrating public policies and technical standards, to guide both public and private stakeholders in the transition to Agri-Work 4.0. In this context, the integration between public and private institutions becomes essential to create a dynamic and responsive regulatory environment. Collaborative governance enables the incorporation of market-driven insights into policy design while leveraging the technical expertise and innovation capacity of the private sector. This partnership fosters regulatory flexibility, accelerates the adoption of emerging technologies, and enhances the responsiveness of institutions to evolving challenges in Agriculture 4.0 (Klerkx et al. 2019).

3.2 *Multiactor Governance: Public, Private, and Civil Society Engagement*

Multi-actor governance involves the participation of diverse stakeholders in the agri-food system, including private and public entities, academic institutions, government agencies, farmers, and nongovernmental organizations (NGOs). While each stakeholder holds distinct interests across different stages of the agri-food system, coordinated and collaborative actions among them are essential to enabling a successful transition toward Agriculture 4.0 (Bayala et al. 2021). Partnerships between private initiatives and the academic community are well established in fields such as medicine, computer science, and engineering, where private corporations often provide financial support for the development of innovative products and technologies. In agriculture, similar collaborations are increasingly important, particularly when they include active participation from farming communities. Farmers contribute valuable field data and local knowledge, helping to guide research and innovation toward practical solutions that address real-world challenges and promote technology adoption on the ground (Hansen et al. 2022). Furthermore, the outcomes of academic research are often made publicly accessible, providing data and tools that benefit the entire agri-food system by supporting evidence-based decision-making, informing the development of new technologies, and guiding the formulation of public policies for farmers, including both large and small-scale producers.

Despite these advantages, several challenges persist in ensuring the equitable and inclusive expansion for building the Agri-Work 4.0 ecosystem. For instance, smallholder and family farmers remain dependent on conventional agricultural practices and face structural limitations such as low digital literacy, limited internet access, and reduced institutional support (Jellason et al. 2021). Conflicts of interest among stakeholders may also hinder effective cooperation, particularly in localities with low financial relevance or political visibility. These conflicts can arise, for example, when private sector priorities do not align with the socioenvironmental needs of rural communities or when local populations exhibit resistance to adopting unfamiliar technologies due to mistrust, lack of information, or perceived risks (Nnanna et al. 2025). In these contexts, public agencies play a pivotal role in mediating relationships by acting as institutional bridges between stakeholders, ensuring that all interests are represented. Additionally, government entities can help reduce asymmetries in access to information and resources, thereby supporting the development of a more integrated, fair, and sustainable governance framework for Agriculture 4.0.

3.3 Capacity Building and Knowledge Networks

Educational level and technical knowledge remain significant barriers to the adoption of Agriculture 4.0, particularly among older adults and low-income populations, who often rely on conventional agricultural practices (Kondoyanni et al. 2024). The lack of digital and technical skills contributes to low confidence in adopting emerging technologies, thereby limiting participation in innovation-driven initiatives. For instance, private and public institutions offer services involving IoT (Internet of Things), remote sensing, mobile applications, and Artificial Intelligence

(AI) chatbots. These technologies rely on a large volume of data that is stored and processed through cloud computing, which raises concerns about the handling of confidential and sensitive information. Moreover, many farmers, particularly older ones, still struggle to interpret and effectively use digital data (Kondoyanni et al. 2024). Consequently, stakeholders such as private institutions tend to redirect their investments toward audiences perceived as more technologically prepared or economically viable. In this context, developing targeted training programs, digital literacy initiatives, and inclusive extension services is essential to ensuring that all segments of the farming population can benefit from technological advancements (Nettle et al. 2025).

Multi-actors' coordination is essential to establishing an effective capacity-building and qualification ecosystem. Private entities and agricultural extension services may provide technological training aimed at enhancing farmers' understanding of innovations and the tangible benefits of their adoption. In addition to introductory training, these actors can facilitate the transition to Agri-Work 4.0 by offering capacity-building initiatives focused on the management, operation, and maintenance of new technologies, ensuring continuity in the learning process and promoting long-term adoption. Moreover, farming communities, in collaboration with NGOs, can actively contribute by sharing local knowledge and experiences, thereby promoting horizontal knowledge exchange and supporting the development of decentralized learning networks in rural regions. Simultaneously, to prepare the next generation of farmers for the Agri-Work 4.0 ecosystem, educational institutions should incorporate emerging technologies into their curricula and offer experiential learning opportunities, including simulations and field-based applications, to strengthen practical skills and innovation readiness. For instance, within the European Union, the Twin Transition Schools initiative aims to transform the educational system by equipping the next generation of farmers with digital and green skills. This program provides experimental learning and skill-building through collaboration among different stakeholders, including farmers, startups, and researchers (European Commission 2025).

Although each stakeholder plays a distinct role within the capacity-building and qualification ecosystem, several structural barriers hinder the effective dissemination of knowledge and skills. One significant limitation is the insufficient number of qualified professionals available to deliver technical training, including within formal educational institutions (Soma and Nuckchady 2021). This shortage is further exacerbated by the rapid pace of technological advancement, which often outpaces the ability of professionals to stay updated and adequately transfer knowledge. As a result, many training programs struggle to keep up with emerging tools and practices, creating gaps between technological availability and actual on-the-ground implementation. In this context, strengthening relationships among stakeholders is essential to bridge these gaps, as horizontal learning networks and multiactor collaboration can foster the exchange of practical knowledge and promote more adaptive and decentralized capacity-building and qualification approaches.

3.4 Inclusive Policy Design and Social Protection Mechanisms

Regulatory frameworks, multiactor governance, and capacity-building environments are fundamental pillars for enabling the transition to Agriculture 4.0. However, this process often encounters barriers such as conflicting interests among stakeholders, which can hinder coordination and slow progress. In this context, inclusive and participatory policy design plays a critical role in reconciling these interests, ensuring that the diverse needs of all agri-food systems are addressed, and creating the institutional conditions necessary for the development and adoption of Agriculture 4.0 innovations (Islam et al. 2024). Moreover, local public policies must be context-sensitive, as many rural regions still lack the necessary infrastructure for farmers to adopt emerging technologies. Structural barriers—such as limited internet connectivity and low levels of digital literacy among farmers—demand targeted interventions, including regionally tailored public policies, capacity-building initiatives, and financial incentives. These localized efforts are crucial to ensure that the benefits of Agriculture 4.0 are equitably distributed and accessible to all farming communities, regardless of their socioeconomic or geographic conditions (Bayala et al. 2021; Jellason et al. 2021).

In many sectors, the interests of private and public institutions are typically prioritized; however, in the context of the transition to Agri-Work 4.0, it is essential that the voices, challenges, and specific needs of farming communities are also addressed. Given their direct involvement in agricultural production and their varying capacities to adopt new technologies, farmers' needs, particularly smallholders, must be actively included in the policy-making process. This approach is fundamental for designing policies that not only support technological innovation but also guide other stakeholders in aligning their actions with the real demands of rural communities. To foster an inclusive ecosystem for Agriculture 4.0, legal frameworks must incorporate provisions that guarantee financial support mechanisms (e.g., subsidized credit, tax incentives), accessible and continuous agricultural extension services, technical assistance for digital tool implementation, and equitable access to education and training opportunities. Financial mechanisms, for example, can provide subsidies or loans for the acquisition of emerging technologies, while rural extension services can act as intermediaries in translating complex digital tools into practical solutions tailored to local contexts. Such measures reduce structural inequalities, prevent the exclusion of vulnerable groups from technological benefits, and ensure that the transition towards Agri-Work 4.0 contributes to sustainable and socially just rural development.

3.5 Approaches to Fostering Agri-Work 4.0 Development

Despite significant technological advancements within the agri-food system, persistent barriers to their effective application continue to impede the consolidation of a fully functional Agri-Work 4.0 ecosystem. Cooperation among stakeholders through institutional arrangements plays a crucial role in driving this transformation forward. Well-structured and strategic approaches to technology adoption are essential to overcome these challenges and foster innovation across the sector. Table 2 presents a set of practical and adaptable suggestions that can guide stakeholders in different countries to design and implement action plans tailored to their local realities. These measures are not only actionable but also capable of generating tangible results in the short and medium term. Given the accelerating pace of technological change and the growing demand for more resilient, efficient, and inclusive agri-food systems, the implementation of these actions has become increasingly urgent.

At the center of this transformation is the farmer, who remains the cornerstone of the agri-food system. As production processes evolve and become increasingly dependent on emerging technologies, farmers must be empowered—not displaced—through accessible training, targeted support, and tools that respond to their real-world needs. The suggestions outlined in Table 2 aim precisely to support this transition. By organizing key interventions around capacity building, institutional coordination, and policy alignment, these approaches provide a practical pathway to foster Agri-Work 4.0. Importantly, they reinforce the idea that investing in people—especially those working directly on the land—is fundamental to building a more resilient, inclusive, and future-oriented agri-food system.

4 Final Considerations

Agriculture 4.0 is driving a profound transformation in work within the agri-food system, fueled by the intensive adoption of emerging technologies. These changes are not neutral—although they bring significant advances in productivity, sustainability, and efficiency, they pose real risks of social and economic exclusion, especially for workers across different segments of the agri-food system who face significant difficulties accessing adequate professional training, modern technological infrastructure, and digital resources essential to keep pace with this technological revolution.

In this context, the concept of Agri-Work 4.0 becomes fundamental to understanding the application of Agriculture 4.0 technologies to working conditions in the agri-food system, highlighting the need to reorganize labor relations, professional profiles, and employment dynamics in light of the challenges and opportunities brought by this new technological revolution. Importantly, Agri-Work 4.0 does not refer solely to the work performed by farmers; it encompasses all actors involved in the Agriculture 4.0 ecosystem, including consultants, technical staff, rural workers, and AgTech professionals. To this end, it is essential to strengthen institutional arrangements that promote digital inclusion, adequate professional training, and effective social protection for all stakeholders involved in the agri-food system. This is crucial because without these foundations, the adoption of emerging technologies may exacerbate existing inequalities, exclude vulnerable

Table 2 Suggestions of approaches with objectives and expected impact to foster the Agri-Work 4.0 development

Approaches	Objectives	Short and long-term impacts
Development of free online platforms for knowledge networks, capacity building, and qualification.	Develop an online platform for stakeholders to exchange technological experiences, enhance data management practices, and access tutorials for capacity building and qualification.	Short: Facilitate free and widespread access to training programs, empowering stakeholders to better engage with emerging technologies Long: Promote the decentralization of responsibilities for technological training across the Agri-food system, fostering a more collaborative innovation ecosystem.
Creation of rural collaborative innovation laboratories for multiactors' technology development.	Establish experimental spaces where universities, startups, farmers, and government can codesign, test, adapt, and develop technology solutions tailored to agri-food system needs.	Short: Foster knowledge exchange and strategic collaboration among diverse stakeholders within the agri-food system. Long: Establish a dynamic and inclusive innovation ecosystem in which all stakeholders actively contribute to the cocreation and continuous advancement of agricultural technologies.
Rural technical schools focus on an intergenera-tional knowledge network.	Create inclusive learning environments where both young and experienced farmers can collaboratively exchange knowledge on emerging technologies and practical field-based agricultural experiences.	Short: Reduce digital exclusion and increase farmers' confidence and readiness in adopting emerging technologies. Long: Foster a sustainable culture of innovation in rural communities by strengthening generational continuity, ensuring long-term knowledge retention, and encouraging lifelong learning across the agricultural workforce.
Development of specific regulatory frameworks for new jobs on Agri-Work 4.0.	Formalize and regulate new Agri-Work 4.0 occupations, such as agricultural drone operators, rural data analysis specialists, and field robotics technicians, among others.	Short: Create a national professional certification system with standardized curriculum guidelines and recognized credentials. Long: Ensure institutional recognition of emerging Agri-Work 4.0 professions, fostering job creation, improving workforce mobility, and strengthening the professionalization of the agri-food system.
Creation of multilevel participatory comanagement platforms for the development of public policies.	Promote stakeholders' collaboration within institutional spaces to co-design regional public policies that address the needs of the entire agri-food system, with particular attention to local farmers with limited economic resources.	Short: Address infrastructure and technology access gaps affecting economically disadvantaged stakeholders. Long: Foster inclusive development across the agri-food system by reducing structural disparities, ensuring that all stakeholders can benefit from and contribute to technological advancement and sustainable growth.

(continued)

Table 2 (continued)

Approaches	Objectives	Short and long-term impacts
Launch national "Agri-Work 4.0 mentoring" programs to scale the adoption of emerging technologies.	Train and distribute skilled professionals who act as technical knowledge multipliers, supporting specific needs for each stakeholder within the agri-food system.	Short: Accelerate the dissemination and practical understanding of emerging technologies by embedding trained mentors within local agri-food contexts. Long: Establishment of an ecosystem of mentors with skills, delivering localized technical services adapted to the unique challenges and opportunities of each territory.
Design national technical assistance task forces to remove barriers to Agriculture 4.0 adoption.	Establish coordinated teams composed of representatives from the institutional arrangement to diagnose territorial challenges and implement targeted, context-specific solutions for technology adoption in agriculture.	Short: Mitigate social, infrastructural, and informational obstacles preventing smallholders and agri-food enterprises from adopting emerging technologies. Long: Enable context-sensitive, inclusive innovation by embedding support structures that adapt to local dynamics and sustain long-term digital integration in agriculture.
Creation of a national financial incentive program for inclusive technology adoption.	Provide tiered financial mechanisms—such as grants, subsidized credit, and tax incentives—supporting stakeholders in implementing emerging Agri-Work 4.0 technology.	Short: Lower financial barriers to the adoption of Agri-Work 4.0 technologies, particularly for economically disadvantaged stakeholders. Long: Establishment of strategies that foster long-term investment capacity, reduce regional technological disparities, and enable structural transformation of the agri-food system.
National, regional, and localized Agriculture 4.0 awareness campaign.	Raise awareness about the opportunities, challenges, and benefits of Agriculture 4.0 by engaging the entire agri-food system through targeted communication strategies tailored to each territorial reality.	Short: Improve understanding of emerging technologies, reduce cultural and informational barriers to adoption, and stimulate initial interest in experimentation with digital solutions. Long: Foster a nationwide innovation culture that values technological adoption in agriculture and promotes inclusive, decentralized engagement with Agriculture 4.0 across diverse agri-food contexts.

workers, and compromise the social and economic sustainability of the sector. Strong and coordinated institutions ensure that the benefits of Agriculture 4.0 are distributed fairly, fostering an environment conducive to responsible innovation and inclusive development.

Thus, this chapter calls on institutions, policymakers, the private sector, and other stakeholders to act in a coordinated and integrated manner in building strategies that ensure the broad distribution of the benefits of Agriculture 4.0, minimizing inequalities and promoting sustainable and socially just development throughout the agri-food system. In addition, the chapter contributes to the Agriculture 4.0 literature by emphasizing the critical role of institutional arrangements and by introducing the concept of Agri-Work 4.0, providing a conceptual foundation for understanding labor transformations in the context of the Fourth Agricultural Revolution. Finally, it is important to emphasize that the objective of this chapter is not to present an absolute truth, but rather to foster debate, stimulating new ideas, reflections, and opinions that may enrich understanding and action in the face of the challenges and opportunities of Agriculture 4.0 in the modern agri-food system. In this sense, future research should engage with critical questions such as:

- How can education and vocational training systems be redesigned to ensure equitable access to the technical and digital skills required by Agri-Work 4.0, particularly among smallholders, informal workers, and marginalized rural populations?
- What new forms of labor are emerging in the era of Agriculture 4.0, and how do these technological advances transform employment and workers' agency within the agri-food system?
- How can universities reformulate their curricula and promote interdisciplinarity among fields such as agronomy, engineering, data science, and management to prepare the new generation of graduates with technical and digital skills aligned with the demands of Agriculture 4.0 in the agri-food system?
- What are the roles, skill requirements, and labor dynamics of different actors within the Agri-Work 4.0 ecosystem—such as consultants, technical staff, and AgTech professionals—and how do they shape the functioning and innovation of the Agriculture 4.0 system?

Acknowledgments We appreciate the support of the researchers from the Digital Agriculture Research Center - Semear Digital. This study was financed by the São Paulo Research Foundation (FAPESP), Brazil. Process Number #2023/12215-3 and #2022/09319-9. We also thank the Digital Agriculture Research Center, School of Agricultural Engineering (FEAGRI), University of Campinas (UNICAMP), and the Coordenação de Aperfeiçoamento de Pessoal de Nível Superior (CAPES), Brazil—Process No. 88887.992998/2024-00, for their institutional and financial support.

Competing Interests The authors have no conflicts of interest that are relevant to the content of this chapter.

References

Abdulai AR, Kc KB, Fraser E (2023) What factors influence the likelihood of rural farmer participation in digital agricultural services? Experience from smallholder digitalization in northern Ghana. Outlook Agric 52(1):57–66. https://doi.org/10.1177/00307270221144641

Bampasidou M, Goldgaber D, Gentimis T, Mandalika A (2024) Overcoming 'digital divides': leveraging higher education to develop next generation digital agriculture professionals. Comput Electron Agric 224:109181. https://doi.org/10.1016/j.compag.2024.108993

Bayala J, Ky-Dembélé C, Dayamba S, Somda J, Ouédraogo M, Diakité A, Chabi A, Alhassane A, Bationo A, Buah SS et al (2021) Multi-actors' coimplementation of climate-smart village

approach in west africa: achievements and lessons learnt. Front Sustain Food Syst 5:637007. https://doi.org/10.3389/fsufs.2021.637007
Bellon-Maurel V, Piot-Lepetit I, Lachia N, Tisseyre B (2023) Digital agriculture in europe and in France: which organisations can boost adoption levels? Crop Pasture Sci 74(6):573–585. https://doi.org/10.1071/CP22065
Benegiamo M (2024) The labor of the future,the future of labor? a just transition critique of the digital agriculture utopia. https://www.politicalecologyljubljana.si/wp-content/uploads/2024/12/iCKZ-2024-Rumena-2-ZBRPredavatelji-www-04C.pdfpage=50. Accessed 28 July 2025
Birner R, Daum T, Pray C (2021) Who drives the digital revolution in agriculture? A review of supply-side trends, players and challenges. Appl Econ Perspect Policy 43(4):1260–1285. https://doi.org/10.1002/aepp.13145
Bouteska A, Sharif T, Bhuiyan F, Abedin MZ (2024) Impacts of the changing climate on agricultural productivity and food security: evidence from Ethiopia. J Clean Prod 449:141793. https://doi.org/10.1016/j.jclepro.2024.141793
Carolan M (2024) Who and what gets recognized in digital agriculture: agriculture 4.0 at the intersectionality of (dis)ableism, labor, and recognition justice. Agric Hum Values 41(4):1465–1480. https://doi.org/10.1007/s10460-024-10560-9
da Silveira F, Lermen FH, Amaral FG (2021) An overview of agriculture 4.0 development: systematic review of descriptions, technologies, barriers, advantages, and disadvantages. Comput Electron Agric 189:106405. https://doi.org/10.1016/j.compag.2021.106405
da Silveira F, Da Silva SLC, Machado FM, Barbedo JGA, Amaral FG (2023) Farmers' perception of the barriers that hinder the implementation of agriculture 4.0. Agric Syst 208:103656. https://doi.org/10.1016/j.agsy.2023.103656
da Silveira F, Corrêa RGF, Baierle IC, Landaverde R, Barbedo JGA (2025) Behavioral profile of farmers in the adoption of agriculture 4.0 technologies in the agri-food system: a case study in Brazil. Front Sustain Food Syst 9. https://doi.org/10.3389/fsufs.2025.1624753
Daum T, Villalba R, Anidi O, Mayienga SM, Gupta S, Birner R (2021) Uber for tractors? Opportunities and challenges of digital tools for tractor hire in India and Nigeria. World Dev 144:105480. https://doi.org/10.1016/j.worlddev.2021.105480
Eastwood CR, Renwick A (2020) Innovation uncertainty impacts the adoption of smarter farming approaches. Front Sustain Food Syst 4. https://doi.org/10.3389/fsufs.2020.00024
European Comission (2025) Promoting responsible twin transitions in European agrifood systems through innovation and learning alliances (TWIN-IN). https://ec.europa.eu/info/funding-tenders/opportunities/portal/screen/how-to-participate/orgdetails/914989749/project/101187101/program/43353764/details. Accessed 24 July 2025
Gargiulo J, Lyons N, García S (2022) Optimising profitability and productivity of pasture-based dairy farms with automatic milking systems. Animal 16(9):100605. https://doi.org/10.1016/j.animal.2022.100605
Hachimi CE, Belaqziz S, Khabba S, Sebbar B, Dhiba D, Chehbouni A (2022) Smart weather data management based on artificial intelligence and big data analytics for precision agriculture. Agriculture 13(1):95. https://doi.org/10.3390/agriculture13010095
Hansen BD, Leonard E, Mitchell MC, Easton J, Shariati N, Mortlock M, Schaefer M, Lamb D (2022) Current status of and future opportunities for digital agriculture in Australia. Crop Pasture Sci. https://doi.org/10.1071/CP21594
Hassoun A, Mhlanga D, Rejeb A, Bhat Z, Buheji M, Bigliardi B (2025) The role of Industry 4.0 in global food security: a promising pathway to ending hunger. Smart Agric Technol:100974. https://doi.org/10.1016/j.atech.2025.100974
Hossain MA, Ferdous N, Ferdous E (2024) Crisis-driven disruptions in global waste management: impacts, challenges and policy responses amid COVID-19, Russia-Ukraine war, climate change, and colossal food waste. Environ Chall 14:100807. https://doi.org/10.1016/j.envc.2023.100807
Howard J, Lincoln JM (2023) Future of work in agriculture. https://doi.org/10.1080/1059924X.2022.2137615

Islam MH, Anam MZ, Hoque MR, Nishat M, Bari AM (2024) Agriculture 4.0 adoption challenges in the emerging economies: implications for smart farming and sustainability. J Econ Technol 2:278–295. https://doi.org/10.1016/j.ject.2024.09.002

Jellason NP, Robinson EJ, Ogbaga CC (2021) Agriculture 4.0: is sub-Saharan Africa ready? Appl Sci 11(12):5750. https://doi.org/10.3389/fcomm.2021.762201

Júnior MRB, de Almeida Moreira BR, dos Santos Carreira V, de Brito Filho AL, Trentin C, de Souza FLP, Tedesco D, Setiyono T, Flores JP, Ampatzidis Y et al (2024) Precision agriculture in the United States: a comprehensive meta-review inspiring further research, innovation, and adoption. Comput Electron Agric 221:108993. https://doi.org/10.1016/j.compag.2024.108993

Klerkx L, Rose D (2020) Dealing with the game-changing technologies of Agriculture 4.0: how do we manage diversity and responsibility in food system transition pathways? Glob Food Sec 24:100347. https://doi.org/10.1016/j.gfs.2019.100347

Klerkx L, Villalobos P (2024) Are agrifoodtech start-ups the new drivers of food systems transformation? An overview of the state of the art and a research agenda. Glob Food Sec 40:100726. https://doi.org/10.1016/j.gfs.2023.100726

Klerkx L, Jakku E, Labarthe P (2019) A review of social science on digital agriculture, smart farming and Agriculture 4.0: new contributions and a future research agenda. NJAS Wageningen J Life Sci 90:100315. https://doi.org/10.1016/j.njas.2019.100315

Kondoyanni M, Loukatos D, Arvanitis KG, Lygkoura KA, Symeonaki E, Maraveas C (2024) Adding machine-learning functionality to real equipment for water preservation: an evaluation case study in higher education. Sustainability 16(8):3261. https://doi.org/10.3390/su16083261

Lidicker WZ Jr (2020) A scientist's warning to humanity on human population growth. Glob Ecol Conserv 24:e01232. https://doi.org/10.1016/j.gecco.2020.e01232

Mandal S, Yadav A, Panme FA, Devi KM, SM SK (2024) Adaption of smart applications in agriculture to enhance production. Smart Agric Technol 7:100431. https://doi.org/10.1016/j.atech.2024.100431

Marinoudi V, Sørensen CG, Pearson S, Bochtis D (2019) Robotics and labour in agriculture: a context consideration. Biosyst Eng 184:111–121. https://doi.org/10.1016/j.biosystemseng.2019.06.013

Mendes J, Pinho TM, Neves dos Santos F, Sousa JJ, Peres E, Boaventura-Cunha J, Cunha M, Morais R (2020) Smartphone applications targeting precision agriculture practices—a systematic review. Agronomy 10(6):855. https://doi.org/10.3390/agronomy10060855

Miine LK, Akorsu AD, Boampong O, Bukari S (2023) Drivers and intensity of adoption of digital agricultural services by smallholder farmers in Ghana. Heliyon 9(12). https://doi.org/10.1016/j.heliyon.2023.e23023

Ndege N, Marshall F, Byrne R (2024) Exploring inclusive innovation: a case study in operationalizing inclusivity in digital agricultural innovations in Kenya. Agric Syst 219:104033. https://doi.org/10.1016/j.agsy.2024.104033

Nettle R, Ingram J, Ayre M (2025) Digiwork: how Agriculture 4.0 is changing work for farm advisers. Front Sustain Food Syst 9:1542007. https://doi.org/10.3389/fsufs.2025.1542007

Nnanna U, Mohammed A, Abubakar T, Yunus J, Eleke U, Sennuga S (2025) The role of technology in transforming rural social systems: an insightful review. J Nutr Food Process 8(1):1–8. https://doi.org/10.31579/2637-8914/281

Prause L (2021) Digital agriculture and labor: a few challenges for social sustainability. Sustainability 13(11):5980. https://doi.org/10.3390/su13115980

Rejeb A, Abdollahi A, Rejeb K, Treiblmaier H (2022) Drones in agriculture: a review and bibliometric analysis. Comput Electron Agric 198:107017. https://doi.org/10.1016/j.compag.2022.107017

Saad MHM, Hamdan NM, Sarker MR (2021) State of the art of urban smart vertical farming automation system: advanced topologies, issues and recommendations. Electronics 10(12):1422. https://doi.org/10.3390/electronics10121422

Soma T, Nuckchady B (2021) Communicating the benefits and risks of digital agriculture technologies: perspectives on the future of digital agricultural education and training. Front Commun 6:762201. https://doi.org/10.3389/fcomm.2021.762201

Spykman O, Emberger-Klein A, Gabriel A, Gandorfer M (2022) Autonomous agriculture in public perception-german consumer segments' view of crop robots. Comput Electron Agric 202:107385. https://doi.org/10.1016/j.compag.2022.107385

Trendov NM, Varas S, Zeng M (2019) Digital technologies in agriculture and rural areas. https://openknowledge.fao.org/handle/20.500.14283/ca4985en. Accessed 28 July 2025

Farmers' Mental Health in the Fourth Agricultural Revolution

David Christian Rose

1 Mental Health of Farmers: Current Global Knowledge

In many parts of the world, there is a growing evidence base surrounding the causes of, and solutions to, poor farmer mental health. While not a problem for all farmers, there are serious cases of poor mental health among farming communities worldwide. For example, a survey of 15,000+ farmers in England and Wales found that over 1/3 of farmers displayed signs of probable or possible depression (Wheeler and Lobley 2022). Though the majority of published studies appear to be from the developed world (Daghagh Yazd et al. 2019), similarly poor levels of farmer and worker mental health have been noted in studies in many places, including, but not limited to—Australia (Hossain et al. 2008—many other studies), Canada (Hagen et al. 2020); China (Liu et al. 2018), Finland (Kallioniemi et al. 2016), India (Das 2011), Ireland (Russell et al. 2023), Mexico (Carvajal et al. 2014), New Zealand (Firth et al. 2007; Knook et al. 2024), Nigeria (Olowogbon et al. 2019), Norway (Logstein 2016), Thailand (Hanklang et al. 2016), and the USA (Rudolphi and Berg 2023). Problems appear to be sometimes more prevalent in farming than in comparable sector, although this has not been universally found (Daghagh Yazd et al. 2019). Demographics seem to play a role in mental health risk factors and help-seeking, including age (Rose et al. 2023; Rudolphi and Berg 2023) and gender (Hammersley et al. 2023; Budge and Shortall 2023).

Many research studies have explored the causes of poor mental health among some farmers, with several different categorisations. One way of categorising pressures is to divide them into (see Rose et al. 2022):

D. C. Rose (✉)
Agriculture and Environment, Harper Adams University, Newport, UK
e-mail: drose@harper-adams.ac.uk

F. da Silveira et al. (eds.), *Agriculture 4.0*,
https://doi.org/10.1007/978-3-032-24552-6_6

1. *Social factors*: may affect any human, but some may be exacerbated in agricultural settings, such as struggles with isolation and loneliness, stigma, family conflict, illness, bereavement, sexuality, rural crime, etc.
2. *Business factors*: specific to farming, such as animal and crop disease, poor finances, lack of time off the farm, succession challenges, tenancy issues, inspections, etc.
3. *Policy and public challenges*: criticism from the public, policy change, bureaucracy.

Some studies have tried to compare stressors across countries, including Nyczaj Kyle et al. (2025) who compared the UK and Canada.

Some farmers can find it difficult to seek or gain support for their mental health for reasons that include poor accessibility of empathetic help in rural areas (Hull et al. 2022; Malone et al. 2025), worries over confidentiality (Riethmuller et al. 2025), and stigma associated with the subject (O'Connor et al. 2024). There is still much to learn about how different stressors affect different farmers and how support can be best targeted (Nye et al. 2025).

A scoping review by Hernández García (2024) suggested that there was no literature (at that point) that had looked into how precision technologies affect farmer stress, although this is an arguable point. It is true to say, however, that studies have rarely directly looked at the role of emergent agricultural technologies in improving or worsening farmer mental health, though some studies have begun to draw a link (e.g. Hansen et al. 2020; Carolan 2023, 2024).

2 Agriculture 4.0 Technologies and Mental Health: An Emerging Area of Interest

It has always been the case in the evolution of technological change in farming that new ideas and methods lead to improvements in some areas, but also regression in others. The same will likely be true in the area of mental health, including effects on the specific factors affecting it outlined above. Hereafter, I review some of the early social science research which has explored the possible social impacts of emergent agricultural technology, with a focus on mental health. These existing studies have either directly explored this issue by making an explicit link between technology and mental health, or indirectly by evaluating how technology addresses one of the known factors influencing mental health (e.g. succession, crime, finances, etc.). Since the technologies, and hence related research, are emergent, it is hard to make definitive statements on the directionality of impact between technology and mental health in farming. It is important, though, to conduct on-farm post-adoption research to build the empirical case for if and how technologies are changing life on the farm, particularly in order to critically interrogate dominant narratives of a "positive, game-changing" revolution. Still to date, much research on farmer technology adoption stops at the point of self-confessed uptake (de Oca Munguia et al. 2021). This prevents us from knowing what happened next on the farm—did the farmer continue to use the tool and did it have overall positive or negative impacts? Did the farmer stop using the tool and why? How did life on the farm change and how do farmers, workers, families, and advisors feel about it? These are all key questions for the next postadoption phase of social science work on the so-called fourth agricultural revolution.

In the following sections, the discussion involves different digital agricultural technologies, some of which may not be typically associated with the so-called agriculture 4.0. Though a clear definition of agriculture 4.0 remains elusive (Klerkx and Rose 2020), various emergent technologies, such as artificial intelligence, robotics, drones, the internet of things, gene editing, and alternative technologies like hydroponics and cultured meat, are generally seen as in scope. Other digital technologies, such as mobile phone apps, some older wearable technologies for livestock, autosteer, or the internet in general, are generally considered to be "older" technologies and are rarely described as revolutionary. However, it is important to note that technology adoption in farming can be slow and non-linear (Rose et al. 2022). In many rural areas of the world, user readiness for digital technologies like smartphone apps or the internet can be low (McCampbell et al. 2023). Furthermore, although some agriculture 4.0 technologies like generative AI might appear to be very different to those that have come before it, it is still likely that many of the benefits and costs of previous digital technologies outlined in the social science literature will be applicable. Therefore, this chapter widens the scope of what an "agriculture 4.0" technology might be and focuses specifically on digital technologies, which may or may not be emergent in specific parts of the world.

3 The Effects of Automation on Lifestyles: Freed Time or Increased Technostress?

Some agriculture 4.0 technologies, particularly robotics and automation, promise to reduce drudgery by taking over dull, dangerous, and time-consuming jobs on farm. In turn, this has the potential to free up time for farmers to do other things, including spending more time with their family and spending more time off the farm, for example on vacation. Research in agriculture illustrates a clear link between time spent off the farm (on nonagricultural pursuits) and farmer mental health (Wheeler and Lobley 2022). Those farmers who do not (or cannot) take holidays or days off are more likely to suffer from mental health problems. Therefore, in principle, technologies that are designed to lighten the load and reduce time-consuming jobs could be beneficial.

Robotic milking systems are one example of an automated technology that can act as a proxy for the likely impacts of other emergent technologies on-farm. These devices automate the process of milking. Instead of the farmer and workers having to herd cattle several times a day into the milking parlour and spend hours (depending on herd size) manually connecting machines to teats, cleaning, and monitoring individual animals, cows come to a robot on their own at a time of their choosing. The robot recognises each individual, checks if they are ready to be milked, automatically connects the milking device to teats, cleans, and lets the cow out afterwards. This process also collects data on individual animals, which can be used to monitor production, health, and welfare parameters.

There is some evidence in the literature that Automated Milking Systems (AMS) have improved the lifestyles and mental health of some farmers, but not for all. In Norway, for example, Hansen et al. have conducted a number of studies exploring the link between adoption of AMS and farmer job satisfaction and well-being. In a 2020 study, they found that it had improved the lifestyles and mental health of some farmers by freeing up time and reducing the need to manually milk at unsociable hours. A positive outcome from AMS was particularly noted if farmers received training and good extension advice and had a successor to the business.

However, not all farmers received a boost to their well-being as a result of AMS adoption. For some, it did not reduce work, but rather changed it. Instead of spending time milking cows, farmers and workers spent time interpreting data collected by the AMS and reacting to alerts. The potential for data-driven farm technologies to increase technostress or cognitive burnout or overload, particularly for people who do not always have advanced computer literacy skills, has been noted in several other studies (e.g. Barrett and Rose 2022; Carolan 2023; King et al. 2023). In addition, a review of how AMS changes the nature of farm work has noted that the dominant discourse of automated technologies reducing work is oversimplistic (Martin et al. 2022). Rather, as Hansen et al. noted, it simply changes the type of work on farm. While this may be attractive to some people, it can create stress for others. Interestingly, the study Hansen et al. (2020) indicated that the effects of new farm technology on mental health can have a temporal element. While there may be an initial increase in stress as a result of having to learn a new technology, adjust work routines, and respond to "teething issues", after this transition period, there can be longer-term benefits if the technology performs as expected.

Other studies have explored how precision or automated technologies could change life on the farm and influence farmer and worker identity, exploring the potential impacts of this on job satisfaction and mental health. As stated above, new technologies could take over tasks that are dangerous to the physical health of humans. For example, the Thorvald modular robot is now commercially available to autonomously treat powdery mildew in strawberry crops instead of humans having to put on protective clothing at night (when the mildew needs to be treated) and use dangerous pesticides (see Rose and Bhattacharya 2023).

The research shows a direct link between good physical health and good mental health (Ohrnberger et al. 2017); therefore, if emergent technologies can reduce or eliminate dull and dangerous farm work, mental health could improve. Some studies have explored how autosteer technologies on tractors may help to reduce stress and fatigue, allowing farmers to undertake other tasks at the same time. For example, a study in Finland by Haapala et al. (2025) used heart rate monitors to compare the well-being of operators using or not using autosteer technologies. The study found that using autosteer technologies on tractors can reduce farmer stress. A similar finding was made in a study of automated spraying in the USA, with a claim that automated steering reduced stress for the operator by 48% (Burgers and Vanderwerff 2022). Furthermore, if farmers are stressed by the inability to recruit labour, which is a problem facing many sectors globally (see Rose and Bhattacharya 2023), labour-saving technologies represent an opportunity to address this stressor. Additionally, autonomous robots could reduce input costs and lead to higher profitability, including for small farms (Lowenberg-DeBoer 2024). Since poor financial outcomes are a key stressor on farmer mental health, if agriculture 4.0 technologies are able to improve profitability, then a key stress factor will be mitigated against. However, for emergent technologies associated with agriculture 4.0, some of which are still in development or early scaling, it is too early to judge what impact they will have on profitability for different types of farm. We are yet to see how "game-changing" so-called agriculture 4.0 technologies really are, and for whom (Klerkx and Rose 2020). New technologies could lead to de-skilling and labour displacement which could negatively impact mental health (also see Sect. 6).

4 Agriculture 4.0: Improving Attractiveness of Farming?

Generational renewal in farming continues to face mounting challenges around the world, including within the European Union where just 11.9% of farm managers are under 40.[1] Attracting young people into the farming industry is, therefore, crucial and many reports on the subject see emergent technologies as part of the solution to attract younger, skilled workers into the sector. In the mental health literature, there is a clear link between job satisfaction and mental health. For some, adopting the latest technologies may be attractive and younger farmers in particular may seek work on farms that have the most cutting-edge equipment. However, for others, the way in which labour-saving technologies change life on the farm may not be desirable. Taking an everyday approach to understanding how technology changes life on the farm is important (Rose et al. 2023). As stated above, new technologies do change farm workflows. Precision livestock technologies, including wearable sensors or artificial intelligence-enabled cameras, can monitor individual animals 24/7, potentially reducing the time needed for humans to check each animal (sometimes in large herds) for health and welfare issues. The effect of this could be that farmers spend less time with their animals, as has been noted in assessments of how AMS changes farm workflows (Beara and Holloway 2019). Automated crop technologies, such as robots or drones, could again reduce the amount of time spent on the land for a farmer, while the increasing move to digital recording and

[1] https://ec.europa.eu/eurostat/statistics-explained/index.php?title=Farmers_and_the_agricultural_labour_force_-_statistics#:~:text=Slightly%20more%20than%20two%20thirds,the%20age%20of%2040%20years

administration can lead a farmer to spend more time indoors on the computer (Rose et al. 2018).

A forward-looking video by John Deere produced in 2012[2] features a farmer called "Terry". It imagines a world in which automated technologies have arrived on the farm. Terry wakes up in a morning and makes coffee, checking a computer projection on his wall for the latest weather forecast. He sits down at his office table, placing the mug onto it, which reveals computer-generated images of his farm, as if he were playing a computer game. From this console, he can control his farm, or allow a team of advisers to control his farm. Only at the end of the day, as dark clouds gather and melancholic background music plays, does Terry decide to go and look at his crops, using a phone app to make diagnostics on his phone.

Here, it is important to acknowledge Carolan's (Carolan 2024, recalling Haraway) point that knowledge is situated; in other words, views towards, and impacts of, agriculture 4.0 technologies on mental health will differ depending on context—the circumstances surrounding the farm, farmers, workers, and families. For some farmers, the John Deere video may be an attractive vision of the future farm. Drawing on farmer interviews in Iowa and Minnesota, Carolan (2023) presents quotes from a farmer called "Beth". As Carolan (2023, 46) writes "digital agriculture came into the picture for Beth because of its role as a vehicle to make dairy farming satisfying to her". Beth said to him (quote directly taken from Carolan 2023):

> I faced a lot of pressure to take over the farm. But I also knew I wouldn't have been happy doing what my parents did. Dairy farming's a 365 days-a-year, twice-a-day, job, for Christ's sake. I'm proud of what we've built, a 21st-century operation that's probably more high-tech, data-oriented as what any of my friends from my high school days are doing. [...] They're off working in cities like Chicago, but this is the real deal; the type of Cloud-based, IT-savvy job people from my generation want.

Another farmer, Lynn, said, *"Forget working harder. I'm all about working smarter, which, yes, in my book means working less"* (from Carolan 2023). For Beth and Lynn, therefore, new technology had been a way of improving job satisfaction, reducing the drudgery of farm work, and improving mental health. For some other younger farmers, agriculture 4.0 technologies could also be seen as improving the attractiveness of the industry and securing their interest in making a succession to the family farm. Since worries over a lack of succession are a major source of anxiety and mental health problems (Rose et al. 2022), anything that attracts farmers' children to take over the farm would likely improve mental health within farming families.

But, Beth was a younger, IT-savvy farmer who was keen to, and able to, embrace new technologies. Based on another set of interviews of farmers across Arizona, California, Colorado, New Mexico, and Wyoming, Carolan (2024) also explored the potential of digital technologies to increase inclusivity and diversity of people in agriculture. He focused particularly on speaking to farmers with disabilities. He quotes farmers who said that digital technologies had enabled them to keep

[2] https://www.youtube.com/watch?v=t08nOEkrX-I&pp=0gcJCdgAo7VqN5tD

working in agriculture despite their disability. One farmer, for example, who was paralysed said that "tractors, drones, e-collars, robotic milkers are technologies that enable me to remain in farming" (quote taken from Carolan 2024). This farmer recalled how his parents' generation had toiled on the farm well past the point of having severe physical ailments. Other farmers had other disabilities, but had found technology a vital route towards continuation in the industry; for example, an autistic farmer recalled finding calm in the routine-based, ordered world of vertical farming, while another with multiple sclerosis could not have worked in the industry without technological aides. Reminiscent of the Hansen et al. (2020) study on robotic milking, other farmers in Carolan's (2024) study argued that technology had freed up time for them to spend on nonfarming activities; enabling them to follow more-than-farmer identities—such as "mothers", "fathers", "husbands", "mothers", "partners", etc". Again, the beneficial impacts of technology on physical health, to succession activities for some disabled individuals, and the potential to free up time, are likely to have a positive knock-on impact on mental health.

For others, the John Deere vision of the future may not be so attractive. Some farmers may want to spend more time with their animals, on their land connected to nature, and to be free from new forms of technostress (Barrett and Rose 2022; Schillings et al. 2021). As Carolan (2023) argues, there are many farmers priced out of adoption of new technologies, a trend noted by many social science papers that find uneven distribution of agriculture 4.0 benefits (see e.g. Bronson 2019). Some farmers may feel that they are being left behind with the rise of agriculture 4.0. There are farmers concerned about data ownership and privacy (Ruder and Wittman 2025), digital divides (Rotz et al. 2019; Mehrabi et al. 2021), the rise of "algorithmic rationality" (Miles 2019; Gardezi and Stock 2021), the reduction of farmer autonomy (Brooks 2021), and the rise of corporate power and land assetisation (Duncan et al. 2024; Fairbairn et al. 2025). There are farmers who are investing in new technology, but who may feel that they do not get the most benefit out of it compared to larger companies on whom they are becoming increasingly reliant for repairs and upgrades (Clapp and Ruder 2020). There has been an intense debate among scholars about the link between biotechnology and farmer suicides in India, with critics claiming that seed monopolisation has been a causative factor in the high rates of suicide (Thomas and De Tavernier 2017). While Thomas and De Tavernier (2017) found the association between seed monopolisation and farmer suicides to be questionable, there may be a link between those farmers who were economically disadvantaged as a result of the changes associated with biotechnology adoption (or non-adoption) and poor mental health outcomes. Therefore, for those farmers who cannot adopt due to digital divides, or do not want to adopt (but may be forced into doing so through assurance schemes or retailer relationship), the march towards agriculture 4.0 may be a source of stress and anxiety.

5 Digital Technologies as Part of the "Landscape of Support"

The literature on farmer mental health is exploring how 'landscapes of support' (Shortland et al. 2023) could be improved. Depending on context, support landscapes typically include a mix of formal providers, such as professional healthcare, mental health charities, and faith groups, and informal help provided by family, friends, peers, rural communities, and advisors who regularly see farmers. Increasingly, digital technologies are becoming part of the landscape of support for farmer mental health. To overcome challenges of isolation and poor accessibility of mental healthcare in rural areas (see Shortland et al. 2023), online counselling support for farmers may hold promise. Through various schemes, such as "Make the Moove" in Ireland (Macra Na Feirme, see Hammersley et al. 2025), or studies investigating the potential of, and demand for, online support in Scotland (Lamont et al. 2024) and Australia (Gunn et al. 2021), it has been found that support provided through digital technology can support farmers with their mental health. For example, the potential benefits of online forums for farmer well-being have been explored by Carlisle et al. (2024), while Gunn et al. have explored both what farmers want from online well-being interventions (Gunn et al. 2021) and have conducted initial evaluations of online resources that have been provided (Gunn et al. 2023). Online education delivery around mental health to agricultural students has also been considered by Knook et al. (2025). These studies have generally concluded that online interventions to support farmer mental health can be effective; the Gunn et al. (2023, 378) study found that the "ifarmwell" initiative was "likely to help farmers reduce their levels of distress and improve their mental wellbeing, by improving their psychological flexibility, ability to focus on the present and accept things beyond their control, as well as by reducing the extent to which they believe unhelpful thoughts". However, the risks of online forums are also currently being explored (Howse et al. 2026). There may be perils in directing farmers to the wrong type of online forum or social media, as we know that such spaces can be sources of conflict and public criticism of farming, which is a key mental health stressor (Rose et al. 2022).

Other nascent initiatives are starting to explore the potential for virtual reality to deliver psychological therapies to farmers (e.g. Rural Minds project in the UK[3]) or how digital technologies can be used to monitor the impacts of support interventions for farmers (Stynes et al. 2025). Some existing digital tools can also help to address isolation and social connectivity, which play a key role in mental health. GPS trackers and other digital systems can monitor the location and performance of farmers and workers on the farm. This can contribute to positive health and safety, particularly if there is an accident on the farm and emergency services/first responders can pinpoint the location of the patient more easily across vast and

[3] https://www.farminguk.com/news/immersive-vr-training-to-help-understand-mental-health-in-farming_67421.html

isolated agricultural landscapes.[4] Again, if physical health is better looked after on-farm, there should be a link with improved mental health.

6 Digital Farm Technologies: The Promises and Perils of Surveillance

However, research has also indicated the mental stress and anxiety that greater digital surveillance of the farm workforce can have. While research has widely pointed out the potential displacement of labour from automated technologies (e.g. Rotz et al. 2019), there have been less studies on how digital technologies could affect the current workforce on-farm. Drones may be used to monitor the performance of workers doing tasks in fields or the picking rate of fruit pickers can be evaluated 24/7 (Prause 2025). Using a case study of the fruit industry in South Africa, Prause (2025) finds that digital systems track worker progress, how quickly they are picking, and how well they are performing tasks. These data can be used to set targets, often unrealistic targets, and can be used as a way of disciplining workers, firing workers, or rewarding the most successful workers. There is a real mental health risk of unchecked forms of surveillance capitalism in which workers constantly feel watched and under pressure to perform (Ruder 2024). I have too witnessed technology being used in this way to monitor workers on a fruit farm in the UK, apparently to support the workforce, but clearly also to reward, disincentivise, and/or punish failure.

7 Shiny, New Technologies: A Magnet for, or a Deterrent to, Rural Crime?

Rural crime is a major cause of poor farmer mental health (Smith 2020). With increasingly more sophisticated and expensive technologies, the proceeds of criminal activity also have the potential to rise. Theft of expensive technologies is on the rise. For example, statistics cited in the UK Parliament in 2025 stated that GPS theft had surged by 137%. Thefts of quad bikes and all-terrain vehicles rose to over £3 million. Drones, autonomous robots, and other technologies associated with agriculture 4.0 are expensive. There is the potential for significant mental health challenges if a farmer suffers loss of a new, expensive technology to rural crime. Despite this potential, there are also ways in which new and existing digital and other farm technologies can also deter crime (e.g. digital trackers, tracking paint, an

[4] https://digital-strategy.ec.europa.eu/en/library/video-how-digital-tools-can-support-wellbeing-and-safety-farmers

engineered farm environment to deter theft, and low-tech methods [though uptake of technology-based measures may be low]) (Smith 2025).

8 Farm Extension and Communication in an AI-World: Opportunity or Risk?

Generative AI, specifically through the use of Large Language Models (LLMs), is being increasingly used in farm extension activities. Programmes such as ChatGPT can analyse large volumes of data and condense instructions to farmers or advisors into simple, accessible outputs. There are several farmer chatbots available or in development (e.g. Farmer.Chat by Digital Green), which offer the opportunity to provide extension advice quickly to isolated farmers, who may not always be able to access human advisers, including ones who are able to communicate in their native language (Tzachor et al. 2023). Since this is an emergent area of scholarship, with social scientists beginning to explore opportunities and risks of LLMs in farm extension, there has, as yet, been no explicit consideration of how it could impact farmers' mental health. This short section, therefore, is somewhat speculative, attempting to map the early scholarship on LLMs against its impact on known stressors of farmer mental health.

There are both opportunities and risks of large language models (Silva et al. 2023; Tzachor et al. 2023; De Clercq et al. 2024; Ibrahim et al. 2024; Marinoudi et al. 2024). As stated, they offer the potential to give quick advice to farmers, utilising large datasets and other sources of information, potentially overcoming the data-action gap on-farm. They could help to give extension advice to those who cannot access advice. However, there are worries over data ownership, digital divides, and misinformation, which could lead to bad advice being given to a farmer. It may also be worth exploring the unintended consequences of digital extension, including through the use of LLMs. Research on farmer mental health indicates that those humans who go up the farm track to see farmers, such as agronomists, accountants, land agents, and more, can play an "accidental counselling" role (Shortland et al. 2023). Since farmers can be isolated from sources of professional support, those individuals who do encounter farmers personally on farm visits could play a key role in spotting signs of distress and signposting towards support. While connections can be made digitally, if extension were to become increasingly virtual, there may be consequences from the loss of personal visits.

It is too early to judge how the increasing use of LLMs in agricultural extension will affect farmers' mental health, but the stakes associated with giving advice are high. If farm management decisions and finances improve or suffer as a result of good or bad advice, it follows that mental health may improve of worse. It also remains to be seen how the rise of generative AI affects the farmer-adviser relationship, which can be a valued and trusted relationship from both perspectives.

9 Concluding Remarks

Digital technologies, including emergent ones associated with agriculture 4.0, are not a panacea for sustainability. The question of how existing digital, or emergent technologies, affect life on the farm is dependent not only on what they can do, but also on how policy instruments (such as grants, regulations, training, infrastructure, etc.) influence the nature of scaling and use, and the ways in which this interacts with farmers' beliefs and worldviews. The impact of technology on mental health will be situated; in other words, depending on farmer circumstances, belief systems, and ability to adapt, it may improve or worsen lifestyles and conditions on the farm, which will have an impact on mental health. The individuality of impacts is worth reiterating; what one farmer considers an improved lifestyle is a view that may not be shared by others. It has always been the case in the evolution of technological change in farming that new ideas and methods lead to improvements in some areas, but also regression in others. This chapter in some ways was difficult to write because of the lack of explicit scholarship on the impacts of emergency agricultural technologies on mental health. A clear message, therefore, is to promote the need to carry out such studies, rooted in farmers' everyday perspectives, from a variety of different demographic and sector viewpoints, and exploring farmers' postadoption experiences. Such scholarship needs to explicitly examine the link between digital technologies and mental health, including examining the effect of specific technology types across farm and farmer types and geographical contexts, sharing learnings between teams of researchers around the world. These studies will need to embrace mixed methodologies, including utilising digital technologies themselves in the evaluation of farmer mental health (see e.g. Stynes et al. 2025), and, where possible, offer everyday, postadoption, longitudinal perspectives into how technology changes life on the farm (for the good or bad).

Competing Interests The author(s) has no competing interests to declare that are relevant to the content of this manuscript.

References

Barrett H, Rose DC (2022) Perceptions of the fourth agricultural revolution: what's in, what's out, and what consequences are anticipated? Sociol Rural 62(2):162–189. https://doi.org/10.1111/soru.12324

Beara C, Holloway L (2019) Beyond resistance: geographies of divergent more-than-human conduct in robotic milking. Geoforum 104:212–221. https://doi.org/10.1016/j.geoforum.2019.04.030

Bronson K (2019) Looking through a responsible innovation lens at uneven engagements with digital farming. NJAS Wageningen J Life Sci 90–91:100294. https://doi.org/10.1016/j.njas.2019.03.001

Brooks S (2021) Configuring the digital farmer: a nudge world in the making? Econ Soc 50(3):374–396. https://doi.org/10.1080/03085147.2021.1876984

Budge H, Shortall S (2023) Agriculture, COVID-19 and mental health: does gender matter? Sociol Rural 63(S1):82–94. https://doi.org/10.1111/soru.12408
Burgers TA, Vanderwerff KJ (2022) Vision and radar steering reduces agricultural sprayer operator stress without compromising steering performance. J Agric Saf Health 28(3):163–179. https://doi.org/10.13031/jash.15060
Carlisle K, Kamstra P, Carlisle E, McCosker A, De Cotta T et al (2024) A qualitative exploration of online forums to support resilience of rural young people in Australia. Front Public Health 12. https://doi.org/10.3389/fpubh.2024.1335476
Carolan M (2023) Digital agriculture killjoy: happy objects and cruel quests for the good life. Sociol Rural 63(S1):37–56. https://doi.org/10.1111/soru.12398
Carolan M (2024) Who and what gets recognized in digital agriculture: agriculture 4.0 at the intersectionality of (Dis)Ableism, labor, and recognition justice. Agric Hum Values 41:1465–1480. https://doi.org/10.1007/s10460-024-10560-9
Carvajal S, Kibor C, McClelland D, Ingram M, Zapien J et al (2014) Stress and sociocultural factors related to health status among US–Mexico border farmworkers. J Immigr Minor Health 16:1176–1182. https://doi.org/10.1007/s10903-013-9853-1
Clapp J, Ruder S-L (2020) Precision technologies for agriculture: digital farming, gene-edited crops, and the politics of sustainability open access. Glob Environ Politics 20(3):49–69. https://doi.org/10.1162/glep_a_00566
Daghagh Yazd S, Wheeler SA, Zuo A (2019) Key risk factors affecting farmers' mental health: a systematic review. Int J Environ Res Public Health 16(23):4849. https://doi.org/10.3390/ijerph16234849
Das A (2011) Farmers' suicide in India: implications for public mental health. Int J Soc Psychiatry 57:21–29. https://doi.org/10.1177/0020764009103645
De Clercq D, Nehring E, Mayne H, Mahdi A (2024) Large language models can help boost food production, but be mindful of their risks. Front Artif Intell 7. https://doi.org/10.3389/frai.2024.1326153
de Oca Munguia OM, Pannell DJ, Llewellyn R, Stahlmann-Brown P (2021) Adoption pathway analysis: representing the dynamics and diversity of adoption for agricultural practices. Agric Syst 191:103173. https://doi.org/10.1016/j.agsy.2021.103173
Duncan E, Rotz S, Magnan A, Bronson K (2024) Disciplining land through data: the role of agricultural technologies in farmland assetisation. Sociol Rural 62(2):231–249
Fairbairn M, Faxon HO, de Wit MM, Bronson K, Kish Z et al (2025) Digital agriculture will perpetuate injustice unless led from the grassroots. Nat Food 6:312–315. https://doi.org/10.1038/s43016-025-01137-8
Firth HM, Williams SM, Herbison GP, McGee RO (2007) Stress in New Zealand farmers. Stress Health 23:51–58. https://doi.org/10.1002/smi.1119
Gardezi M, Stock R (2021) Growing algorithmic governmentality: interrogating the social construction of trust in precision agriculture. J Rural Stud 84:1–11. https://doi.org/10.1016/j.jrurstud.2021.03.004
Gunn KM, Barrett A, Hughes-Barton D, Turnbull D, Short CE et al (2021) What farmers want from mental health and wellbeing-focused websites and online interventions. J Rural Stud 86:298–308. https://doi.org/10.1016/j.jrurstud.2021.06.016
Gunn KM, Skaczkowski G, Dollman J, Vincent AD, Brumby S, Short CE, Turnbull D (2023) A self-help online intervention is associated with reduced distress and improved mental wellbeing in Australian farmers: the evaluation and key mechanisms of www.ifarmwell.com.au. J Agromedicine 28(3):378–392. https://doi.org/10.1080/1059924X.2022.2156642
Haapala H, Kalmari J, Sarvela K, Linna P, Appelgren I (2025) Economic, environmental, and human benefits of GPS-guided autosteer in grain seeding. In: 2025 ASABE annual international meeting, p 2501522. https://doi.org/10.13031/aim.202501522
Hagen BNM, Albright A, Sargeant J, Winder CB, Harper SL, O'Sullivan TL, Jones-Bitton A (2020) Tailored mental health literacy training improves mental health knowledge and confidence among Canadian farmers. Int J Environ Res Public Health 17(11). https://doi.org/10.3390/ijerph17113807

Hammersley C, Meredith D, Richardson N, Carroll P, McNamara J (2023) Mental health, societal expectations and changes to the governance of farming: reshaping what it means to be a 'man' and 'good farmer' in rural Ireland. Sociol Rural 63(S1):57–81. https://doi.org/10.1111/soru.12411

Hammersley C, Richardson N, Meredith D, McNamara J, Carroll P, Jenkins P (2025) Analysing the effectiveness of a bespoke farmer mental health training programme targeted at farm advisors in Ireland. J Agric Educ Ext 31(1):90–118. https://doi.org/10.1080/1389224X.2024.2339801

Hanklang S, Kaewboonchoo O, Morioka I, Plernpit S-A (2016) Gender differences in depression symptoms among rice farmers in Thailand. Asia Pac J Public Health 28:83–93. https://doi.org/10.1177/1010539515620631

Hansen BG, Bugge CT, Skibrek PK (2020) Automatic milking systems and farmer wellbeing–exploring the effects of automation and digitalization in dairy farming. J Rural Stud 80:469–480. https://doi.org/10.1016/j.jrurstud.2020.10.028

Hernández García D (2024) A scoping review on how precision farming technology affects farmers' mental Well-being. Student Thesis, University of Twente. https://essay.utwente.nl/101024/1/Hern%C3%A1ndezGarc%C3%ADaDaniel_BA_EEMCS.pdf

Hossain D, Eley R, Coutts J, Gorman D (2008) Mental health of farmers in southern Queensland: issues and support. Aust J Rural Health 16:343–348. https://doi.org/10.1111/j.1440-1584.2008.01014.x

Howse J, Rose DC, Watson K, Maye D, Morgans L (2026) A SWOT analysis of an online farming forum as a space for supporting farmer well-being. https://www.sciencedirect.com/science/article/pii/S074301672600029X

Hull MJ, Gunn KM, Smith AE, Jones M, Dollman J (2022) "We're lucky to have doctors at all"; a qualitative exploration of Australian farmers' barriers and facilitators to health-related help-seeking. Int J Environ Res Public Health 19(17):11075. https://doi.org/10.3390/ijerph191711075

Ibrahim A, Senthilkumar K, Saito K (2024) Evaluating responses by ChatGPT to farmers' questions on irrigated lowland rice cultivation in Nigeria. Sci Rep 14:1–8. https://doi.org/10.1038/s41598-024-53916-1

Kallioniemi MK, Simola A, Kaseva J, Kymäläinen H-R (2016) Stress and burnout among Finnish dairy farmers. J Agromedicine 21:259–268. https://doi.org/10.1080/1059924X.2016.1178611

King E, Lamont K, Wendelboe-Nelson C, Williams C, Stark C (2023) Engaging the agricultural community in the development of mental health interventions: a qualitative research study. BMC Psychiatry 23:399. https://doi.org/10.1186/s12888-023-04806-9

Klerkx L, Rose DC (2020) Dealing with the game-changing technologies of Agriculture 4.0: how do we manage diversity and responsibility in food system transition pathways? Glob Food Sec 24:100347. https://doi.org/10.1016/j.gfs.2019.100347

Knook J, Carver D, Gear K, Eastwood C (2024) Assessing rural wellbeing through design of a farmer-focused survey methodology: the importance of cultural health, ease of use and farming specific questions. J Rural Stud 111:103412. https://doi.org/10.1016/j.jrurstud.2024.103412

Knook J, Stanley-Clarke N, Maris R, Hay A, Andrews C, Winder L (2025) A mixed methods assessment of an online mental health and resilience program for agricultural sector students. J Agromedicine 30(1):95–104. https://doi.org/10.1080/1059924X.2024.2421982

Lamont K, van Woerden HC, King E, Wendelboe-Nelson C, Humphry RW et al (2024) Improving the mental health of farmers: what types of remote support are acceptable, feasible, and improve outcomes? A feasibility RCT. Discov Ment Health 4:4. https://doi.org/10.1007/s44192-023-00054-1

Liu B-P, Qin P, Liu Y-Y, Yuan L, Gu L-X, Jia C-X (2018) Mental disorders and suicide attempt in rural China. Psychiatry Res 261:190–196. https://doi.org/10.1016/j.psychres.2017.12.087

Logstein B (2016) Farm-related concerns and mental health status among Norwegian farmers. J Agromedicine 21:316–326. https://doi.org/10.1080/1059924X.2016.1211055

Lowenberg-DeBoer J (2024) Economics of agricultural robotics. In: van Henten E, Edan Y (eds) Advances in agri-food robotics. Burleigh Dodds Science Publishing, Cambridge, UK

Malone S, Firnhaber J, O'Hagan AD, O'Keefe S, McNamara J, O'Connor S (2025) "Lacking the rural empathy"; Irish farmers' and stakeholders' opinions on current mental health services and preferences for support. J Rural Stud 114:103508. https://doi.org/10.1016/j.jrurstud.2024.103508
Marinoudi V, Benos L, Villa CC, Kateris D, Berruto R et al (2024) Large language models impact on agricultural workforce dynamics: opportunity or risk? Smart Agric Technol 9:100677. https://doi.org/10.1016/j.atech.2024.10067
Martin T, Gasselin P, Hostiou N, Feron G, Laurens L et al (2022) Robots and transformations of work in farm: a systematic review of the literature and a research agenda. Agron Sustain Dev 42:66. https://doi.org/10.1007/s13593-022-00796-2
McCampbell M, Adewopo J, Klerkx L, Leeuwis C (2023) Are farmers ready to use phone-based digital tools for agronomic advice? Ex-ante user readiness assessment using the case of Rwandan banana farmers. J Agric Educ Ext 29(1):29–51. https://doi.org/10.1080/1389224X.2021.1984955
Mehrabi Z, McDowell MJ, Ricciardi V, Levers C, Martinez JD, Mehrabi N, Wittman H, Ramankutty N, Jarvis A (2021) The global divide in data-driven farming. Nat Sustain 4:154–160. https://doi.org/10.1038/s41893-020-00631-0
Miles C (2019) The combine will tell the truth: on precision agriculture and algorithmic rationality. Big Data Soc 6(1). https://doi.org/10.1177/2053951719849444
Nyczaj Kyle S, Barkus E, Dunne S, Jones A (2025) A comparison of farm stressors in the UK and Canada. The Round Table 114(3):266–285. https://doi.org/10.1080/00358533.2025.2498161
Nye C, Wheeler R, Rose DC, Becot F, Holton M et al (2025) Mental health, well-being and resilience in agricultural areas: a research agenda for the Global North. J Rural Stud 114:103506. https://doi.org/10.1016/j.jrurstud.2024.103506
O'Connor S, O'Hagan AD, Firnhaber J, O'Shaughnessy BR, McNamara J et al (2024) Sowing seeds of awareness: a cross-sectional analysis of mental health literacy and help-seeking in Irish farmers. J Occup Med Toxicol 19:47. https://doi.org/10.1186/s12995-024-00446-1
Ohrnberger J, Fichera E, Sutton M (2017) The relationship between physical and mental health: a mediation analysis. Soc Sci Med 195:42–49. https://doi.org/10.1016/j.socscimed.2017.11.008
Olowogbon TS, Yoder AM, Fakayode SB, Falola AO (2019) Agricultural stressors: identification, causes and perceived effects among Nigerian crop farmers. J Agromedicine 24:46–55. https://doi.org/10.1080/1059924X.2018.1538915
Prause L (2025) The farm as digital factory: controlling labour and nature in digital agriculture. J Peasant Stud:1–19. https://doi.org/10.1080/03066150.2024.2443667
Riethmuller ML, McEvoy PM, Newnham EA (2025) Farmers' access to mental health services and community supports: a mixed methods analysis of service preferences, barriers and attitudes. J Rural Stud 117:103657. https://doi.org/10.1016/j.jrurstud.2025.103657
Rose DC, Barkemeyer A, de Boon A, Price C, Roche D (2023) The old, the new, or the old made new? Everyday counter-narratives of the so-called fourth agricultural revolution. Agric Hum Values 40:423–439. https://doi.org/10.1007/s10460-022-10374-7
Rose DC, Bhattacharya M (2023) Adoption of autonomous robots in the soft fruit sector: grower perspectives in the UK. Smart Agric Technol 3:100118. https://doi.org/10.1016/j.atech.2022.100118
Rose DC, Morris C, Lobley M, Winter M, Sutherland WJ, Dicks LV (2018) Exploring the spatialities of technological and user re-scripting: the case of decision support tools in UK agriculture. Geoforum 89:11–18. https://doi.org/10.1016/j.geoforum.2017.12.006
Rose DC, Shortland F, Hall J, Hurley P, Little R, Nye C, Lobley M (2022) The impact of COVID-19 on farmers' mental health: a case study of the UK. J Agromedicine 28(3):346–364. https://doi.org/10.1080/1059924X.2022.2137616
Rotz S, Gravely E, Mosby I, Duncan E, Finnis E, Horgan M et al (2019) Automated pastures and the digital divide: how agricultural technologies are shaping labour and rural communities. J Rural Stud 68:112–122. https://doi.org/10.1016/j.jrurstud.2019.01.023

Ruder S-L, Wittman H (2025) Agricultural data governance from the ground up: exploring data justice with agri-food movements. Big Data Soc 12(1). https://doi.org/10.1177/20539517251330182

Ruder S-L (2024) The 'terms and conditions' of surveillance capitalism: theorizing agricultural data policy and governance. J Peasant Stud. https://doi.org/10.1080/03066150.2024.2429480

Rudolphi JM, Berg RL (2023) Mental health of agricultural adolescents and adults: preliminary results of a five-year study. Front Public Health 11. https://doi.org/10.3389/fpubh.2023.1056487

Russell T, Stapleton A, Markey A, McHugh L (2023) Dying to farm: developing a suicide prevention intervention for farmers in Ireland. http://hdl.handle.net/10147/642916

Schillings J, Bennett R, Rose DC (2021) Animal welfare and other ethical implications of Precision Livestock Farming technology. CABI Agric Biosci 2:17. https://doi.org/10.1186/s43170-021-00037-8

Shortland F, Hall J, Hurley P, Little R, Nye C, Lobley M, Rose DC (2023) Landscapes of support for farming mental health: adaptability in the face of crisis. Sociol Rural 63(S1):116–140. https://doi.org/10.1111/soru.12414

Silva B, Nunes L, Esteva R, Aski V, Chandra R (2023) GPT-4 as an agronomist assistant? Answering agriculture exams using large language models. *arXiv* [preprint] arXiv.2310.06225. https://doi.org/10.48550/arXiv.2310.06225

Smith K (2020) Desolation in the countryside: how agricultural crime impacts the mental health of British farmers. J Rural Stud 80:522–531. https://doi.org/10.1016/j.jrurstud.2020.10.037

Smith K (2025) Crime and safety in rural England. In: Ceccato V, Harnkess A (eds) Crime, peoples and places: perspectives on rural safety and justice. Routledge, London

Stynes G, Stapleton A, Moore B, Russell T, O'Connor M, Richardson N, Ruiz FJ, McHugh L (2025) Effectiveness of a process-based approach to farmer wellbeing: a randomized multiple baseline single-case experimental design. J Contextual Behav Sci 37:100932. https://doi.org/10.1016/j.jcbs.2025.100932

Thomas G, De Tavernier J (2017) Farmer-suicide in India: debating the role of biotechnology. Life Sci Soc Policy 13:8. https://doi.org/10.1186/s40504-017-0052-z

Tzachor A, Devare M, Richards C, Pypers P, Ghosh A et al (2023) Large language models and agricultural extension services. Nat Food 4:941–948. https://doi.org/10.1038/s43016-023-00867-x

Wheeler R, Lobley M (2022) Health-related quality of life within agriculture in England and Wales: results from a EQ-5D-3L self-report questionnaire. BMC Public Health 22:1395. https://doi.org/10.1186/s12889-022-13790

Drivers and Barriers to the Adoption of Agriculture 4.0 Technologies: A Microeconomic and Political Perspective

Zdeňka Žáková Kroupová, Lenka Rumánková, Renata Aulová, and Lukáš Čechura

1 Introduction

The twenty-first century has introduced a range of complex challenges to agricultural production, including population growth, depletion of natural resources, land degradation, biodiversity loss driven by environmental pollution, and climate change—particularly global warming. As a result, agricultural systems are under unprecedented pressure to increase productivity while mitigating negative environmental externalities. In response, advanced technological innovations are increasingly being developed to address these challenges, and policymakers are formulating strategies to support their adoption, particularly through the integration of new technologies into agricultural practices (Feder et al. 1985).

Agriculture 4.0 offers a range of social, economic, and environmental benefits. By transforming traditional farming practices, it enhances efficiency, productivity, and sustainability in agriculture, aligning with several of the United Nations' Sustainable Development Goals.

The social dimension plays a very important role in the adoption of Agriculture 4.0 technologies. Farmers' attitudes and perceptions, educational background, and access to information significantly influence whether and how these innovations are adopted. With increasing digitalization, it is also important to consider issues of trust regarding the processing of private data, as well as shifts in social identity and authority resulting from the automation of decision-making processes (Boyer et al. 2024). One notable social benefit of Agriculture 4.0 is its potential to attract

Z. Ž. Kroupová (✉) · L. Rumánková · R. Aulová · L. Čechura
Department of Economics, Faculty of Economics and Management, Czech University of Life Sciences Prague, Prague, Czech Republic
e-mail: kroupovaz@pef.czu.cz

F. da Silveira et al. (eds.), *Agriculture 4.0*,
https://doi.org/10.1007/978-3-032-24552-6_7

younger generations to the farming sector by modernizing the sector and making decision-making more efficient (Doornbos et al. 2024).

Peer-reviewed scholarship frequently reports cost savings in key operational areas, including fertilizer, seed, fuel, or irrigation costs as one of the most common economic benefits of using Agriculture 4.0 technologies (Martinez-Guanter et al. 2020; Tahmasebi et al. 2023). More efficient resource use enabled by these technologies leads to higher yields, improved productivity, and increased profitability–central economic incentives for adoption (Kazlauskas et al. 2022; Quan et al. 2023).

Reducing the environmental burden is another major advantage associated with the use of these technologies. Reported benefits of the adoption of Agriculture 4.0 technologies include water conservation, decreased use of chemical inputs due to more controlled application, reduced chemical residues in groundwater, enhanced nutrient recycling, improved soil quality, and reduced CO_2 emissions (Lavrenko et al. 2021; Jovarauskas et al. 2021; Denora et al. 2023).

Despite these benefits, the adoption rate of Agriculture 4.0 technologies remains suboptimal (Amoussouhoui et al. 2024). To understand the barriers to adoption, it is crucial to examine the decision-making processes of farmers and the various factors that influence their choices. A comprehensive understanding of these drivers and barriers is crucial from both microeconomic and policy perspectives. Such insights can help the industry design technologies that are more likely to be adopted by farmers and develop more effective marketing strategies (Degieter et al. 2023).

From a policymaking perspective, identifying the key factors behind adoption decisions is critical for designing policies that effectively support the widespread use of advanced agricultural technologies (Tey and Brindal 2022). Without such understanding, policy efforts to promote technological innovation may fail to be effective (Weersink and Fulton 2020), ultimately hindering the agricultural sector's transition toward greater productivity and sustainability.

2 Adoption of Agriculture 4.0 Technologies

2.1 Technology Adoption Decision-Making Process

From a standard microeconomic perspective, the adoption of Agriculture 4.0 technologies can be viewed as a process of profit maximization, or more broadly as expected profit or utility maximization. The profit maximization model compares the net returns with and without the new technology, requiring data on both the costs and the expected impact on production. However, because these outcomes are not known at the time the farmer makes an investment decision, the simple net return model is often extended to account for the risks associated with the new technology. This results in the development of expected profit or expected utility maximization models, which incorporate the uncertainty associated with adopting new technologies (Weersink and Fulton 2020).

The assumption that profit maximization drives agricultural producers' decisions predominates in Agriculture 4.0 adoption research. However, insights from behavioral economics and reviews of adoption determinants reveal that factors beyond profit maximization also influence farmers' choices. These factors are explained by adoption theory, which understands technology adoption as a form of behavior change. Among the most well-known theories are the Theory of Reasoned Action (TRA) (Fishbein and Ajzen 1975), the Technology Acceptance Model (TAM) (Davis et al. 1989), the Theory of Planned Behavior (TPB) (Ajzen 1991), and the Unified Theory of Acceptance and Use of Technology (UTAUT) (Venkatesh et al. 2003).

According to the TRA, an individual's behavior can be predicted by their beliefs and attitudes, which form a causal chain leading to intentions and, ultimately, behavior. The theory assumes that individuals behave rationally, evaluating the likelihood of a particular outcome. If the expected result is perceived as desirable or valuable, they are more likely to develop a positive attitude toward the behavior. However, intentions are also shaped by subjective norms (i.e., social pressure), which stem from the individual's perception of how others will view their actions. In this model, personal attitudes and social norms directly influence behavioral intentions, making these intentions the strongest predictor of actual behavior. All other environmental factors affect behavior only indirectly, through their influence on attitudes and subjective norms (Tsai et al. 2012). This model has been empirically applied in the context of Agriculture 4.0 adoption by Hüttel et al. (2020), who examined farmers' intentions to use sustainable digital fertilization methods in Germany.

Ajzen's TPB extends the TRA to account for situations in which individuals do not have full voluntary control over their actions. The TPB proposes that behavioral intention is influenced by three key factors: attitude, subjective norms, and perceived behavioral control–reflecting an individual's perception of internal and external constraints on behavior (Granić 2024). This framework has been applied to the study of the adoption of Agriculture 4.0 by Feisthauer et al. (2024) and Yang et al. (2024).

Davis's TAM was specifically developed to explain how individuals adopt information technologies. It emphasizes two key factors influencing technology acceptance: perceived usefulness and ease of use, both of which are based on the technology's characteristics. Perceived usefulness refers to the degree to which users believe a technology will enhance their performance or efficiency, while perceived ease of use relates to the anticipated effort required to use the technology. TAM suggests that users are less likely to adopt technologies that require significant effort, and that acceptance depends on the technology's ability to effectively fulfill its intended purpose (Davis et al. 1989). In the original model, perceived usefulness and perceived ease of use shape an individual's attitude toward using the technology, which then influences behavioral intention and actual usage. However, later studies have found that both factors can directly influence behavioral intention and actual usage, without being mediated by attitude, reducing the need to consider attitude as a mediating factor.

Subsequent research has shifted focus to exploring how various individual and contextual factors influence perceived usefulness and perceived ease of use (Venkatesh and Davis 2000). This is evident in TAM studies on agriculture, where self-efficacy—i.e., farmers' awareness of their ability to carry out a particular activity—is one of the most frequently examined factors influencing both constructs. Self-efficacy reflects how individuals evaluate their skills, experience, knowledge, and confidence in using new technologies (Sara et al. 2024). In Agriculture 4.0 studies, TAM has been used to explain individuals' intentions by Caffaro et al. (2020), Michels et al. (2021), and Sara et al. (2024). Across these studies, the relationship between perceived usefulness, perceived ease of use, and the intention to adopt has varied: in some cases, all factors are equally influential, while in others, only perceived usefulness significantly affects farmers' intentions.

Finally, Venkatesh's UTAUT integrates insights from previous theories and outlines four core components: performance expectancy, effort expectancy, social influence, and facilitating conditions. The first three factors shape user intentions to adopt the technology, while the fourth directly affects actual usage behavior. Performance expectancy refers to the degree to which an individual believes the technology will improve their performance, similar to perceived usefulness in TAM. Effort expectancy, comparable to perceived ease of use in TAM, reflects the perceived ease of using the technology. Social influence captures perceived pressure from others to adopt the technology. Lastly, facilitating conditions include external factors and resources that either enable or hinder the adoption of the technology. The model also proposes that gender, age, experience, and voluntariness of use moderate the impact of these factors on intentions and behavior. The UTAUT model demonstrates greater explanatory power than earlier models by incorporating task-technology fit, social influence, and resource availability within facilitating conditions (Dissanayake et al. 2022). As a result, UTAUT is currently considered one of the most influential frameworks for understanding technology acceptance and utilization (Piancharoenwong and Badir 2024).

In Agriculture 4.0 adoption research, UTAUT has been used as a theoretical model by scholars such as Li et al. (2020), Schukat and Heise (2021), and Giua et al. (2022). Li et al. (2020) adapted the UTAUT model by adding perceived risks—financial, technical, production, and managerial—that negatively affect adoption intentions. In contrast, Schukat and Heise (2021) extended the UTAUT2 model (developed by Venkatesh et al. 2012), which builds on the original UTAUT by incorporating three additional constructs: hedonic motivation, price value, and habit. Schukat and Heise (2021) further enriched the model by incorporating trust and technology readiness, and highlighted trust as a critical factor in agricultural business relationships noting that without it, farmers are reluctant to adopt new technologies despite their potential benefits. Trust is also central to data ownership, privacy, and security concerns, as it can shape farmers' risk perceptions associated with a given technology (Bekee et al. 2024).

Technology readiness, the second factor, refers to an individual's preparedness to implement technology. It encompasses personal competence and control beliefs, reflecting confidence in one's ability to use a given technology (Schukat and Heise 2021). Building on this framework, Yeo and Keske (2024) examined barriers to digital agricultural adoption in California. Their findings were consistent with those of Schukat and Heise (2021) and indicated that economic profitability is the dominant factor influencing adoption decisions, while trust in both technology's performance and in those recommending it acts as a moderating factor, either facilitating or hindering adoption.

While the models mentioned above focus on individual-level technology adoption, organizational-level adoption is addressed by theories such as the Technology, Organization, and Environment (TOE) Framework (Tornatzky and Fleischer 1990), which is among the most well-known frameworks in the field (Granić 2024). The TOE framework identifies three key dimensions that influence adoption decisions: technological, organizational, and environmental. The technological dimension primarily includes the basic technological conditions available (e.g., technological infrastructure, capabilities) and the characteristics of technology use (e.g., relative advantages, compatibility, cost, and complexity). The organizational dimension refers to the farm's characteristics and resources, such as the quality of human capital, farm size, and availability of slack resources. The environmental dimension encompasses factors like industry and market structure, the presence or absence of technology service providers, and government regulations (Li et al. 2024; Baker 2012). In Agriculture 4.0 adoption research, this framework has been applied by Bahari et al. (2024) to investigate the factors shaping Internet of Things (IoT) adoption in Malaysia as well as by Yoon et al. (2020) to analyze the adoption of smart farming in Korea. Both papers reached a similar conclusion regarding the importance of compatibility for technology adoption.

These theories and models are continually modified and extended to better capture the complexity of the adoption process. In addition to developing new models, there is also ongoing theoretical synthesis within the analysis at the individual and organizational level, as well as across these levels. For example, Li et al. (2024) combine theories explaining individual behavior with the TOE framework.

Based on the aforementioned theories, empirical evaluations of the adoption patterns of Agriculture 4.0 technologies can be carried out through either ex-ante or ex-post approaches. Ex-ante models investigate the potential acceptance of emerging technologies, while ex-post models analyze the factors determining the current level of adoption. These approaches are often combined in studies analyzing technology diffusion, which typically exhibit a sigmoidal or S-shaped trajectory (Weersink and Fulton 2020). The corresponding models are based on diffusion theory, which addresses a macrolevel perspective of the diffusion of innovation over time.

Rogers's Innovation Diffusion, one of the most well-known frameworks, identifies five key characteristics that influence the adoption of innovations: relative advantage, compatibility, observability, trialability, and complexity (Rogers 2003).

Relative advantage refers to the perceived benefits of an innovation compared to its predecessors. Compatibility assesses how well the innovation aligns with potential adopters' attitudes and norms. Observability reflects the extent to which the results of an innovation are visible to others. Trialability allows potential adopters to experiment with the innovation before committing to full adoption. Complexity represents how difficult the innovation is perceived to be in terms of understanding and implementation (Rogers 2003). These characteristics are also commonly featured in adoption studies using the TOE framework.

2.2 *Drivers and Barriers to the Adoption of Agriculture 4.0 Technologies*

The adoption of new agricultural technologies is generally influenced by a wide range of factors, including the characteristics of the technology itself; the agroecological and biophysical conditions of implementation; producers' attitudes toward innovation; risk aversion; and organizational inertia. Table 1 summarizes the most commonly cited factors in empirical studies on Agriculture 4.0 technology adoption, whether in terms of intention to adopt or actual adoption behavior.

For clarity, these factors are grouped into five categories: economic, organizational (including management and human resources), technical and technological, policy-related, and environmental and social responsibility. Economic factors include financial considerations such as investment costs, resource availability, uncertainty around return on investment, and market dynamics (e.g., competitive pressure). Organizational factors refer to the human and structural components of farm operations, such as management characteristics (age, gender, education), levels of knowledge and skill, training availability, organizational culture, leadership support, and farm size. Technical and technological factors capture the key traits of Agriculture 4.0 technologies, including their usability, compatibility, complexity, and expected benefits. Policy factors encompass government support schemes, regulations, and certification requirements that either enable or hinder technology adoption. Finally, environmental and social responsibility factors relate to sustainability concerns and the adoption of practices aimed at minimizing environmental impact while maintaining farm viability.

This categorization represents one of several possible classification frameworks. The literature (e.g., Tey and Brindal 2012; da Silveira et al. 2021; Abbasi et al. 2022; Piancharoenwong and Badir 2024) offers diverse approaches to organizing the determinants of Agriculture 4.0 adoption. The classification shown in Table 1 aligns most closely with that proposed by da Silveira et al. (2021). Fragomeli et al. (2024) further refine this framework, distinguishing organizational factors into three subcategories: individual factors (e.g., age, gender, education, digital skills, experience, and innovativeness), psychological factors (e.g., interest, skepticism, trust, risk perception, and social influence—including collaboration amongst farm-

Table 1 Drivers and barriers to Agriculture 4.0 adoption

Factors	Description	Type	Examples of studies
Economic factors			
High investment cost	High investment costs and time requirements reduce the willingness to adopt Agriculture 4.0. Adoption may be discouraged if the expected benefits are insufficient to offset the initial investment	Barrier	Yoon et al. (2020), Bahari et al. (2024), Ammann et al. (2022), Makinde et al. (2022), Troiano et al. (2023), Vecchio et al. (2022), Pandeya et al. (2025), Kendall et al. (2022), Wang et al. (2023)
Lack of resources	A shortage of resources, including financial constraints, hinders the adoption of Agriculture 4.0	Barrier	Aubert et al. (2012), Bahari et al. (2024), Kendall et al. (2022)
Return on investment	Uncertainty about the return on investment can impede the adoption of Agriculture 4.0, as farmers may hesitate to invest without a clear understanding of the financial benefits or long-term profitability	Barrier	Makinde et al. (2022), Pandeya et al. (2025)
Competitive pressures (cost, productivity, and quality)	The pressure farms face from competitors encourages the adoption of innovative solutions to stay competitive in the market	Driver	Yoon et al. (2020), Bahari et al. (2024),
Organizational factors, management, and human resources			
Manager age	Older farmers are less willing to adopt and utilize new technologies, whereas younger farmers demonstrate greater enthusiasm and a stronger inclination to embrace modern technologies.	Barrier	Michels et al. (2020), Vecchio et al. (2022), Pandeya et al. (2025), Aubert et al. (2012), Kendall et al. (2022)
Manager gender	Male managers are generally more willing to adopt Agriculture 4.0 than female managers, as studies suggest men may be more achievement-oriented in this context.	Barrier	Michels et al. (2020), Pandeya et al. (2025), Kendall et al. (2022)
Lack of knowledge	Limited knowledge of Agriculture 4.0 hinders adoption and effective use.	Barrier	Ammann et al. (2022), Aubert et al. (2012), Kendall et al. (2022),
Lack of skilled human resources	A lack of skilled employees can impede the adoption of Agriculture 4.0.	Barrier	Yoon et al. (2020), Giorgio et al. (2024), Vecchio et al. (2022), Aubert et al. (2012), Kendall et al. (2022)

(continued)

Table 1 (continued)

Factors	Description	Type	Examples of studies
Lack of training and education	Limited training and education hinder the adoption of Agriculture 4.0 by failing to equip workers with the skills needed to implement and utilize these technologies effectively. This knowledge deficit creates resistance, reduces system efficiency, and introduces errors that undermine the perceived benefits of Agriculture 4.0 solutions	Barrier	Giorgio et al. (2024), Kendall et al. (2022)
Organizational inertia and risk-averse culture	Farms inclined to maintain the status quo may resist adopting new technologies due to perceived risks such as disruptions, financial losses, or potential failure	Barrier	Patil and Veettil (2024)
Farm size	Large farms are more likely to adopt new technologies due to greater flexibility and risk absorption. In contrast, small farms lag due to economies of scale, productivity differences, input prices, and higher transaction costs	Driver	Bahari et al. (2024), Michels et al. (2020), Ammann et al. (2022), Vecchio et al. (2022), Pandeya et al. (2025), Aubert et al. (2012), Kendall et al. (2022)
Labor intensity	Greater pressure to increase efficiency and reduce labor costs in high-labor-intensity production drives the adoption of technologies that reduce manual labor and enhance productivity	Driver	Vecchio et al. (2022)
Manager education	Higher education of the manager is associated with a greater willingness to adopt Agriculture 4.0	Driver	Michels et al. (2020), Vecchio et al. (2022), Pandeya et al. (2025), Aubert et al. (2012), Kendall et al. (2022)
Management support	Management support refers to the commitment of senior leadership to invest time in evaluating plans, forecasting outcomes, and addressing managerial challenges. It is acknowledged as a critical factor in successfully adopting new technologies	Driver	Bahari et al. (2024)
Technical and technological factors			
Usability	Ease of use describes how simple and intuitive a technology is for users, emphasizing the effort required for effective use. Technologies demanding significant effort or specialized skills are often avoided	Barrier	Pandeya et al. (2025), Aubert et al. (2012)

(continued)

Table 1 (continued)

Factors	Description	Type	Examples of studies
Complexity	High complexity creates barriers to adoption, as systems that are difficult to understand or implement may deter potential users	Barrier	Yoon et al. (2020), Bahari et al. (2024), Ammann et al. (2022), Makinde et al. (2022), Giorgio et al. (2024), Vecchio et al. (2022), Pandeya et al. (2025), Wang et al. (2023)
Lack of compatibility	Compatibility refers to how well new technologies integrate with existing systems, equipment, practices, needs, policies, and legal requirements. Low compatibility hinders the adoption of novel technologies	Barrier	Yoon et al. (2020), Bahari et al. (2024), Ammann et al. (2022), Aubert et al. (2012), Giorgio et al. (2024), Kendall et al. (2022), Wang et al. (2023),
Lack of scalability and interop- erability	Interoperability refers to the ability of systems to work together, exchange information, and integrate solutions. Limited interoperability and scalability among agricultural technologies disrupt workflows, hinder data exchange, and create bottlenecks. This increases costs and reduces farmers' willingness to adopt innovations	Barrier	Giorgio et al. (2024)
Lack of physical and IT infrastructure	Insufficient or outdated infrastructure significantly limits the implementation and effectiveness of Agriculture 4.0 in farming operations	Barrier	Ammann et al. (2022), Bahari et al. (2024), Wang et al. (2023)
Distrust and data security concerns	Lower trust in a technology and/or the entities responsible for its management and operation, particularly regarding data security concerns, reduces the likelihood of adoption	Barrier	Ammann et al. (2022), Pandeya et al. (2025), Idowu et al. (2023), Wang et al. (2023)
Relative advantage	Measurable benefits of Agriculture 4.0, such as reduced costs, streamlined operations, and better resource utilization, make these solutions attractive for farms aiming to increase productivity and efficiency	Driver	Yoon et al. (2020), Bahari et al. (2024), Ammann et al. (2022), Troiano et al. (2023), Vecchio et al. (2022), Aubert et al. (2012), Kendall et al. (2022)
Policy factors			
Government support	Government support includes administrative assistance and financial incentives for farms to implement and utilize new technologies	Driver	Yoon et al. (2020), Bahari et al. (2024), Ammann et al. (2022), Kendall et al. (2022),

(continued)

Table 1 (continued)

Factors	Description	Type	Examples of studies
Regulations and certifications	Regulatory frameworks and certification requirements that mandate or incentivize the monitoring and reporting of agricultural inputs and practices, which create compliance pressures that drive farmers to adopt Agriculture 4.0 for accurate data collection, efficient resource management, and transparent documentation of environmental performance	Driver	Ammann et al. (2022), Aubert et al. (2012), Kendall et al. (2022)
Environmental factors and social responsibility			
Sustainability	Societal pressure to pursue sustainability, combined with the belief that Agriculture 4.0 can conserve resources, reduce ecological footprints, and ensure long-term economic viability drives adoption	Driver	Troiano et al. (2023)
Environmental risk concerns	Changes in natural systems influence farmers' risk preferences and their adoption of new technologies. Agriculture 4.0 can help manage environmental risks, increasing farmers' willingness to adopt them	Driver	Kendall et al. (2022)

Source: own processing

ers), and farm-related factors (e.g., farm size, type of production, and degree of specialization).

Piancharoenwong and Badir (2024) offer a simpler structure, categorizing adoption factors into two broad groups: internal and external. Internal factors are tied to business imperatives such as improving efficiency and management control and are influenced by farmers' knowledge, skills, attitudes, age, and preferences. External factors relate to the accessibility and affordability of Agriculture 4.0 technologies and the provision of supporting and complementary programs.

Additionally, adoption factors can be further categorized into drivers and barriers (see Table 1). Systematic reviews consistently identify the primary barriers to adoption as high investment costs, limited compatibility, and insufficient skills and knowledge. Conversely, perceived profitability (i.e., relative advantage) and government support are considered among the strongest drivers of Agriculture 4.0 adoption.

Acquiring Agriculture 4.0 technologies is associated with high investment costs and often longer payback periods, particularly for large-scale systems. The expected return on investment plays a critical role in determining whether a farm can afford to adopt such technologies. These costs are especially burdensome for Small and Medium-Sized Farms (SMFs), which typically face constrained access to financial resources and limited liquidity that restrict their ability to adopt new technologies (Yoon et al. 2020).

Technological compatibility, interoperability, and scalability are crucial from both technological and economic perspectives. Farmers tend to avoid adopting technologies that are incompatible with existing practices, regulations, or legal frameworks (Makinde et al. 2022). A comprehensive review of Agriculture 4.0 by da Silveira et al. (2021) identifies the lack of technological compatibility as one of the most significant barriers to adoption alongside concerns about reliability and complexity. Within Agriculture 4.0, interoperability and scalability are equally critical. For optimal use, devices must be able to communicate with one other. When interoperability is limited, workflows are disrupted, data exchange becomes difficult, and performance bottlenecks emerge—leading to higher implementation and maintenance costs. Similarly, poor scalability shortens the economic lifespan of investments and restricts expansion potential, thereby reducing farmers' willingness to adopt Agriculture 4.0 systems more broadly.

Despite the costs, farmers consider the expected benefits in their decision-making, as these benefits naturally represent key drivers of adoption. Numerous studies have found that Agriculture 4.0 can increase farm productivity and product quality while reducing labor, energy, and fertilizer use. These improvements can provide farms with a significant competitive advantage (Yoon et al. 2020). Ammann et al. (2022), using the Delphi method in their study on precision farming in Swiss outdoor vegetable production, identified cost savings, increased revenues, and labor/time efficiency as dominant adoption drivers. In this context, profitability remains the core determinant in adoption decisions (Lee et al. 2021; Fragomeli et al. 2024).

Knowledge and skills are also critical prerequisites for adopting Agriculture 4.0 technologies. Due to their complexity, these technologies require a high level of human capital. Insufficient knowledge or digital proficiency can significantly hinder both adoption and effective use (Tey and Brindal 2012). Therefore, more highly educated farmers are generally better equipped to meet the demands of Agriculture 4.0. Targeted training and experience-building initiatives can help bridge skill gaps and encourage broader adoption (Gerli et al. 2022).

Age also plays a role in adoption. Research consistently shows that younger, more educated farmers—especially those with training in digital tools—are more enthusiastic and less risk-averse, making them early adopters. Older farmers, by contrast, often have shorter planning horizons, fewer incentives to change, and rely more heavily on established practices and experience, which results in avoidance of adopting highly capital-intensive innovations (Tey and Brindal 2012).

Larger farms have a significant advantage over SMFs in adopting capital-intensive technologies. Their access to highly skilled labor, internal capital reserves, data for decision-making, and external financing–as well as economies of scale—enables more efficient investment in Agriculture 4.0 (Barnes et al. 2019).

Farm and crop specialization may also shape adoption. While less frequently discussed in the literature, indicators such as income share from specialized activities, unpaid farm labor, land ownership, and farm location all influence likelihood of adoption (Barnes et al. 2019). Specialized farms benefit from economies of scale, which helps overcome cost barriers associated with adopting capital-intensive innovations. Their managerial focus also enhances their ability to integrate and operate advanced technologies effectively and provides them with a stronger motivation to adopt Agriculture 4.0 tools that mitigate risks inherent in specialized systems (Watcharaanantapong et al. 2014).

Land tenure influences adoption patterns. Farmers who own land tend to invest more readily in new technologies than those who rent land, due to lower risk aversion and stronger motivation to enhance long-term productivity. Secure land tenure encourages longer investment horizons, making high-cost technologies more appealing (Tey and Brindal 2012).

The location of a farm is highly connected to its soil quality, which also influences adoption patterns. Agricultural producers operating on high-quality land often make use of Agriculture 4.0 technologies to preserve soil productivity over time (Paxton et al. 2011).

Growing environmental concerns are also influencing adoption. Climate change-related risks such as increased yield variability and greater climatic fluctuations as well as societal pressure for sustainable practices may motivate producers to adopt technologies that enable more precise resource management and reduce environmental harm (Lee et al. 2021).

2.3 Policy Implications

The literature highlights that the high investment costs of Agriculture 4.0 technologies are a significant barrier to widespread adoption. Supportive policies for these investments can play a crucial role in influencing farmers' decisions regarding adoption. Ozcelik and Güldal (2024) posit that support mechanisms, such as direct payments or subsidized loans, can significantly shift farmers' intentions towards adopting Agriculture 4.0 technologies. Such support is particularly impactful for SMFs, as they often face considerable constraints (Yoon et al. 2020). There are also significant public benefits associated with Agriculture 4.0, including their contributions to more sustainable and resilient agricultural systems, which further support and justify adoption.

Many of the Agriculture 4.0 applications that have the greatest societal benefits are often not economically viable for farmers, which leads to low adoption rates. This discrepancy highlights the need for policy interventions to bridge the gap between societal benefits and individual economic incentives. Moreover, support for digital innovation can serve as a means to achieve policy goals including reducing environmental externalities, increasing resilience, and improving animal welfare cost-efficiently (Barnes et al. 2019; Finger 2023).

Agriculture 4.0 innovations are both developed and enter the market relatively quickly. Thus, the most efficient technology may change quickly. With this consideration, Finger (2023) finds that policy interventions should not focus on specific Agriculture 4.0 technologies. Instead, policymakers should create an enabling environment and design incentive schemes that promote developing and implementing the most cost-efficient strategies to achieve broader policy goals.

These interventions can take various forms such as financial incentives, economic subsidies, and regulatory obligations. Previous scholarship finds that different types of policy instruments can have varying impacts on adoption rates, with economic incentives generally providing greater motivation than traditional regulations, such as environmental rules. Additionally, results-based payments, such as compensating farmers per kg of nitrogen reduction, support the adoption of Agriculture 4.0 more effectively than area-based payments (Kerr and Newell 2003).

Finger (2023) suggests designing policy interventions that combine sufficient IT infrastructure, frameworks for data use and exchange, promote education and information, encourage collaboration and resource sharing, and implement economic instruments such as taxes on inputs like fertilizers and pesticides, as well as targeted subsidies and direct payments. This aligns with the findings of empirical studies analyzing the adoption of Agriculture 4.0 (Huber et al. 2023).

Literature confirms that SMFs face significant resource constraints hindering their adoption of Agriculture 4.0 (Kendall et al. 2022; Vecchio et al. 2022). Economic incentives and public subsidies remain indispensable for these farms to overcome the substantial upfront investments required for Agriculture 4.0 technologies (Yoon et al. 2020). A number of initiatives have been developed to reduce individual costs while facilitating knowledge transfer for SMFs including targeted subsidies for small agricultural businesses, collaborative approaches (e.g., joint investments), equipment-sharing arrangements, farmer networks, and formal cooperatives.

Previous studies reveal that the adoption of Agriculture 4.0 is significantly influenced by the knowledge and skills of the farmer as well as broader social and cultural factors. To enhance the knowledge and skills of the farmer, it is recommended that educational and outreach initiatives be developed, including cooperation with universities. In addition, training and technical support policies can be introduced to ensure the proper use of technology. These educational instruments should not be limited to a period of time but should instead support farmers throughout their careers and should serve to enrich their understanding on established and emerging smart technologies (Gerli et al. 2022).

Although educational programs and formal educators play a key role in developing digital skills in the agribusiness sector, Gerli et al. (2022) highlight the important contribution of informal sources in shaping positive perceptions and building trust in smart farming technologies. Informal sources include word-of-mouth and peer observation. It is important to note that these informal sources do not replace formal training, but rather complement it by addressing the psychological barriers to adoption that formal education alone cannot overcome. For example, governments can support disseminating information about Agriculture 4.0 technologies by presenting examples of good practices through storytelling solutions. Carefully designed narratives can influence potential adopters' cognitive and emotional decision-making processes. Gerli et al. (2022) also highlight the role of farmers' unions in establishing online platforms and organizing events where members can discuss and compare their experiences with Agriculture 4.0 technologies.

In addition, empirical studies show that moral norms play a crucial role in shaping farmers' attitudes towards Agriculture 4.0 (e.g., Feisthauer et al. 2024). These perceived moral obligations and norms evolve and cannot be imposed through government regulation. However, government campaigns that raise awareness of the environmental benefits of Agriculture 4.0 can help motivate farmers recognize and actively contribute to this development (Feisthauer et al. 2024).

Empirical studies also highlight the importance of compatibility and interoperability that, from a policy perspective, standardization could support. Policymakers could prioritize developing open technical standards to ensure interoperability between Agriculture 4.0 components–from IoT sensors and autonomous machinery to data analytics platforms and farm management systems. Such standardization would accelerate technology adoption as well as help prevent market concentration (Aubert et al. 2012).

Moreover, concerns regarding data security may hinder the adoption of these technologies. According to Abbasi et al. (2022), cybersecurity is a critical challenge that must be tackled in Agriculture 4.0, involving various privacy-preserving mechanisms and federated learning techniques. Policies could be developed to protect farmer data, helping to bridge the trust gap between farmers and service providers.

3 Conclusion

The adoption of Agriculture 4.0 provides an opportunity for transforming agricultural practices. Advancements in precision agriculture as well as the integration of the IoT, robotics, artificial intelligence, and Big Data provide promising solutions to global challenges such as food security, resource management, and environmental sustainability. Therefore, economists and policymakers are increasingly focusing on the diffusion and adoption of these technologies.

Since the introduction of Rogers' Diffusion of Innovations Theory, the question of innovation adoption has attracted attention in academic circles. Theoretical models undergo continuous development and refinement, expanding factors considered as the determinants of adoption behavior. The synthesis of theoretical models and empirical results illustrates that factors at both the individual and organizational level can influence technology adoption.

Empirical results confirm that despite substantial benefits across economic, agronomic, social, and environmental dimensions, the adoption of Agriculture 4.0 technology is hindered by various factors. These include high initial costs, technological complexity, social resistance, and interoperability challenges. Addressing these barriers requires a collective effort consisting of governments, the private sector, and local communities to ensure that the technologies are accessible, affordable, and tailored to the needs of farmers. The successful integration of Agriculture

4.0 depends on overcoming all these challenges and fostering a collaborative and inclusive approach. With an appropriate policy framework, technical support, and infrastructural investments, Agriculture 4.0 has the potential to improve farming practices and contribute to a more sustainable and food-secure future.

Acknowledgments This work was supported by the Ministry of Agriculture of the Czech Republic, under the ZEME program [Project No. QK 23020058].

Competing Interests The author(s) has no competing interests to declare that are relevant to the content of this manuscript.

References

Abbasi R, Martinez P, Ahmad R (2022) The digitization of agricultural industry – a systematic literature review on agriculture 4.0. Smart Agric Technol 2:100042. https://doi.org/10.1016/j.atech.2022.100042

Ajzen I (1991) The theory of planned behavior. Organ Behav Hum Decis Process 50(2):179–211. https://doi.org/10.1016/0749-5978(91)90020-T

Ammann J, Umstätter C, El Benni N (2022) The adoption of precision agriculture enabling technologies in Swiss outdoor vegetable production: a Delphi study. Precis Agric 23:1354–1374. https://doi.org/10.1007/s11119-022-09889-0

Amoussouhoui R, Arouna A, Ruzzante S, Banout J (2024) Adoption of ICT4D and its determinants: a systematic review and meta-analysis. Heliyon 10(9):e30210. https://doi.org/10.1016/j.heliyon.2024.e30210

Aubert BA, Schroeder A, Grimaudo J (2012) IT as an enabler of sustainable farming: an empirical analysis of farmers' adoption decision of precision agriculture technology. Decis Support Syst 54(1):510–520. https://doi.org/10.1016/j.dss.2012.07.002

Bahari M, Arpaci I, Der O, Akkoyun F, Ercetin A (2024) Driving agricultural transformation: unraveling key factors shaping IoT adoption in smart farming with empirical insights. Sustainability 16(5):2129. https://doi.org/10.3390/su16052129

Baker J (2012) The technology–organization–environment framework. In: Dwivedi Y, Wade M, Schneberger S (eds) Information systems theory, Integrated series in information systems, vol 28. Springer, New York, pp 231–245. https://doi.org/10.1007/978-1-4419-6108-2_12

Barnes AP, Soto I, Eory V, Beck V, Balafoutis A, Sánchez B, Vangeyte J, Fountas S, van der Wal T, Gómez-Barbero M (2019) Exploring the adoption of precision agricultural technologies: a cross-regional study of EU farmers. Land Use Policy 80:163–174. https://doi.org/10.1016/j.landusepol.2018.10.004

Bekee B, Segovia MS, Valdivia C (2024) Adoption of smart farm networks: a translational process to inform digital agricultural technologies. Agric Hum Values 41:1573–1590. https://doi.org/10.1007/s10460-024-10566-3

Boyer CN, Cavasos KE, Greig JA, Schexnayder SM (2024) Influence of risk and trust on beef producers' use of precision livestock farming. Comput Electron Agric 218:108641. https://doi.org/10.1016/j.compag.2024.108641

Caffaro F, Cremasco MM, Roccato M, Cavallo E (2020) Drivers of farmers' intention to adopt technological innovations in Italy: the role of information sources, perceived usefulness, and perceived ease of use. J Rural Stud 76:264–271. https://doi.org/10.1016/j.jrurstud.2020.04.028

da Silveira F, Lermen FH, Amaral FG (2021) An overview of agriculture 4.0 development: systematic review of descriptions, technologies, barriers, advantages, and disadvantages. Comput Electron Agric 189:106405. https://doi.org/10.1016/j.compag.2021.106405

Davis F, Bogozzi R, Warshaw P (1989) User acceptance of computer technology: a comparison of two theoretical models. Manag Sci 35:982–1003

Degieter M, De Steur H, Tran D, Gellynck X, Schouteten JJ (2023) Farmers' acceptance of robotics and unmanned aerial vehicles: a systematic review. Agron J 115:2159–2173. https://doi.org/10.1002/agj2.21427

Denora M, Candido V, D'Antonio P, Perniola M, Mehmeti A (2023) Precision nitrogen management in rainfed durum wheat cultivation: exploring synergies and trade-offs via energy analysis, life cycle assessment, and monetization. Precis Agric 24(6):2566–2591. https://doi.org/10.1007/s11119-023-10053-5

Dissanayake CAK, Jayathilake W, Wickramasuriya HVA, Dissanayake U, Kopiyawattage KPP, Wasala WMCB (2022) Theories and models of technology adoption in agricultural sector. Hum Behav Emerg Technol 2022:9258317. https://doi.org/10.1155/2022/9258317

Doornbos J, Bennin KE, Babur Ö, Valente J (2024) Drone technologies: a tertiary systematic literature review on a decade of improvements. IEEE Access 12:23220–23239. https://doi.org/10.1109/ACCESS.2024.3364676

Feder G, Just R, Zilberman D (1985) Adoption of agricultural innovations in developing countries: a survey. Econ Dev Cult Chang 33(2):255–298. https://doi.org/10.1086/451461

Feisthauer P, Hartmann M, Börner J (2024) Behavioral factors driving farmers' intentions to adopt spot spraying for sustainable weed control. J Environ Manag 353:120218. https://doi.org/10.1016/j.jenvman.2024.120218

Finger R (2023) Digital innovations for sustainable and resilient agricultural systems. Eur Rev Agric Econ 50(4):1277–1309. https://doi.org/10.1093/erae/jbad021

Fishbein M, Ajzen I (1975) Belief, attitude, intention, and behaviour: an introduction to theory and research. Reading, Addison-Wesley

Fragomeli R, Annunziata A, Punzo G (2024) Promoting the transition towards Agriculture 4.0: a systematic literature review on drivers and barriers. Sustainability 16(6):2425. https://doi.org/10.3390/su16062425

Gerli P, Clement J, Esposito G, Mora L, Crutzen N (2022) The hidden power of emotions: how psychological factors influence skill development in smart technology adoption. Technol Forecast Soc Chang 180:121721. https://doi.org/10.1016/j.techfore.2022.121721

Giorgio A, Penate Lopez LP, Bertoni D, Cavicchioli D, Ferrazzi G (2024) Enablers to digitalization in agriculture: a case study from Italian field crop farms in the Po River valley, with insights for policy targeting. Agriculture 14(7):1074. https://doi.org/10.3390/agriculture14071074

Giua C, Materia VC, Camanzi L (2022) Smart farming technologies adoption: which factors play a role in the digital transition? Technol Soc 68:101869. https://doi.org/10.1016/j.techsoc.2022.101869

Granić A (2024) Technology adoption at individual level: toward an integrated overview. Univ Access Inf Soc 23:843–858. https://doi.org/10.1007/s10209-023-00974-3

Huber R, Späti K, Finger R (2023) A behavioural agent-based modelling approach for assessing policies supporting precision agriculture. Ecol Econ 212:107936. https://doi.org/10.1016/j.ecolecon.2023.107936

Hüttel S, Leuchten MT, Leyer M (2020) The importance of social norm on adopting sustainable digital fertilisation methods. Organ Environ 35(1):79–102. https://doi.org/10.1177/1086026620929074

Idowu AR, Wachenheim C, Hanson E, Sickler A (2023) The disposition of data from precision agricultural technologies: what do young agriculturalists think? Technol Soc 75:102389. https://doi.org/10.1016/j.techsoc.2023.102389

Jovarauskas D, Steponavičius D, Kemzūraitė A, Zinkevičius R, Venslauskas K (2021) Comparative analysis of the environmental impact of conventional and precision spring wheat fertilization under various meteorological conditions. J Environ Manag 296:113150. https://doi.org/10.1016/j.jenvman.2021.113150

Kazlauskas M, Šarauskis E, Lekavičienė K, Naujokienė V, Romaneckas K, Bručienė I, Buragienė S, Steponavičius D (2022) The comparison analysis of uniform-and variable-rate fertilizations on winter wheat yield parameters using site-specific seeding. PRO 10(12):2717. https://doi.org/10.3390/pr10122717

Kendall H, Clark B, Li W, Jin S, Jones GD, Chen J, Tayor J, Li Z, Frewer LJ (2022) Precision agriculture technology adoption: a qualitative study of small-scale commercial "family farms" located in the North China Plain. Precis Agric 23:319–351. https://doi.org/10.1007/s11119-021-09839-2

Kerr S, Newell RG (2003) Policy-induced technology adoption: evidence from the U.S. lead phasedown. J Ind Econ 51:317–343. https://doi.org/10.1111/1467-6451.00203

Lavrenko SO, Lavrenko NM, Maksymov DO, Maksymov M, Didenko NO, Islam KR (2021) Variable tillage depth and chemical fertilization impact on irrigated common beans and soil physical properties. Soil Tillage Res 212(August):105024. https://doi.org/10.1016/j.still.2021.105024

Lee CL, Strong R, Dooley K (2021) Analyzing precision agriculture adoption across the globe: a systematic review of scholarship from 1999–2020. Sustainability 13(18):10295. https://doi.org/10.3390/su131810295

Li W, Clark B, Taylor JA, Kendall H, Jones G, Li Z, Jin S, Zhao C, Yang G, Shuai C, Cheng X, Chen J, Yang H, Frewer LJ (2020) A hybrid modelling approach to understanding adoption of precision agriculture technologies in Chinese cropping systems. Comput Electron Agric 172:105305. https://doi.org/10.1016/j.compag.2020.105305

Li L, Min X, Guo J, Wu F (2024) The influence mechanism analysis on the farmers' intention to adopt Internet of Things based on UTAUT-TOE model. Sci Rep 14:15016. https://doi.org/10.1038/s41598-024-65415-4

Makinde A, Islam MM, Wood KM, Conlin E, Williams M, Scott SM (2022) Investigating perceptions, adoption, and use of digital technologies in the Canadian beef industry. Comput Electron Agric 198:107095. https://doi.org/10.1016/j.compag.2022.107095

Martinez-Guanter J, Agüera P, Agüera J, Pérez-Ruiz M (2020) Spray and economics assessment of a UAV-based ultra-low-volume application in olive and citrus orchards. Precis Agric 21(1):226–243. https://doi.org/10.1007/s11119-019-09665-7

Michels M, Fecke W, Feil J-H, Musshoff O, Lülfs-Baden F, Krone S (2020) Anytime, anyplace, anywhere - a sample selection model of mobile internet adoption in German agriculture. Agribusiness 36:192–207. https://doi.org/10.1002/agr.21635

Michels M, von Hobe CF, Weller von Ahlefeld PJ, Musshoff O (2021) The adoption of drones in German agriculture: a structural equation model. Precis Agric 22:1728–1748. https://doi.org/10.1007/s11119-021-09809-8

Ozcelik A, Güldal HT (2024) From conventional to smart: farmers' preferences under alternative policy scenarios. New Medit. https://10.30682/nm2401a.

Pandeya S, Gyawali BR, Upadhaya S (2025) Factors influencing precision agriculture technology adoption among small-scale farmers in Kentucky and their implications for policy and practice. Agriculture 15(2):177. https://doi.org/10.3390/agriculture15020177

Patil V, Veettil PC (2024) Farmers' risk attitude, agricultural technology adoption and impacts in Eastern India. Agric Food Secur 13:50. https://doi-org.infozdroje.czu.cz/10.1186/s40066-024-00497-x

Paxton KW, Mishra AK, Chintawar S, Roberts RK, Larson JA, English BC, Lambert DM, Marra MC, Larkin SL, Martin SW (2011) Intensity of precision agriculture technology adoption by cotton producers. Agric Resour Econ Rev 40(1):133–144. https://doi.org/10.1017/S1068280500004561

Piancharoenwong A, Badir YF (2024) IoT smart farming adoption intention under climate change: the gain and loss perspective. Technol Forecast Soc Chang 200:123192. https://doi.org/10.1016/j.techfore.2023.123192

Quan X, Guo Q, Ma J, Doluschitz R (2023) The economic effects of unmanned aerial vehicles in pesticide application: evidence from Chinese grain farmers. Precis Agric 24(5):1965–1981. https://doi.org/10.1007/s11119-023-10025-9

Rogers E (2003) Diffusion of innovations, 5th edn. Free Press. ISBN 9780743222099

Sara G, Todde G, Pinna D, Caria M (2024) Investigating the intention to use augmented reality technologies in agriculture: will smart glasses be part of the digital farming revolution? Comput Electron Agric 224:109252. https://doi.org/10.1016/j.compag.2024.109252

Schukat S, Heise H (2021) Towards an understanding of the behavioral intentions and actual use of smart products among German farmers. Sustainability 13(12):6666. https://doi.org/10.3390/su13126666

Tahmasebi M, Gohari M, Sharifi Malvajerdi A, Hedayatipoor A (2023) Development and field evaluation of a variable-depth tillage tool based on a horizontal pneumatic sensor measurement. J Agric Mach 13(1):85–99. https://doi.org/10.22067/jam.2023.79231.1128

Tey YS, Brindal M (2012) Factors influencing the adoption of precision agricultural technologies: a review for policy implications. Precis Agric 13:713–730. https://doi.org/10.1007/s11119-012-9273-6

Tey Y, Brindal M (2022) A meta-analysis of factors driving the adoption of precision agriculture. Precis Agric 23(2):353–372. https://doi.org/10.1007/s11119-021-09840-9

Tornatzky LG, Fleischer M (1990) The processes of technological innovation. Lexington Books, Lexington

Troiano S, Carzedda M, Marangon F (2023) Better richer than environmentally friendly? Describing preferences toward and factors affecting precision agriculture adoption in Italy. Agric Econ 11:16. https://doi.org/10.1186/s40100-023-00247-w

Tsai MT, Chen KS, Chien JL (2012) The factors impact of knowledge sharing intentions: the theory of reasoned action perspective. Qual Quant 46:1479–1491. https://doi.org/10.1007/s11135-011-9462-9

Vecchio Y, De Rosa M, Pauselli G, Masi M, Adinolfi F (2022) The leading role of perception: the FACOPA model to comprehend innovation adoption. Agric Econ 10:5. https://doi.org/10.1186/s40100-022-00211-0

Venkatesh V, Davis FD (2000) A theoretical extension of the technology acceptance model: four longitudinal field studies. Manag Sci 46(2):186–204. https://doi.org/10.1287/mnsc.46.2.186.11926

Venkatesh V, Morris M, Davis G, Davis F (2003) User acceptance of information technology: toward a unified view. MIS Q 27(3):425–478. https://doi.org/10.2307/30036540

Venkatesh V, Thong JYL, Xu X (2012) Consumer acceptance and use of information technology: extending the unified theory of acceptance and use of technology. MIS Q 36:157–178. https://doi.org/10.2307/41410412

Wang T, Jin H, Sieverding HL (2023) Factors affecting farmer perceived challenges towards precision agriculture. Precis Agric 24:2456–2478. https://doi.org/10.1007/s11119-023-10048-2

Watcharaanantapong P, Roberts R, Lambert D, Larson J, Velandia M, English B, Rejesus R, Wang C (2014) Timing of precision agriculture technology adoption in US cotton production. Precis Agric 15(4):427–446. https://doi.org/10.1007/s11119-013-9338-1

Weersink A, Fulton M (2020) Limits to profit maximization as a guide to behavior change. Appl Econ Perspect Policy 42:67–79. https://doi.org/10.1002/aepp.13004

Yang Q, Al Mamun A, Masukujjaman M, Makhbul ZKM, Zhong Y (2024) Adoption of internet of things-enabled agricultural systems among Chinese agro-entreprises. Precis Agric 25:2477–2504. https://doi.org/10.1007/s11119-024-10182-5

Yeo ML, Keske CM (2024) From profitability to trust: factors shaping digital agriculture adoption. Front Sustain Food Syst 8:1456991. https://doi.org/10.3389/fsufs.2024.1456991

Yoon C, Lim D, Park C (2020) Factors affecting adoption of smart farms: the case of Korea. Comput Hum Behav 108:106309. https://doi.org/10.1016/j.chb.2020.106309

Economic Viability of Agriculture 4.0 Technologies in the Agrifood System

Ricardo Gonçalves de Faria Corrêa, Miguel Afonso Sellitto, Ismael Cristofer Baierle, Rodrigo Pessotto Almeida, and Joana Siqueira de Souza

1 Introduction

The agri-food sector faces major challenges due to increasing food demand driven by population growth and shifting consumption patterns (FAO 2020a, b; van Dijk et al. 2020). Simultaneously, climate change, resource scarcity, and sustainability pressures intensify the need for innovative agricultural practices (Rockström et al. 2020; Basso and Antle 2020). Agriculture 4.0 has emerged as a strategic response, integrating digital technologies—such as Internet of Things (IoT), Big Data, Artificial Intelligence (AI), robotics, and drones—to enhance the efficiency, sustainability, and productivity of food systems (Klerkx and Rose 2020).

These technologies enable precise resource management, real-time monitoring, and data-driven decision-making, which generate economic, environmental, and social benefits (Kamilaris et al. 2017; Farooq et al. 2020; Lioutas et al. 2021). However, evaluating their adoption requires more than technical or financial analysis. We need integrated frameworks that include not only financial metrics but also sociotechnical and environmental perspectives (Basso and Antle 2020; Beier

R. G. de Faria Corrêa (✉)
Federal University of Rio Grande (FURG), Santo Antônio da Patrulha, RS, Brazil
e-mail: ricardofariacorrea@furg.br

M. A. Sellitto
Univesity of Vale do Rio dos Sinos (UNISINOS), Porto Alegre, RS, Brazil
e-mail: sellitto@unisinos.br

I. C. Baierle
Federal University of Rio Grande (FURG), Santo Antônio da Patrulha, RS, Brazil
e-mail: ismaelbaierle@furg.br

R. P. Almeida · J. S. de Souza
Federal University of Rio Grande do Sul (UFRGS), Santo Antônio da Patrulha, RS, Brazil
e-mail: rodrigo.pessotto@ufrgs.br; joana.souza@ufrgs.br

F. da Silveira et al. (eds.), *Agriculture 4.0*,
https://doi.org/10.1007/978-3-032-24552-6_8

et al. 2020; Filmer and Jackson 2021). Lezoche et al. (2020) emphasize Agriculture 4.0's potential to increase output, reduce environmental impact, and promote social well-being. These technologies also align with the UN Sustainable Development Goals—especially Zero Hunger, Climate Action, and Responsible Consumption (FAO 2020a, b). Thus, understanding the key technologies and their economic implications is essential.

Despite these benefits, small and medium-sized producers still face barriers—such as high implementation costs, infrastructure gaps, and limited technical skills—that restrict widespread adoption (Kamilaris et al. 2017; Lezoche et al. 2020). While some studies confirm the long-term gains in productivity and cost reduction (FAO 2020a, b), few provide standardized tools to assess economic viability in the diverse and complex agri-food landscape (Filmer and Jackson 2021). This gap limits evidence-based decision-making for farmers, investors, and policymakers. Without robust, multidimensional assessment tools, it becomes difficult to scale Agriculture 4.0 sustainably (Aubert et al. 2012; Basso and Antle 2020). Researchers must therefore develop tailored investment models that reflect territorial specificities and systemic complexity.

To address this challenge, this chapter expands the investment assessment framework proposed by Almeida et al. (2022). We adapt it to include environmental considerations and align its sociotechnical dimension with the agri-food sector. Grounded in Innovation Diffusion Theory (Rogers 2003), the framework guides decision-makers through the early stages of adoption and supports comprehensive evaluation of Agriculture 4.0 technologies.

2 An Integrated Framework to Assess Agricultural 4.0 Technologies Investments

To assess Agriculture 4.0 technologies, it is essential to understand the decision-making process that underpins their adoption. Rogers (2003), through the Innovation Diffusion Theory, outlines five stages in the innovation-decision process: knowledge, persuasion, decision, implementation, and confirmation. This chapter focuses on the first three stages, which support preadoption evaluation. Figure 1 presents the proposed framework to assess agricultural 4.0 technologies investments.

Adopting digital technologies in agriculture is not an immediate choice but a structured journey. The process starts with the *Knowledge stage*, where producers or agrifood system managers become aware of a specific innovation—such as precision irrigation, autonomous tractors, or blockchain-based traceability—and gain a basic understanding of its purpose, function, and needs. Next, during the *Persuasion stage*, decision-makers analyze the technology from economic, financial, sociotechnical, and environmental perspectives. They seek detailed information, weigh potential benefits—like higher yields or greater efficiency—against costs, risks, and oper-

Fig. 1 Integrated framework for Agriculture 4.0 technology investment assessment

ational challenges, and often rely on peer insights and real-world cases. Finally, the *Decision stage* involves a firm commitment to adopt or reject the technology, frequently through resource allocation to pilot tests or small-scale trials. This step marks the initial operational integration of the technology, based on its perceived economic promise and strategic relevance. The next sections detail these stages.

2.1 Stage 1: Knowledge: Identifying Potential Technologies

The adoption of Agriculture 4.0 technologies begins with identifying relevant innovations, understanding their requirements, and assessing their suitability for each farm's specific characteristics. Farms vary widely in their technological maturity, workforce structure, and management style. These factors significantly influence the feasibility and success of technology adoption. Before selecting a technology, producers must understand both the internal context of their operation and the external requirements of the innovation. For example, implementing a dairy welfare sensor demands not only hardware and connectivity but also data storage, analytics tools, and a cultural shift in farm operations. Devices such as computers and smartphones become essential to manage the new information flow. Each technology presents its own prerequisites and benefits. The following sections summarize key Agriculture 4.0 technologies and their applications:

Internet of Things (IoT): IoT connects physical devices through sensors and networks, enabling real-time data collection and exchange. In agriculture, IoT supports:

- Soil and climate monitoring: sensors measure variables like moisture and nutrients, improving irrigation and fertilization efficiency (Lezoche et al. 2020).
- Machinery management: IoT tracks equipment usage and maintenance, reducing downtime and operational costs (Farooq et al. 2020; Lima et al. 2020).
- Animal welfare: wearable sensors monitor livestock health and behavior, enabling rapid response to stress or illness (Lezoche et al. 2020). When combined with AI and Big Data, IoT technologies enhance the intelligence and sustainability of farming systems.

Big Data and Data Analytics: agriculture produces vast data from sensors, drones, satellites, and machines. Big Data analytics transforms these datasets into actionable insights. Machine learning models, for instance, can predict harvest times or improve logistics by optimizing storage and transport (Lezoche et al. 2020; FAO 2020a, b).

Artificial Intelligence (AI) and Machine Learning (ML): AI automates complex tasks and improves decision-making. Algorithms analyze imagery to detect early plant stress, allowing timely interventions that reduce losses and input use (Lezoche et al. 2020). AI systems can also fine-tune resource applications to specific field conditions, improving sustainability (FAO 2020a, b).

Robotics and automation: robots automate labor-intensive tasks such as planting, weeding, and harvesting. For example, smart harvesters identify and pick ripe fruits delicately, reducing damage. Automated weed removal decreases herbicide use, improving both efficiency and environmental outcomes (Lezoche et al. 2020).

Drones and remote sensing: Unmanned Aerial Vehicles (UAVs) equipped with cameras and sensors offer high-resolution aerial data. They detect plant stress, assess moisture levels, and identify pests or diseases. This enables timely, precise interventions. Drones can also apply fertilizers or pesticides accurately and safely, even in hard-to-reach areas (FAO 2020a, b).

2.2 Stage 2: Persuasion: Evaluating Technologies from Different Lens

To determine whether a technology should be adopted, it must be evaluated from different perspectives. In the case of an Agriculture 4.0 technology, economic and financial aspects alone are not enough. Nonmonetary benefits or losses can be hidden in a purely financial assessment. Therefore, the sociotechnical and environmental perspectives cover these aspects that are not captured. Each of these three perspectives is presented below.

2.2.1 Economic and Financial Perspectives

While this chapter argues that a purely financial analysis is insufficient for evaluating Agriculture 4.0 technologies, it remains the non-negotiable foundation of any investment decision. A project that is not economically viable cannot be sustainable in the long run. This perspective is best understood through two distinct but complementary lenses: the economic lens, which focuses on long-term profitability and value creation, and the financial lens, which addresses liquidity and the risk associated with the time required to recover capital (Brealey et al. 2020). The economic lens evaluates an investment's ability to generate returns above its cost of capital, considering the time value of money. The primary tool for this is the

Net Present Value (NPV), which calculates the present value of all future cash flows (FC_t) generated by an investment, discounted by the minimum attractive rate of return (MARR), and subtracts the initial investment cost (FC_0) (Eq. 1).

$$\text{NPV} = \sum_{t=1}^{m} \frac{\text{FC}_t}{(1 + \text{MARR})^t} + \text{FC}_0 \tag{1}$$

A positive NPV indicates that the project is expected to generate more value than its cost, making it economically attractive (Ross et al. 2019). In the context of Ag 4.0, estimating these cash flows is a significant challenge:

- *Initial investment (FC_0)*: goes beyond the hardware price (e.g., a drone, a robot). It must include costs of installation, system integration, and, crucially, the substantial cost of initial operator training and process re-engineering.
- *Future cash flows (FC_t)*: these are a mix of cost savings and revenue enhancements. For instance, precision spraying technologies reduce expenditure on agrichemicals, while soil moisture sensors can lead to yield increases by optimizing irrigation. However, many benefits, such as improved decision quality from data analytics, are difficult to quantify precisely, often leading to conservative and underestimated NPVs—a well-documented challenge in the evaluation of advanced manufacturing technologies (Kaplan 1986).

To compare projects of different scales, the Internal Rate of Return (IRR) is used. The IRR is the discount rate at which the NPV of a project equals zero. A project is considered viable if its IRR is higher than the firm's MARR. While NPV and IRR measure profitability, the *Payback Period* measures a different dimension: risk and liquidity. It calculates the time required for a project's cumulative discounted cash flows to equal the initial investment.

For agricultural producers, who often face tight cash flow cycles and high market uncertainty, a short payback period can be psychologically and financially more compelling than a high NPV with a long-term return. A farmer might rationally choose a technology with a 2-year payback over one with a higher NPV but a 5-year payback, simply because the risks associated with commodity prices, weather, and technology obsolescence increase dramatically over longer time horizons (Boehlje and Lins 1998). This highlights the importance of the financial perspective in managing the producer's risk appetite.

The viability of any project is highly sensitive to the discount rate (MARR). A common method to determine the MARR is the *weighted average cost of capital (WACC)*, which represents the blended cost of a firm's equity and debt (Modigliani and Miller 1963) (Eq. 2):

$$\text{WACC} = \frac{E}{(E + D)} R_{\text{e}} + \frac{D}{(E + D)} R_{\text{D}} (1 - T) \tag{2}$$

Where:

- E is the market value of equity; D is the market value of debt.
- R_E is the cost of equity (the return shareholders expect).
- R_D is the cost of debt (the interest rate on loans).
- T is the corporate tax rate.

In agriculture, the WACC is particularly relevant. The cost of equity (R_E) is often high due to the inherent risks of the sector (e.g., climate volatility, pests, price fluctuations). Therefore, government policies that provide subsidized financing or "green loans" for sustainable technologies can directly lower the cost of debt (R_D), reducing the overall WACC (Patrick 1972). This makes capital-intensive Ag 4.0 projects, especially those with environmental benefits, significantly more viable. A lower MARR makes future cash flows more valuable in present terms, favoring investments with longer-term returns. Beyond firm-level decisions, public policies and data governance frameworks play a crucial role in shaping the economic feasibility of Agriculture 4.0 technologies. Subsidies, tax incentives, and credit programs—such as green loans—can significantly lower the WACC, accelerating adoption by mitigating upfront costs. Furthermore, data governance—including security, privacy, and ownership rules—affects producers' willingness to share and use digital data, influencing collaborative platforms and technology-enabled business ecosystems. Establishing transparent governance standards ensures trust, protects producers' rights, and facilitates equitable value distribution across the agri-food chain.

Ultimately, while these quantitative tools are essential for establishing a baseline of financial viability, their limitations in capturing intangible benefits and navigating the high uncertainty of agriculture underscore the need to enrich the analysis. The following sections will demonstrate how incorporating sociotechnical and environmental perspectives creates a more robust and realistic framework for decision-making.

2.2.2 Social Technical Perspective

Successfully adopting advanced technologies in digital transformation requires a sociotechnical approach that integrates technical, organizational, social, and environmental elements (Beier et al. 2020; Gillani et al. 2020). Cherns' Sociotechnical Theory (1976) views productive systems as the result of interactions between technical and social subsystems, whose effectiveness depends on their joint optimization. In Industry 4.0, this perspective becomes even more critical due to the high degree of integration between digital technologies—such as IoT, AI, Big Data, and cloud computing—and the physical, organizational, and human environments where they operate (Masood and Egger 2019; Sony and Naik 2020). This complexity demands evaluation beyond purely financial criteria. In this context, Almeida et al. (2022) proposed a framework with 12 sociotechnical criteria grouped into four dimensions: technology, organization, external environment, and social aspects. Developed through a systematic literature review and validated by academic and industry experts, the model supports broader and more sustainable investment decisions.

The technological dimension evaluates maturity, implementation complexity, and compatibility with existing infrastructure. The organizational dimension examines process readiness, structural adequacy, and strategic alignment. The external environment includes competitive pressure, regulatory requirements, and the capacity of external partners. The social dimension assesses working conditions, employee acceptance, and workforce skills. This framework provides a holistic view of Industry 4.0 investments, allowing decision-makers to consider not only economic returns but also sociotechnical risks and benefits. It promotes more effective and sustainable technology implementation. The same logic applies to Agriculture 4.0, which integrates digital technologies into farming to boost productivity, sustainability, and traceability. As agribusiness increasingly adopts sensors, drones, management systems, and predictive tools, decision-making becomes more complex. As in industry, evaluating digital transformation in rural settings requires a sociotechnical perspective.

Sociotechnical Criteria

Adopting technologies in the agro-industrial sector requires more than technical or economic feasibility. The sector's complexity—shaped by climate variability, geographic dispersion, diverse organizational cultures, uneven infrastructure, and varying workforce skills—demands a sociotechnical approach to assess readiness and maturity. Almeida et al. (2022) offer a framework of 12 sociotechnical criteria grouped into four dimensions: technological, organizational, external environment, and social. This model can be adapted to the realities of the agro-industrial context.

In the technological dimension, technological maturity considers the digitalization level of farms and value chains, including sensor use, machinery automation, and rural connectivity—often uneven across small and large operations. Implementation complexity includes rural challenges like remote access, weak signals, and limited maintenance services. Technological compatibility refers to the integration with heterogeneous equipment and existing systems such as farm-specific Enterprise Resource Planning (ERPs) or digital field logs.

In the organizational dimension, organizational structure reflects common setups—family-based businesses, low professionalization, centralized decisions, and conservative cultures. Process readiness assesses the maturity of production, logistics, and administrative processes, often influenced by seasonality and low standardization. Strategic coherence evaluates alignment between the technology and business goals, such as productivity, sustainability, traceability, and export compliance.

The external environment includes institutional and regulatory pressure, such as pesticide regulations, sustainability certifications, and environmental policies. Competitive pressure reflects global market demands for quality, logistics, and sustainability. Partner maturity evaluates the capacity of suppliers, cooperatives, and logistics providers to integrate with digital solutions—often a challenge in chains with varying levels of digitalization. The social dimension includes working conditions, focused on ergonomics, safety, and rural workers' well-being. Technology acceptance measures how managers and workers—often skeptical and distant from tech hubs—respond to innovation. Competency adequacy examines the workforce's readiness, from operating drones and interpreting satellite images to basic digital literacy in low-education contexts.

Adapting these criteria to agro-industrial contexts helps diagnose technological, organizational, and cultural readiness more accurately. This reduces implementation risks, avoids misaligned investments, and highlights the role of people and social structures in the digital transformation of agribusiness. This systemic lens also helps prioritize initiatives that better match rural realities, lowering the risk of rejection or inefficiency. In a sector exposed to external shocks and constant demands for efficiency, productivity, and regulatory compliance, sociotechnical criteria are no longer optional—they are a strategic imperative.

These criteria are assessed and weighted based on their relevance to the firm. We begin by scoring the technological impact of each criterion: high and very high impacts receive scores of 4 and 5; low or nonexistent impacts receive 1 or 2. Negative impacts use the same scale but are converted using the inverse (1/score). We calculate the final index by multiplying each score by its respective weight. Criteria can be weighted using multicriteria analysis methods such as Analytic Hierarchy Process (AHP) (Saaty 2008). The same process applies to the environmental index. If the resulting index is greater than 1, the overall weighted impact is positive; the lower the value, the more negative the outcome.

A critical aspect of digital transformation in agriculture is bridging the gap between smallholders and large-scale farmers. Smaller producers often face disproportionate challenges, such as lack of infrastructure and limited digital literacy, making individual adoption economically unfeasible. Cooperatives and collective arrangements, including digital platforms for shared equipment and data services, can reduce individual costs and democratize access to Agriculture 4.0 technologies. These arrangements foster economies of scale and collective learning, contributing to more inclusive rural development.

2.2.3 Environmental Perspective

Beyond the traditional metrics of economic, financial, and sociotechnical analysis, the environmental perspective stands as a crucial, and arguably defining, element in assessing the viability of Agriculture 4.0 technologies. While other industries may treat environmental factors as externalities, in the agri-food sector, the environment is the very foundation of production. Thus, technologies must not only meet financial hurdles but also satisfy minimum environmental performance levels. Conversely, projects with a low net present value (NPV) or a long payback period may be justified by a substantial environmental return, which can unlock new forms of value.

Quantifying Environmental Value: Direct Financial Pathways

While many environmental benefits are long-term and difficult to price, certain outcomes enabled by Ag 4.0 technologies have a clear path to financial realization:

- *Regulatory compliance and risk mitigation*: technologies like sensors and blockchain generate auditable, tamper-proof data that prove compliance with strict regulations on water use, agrichemicals, and waste. This reduces regulatory risk and boosts credibility with investors.
- *Access to premium markets*: satellite imagery and on-farm sensors can verify regenerative practices, while traceability tools communicate sustainability to consumers. This opens premium markets, increases product value, and strengthens brand loyalty.
- *New revenue from ecosystem services*: verified environmental data creates new market opportunities. For example, IoT sensors that track gains in Soil Organic Carbon (SOC) enable farms to sell carbon credits, turning healthy soil into a financial asset (Lal 2020).

Intrinsic Value and Long-Term Resilience

Other environmental benefits, while not always directly monetizable, are fundamental to the long-term resilience and legitimacy of the farming operation:

- *Resource use efficiency*: smart technologies enable the core principle of "doing more with less." Variable rate technology, for example, optimizes fertilizer application, which provides the dual benefit of reducing input costs and mitigating environmental damage, such as the eutrophication of waterways (Schimmelpfennig 2016). This directly aligns with the UN's Sustainable Development Goal of Responsible Consumption and Production (FAO 2020a, b). This concept is also applicable to water and soil use, promoting their efficient use.
- *Climate action and footprint reduction*: Ag 4.0 technologies are key to climate change mitigation and adaptation. Route optimization for machinery and the use of lightweight drones instead of agricultural aircraft can drastically reduce fossil fuel consumption per hectare, lowering the farm's carbon footprint and enhancing operational efficiency (FAO 2020a, b).
- *Biodiversity and ecosystem health*: precision technologies foster a healthier ecosystem. "See-and-spray" systems that target individual weeds can reduce herbicide use by over 90%, minimizing pesticide drift into sensitive buffer zones. This protects vital pollinators and beneficial insects, creating a positive feedback loop that supports the productivity of the entire agricultural system.

In essence, the implementation of environmental sensing and management technologies transforms the farm from a passive subject of environmental conditions into an active generator of valuable data. This data qualifies the operation for new business models based on ecosystem services, positioning sustainability as a core driver of future profitability.

2.3 Stage 3: Decision: Integrating Perspectives to Determine the Technology Adoption

To facilitate a comprehensive assessment, these four dimensions are operationalized into a set of quantitative and qualitative criteria, a process central to building robust performance frameworks (Kaplan and Norton 1996). These criteria are subsequently measured objectively and synthesized into a single Composite Index (CI), enabling a direct comparison between disparate technologies. This aggregation is often achieved through established methodologies like the Analytic Hierarchy Process (AHP) or other scoring models (Saaty 2008). Therefore, this proposed framework extends beyond traditional economic metrics by leveraging a Multi-Criteria Decision-Making approach, a method widely recognized as essential for evaluating complex technology investments where financial metrics alone are insufficient (Chan et al. 2001; Almeida et al. 2022).

First, to compare heterogeneous indicators (e.g., years, monetary values, Likert scores), it is necessary to normalize each perspective's index (I) onto a common, dimensionless scale. For this purpose, a -1-to-1 scale is used, where -1 represents the least desirable outcome and 1 represents the most desirable. This requires defining a maximum performance threshold (I_{Max})for each indicator, typically based on expert judgment, strategic goals, or historical data.

To calculate the composite index, it is necessary to normalize all perspective indexes (I) into a common scale, ranging from -1 to 1, where 1 represents an excellent level. Para esse propósito, foi definida uma máxima performance (I_{Max}) para cada perspectiva, que após a parametrização representaria a nota 1, se alcançada. A forma de cálculo varia conforme o tipo de orientação do index (maior-melhor ou menor-melhor). The normalization of the economic indicator is obtained by the ratio of the index to its maximum performance. For the financial indicator, the normalization consists of the same procedure, but with the outcome being subtracted from 1. The other indices are normalized (I_{N}) using Eqs. 3 and 4:

$$I_{\text{N}} = \frac{I - 1}{I_{\text{max}} - 1}; \; I \geq 0 \tag{3}$$

$$I_{\text{N}} = -\frac{1/(I - 1)}{I_{\text{max}} - 1}; \; I < 0 \tag{4}$$

Once normalized, the scores are aggregated into the final *Composite Index (CI)* through a weighted average. This step allows the framework to be tailored to the specific context and strategic priorities of the farm. These weights are not fixed. This contextual flexibility enhances the robustness and applicability of the framework, as the weights can be customized depending on the farm's strategic orientation. For instance, an organic farm may assign greater weight to the environmental perspective, while a commodity-oriented farm might prioritize financial and economic returns. This reinforces the notion that technology adoption in agriculture is not solely a question of return on investment, but rather a multidimensional decision shaped by economic realities, environmental pressures, and human factors.

Ultimately, the final composite index serves as a powerful decision support indicator. It enables farmers, advisors, and policymakers to rank technologies, compare scenarios, and make informed choices aligned with both short-term feasibility and long-term sustainability goals.

Traditional financial indicators such as NPV, IRR, and Payback remain essential, but complementary metrics can enrich investment assessment. Resilience indicators—for example, how stable the farm's operations are during climate or market shocks, as shown by variability in profitability—capture risk management benefits. Similarly, environmental sustainability indicators (e.g., carbon footprint reduction, water-use efficiency, biodiversity conservation) provide a more holistic view of technology performance. Integrating these metrics supports a transition toward long-term, regenerative agricultural models.

3 Application of the Framework: A Case Study

To illustrate the practical application of the framework, this section analyses a hypothetical case study. Consider a diversified farm located in the Pampa biome of Southern Brazil, an enterprise specializing in both rice cultivation and beef cattle ranching. The management is currently evaluating the acquisition of an Unmanned Aerial Vehicle (UAV), or drone, for two primary functions: crop monitoring and livestock counting. It is noteworthy that, due to the region's characteristics, such operations are typically classified as large-scale enterprises by the financial institutions that provide investment loans. This case will be used to demonstrate the framework's functionality in assessing a technology adoption decision.

3.1 Knowledge Stage

During the knowledge stage, the farm progresses from initial awareness to a more nuanced understanding of the technology. A key realization is that the drone is not a recreational device but a piece of professional-grade equipment, whose safe and effective deployment necessitates specialized operator training. This prompts an internal capability assessment to evaluate the aptitude and willingness of the existing workforce to become certified operators. The analysis concludes that the farm meets the foundational prerequisites for successfully integrating this technology into its operations.

3.2 *Persuasion Stage*

At this stage, the farm evaluates the viability of adopting the drone from three distinct perspectives, as proposed by the framework.

3.2.1 Economic and Financial

Adopting the drone as a technology involves more than just the initial acquisition cost; it also encompasses operator training, maintenance, and the potential need for technical support contracts. Over time, maintenance costs are expected to increase. Conversely, the benefit stream is not static; technological advancements and future integration possibilities are projected to yield greater returns over the technology's lifespan. For instance, a drone that was recently used solely for crop inspection has now gained the additional functionality of animal counting. It is anticipated that future updates will enable weight estimation and, eventually, individual animal identification and tracking. These evolving capabilities can be translated into quantifiable benefits, such as labor savings and loss reduction.

Therefore, the cash inflows for this technology acquisition project are based on: savings in manual inspection labor, productivity gains from the faster diagnosis of pests and crop failures, and improved herd-counting accuracy. Furthermore, the drone's useful life is estimated to be 8 years. The resulting cash flow projection for this technology adoption is presented in Table 1.

The initial investment, covering both the acquisition and training, amounts to $5000.00. Beginning in the second year, annual maintenance costs are incurred, starting at $100.00 and increasing by $10.00 each subsequent year. Conversely, the project generates cost savings starting at $1200.00 in the first year, and these savings

Table 1 Free cash flow from the adoption of drones as an Ag4.0 technology

Year (n)	Costs (US$)	Revenues/ economies (US$)	Free cash flow (FCF)	FCF cumulative	Discounted cash flow (DFC)	DCF cumulative
0	5000.00		−5000.00	−5000.00	−5000.00	−5000.00
1	0.00	1200.00	1200.00	−3800.00	1062.75	−3937.25
2	100.00	1250.00	1150.00	−2650.00	901.98	−3035.28
3	110.00	1300.00	1190.00	−1460.00	826.59	−2208.68
4	120.00	1350.00	1230.00	−1460.00	756.66	−1452.03
5	130.00	1400.00	1270.00	−230.00	691.90	−760.13
6	140.00	1450.00	1310.00	1040.00	632.06	−128.06
7	150.00	1500.00	1350.00	2350.00	576.86	448.80
8	160.00	3050.00	2890.00	3700.00	1093.67	1542.47

grow by an additional $50.00 annually. The cash flow projection concludes with the drone's salvage value of $1500.00 in the final year.

To calculate the project's NPV, the company's capital structure was assumed to be composed of 40% debt, with a pre-tax cost of debt (R_D) of 13.5% per year and a corporate tax rate (T) of 27.5%, while the cost of equity (R_E) is 15% per year. These figures are based on the current interest rate environment in Brazil.[1] Consequently, the Weighted Average Cost of Capital (WACC), which serves as the minimum attractive rate of return (MARR) for this analysis, is calculated to be 12.92% per year (check using Eq. 2).

Using this MARR, the project's NPV is $1542.47 (according to Eq. 1), which indicates that the project is economically viable. Furthermore, its internal rate of return (IRR) of 20.47% per year exceeds the MARR, further confirming its economic feasibility. Moreover, the project's Profitability Index (PI)—calculated by dividing the present value of future cash flows by the initial investment—is 1.31. This means that for every $1.00 invested, the project is expected to generate $1.31 in present value. PI becomes more relevant when it is necessary to compare different technologies with a great amplitude of the investment amount. Therefore, from an economic standpoint, the project is considered advantageous.

From a financial perspective, however, the project's Discounted Payback Period is 6.22 years, which reveals a less favorable profile. The initial investment will only be recovered after 78% of the equipment's useful life has elapsed. If the agricultural producer requires a liquidity horizon of 5 years or less, the project would be deemed unattractive from a financial standpoint.

This analysis of the drone investment vividly illustrates a classic strategic dilemma: a project that is attractive from a long-term economic perspective, as indicated by its positive NPV and robust IRR, can simultaneously appear unattractive from a short-term financial standpoint due to its extended payback period. This divergence underscores that a decision based solely on traditional financial metrics is incomplete. It is precisely in such situations of financial ambiguity that the sociotechnical and environmental perspectives become decisive, providing the additional criteria necessary to make a well-rounded and strategically sound investment choice.

3.2.2 Sociotechnical

The sociotechnical analysis provides crucial insights beyond the economic assessment. Ensuring strong sociotechnical performance facilitates a smoother implemen-

[1] In 2025, Brazil faces a sharp rise in its benchmark interest rate, reaching 15% per year by July. Although the government maintains incentives for agribusiness investments, higher borrowing costs have directly impacted rural producers, increasing the financial burden of agricultural credit lines. Source: https://globorural.globo.com/economia/noticia/2025/06/selic-puxara-alta-dos-juros-no-plano-safra.ghtml.

Table 2 Sociotechnical Index calculation for drone adoption

Dimension	Criteria	Score	Weight	Weighted score
Technology	Technological maturity	2	3%	0.06
	Technological complexity	0.33	8%	0.03
	Technological compatibility	3	1%	0.03
Organization	Production system readiness	2	3%	0.06
	Organizational structure	2	4%	0.08
	Strategic coherence	5	14%	0.70
External environment	Competitive pressure	1	0%	0.00
	Regulatory compliance	2	2%	0.04
	Supplier conditions	1	2%	0.02
Social	Working conditions	5	5%	0.25
	People's acceptance of new technologies	0.25	30%	0.08
	Workforce skills and competences	3	28%	0.84
Total			**100%**	**2.18**

Note: A score of 5 represents the most favorable outcome. Scores below 1 indicate that perceived drawbacks outweigh benefits for that specific criterion

tation and utilization of the technology, which in turn can lead to improved economic outcomes (Gillani et al. 2020). Furthermore, technology adoption has both positive and negative sociotechnical impacts. Accordingly, the drone adoption was evaluated across four dimensions and their respective criteria (Table 2).

The Technology dimension shows that the drone improves the farm's technological maturity and integrates well with existing infrastructure, but it introduces operational complexity. In the organizational dimension, the drone aligns strongly with the farm's strategy (score 5), contributing significantly to the overall score (14% weight). Although the farm's family-based structure and informal processes may require external technical support and system adjustments, these challenges are manageable and do not hinder adoption.

The external environment dimension has limited influence due to its low weight. As a commodity-driven operation with mature partners, the farm experiences little competitive pressure. However, regulatory compliance is becoming more relevant, and the drone can support this indirectly by generating valuable operational data.

The social dimension carries the most weight (63%) and plays a decisive role. The key concern—people's acceptance of new technologies—scored just 0.25, revealing strong resistance, especially from staff involved in traditional livestock handling. On the other hand, the drone is expected to improve working conditions by reducing long-distance manual crop inspections and by contributing to workforce upskilling. Notably, a young, digitally inclined employee interested in operating the drone helps mitigate this resistance. The composite sociotechnical score is 2.18 out of 5, suggesting a slightly above-low impact. Despite strong strategic alignment, the

Table 3 Environmental index calculation for drone adoption

Dimension	Criteria	Score	Weight	Weighted score
Financial pathway	Regulatory compliance and risk mitigation	3.00	24%	0.72
	Access to premium market	1.00	6%	0.06
	New revenue stream from ecosystem services	2.00	3%	0.06
Intrinsic value	Resource use efficiency	5.00	13%	0.65
	Climate action and footprint reduction	5.00	27%	1.35
	Biodiversity and ecosystem health	2.00	27%	0.54
Total			**100%**	**3.38**

Note: a score of 5 represents the most favorable outcome

adoption faces major social challenges, particularly around user acceptance. The next step in the evaluation focuses on the environmental perspective, assessing the technology's interaction with the natural ecosystem.

3.2.3 Environmental

The adoption of the drone offers direct contributions to the farm's sustainability goals, as quantified in Table 3.

Within the *Financial Pathway* dimension, the technology provides moderate support (a score of 3) for regulatory compliance and operational risk mitigation, primarily through image capture and early threat identification. Its potential to facilitate new revenue streams from ecosystem services (by providing data that attests to responsible resource management) is present but currently limited (a score of 2). Particularly, the technology does not enable access to new premium markets, resulting in a score of 1 for that criterion.

The *Intrinsic Value* dimension holds the greatest weight (67%) in the environmental perspective, and it is here that the drone's benefits are most pronounced. It has a high positive impact (a score of 5) on *Resource Use Efficiency*, as aerial pest identification allows for more targeted agrichemical application. Similarly, it makes a significant contribution (a score of 5) to *Climate Action* by reducing the need for heavy vehicles for inspection, thereby lowering CO_2 emissions. Finally, it offers a modest contribution (a score of 2) to Biodiversity and Ecosystem Health, enabling the monitoring of on-farm conservation areas and the early detection of threats such as wildfires. Notably, the assessed technology presents no discernible negative environmental impacts; on the contrary, its contributions are clearly positive. The quantified and weighted assessment results in a final environmental index of 3.38 out of a possible 5, representing a moderately strong positive impact. The next stage combines the three perspectives.

Table 4 Composite index calculation for drone adoption

	Perspective				
	Financial	Economic	Sociotechnical	Environmental	**Total**
Index	6.22	31%	2.18	3.38	
Unit	Year	ROI	Likert	Likert	
Maximum performance	8	100%	5	5	
Orientation	Lower is better	Higher is better	Higher is better	Higher is better	
Normalized index	**0.22**	**0.31**	**0.30**	**0.60**	
Perspective weight	0.2	0.3	0.25	0.25	1.00
Weighted normalized index	**0.04**	**0.09**	**0.07**	**0.15**	**0.36**

Table 5 Monetization of nonfinancial perspectives

	Perspective			
Monetization ($)	Economic	Sociotechnical	Environmental	Total
NPV	1542.47			
NPV/weighted normalized index	16,666.67			
Adjusted perspective NPV	**1542.47**	**1230.90**	**2479.95**	**5253.32**

3.3 Decision Stage

With the individual analyses complete, the decision-maker is now faced with a multifaceted and conflicting picture: a project that is economically viable (positive NPV) but financially risky (long payback), sociotechnically challenging (low user acceptance), and environmentally beneficial. The framework resolves this complexity by synthesizing these disparate measures into a single, weighted composite index, providing a holistic and quantifiable basis for the ultimate adoption decision, as detailed in Table 4.

For this case, the decision model assigns a 50% combined weight to the traditional economic and financial criteria, while the sociotechnical and environmental perspectives each receive a 25% weight. After normalization and weighting, the resulting composite index is 0.36. This indicates that the technology's overall performance across all categories is 36% of the ideal maximum. An analysis of the weighted scores reveals that the environmental perspective (0.15) is the primary contributor to this positive result, followed by the economic (0.09) and sociotechnical (0.07) perspectives. To provide a more tangible interpretation of these nonfinancial contributions, it is possible to monetize their impact relative to the project's NPV. By establishing a linear relationship between the economic NPV and its weighted normalized index, we can estimate the equivalent monetary value of the sociotechnical and environmental benefits, as presented in Table 5.

The drone investment assigns 50% of the weight to the economic-financial perspective and 25% each to the sociotechnical and environmental perspectives. After weighting, the composite index reaches 0.36, indicating a positive outcome equal to 36% of the maximum expected performance across all categories. Among the weighted scores, the environmental perspective contributes the most (0.15), followed by the economic (0.09) and sociotechnical (0.07) dimensions. Assuming a linear relationship between NPV and the weighted index, we can estimate the monetary value of each perspective's contribution—as shown in Table 5.

This monetization concludes that adopting the drone is highly advantageous. When translated into monetary terms, the environmental benefits (2479.95) significantly exceed the direct economic gains, and the sociotechnical (1230.90) are comparable in value. The total adjusted NPV of $5253.32—which now incorporates the quantified value of nonfinancial factors—is substantially greater than the initial investment. However, a crucial caveat remains: these estimated benefits are contingent upon the farm's ability to implement and effectively utilize the technology.

4 Discussion

This chapter proposed and operationalized an integrated framework for evaluating the economic viability of Agriculture 4.0 technologies, addressing the limitations of purely financial approaches criticized by scholars such as Filmer and Jackson (2021). Its application to a case study reveals key insights into technology adoption in the agri-food sector.

First, the analysis highlights the tension between value perspectives. The drone investment, while showing a positive NPV, raised financial concerns due to its long payback period. This supports Kaplan's (1986) critique that traditional discounted cash flow methods often miss the strategic value of innovation. It also aligns with Boehlje and Lins (1998), who emphasize that under uncertainty, liquidity and risk often outweigh long-term profitability in agricultural decision-making. Our framework does not resolve these trade-offs but makes them explicit. By structuring the evaluation across multiple dimensions, it helps expose and manage adoption barriers (da Silveira et al. 2021).

Second, the environmental dimension played a decisive role in reframing the investment. In our case, environmental gains outweighed direct financial returns, transforming a moderately attractive project into a strategic opportunity. This underscores the chapter's key argument: in agriculture, sustainability is not an external benefit—it is central to value creation. Quantifying and integrating these impacts responds to the FAO's (2020a, b) call for sustainable agricultural systems, as advocated by Basso and Antle (2020).

Another implication of digitalization is the emergence of new business models. Servitization—where technologies are offered as services rather than products—reduces upfront investment and facilitates continuous upgrades. Shared-use models, such as machinery rings or digital marketplaces, allow producers to access advanced equipment and analytics on demand. These models redistribute risk, lower barriers to entry, and make Agriculture 4.0 solutions accessible to a broader range of producers.

Finally, the composite index—while insightful—relies on assumptions subject to uncertainty. Agricultural investments operate in volatile conditions. Thus, static analyses fall short. Integrating scenario analysis improves strategic planning by assessing the investment under pessimistic, likely, and optimistic futures. This enhances agility and foresight, helping decision-makers prepare for multiple outcomes.

Ultimately, the framework's value lies not in providing a single "correct" answer but in enabling a more strategic, informed discussion about the multidimensional impacts of digital technologies in agriculture.

5 Conclusion

The growing complexity of challenges in the agri-food sector demands integrated and innovative approaches to evaluate the adoption of digital technologies. This chapter proposed and operationalized an analytical framework to assess the economic viability of Agriculture 4.0 solutions, integrating economic-financial, sociotechnical, and environmental dimensions. Grounded in Innovation Diffusion Theory (Rogers 2003) and multicriteria evaluation models, the framework was adapted to the specific conditions of the agro-industrial sector, accounting for variations in scale, infrastructure, organizational culture, and environmental sensitivity.

The analysis reinforces that evaluating Agriculture 4.0 technologies requires more than financial metrics. Intangible costs and benefits—such as workforce training, social acceptance, and environmental impact—play a decisive role in adoption success and sustainability. Incorporating a sociotechnical perspective tailored to rural realities and treating environmental aspects as central strengthens both the robustness and legitimacy of decision-making.

The case study on drone adoption illustrated the framework's ability to provide objective and comparative insights to support decision-making. The use of three future scenarios emphasized the importance of addressing uncertainty and operational variability, promoting more resilient choices under diverse conditions.

Ultimately, this chapter offers both a conceptual and practical tool for producers, managers, policymakers, and researchers seeking to drive the digital transformation of the agri-food system with economic, social, and environmental responsibility. By addressing a methodological gap in the literature, the proposed framework supports more strategic, sustainable, and context-sensitive decisions for agro-industrial development.

Future research should move beyond single-technology assessments and adopt comparative and systemic approaches that capture the economic, social, and environmental impacts of multiple digital solutions. This would enable policymakers and practitioners to prioritize interventions that maximize collective value for the entire agri-food system.

Competing Interests The author(s) has no competing interests to declare that are relevant to the content of this manuscript.

References

Almeida RP, Ayala NF, Benitez GB, Kliemann Neto FJ, Frank AG (2022) How to assess investments in industry 4.0 technologies? A multiple-criteria framework for economic, financial, and sociotechnical factors. Prod Plan Control 34(16):1583–1602

Aubert BA, Schroeder A, Grimaudo J (2012) IT as enabler of sustainable farming: an empirical analysis of farmers' adoption decision of precision agriculture technology. Decis Support Syst 54(1):510–520

Basso B, Antle J (2020) Digital agriculture to design sustainable agricultural systems. Nat Sustain 3:254–256. https://doi.org/10.1038/s41893-020-0510-0

Beier G, Ullrich A, Niehoff S, Reißig M, Habich M (2020) Industry 4.0: how it is defined from a sociotechnical perspective and how much sustainability it includes—a literature review. J Clean Prod 259:120856. https://doi.org/10.1016/j.jclepro.2020.120856

Boehlje MD, Lins DA (1998) Risks and risk management in industrialized agriculture. Agric Finance Rev 58(1):1–16

Brealey RA, Myers SC, Allen F (2020) Principles of corporate finance, 13th edn. McGraw-Hill Eduction

Chan FTS, Chan MH, Lau RWL (2001) Investment appraisal techniques for advanced manufacturing technology (AMT): a literature review. Integr Manuf Syst 12(1):35–47

da Silveira F, Lermen FH, Amaral FG (2021) An overview of agriculture 4.0 development: systematic review of descriptions, technologies, barriers, advantages, and disadvantages. Comput Electron Agric 189. https://doi.org/10.1016/j.compag.2021.106405

FAO (2020a) Digital agriculture: farmers in the driver's seat. Food and Agriculture Organization of the United Nations

FAO (2020b) The State of Food and Agriculture 2020—overcoming water challenges in agriculture. Food and Agriculture Organization of the United Nations, Rome. https://doi.org/10.4060/cb1447en

Farooq MS, Riaz S, Abid A, Umer T, Zikria YB (2020) Role of IoT technology in agriculture: a systematic literature review. Electronics 9(2):319. https://doi.org/10.3390/electronics9020319

Filmer D, Jackson S (2021) Measuring the return on investment of digital agriculture: a critical challenge. World Bank Policy Briefs

Gillani F, Chatha KA, Jajja MSS, Farooq S (2020) Implementation of digital manufacturing technologies: antecedents and consequences. Int J Prod Econ 229:107748. https://doi.org/10.1016/j.ijpe.2020.107748

Kamilaris A, Kartakoullis A, Prenafeta-Boldú FX (2017) A review on the practice of big data analysis in agriculture. Comput Electron Agric 143:23–37. https://doi.org/10.1016/j.compag.2017.09.037

Kaplan RS (1986) Must CIM be justified by faith alone? Harv Bus Rev 64(2):87–95

Kaplan RS, Norton DP (1996) The balanced scorecard: translating strategy into action. Harvard Business Press, Boston

Klerkx L, Rose D (2020) Dealing with the game-changing technologies of Agriculture 4.0: how do we manage diversity and responsibility in food system transition pathways? Glob Food Sec 24:100347. https://doi.org/10.1016/j.gfs.2019.100347

Lal R (2020) Soil organic matter and soil quality. In: No-till farming systems for sustainable agriculture. Springer, Cham, pp 3–21

Lezoche M, Hernandez JE, Díaz MD, Panetto H, Kacprzyk J (2020) Agri-food 4.0: a survey of the supply chains and technologies for the future agriculture. Comput Ind 117:103187

Lima GCF, Barbieri FL, Seki AE, Jorge A (2020) Agro 4.0: enabling agriculture digital transformation through IoT. Rev Ciênc Agron 51(spe) https://www.scielo.br/j/rca/i/2020.v51nspe/

Lioutas ED, Charatsari C, De Rosa M (2021) Digitalization of agriculture: a way to solve the food problem or a trolley dilemma? Technol Soc 67:101744. https://doi.org/10.1016/j.techsoc.2021.101744

Masood T, Egger J (2019) Augmented reality in support of industry 4.0– implementation challenges and success factors. Robot Comput Integr Manuf 58:181–195. https://doi.org/10.1016/j.rcim.2019.02.003

Modigliani F, Miller MH (1963) Corporate income taxes and the cost of capital: a correction. Am Econ Rev 53(3):433–443

Patrick GF (1972) Effects of federal income tax laws on the distribution of benefits of farm programs. Am J Agric Econ 54(5):903–908

Rockström J, Williams J, Daily G, Noble A, Matthews N, Gordon L et al (2020) Sustainable intensification of agriculture for human prosperity and global sustainability. Ambio 46(1):4–17. https://doi.org/10.1007/s13280-016-0793-6

Rogers EM (2003) Diffusion of Innovations, 5th edn. Free Press, New York

Ross SA, Westerfield RW, Jaffe J (2019) Corporate finance, 12th edn. McGraw-Hill Education

Saaty TL (2008) Decision making with the analytic hierarchy process. Int J Serv Sci 1(1):83–98

Schimmelpfennig D (2016) The adoption of precision agriculture. USDA Economic Research Service

Sony M, Naik S (2020) Industry 4.0 integration with socio technical systems theory: a systematic review and proposed theoretical model. Technol Soc 61:101248. https://doi.org/10.1016/j.techsoc.2020.101248

van Dijk M, Morley T, Rau ML, Saghai Y (2020) A meta-analysis of projected global food demand and population growth. Nat Sustain 4:233–239. https://doi.org/10.1038/s41893-020-00617-2

Servitization in Agriculture 4.0

Guilherme Sales Smania, Néstor Fabián Ayala, and Glauco Henrique de Sousa Mendes

1 Introduction

Over the past decades, large manufacturing firms have progressively adopted service-oriented business models (Kowalkowski et al. 2017; Baines et al. 2020; Favoretto et al. 2022). The integration of services into product offerings was initially defined as servitization by Vandermerwe and Rada (1988). Currently, servitization is understood as the transition or transformation of manufacturing firms toward business models focused on services (Kowalkowski et al. 2017; Martinez et al. 2017). This shift involves a process of organizational restructuring, driven by the increasing strategic relevance of services, which leads to an expanded service portfolio and greater alignment with service-oriented strategies (Baines 2015; Kowalkowski et al. 2017; Martinez et al. 2017). As a result, servitized firms have started to deliver integrated product and service offerings, known as Product-Service Systems (PSS) (Tukker 2015). In developed countries, approximately two-thirds of manufacturers have adopted servitization strategies, with services representing, on average, 30% of total revenues (Vendrell-Herrero et al. 2017).

Servitization strategies have been implemented across multiple industries, including agriculture (Smania et al. 2022; Van Dyck et al. 2024), energy (Lütjen et al. 2019), aerospace (Grubic and Jennions 2018; Korkeamäki et al. 2021), telecommunications (Kamalaldin et al. 2020; Sjödin et al. 2020), mining (Sjödin et al. 2020; Korkeamäki et al. 2021), and automotive (Genzlinger et al. 2020;

G. S. Smania (✉) · G. H. de Sousa Mendes
Department of Production Engineering, Federal University of São Carlos, São Carlos, SP, Brazil

N. F. Ayala
Organizational Engineering Group, Department of Industrial Engineering, Federal University of Rio Grande do Sul, Porto Alegre, RS, Brazil

School of Engineering and Sciences, Tecnológico de Monterrey, Mexico City, Mexico

F. da Silveira et al. (eds.), *Agriculture 4.0*,
https://doi.org/10.1007/978-3-032-24552-6_9

Bohnsack et al. 2021). In the agricultural sector, manufacturers have made significant progress in implementing PSS and developing service-based offerings (Van Dyck et al. 2024). These initiatives reflect a broader strategic shift among leading agricultural machinery firms toward the provision of smart farming services, supported by Agriculture 4.0 technologies and digital platforms (Smania et al. 2022).

The evolution of agriculture has compelled farmers to increase investments in technology to meet food industry quality standards (Vidickiene and Gedminaite-Raudone 2018), enhance production efficiency in a global market (Matheny and Leahy 2007), and address uncertainties related to climate change (FAO et al. 2018). In response, leading agricultural machinery manufacturers have adopted Agriculture 4.0 technologies to integrate tractors and harvesters, enabling the delivery of smart services (Smania et al. 2022). These firms are moving toward digital servitization by offering services such as remote monitoring, predictive maintenance, and operational management (Porter and Heppelmann 2014).

Two main drivers support this transition. The first is Agriculture 4.0, which involves applying technologies to collect, process, and analyze high-resolution, multisource data to support crop production decisions (Kaňovská and Tomášková 2018; Ozdogan et al. 2017). The second driver stems from the evolving needs of farmers: (i) to improve productivity and remain competitive in markets with less restrictive regulations and lower costs (Matheny and Leahy 2007); (ii) to comply with new environmental regulations (Lichtenberg 2019); and (iii) to reduce the inherent uncertainties of agricultural activity (Kaňovská and Tomášková 2018).

As a result of both technology-push and demand-pull forces, machinery manufacturers have begun offering advanced, tailored solutions (Kaňovská and Tomášková 2018). From an equipment perspective, tractors and harvesters are now equipped with sensors and data systems that support optimization, remote diagnostics, predictive maintenance, and simulation-based prototyping (Smania et al. 2022; Van Dyck et al. 2024). From an operational standpoint, smart PSSs enable real-time harvest monitoring and the prediction of input consumption (e.g., seeds, fuel, fertilizers) using historical data. John Deere, for instance, offers a management system as a platform that connects its fleet, enabling farmers to monitor and control operations remotely (Porter and Heppelmann 2014; Kaňovská and Tomášková 2018). These platforms also support equipment maintenance and performance tracking.

In summary, Agriculture 4.0 requires manufacturing firms to transition toward service-oriented business models. For this transformation to occur at the necessary pace and scale across both developed and developing countries, a broad understanding of the servitization concept is essential. This chapter contributes to that goal by presenting the key foundations of servitization, exploring specific opportunities within Agriculture 4.0, and addressing the main challenges that firms may face in this transformation process. Furthermore, it outlines emerging trends that can support agricultural machinery manufacturers in advancing toward digital servitization and proposes avenues for future research to further expand scholarly understanding of servitization in Agriculture 4.0.

2 Type of Service Related to Products and the Role of Digital Technologies

Cusumano et al. (2015) classify services offered by product-oriented firms into three main types, based on their strategic roles: smoothing, adapting, and substituting services. Smoothing services are loosely coupled with the core product and aim to facilitate its sale or use without modifying its essential functions. Examples include basic maintenance, financing, or standard training. In contrast, adapting services are more integrated and seek to expand the product's functionality or adapt it to specific customer contexts, requiring more intensive data exchange and interaction between provider and user. Finally, substituting services eliminates the need for product ownership, enabling customers to access the product's function through service delivery. These align with use- and result-oriented Product-Service Systems (Tukker 2004) and the notion of advanced services (Baines and Lightfoot 2013).

Beyond classification, Cusumano et al. (2015) argue that the relevance of each type of service shifts along the industry lifecycle. In early stages, adapting services dominate, helping firms respond to uncertainty and customer-specific needs. As markets mature, smoothing services gain prominence by supporting standardization and operational efficiency. Substituting services, although occasionally emerging in early phases under extreme uncertainty, typically become more viable in mature stages as firms explore new customer segments.

In recent years, however, digitalization has reshaped the delivery and potential of all these service types. Technologies associated with Industry 4.0, such as the Internet of Things, big data analytics, cloud computing, and digital platforms, have enabled more efficient and scalable deployment of services across all three categories (Frank et al. 2019). Smoothing services benefit from automation and remote diagnostics; adapting services are strengthened by advanced analytics and real-time monitoring that personalize the offering; and substituting services become more viable through digital infrastructures that support usage-based access and outcome-focused contracts.

As highlighted by Frank et al. (2019), the interplay between servitization and Industry 4.0 technologies leads to new configurations of service-based business models. This convergence allows manufacturers to bundle digital capabilities with physical assets, transforming not only how services are delivered but also how value is co-created with customers. Hence, companies are increasingly combining multiple types of services, enabled by digital technologies, to respond dynamically to market demands, customer preferences, and operational contexts. This multilayered service strategy reflects a shift from linear servitization paths to more modular and adaptive configurations, where digitalization acts as a critical enabler of scalability, customization, and performance optimization.

3 Smart Products and Digital Servitization

As seen, digital technologies and smart products are the foundation for the delivery of new services. Smart products are physical goods enhanced with embedded intelligence and connectivity. According to Porter and Heppelmann (2014, 2015), these products combine three core elements: physical components (such as mechanical and electrical structures), smart components (including sensors, microprocessors, and software), and connectivity components (which enable data exchange through communication networks). This configuration allows the product to monitor its condition and environment, control its operations, optimize performance using algorithms, and, in some cases, act autonomously.

By continuously collecting and transmitting data, smart products enable new forms of interaction between manufacturers and users that extend beyond the point of sale. These interactions occur throughout the product's lifecycle and support the delivery of advanced services, such as remote diagnostics, predictive maintenance, and performance monitoring. As a result, the product evolves into a platform for service provision, generating value over time and reinforcing the shift toward service-oriented business models. In Agriculture 4.0, this evolution is particularly relevant. Smart machinery facilitates data-driven decision-making on the farm and plays a strategic role in improving productivity and resource efficiency.

The transformation of agricultural machinery through digital technologies is laying the foundation for what is known as "*digital servitization.*" This concept refers to the convergence of two major developments: servitization, which describes the shift from product-based to service-oriented business models (Kowalkowski et al. 2017; Baines et al. 2020), and digitalization, understood as the increasing use of digital technologies to connect people, systems, and assets, thereby enabling new offerings and transforming business processes (Verhoef et al. 2021). Digital servitization captures the ongoing interaction between these dynamics and underscores the role of technologies such as the Internet of Things, Big Data, Digital Twins, and Artificial Intelligence in shaping innovative services and reconfiguring how manufacturing firms create value (Paschou et al. 2020; Gebauer et al. 2021; Favoretto et al. 2022). Building on these definitions of digital servitization, we define Servitization in Agriculture 4.0 as "*the process of transformation whereby actors in the agricultural ecosystem create, deliver, and capture value through service-oriented business models enabled by Agriculture 4.0 technologies.*"

Indeed, the shift toward digital servitization is already evident in agriculture. Manufacturers traditionally focused on improving the performance of individual machines such as tractors and harvesters. Today, companies like John Deere and AGCO are moving toward optimizing the performance of a broader set of connected equipment, including planters, sprayers, and irrigation systems. These machines are linked through digital platforms that allow for the coordination of operations across the entire production system, generating value beyond the physical functionality of each asset (Porter and Heppelmann 2014). As a result, the industry evolves from

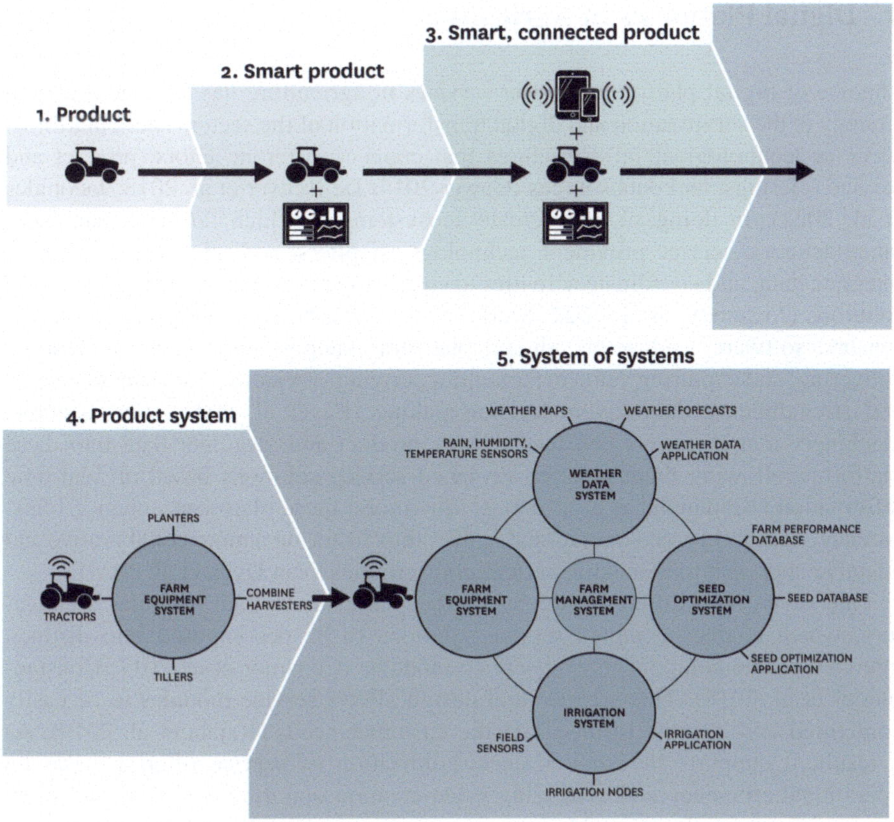

Fig. 1 From smart products to system of systems. (Porter and Heppelmann 2014)

producing stand-alone machinery to delivering integrated agricultural solutions based on real-time data.

This evolution is progressing toward what is known as "systems of systems" (see Fig. 1). In this model, agricultural machinery is integrated with external data sources such as weather forecasts derived from satellites, soil condition sensors, and market information on crop prices and commodity futures. Leading agricultural machinery manufacturers are actively connecting their equipment to agronomic data platforms that help farmers make informed decisions. The goal is to optimize the full farming cycle, from planting to harvesting, by coordinating resources and operations through adaptive, data-driven processes (Markfort et al. 2022; Porter and Heppelmann 2014).

4 Digital Platforms in Agriculture

The use of digital platforms to offer services in agriculture has become a central strategy in the servitization and digital transformation of the sector. These platforms serve as technological infrastructures that connect different actors, product and service offerings, and data sources (Gawer 2014; De Reuver et al. 2018; Jacobides et al. 2024). In doing so, they create ecosystems in which farmers, machinery manufacturers, service providers, technology suppliers, and other actors interact, integrate data, and coordinate activities to enable the provision of smart agricultural solutions (Jovanovic et al. 2022; Markfort et al. 2022). By connecting machines, sensors, software, and actors, digital platforms support new ways of creating, delivering, and capturing value in agriculture, driving service-based business models and strengthening data-driven decision-making (Fu et al. 2022). For example, machinery manufacturers can feed diverse product and customer data into these platforms, allowing them to offer advanced service solutions based on real-time information (Kohtamäki et al. 2019). Furthermore, these platforms enhance transparency, traceability, and operational agility, thus fostering a more collaborative and adaptive approach to managing agricultural activities (Van Dyck et al. 2024).

One of the key differentiators of digital platforms is their modular architecture, which enables complex service solutions to be decomposed into distinct, interchangeable components and service modules (Cenamor et al. 2017; Constantinides et al. 2018). This architectural design allows service modules to be easily configured and adapted to meet specific customer needs (Rajala et al. 2019). As a result, it supports the large-scale customization of service offerings, driving operational efficiency and expanding value-creation opportunities (Cenamor et al. 2017; Eloranta et al. 2021). Modules such as crop monitoring, asset management, climate analysis, traceability, and agronomic diagnostics can be added, updated, or removed as needed—without necessarily compromising the overall platform stability. Additionally, modularity enables the agile development of solutions tailored to farmers' specific needs and local conditions. This technical feature fosters continuous innovation and facilitates the integration and operation of multiple external providers within the digital ecosystem (Eloranta and Turunen 2016; Sjödin et al. 2020).

Through these platforms, machinery manufacturers and other providers can offer smart solutions that span a wide range of services, addressing both the machinery itself and farmers' agricultural operations (Van Dyck et al. 2024). At the machinery level, services focus on optimizing the overall effectiveness of agricultural equipment and may include telemetry and remote monitoring, predictive maintenance, remote calibration, navigation systems, fleet management, and over-the-air updates. At the operational level, platforms provide functionalities aimed at enhancing farmers' performance by supporting more efficient planning and execution of activities such as planting, fertilization, irrigation, and harvesting. These services include crop planning, input management, profitability analysis, weather forecasting, crop monitoring, soil mapping, and variable rate application (Markfort et al. 2022).

By combining product-oriented and customer-oriented services, digital platforms strengthen their overall value proposition, enabling farmers to improve efficiency, sustainability, and competitiveness in their operations.

5 Challenges of Digital Servitization in Agriculture

Despite its potential benefits for actors within the agricultural ecosystem, digital servitization presents some challenges that must be addressed to ensure the successful delivery of smart solutions. One critical challenge is gaining access to data generated by machines and other physical assets. In digital servitization, the development and provision of advanced services—such as predictive maintenance, agricultural consulting, and productivity analytics—depend heavily on data collected on-farm (Dalenogare et al. 2022; Frank et al. 2022). However, many farmers still resist sharing their data with machinery manufacturers and other service providers due to concerns over data leakage (Eggert et al. 2022). In particular, they fear that sensitive business information such as productivity and planting data may be disclosed to competitors or used by providers as leverage to increase their bargaining power. Another reason is the low perceived value of smart services. Many farmers continue to exhibit cultural resistance and fail to recognize the potential benefits that smart services can bring to their operations (Schäfer et al. 2023). Finally, the lack of clear definitions regarding data ownership may impede data access agreements between farmers and service providers, thereby discouraging farmers from sharing their data (Langley 2022).

Another significant challenge is the lack of interoperability among agricultural machines, devices, and systems. On modern farms, data are collected from a variety of sources, including machines (such as tractors, harvesters, seeders, planters, and sprayers) from different brands, sensors, drones, and farm management software (Markfort et al. 2022; Van Dyck et al. 2024). These machines and devices must be able to exchange data to provide farmers with reliable information for making more accurate operational decisions (Langley 2022). For instance, data exchange among machines is crucial to prevent overlapping fertilizer applications. Nonetheless, many agricultural machinery manufacturers are reluctant to ensure data interoperability, often adopting proprietary systems that restrict data sharing to equipment within their own brand (Ferreira and Lind 2023; Momeni et al. 2023). In such cases, limiting interoperability enables manufacturers to offer exclusive services to farmers, creating lock-in effects and reducing competition within the ecosystem. Furthermore, the lack of data compatibility or standardization increases the costs of manual data conversion, or even prevents data integration altogether (Smania et al. 2024). This can lead to data silos, ultimately hindering the development and delivery of smart solutions.

Data overload also emerges as a significant challenge for digital servitization in agriculture. Indeed, the abundance of data presents an inherently ambiguous situation for both service providers and farmers. On the one hand, as previously discussed, data serves as a key input for smart services, providing reliable insights for the development of tailored solutions for customers (Eggert et al. 2022; Frank et al. 2022). Thus, data play a critical role in digital servitization. On the other hand, many small and medium-sized farmers lack the appropriate digital technologies and systems needed to process the large volumes of data generated by smart machines and services (Schäfer et al. 2023). In addition, they often do not have access to specialized personnel capable of interpreting this information and acting on real-time data (Chávez et al. 2023). As a result, rather than enabling value creation, the abundance of data can lead to confusion, ultimately hindering farmers from capturing the potential value that data-driven services can offer (Smania et al. 2024).

Finally, another key challenge is the interdependence among actors within the agricultural ecosystem. Delivering smart solutions to farmers requires the effective integration of various actors, including agricultural machinery manufacturers, telecommunications companies, technology suppliers, drone providers, meteorological service providers, software developers, input suppliers, digital platform providers, service intermediaries, and government agencies (Markfort et al. 2022; Van Dyck et al. 2024; Rodríguez et al. 2025). Therefore, building trust and fostering mutual commitment among these actors is essential. Otherwise, the lack of engagement from a single actor can undermine the digital servitization efforts of others (Dalenogare et al. 2023). For example, the successful provision of smart solutions depends on dealer commitment to clearly communicate their benefits to farmers and ensure effective service delivery. Moreover, power dynamics within the ecosystem may shift as a result of this interdependence. In particular, technology suppliers and platform providers can assume a central role by controlling critical digital resources required for smart service provision, potentially challenging the traditional dominance of machinery manufacturers (Ferreira and Lind 2023; Momeni et al. 2023). As such, effectively managing the interdependencies among actors is crucial to prevent conflicts and tensions within the ecosystem.

6 Future Trends for Servitization in Agriculture 4.0

Future trends in servitization within Agriculture 4.0 primarily involve the transformation of the agriculture into an increasingly data-driven sector. As discussed in this chapter, one particularly strong and steadily growing trend is platformization. Leading agricultural machinery manufacturers—such as John Deere, AGCO, CNH Industrial, and Mahindra—have implemented digital platforms to connect various farming machines and physical assets, integrating vast amounts of data to enable the delivery of connected services based on real-time farm information (Van Dyck et al. 2024). These platforms expand digital ecosystems by allowing third-party service providers, agricultural consulting firms, and technology suppliers to offer services directly to farmers (Markfort et al. 2022). Another trend linked to platformization is the rise of integrative platforms, such as Bayer's Climate FieldView, which facilitates the connection of machines from different manufacturers through interoperability protocols. This enables data integration into a single portal, thus enhancing data visualization and supporting the provision of connected services.

With the advancement of Artificial Intelligence (AI) and machine learning technologies, another key trend is the use of AI tools to deliver predictive and data-driven services. By applying machine learning algorithms, AI can analyze diverse datasets collected on the farm and identify patterns related to crop health, soil conditions, weather variability, and equipment performance. This enables the delivery of predictive services that offer improved recommendations for fertilization, planting, and harvesting based on real-time and specific conditions, thereby enhancing farmers' decision-making (Sjödin et al. 2021; Ayala et al. 2025). AI also facilitates the provision of customized smart agricultural solutions, as it can learn from historical farm data and generate actionable insights tailored to local contexts (Kohtamäki et al. 2025). Overall, AI holds strong potential to support the provision of services that make agricultural systems more responsive and efficient (Sjödin et al. 2023).

A third and final emerging trend in the servitization of agriculture is the development of digital services that support sustainable farming practices. As regulatory pressures, certification requirements, and consumer demands for environmentally responsible agriculture continue to grow, agricultural machinery manufacturers and technology providers are increasingly offering services that help farmers monitor and reduce the environmental impact of their operations (Kolling et al. 2022). These services include, for instance, solutions for tracking carbon footprint, optimizing input usage, monitoring biodiversity, and ensuring compliance with international sustainability standards. By collecting and analyzing operational and environmental data, such services not only help farmers meet regulatory requirements but also create added value by enabling access to more demanding markets and promoting greener practices (Camel et al. 2025; Vidickienė and Lankauskienė 2025). As a result, sustainability is no longer merely a regulatory obligation but becomes a strategic differentiator enabled by digital services in agribusinesses.

7 Research Agenda on Servitization in Agriculture 4.0

Considering the emerging topics in research on servitization in Agriculture 4.0, we propose some avenues for scholars to explore in order to advance knowledge in the field. First, while there is extensive research on enabling technologies for Agriculture 4.0 (e.g., Klerkx et al. 2019; Liu et al. 2020; Javaid et al. 2022), little is known about which of these technologies drive servitization within the agri-food ecosystem and how they shape servitized business models. We therefore believe that further examining the adoption of Agriculture 4.0 technologies in servitization, along with the barriers to their implementation and the ways in which such technologies can create new trajectories for service innovation, can both enrich the knowledge on servitization in Agriculture 4.0 and provide valuable insights for practitioners within the agri-food ecosystem.

Second, a highly relevant research topic in servitization concerns the development of capabilities for delivering product-service offerings. Prior studies have examined this topic extensively, including digitalization capabilities for value co-creation (Lenka et al. 2017), capabilities for orchestrating ecosystems for service innovation (Lütjen et al. 2019), capabilities for implementing digital servitization in service triads (Marcon et al. 2022), capabilities for leveraging AI throughout back- and front-office service provision (Ayala et al. 2025), and capabilities for unlocking data value through digital platforms (Jia et al. 2024). Future research could build on these insights by investigating such capabilities more deeply within the agri-food ecosystem. In particular, it could explore how manufacturers, farmers, and other actors develop and orchestrate these capabilities to create and capture value through servitized business models.

Third, as agricultural machinery manufacturers and other service providers transition to servitized business models, they are reshaping their traditional value chains into broader ecosystems that foster innovative product-service offerings (Markfort et al. 2022; Van Dyck et al. 2024). This transformation requires integrating new partners, such as startups, technology providers, and data companies, to incorporate additional Agriculture 4.0 solutions into their offerings (Smania et al. 2022). We therefore encourage scholars to further explore the ecosystem transformation toward servitization in Agriculture 4.0, investigating how companies build new partnerships and orchestrate their ecosystems to advance new product-service solutions. Such insights can enrich theory by deepening our understanding of servitized business models in complex agri-food ecosystems and inform practice by guiding industry actors in adopting effective service-oriented strategies that leverage the broader ecosystem.

Finally, recent studies have increasingly addressed the sustainability benefits of data-driven services in the agri-food ecosystem (e.g., Kolling et al. 2022; Ciasullo et al. 2024; Camel et al. 2025; Vidickienė and Lankauskienė 2025). However, further research is needed to better understand how servitization fosters sustainable practices within this ecosystem. Scholars could investigate the economic, environmental, and social impacts of servitization practices in agribusinesses. Furthermore, future studies could examine how Agriculture 4.0 technologies unlock opportunities for new service offerings that deliver sustainable outcomes, and how sustainable practices are coordinated among multiple actors in the agri-food ecosystem, thereby revealing practices that actors can adopt to implement more responsible service strategies.

8 Conclusion

In this chapter, we present how manufacturers have progressed in implementing servitization within the agricultural industry. Servitization refers to the shift from product-oriented to service-oriented business models. Within the domain of servitization, manufacturers can offer three types of product-related services: smoothing, adapting, and substituting services. These service types reshape the customer's relationship with the product in distinct ways, ranging from support (smoothing), to customization (adapting), to full replacement (substituting). We also highlight the rise of digital servitization in agriculture, emphasizing the role of digital technologies in connecting machines, sensors, systems, and platforms to integrate diverse data and enable the delivery of smart, advanced services. Finally, we uncover the main challenges facing digital servitization in agriculture and discuss future trends, illustrating how digital services can drive progress in the sector. Therefore, this chapter contributes to the emerging field of servitization in Agriculture 4.0, which remains largely underexplored.

Competing Interests The author(s) has no competing interests to declare that are relevant to the content of this manuscript.

References

Ayala NF, da Silva JR, Tinoco MAC, Saccani N, Frank AG (2025) Artificial intelligence capabilities in digital servitization: identifying digital opportunities for different service types. Int J Prod Econ 284:109604

Baines T (2015) Exploring service innovation and the servitization of the manufacturing firm. Res Technol Manag 58(5):9–11

Baines T, Lightfoot HW (2013) Servitization of the manufacturing firm: exploring the operations practices and technologies that deliver advanced services. Int J Oper Prod Manag 34(1):2–35

Baines T, Bigdeli AZ, Sousa R, Schroeder A (2020) Framing the servitization transformation process: a model to understand and facilitate the servitization journey. Int J Prod Econ 221:107463

Bohnsack R, Kurtz H, Hanelt A (2021) Re-examining path dependence in the digital age: the evolution of connected car business models. Res Policy 50(9):104328

Camel A, Belhadi A, Kamble S, Wetzels M, Touriki FE (2025) Servitizing for sustainability: leveraging technology-enabled platforms and service innovation for carbon reduction in Africa's Agri-Food Sector: a dynamic capabilities perspective. J Bus Res 189:115166

Cenamor J, Sjödin DR, Parida V (2017) Adopting a platform approach in servitization: leveraging the value of digitalization. Int J Prod Econ 192:54–65

Chávez CAG, Unamuno G, Despeisse M, Johansson B, Romero D, Stahre J (2023) Analyzing the risks of digital servitization in the machine tool industry. Robot Comput Integr Manuf 82:102520

Ciasullo MV, Savastano M, Douglas A, Ferrara M, Fiorentino S (2024) How sustainable is smart farming? The contribution of service platforms to innovate Italian agribusinesses. Sinergie 42(1):45–69

Constantinides P, Henfridsson O, Parker GG (2018) Introduction—platforms and infrastructures in the digital age. Inf Syst Res 29(2):381–400

Cusumano MA, Kahl SJ, Suarez FF (2015) Services, industry evolution, and the competitive strategies of product firms. Strateg Manag J 36(4):559–575

Dalenogare LS, Le Dain MA, Benitez GB, Ayala NF, Frank AG (2022) Multichannel digital service delivery and service ecosystems: the role of data integration within Smart Product-Service Systems. Technol Forecast Soc Chang 183:121894

Dalenogare LS, Le Dain MA, Ayala NF, Pezzotta G, Frank AG (2023) Building digital servitization ecosystems: an analysis of inter-firm collaboration types and social exchange mechanisms among actors. Technovation 124:102756

De Reuver M, Sørensen C, Basole RC (2018) The digital platform: a research agenda. J Inf Technol 33(2):124–135

Eggert CG, Winkler C, Volkmann A, Schumann JH, Wünderlich NV (2022) Understanding intra- and interorganizational paradoxes inhibiting data access in digital servitization. Ind Mark Manag 105:404–421

Eloranta V, Turunen T (2016) Platforms in service-driven manufacturing: leveraging complexity by connecting, sharing, and integrating. Ind Mark Manag 55:178–186

Eloranta V, Ardolino M, Saccani N (2021) A complexity management approach to servitization: the role of digital platforms. Int J Oper Prod Manag 41(5):622–644

FAO et al (2018) The state of food security and nutrition in the world 2018. In: Building climate resilience for food security and nutrition. Food and Agricultural Organization of the United Nations

Favoretto C, Mendes GH, Oliveira MG, Cauchick-Miguel PA, Coreynen W (2022) From servitization to digital servitization: how digitalization transforms companies' transition towards services. Ind Mark Manag 102:104–121

Ferreira CC, Lind F (2023) Supplier interfaces in digital transformation: an exploratory case study of a manufacturing firm and IoT suppliers. J Bus Ind Mark 38(6):1332–1344

Frank AG, Mendes GH, Ayala NF, Ghezzi A (2019) Servitization and Industry 4.0 convergence in the digital transformation of product firms: a business model innovation perspective. Technol Forecast Soc Chang 141:341–351

Frank AG, Mendes GH, Benitez GB, Ayala NF (2022) Service customization in turbulent environments: service business models and knowledge integration to create capability-based switching costs. Ind Mark Manag 100:1–18

Fu W, Zhang M, Zhao X, Jia F (2022) Interplay between servitization and platforms: a longitudinal case study. Int J Oper Prod Manag 42(4):471–499

Gawer A (2014) Bridging differing perspectives on technological platforms: toward an integrative framework. Res Policy 43(7):1239–1249

Gebauer H, Paiola M, Saccani N, Rapaccini M (2021) Digital servitization: crossing the perspectives of digitization and servitization. Ind Mark Manag 93:382–388

Genzlinger F, Zejnilovic L, Bustinza OF (2020) Servitization in the automotive industry: how car manufacturers become mobility service providers. Strateg Chang 29(2):215–226

Grubic T, Jennions I (2018) Do outcome-based contracts exist? The investigation of power-by-the-hour and similar result-oriented cases. Int J Prod Econ 206:209–219

Jacobides MG, Cennamo C, Gawer A (2024) Externalities and complementarities in platforms and ecosystems: from structural solutions to endogenous failures. Res Policy 53(1):104906

Javaid M, Haleem A, Singh RP, Suman R (2022) Enhancing smart farming through the applications of Agriculture 4.0 technologies. Int J Intell Netw 3:150–164

Jia Y, Cui L, Su J, Wu L, Akter S, Kumar A (2024) Digital servitization in digital enterprise: leveraging digital platform capabilities to unlock data value. Int J Prod Econ 278:109434

Jovanovic M, Sjödin D, Parida V (2022) Co-evolution of platform architecture, platform services, and platform governance: expanding the platform value of industrial digital platforms. Technovation 118:102218

Kamalaldin A, Linde L, Sjödin D, Parida V (2020) Transforming provider-customer relationships in digital servitization: a relational view on digitalization. Ind Mark Manag 89:306–325

Kaňovská L, Tomášková E (2018) Drivers for smart servitization in manufacturing companies. AGRIS Online Pap Econ Inform 10(3):57–68

Klerkx L, Jakku E, Labarthe P (2019) A review of social science on digital agriculture, smart farming and agriculture 4.0: new contributions and a future research agenda. NJAS Wageningen J Life Sci 90:100315

Kohtamäki M, Parida V, Oghazi P, Gebauer H, Baines T (2019) Digital servitization business models in ecosystems: a theory of the firm. J Bus Res 104:380–392

Kohtamäki M, Brekke T, Naeem R, Sjödin D, Parida V (2025) Managerial practices shaping the transformation towards artificial intelligence-enabled product–service systems. Int J Prod Econ 286:109658
Kolling C, de Medeiros JF, Ribeiro JLD, Morea D (2022) A conceptual model to support sustainable Product-Service System implementation in the Brazilian agricultural machinery industry. J Clean Prod 355:131733
Korkeamäki L, Kohtamäki M, Parida V (2021) Worth the risk? The profit impact of outcome-based service offerings for manufacturing firms. J Bus Res 131:92–102
Kowalkowski C, Gebauer H, Kamp B, Parry G (2017) Servitization and deservitization: overview, concepts, and definitions. Ind Mark Manag 60:4–10
Langley DJ (2022) Digital product-service systems: the role of data in the transition to servitization business models. Sustainability 14(3):1303
Lenka S, Parida V, Wincent J (2017) Digitalization capabilities as enablers of value co-creation in servitizing firms. Psychol Mark 34(1):92–100
Lichtenberg E (2019) Conservation and the environment in US farm legislation. EuroChoices 18(1):49–55
Liu Y, Ma X, Shu L, Hancke GP, Abu-Mahfouz AM (2020) From industry 4.0 to agriculture 4.0: current status, enabling technologies, and research challenges. IEEE Trans Industr Inform 17(6):4322–4334
Lütjen H, Schultz C, Tietze F, Urmetzer F (2019) Managing ecosystems for service innovation: a dynamic capability view. J Bus Res 104:506–519
Marcon É, Marcon A, Ayala NF, Frank AG, Story V, Burton J et al (2022) Capabilities supporting digital servitization: a multi-actor perspective. Ind Mark Manag 103:97–116
Markfort L, Arzt A, Kögler P, Jung S, Gebauer H, Haugk S et al (2022) Patterns of business model innovation for advancing IoT platforms. J Serv Manag 33(1):70–96
Martinez V, Neely A, Velu C, Leinster-Evans S, Bisessar D (2017) Exploring the journey to services. Int J Prod Econ 192:66–80
Matheny G, Leahy C (2007) Farm-animal welfare, legislation, and trade. Law Contemp Probl 70:325
Momeni K, Raddats C, Martinsuo M (2023) Mechanisms for developing operational capabilities in digital servitization. Int J Oper Prod Manag 43(13):101–127
Ozdogan B, Gacar A, Aktas H (2017) Digital agriculture practices in the context of agriculture 4.0. J Econ Finance Account 4(2):186–193
Paschou T, Rapaccini M, Adrodegari F, Saccani N (2020) Digital servitization in manufacturing: a systematic literature review and research agenda. Ind Mark Manag 89:278–292
Porter ME, Heppelmann JE (2014) How smart, connected products are transforming competition. Harv Bus Rev 92(11):64–88
Porter ME, Heppelmann JE (2015) How smart, connected products are transforming companies. Harv Bus Rev 93(10):96–114
Rajala R, Brax SA, Virtanen A, Salonen A (2019) The next phase in servitization: transforming integrated solutions into modular solutions. Int J Oper Prod Manag 39(5):630–657
Rodríguez MA, Lepratte L, Yoguel G, Rabetino R (2025) Digital Service Innovation as a co-production of socio-technical assemblages oriented toward servitization. J Enterp Inf Manag
Schäfer F, Gebauer H, Gröger C, Gassmann O, Wortmann F (2023) Data-driven business and data privacy: challenges and measures for product-based companies. Bus Horiz 66(4):493–504
Sjödin D, Parida V, Kohtamäki M, Wincent J (2020) An agile co-creation process for digital servitization: a micro-service innovation approach. J Bus Res 112:478–491
Sjödin D, Parida V, Palmié M, Wincent J (2021) How AI capabilities enable business model innovation: scaling AI through co-evolutionary processes and feedback loops. J Bus Res 134:574–587
Sjödin D, Parida V, Kohtamäki M (2023) Artificial intelligence enabling circular business model innovation in digital servitization: conceptualizing dynamic capabilities, AI capacities, business models and effects. Technol Forecast Soc Chang 197:122903

Smania GS, Mendes GHS, Godinho Filho M, Osiro L, Cauchick-Miguel PA, Coreynen W (2022) The relationships between digitalization and ecosystem-related capabilities for service innovation in agricultural machinery manufacturers. J Clean Prod 343:130982

Smania GS, Ayala NF, Coreynen W, Mendes GH (2024) Data-related tensions in digital servitization ecosystems: a systematic literature review. Ind Mark Manag 123:31–48

Tukker A (2004) Eight types of product–service system: eight ways to sustainability? Experiences from SusProNet. Bus Strateg Environ 13(4):246–260

Tukker A (2015) Product services for a resource-efficient and circular economy–a review. J Clean Prod 97:76–91

Van Dyck M, Lüttgens D, Diener K, Piller F, Pollok P (2024) From product to platform: how incumbents' assumptions and choices shape their platform strategy. Res Policy 53(1):104904

Vandermerwe S, Rada J (1988) Servitization of business: adding value by adding services. Eur Manag J 6(4):314–324

Vendrell-Herrero F, Bustinza OF, Parry G, Georgantzis N (2017) Servitization, digitization and supply chain interdependency. Ind Mark Manag 60:69–81

Verhoef PC, Broekhuizen T, Bart Y, Bhattacharya A, Dong JQ, Fabian N, Haenlein M (2021) Digital transformation: a multidisciplinary reflection and research agenda. J Bus Res 122:889–901

Vidickiene D, Gedminaite-Raudone Z (2018) Challenges for agricultural policy in the service-driven economic system. Ekon Poljopriv (Econ Agric) 65(4):1545–1555

Vidickienė D, Lankauskienė R (2025) Sustainability through the prism of innovative service-oriented business model in farming. Discov Sustain 6(1):286

Perspectives on Agricultural Technologies for Sustainable Transitions. Their Role in Farming System Innovation Toward Sustainable Agriculture

Davide Rizzo

1 Introduction and Conceptual Framework

Various types of technologies mediate farmers' decision-making for the management of crops, livestock, and resources. Tools and equipment materialize the related technical evolutions, and mark the major steps in the so-called agricultural revolutions. First, there was the introduction of tools to facilitate farmers' actions (Sigaut 1989), followed by the use of different energy sources, from animal to fossil (Harchaoui and Chatzimpiros 2019). Finally, the deployment of global navigation satellite systems eased the geolocation of equipment. The latest stage is the digitalization and automation of equipment for farming practices, also enabling the development of precision agriculture.

Agricultural robotics is at the crossroads of these evolutions. On the one hand, there is machinery automation through mobile tool carriers. On the other hand, there is the development of a data-driven approach to agriculture involving intensive data collection. Although agricultural robotics is still seen as a mere replacement for traditional tractors, its potential extends far beyond this role if framed within a system agronomy approach.

Robotics is the tip of the massive iceberg of agricultural digitalization. Digitalization itself opens new opportunities but also raises questions about farmers' mastery of it. Indeed, digitalization allows for increased equipment connectivity and data collection. However, its intangible dimensions can reduce farmers' control and introduce new actors in the knowledge production systems (Legun et al. 2023) and related value chains (Isaac and Pouyat 2018).

D. Rizzo (✉)
LISAH, Univ Montpellier, AgroParisTech, INRAE, Institut Agro, IRD, Montpellier, France
e-mail: davide.rizzo@ird.fr

F. da Silveira et al. (eds.), *Agriculture 4.0*,
https://doi.org/10.1007/978-3-032-24552-6_10

This chapter provides an overview of the issues and opportunities related to current agricultural technologies for sustainable transitions, adopting a multilevel perspective on both the agricultural system and the technological systems. The aim is to critically examine the interactions, actors, and new interfaces as a background for evaluating the contribution of Agriculture 4.0 technologies to the Sustainable Development Goals (SDGs), and to offer a nuanced and practice-oriented perspective with a focus on field robotics, mainly from a Western world (European) perspective. To this end, this chapter deliberately draws upon an expert-based synthesis of selected trends that allow for questioning agricultural technologies from a sustainable transitions perspective.

Similar topics have been recently addressed in great detail in the book edited by Dörr and Nachtmann (2022), where multiple authors covered all aspects of digital transformation for sustainable agriculture, with several examples from Germany and Europe. Indeed, technologies in agriculture are in a period of full bloom, making it difficult to grasp all the nuances across different contexts. Rather than providing a formal systematic review, this chapter draws on the author's interdisciplinary expertise and engagement with recent research, projects, and stakeholder networks to identify key developments, tensions, and pathways in technology and farming system innovation. However, this approach inherently carries the limitations of a selective perspective, which are partly mitigated through cross-references to complementary chapters in the present volume.

The rest of the chapter is organized as follow. First, it addresses the conceptualization of technologies in the agricultural context (Sect. 2). Then, it expands on the implications of innovations both for the task execution and the farming and sociotechnical system levels (Sect. 3), illustrated the different types of field demonstrations (Sect. 4). On these bases, the chapter then outlines the specific challenges emerging for sustainability, comparing them to the alternative transition pathways (Sect. 5), and concludes by discussing the next steps for agricultural technologies in sustainable transitions (Sect. 6).

2 Understanding Technologies in the Agricultural Context

> Do I think today's farmers need a robot? I think today's robots need a farmer! — Rod Karter, an Australian cattle farmer, commenting on a test of a four-legged herding robot by Professor Sukkarieh's team (2022)

Farmers make decisions and implement their actions, mainly using equipment. In this regard, technologies should not be considered *the* solution, but rather another tool. Indeed, the adoption and use of agricultural technologies cannot be considered alone. Technologies are embedded within complex sociotechnical systems shaped by local and wider contexts, which also frame farmers' decision-making processes and the execution of farming activities. As such, the notion of technology adoption or acceptability is insufficient to understand actual technology transitions in agriculture.

Two materialist philosophy approaches frame this chapter. On the one hand, Leroi-Gourhan's philosophy on the *technical fact*, defined as the observable results of a long-term evolutionary process (i.e., the technical tendency) of adapting tools to execute a task in response to physical constraints (Leroi-Gourhan 1943, 1945), thus framing the comparison of technologies across time (Audouze 2002). This philosophical and ethnological approach was subsequently applied to agronomy (Gras et al. 1989) to explore farmers' decision-making mediated by technical artifacts. On the other hand, the *multi-level perspective* elaborated by Geels (2002) explains technology transitions within the complex social configurations that structure the interactions among relevant actors. These levels encompass symbolic meanings, financial and governance rules, operational networks for production, distribution, and maintenance, as well as all the other networks needed to build and use a given technology. Bridging these two frameworks provides a way to examine the role of current technologies in farming system innovation, considering how the intangible dimensions of digitalization challenge their adoption and use.

This section comprises of three parts. First, it synthetizes of the two theoretical frameworks and their contribution to understanding the mismatch induced by digitalization. Second, it focuses on measured, data-driven agriculture that goes beyond the simple improvement of task execution precision. Finally, it proposes a classification of agricultural technologies in light of these premises.

2.1 Agricultural Actions from a Technological Perspective

In sociotechnical studies, there is a return to pragmatism (Miettinen et al. 2009) that helps in approaching farmers' practices in terms of observable effects within the interactions between the farmers' decision-making process and the everyday contextual materiality. This echoes the use of pragmatism to strengthen the systems approach, with implications for situated ontology and methodological multiplicity (Zhu 2025). For agriculture, the reference frame can be the technical fact as interpreted by Gras et al. (1989), drawing upon the ethnomethodological notion of the *technical fact* theorized by Leroi-Gourhan.

From an agronomic perspective, the technical fact in agronomy (Fig. 1) places the farmer's decision-making at the core, under the influence of both the biophysical environment (e.g., soils, climate, and topography) and the socioeconomic setting (e.g., policy, markets, and social norms). Farmers' decision-making intervenes through technical facts to change the state of targets and achieve production goals. In this context, the progressive development and availability of measurement tools inform and refine decision-making, creating new pathways for learning but also new dependencies.

The key idea underpinning the notion of technical fact is that technology acts as a mediator between the biophysical and socioeconomic environments, as perceived

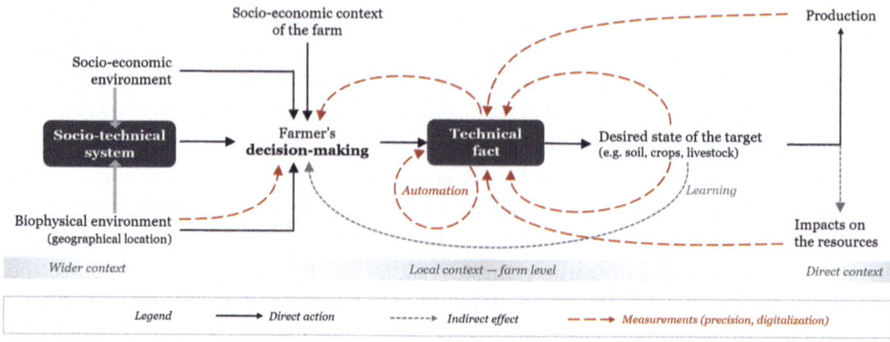

Fig. 1 Agronomic perspective on the technical fact. The scheme represents the conceptual connections around the farmer's decision-making and the technical fact, from the sociotechnical system to the output of on-farm actions, considering the contexts at different levels. (*Source*: translated and largely adapted from Gras et al. (1989, fig. 1))

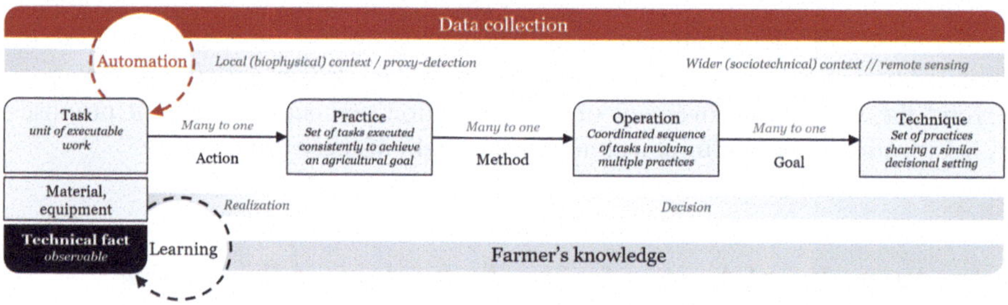

Fig. 2 Ontological representation of farming activities. The scheme represents the relationships linking the unit of work realized by each technical fact to the most structured decisions regarding farming techniques. Data collection and digitalization (in red) add a new layer to the farmer's knowledge

by farmers in their decision-making. In this sense, the effects of the farmers' actions, mediated by technologies, make the decision-making process observable. This framework has been used, for example, to describe and understand soil and water management practices (Smit et al. 2017; Rizzo et al. 2025), as well as the tinkering of agricultural machinery to improve soil management (Kefi et al. 2024).

The technical fact is the observable manifestation of the farmer's decision. From the perspective of a situated ontology (i.e., contextualized), a technical fact can be linked to a task, defined as a unit of executable work (Fig. 2). A set of tasks executed consistently to achieve an agricultural goal constitutes a practice. Multiple practices can be coordinated into agricultural operations, which result from the local combinations of farmers' knowledge, available materials, and goals (Landais and Deffontaines 1988), and collectively comprises various farming techniques (Fig. 2). Agronomic research studies farming practices and operations to reduce impacts on resources and trade-offs but also to increase, enhance sustainability, or and account for innovations in materials, equipment, and other available technologies (Huttunen 2019; Naim et al. 2025a).

This task-method-goal continuum is an ontological breakdown that provides a conceptual bridge between farmers' knowledge and the technical facts. Here, automation and data collection intervene at different levels—automating tasks, informing operations, and ultimately reconfiguring techniques. The diagram in Fig. 2 emphasizes that innovation in agricultural technologies is about reshaping how knowledge, action, and context interact over time beyond precise task automation.

2.2 Beyond Precision: Measured (Data-Driven) Agriculture

> Precision farming is a management approach that focuses on observing, measuring, and responding to variability in crops, fields, and animals in near-real time. (EU CAP Network 2025)

Digital data collection dramatically increases the quantity of measurements, which then translates into more accurate task preparation and execution. In this sense, agricultural digitalization frequently falls under the label of precision agriculture. However, the focus on precision overlooks the real novelty of the related technologies, which lies in the collection and use of data. The definition of precision farming by the European Union Common Agricultural Policy Network (2025) clearly emphasizes the role of measurements in management, similar to the definition elaborated by the International Society of Precision Agriculture (ISPA 2018). In this regard, Gilbert Grenier (2018) proposes using the term "measured agriculture" for its double meaning of (1) a data-driven agriculture and (2) carefully weighed input usage. The latter specifically highlights the role of data in evaluating practices and optimizing resources with greater attention. Compared to the more technocentric notion of *precision*—focused on refining resource application only where and when needed, minimizing waste while maximizing yield—the notion of *measured* is more explicit about the implications for sustainability. This switches the focus from basic quantitative measurement to a controlled qualitative optimization.

Besides semantic nuances, the key idea is to place digitalization within the agricultural perspective, focusing on its contribution to farming practices rather than on the potentials and impacts of the technologies. The aim is to set the scene for a description of technologies from a farming system perspective as simply another tool in the growing array of available equipment. Indeed, studies on impact and adoption try to bridge a mismatch between technology and agricultural research. Instead, an agriculture-centered perspective could allow reframing the technologies as part of the more complex farmers' decision-making system (Higgins et al. 2023).

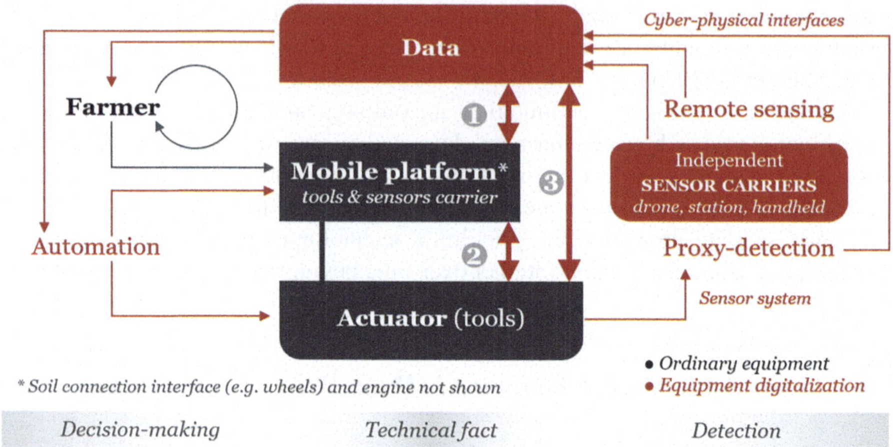

Fig. 3 Digitalization of agricultural equipment. The three critical components of agricultural equipment, from ordinary (black) to digitalized (red), highlight the emergence of new cyber-physical interfaces. Numbers exemplify these interfaces, such as the connected equipment (1), the tractor-implement management (2), or the data-driven actuator directly controlled through on-board proxy detection (3)

2.3 *Digitalization of the Agricultural Technical Fact*

The most critical piece of technology in previous agricultural evolutions has been machinery. The transition to Agriculture 4.0 and beyond is marked by its digitalization as part of a more pervasive presence of data. Ordinary equipment, at its core, is composed of a mobile platform and an actuator: a tool that allows for the execution of the desired task that realizes the farmer's decision and produces a technical fact (Fig. 3).

In the context of mechanized agriculture, the availability of engines has pushed the distinction between the mobile platform and the actuator that realizes the task. The tractor is the paradigmatic example of this mobile platform that provides energy through the engine to different types of tools and implements adapted to several tasks. Even in the case of specialized machines, such as combines, planters, or sprayers, the technologies follow different pathways for the engine and for the actuators, the latter needing to meet specifically agronomic targets.

Beyond this initial distinction, digitalization introduces data as a third element that is changing the ordinary configuration (Fig. 3). First, the platform began carrying various types of sensors, initially for simple driving assistance and safety, and ultimately for automation. Then, automation prompted the emergence of new cyber-physical interfaces that have driven the current technological revolution in agriculture.

Data flows bidirectionally between platforms, tools, sensor carriers (such as drones and handheld devices), and centralized data infrastructures. This architecture highlights three interlinked processes that revolve around the technical fact: detection (via remote sensing or proxy detection), decision-making (increasingly data-driven), and action (through actuation tools). It encapsulates how digital agriculture functions as a hybrid system where farmers, algorithms, and machines are co-actors in producing technical outcomes.

Robots represent a unique aspect of the agricultural technology evolution. Besides being highly digitalized machinery, they are also a tangible part of digitalization. The key novelty compared to digitalized machinery is their capacity to execute physical tasks autonomously (Daum 2021) with possible adaptation to the environment (Coiffet 2007). In this sense, robots repurpose the brain–body duality, embodying the digitalized knowledge (Dario et al. 2001; van Velden et al. 2024). Ultimately, agricultural robots should be evaluated beyond their capacity to replace humans for specific tasks, particularly regarding their potential of becoming a farmer's peer teammate (Puig-Pey et al. 2022; Legun et al. 2023), comparable to the development of the robot companion approach in human assistance and other sectors (Dario et al. 2011). In this regard, it is critical to remember that agricultural robots work with a large variety of living beings beyond humans (Ayris et al. 2024).

3 Implications of Agricultural Technologies for Sustainable Transitions

Technologies that mediate between farmers' decision-making processes and the technical facts lie at the intersection of two innovation systems: farming and sociotechnical systems. This section explores the key features that influence the promises and anchoring of technologies in agriculture (Elzen et al. 2017; Martin and Schnebelin 2024), following a multi-level perspective on sociotechnical transitions (Geels 2002).

Using a metaphor by Gaëtan Sevrac, co-founder of Naïo Technologies (at FIRA, 2021), an agricultural robot is like a washing machine: you cannot simply push a button to make it work correctly. Instead, you need to understand how it functions, at least to pick a preset programs and understand how to prepare your laundry to utilize it. Beyond the metaphor, a new piece of technology like an agricultural robot has two fronts: (1) its own functioning and (2) its interactions with the existing system. At the level of the technical fact, every new piece of technology either emerges to meet an existing need or must be questioned regarding its contribution to solving existing issues.

3.1 Innovation of the Physical Part: Platforms and Actuators

The basic contribution of technology to farming systems come from the tangible components, including the carrier and an actuator. For example, in the case of field robots, what is interesting from a farming system perspective is their specific size and autonomy, both of which increase the operational window for time-sensitive operations like weeding or crop protection.

Firstly, robots benefit from lighter mobile platforms compared to ordinary equipment, thus expanding the operational window for given soil trafficability conditions (Naim et al. 2025a). This advantage is further increased by the absence of a driver cab, particularly for autonomous tractors, which allows for a more efficient load distribution, resulting in higher traction capacity compared to cabbed machinery of the same weight.

Secondly, autonomy enables tasks to be executed at any agronomically relevant time, thus allowing for more timely use compared to equipment with a driver. In summary, the tasks can and should be redefined to consider the operational specificities of the new piece of equipment. In the example of field robots, in the short term, this includes re-evaluating the timing of task execution, drawing upon the larger *fieldability*. In the long term, the change in task planning can enable the complete redesign of operations like weeding or crop protection. For instance, the possibility of executing weeding from the earlier stages of weed development could enable a more effective switch from chemical to mechanical control.

In the search for a pertinent business model, some manufacturers have opted to split the offering. Besides the direct purchase of a robot, they also propose the robot as a service (Fourati-Jamoussi et al. 2023). As such, farming, weeding, or any other task the robot can execute is sold as a "solution" (Migende et al. 2022, pp. 196–197). Since most field robots are still in their development stage, this also gives customers the advantage of benefiting from the close support of the manufacturer and provides the manufacturer with new field data to inform development. In summary, *servitization* can make advanced technology more accessible (cf. chapter "Agriculture 5.0: Building Resilience by Integrating Social, Technological and Economic Innovations").

3.2 Role of the Digital Part

> [...] We're being asked to put our trust not just in other people, but in computers and algorithms. Algorithms most of us don't fully understand. Whose workings might be a mystery even to those who use them to run their businesses. (Vestager 2017)

The specificity of current agricultural technologies, as seen above, are the introduction of a massive amount of intangible data that collects information and knowledge about farming practices and their context of setup and execution. In this sense, their adoption or use questions the knowledge underpinning the farmer's decision-making process prior to the technical facts. Accordingly, a given piece of technology will be a "solution" if it meets the farmers' goals or fits their farming knowledge management system. The risk is instead an information and knowledge overload (infobesity) for the farmer (Hammond et al. 2023), who must face the integration of over-detailed data across fragmented data sources, with possible negative effects on their mental health (see chapter "Agriculture 4.0 and Transformation in Smallholder Farming Systems" in this book).

Data-driven or evidence-based choices present a great variety of advantages, from the different/higher control of input use to the potential to inform a complete redesign of farming systems (cf. Sect. 3.4). In addition, data collection can serve for certification in both regulatory and value chain contexts. However, farmers are in a weak position regarding data pooling and the risks associated with intellectual property protection (Bronson and Knezevic 2016; Jakku et al. 2018; MacPherson et al. 2022; Burch et al. 2023), further weakened by the disintermediation typical of digitalization (Martin and Schnebelin 2024).

Beyond the single problem patch or individual solution, the digital component of current agricultural technologies requires integration into wider knowledge management. This can be done at the farm level, within Farm Management Information Systems (FMIS), or more broadly by redefining Digital Agricultural Knowledge Information Systems (DAKIS) capable of bridging farms, extension services, and land managers (Mouratiadou et al. 2023).

A final interesting contribution of the purely digital data to transition pathways is the opening of exploration possibilities, as seen in digital games. Farming Simulator is a peculiar case because it reproduces a virtual farming reality and can eventually shape the perception of the farmer's duties for the future generation of farmers (Foureaux and Daum 2025). In the same vein, multiple serious games are emerging, allowing different actors to explore alternative pathways before acting in the field (Dernat et al. 2025).

3.3 Specific Challenges of the New Cyber-Physical Interfaces

Next, there is the evaluation of the cyber-physical interface. In this regard, some studies are exploring the influence of data-driven technologies on the physical health of operators (farmers or workforce). For instance, auto-steering can reduce the tractor driver's physical and/or mental workload during weeding (Gras et al. 2025), seeding (Sarvela et al. 2024), spraying (Burgers and Vanderwerff 2022), or harvesting (MacEachern et al. 2023) while keeping the same quality of the results. Despite the variety of tasks, new operators appear to benefit more from automation compared to skilled operators, thus confirming that these technologies might help in addressing the skilled labor shortage. To improve the analysis of this preliminary evidence, the authors also highlight the need to include engine data (Sarvela et al. 2024). Indeed, CAN bus data enable a detailed analysis and understanding of equipment functioning that can strengthen its finer mastery and optimization (Mattetti et al. 2022; Varani et al. 2025).

Agriculture 4.0 technologies, such as robots, have raised specific challenges in terms of certifications, partly overlapping with autonomous vehicles. One key need is compliance with established safety standards such as ISO 18497, *Safety of highly automated agricultural machines*, and ISO 25119 about the functional *Safety-related parts of control systems* (Lowenberg-DeBoer et al. 2022). These standards require robust obstacle detection, emergency stop functions, speed limitations, and thorough risk analysis prior to market deployment. Engineers evaluating existing prototypes found that only around 50 percent of functionalities complied with ISO 18497 and that traditional risk-assessment tools (like ISO 12100) often fall short when data on novel systems is lacking (Aby and Issa 2023). Regulatory frameworks remain inconsistent and frequently outdated. In the European Union, the *Machinery Directive* 2006/42/EC governs farm machinery, but does not specifically address autonomous field robots; emerging robotics functions can thus fall into legal gray zones (Basu et al. 2020). In the United States, certain states like California require an on-site operator even when robots operate autonomously, limiting fully driverless deployment (FIRA Team 2020).

Stakeholders also emphasize the role of harmonized certification, especially to support trade and mutual recognition across jurisdictions. Industry organizations, such as CEMA—the European Agricultural Machinery Industry Association—and the OECD advocate for coordinated certification schemes akin to those used for tractors, where compliance tested in one country is acknowledged elsewhere. This requires the definition of common frameworks to standardize risk analysis and functional architecture across use cases (Happich et al. 2025).

Lastly, third-party safety evaluation services have emerged to assist manufacturers in preparing compliant safety dossiers that align with ISO standards and pending European Machinery Regulation 2027/1230. However, developers still face considerable complexity in translating technical performance into legally recognized certification, and the sector remains characterized by fragmented rules, evolving definitions of operator responsibility, and a lack of tested precedents for autonomous, unmanned deployment (Basu et al. 2020). All of this is prompting the construction of dedicated benchmarking platforms (Imbert et al. 2016), such as the AgroTechnoPôle in France (https://www.agrotechnopole.fr/en/), and the activities of sector associations like RobAgri to work on shared normative frameworks.

From a different perspective, it is interesting to note that some robot manufacturers have opted to propose both autonomous machines and hitched implement versions. This aims to smooth the transition between the adoption of fully autonomous vehicles and the development of robotic implementations, such in the case of Ecorobotix.

Example Ecorobotix (https://ecorobotix.com/) is an example of a manufacturer that has developed a dual-version approach. On the one hand, the robot AVO, derived from the D-Arm iconic solar-powered weeder (Fig. 4). On the other hand, the ARA implement version. Interestingly, ARA was developed based on a version

Fig. 4 Ecorobotix robots. From the early prototype (left), showcased in a sugar beet field at Techniday UNEAL (France, 20 June 2018), to the later AVO version (right) exposed at the international forum of agricultural robots (FIRA, Toulouse, 11 November 2019). (Source: © Davide Rizzo)

of the robot that could be hitched to a tractor for road transportation. This approach allows for exploiting automation and advanced technologies while reducing costs for the user (Honcoop 2024).

3.4 Interactions with the Farming System

The interactions between agricultural technologies and farming systems start at the target level, namely the plant, animal, or soil. This could concern, for instance, redefining plant-breeding criteria for the adaptation of plant architecture to the interaction with automatic actuators (e.g., visibility and fruit picking resistance). The same technologies will indeed have different relevance and applicability depending on the specific farming system (e.g., arable or mixed cropping, highly diversified or specialized), targets (annual crops, vegetables, and trees), farmland spatial configurations, or regional agricultural governance (e.g., national or sectoral policies). As such, the adoption and use of new technologies require the acquisition of relevant agronomic references to establish their use in various agropedoclimatic contexts. These biophysical interactions intersect with socioeconomic dependencies, exemplified by data portability risks.

This also raises the question of specificity versus flexibility of a given technology for a single task (e.g., dedicated robots or decision support tools) versus modular or flexible technologies (e.g., tool carriers, or digital tool integration with farm management information systems). Insofar, Wiki AgriTech is the only available platform enabling a quite exhaustive comparison of the different agricultural technologies—1916 at the time this chapter was written—from a farming system perspective, with every piece of innovation framed within the farmer' goals or faming system type (https://www.wiki-agri-tech.com/).

Fig. 5 The patchCROP project carried out by ZALF in Steinhöfel, Germany. (Source: © Davide Rizzo, 20 September 2024)

The use of small, data-driven equipment can enable the diversification of intra-field organization, such as pixel farming (Ditzler and Driessen 2022). Another example is the strip cropping—including soy, maize, spring barley, and field beans—being tested by the KAFE challenge on 40 ha (Agreenculture 2025). It builds upon the previous CENTEOL challenge, carried out in 2018 on 50 ha of maize, for which a fleet of robots executed soil preparation, seeding, and the first weeding (2019). Finally, two projects that are even more ambitious explore the use of automated machine on an entire farm of 35 ha or even the complete rearrangement of crop management units based on soil and biodiversity potential, then managed with data-driven equipment (Fig. 5). The former is the Hand Free farm project (https://www.handsfree.farm/), coordinated by the Harper Adams University (Franklin 2024) and drawing upon the previous Hands Free Hectare experiment (2019). The latter is the patchCROP project (https://comm.zalf.de/sites/patchcrop/SitePages/Homepage.aspx) coordinated by ZALF (Donat et al. 2022; Grahmann et al. 2024). These two projects also address a current limitation of agricultural field robots, which is access to the field.

One should remember that, at least in Europe, autonomous machinery is still limited in on-road movement. Therefore, additional logistics still have to be considered when using robots (Lowenberg-DeBoer et al. 2022). In this sense, specific solutions are emerging, such as the vehicle transport system by Krone and Lemken "Combined power," an autonomous tool carrier (https://combined-powers.com/) that allows using a tractor to move the machine instead of a truck with a skilled driver. A more extreme solution is being developed by NextAT with a modular implement carrier (https://www.nexat.de/en/the-system/) that has a convertible structure allowing the cab position to be changed to drive on the road or in the field. These technologies aim to reduce the time an operator is exclusively dedicated to robot movement.

In summary, at the farming system level, current agricultural technologies create new dependencies, such as the need for connectivity that possibly hampers data logging and the correct functioning of the highly digitalized equipment. It also offers freedoms, particularly wider operational windows for mechanical interventions (e.g., weeding), which can be further improved with historical data analysis and real-time proxy detection. However, many of these technologies are still in their initial development stages and thus unstable; their rapid obsolescence casts farmers more as beta-testers than users. In addition, the lack of updated and adapted agronomic references reduces the possibility of exploiting the technologies to their fullest potential. In this regard, dealerships bridge the knowledge gap and are re-configuring as temporary training centers for the adoption and set up of the most advanced technologies they distribute (Rizzo et al. 2021). The role of dealerships can also be crucial to overcome the lack of interoperability between different implementations and machinery.

3.5 Sociotechnical System Reconfigurations for Sustainability

Technologies are embedded in complex systems. As such, their uptake requires changes in the system configuration and cannot be evaluated alone. In this sense, the question of agricultural technology adoption requires reformulation to address the system level, and it could be: *What are the changes in the system that allow for the introduction of new technologies?* These changes entail a reconfiguration of the system (Geels 2002) that can be triggered or blocked by several factors. However, the adoption and use of technologies for sustainable transitions intercept concurrent, deep changes in the agricultural sector. To tackle these systemic changes, new arenas emerge, such as the French association for Agricultural Robotics, RobAgri (https://robagri.fr/) (Fig. 6).

Example RobAgri. It is a French-based non-profit association founded in 2017. It was created to meet one of the priorities identified in two of the series of programmatic reports commissioned by the French government to define the strategies for agriculture in several sectors, including equipment manufacturing (Bournigal 2014) and environmentally and economically sustainable innovation for agriculture (Bournigal et al. 2015). Two major actors facilitated the configuration and birth of the association: the French Association for agricultural equipment manufacturers

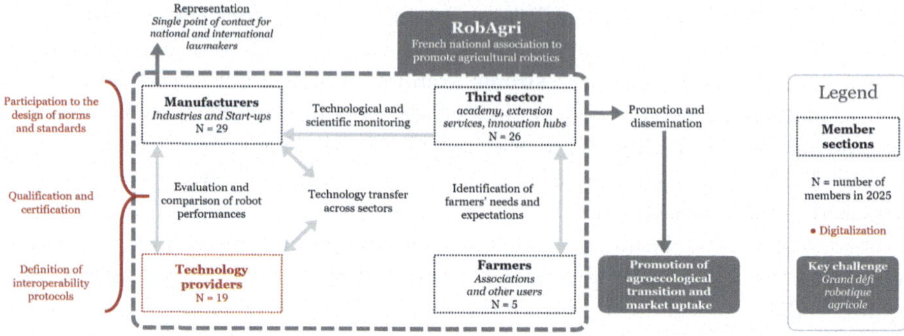

Fig. 6 Schematic representation of the RobAgri association. An example of the configuration of a technology innovation system

and agricultural environment providers AXEMA and the robot manufacturer Naïo Technologies (founded in 2011). Since 2018, the organization of the yearly scientific conference (e.g., Lenain et al. 2021) at FIRA (see Sect. 4.3) has been an example of RobAgri's technological monitoring and transfer.

Some technologies are already reconfiguring the systems. Nonetheless, the emerging actors, namely start-ups, remain quite fragile, especially in their search for successful business models (Fourati-Jamoussi et al. 2023). For the more solid ones, a solution lies in building strategic partnerships (Honcoop 2025). For instance, Naïo Technologies, a French-based robot manufacturer established in 2011 with hundreds of robots already operating in the field worldwide, is currently obliged to redefine its structure to stay in the market (Guillaumot 2025). Altogether, this increases uncertainties for farmers regarding both the durability of the technology provider and data portability in case of lack of interoperability, with a serious risk of data lock-in.

A pioneering study by Léa Stiefel (2023) explored at a country level the risks and advantages of a system reconfiguration for digital data governance. She compared two alternative models: a completely centralized data management, easier for operators but with a risk of creating an external dependency, versus a decentralized management, more secure for farmers but requiring a higher data literacy. In perspective, such a comparison opens up reflection on the integration of current technologies within existing agricultural knowledge information systems and the redefinition of the roles for extension services, distributors, technology providers, and data infrastructure managers (Fielke et al. 2019; Schnebelin et al. 2021). This eventually requires a revision of educational and training programs to increase the data literacy of all the concerned actors (Rizzo et al. 2019) (cf. chapter "Science Center for Development in Digital Agriculture").

4 Demonstrations and Challenges to Facilitate the Adoption

Changes in sociotechnical configurations can be triggered by the emergence of new interfaces among actors of the agricultural technological sector, such as exhibitions and field demonstrations, which create opportunities for new coordination and partnerships to emerge.

First, the preparation of these events generally requires the creation of a consortium among different actors, thus facilitating their interactions. Then, during the event, different stakeholders can meet to discuss and share viewpoints and technology demonstrations. Finally, the event itself can enable the creation of durable connections once the transactional costs have been paid.

4.1 In-field Demonstrations

The first and most classical interfaces are field demonstrations of new technologies (Sutherland and Marchand 2021). Field demonstrations create room for the technology providers to showcase their products and solutions, and possibly for peer-to-peer exchanges about use cases in comparable contexts (Ingram et al. 2018). Taking the example of robots again, one can observe two approaches.

Firstly, simple exhibitions in field contexts, copying what has been done for ordinary agricultural machinery for decades. Despite being a well-known format, with several historical events already included in seasonal planning in many agricultural regions, the inclusion of field robots is still very recent. For instance, the first field demonstration of agricultural robots was held in France in 2021 at the Innov'Agri yearly fair (Fig. 7).

Example The organization of the first field robot demonstration in France. It was planned in France in 2020, even though delayed to 2021 because of the pandemic. The *Robot camp—Village robotique* in French—was hosted by the Innov'Agri agricultural fair, organized by the Groupe France Agricole (GFA), editor of the eponymous magazine and part of *Nouvelle Génération de Presse Agricole* NGPA, owned by Isagri, an important software editor for agriculture. The fair generally showcases innovative topics as "camps," which are dedicated spaces that include field demonstrations and a series of pitches and conferences on a three-day program (Fig. 7e). The authors of this chapter, François Chéry (GFA) and Benoît Egon (NPGA group), defined the format and selected the robots to display in the field. Despite some start-ups and manufacturers having been around for almost ten years, only a few had a robot ready for live field demonstrations with farmers. Besides, a couple of them opted to be at the fair but in other sections. In the end, the camp was one of the key features of the entire fair and attracted most of the visitors, including the French Ministry of Agriculture at the time, Mr. Julien Denormandie. It was the first field demonstration for two autonomous tractors/tool carriers—Robotti by AgroIntelli (Fig. 7c) and Agbot by AgXeed (Fig. 7f)—along with a data scouting

Fig. 7 The first in-field demonstration of agricultural robots in France at the Robot camp, during Innov'Agri, in September 2021. (Source: © Davide Rizzo)

rover (SentiV by Meropy, Fig. 7b) and the already distributed autonomous seeder and weeder FD20 by FarmDroid (Fig. 7d). On the one hand, most of the farmers were confronted for the first time with these new technologies. The fair also gave other machinery manufacturer and technology providers the opportunity to meet the robot manufacturers and talk directly to the research and development teams.

Finally, many students in agriculture, both from high schools and colleges, were introduced to agricultural robotics.

4.2 Challenges and Competitions

Challenges are another type of event relevant to creating new interfaces for the adoption of agricultural technologies. These are problem-focused competitions where teams (often students, researchers, or start-ups) deploy technology in outdoor fields to perform defined tasks. These events borrow the incentive model of innovation prizes: they set concrete goals (e.g., autonomous weeding, crop mapping, harvesting) and reward the most effective solutions. For example, since 2003, the Field Robot Event annual international contest originating at Wageningen University serves as a motivating, project-based learning challenge to build small autonomous robots to perform field tasks (Oksanen et al. 2009), similarly to other events dedicated to student teams (Rizzo et al. 2018). On a more structured level, the Europe's Agri-food Competition for Robot Evaluation (ACRE) initiative explicitly "adopts the concept of benchmarking through competitions" to serve as a real-world testbed for advanced farm robots https://metricsproject.eu/agri-food/3rd-acre-workshop/. Likewise, the U.S. Farm Robotics Challenge is framed as "a transformative experience" to solve real-world farm problems and advance sustainable agriculture (UC Agriculture and Natural Resources 2023). The emphasis is on the novel use of technology in a farm setting, with teams that can mix commercial gear and custom builds, but must solve the designated tasks *in situ*.

Finally, there are episodic and limited in scope challenges organized by constructors to showcase the potential of new technologies in the field. For instance, in France, the implement manufacturer Kuhn and the automation company and robot manufacturer Agreenculture have collaborated with other technical partners to manage an entire cropping season—Challenge Centéol, 2018 and KAFE, 2025 (see Sect. 3.4)—or the three yearly edition of the Softirob challenge (2023, 2024, 2025) to compare the seeding performance of several field robots (Ladevèze 2025).

4.3 New Dedicated Fairs and Exhibitions

In-field challenge events and demonstrations play a critical role in increasing farmers' confidence and intention to adopt innovations in agricultural technology knowledge by making potential benefits visible (*observability*) and allowing hands-on trials (*trialability*) (Hayes et al. 2015; Ingram et al. 2018). However, these events are episodic and resource-intensive, often lacking systematic follow-up to assess long-term impacts, and their benefits may not generalize beyond the specific context of the demonstration (Hussain and Maharjan 2025). This is where the evolution of fairs and the emergence of dedicated exhibitions come into play, such as FIRA

Fig. 8 Overview of different editions of the International Forum of Agricultural Robots (FIRA) from the indoor edition (e.g., 2019 and 2021, on the top) to the field event (e.g., 2024, on the bottom). (Source: © Davide Rizzo)

(Fig. 8), to account for scalability, continuity, and context-specific tailoring to ensure sustained technology integration.

Example The International Forum for Agricultural Robotics—Forum International de la Robotique Agricole, FIRA https://www.agricultural-robotics.com/fira in French—is an example of a pioneering yearly event in the sector of agricultural technologies. Its evolution reflects dynamics that are useful to understand the opportunities for coordination among actors. It started in 2016 in Toulouse (southern France) as an indoor event to bring together key players in the sector, with around 200 participants (see the video summary, 2017). It was initially promoted as a spin-off of Naïo Technologies, to display their robots and promote the structuring of the sector. In 2019, the Global Organization for Agricultural Robotics—GOFAR, a French-Based non-profit association—took over the organization and, since 2022, launched FIRA USA in California, accordingly rebranding the original event as World FIRA. Both events regularly attract robot manufacturers, agricultural businesses, producers, investors, suppliers, entrepreneurs, and researchers from all over the world. Besides this event, GOFAR also participates in other promotional fairs, often helping existing machinery events to set up new spaces dedicated to robotics, supporting networking, and promoting studies and technical consulting. Starting from the Open Day in May 2022, FIRA evolved from an indoor event to a coupled conference program and in-field robot demonstrations (Fig. 8). This was made possible by tight collaboration with the French federation of the cooperatives for the use of agricultural machinery (FNCUMA), thus enabling closer interaction with farmers' lead entities.

5 Agricultural Technologies and Sustainable Transitions

Societal expectations and farmers' awareness and expectations for reducing trade-offs of agricultural activities and improving the sustainability of resource usage are opening opportunities for the sociotechnical reconfigurations, which are needed for a technological transition (Geels 2019).

5.1 Sustainability of Agricultural Technologies

Agricultural digital technologies, especially robots, potentially contribute to various Sustainable Development Goals (SDGs) by enabling the transition of agricultural practices toward greater efficiency, productivity, and environmental sustainability. This advancement comes with inherent synergies and trade-offs that need careful consideration.

Existing literature on the sustainability of agricultural technologies, especially for robots, mainly focuses on analyzing environmental efficiency (e.g., input reduction, soil health). For example, lifecycle studies highlight the scale discrepancies between benefits observed locally and trade-offs that can affect the wider context, such as rare element extraction or workforce marginalization. However, there is a geographical bias, with research mostly concentrated in Western countries, China, India, and Brazil, where technologies are more easily adopted compared to low-income countries. In terms of geographical tensions, the most advanced technologies also raise concerns about e-waste generation impacting extraction sites and potentially being overlooked at usage sites. However, early lifecycle research pinpoint that robots and the latest tractor models might have comparable trade-offs (Pradel et al. 2022).

Fewer studies address socioeconomic dimensions due to the early stage of technology development and use. The most critical trade-offs concern social equity impacts, labor displacement, and the risk of exacerbating the digital divide (Table 1).

Table 1 Overview of the potential contribution of current agricultural technologies to sustainable development goals (SDGs, https://www.un.org/sustainabledevelopment), with a focus on robotics

SDG	Contribution	Challenges/trade-offs
1 NO POVERTY	No relevant contribution	Displacement of farm workforce (Rose et al. 2023), and a known risk of wealth accumulation (Guenat et al. 2022)
2 ZERO HUNGER	In a technocentric perspective, increase of productivity (target 2.3) through the optimization of resource usage (target 2.4) (Naim et al. 2025a), and even the replacement of labor shortage. Data collection can inform decision-making and farm management, leading to more robust and accountable strategic and tactical actions	Reinforcement of unsustainable resource-intensive farming models, and marginalization of smallholders. Exacerbation of the unequal distribution of gains through the value chain (Haidegger et al. 2023)
3 GOOD HEALTH AND WELL-BEING	Indirectly contributes via safer, cleaner food. By reducing agrochemical use (via precision spraying) and improving animal health monitoring, field robots can lower exposure to pesticides and reduce food contaminants. Improvement of farmers' health and stress reduction (Burgers and Vanderwerff 2022; MacEachern et al. 2023; Sarvela et al. 2024; Gras et al. 2025)	Mental health burden due to uncertainties and dependencies created by new technologies
4 QUALITY EDUCATION	In agricultural education, the presentation and mastery of current technologies can support quality technical, vocational, and tertiary education (target 4.3) and facilitate the development of relevant skills for decent jobs (target 4.4), especially in digital literacy (target 4.6). In other sectors, experimenting with farmbots can increase the attractiveness of agriculture	Widening of the digital divide and educational gaps among countries and social groups. Engagement programs are needed to include marginalized groups (Ayris et al. 2024)
5 GENDER EQUALITY	Automation of repetitive, strenuous, or hazardous tasks can free women farmers and create opportunities in tech-driven agriculture, freeing time for education or diverse work (target 5.5)	Increase the gender divide already existent in STEM studies

(continued)

Table 1 (continued)

SDG	Contribution	Challenges/trade-offs
6 CLEAN WATER AND SANITATION	Field robots can optimize irrigation (e.g., AI-controlled sprayers, soil-moisture sensors), supporting efficient water use (target 6.4)	Displacement of pollution from the field (where technologies are utilized) to the place of manufacturing and resource extraction
7 AFFORDABLE AND CLEAN ENERGY	Dedicated robots can ease the spread of renewable energies, either directly with the switch from thermal to electric or solar-powered engines (target 7.2) (Naim et al. 2025b), or by showcasing their use	Without a careful lifecycle design, displacement of energy consumption and even an increase due to the use of rare resources (Pradel et al. 2022)
8 DECENT WORK AND ECONOMIC GROWTH	Agricultural technologies directly contribute to the upgrading and innovation (target 8.2), possibly reducing heavy tasks and enabling creativity and innovation (target 8.3). A proper use could even reduce environmental degradation (target 8.4) and facilitate inclusiveness (target 8.5), and increase attractiveness of agriculture to youth (target 8.6)	Significant risk of job displacement and marginalization of some groups in the society, and their generated incapability if the digitalization is too radical
9 INDUSTRY, INNOVATION AND INFRASTRUCTURE	Directly fosters innovation (target 9.5) by driving R&D in robotics, AI, and digital agriculture. Automated farms and smart supply chains are examples of resilient infrastructure (9.a). Investments in robotics lead to new industries (agritech) and infrastructure (data networks for IoT in farming). Promotion of retrofitting of existing equipment and creation of new business opportunities (target 9.4)	Creation of new dependencies on digital infrastructures and an increase of the digital divide (risk for target 9.c) Uptake of small-scale enterprises by regime (sensu Geels 2002) manufacturers and industries. Lack of investment in dedicated research in low- and medium-income countries (risk for target 9.5) with a consequent lack of support for domestic technology development (9.b)

(continued)

Table 1 (continued)

SDG	Contribution	Challenges/trade-offs
10 REDUCED INEQUALITIES	Potentially reduces inequalities by raising smallholder productivity (e.g., low-cost robots targeted at small farms) and improving food access (Haidegger et al. 2023). By lowering food costs, robotics can help poorer communities (target 10.1). Global partnerships on agricultural technologies can support development in low- and medium-income countries	Marginalization of low- and medium-income countries through the concentration of critical technologies (e.g., satellites), lack of inclusive policies (Guenat et al. 2022), and displacement of trade-offs, such as with rare element extraction for Western localized industries. Increase in rural–urban disparities for remotely controlled technologies. Further marginalization of the low-skilled migrant workforce
11 SUSTAINABLE CITIES AND COMMUNITIES	Supports urban agriculture (vertical farms, rooftop robots) and better food recycling. Robots sort food waste and supply local markets, making cities more self-reliant (target 11.6 on waste)	Over-reliance on tech could exclude informal urban farmers; planning must remain participatory (see goal 16)
12 RESPONSIBLE CONSUMPTION AND PRODUCTION	Precision farming cuts overuse of inputs (target 12.2), and food-processing robots sort and salvage imperfect produce, reducing waste (target 12.3)	Lifecycle impacts: digital technologies and robot hardware require rare minerals and create e-waste (Pradel et al. 2022)
13 CLIMATE ACTION	Current agricultural technologies are expected to enable climate-smart agriculture (target 13.2). Precision planting and tillage could reduce greenhouse gases from fuel and fertilizers	Early-stage adoption can increase emissions before realizing cuts, due to scale-up and energy use
14 LIFE BELOW WATER	Indirect benefits: By minimizing runoff of fertilizers and pesticides into waterways (via precision application), agricultural robots help reduce water pollution (target 14.1)	Indirect risks include e-waste leaching into water if not managed

(continued)

Table 1 (continued)

SDG	Contribution	Challenges/trade-offs
15 LIFE ON LAND	Lightweight or unmanned aerial vehicles lessen soil compaction, preventing erosion (target 15.3). Technologies that enable mechanical weeding and precise use of phytochemicals can help protect biodiversity (15.9). Data collection can also contribute to monitoring and managing soil health (Naim et al. 2025a)	Risk of favoring intensive, simplified farming systems. In particular, technologies enabling large-scale homogenized farming may worsen habitat and biodiversity loss
16 PEACE, JUSTICE AND STRONG INSTITUTIONS	Consistent and robust data collection can increase accountability and transparency of agricultural policies (target 16.6)	Data concentration and risk of leakage or misuse, in addition to the exclusive control by foreign agritech companies, challenging transparency (target 16.10) (Stiefel 2023)
17 PARTNERSHIPS FOR THE GOALS	The need for harmonization and policy-making requires the creation of international consortia that could ease technology transfer and help spread innovation (targets 17.7 and 17.8)	Lack of frameworks and references for the evaluation of actual technology sustainability Over-reliance on proprietary tech that limits access for poorer countries

The literature references are only examples and are absolutely nonexhaustive. *The content of this publication has not been approved by the United Nations and does not reflect the views of the United Nations, its officials, or Member States*

5.2 Alternative Transition Pathways

> Precision farming is thus, somewhat ironically, a high-technology means for rationalizing the use of old—if not obsolete—farming inputs. It can be seen as part of a process of restructuring of agricultural practices [. . .]. (Wolf and Buttel 1996, p. 1274)

While high-tech solutions promise precise control and automation, alternative pathways harness ecological processes and local knowledge to support the transitions toward sustainable agriculture. For instance, biofertilization addresses the reduction of external inputs by exploring the natural boosting of soil health through microbial formulations. Agroforestry, intercropping, and regenerative agriculture work instead on redesigning land management systems to improve biodiversity, soil structure, and carbon sequestration more intensively than the ordinary diversification of crop sequences.

Most of these practices are framed within the agroecological practices, even though their development and adoption can be easily included in a larger variety

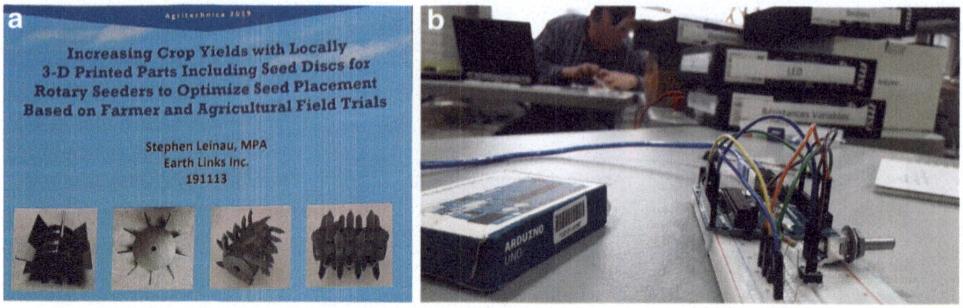

Fig. 9 Example of low-tech technology development. Some 3-D printed parts of a low-cost weeder presented at AgriTechnica (Stuttgart, Germany) by Stephen Leinau for EarthLinks https://earthlinksinc.org/about/ (left). A sensor being developed in the agricultural FabLab ® AgriLab in UniLaSalle Beauvais, France (right). (Source: © Davide Rizzo, 2019)

of sustainable transition pathways. Research and development should explore further how digital agricultural technologies and Agriculture 4.0 can support their development (Naim et al. 2025a) or help evaluate their actual effects and relevance for farmers' and societal goals.

Low-tech approaches also offer a middle path. Farmers need machines that they can repair themselves to master their availability, especially for time-sensitive tasks such as soil preparation, seeding, and harvesting. For instance, this drives a taste for traditional internal-combustion engines over sealed electric motors. Such "reparability" extends tool lifespan and reduces waste. It also lowers the environmental footprint by limiting resource-intensive replacements. Sobriety—designing robots for minimal energy use and simple maintenance—thus becomes a key criterion in the early adoption and use of agricultural robots (Naim et al. 2025b).

Parallel to production-focused innovations, grassroots innovation ecosystems foster inclusive transitions. FabLabs and maker spaces exemplify this by enabling open-source co-creation of agricultural tools, democratizing technology access via a community of practice, and dedicated equipment (Salembier et al. 2020; Rizzo et al. 2022). This can support, for instance, the development of 3D-printed repair parts or simple tools, or modular sensors for low-cost data collection (Fig. 9).

Institutional efforts reinforce this trend. The French National Federation of Agricultural Machinery Cooperatives (FNCUMA) advocates for "responsible mechanization," emphasizing shared ownership, local servicing, and ecological design. Such low-tech solutions bridge accessibility gaps while challenging the high-consumption model of mainstream digital agriculture.

Collectively, these pathways—agroecology, low-tech mechanization, and grassroots innovation—contribute to a sovereignty-centered framework for sustainable transitions. They prioritize modularity, reduced resource intensity, and community self-determination, offering scalable alternatives to top-down technological fixes, essential elements for a just and durable transition.

5.3 Contextual Challenges for Agricultural Technologies

Agricultural technologies and farming systems evolve on different timelines and within separate systems. Their main points of contact—farm equipment and farmers' decisions—are often where these mismatches create the most tension. Agricultural Knowledge and Innovation Systems (AKIS) can help reduce this stress through support and learning. However, they may also worsen it if they fail to adapt quickly. Sustainability goals can offer a common frame to align these changes. Framed by these goals, life-cycle assessment—the systematic analysis of environmental impacts from raw-material extraction through disposal—becomes essential to compare and prioritize emerging tools (Pradel et al. 2022) and to quantify the broader ecological footprint of digital agriculture (Tzachor et al. 2022). For instance, life-cycle assessment research exposes how rare earth extraction for precision sensors and energy use for data centers may offset input-reduction benefits, challenging SDG alignment.

To address these systemic trade-offs, technology development must embed *Responsible Research and Innovation* principles (Fielke et al. 2023). It is an approach that embeds anticipation of impacts, inclusion of stakeholders, reflexivity on ethics, and responsiveness to societal needs to ensure that innovations meet societal needs and avoid unintended harms (Stilgoe et al. 2013; Eastwood et al. 2022).

Indeed, the highly digitalized agricultural technologies come with multiple environmental and socioeconomic drawbacks (Prutzer et al. 2023). Beyond environmental framing and governance, external pressures such as climate change and geopolitical competition for critical minerals (e.g., lithium and rare earths) increasingly shape both the design and deployment of field robotics and precision implements. The introduction of highly digitalized technologies and automation can displace routine labor, yet under-delivers on attracting younger workers seeking meaningful, digitally augmented roles—a workforce gap undermining SDG 8 for decent work (Glover and Sumberg 2025). Finally, the rapid "datafication" and automation of farming operations raise ethical questions around data ownership, algorithmic transparency, and fair benefit-sharing (Bronson 2019; Carolan 2020; Alexander et al. 2024; Steinke et al. 2024).

The European Union-funded DESIRA project, by addressing the development and role of socio-cyber-physical systems, has demonstrated that participatory digital co-design can bridge these divides, yielding resilient, low-carbon tools that support sustainable transitions (Debryune et al. 2021). So, it can exemplify responsible research innovation in action, co-designing technologies with farmers to balance innovation with equity and ecological resilience.

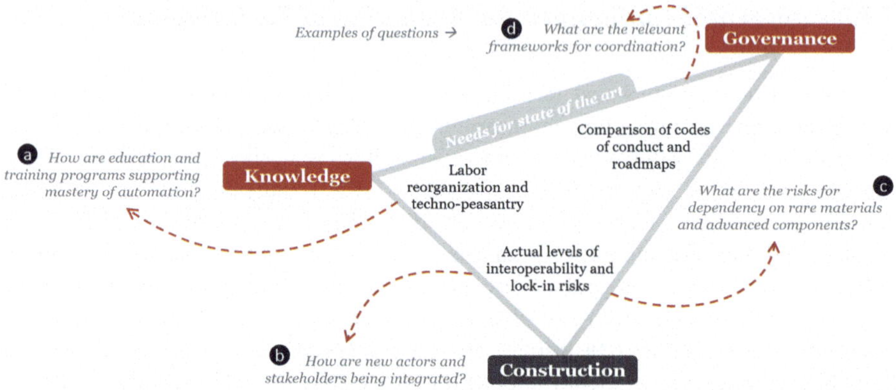

Fig. 10 Perspectives for research and public policy. Needs for state-of-the-art and examples of questions (a to d) on three main poles: knowledge, governance, and construction

6 Perspectives and Suggestions for Research and Public Policy

> The problem is that we have Paleolithic emotions, medieval institutions, and godlike technology. And it's terrifically dangerous.—Edward O. Wilson (2009)

Building on the ethical innovation approach outlined above, this section presents some long-term perspectives for advancing research on agricultural technologies that support sustainable transitions. The field continues to evolve, with frequent reviews of technical development (Spagnuolo et al. 2025; Vijayakumar et al. 2025), research challenges (U.S. National Science Foundation 2024), and ethical considerations (Dara et al. 2022). Here, agricultural robotics and automation are analyzed as components of broader farming system innovation, extending beyond short-term technical recommendations to address the systemic questions they raise. This includes examining how individual automated tasks are embedded within the wider sociotechnical context that enable or constrain them.

Framing the development of automation and robots for smallholders, especially in low- and middle-income countries, requires shifting innovation priorities away from narrow output maximization toward approaches grounded in people inclusion, holistic sustainability, and equitable resource management (Ayris et al. 2024). To this end, the perspectives and recommendations for research and public policy are organized around three connected axes: knowledge, construction, and governance (Fig. 10). The goal is to highlight gaps in the current state of the art and to propose questions that could strengthen the assessment of the long-term implications of adopting robots and digital technologies.

This list of recommendations clarifies immediate priorities and encourages interdisciplinary collaboration to tackle the complexities of agricultural robotics and automation, prioritizing equal distribution of benefits while evaluating threats to sustainability and social structures.

6.1 Knowledge: Farmers' Agency, Skills, and Ethical Considerations

The knowledge axis refers to the essential role of farmers and related actors in defining farming practices that incorporate automation and robotics. Three main perspectives can be highlighted: preserving farmer agency, managing the consequences of labor reorganization, and crosscutting ethical considerations.

6.1.1 Agency

Preserving farmers' agency is essential to ensure that automated systems are fully mastered and not treated as an opaque "oracle," thereby developing something farmers genuinely want (Sangwa and Mutabazi 2025).

The primary recommendation is to build trust through knowledge circulation and training, enabling farmers to comprehend and negotiate the development of relevant technologies.

Research and public institutions should lead in creating agronomic references for these technologies, making development processes transparent and inclusive. This approach can foster trust by demonstrating local adaptations, which are essential for disseminating agricultural innovation, and evaluating the attitude of local communities (Zeddies et al. 2024).

6.1.2 Skills

The introduction of new machinery and automation advancements leads to labor reorganization in high-income countries, where the agricultural workforce is declining. However, in low- and middle-income countries, adopting these technologies without robust capacity-building support risks fostering a sort of agricultural "techno-peasantry" (Black 1983; Shvyrkov 2025).

- Research should investigate how education and training programs are addressing the full mastery of automation technologies to redesign farming systems (Ritz et al. 2019; Hetman et al. 2024).
- For instance, what knowledge networks support skills related to agricultural technologies?

6.1.3 Ethical Considerations

Critical research must address the socioeconomic and cultural impacts of robotics and automation on specific rural communities (Rotz et al. 2019).

Study on the adoption of automated technologies should include the ethical and spiritual tensions that emerge when automated algorithmic agriculture displaces traditional agrarian ethics (Holm et al. 2024) and the ways these technologies are embodied by farmers (van Velden et al. 2024).

6.2 Construction: Interoperability and Dependencies

The construction axis refers to perspectives related to technology providers, encompassing both their market role and the tangible construction of automation technologies and robots. As the sector remains highly exploratory, frequently led by new actors or newcomers in the agricultural sector (Rizzo et al. 2019), research should compare actual levels of interoperability and assess the lock-in risks.

6.2.1 Interoperability: Lock-in and Business Models

Beyond the obvious technical issues of interoperability, there is a need to integrate the technology providers in the farming system design and functioning.

- A sociotechnical comparison of their trajectories, as well as of the definition of their business model, can clarify how they negotiate their roles (Shockley et al. 2021; Fourati-Jamoussi et al. 2023).
- A pertinent example is Naïo (Technologies), a successful agricultural robotics start-up founded in 2011: Despite extensive experience, international expansion from France, and over 300 operational robots, it faced financial difficulties resolved only recently through a joint venture involving public and regional investment funds. This example, among others, illustrates that stable business models in agricultural robotics are still evolving (Alarcon et al. 2021) and need to be carefully analyzed by research and policy.

6.2.2 Dependencies and Geopolitical Risks

Dependency poses risks during construction—such as reliance on rare materials and advanced components (e.g., semiconductors and microchips)—and during use, through lock-ins to proprietary clouds, or data formats.

- The emerging applications of life-cycle analysis in the agricultural robotics and automation industry (e.g., Pradel et al. 2022) should consider the impact of dependency on outsourced materials. For instance, the disruption of global logistics during the 2020 pandemic slowed or halted tractor production due to a shortage of microchips. The more advanced the technology, the more these dependencies can disrupt the production cycle.

- In the background, this may involve mapping and evaluating the creation and maintenance of open and shared digital infrastructures to promote interoperability.
- From a broader viewpoint, research and policy should assess how these dependencies generate new risks tied to geopolitical trends in accessing raw materials or data storage (Atkinson et al. 2025).
- Additionally, dependence on external providers and potential erosion of traditional knowledge systems threaten local agricultural autonomy and cultural practices (Bianchi et al. 2025; Manzoor et al. 2025), thus requiring dedicated evaluation.

6.3 Governance: Data and Liability

The third axis for research and policy involves addressing the governance of automation transitions in agriculture. Technologies advance far quicker than legal frameworks, creating uncertainties for stakeholders, particularly around data management and liability for autonomous machinery.

6.3.1 Data Governance

Data generated, collected, and managed by automation technologies expose intangible actors' knowledge, as seen in the previous sections, raising fundamental ownership questions.

- Research should examine the specificities of low- and middle-income countries in technological transitions to identify necessary context-sensitive adaptations.
- Scholarly effort must be directed toward understanding the views of diverse stakeholders (including farmers and workers) to contextualize academic perspectives, particularly regarding the ownership of data collected by robots (see the pilot study for the United Kingdome by Ayris et al. 2024).
- Stiefel's work (2023) on Switzerland's digital transition illustrates the issues at stake between centralized data systems—efficient but potentially controlled by private vendors—and decentralized approaches, which promote capacity building but may lag in performance. This example shows the convergence with long-known questions about the overall digital transition pathways at multiple levels (Black 1983; Fielke et al. 2019).

6.3.2 Liability

Assigning responsibility for autonomous machinery is complex due to overlaps among agricultural production, labor regulation, automation, and digitalization (Holm et al. 2024). The rapid evolution of artificial intelligence demonstrates the risks of regulatory lag, with some sectors moving toward what Abiri (2025) describes as a "mutually assured deregulation," with policy delegating technology developers to lead the way in an uncharted territory. This is particularly urgent as the boundaries between digitalization, automation, and autonomous AI systems become increasingly blurred (Balkin 2017), and as accountability for algorithmic actions remains unsettled (Alexander et al. 2024).

Industry-developed codes of practice play an important preparatory role by establishing operational and organizational benchmarks to reduce uncertainty. Despite not being legally binding, these types of documents help shape the professional standards and can inspire future legal frameworks. Available examples allow identifying two domains to be integrated by research and policy design:

- *Operational safety*, with documents that provide practical guidance for safe deployment: examples are the British Standards Institution's Code of Practice (BS 8646:2023) and the Australian industry-led code of practice about agricultural mobile field machinery (Grain Central 2021).
- *Data governance*, covered for instance by the European code of conduct on agricultural data sharing by contractual agreement (COPA and COGECA 2018), and two farmer-led initiatives to certify the respect of agricultural data use based on predefined principles, in France (FNSEA and Jeunes Agriculteurs 2019) and the United States (American Farm Bureau Federation 2024).
- The codes of conduct and practices articulate high-level principles and call research and policy for improving standards for contractual readability to close the critical "trust gap" in digitalization between farmers and technology providers. Research and policy should target the actual user readability and the trends in responsibility assignment in contractual agreements.
- Comparative research on existing codes of conduct will also be essential to evaluate private-sector commitments and to support public policy in developing effective legal frameworks.

7 Conclusion: What Is Next for Technology in Agricultural Transition Towards Sustainability?

The adoption and use of Agriculture 4.0 technologies in farming practices represent a disruption in how farming systems operate and evolve. As the previous sections have explored, the high digitalization of knowledge and the cyber-physical interfaces in field robotics and other new technologies promise significant potential for enhancing productivity, efficiency, and sustainability in agriculture. However, this transition is not without its challenges. The intangible dimensions of digitalization, such as data management and connectivity, introduce new complexities and dependencies that providers must tackle to ensure they align with the broader goals of sustainable development and farmers' needs and mastery.

Regardless of the specific technology or sociotechnical configuration, this chapter focused on digitalization as a new component of farming systems and decision-making processes, to highlight challenges to farmers' mastery of intangible knowledge being produced and mobilized. Adopting the technical fact as the main theoretical framework helped stress the mismatch between the materiality of farming practices and the intangibility of digitalization. Building on the "technical fact" framework, future research should empirically assess how digital and mechanical innovations reshape farming practices across diverse farming systems, accounting for both material and informational dimensions.

The rapid pace of technological advancement necessitates continuous learning and adaptation. Educational programs and training initiatives must evolve to keep pace with these changes, ensuring that farmers are not left behind in this digital revolution. The role of agricultural knowledge and innovation systems (AKIS) is crucial in this regard, providing the support and resources needed for farmers to evaluate and adapt these technologies. Moreover, there is a pressing need to integrate complementary systemic analyses—such as lifecycle assessments and Responsible Research and Innovation metrics—into technology design cycles from scratch, so that sustainability is considered *by design* rather than as an ancillary benefit.

In conclusion, the future of agricultural technologies lies in their ability to be seamlessly integrated into farming systems, enhancing rather than complicating the decision-making processes of farmers. As Agriculture 4.0 moves forward, it is essential to focus on creating and developing technologies that are not only advanced but also accessible, sustainable, and aligned with the broader goals of agricultural sustainability.

Acknowledgments Thanks to Gautier Moray (EXXACT Robotics) for the updates and insights about trends in the agricultural robot sector, and to David Christian Rose (Harper Adams University) for the inspiring reading suggestions and copious references about the responsible research innovation, which I have been able to incorporate in the text far less than what I would have desired.

Funding The activities of Davide Rizzo are funded by the French National Research Agency (project no. ANR-22-CPJ1-0050-01), the IRD, and the University of Montpellier via the I-SITE MUSE excellence program in the framework of the junior professorship tenure track in landscape agronomy.

Declaration of Generative AI and AI-Assisted Technologies in the Writing Process During the preparation of this work, the author used different online, free generative AI services to check for English Language consistency and to improve fluidity and clarity of the text. After using these services, the author carefully reviewed and edited the content as needed and takes full responsibility for the content of the publication.

Competing Interests The author(s) has no competing interests to declare that are relevant to the content of this manuscript.

References

Abiri G (2025) Mutually assured deregulation (arXiv:2508.12300). arXiv. https://doi.org/10.48550/arXiv.2508.12300

Aby GR, Issa SF (2023) Safety of automated agricultural machineries: a systematic literature review. Safety 9:13. https://doi.org/10.3390/safety9010013

Agreenculture (2025) KAFE challenge. In: LinkedIn. https://web.archive.org/web/20250718201709/https://www.linkedin.com/posts/agreenculture_the-kafe-challenge-is-gaining-momentum-activity-7331282949130108929-W1ZJ/. Accessed 18 July 2025

Alarcon P, Marco-Jimenez F, Horigan V et al (2021) A review of cleaning and disinfection guidelines and recommendations following an outbreak of classical scrapie. Prev Vet Med 193. https://doi.org/10.1016/j.prevetmed.2021.105388

Alexander CS, Yarborough M, Smith A (2024) Who is responsible for 'responsible AI'?: navigating challenges to build trust in AI agriculture and food system technology. Precis Agric 25:146–185. https://doi.org/10.1007/s11119-023-10063-3

American Farm Bureau Federation (2024) Ag Data Core Principles. Ag Data Transparent. https://www.agdatatransparent.com/principles

Atkinson RD, Ostertag M, Long T (2025) A time to act: policies to strengthen the US robotics industry. Information Technology & Innovation Foundation (ITIF), p 27. https://itif.org/publications/2025/07/18/time-to-act-policies-to-strengthen-us-robotics-industry/

Audouze F (2002) Leroi-Gourhan, a philosopher of technique and evolution. J Archaeol Res 10:277–306. https://doi.org/10.1023/A:1020599009172

Ayris K, Jackman A, Mauchline A, Rose DC (2024) Exploring inclusion in UK agricultural robotics development: who, how, and why? Agric Hum Values 41:1257–1275. https://doi.org/10.1007/s10460-024-10555-6

Balkin JM (2017) The three laws of robotics in the age of big data. 016 Sidley Austin distinguished lecture on big data law and policy. Ohio State Law J 78(5):1217–1241. https://openyls.law.yale.edu/server/api/core/bitstreams/08fc8025-26f0-4725-88da-bdc2d794df18/content

Basu S, Omotubora A, Beeson M, Fox C (2020) Legal framework for small autonomous agricultural robots. AI Soc 35:113–134. https://doi.org/10.1007/s00146-018-0846-4

Bianchi A, Corbo C, Parigi F et al (2025) Assessing the social impacts of Digital Agriculture Technology Solutions: a practical tool. Bio-Based Appl Econ. https://doi.org/10.36253/bae-16929

Black ER (1983) Politics on a microchip. Can J Polit Sci Rev Can Sci Polit 16:675–690

Bournigal J-M (2014) Définir ensemble le futur du secteur des agroéquipements [Rapport de la mission agroéquipements, Irstea]. IRSTEA - Institut national de recherche en sciences et technologies pour l'environnement et l'agriculture, p 149. http://agriculture.gouv.fr/ministere/definir-ensemble-le-futur-du-secteur-des-agroequipements

Bournigal J-M, Houiller F, Lecouvey P, Pringuet P (2015) Agriculture – Innovations 2025: 30 projets pour une agriculture compétitive & respectueuse de l'environnement. [Rapport aux Ministres en charge de l'agriculture et en charge de la recherche]. Ministère de l'Agriculture, de l'Agroalimentaire et de la Forêt, p 135. http://agriculture.gouv.fr/agriculture-innovation-2025-des-orientations-pour-une-agriculture-innovante-et-durable

Bronson K (2019) Looking through a responsible innovation lens at uneven engagements with digital farming. NJAS - Wagening J Life Sci. https://doi.org/10.1016/j.njas.2019.03.001

Bronson K, Knezevic I (2016) Big Data in food and agriculture. Big Data Soc 3:2053951716648174. https://doi.org/10.1177/2053951716648174

Burch KA, Nafus D, Legun K, Klerkx L (2023) Intellectual property meets transdisciplinary co-design: prioritizing responsiveness in the production of new AgTech through located response-ability. Agric Hum Values 40:455–474. https://doi.org/10.1007/s10460-022-10378-3

Burgers TA, Vanderwerff KJ (2022) Vision and radar steering reduces agricultural sprayer operator stress without compromising steering performance. J Agric Saf Health 28:163–179. https://doi.org/10.13031/jash.15060

Catalyst (Producer) (2022, July 12) Farmer needs a robot [Video recording]. Australian Broadcasting Corporation. https://www.abc.net.au/contentsales/programsandgenres/farmer-needs-robot/13970184

Carolan M (2020) Acting like an algorithm: digital farming platforms and the trajectories they (need not) lock-in. Agric Hum Values 37:1041–1053. https://doi.org/10.1007/s10460-020-10032-w

Coiffet P (2007) Robots industriels: Concepts, définitions et classifications. In: Automatique et ingénierie systéme. Éditions Techniques de l'Ingénieur, p S7700. https://doi.org/10.51257/a-v1-s7700

COPA, COGECA (2018) Code of Conduct on agricultural data sharing. https://www.fao.org/family-farming/detail/en/c/1370911/

Dara R, Hazrati Fard SM, Kaur J (2022) Recommendations for ethical and responsible use of artificial intelligence in digital agriculture. Front Artif Intell 5. https://doi.org/10.3389/frai.2022.884192

Dario P, Murphy R, Bonasso P et al (2001) Sentience in robots: applications and challenges. IEEE Intell Syst 16:66–69. https://doi.org/10.1109/MIS.2001.956083

Dario P, Verschure PFMJ, Prescott T et al (2011) Robot companions for citizens. Procedia Comput Sci 7:47–51. https://doi.org/10.1016/j.procs.2011.12.017

Daum T (2021) Farm robots: ecological utopia or dystopia? Trends Ecol Evol 36:774–777. https://doi.org/10.1016/j.tree.2021.06.002

Debryune L, Brunori G, Casares B, Nieto E (2021) Key digital game changers in shaping the future of agriculture in 2040. View's from DESIRA's rural digitalisation forum experts (Long-Term Vision for Rural Areas, p 10). DESIRA EU Project. https://desira2020.eu/contribution-ltvra-agriculture/

Dernat S, Grillot M, Andreotti F, Martel G (2025) A sustainable game changer? Systematic review of serious games used for agriculture and research agenda. Agric Syst 222:104178. https://doi.org/10.1016/j.agsy.2024.104178

Ditzler L, Driessen C (2022) Automating agroecology: how to design a farming robot without a monocultural mindset? J Agric Environ Ethics 35:2. https://doi.org/10.1007/s10806-021-09876-x

Donat M, Geistert J, Grahmann K et al (2022) Patch cropping- a new methodological approach to determine new field arrangements that increase the multifunctionality of agricultural landscapes. Comput Electron Agric 197:106894. https://doi.org/10.1016/j.compag.2022.106894

Dörr J, Nachtmann M (eds) (2022) Handbook digital farming: digital transformation for sustainable agriculture. Springer Berlin Heidelberg, Berlin, Heidelberg

Eastwood CR, Dela Rue B, Edwards JP, Jago J (2022) Responsible robotics design-a systems approach to developing design guides for robotics in pasture-grazed dairy farming. Front Robot AI 9. https://doi.org/10.3389/frobt.2022.914850

Elzen B, Janssen A, Bos AP (2017) Portfolio of promises: designing and testing a new tool to stimulate transition towards sustainable agriculture. In: AgroEcological transitions. Wageningen University & Research, pp 143–161. Wageningen, The Netherlands

EU CAP Network (2025) Precision farming, livestock farming, forestry and food processing. In: Innov. Knowl. Exch. EIP-AGRI. https://eu-cap-network.ec.europa.eu/support/innovation-knowledge-exchange-eip-agri/digitalisation/digital-tools-automation-and-platforms/precision-farming-livestock-farming-forestry-and-food-processing_en. Accessed 18 July 2025

Fielke SJ, Garrard R, Jakku E et al (2019) Conceptualising the DAIS: implications of the 'digitalisation of agricultural innovation systems' on technology and policy at multiple levels. NJAS - Wageningen J Life Sci 90–91:100296. https://doi.org/10.1016/j.njas.2019.04.002

Fielke SJ, Lacey J, Jakku E et al (2023) From a land 'down under': the potential role of responsible innovation as practice during the bottom-up development of mission arenas in Australia. J Responsible Innov 10:2142393. https://doi.org/10.1080/23299460.2022.2142393

FIRA Forum (Director). (2019, January 28). FIRA 2018: Prèsentation des challenges CENTEOL 2018 & Projet Hands Free Hectare [Video recording]. https://www.youtube.com/watch?v=hTfh7ouAu9E

FIRA Team (2020) Regulation and standardization of agricultural robots: where are we now in Europe, and in the world? In: Agric. Robot. https://web.archive.org/web/20250000000000*/https://www.agricultural-robotics.com/news/regulation-standardization-agricultural-robots. Accessed 20 July 2025

FNSEA, Jeunes Agriculteurs (2019) Charte sur l'utilisation des données agricoles Data Agri. Valoriser et sécuriser les données des exploitations agricoles dans les contrats

Fourati-Jamoussi F, Dantan J, Dubois MJF, Rizzo D (2023) Impact of digital technologies on farms' business models. Cahiers COSTECH - Cahiers Connaissance, Organisation et Systèmes Techniques 6:21. https://doi.org/10.34746/cahierscostech155

Foureaux S, Daum T (2025) "But don't think it is a game": agricultural videogames and "good farming". J Rural Stud 117:103686. https://doi.org/10.1016/j.jrurstud.2025.103686

Franklin K (2024) Challenges and experiences from the hands free farm project. Key note. In: Advances and innovations in agricultural engineering. NJF - EurAgEng, Herning, pp 15–16

Geels FW (2002) Technological transitions as evolutionary reconfiguration processes: a multi-level perspective and a case-study. Res Policy 31:1257–1274. https://doi.org/10.1016/S0048-7333(02)00062-8

Geels FW (2019) Socio-technical transitions to sustainability: a review of criticisms and elaborations of the multi-level perspective. Curr Opin Environ Sustain 39:187–201. https://doi.org/10.1016/j.cosust.2019.06.009

Glover D, Sumberg J (2025) Skill and deskilling in an uncertain agricultural world: introduction to the special issue. Outlook Agric:00307270251337112. https://doi.org/10.1177/00307270251337112

Grahmann K, Reckling M, Hernández-Ochoa I et al (2024) Co-designing a landscape experiment to investigate diversified cropping systems. Agric Syst 217:103950. https://doi.org/10.1016/j.agsy.2024.103950

Grain Central (2021) Code of practice. Agricultural mobile field machinery with autonomous functions in Australia. Grain Central, Bulimba Qsd, Australia. https://www.graincentral.com/ag-tech/code-of-practice-drives-the-future-of-autonomous-farm-machinery/

Gras R, Benoît M, Deffontaines J-P et al (1989) Le Fait technique en agronomie: activité agricole, concepts et méthodes d'étude. Institut National de la Recherche Agronomique. Édition L'Harmattan, Paris, France

Gras J-P, Valloo Y, Tisseyre B (2025) Evaluating the benefits of auto-steering on driving conditions in viticulture. Precis Agric 25:163–170. https://doi.org/10.1163/9789004725232_019

Grenier G (2018) Agriculture de précision. Editions France Agricole, Paris, France

Guenat S, Purnell P, Davies ZG et al (2022) Meeting sustainable development goals via robotics and autonomous systems. Nat Commun 13:3559. https://doi.org/10.1038/s41467-022-31150-5

Guillaumot M (2025) The End of Naïo Technologies Era !? Look back at this adventure within this Ag Robotics Pioneer. In: Agtech Mark. https://www.agtechmarket.net/news/naiolastnews. Accessed 18 July 2025

Haidegger T, Mai V, Mörch CM et al (2023) Robotics: enabler and inhibitor of the sustainable development goals. Sustain Prod Consum 43:422–434. https://doi.org/10.1016/j.spc.2023.11.011

Hammond J, Pagella T, van Etten J et al (2023) Editorial: Agile data-oriented research tools to support smallholder farm system transformation. Front Sustain Food Syst 7. https://doi.org/10.3389/fsufs.2023.1128513

Harvard Museum of Natural History (Director) (2009) Looking back looking forward: a conversation with James D Watson and Edward O Wilson [Video recording]. https://www.youtube.com/watch?v=N8_W2cBAO7s

Happich G, Grever A, Schöning J (2025) Agricultural Industry Initiatives on Autonomy: how collaborative initiatives of VDMA and AEF can facilitate complexity in domain crossing harmonization needs (arXiv:2501.17962). arXiv. https://doi.org/10.48550/arXiv.2501.17962

Harchaoui S, Chatzimpiros P (2019) Energy, nitrogen, and farm surplus transitions in agriculture from historical data modeling. France, 1882–2013. J Ind Ecol 23:412–425. https://doi.org/10.1111/jiec.12760

Hayes KJ, Eljiz K, Dadich A et al (2015) Trialability, observability and risk reduction accelerating individual innovation adoption decisions. J Health Organ Manag 29:271–294. https://doi.org/10.1108/JHOM-08-2013-0171

Hetman M, Giovanelli V, Medici M (2024) An overview of higher education AgTech in Europe: general features, sustainability content, and life-long learningpathways. In: LATEST - local-focused AgTech education for successful agricultural transitions Erasmus+ project Deliverable PR1, https://latest-project.eu/project-result-1

Higgins V, van der Velden D, Bechtet N et al (2023) Deliberative assembling: tinkering and farmer agency in precision agriculture implementation. J Rural Stud 100:103023. https://doi.org/10.1016/j.jrurstud.2023.103023

Holm S, Pedersen SM, Tamirat TW (2024) Robots in agriculture – a case-based discussion of ethical concerns on job loss, responsibility, and data control. Smart Agric Technol 9:100633. https://doi.org/10.1016/j.atech.2024.100633

Honcoop P (2024) The shift from single-task autonomous solutions to smart implements. In: LinkedIn. https://web.archive.org/web/20250718211646/https://www.linkedin.com/pulse/shift-from-single-task-autonomous-solutions-smart-patrick-honcoop-f8vpf/. Accessed 18 July 2025

Honcoop P (2025) Unlocking the power of strategic partnerships in AgTech | LinkedIn. In: AgTech Edge. https://web.archive.org/web/20250718220319/https://www.linkedin.com/pulse/unlocking-power-strategic-partnerships-agtech-patrick-honcoop-gf1he/. Accessed 19 July 2025

Hussain N, Maharjan KL (2025) Impact of on-farm demonstrations on technology adoption, yield, and profitability among small farmers of wheat in Pakistan—an experimental study. Agriculture 15:214. https://doi.org/10.3390/agriculture15020214

Huttunen S (2019) Revisiting agricultural modernisation: interconnected farming practices driving rural development at the farm level. J Rural Stud 71:36–45. https://doi.org/10.1016/j.jrurstud.2019.09.004

Imbert P, Petit N, Siben C (2016) Plateaux techniques pour agroéquipements. https://agriculture.gouv.fr/plateaux-techniques-pour-lagroequipement

Ingram J, Chiswell HM, Mills J et al (2018) Enabling learning in demonstration farms: a literature review. Int J Agric Ext 2018:29–42. https://doi.org/10.5281/zenodo.6619865

Isaac H, Pouyat M (eds) (2018) La valeur des données agricoles. Renaissance Numèrique, Paris, France. http://www.renaissancenumerique.org/publications/la-valeur-des-donnees-agricoles

ISPA (2018) Precision Ag definition. In: Int. Soc. Precis. Agric. https://web.archive.org/web/20211007061405/https://www.ispag.org/about/definition. Accessed 26 Mar 2021

Jakku E, Taylor B, Fleming A et al (2018) "If they don't tell us what they do with it, why would we trust them?" Trust, transparency and benefit-sharing in smart farming. NJAS - Wagening J Life Sci:100285. https://doi.org/10.1016/j.njas.2018.11.002

Kefi S, Dubois MJ-F, Rizzo D (2024) Tinkering for soil care. Cah COSTECH 7:198. https://zenodo.org/records/13923965

Ladevèze T (2025) Gabaston et Espéchède : le Softirob Challenge revient pour sa 3e édition ce 25 juin. SudOuest.fr. https://web.archive.org/web/20250720134133/. https://www.sudouest.fr/pyrenees-atlantiques/gabaston/gabaston-et-espechede-le-softirob-challenge-revient-pour-sa-3e-edition-ce-25-juin-24922982.php

Landais E, Deffontaines J-P (1988) Les pratiques des agriculteurs. Point de vue sur un courant nouveau de la recherche agronomique. Études Rural 109:125–158. https://doi.org/10.3406/rural.1988.3226

Legun K, Burch KA, Klerkx L (2023) Can a robot be an expert? The social meaning of skill and its expression through the prospect of autonomous AgTech. Agric Hum Values 40:501–517. https://doi.org/10.1007/s10460-022-10388-1

Lenain R, Peyrache J, Savary A, Séverac G (2021) Agricultural robotics: part of the new deal? FIRA 2020 conclusions - With 27 agricultural robot information sheets. Quae

Leroi-Gourhan A (1943) Évolution et techniques. Tome 1 : L'Homme et la Matière. Albin Michel

Leroi-Gourhan A (1945) Évolution et techniques. Tome 2 : Milieu et techniques. Albin Michel

Lowenberg-DeBoer J, Behrendt K, Ehlers M-H et al (2022) Lessons to be learned in adoption of autonomous equipment for field crops. Appl Econ Perspect Policy 44:848–864. https://doi.org/10.1002/aepp.13177

MacEachern CB, Esau TJ, Zaman QU, Farooque AA (2023) Assessing the effect of machine automation on operator heart and breathing rate during mechanical harvesting of wild blueberries. Smart Agric Technol 4:100171. https://doi.org/10.1016/j.atech.2023.100171

MacPherson J, Voglhuber-Slavinsky A, Olbrisch M et al (2022) Future agricultural systems and the role of digitalization for achieving sustainability goals. A review. Agron Sustain Dev 42:70. https://doi.org/10.1007/s13593-022-00792-6

Manzoor F, Wei L, Siraj M et al (2025) Digital agriculture technology adoption in low and middle-income countries—a review of contemporary literature. Front Sustain Food Syst 9. https://doi.org/10.3389/fsufs.2025.1621851

Martin T, Schnebelin É (2024) Agriculture numérique : une promesse au service d'un nouvel esprit du productivisme. Nat Sci Sociétés. https://doi.org/10.1051/nss/2023046

Mattetti M, Medici M, Canavari M, Varani M (2022) CANBUS-enabled activity-based costing for leveraging farm management. Comput Electron Agric 194:106792. https://doi.org/10.1016/j.compag.2022.106792

Miettinen R, Samra-Fredericks D, Yanow D (2009) Re-turn to practice: an introductory essay. Organ Stud 30:1309–1327. https://doi.org/10.1177/0170840609349860

Migende J, Sonnen J, Schauff S et al (2022) Agronomy perspective. In: Dörr J, Nachtmann M (eds) Handbook digital farming: digital transformation for sustainable agriculture. Springer, Berlin, Heidelberg, pp 191–275. https://doi.org/10.1007/978-3-662-64378-5_4

Mouratiadou I, Lemke N, Chen C et al (2023) The Digital Agricultural Knowledge and Information System (DAKIS): employing digitalisation to encourage diversified and multifunctional agricultural systems. Environ Sci Ecotechnology 16:100274. https://doi.org/10.1016/j.ese.2023.100274

Naim M, Rizzo D, Sauvée L, Medici M (2025a) Advancing agroecology and sustainability with agricultural robots at field level: a scoping review. Comput Electron Agri 237:110650. https://doi.org/10.1016/j.compag.2025.110650

Naim M, Rizzo D, Sauvée L, Medici M (2025b) A systematic classification of agrobots to inform farmers' choice and clarify market development. In: Stafford JV (ed) Prècision agriculture '25. Wageningen Academic, pp 991–997. https://doi.org/10.1163/9789004725232_130

Naïo Technologies (Director) (2017, January 11) FIRA 2016: Retour sur le Forum International de la Robotique Agricole [Video recording]. https://www.youtube.com/watch?v=wmBEBmkGq1o

Oksanen T, Tiusanen J, Kostamo J (2009) Probabilistic robotics. In: van Henten EJ, Goense D, Lokhorst C (eds) Precision agriculture'09. Brill, pp 849–856. https://doi.org/10.3920/9789086866649_102

Pradel M, de Fays M, Seguineau C (2022) Comparative life cycle assessment of intra-row and inter-row weeding practices using autonomous robot systems in French vineyards. Sci Total Environ 838:156441. https://doi.org/10.1016/j.scitotenv.2022.156441

Prutzer E, Gardezi M, Rizzo DM et al (2023) Rethinking 'responsibility' in precision agriculture innovation: lessons from an interdisciplinary research team. J Responsible Innov 10:2202093. https://doi.org/10.1080/23299460.2023.2202093

Puig-Pey A, Sanfeliu A, Leroux C et al (2022) A new methodological approach to analyze human roles in human-robot interaction scenarios. In: 2022 IEEE international conference on advanced robotics and its social impacts (ARSO), pp 1–6. https://doi.org/10.1109/arso54254.2022.9802978

Ritz S, Rizzo D, Fourati F et al (2019) Training in agricultural technologies: a new prerequisite for smart farming. In: Sustainable agriculture: an opportunity for innovation in machinery and systems. AXEMA - EurAgEng, Villepinte (FRA), 36–50. https://doi.org/10.5281/zenodo.20369541

Rizzo D, Jaber M, Andrii Y et al (2018) A robot from the scratch in 5 months. How agronomy students could master agricultural machinery innovation. In: Farming systems: facing uncertainties and enhancing opportunities. GRC, Chania, p 11. https://doi.org/10.5281/zenodo.20369376

Rizzo D, Ritz S, Combaud A (2019) Mastering AgTech again: emerging of new training offers and intermediate players. In: Proceedings of the international scientific conference "rural development". Kaunas, Lithuania, pp 57–61. https://doi.org/10.15544/RD.2019.044

Rizzo D, Bach A, Combaud A (2021) Distribution of precision agriculture technologies: a first survey of French dealerships. In: The conference of innovations and solutions for efficient and sustainable agriculture. APS Axema Promotion & Services, Nantes, pp 235–244. https://doi.org/10.5281/zenodo.5602074

Rizzo D, Combaud A, Schnuriger N et al (2022) Is farming technology innovation locus dependent? Making of an agricultural fablab. In: Farming systems facing climate change and resource challenges. IFSA International Farming System Association, Évora, pp 623–637. https://doi.org/10.5281/zenodo.6420888

Rizzo D, Vinatier F, Jacob F et al (2025) A framework for the sustainable maintenance of permanent runoff management structures in rainfed agriculture under climate change. J Environ Manag 377:124718. https://doi.org/10.1016/j.jenvman.2025.124718

Rose DC, Barkemeyer A, de Boon A et al (2023) The old, the new, or the old made new? Everyday counter-narratives of the so-called fourth agricultural revolution. Agric Hum Values 40:423–439. https://doi.org/10.1007/s10460-022-10374-7

Rotz S, Gravely E, Mosby I et al (2019) Automated pastures and the digital divide: how agricultural technologies are shaping labour and rural communities. J Rural Stud. https://doi.org/10.1016/j.jrurstud.2019.01.023

Salembier C, Segrestin B, Sinoir N et al (2020) Design of equipment for agroecology: coupled innovation processes led by farmer-designers. Agric Syst 183:102856. https://doi.org/10.1016/j.agsy.2020.102856

Sangwa S, Mutabazi P (2025) Artificial intelligence in agriculture: ethical stewardship, responsible innovation, and governance for sustainable food systems. Intell Agric 1:25–45. https://doi.org/10.54963/ia.v1i2.1684

Sarvela K, Kalmari J, Haapala H (2024) GIS-based assessment of operator stress: heart rate analysis in smart vs. manual field operations. NJF - EurAgEng, Herning

Schnebelin É, Labarthe P, Touzard J-M (2021) How digitalisation interacts with ecologisation? Perspectives from actors of the French Agricultural Innovation System. J Rural Stud 86:599–610. https://doi.org/10.1016/j.jrurstud.2021.07.023

Shockley J, Dillon C, Lowenberg-DeBoer J, Mark T (2021) How will regulation influence commercial viability of autonomous equipment in US production agriculture? Appl Econ Perspect Policy n/a:1–14. https://doi.org/10.1002/aepp.13178

Shvyrkov A (2025) Artificial intelligence, the death of society, and the birth of human. https://doi.org/10.2139/ssrn.5574258

Sigaut F (1989) La naissance du machinisme agricole moderne. Anthropol Sociétés 13:79–102. https://doi.org/10.7202/015078ar

Smit H, Muche R, Ahlers R, van der Zaag P (2017) The political morphology of drainage—how gully formation links to state formation in the Choke Mountains of Ethiopia. World Dev 98:231–244. https://doi.org/10.1016/j.worlddev.2017.04.031

Spagnuolo M, Todde G, Caria M et al (2025) Agricultural robotics: a technical review addressing challenges in sustainable crop production. Robotics 14:9. https://doi.org/10.3390/robotics14020009

Steinke J, Schumann C, Langan S et al (2024) Fostering social inclusion in development-oriented digital food system interventions. Agric Syst 215:103882. https://doi.org/10.1016/j.agsy.2024.103882

Stiefel L (2023) Politiques des architectures numériques. Cheminements ethnographiques dans la conception d'alternatives à la centralisation des données. PhD Thesis, Université de Lausanne

Stilgoe J, Owen R, Macnaghten P (2013) Developing a framework for responsible innovation. Res Policy 42(9):1568–1580. https://doi.org/10.1016/j.respol.2013.05.008

Sutherland L-A, Marchand F (2021) On-farm demonstration: enabling peer-to-peer learning. J Agric Educ Ext 27:573–590. https://doi.org/10.1080/1389224X.2021.1959716

Tzachor A, Devare M, King B et al (2022) Responsible artificial intelligence in agriculture requires systemic understanding of risks and externalities. Nat Mach Intell 4:104–109. https://doi.org/10.1038/s42256-022-00440-4

U.S. National Science Foundation (2024) Joint National Science Foundation and United States Department of Agriculture National Institute of Food and Agriculture Funding Opportunity: supporting Foundational Research in Robotics (FRR).

UC Agriculture and Natural Resources (2023) Farm Robotics Challenge launches 2024 college competition. https://web.archive.org/web/20250720125726/https://ucanr.edu/blog/anr-news-releases/article/farm-robotics-challenge-launches-2024-college-competition. Accessed 20 July 2025

van Velden D, Klerkx L, Dessein J, Debruyne L (2024) Cyborg farmers: embodied understandings of precision agriculture. Sociol Rural 64:3–21. https://doi.org/10.1111/soru.12456

Varani M, Molari G, Annesi G, Mattetti M (2025) Automated activity-based costing for large tractor fleets: a scalable CANBUS framework for farm economics. Smart Agric Technol 12:101171. https://doi.org/10.1016/j.atech.2025.101171

Vestager M (2017) Clearing the path for innovation. Web Summit. https://www.youtube.com/watch?v=r0fCq9Wg5PA

Vijayakumar S, Murugaiyan V, Ilakkiya S et al (2025) Opportunities, challenges, and interventions for agriculture 4.0 adoption. Discov Food 5:265. https://doi.org/10.1007/s44187-025-00576-3

Wolf SA, Buttel FH (1996) The political economy of precision farming. Am J Agric Econ 78:1269–1274. https://doi.org/10.2307/1243505

Zeddies HH, Busch G, Qaim M (2024) Positive public attitudes towards agricultural robots. Sci Rep 14:15607. https://doi.org/10.1038/s41598-024-66198-4

Zhu Z (2025) Pragmatic ontology—enhancing the philosophical foundation of critical systems thinking/practice. Syst Res Behav Sci 42:83–97. https://doi.org/10.1002/sres.3089

Part II
Structuring the New Ecosystem

Agriculture 4.0 Ecosystems in the Agri-Food Sector

Franco da Silveira, Rafael Landaverde, Filipe Molinar Machado, Everton Castelão Tetila, and Jayme Garcia Arnal Barbedo

1 Toward an Agriculture 4.0 Ecosystem

The twenty-first-century agri-food system is undergoing profound transformations, driven by complex and interconnected challenges. Among these are exponential population growth (Ghosh et al. 2024), global climate change and its impacts (Zurek et al. 2022), geopolitical conflicts disrupting production chains and markets (El Bilali and Ben Hassen 2024), and growing environmental degradation compromising essential natural resources (Midler 2022). These factors threaten global food security and reveal the limits of conventional production, distribution, and consumption models (Webb et al. 2020). In this context, seeking innovative, robust, and integrated solutions is imperative—solutions capable of efficiently and sustainably addressing the agri-food system's challenges (Klerkx and Rose 2020).

F. da Silveira (✉) · J. G. A. Barbedo
EMBRAPA Digital Agriculture, Campinas, São Paulo, Brazil

R. Landaverde
Texas A&M University, College Station, TX, USA

F. M. Machado
Regional Integrated University of Alto Uruguay and the Missões, Santo Ângelo, Brazil

E. C. Tetila
Federal University of Grande Dourados, Dourados, Brazil

F. da Silveira et al. (eds.), *Agriculture 4.0*,
https://doi.org/10.1007/978-3-032-24552-6_11

In this context, Agriculture 4.0[1] has gained prominence as a transformative approach capable of systemically revolutionizing the agri-food system. Its impact goes beyond on-farm production, encompassing storage, logistics, distribution, marketing, and food consumption (Ting et al. 2011). This transformation relies on integrating emerging technologies such as IoT, AI, Blockchain, agricultural robotics, drones, smart sensors, Big Data, and Chatbots (da Silveira et al. 2023b; Majumdar et al. 2025; Sai et al. 2025). Applied in a coordinated manner, these technologies optimize production processes, resource use, supply chain traceability, and strategic decision-making based on real-time data (da Silveira et al. 2021; Latino et al. 2022). By reconfiguring operational and informational dynamics, Agriculture 4.0 advances toward a more efficient, resilient, inclusive, and environmentally sustainable model aligned with SDGs, particularly SDG 2, SDG 3, SDG 12, SDG 13, and SDG 15 (Fragomeli et al. 2024; Perçin et al. 2025). Thus, it enhances productivity while contributing to fairer and more equitable food systems (Carolan 2023; Kusakina and Sokolov 2022).

However, its benefits are not yet fully realized in practice (da Silveira et al. 2025). Progress is often hindered by overly optimistic expectations, technical limitations, socioeconomic barriers, and inequalities in access and capacity—especially among small- and medium-sized farmers (Eastwood et al. 2021; Abbasi et al. 2022; Fragomeli et al. 2024; Bissadu et al. 2025). Adoption is more limited in developing countries due to high costs, insufficient digital infrastructure, low technical training, and cultural resistance (da Silveira et al. 2023a; Islam et al. 2024; Mihret et al. 2025). Developed countries generally have more resources, advanced infrastructure, and supportive policies, though challenges remain in ensuring full, inclusive, and equitable adoption (Khan 2025).

Agriculture 4.0 is more than adopting isolated technologies. It is a complex, systemic phenomenon, characterized by dynamic interactions among multiple interdependent elements forming an evolving technological, social, and institutional ecosystem (da Silveira et al. 2023b). This ecosystem relies on technological infrastructures, generally grouped into physical and digital components. However, technology is only one dimension. Agriculture 4.0 also involves diverse social, economic, and institutional actors, including farmers, cooperatives, technology companies, agri-foodtech startups, educational and research institutions, regulatory bodies, policymakers, and consumers. These actors collaborate across multiscale and cross-sectoral networks, fostering knowledge exchange, co-creation, and strategies for innovation, sustainability, and adaptation (Maria et al. 2021; Charatsari et al. 2022; Lioutas and Charatsari 2022; Cosby et al. 2024; Klerkx and Villalobos 2024).

[1] This chapter adopts the broad definition of Agriculture 4.0 proposed by da Silveira et al. (2021): "agriculture 4.0 is the implementation of emerging technologies and innovative services on the agriculture, that requires a cultural and behavioral change in all actors involved in the agricultural production chain, to increase their productivity and efficiency, and support a more sustainable agriculture, using precise and momentary of information that will help make strategic *decisions*."

Furthermore, the Agriculture 4.0 ecosystem encompasses institutional, social, environmental, and ethical dimensions that influence its structuring and operation. These include governance of data and technology, information privacy and security, digital inclusion across territories, sustainable production processes, and social equity in access to and benefits from innovations (Ingram et al. 2022; Martens and Zscheischler 2022; Carolan 2023). Understanding Agriculture 4.0 as an ecosystem implies recognizing that effective implementation depends on the synergy between technologies, people, institutions, and values. This perspective broadens analyses beyond technical transformation to include socioeconomic impacts, structural challenges, and political-institutional conditions for broad, inclusive, and sustainable adoption (da Silveira et al. 2021).

The complexity and interdependence of the various elements that make up the Agriculture 4.0 ecosystem render the process of expanding and disseminating these technologies within the agri-food system a significant challenge (da Silveira et al. 2025). These dynamic demands include integrated approaches, effective public policies, technical training, and governance models that promote inclusion and sustainability. Although fundamental for the sector's advancement, this ecosystem remains underexplored in academic literature, revealing an important gap in understanding its dynamics, challenges, and interrelationships. To address this need, the objective of this chapter is to discuss the main key dimensions that form the Agriculture 4.0 ecosystem within the agri-food system, contributing to a broader understanding of the conditions necessary to enable its implementation and expand its adoption. The information presented in this chapter reflects researchers' perspectives and interpretations on the subject, serving as an exploratory basis for debate, without intending to establish absolute or definitive truths.

2 Agriculture 4.0 Ecosystem: Key Elements

This section presents the main aspects that shape the Agriculture 4.0 ecosystem, focusing on its technological, social, political, economic, and environmental dimensions. This selection is not exhaustive or definitive but aims to provide a representative overview of the most relevant components in this context. Additionally, for each element addressed, specific challenges are identified that must be overcome to foster its development and strengthen the ecosystem.

2.1 Infrastructural Elements

2.1.1 Physical Infrastructure

Physical infrastructure constitutes the essential structural foundation for the development and operationalization of Agriculture 4.0. It encompasses the tangible elements that enable the functioning of technologies in rural areas, such as telecommunications towers, fiber optic cables, gateways, routers, communication satellites, and energy sources—both renewable and conventional. However, one of the main challenges for consolidating this infrastructure within the Agriculture 4.0 ecosystem is the territorial inequality in access to these resources. In many rural regions, especially the most remote ones, technical limitations, high installation costs, and lack of public and private incentives persist. These factors hinder the creation of environments adequately prepared for the large-scale adoption of digital technologies in agriculture, deepening the structural asymmetries within the agri-food system.

2.1.2 Digital Infrastructure

Digital infrastructure is the invisible technological foundation that supports the intelligence and connectivity of the modern agri-food system. Comprised of systems, platforms, and networks capable of storing, processing, and transmitting large volumes of data generated in the field, this infrastructure includes high-capacity data centers that ensure secure storage and efficient processing of agricultural information, supporting advanced applications such as predictive analytics, artificial intelligence, and machine learning. Mobile networks—including 3G, 4G, and 5G—provide fast and reliable communication among connected devices, machinery, and digital platforms. Complementing these networks, LPWAN (Low-Power Wide-Area Network) technologies, such as LoRaWAN and NB-IoT, enable efficient connection of sensors distributed over large areas with low energy consumption and long range. Together, these digital layers enable process automation, remote monitoring, and data-driven decision-making in real time, transforming the field into a connected and intelligent environment. However, the complexity of integrating diverse systems, platforms, and digital devices, coupled with the lack of unified standards, poses a significant challenge, hindering interoperability and limiting the scalability of Agriculture 4.0 solutions.

2.2 Technologies

2.2.1 Basic Technologies

Basic technologies constitute the established foundation of precision agriculture, forming the cornerstone upon which the digitalization of the agri-food system is developed, which in turn underpins Agriculture 4.0 (da Silveira and Amaral 2023). Tools such as drones, IoT sensors, GPS, digital platforms, and autopilot farming systems integrate to provide accurate, real-time insights into agricultural operations. They enable continuous collection of environmental and production data, facilitate georeferenced field mapping, and automate routine tasks, increasing efficiency and reducing waste. The synergy among these elements creates a conducive environment for the application of simple yet effective digital solutions that are already significantly transforming the field. A significant challenge lies in the insufficient technical training of producers and operators to effectively utilize these tools, coupled with a lack of ongoing technical support to ensure proper data interpretation and informed decision-making.

2.2.2 Advanced Technologies

Advanced technologies are driving the transformation of the modern agri-food system, elevating its processes to higher levels of intelligence, autonomy, and integration. These innovations turn traditional production chains into complex, adaptive, and interconnected systems. Artificial intelligence and big data analytics process large volumes of data to generate strategic insights, enabling demand forecasting, logistics optimization, and more efficient decision-making. Autonomous robots perform specialized operations with high precision and minimal human intervention, while blockchain technology ensures transparency, security, and traceability across the supply chain. Augmented and virtual reality tools facilitate immersive training and operational support, enhancing the interface between operators and digital systems. Intelligent systems and chatbots improve communication and automate administrative processes, and 3D printing enables the production of innovative foods, such as 3D-printed vegan steaks, to meet new dietary habits, offering sustainable and customized solutions for protein production within the agri-food system (Watkins et al. 2022). A significant challenge in adopting these advanced technologies is ensuring that investments in innovation translate into tangible and accessible benefits for all stakeholders within the agri-food system. Furthermore, the technical complexity of integrating multiple platforms and devices, combined with a lack of specialized training, limits the scalability and effectiveness of these solutions across diverse contexts.

2.2.3 Emerging Technologies

Emerging technologies represent the state of the art in innovation within the agri-food system, combining disruptive advances with great potential to transform production systems (Taha et al. 2025). Cooperative robotics systems and autonomous drones promise greater coordination and efficiency in the field. Nanosensors offer unprecedented levels of precision in environmental and phytosanitary detection. Digital twins create virtual replicas of crops for real-time simulation and optimization. Quantum computing and digital biotechnology, including CRISPR gene editing, integrated with artificial intelligence, point to a future where the agri-food system will be highly personalized, efficient, and sustainable (Lockie 2021). These emerging technologies complement each other by providing integrated solutions to complex challenges. The greatest challenge for these technologies lies in their technological maturity still under development, in addition to regulatory and ethical issues, and the need for robust investments in research and infrastructure for their effective adoption.

2.3 *Institutional Framework*

The institutional framework of Agriculture 4.0 consists of an integrated set of governance mechanisms operating at multiple levels—local, regional, national, and international—to guide, regulate, and promote a modern, innovative, and sustainable agri-food system. This framework encompasses innovative public policies, specific regulations, adapted laws, updated technical standards, risk mitigation strategies, and initiatives to foster digital inclusion, access to credit, financing, and technical training (Ehlers et al. 2022). Additionally, it involves coordination among the public sector, private industry, academia, and civil society to ensure the creation of an environment conducive to innovation, sustainability, and equity in the agricultural sector. Ensuring convergence across multiple governance spheres and sectors is a crucial challenge for the institutional framework of Agriculture 4.0, requiring the elimination of fragmented regulatory barriers to create an agile, integrated, and innovation-friendly normative environment capable of rapidly responding to the demands of the modern agri-food system.

2.4 *Stakeholders*

2.4.1 Users and Implementers

These are farm machinery operators, rural producers, their families, and entrepreneurs who directly engage in the implementation and operation of Agriculture 4.0 technologies within the agri-food system (da Silveira et al. 2023a). These stakeholders are at the forefront of technological adoption, integrating digital solutions into daily production practices, and possess practical knowledge of the local environmental, economic, and social conditions that impact the use of these technologies. However, they face significant barriers to effective adoption, such as cultural resistance to change, low familiarity with digital tools, and limited technical training.

2.4.2 Technology Developers and Providers

This group includes software companies, sensor and automation firms, AgriFoodTech startups, and agribusiness multinationals responsible for developing, customizing, and supplying advanced technological solutions for the agri-food system (Zambon et al. 2019; Klerkx and Villalobos 2024; Vlăduț and Ungureanu 2024). Their role involves research, development, field testing, and technical support, focusing on innovation that meets the specific demands of Agriculture 4.0. Nonetheless, adapting technologies to heterogeneous local realities, the complexity of integrating with existing systems, and the ongoing need for investment in research and development limit the availability and accessibility of these solutions.

2.4.3 Knowledge and Talent Builders

These comprise public and private universities, technical education institutions, digital training platforms, and research and innovation centers that promote the training of qualified professionals and the generation of applied knowledge in Agriculture 4.0 within the agri-food system (Soma and Nuckchady 2021). They are essential for aligning education and research with the field's demands, ensuring the update and development of necessary technical skills. However, there is a discrepancy between traditional academic curricula and the dynamic demands of the digital agri-food system, aggravated by insufficient articulation between educational institutions, research centers, and the productive sector, resulting in gaps in the training of professionals with practical and innovative skills (Bampasidou et al. 2024). Additionally, limited infrastructure for continuous training and the low incorporation of active learning methodologies hinder the rapid updating and qualification needed for effective technology transfer and adoption.

2.4.4 Local Facilitators and Articulators

This group includes producer associations, cooperatives, unions, NGOs, regional innovation hubs, and consultants who act as bridges between Agriculture 4.0 technologies and local realities within the agri-food system (Ciruela-Lorenzo et al. 2020; Stojanova et al. 2022). They facilitate access to resources, promote stakeholder coordination, and assist in adapting technological solutions to the socioeconomic and cultural conditions specific to rural communities. However, institutional fragmentation, limited coordination capacity, and scarce resources for local action compromise the effectiveness of these facilitators in disseminating and adopting technologies.

2.4.5 Public and Institutional Agents

These comprise municipal, state, and federal public agencies, as well as multilateral organizations and international cooperation bodies responsible for policy formulation, regulations, and programs to foster technological innovation for Agriculture 4.0 in the agri-food system (Rotz et al. 2019; da Silveira et al. 2021). They create the regulatory environment, encourage investments, promote infrastructure, and support strategic projects aimed at sustainable development and competitiveness of the agri-food system. However, the effectiveness of these actions is frequently limited by fragmented governance, lack of integrated and long-term strategic policies, and difficulties in aligning public incentives with the real technological needs of producers and other stakeholders in the agri-food system.

2.5 Sustainability, Inclusion, and Rural Well-Being

2.5.1 Environmental Goals

Within the agri-food system, Agriculture 4.0 offers concrete opportunities to reduce the environmental impacts associated with food production. By enabling more precise management of resources such as water, soil, and inputs, it supports the rational use of fertilizers and pesticides, reducing contamination and waste. It also allows for continuous monitoring of greenhouse gas emissions, carbon footprints, and the integrity of agricultural ecosystems, while fostering the transition toward regenerative practices and renewable energy sources. However, one of the main challenges is to prevent environmental sustainability from becoming merely a rhetorical add-on to innovation, without real commitment to structural changes in the dominant agri-food model (Bellon-Maurel et al. 2022).

2.5.2 Social Goals

From a social perspective, Agriculture 4.0 should be guided by principles of equity, territorial justice, and active inclusion within the agri-food system. This entails expanding access for smallholders, women, rural youth, and traditional communities to the resources, knowledge, and opportunities generated by technological transformation in the countryside (Xie et al. 2021; Daum et al. 2022; Gumbi et al. 2023; Thakur 2023; Sun and Li 2025). At the same time, it is essential to recognize and mitigate the negative impacts of digitalization, such as social isolation, stress caused by hyperconnectivity, and the weakening of community ties. Promoting rural well-being requires creating conditions that respect local ways of life, harmonizing productivity, social relations, and cultural identity. The most pressing challenge is to prevent the deepening of social inequalities in rural areas, where modernization tends to benefit only those groups with greater technical and economic capacity, leaving significant portions of the population marginalized amid ongoing changes.

2.6 International Cases

Several international projects have been driving the development of the Agriculture 4.0 ecosystem, integrating technological innovation, sustainability, and social inclusion into global agri-food systems. In Australia, Digiscape (https://research.csiro.au/digiscape/), a CSIRO initiative, harnesses the digital revolution to create smart tools that increase productivity and sustainability for farmers and land managers, generating expected economic impacts exceeding 400 million dollars annually by 2028. In Europe, the DESIRA project (https://desira2020.agr.unipi.it/the-project/), coordinated by the University of Pisa, employs a network of 20 rural Living Labs to assess the socioeconomic effects of digitalization, promoting policies that foster inclusive and responsible digital transformation in rural areas. The DEMETER program (https://h2020-demeter.eu/), funded by Horizon 2020, drives the large-scale adoption of interoperable IoT platforms across 18 countries, involving thousands of farmers to optimize decision-making and strengthen the sustainability of the European agri-food sector. In Brazil, Semear Digital (https://www.semear-digital.cnptia.embrapa.br/) establishes Agrotechnological Districts in strategic regions to identify connectivity bottlenecks and develop digital solutions aligned with local needs, strengthening collaboration among producers, startups, public and private institutions, and fostering sustainable technological capacity building. Complementing these efforts, the European project TWIN-IN (https://ec.europa.eu/info/funding-tenders/opportunities/portal/screen/how-to-participate/org-details/914989749/project/101187101/program/43353764/details) acts at the convergence of the digital and green transition ("twin transition") by developing and implementing innovative education and training models, empowering various actors in the agri-food chain to face the challenges of sustainability and integrated digitalization. Beyond these highlighted examples, many other cases worldwide also significantly contribute to the consolidation of technological, resilient, and socially inclusive ecosystems for Agriculture 4.0. The importance of these initiatives lies in their contribution to building more efficient, sustainable, and inclusive agri-food systems. By integrating technology, public policies, and capacity building, these projects strengthen the ability of agri-food systems to address critical global challenges such as food security, climate change, and balanced rural development. Thus, they advance the creation of innovative and resilient agri-food systems, prepared to respond to the complex and dynamic demands of the twenty-first century.

Understanding the Agriculture 4.0 ecosystem is essential to drive the transformation of the agri-food system into a more sustainable, resilient, and inclusive model capable of meeting the growing global demand for quality food while preserving vital natural resources (Wolfert et al. 2023). Therefore, it is fundamental that researchers, public managers, entrepreneurs, and producers deepen their understanding of the multiple dimensions and interdependencies present in this ecosystem. Such comprehensive knowledge enables the development and implementation of more effective strategies, policies, and technologies, fostering innovation and ensuring that Agriculture 4.0 is adopted equitably and at scale. Figure 1 aims to represent these complex relationships and key elements, serving as a conceptual map that guides various stakeholders to navigate the challenges and opportunities inherent in this transformative process. Given its complexity and relevance, the Agriculture 4.0 ecosystem deserves to be more thoroughly explored

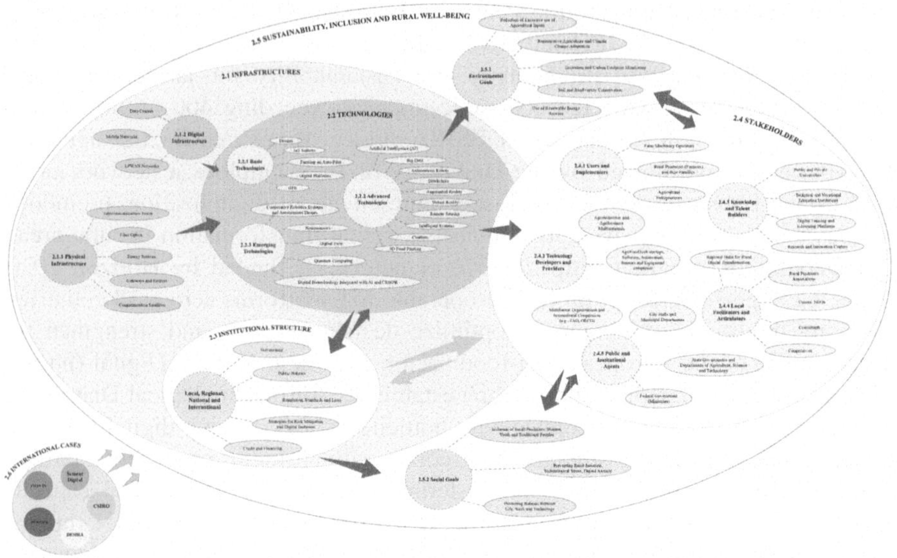

Fig. 1 A conceptual overview of the Agriculture 4.0 ecosystem in the agri-food system

and better understood, as only then will it be possible to maximize its positive impact and promote significant advances in the modern agri-food system.

The lack of a systemic understanding of the Agriculture 4.0 ecosystem can lead to several negative consequences for the technological and institutional development of the agri-food sector. Fragmented adoption of technologies—without considering local contexts, the capacities of involved stakeholders, and the interdependence among infrastructure, governance, and training—can result in ineffective solutions, wasted public and private resources, and growing inequalities in rural areas. Moreover, overlooking the social and environmental dimensions of the ecosystem may exacerbate digital exclusion, concentrating on benefits, and contribute to environmental degradation. A lack of coordination between public policies, value chains, and innovation systems also undermines the scalability and sustainability of digital initiatives. Therefore, without an integrated and strategic vision of this ecosystem, not only is technological progress in agriculture at risk, but so are the broader goals of modernization, inclusion, and sustainability within the agri-food system.

3 Proposals for Strengthening the Agriculture 4.0 Ecosystem

For Agriculture 4.0 to effectively and sustainably contribute to the transformation of the agri-food system, it is essential to reinforce the foundational pillars of its ecosystem. This requires integrated initiatives that articulate public policies, technological innovation, social inclusion, infrastructure development, and participatory governance (Steinke et al. 2024). Below are strategic proposals aimed at consolidating this ecosystem as a robust basis for the digital and intelligent transition of the agri-food system.

- Strengthening digital governance and institutional frameworks:

The advancement of Agriculture 4.0 depends on solid digital governance capable of coordinating diverse interests, establishing technical standards, and ensuring ethical and inclusive use of technologies within the agri-food system. It is necessary to create dialogue platforms among governments, producers, companies, researchers, and civil society to define guidelines on data usage, interoperability, cybersecurity, and equitable access to innovations. Institutional strengthening should prioritize transparent and participatory mechanisms that ensure the benefits of digitalization reach all links of the production chain, especially the most vulnerable.

- Support for living labs and collaborative experimentation environments:

The creation of local innovative ecosystems, such as living labs, allows the development, testing, and validation of Agriculture 4.0 technologies in real-world conditions, with active participation from farmers, companies, researchers, technicians, and policymakers. These environments foster co-innovation, accelerate learning cycles, and strengthen the adaptive capacity of rural territories. Encouraging public–private partnerships and territorial innovation networks is vital to expand the reach and effectiveness of these initiatives.

- Structured investment in digital training and technological literacy in rural areas:

Human capital is one of the main pillars of any technological ecosystem. In the case of Agriculture 4.0, this implies significantly expanding investments in technical education, digital extension, and technological literacy aimed at farmers, rural youth, women, and traditional communities. More than training for tool usage, it involves developing skills for data analysis, evidence-based decision-making, and strategic use of technology in inclusive and sustainable production models.

- Promotion of open, interoperable, and accessible digital platforms:

The diversity of technologies and providers in agriculture demands solutions that can communicate with each other. The development of open and interoperable digital platforms is essential to avoid ecosystem fragmentation, reduce barriers for small producers, and broaden access to technological solutions. Moreover, these platforms should be designed with a focus on usability, social inclusion, and adaptation to different scales and regional contexts, promoting transparency and technological sovereignty.

These guidelines should not be seen as isolated actions but as interdependent components of a broader transformation architecture. Strengthening the Agriculture 4.0 ecosystem requires a strategic, systemic, and inclusive vision—integrating knowledge, infrastructure, innovation, and public policies—to drive the evolution of a more resilient, intelligent, and equitable agri-food system.

4 Final Considerations

Agriculture 4.0 should be understood as a complex and integrated ecosystem, where physical and digital infrastructure, emerging technologies, social stakeholders, and institutions interact to transform the agri-food system in a sustainable and inclusive manner. This chapter, by discussing the main elements that structure this ecosystem, contributes to expanding the understanding of the conditions necessary for its implementation and to promoting broad, inclusive, and sustainable adoption.

Given the diverse contexts, especially between developed and developing countries, consolidating this ecosystem requires integrated and flexible approaches capable of overcoming challenges related to unequal access, varying technical capacities, and the development of effective and inclusive public policies. To support this consolidation, three priority lines of future research are highlighted:

- Strategies to consolidate multisectoral collaborative networks that integrate producers, companies, research institutions, government, and civil society, promoting co-creation and knowledge sharing within the Agriculture 4.0 ecosystem. For example, creating regional innovation hubs where startups, universities, and farmer cooperatives collaborate on AI-based precision farming solutions.
- Adaptive and inclusive governance models ensure effective participation of all stakeholders, with a special focus on digital inclusion and the capacity building of small and medium producers, enabling the sustainable dissemination of emerging technologies. For instance, implementing public–private training programs that equip rural technicians with IoT and drone operation skills for smallholder farms.
- Approaches to promote effective articulation among different stakeholders, institutions, and governance levels, strengthening coordination and cooperation necessary to scale emerging technologies. This should be accompanied by the development of physical and digital infrastructure capable of adapting to regional and cultural specificities, ensuring continuous, efficient, and scalable operation of the Agriculture 4.0 ecosystem. As an example, establishing interoperable data platforms that connect government extension services, AgTech companies, and local cooperatives for real-time monitoring of crop health, food production, and supply chain logistics across the agri-food system.

Thus, this chapter presents a fundamental conceptual framework that serves as a starting point for future discussions on the global implementation of Agriculture 4.0 and its implications for agri-food systems, stimulating the advancement of innovative and systemic solutions to address current challenges of food security, sustainability, and equity.

Acknowledgments We appreciate the support of the researchers from the Digital Agriculture Research Center—Semear Digital. This study was financed by the São Paulo Research Foundation (FAPESP), Brazil. Process Number #2023/12215-3, #23/03870-8, and #2022/09319-9.

Competing Interests The authors have no conflicts of interest that are relevant to the content of this chapter.

References

Abbasi R, Martinez P, Ahmad R (2022) The digitization of agricultural industry–a systematic literature review on agriculture 4.0. Smart Agric Technol 2:100042. https://doi.org/10.1016/j.atech.2022.100042

Bampasidou M, Goldgaber D, Gentimis T, Mandalika A (2024) Overcoming 'digital divides': leveraging higher education to develop next generation digital agriculture professionals. Comput Electron Agric 224:109181. https://doi.org/10.1016/j.compag.2024.109181

Bellon-Maurel V, Lutton E, Bisquert P, Brossard L, Chambaron-Ginhac S, Labarthe P et al (2022) Digital revolution for the agroecological transition of food systems: a responsible research and innovation perspective. Agric Syst 203:103524. https://doi.org/10.1016/j.agsy.2022.103524

Bissadu KD, Sonko S, Hossain G (2025) Society 5.0 enabled agriculture: drivers, enabling technologies, architectures, opportunities, and challenges. Inf Process Agric 12(1):112–124. https://doi.org/10.1016/j.inpa.2024.04.003

Carolan M (2023) Digital agriculture killjoy: happy objects and cruel quests for the good life. Sociol Rural 63:37–56. https://doi.org/10.1111/soru.12398

Charatsari C, Lioutas ED, Papadaki-Klavdianou A, Michailidis A, Partalidou M (2022) Farm advisors amid the transition to agriculture 4.0: professional identity, conceptions of the future and future-specific competencies. Sociol Rural 62(2):335–362. https://doi.org/10.1111/soru.12364

Ciruela-Lorenzo AM, Del-Aguila-Obra AR, Padilla-Meléndez A, Plaza-Angulo JJ (2020) Digitalization of agri-cooperatives in the smart agriculture context. proposal of a digital diagnosis tool. Sustainability 12(4):1325. https://doi.org/10.3390/su12041325

Cosby A, Sullivan MA, Manning J, Harreveld B (2024) Tech-ready teachers for Agriculture 4.0: a teacher–industry partnership case study. Educ Train 66(6):668–691. https://doi.org/10.1108/ET-05-2023-0166

da Silveira F, Amaral FG (2023) Agriculture 4.0. In: Zhang Q (ed) Encyclopedia of smart agriculture technologies. Springer, Cham. https://doi.org/10.1007/978-3-030-89123-7_207-3

da Silveira F, Lermen FH, Amaral FG (2021) An overview of agriculture 4.0 development: systematic review of descriptions, technologies, barriers, advantages, and disadvantages. Comput Electron Agric 189:106405. https://doi.org/10.1016/j.compag.2021.106405

da Silveira F, da Silva SLC, Machado FM, Barbedo JGA, Amaral FG (2023a) Farmers' perception of the barriers that hinder the implementation of agriculture 4.0. Agric Syst 208:103656. https://doi.org/10.1016/j.agsy.2023.103656

da Silveira F, Barbedo JGA, da Silva SLC, Amaral FG (2023b) Proposal for a framework to manage the barriers that hinder the development of agriculture 4.0 in the agricultural production chain. Comput Electron Agric 214:108281. https://doi.org/10.1016/j.compag.2023.108281

da Silveira F, Corrêa RGF, Baierle IC, Landaverde R, Barbedo JGA (2025) Behavioral profile of farmers in the adoption of agriculture 4.0 technologies in the agri-food system: a case study in Brazil. Front Sustainable Food Syst 9:1624753. https://doi.org/10.3389/fsufs.2025.1624753

Daum T, Adegbola PY, Adegbola C, Daudu C, Issa F, Kamau G et al (2022) Mechanization, digitalization, and rural youth-stakeholder perceptions on three mega-topics for agricultural transformation in four African countries. Glob Food Sec 32:100616. https://doi.org/10.1016/j.gfs.2022.100616

Eastwood CR, Edwards JP, Turner JA (2021) Anticipating alternative trajectories for responsible Agriculture 4.0 innovation in livestock systems. Animal 15:100296. https://doi.org/10.1016/j.animal.2021.100296

Ehlers MH, Finger R, El Benni N, Gocht A, Sørensen CA, Gusset M et al (2022) Scenarios for European agricultural policymaking in the era of digitalisation. Agric Syst 196:103318. https://doi.org/10.1016/j.agsy.2021.103318

El Bilali H, Ben Hassen T (2024) Disrupted harvests: how Ukraine–Russia war influences global food systems–a systematic review. Policy Stud 45(3–4):310–335. https://doi.org/10.1080/01442872.2024.2329587

Fragomeli R, Annunziata A, Punzo G (2024) Promoting the transition towards agriculture 4.0: a systematic literature review on drivers and barriers. Sustainability 16(6):2425. https://doi.org/10.3390/su16062425

Ghosh A, Kumar A, Biswas G (2024) Exponential population growth and global food security: challenges and alternatives. In: Bioremediation of emerging contaminants from soils. Elsevier, pp 1–20. https://doi.org/10.1016/B978-0-443-13993-2.00001-3

Gumbi N, Gumbi L, Twinomurinzi H (2023) Towards sustainable digital agriculture for smallholder farmers: a systematic literature review. Sustainability 15(16):12530. https://doi.org/10.3390/su151612530

Ingram J, Maye D, Bailye C, Barnes A, Bear C, Bell M et al (2022) What are the priority research questions for digital agriculture? Land Use Policy 114:105962. https://doi.org/10.1016/j.landusepol.2021.105962

Islam MH, Anam MZ, Hoque MR, Nishat M, Bari AM (2024) Agriculture 4.0 adoption challenges in the emerging economies: implications for smart farming and sustainability. J Econ Technol 2:278–295. https://doi.org/10.1016/j.ject.2024.09.002

Khan F (2025) Agriculture 4.0 as a way forward to sustainable agriculture in Australia. JSFA Rep 5(8):337–358. https://doi.org/10.1002/jsf2.70014

Klerkx L, Rose D (2020) Dealing with the game-changing technologies of Agriculture 4.0: how do we manage diversity and responsibility in food system transition pathways? Global Food Secur 24:100347. https://doi.org/10.1016/j.gfs.2019.100347

Klerkx L, Villalobos P (2024) Are AgriFoodTech start-ups the new drivers of food systems transformation? An overview of the state of the art and a research agenda. Glob Food Sec 40:100726. https://doi.org/10.1016/j.gfs.2023.100726

Kusakina ON, Sokolov SV (2022) Human capital development in rural areas from the standpoint of the Sustainable Development Goals (SDGs). In: International conference on "advances in management, business and technology toward sustainable development". Springer, Cham, pp 171–179. https://doi.org/10.1007/978-3-031-20803-4_18

Latino ME, Menegoli M, Lazoi M, Corallo A (2022) Voluntary traceability in food supply chain: a framework leading its implementation in Agriculture 4.0. Technol Forecast Soc Chang 178:121564. https://doi.org/10.1016/j.techfore.2022.121564

Lioutas ED, Charatsari C (2022) Innovating digitally: the new texture of practices in agriculture 4.0. Sociol Rural 62(2):250–278. https://doi.org/10.1111/soru.12356

Lockie S (2021) Agriculture and the future of food through digital and other novel technologies. In: Digital transformation in a post-covid world. CRC Press, pp 307–322

Majumdar P, Mitra S, Bhattacharya D, Bhushan B (2025) Enhancing sustainable 5G powered agriculture 4.0: summary of low power connectivity, internet of UAV things, AI solutions and research trends. Multimed Tools Appl 84(17):17389–17433. https://doi.org/10.1007/s11042-024-19728-1

Maria K, Maria B, Andrea K (2021) Exploring actors, their constellations, and roles in digital agricultural innovations. Agric Syst 186:102952. https://doi.org/10.1016/j.agsy.2020.102952

Martens K, Zscheischler J (2022) The digital transformation of the agricultural value chain: discourses on opportunities, challenges and controversial perspectives on governance approaches. Sustainability 14(7):3905. https://doi.org/10.3390/su14073905

Midler, E (2022) Environmental degradation: impacts on agricultural production. https://ieep.eu/wp-content/uploads/2022/12/Policy-brief_Environmental-degradation.-Impacts-on-agricultural-production_IEEP-2022.pdf. Accessed on 30 July 2025

Mihret YC, Takele MM, Mintesinot SM (2025) Advancements in agriculture 4.0 and the needs for introduction and adoption in Ethiopia: a review. Adv Agric 2025(1):8828400. https://doi.org/10.1155/aia/8828400

Perçin S, Bayraktar Y, Kumar V (2025) Evaluating the Adoption of Industry 4.0, Sustainability and Circular Economy Drivers to Achieve Sustainable Development Goals–Oriented Agri-Food Supply Chains. Bus Strateg Environ 34(4):4871–4895. https://doi.org/10.1002/bse.4221

Rotz S, Duncan E, Small M, Botschner J, Dara R, Mosby I, Fraser ED (2019) The politics of digital agricultural technologies: a preliminary review. Sociol Rural 59(2):203–229. https://doi.org/10.1111/soru.12233

Sai S, Kumar S, Gaur A, Goyal S, Chamola V, Hussain A (2025) Unleashing the power of generative AI in agriculture 4.0 for smart and sustainable farming. Cogn Comput 17(1):63. https://doi.org/10.1007/s12559-025-10420-6

Soma T, Nuckchady B (2021) Communicating the benefits and risks of digital agriculture technologies: perspectives on the future of digital agricultural education and training. Front Commun 6:762201. https://doi.org/10.3389/fcomm.2021.762201

Steinke J, Schumann C, Langan S, Müller A, Opola FO, Ortiz-Crespo B, van Etten J (2024) Fostering social inclusion in development-oriented digital food system interventions. Agric Syst 215:103882. https://doi.org/10.1016/j.agsy.2024.103882

Stojanova S, Cvar N, Verhovnik J, Božić N, Trilar J, Kos A, Stojmenova Duh E (2022) Rural digital innovation hubs as a paradigm for sustainable business models in europe's rural areas. Sustainability 14(21):14620. https://doi.org/10.3390/su142114620

Sun Y, Li S (2025) The impact of digital development on non-agricultural employment of rural women: evidence from the broadband China strategy. Appl Econ 57(18):2224–2240. https://doi.org/10.1080/00036846.2024.2322584

Taha MF, Mao H, Zhang Z, Elmasry G, Awad MA, Abdalla A et al (2025) Emerging technologies for precision crop management towards agriculture 5.0: a comprehensive overview. Agriculture 15(6):582. https://doi.org/10.3390/agriculture15060582

Thakur N (2023) Women farmers and technologies in agriculture: a review of current practices. In: Munshi S, Singh M (eds) Women farmers: unheard being heard. Sustainability sciences in Asia and Africa. Springer, Singapore. https://doi.org/10.1007/978-981-19-6978-2_12

Ting KC, Abdelzaher T, Alleyne A, Rodriguez L (2011) Information technology and agriculture: global challenges and opportunities. Bridge 41(3):6–13

Vlăduț NV, Ungureanu N (2024) Beyond agriculture 4.0: design and development of modern agricultural machines and production systems. Agriculture 14(7):991. https://doi.org/10.3390/agriculture14070991

Watkins P, Logan A, Bhandari B (2022) Three-dimensional (3D) food printing—an overview. In: Food Engineering Innovations Across the Food Supply Chain, pp 261–276. https://doi.org/10.1016/B978-0-12-821292-9.00003-0

Webb P, Benton TG, Beddington J, Flynn D, Kelly NM, Thomas SM (2020) The urgency of food system transformation is now irrefutable. Nat Food 1(10):584–585. https://doi.org/10.1038/s43016-020-00161-0

Wolfert S, Verdouw C, van Wassenaer L, Dolfsma W, Klerkx L (2023) Digital innovation ecosystems in agri-food: design principles and organizational framework. Agric Syst 204:103558. https://doi.org/10.1016/j.agsy.2022.103558

Xie L, Luo B, Zhong W (2021) How are smallholder farmers involved in digital agriculture in developing countries: a case study from China. Land 10(3):245. https://doi.org/10.3390/land10030245

Zambon I, Cecchini M, Egidi G, Saporito MG, Colantoni A (2019) Revolution 4.0: industry vs. agriculture in a future development for SMEs. Processes 7(1):36. https://doi.org/10.3390/pr7010036

Zurek M, Hebinck A, Selomane O (2022) Climate change and the urgency to transform food systems. Science 376(6600):1416–1421. https://doi.org/10.1126/science.abo2364

Digital Agricultural Innovation: A Multiple Contexts Perspective

Evagelos D. Lioutas and Chrysanthi Charatsari

1 Introduction

As the introduction of digital technologies in agriculture shows a considerable momentum in recent years, research interest has turned to the factors that catalyze the innovation process in digital agricultural innovation systems (DISs). Such a systems perspective has its roots in one of the most vivid streams of social science of innovation: agricultural innovation systems scholarship. Largely built upon the theoretical underpinnings of innovation systems that emphasize the systemic and interactive nature of innovation process (Lundvall et al. 1988; Nelson and Nelson 2002; Smits and Kuhlmann 2004), and following the highly influential work of key scholars in the field of agricultural innovation (Spielman et al. 2008, 2009; Klerkx et al. 2010, 2012), a rich corpus of knowledge on agricultural innovation systems has been gradually developed.

Today, there is consensus among researchers that innovation systems, including those operating in agricultural sectors, are open, socio-technical systems, in the sense that they combine a technical dimension, referring to the technologies incorporated into them, and a social dimension, which is formed by the sum of social entities participating in the innovation process (Kline and Rosenberg 2010). The social actors within the system pursue innovation goals by interacting and fueling the coevolution of innovation (Klerkx et al. 2012).

The emergence of digital technologies has led to the inevitable digitalization of agricultural innovation systems, transforming them into Digital Innovation Systems

E. D. Lioutas (✉)
Department of Supply Chain Management, International Hellenic University, Katerini, Greece
e-mail: evagelos@agro.auth.gr

C. Charatsari
School of Agriculture, Aristotle University of Thessaloniki, Thessaloniki, Greece

F. da Silveira et al. (eds.), *Agriculture 4.0*,
https://doi.org/10.1007/978-3-032-24552-6_12

(DISs) (Fielke et al. 2019). Adopting a socio-technical lens and following the line of thinking outlined in agricultural innovation systems research, a DIS can be defined as an open system in which heterogeneous actors collaboratively develop and utilize digital innovation through mobilizing resources, cocreating knowledge, and adapting to the new conditions shaped by the digitalization of agriculture.

However, DISs are not isolated islands. As it happens in all types of innovation systems, they are affected by a variety of contextual forces, which determine their ability to deliver positive impacts on innovators and society, as Rajalahti et al. (2008) have explained in one of the earliest reviews of agricultural innovation systems. Relevant work indicates that DISs and the innovation trajectories they are following are affected by a range of contexts, including social, cultural, political/regulatory, and technological (Fielke et al. 2019; Jakku et al. 2019; Bryant and Higgins 2021; da Silveira et al. 2021; Schnebelin et al. 2021; Issa et al. 2022; Pauschinger and Klauser 2022; De Rosa et al. 2024; Charatsari et al. 2024b; Dou et al. 2025). In this conceptual chapter, we aim to critically examine the impacts that these contexts may have on the process of digital agricultural innovation, while also presenting their main characteristics.

2 Contexts That Catalyze Digital Agricultural Innovation

2.1 Social Contexts

The social context of a DIS encompasses an internal (or micro) and an external (or macro) layer, which define its character and behavior, as Katz (2006) puts it. The first refers to the actors directly linked to the system, forming the core "innovation community" (Fielke et al. 2019). Beyond farmers who adopt digital technologies, their workforce, AgTech companies and technology suppliers, in this layer are included farm advisors, innovation brokers, research and university institutes, investment firms, startups and other "technology natives" like developers of software applications, data scientists, and engineers (Regan 2019; Lioutas and Charatsari 2022; Guthman and Butler 2023; Klerkx and Villalobos 2024; Nettle et al. 2025), and also actors operating in the borders of the system, such as owners of agricultural media (Paulus et al. 2024) and agriculture-related public organizations (Schnebelin et al. 2021). Notably, the social layer of innovation systems is in constant flux due to their openness, which allows new actors to enter the innovation process (Wang et al. 2018; Markow et al. 2023).

The high heterogeneity of the actors participating in the innovation community, owing to the divergent backgrounds, fields of expertise, and capacities they have, generates a complex social web within which varying cognitive schemes coexist. As early as 2000, Nooteboom noticed the need to consider such a potential "cognitive distance" between actors participating in innovation systems. In his view, this distance sets a scene allowing novel ideas and, consequently, innovation to emerge. Nevertheless, it has the significant disadvantage of incomprehensibility (Nooteboom

2000). Indeed, the involvement of actors with considerably different backgrounds in digital agricultural innovation systems offers opportunities for collaboratively ideating and developing solutions through incorporating various objectives (Mannari et al. 2025), needs (McGrath et al. 2024), and values (Stitzlein et al. 2020), thus opening new, nonenvisioned beforehand innovation paths. However, the cooperation of actors belonging to different spheres of social and economic life is not always smooth. Merging dissimilar or even contrasting priorities, and understandings of digital innovation is often challenging in practice (Lioutas and Charatsari 2022).

In addition, as it always happens in such complex social systems, participating actors have varying and often competitive interests that they may find difficult to give up (Oliver 2004; Botha et al. 2014). When multiple competitive interests are pursued, new power dynamics emerge (Cronin et al. 2024). Actors who own key resources—human, intellectual, artefactual, and natural—concentrate power (Avelino and Rotmans 2009) that may be exercised to advance their agendas to the detriment of co-innovators. Studies converge to show that this behavioral pattern is not uncommon in digital agriculture (Pauschinger and Klauser 2022; Hassim et al. 2024), leading to asymmetric distribution of the benefits and value that digital technologies encompass (Hackfort 2023), and, hence, jeopardizing the sustainability of DISs (Lioutas and Charatsari 2020a).

To avoid shifting the system in directions that serve specific interests, codesign approaches to innovation products and paths have been developed. However, despite its positive results (Ayre et al. 2019; Stitzlein et al. 2020; Kenny and Regan 2021), codesign is not a synonym of mutual decision-making (Jakku et al. 2023), since power dynamics catalyze the innovation process, often leading to distinctions between developers and implementers of innovation (McCampbell et al. 2022).

Irrespective of the power geometries that characterize DISs, social contexts are spaces within which knowledge and skills are produced. While actors innovate digitally, they experience knowledge and skill gaps associated not only with the utilization and exploitation of technologies (Weersink et al. 2018; Gumbi et al. 2023) but also, and more importantly, with the need to reconsider old roles (Charatsari et al. 2023), occupy new ones (Nettle et al. 2025), and add novel elements to their identities (Charatsari et al. 2022). As the experience of transitioning toward digitalization grows, social contexts allow new forms of knowledge to emerge and be distributed across DISs (e.g., knowledge and skills necessary for using digital tools or understanding their operational limits), while—following a commonly observed time lag in their development (Bacon et al. 2020)—tacit knowledge and skills in innovating are gradually built, through interacting with other actors and the digitalization process itself.

However, as noted above, the social context also includes a macro-level, consisting of actors not belonging to the digital innovation system, which can be either entities that form antagonistic economic sectors that are possibly threatened by digital innovation or actors generally described in the literature with terms like "citizens" or "society." Although this external social layer receives limited research attention, these actors can exert varying degrees of influence on the system, ranging from very limited to very intense.

To understand how actors not directly linked to the digital agricultural innovation process itself affect innovation systems, we may need to reflect upon past examples of technological innovation. What these examples teach us is that when a new technology enters society, the level of acceptability can be high due to the expectations that social actors have and/or technology providers cultivate. However, not infrequently, as technologies gain social acceptance, some risks associated with their use come to the fore, leading to social opposition from individuals or groups.

The opposition to a new technology or set of technologies can be weak and nonimpactful for an innovation system, as it happened during and after the first World War in the United States, where the "Horse Association of America," a group consisting of actors (e.g., ferries, horse veterinarians) who saw the introduction of tractors in the country as a threat for their interests expressed their resistance to the upcoming change without any significant outcome (Olmstead and Rhode 1994; Ellenberg 2000). Nevertheless, in other cases, opposition can have a negative impact on innovation. The example of wind power technology adoption is quite illustrative of this pattern: although such technologies emerged (and were promoted) as greener than conventional power plants, gradually a part of the public developed negative perceptions of them, and groups began to oppose their installation and use (Devine-Wright 2005; Toke et al. 2008).

Turning back to digital agriculture, research examining public perceptions of agricultural digitalization yields mixed conclusions. While work examining the acceptance and attitude of consumers/citizens toward digital agriculture shows a relatively positive—albeit not free from concerns—social attitude (Krampe et al. 2021; Pfeiffer et al. 2021), studies focusing on specific clades of digitalization, like farm robotics (Langer and Kühl 2024) and gene editing (Schnack et al. 2024), or niches operating within agrifood systems, like short food supply chains (Charatsari et al. 2024a), expose a degree of skepticism toward digitalization.

Although it is hard to predict if and how public opinion will change in the future, there is always the possibility of a negative shift in the acceptability of digital agriculture. While the media, as major influencers of public attitudes toward technology, emphasize the positive attributes of digital agricultural technologies (Preininger 2024), stressing their potential to enhance environmental sustainability and allow an easier and better farm life (Mohr and Höhler 2023), there are already voices that stress the possible adverse effects of digitalization or even advocate an "anti-smart" way of thinking, which can also be extended to include agricultural modernization. Additionally, situational factors or technology failures may lead citizens to develop a negative attitude toward agricultural digitalization.

2.2 Cultural Contexts

Culture is strongly interconnected with social contexts. It refers to a sum of immaterial attributes, like ideas, emotions, and behaviors, that members of societies or specific social groups share (Harris 1985). The notion that cultural contexts enable or resist technological advancement is not new. As Leidner and Kayworth (2006) observed, cultural settings have significant impacts on how novel technology is developed, adopted, and used.

The existence of an innovation-enabling culture is a sine qua non for supporting digital agriculture (Abdulai et al. 2023). As Ridley (2020) argues, a pivotal characteristic of such a culture is the freedom to exchange ideas and imagine different futures that these ideas create, develop visions and plans to arrive at desirable future states, invest in these visions, experiment, and even fail, thereby refining initial ideas and initiating another innovation journey. Hence, in societies where a democratic and free culture prevails, innovation can flourish.

However, what happens in culture after the emergence of technological innovation? The adoption of a technology by societal actors leads to its gradual integration into dominant cultures, reshaping them and instilling new cultural norms (Alsaleh 2024). The rise of Internet use is just an example of how cultural landscapes (e.g., communication patterns, community building, ideas about reality) are affected by new technologies (Brey 2006). As a novel digital culture emerges, new artifacts are developed to serve it (Gere 2008), further supporting the digitalization of society. Following Lévy's (1997) ideas on what he termed "cyberculture," we can define digital culture as the set of new ways of conceiving things and phenomena, acting, working, valuing, and, finally, existing in a digitalized world.

This turn to a culture that is heavily entangled with technologies (Murphie and Potts 2017) often leads to an uncritically accepted confidence to the capacity of technology to positively transform our world (Huesemann and Huesemann 2011), which promotes a culture of techno-solutionism, i.e., an overoptimism, or even blind faith to the ability of technological tools to solve major societal problems (Morozov 2013). Several studies suggest that agricultural digitalization involves a level of solutionism, emphasizing the potential positive outcomes of going digital and overlooking or underestimating significant adverse side effects (Gras and Cáceres 2020; Donaldson 2022; Guthman and Butler 2023). Hence, although current cultural dynamics facilitate the development of digital agriculture, overconfidence biases can mask the potential unintended effects of digitalization.

Finally, it is important to make the distinction between the general prevailing culture in a sector and/or geographical area and the many coexisting subcultures, i.e., the sum of cultural ideals, values, and practices that are shared among the members of a social cluster and make it behave differently from other, relatively homogeneous social groups (Blackman 2014). Although mainstream culture may facilitate the expansion of digital agricultural innovation, subcultures may view it as a threat to their distinctiveness. Indeed, there are indications that coupling digitalization with agroecological culture may be challenging (Binetsch et al. 2023; Shilomboleni and Schnurr 2025). The same is true for the case of organic agriculture, where technological sophistication is not an essential element of farmers' culture (Tereso et al. 2012), or in alternative food networks, where there is an emphasis on keeping farm practices less industrialized and, perhaps, high-tech free (Lioutas and Charatsari 2020b).

2.3 Policy-Legal Contexts

Policy and legal contexts are well-known catalysts for innovation. Beyond core innovation policies (those policy and legislative frameworks that directly deal with an innovation, which, usually, have as a central aim its promotion through straightforwardly supporting innovation development and its uptake), a series of—not always so obviously relevant—policy axes can positively or negatively impact innovation process: taxation policies, legislations governing venture capital investment, frameworks promoting interactivity and collaboration among actors operating in the public sphere and the private realm, or intellectual property rights policies are only a few examples (Ekboir 2003; Avnimelech and Teubal 2008; Bergek et al. 2010; Dolfsma and Seo 2013).

Characterized by a positive stance toward modernizing farming, current core innovation policy frameworks in many countries support the promotion of digital agriculture, either by emphasizing the need to shift to digitally enabled agrifood production or by setting forth mechanisms that facilitate the digital transition of agrifood systems. For example, the European Union's Common Agricultural Policy encourages Member States to embed agricultural digitalization into their national development policies (Barabanova and Krzysztofowicz 2023), while the Union supports investments in a range of relevant activities, including the development of information and communication technology infrastructures and precision agriculture (European Union 2021).

In the United States, the Department of Agriculture offers loans and funding to increase broadband connectivity in rural areas (Rural Utilities Service 2024) while promoting the adoption of precision agriculture technologies (USDA 2019). In Australia, the government prioritizes the development of digital skills in the agrifood system, the maximization of agricultural data use, the establishment of data standards, the increase of interoperability, the cultivation of a digitalization culture in the sector, and the improvement of relevant infrastructure (Australian Government 2022).

Other countries have also integrated agricultural digitalization into their strategic policy interventions, such as the "Strategy for digital transformation" in Brazil (Centro de Gestão e Estudos Estratégicos 2022), or funding policies, like the AgriInnovate program in Canada, which is a scheme that funds projects aiming to accelerate digital innovation in agriculture (Canadian Agri-food Policy Institute 2025). Following a different approach, as part of its policy initiatives to the direction of supporting agricultural digitalization, Japan developed tools that support agricultural digitalization, such as an information research platform that collects and summarizes agricultural data (Kawamura et al. 2022) and a specialized agricultural knowledge Artificial Intelligence system, to offer farmers relevant information when needed (National Agriculture and Food Research Organization 2024). Beyond the Global North, a series of policy frameworks have been recently initiated in African (Degila et al. 2023) and Asian countries (Montesclaros and Teng 2023).

In general, policy organizations currently adopt a positive stance toward agricultural digitalization. At the same time, they seem to be quite optimistic about the capacity of digital technologies to enhance food productivity and security (Lajoie-O'Malley et al. 2020). However, beyond the enabling policies—and their rhetoric about the "need" to shift to a more digitalized future—there are several policy and legal blocks in the process.

A commonly mentioned issue in the agricultural digitalization literature is the lack of clear policies regulating data ownership, management, and governance (Jakku et al. 2019; Rotz et al. 2019; Atik 2022; Brown et al. 2023; Ruder and Wittman 2025). Although relevant frameworks do exist in some regions—e.g., the European Data Governance Act (European Union 2022) or the Australian Data and Digital Government Strategy (Australian Government 2023)—they offer instead general suggestions or declare their commitment to deliver results shortly. For instance, the European Union emphasizes the need to take action "to increase trust in data sharing by establishing appropriate mechanisms for control by data subjects and data holders over data that relates to them" (European Union 2022, L152/4) and develop a union-wide data governance framework, without offering explicit directions on how to achieve these purposes. Given that the process of launching relevant regulations is usually time consuming, a time lag may exist between the introduction of digital innovation and the development of relevant regulatory mechanisms.

In other cases, old and new policies and legal frameworks are conflicting, thus generating a "neither allowed nor forbidden" situation for digital innovation. To offer an example, in Greece, Article 23 of Law 4036/2012 (Greek Ministry of Rural Development and Food, 2012) explicitly mentions that "aerial spraying is prohibited" in the country, defining, however, aerial spraying as "the application of pesticides from an airplane or helicopter." Since the law came into force well before the expansion of the use of drones in agriculture, it does not include a clear framework governing them. Similar conflicts between old regulatory priorities and new functions of digital technologies exist in many countries—see Frankelius et al. (2019) for a case in Sweden—revealing the disturbing effects of digitalization in the political and legal contexts, as well as the enabling or inhibiting power of legislative and policy mechanisms for digital innovation.

2.4 Technological Contexts

Last but not least among the contexts that determine the emergence and success of digital agricultural innovation is technology itself. The capacity to move on to the digital era—a future in which digital technologies perform critical tasks in agrifood systems—heavily relies on "the present" of technology. There is little doubt that, to be efficient (or even to be produced), digital technologies require a functional infrastructure, consisting of a bundle of baseline technologies that set the foundation for digital tools to operate (Bernhardt et al. 2021; Dwivedi and Sherly 2025; Mansoor et al. 2025).

Beyond this on/off requirement, which refers to the compatibility of digital agricultural technologies with the current technological base of an agricultural system (Lioutas and Charatsari 2020b), the technological context of a DIS refers to the sum of the technologies that operate or are under development at a given time, their interdependencies, and their parallel evolutionary paths. To better understand what this context entails, it is essential to view technologies as combinations of existing artifacts, which are then organized into technological domains (a set of new technologies that share characteristics and functionalities) that generate a new "world of the possible," to use Arthur's words (2009). Simply put, they create opportunities that were not fully open before the emergence of these domains, determining, in this way, what is possible in a given period.

For Arthur (2009), the combinational nature of technology leads to a continuous cycle of innovation; a constant evolution toward new artifacts that are produced and utilized to solve old, current, or possible future problems. However, not all technologies follow the same evolutionary pace. Some progress faster than others, while some, inevitably, have a short life cycle, either because they cannot meet the expectations of potential users or because other artifacts surpass them during the digital innovation race. Indeed, technologies are often competitive, fighting for a position in society. Theories of competitive substitution of technologies utilize the predator–prey example to explain this pattern of technological evolution: new technologies act as predators, aiming to replace older ones (Fisher and Pry 1971). In that case, technology providers constantly produce new artifacts that promise to address current and future societal needs and/or wants in an easier, cheaper, or more convenient way (or in all these ways). The shift from the cassette tape to the compact disc, and then to purely digital music and streaming services, is an example of this pattern.

However, this norm is not the standard. For some periods—which, not rarely, are quite extensive—old and new technologies coexist. Consider, for example, farm irrigation technology. While sophisticated digital systems incorporating Internet of Things sensing techniques and Artificial Intelligence are used (Hammouch et al. 2024; Gaitan et al. 2025), older technologies such as conventional drip irrigation, center pivot systems, and sprinkler irrigation technologies remain in use today. When considering digital innovation in agriculture, the substitution of analog technologies is not always the endpoint.

On the other hand, competition can also be observed between digital technologies, taking the form of symbiotic coexistence (Utterback et al. 2018). In that form, the evolution of a specific technology leads to the parallel evolution of another, thus creating a coevolutionary pattern for both of them (e.g., the coevolution of farm sensors and big data analytics). When such a pattern is followed, technologies undergo reciprocal adaptations, supporting the further development and acceleration of innovation (Coccia and Watts 2020).

3 On the Nature of Contexts

The four contexts presented above are fluid, mutually formative, and—at least partially—rely on their past. Below, we briefly discuss these properties of the multiple contexts that catalyze digital agricultural innovation.

Fluidity concerns the capacity of a context to change forms, as an output of both internal and external forces. Following Nenonen et al. (2014), fluidity can be the outcome of changes that (usually) incrementally alter a context. Although in short periods, social, cultural, political/legal, and technological contexts may exhibit a level of equilibrium, they are not static. Instead, they constantly undergo minor or significant changes that allow new dynamics to emerge (Chandler et al. 2019). New actors enter DISs, altering their internal social structure, while, as explained above, the perceptions and attitudes of the external social context may change over time. The latter form of change is strongly associated with the transformation of cultural contexts and the values they endorse and prioritize. Policies and legal frameworks, and, of course, technologies, are also evolving.

Importantly, these contexts coexist and are interwoven. They are interdependent and mutually formative, in that an alteration in one of the contexts may spur corresponding changes in all the remaining contexts. For instance, as Arthur (2009) suggests, when a new technological domain takes shape, the social and cultural contexts encounter it and, as a result, alter themselves through embedding technology into their routines and practices, and develop new modes of acting, working, thinking, behaving, and living with the new technologies. Nevertheless, not all types of social and cultural changes are positive, leading to the initiation of new policies and the enactment of legislative frameworks that regulate the development and utilization of technology.

Notably, to better understand changes that occur within and between contexts and the responses they produce, a thorough consideration of their historicity is essential. Contexts and their components are, to some extent, reflections of their past cycles, which, as Eastwood et al. (2021) argue, may define their futures. Simply put, the history of social, cultural, political, and technological contexts—or aspects of it—is bequeathed to their present and future states in the form of rules, routines, logics, artifacts (Sannino and Engeström 2018), and ways of adapting to new conditions (Charatsari et al. 2020). The power dynamics that emerge in DISs are partially influenced by past passages of social and cultural contexts to new, technologically advanced states (Abdulai 2022), policies are by nature based on past frameworks and emerge through layering strategies that add new instruments into extant contexts (Daugbjerg and Swinbank 2016), while technologies evolve by connecting new artifacts with existing technological blocks (Coccia 2019; Coccia and Watts 2020) and do have their history even when portrayed as unconnected with past technical foundations (Duncan et al. 2021) and, hence, even when radically novel, they rely on previous evolutionary lines of their predecessors.

4 Discussion and Concluding Thoughts

In this conceptual study, we aimed to outline the nature and impact of four interrelated and interdependent contexts within which DISs operate. Departing from the assertion that digital agricultural innovation takes place in complex socio-technical systems and is catalyzed by internal and external factors (Fielke et al. 2019; Jakku et al. 2019; Lioutas et al. 2019; Pauschinger and Klauser 2022; Daniel et al. 2025; Fielke et al. 2025), we presented how and why social, cultural, political/legal, and technological contexts catalyze DISs.

As our theoretical conceptualization indicates, contexts simultaneously enable and constrain digital agricultural innovation. Although each context can be examined separately, the interdependency among them requires a systemic consideration of their impacts on the innovation process and its outcomes. The fluidity of all four contexts described in the previous section makes any attempt to predict their future alterations extremely challenging. However, historicity can be used as a compass, assisting scholars in anticipating how these contexts—and, consequently, DISs—will likely evolve.

The relatively brief (due to space constraints) analysis of social, cultural, policy/legal, and technological contexts presented herein can be used as a jumping-off point for researchers studying DISs and their future trajectories. Today, there is a growing and insightful literature inquiring into the futures that digitalization of agriculture shapes (e.g., Fleming et al. 2021; Eastwood et al. 2021; Charatsari et al. 2022). Although further conceptualization of our multiple contexts perspective is needed, our framework can be integrated into the body of research focusing on how digitalization is transforming agrifood systems.

Of course, other contexts can also have more or less potent impacts on digital agricultural innovation. Ecological contexts affect the operation and performance of several digital technologies that are used in agriculture. Technology-oriented research indicates that installing digital devices like sensing systems is difficult in some regions (Mansoor et al. 2025), while sensors and agricultural robots may underperform in harsh environments (Fountas et al. 2020). Economic and ethical contexts also play a role in shaping the dynamics of digital agricultural innovation. Future research will likely theorize the influence of these contexts in DISs, thus helping to develop a more complete framework of the factors that set the scene upon which agricultural digitalization unfolds.

To conclude, this study sheds light on the critical role of multiple, coexisting, coevolving, and mutually formative contexts in the process of digital agricultural innovation. However, many relevant questions await answers. How does the heterogeneity of actors who form the digital innovation community shape the dynamics of co-innovation? What is the impact of antagonistic sectors (e.g., legacy technology providers) on digitalization? How will public acceptance of digital agricultural technologies evolve? Is public opposition to agricultural digitalization a possible future scenario? To what extent do digitalization-promoting policies generate a wave of techno-solutionism? What types of technologies are suitable for and compatible with the niches that exist within agrifood systems? What types of policies are needed to promote responsible agricultural digitalization? What are the risks associated with the time lag between the emergence of new technologies and the launching of relevant regulatory frameworks? How will the development of new technology domains affect the social fabric of agrifood systems and the very nature of farming?

Social science on digital agriculture is a rapidly growing field, and scholars from around the world have already incorporated some of these questions into their agendas. Our perspective can possibly point them toward a more comprehensive conceptualization of the forces that catalyze DISs, hopefully opening up a new path of inquiry for scholarly work in agricultural digitalization.

Competing Interests The authors have no conflicts of interest to declare that are relevant to the content of this chapter.

References

Abdulai AR (2022) A new green revolution (GR) or neoliberal entrenchment in agri-food systems? Exploring narratives around digital agriculture (DA), food systems, and development in sub-Sahara Africa. J Dev Stud 58:1588–1604. https://doi.org/10.1080/00220388.2022.2032673

Abdulai AR, Gibson R, Fraser ED (2023) Beyond transformations: zooming in on agricultural digitalization and the changing social practices of rural farming in northern Ghana, West Africa. J Rural Stud 100:103019. https://doi.org/10.1016/j.jrurstud.2023.103019

Alsaleh A (2024) The impact of technological advancement on culture and society. Sci Rep 14:32140. https://doi.org/10.1038/s41598-024-83995-z

Arthur WB (2009) The nature of technology: what it is and how it evolves. Simon and Schuster, New York

Atik C (2022) Towards comprehensive European agricultural data governance: moving beyond the "data ownership" debate. IIC Int Rev Intellect Prop Compet Law 53:701–742. https://doi.org/10.1007/s40319-022-01191-w

Australian Government (2022) Digital foundations for agriculture strategy: driving the development and uptake of digital technologies in the Australian agriculture, fisheries and forestry industry. Available at: https://www.agriculture.gov.au/sites/default/files/documents/digital-foundations-agriculture-strategy.pdf

Australian Government (2023) Data and digital government strategy: the data and digital vision for a world-class APS to 2030. Commonwealth of Australia 2023. Available at: https://www.dataanddigital.gov.au/sites/default/files/2023-12/Data%20and%20Digital%20Government%20Strategy%20v1.0.pdf

Avelino F, Rotmans J (2009) Power in transition: an interdisciplinary framework to study power in relation to structural change. Eur J Soc Theory 12:543–569. https://doi.org/10.1177/1368431009349830

Avnimelech G, Teubal M (2008) From direct support of business sector R&D/innovation to targeting venture capital/private equity: a catching-up innovation and technology policy life cycle perspective. Econ Innov New Technol 17:153–172. https://doi.org/10.1080/10438590701279417

Ayre M, Mc Collum V, Waters W, Samson P, Curro A, Nettle R et al (2019) Supporting and practising digital innovation with advisers in smart farming. NJAS Wageningen J Life Sci 90:100302. https://doi.org/10.1016/j.njas.2019.05.001

Bacon E, Williams MD, Davies G (2020) Coopetition in innovation ecosystems: a comparative analysis of knowledge transfer configurations. J Bus Res 115:307–316. https://doi.org/10.1016/j.jbusres.2019.11.005
Barabanova Y, Krzysztofowicz M (2023) Digital transition: long-term implications for eu farmers and rural communities, publications office of the European Union, Luxembourg, 2023. https://doi.org/10.2760/093463, JRC134571
Bergek A, Jacobsson S, Hekkert M, Smith K (2010) Functionality of innovation systems as a rationale for and guide to innovation policy. In: Smits R, Kuhlmann S, Shapira P (eds) The theory and practice of innovation policy. Edward Elgar Publishing, Cheltenham, pp 117–146
Bernhardt H, Schumacher L, Zhou J, Treiber M, Shannon K (2021) Digital agriculture infrastructure in the USA and Germany. Eng Proc 9(1):1. https://doi.org/10.3390/engproc2021009001
Binetsch J, Bun H, Westpfahl RH, Krug J, Lehmann R, Bravo ES et al (2023) A sustainable future for small-scale farming? Qualitative research on agroecology and digitalization. HCIAS Working Papers on Ibero-America
Blackman S (2014) Subculture theory: an historical and contemporary assessment of the concept for understanding deviance. Deviant Behav 35:496–512. https://doi.org/10.1080/01639625.2013.859049
Botha N, Klerkx L, Small B, Turner JA (2014) Lessons on transdisciplinary research in a co-innovation programme in the New Zealand agricultural sector. Outlook Agric 43:219–223. https://doi.org/10.5367/oa.2014.0175
Brey P (2006) Evaluating the social and cultural implications of the internet. ACM SIGCAS Comput Soc 36:41–48. https://doi.org/10.1145/1195716.1195721
Brown C, Regan Á, Van Der Burg S (2023) Farming futures: perspectives of Irish agricultural stakeholders on data sharing and data governance. Agric Hum Values 40:565–580. https://doi.org/10.1007/s10460-022-10357-8
Bryant M, Higgins V (2021) Securitising uncertainty: ontological security and cultural scripts in smart farming technology implementation. J Rural Stud 81:315–323. https://doi.org/10.1016/j.jrurstud.2020.10.051
Canadian Agri-food Policy Institute (2025) The future is digital: digital agriculture and Canadian agriculture policy. Available at: https://capi-icpa.ca/wp-content/uploads/2025/05/2025-05-20-Digital-Agriculture-EN-CAPI.pdf
Centro de Gestão e Estudos Estratégicos (2022) Estratégia Brasileira para a Transformação Digital (E-Digital): Ciclo 2022–2026. Available at: https://www.gov.br/mcti/pt-br/acompanhe-o-mcti/transformacaodigital/arquivosestrategiadigital/e-digital_ciclo_2022-2026.pdf
Chandler JD, Danatzis I, Wernicke C, Akaka MA, Reynolds D (2019) How does innovation emerge in a service ecosystem? J Serv Res 22:75–89. https://doi.org/10.1177/1094670518797479
Charatsari CD, Lioutas E, De Rosa M, Papadaki-Klavdianou A (2020) Extension and advisory organizations on the road to the digitalization of animal farming: an organizational learning perspective. Animals 10:2056. https://doi.org/10.3390/ani10112056
Charatsari C, Lioutas ED, Papadaki-Klavdianou A, Michailidis A, Partalidou M (2022) Farm advisors amid the transition to agriculture 4.0: professional identity, conceptions of the future and future-specific competencies. Sociol Rural 62:335–362. https://doi.org/10.1111/soru.12364
Charatsari C, Michailidis A, Lioutas ED, Bournaris T, Loizou E, Paltaki A, Lazaridou D (2023) Competencies needed for guiding the digital transition of agriculture: are future advisors well-equipped? Sustainability 15:15815. https://doi.org/10.3390/su152215815
Charatsari C, Lioutas ED, De Rosa M (2024a) Going short and going digital: how do consumers view the impacts of digitalizing short food supply chains? Sustainability 16:11241. https://doi.org/10.3390/su162411241
Charatsari C, Michailidis A, Francescone M, De Rosa M, Aidonis D, Bartoli L et al (2024b) Do agricultural knowledge and innovation systems have the dynamic capabilities to guide the digital transition of short food supply chains? Information 15:22. https://doi.org/10.3390/info15010022

Coccia M (2019) The theory of technological parasitism for the measurement of the evolution of technology and technological forecasting. Technol Forecast Soc Chang 141:289–304. https://doi.org/10.1016/j.techfore.2018.12.012

Coccia M, Watts J (2020) A theory of the evolution of technology: technological parasitism and the implications for innovation management. J Eng Technol Manag 55:101552. https://doi.org/10.1016/j.jengtecman.2019.11.003

Cronin E, Block T, Fosselle S, Rogge E (2024) Whose knowledge counts? Power dynamics in the co-production of knowledge and innovation in agri-food systems. Sci Public Policy 51:1117–1132. https://doi.org/10.1093/scipol/scae050

da Silveira F, Lermen FH, Amaral FG (2021) An overview of agriculture 4.0 development: systematic review of descriptions, technologies, barriers, advantages, and disadvantages. Comput Electron Agric 189:106405. https://doi.org/10.1016/j.compag.2021.106405

Daniel L, Groeger L, Hölzle K (2025) Unpacking smart farming innovation: A systematic literature review on technological change in agriculture. J Eng Technol Manag 77:101898. https://doi.org/10.1016/j.jengtecman.2025.101898

Daugbjerg C, Swinbank A (2016) Three decades of policy layering and politically sustainable reform in the European Union's Agricultural Policy. Governance 29:265–280. https://doi.org/10.1111/gove.12171

De Rosa M, Charatsari C, Lioutas ED, Masi M, Vecchio Y, Francescone M (2024) Contextualising digitalisation through ambidexterity and new territorial proximities. J Rural Stud 109:103322. https://doi.org/10.1016/j.jrurstud.2024.103322

Degila J, Sodedji FAK, Avakoudjo HGG, Tahi SPG, Houetohossou SCA, Honfoga AC et al (2023) Digital agriculture policies and strategies for innovations in the agri-food systems—cases of five West African countries. Sustainability 15:9192. https://doi.org/10.3390/su15129192

Devine-Wright P (2005) Beyond NIMBYism: towards an integrated framework for understanding public perceptions of wind energy. Wind Energy Int J Prog Appl Wind Power Convers Technol 8:125–139. https://doi.org/10.1002/we.124

Dolfsma W, Seo D (2013) Government policy and technological innovation—a suggested typology. Technovation 33:173–179. https://doi.org/10.1016/j.technovation.2013.03.011

Donaldson A (2022) Digital from farm to fork: infrastructures of quality and control in food supply chains. J Rural Stud 91:228–235. https://doi.org/10.1016/j.jrurstud.2021.10.004

Dou X, Zheng Y, Yang D (2025) Promoting digital transformation in facility agriculture: the role of government policies and digital literacy. Front Sustainable Food Syst 9:1575666. https://doi.org/10.3389/fsufs.2025.1575666

Duncan E, Glaros A, Ross DZ, Nost E (2021) New but for whom? Discourses of innovation in precision agriculture. Agric Hum Values 38:1181–1199. https://doi.org/10.1007/s10460-021-10244-8

Dwivedi S, Sherly MA (2025) Multivariate assessment of digital agriculture and irrigation potential: application to India. Agric Syst 227:104328. https://doi.org/10.1016/j.agsy.2025.104328

Eastwood CR, Edwards JP, Turner JA (2021) Anticipating alternative trajectories for responsible Agriculture 4.0 innovation in livestock systems. Animal 15:100296. https://doi.org/10.1016/j.animal.2021.100296

Ekboir JM (2003) Research and technology policies in innovation systems: zero tillage in Brazil. Res Policy 32:573–586. https://doi.org/10.1016/S0048-7333(02)00058-6

Ellenberg GB (2000) Debating farm power: draft animals, tractors, and the United States Department of Agriculture. Agric Hist 74:545–568

European Union (2021) Regulation (EU) 2021/2115 of the European Parliament and of the council of 2 December 2021. Available at: https://eur-lex.europa.eu/legal-content/EN/TXT/PDF/?uri=CELEX:32021R2115

European Union (2022) Off J Eur Union L 152:Regulation (EU) 2022/868 of the European Parliament and of the Council of 30 May 2022 on European data governance and amending Regulation (EU) 2018/1724 (Data Governance Act). Available at: https://eur-lex.europa.eu/legal-content/EN/TXT/PDF/?uri=CELEX:32022R0868

Fielke SJ, Garrard R, Jakku E, Fleming A, Wiseman L, Taylor BM (2019) Conceptualising the DAIS: implications of the 'digitalisation of agricultural innovation systems' on technology and policy at multiple levels. NJAS Wageningen J Life Sci 90:1–11. https://doi.org/10.1016/j.njas.2019.04.002

Fielke S, Fleming A, Jakku E, Stitzlein C, Ricketts K, Cornish G et al (2025) "The end point is a... more appropriate innovation ecosystem" Mission-oriented and responsible innovation in Australian agricultural systems. Agric Syst 227:104359. https://doi.org/10.1016/j.agsy.2025.104359

Fisher JC, Pry RH (1971) A simple substitution model of technological change. Technol Forecast Soc Chang 3:75–88. https://doi.org/10.1016/S0040-1625(71)80005-7

Fleming A, Jakku E, Fielke S, Taylor BM, Lacey J, Terhorst A, Stitzlein C (2021) Foresighting Australian digital agricultural futures: applying responsible innovation thinking to anticipate research and development impact under different scenarios. Agric Syst 190:103120. https://doi.org/10.1016/j.agsy.2021.103120

Fountas S, Espejo-García B, Kasimati A, Mylonas N, Darra N (2020) The future of digital agriculture: technologies and opportunities. IT Prof 22:24–28. https://doi.org/10.1109/MITP.2019.2963412

Frankelius P, Norrman C, Johansen K (2019) Agricultural innovation and the role of institutions: lessons from the game of drones. J Agric Environ Ethics 32:681–707. https://doi.org/10.1007/s10806-017-9703-6

Gaitan NC, Batinas BI, Ursu C, Crainiciuc FN (2025) Integrating artificial intelligence into an automated irrigation system. Sensors 25:1199. https://doi.org/10.3390/s25041199

Gere C (2008) Digital culture. Reaktion Books, London

Gras C, Cáceres DM (2020) Technology, nature's appropriation and capital accumulation in modern agriculture. Curr Opin Environ Sustain 45:1–9. https://doi.org/10.1016/j.cosust.2020.04.001

Gumbi N, Gumbi L, Twinomurinzi H (2023) Towards sustainable digital agriculture for smallholder farmers: a systematic literature review. Sustainability 15:12530. https://doi.org/10.3390/su151612530

Guthman J, Butler M (2023) Fixing food with a limited menu: on (digital) solutionism in the agri-food tech sector. Agric Hum Values 40:835–848. https://doi.org/10.1007/s10460-023-10416-8

Hackfort S (2023) Unlocking sustainability? The power of corporate lock-ins and how they shape digital agriculture in Germany. J Rural Stud 101:103065. https://doi.org/10.1016/j.jrurstud.2023.103065

Hammouch H, El-Yacoubi M, Qin H, Berbia H (2024) A systematic review and meta-analysis of intelligent irrigation systems. IEEE Access 12:128285–128304. https://doi.org/10.1109/access.2024.3421322

Harris M (1985) Culture, people, nature: an introduction to general anthropology, 4th edn. Harper & Row, New York

Hassim OA, Osman I, Awal A, Amin FM, Alam BP (2024) Navigating the path to equitable and sustainable digital agriculture among small farmers in Malaysia: a comprehensive review. Information Management and Business Review 16:173–188

Huesemann M, Huesemann J (2011) Techno-fix: why technology won't save us or the environment. New Society Publishers, Gabriola Island

Issa H, Jabbouri R, Palmer M (2022) An artificial intelligence (AI)-readiness and adoption framework for AgriTech firms. Technol Forecast Soc Chang 182:121874. https://doi.org/10.1016/j.techfore.2022.121874

Jakku E, Taylor B, Fleming A, Mason C, Fielke S, Sounness C, Thorburn P (2019) "If they don't tell us what they do with it, why would we trust them?" trust, transparency and benefit-sharing in smart farming. NJAS Wageningen J Life Sci 90:100285. https://doi.org/10.1016/j.njas.2018.11.002

Jakku E, Fleming A, Espig M, Fielke S, Finlay-Smits SC, Turner JA (2023) Disruption disrupted? Reflecting on the relationship between responsible innovation and digital agriculture research

and development at multiple levels in Australia and Aotearoa New Zealand. Agric Syst 204:103555. https://doi.org/10.1016/j.agsy.2022.103555
Katz JS (2006) Indicators for complex innovation systems. Res Policy 35:893–909. https://doi.org/10.1016/j.respol.2006.03.007
Kawamura T, Katsuragi T, Kobayashi A, Inatomi M, Oshiro M, Eguchi H (2022) Development of an information research platform for data-driven agriculture. Int J Agric Environ Inf Syst 13:1–19. https://doi.org/10.4018/IJAEIS.302908
Kenny U, Regan A (2021) Co-designing a smartphone app for and with farmers: Empathising with end-users' values and needs. J Rural Stud 82:148–160. https://doi.org/10.1016/j.jrurstud.2020.12.009
Klerkx L, Villalobos P (2024) Are AgriFoodTech start-ups the new drivers of food systems transformation? An overview of the state of the art and a research agenda. Glob Food Sec 40:100726. https://doi.org/10.1016/j.gfs.2023.100726
Klerkx L, Aarts N, Leeuwis C (2010) Adaptive management in agricultural innovation systems: the interactions between innovation networks and their environment. Agric Syst 103:390–400. https://doi.org/10.1016/j.agsy.2010.03.012
Klerkx L, Van Mierlo B, Leeuwis C (2012) Evolution of systems approaches to agricultural innovation: concepts, analysis and interventions. In: Darnhofer I, Gibbon D, Dedieu B (eds) Farming systems research into the 21st century: the new dynamic. Springer, Dordrecht, pp 457–483
Kline SJ, Rosenberg N (2010) An overview of innovation. In: Studies on Science and the Innovation Process: Selected Works of Nathan Rosenberg. National Academy Press, Washington, pp 173–203
Krampe C, Serratosa J, Niemi JK, Ingenbleek PT (2021) Consumer perceptions of precision livestock farming—A qualitative study in three European countries. Animals 11:1221. https://doi.org/10.3390/ani11051221
Lajoie-O'Malley A, Bronson K, van der Burg S, Klerkx L (2020) The future (s) of digital agriculture and sustainable food systems: an analysis of high-level policy documents. Ecosyst Serv 45:101183. https://doi.org/10.1016/j.ecoser.2020.101183
Langer G, Kühl S (2024) Perception and acceptance of robots in dairy farming—a cluster analysis of German citizens. Agric Hum Values 41:249–267. https://doi.org/10.1007/s10460-023-10483-x
Leidner DE, Kayworth T (2006) A review of culture in information systems research: toward a theory of information technology culture conflict. MIS Q 30:357–399
Lévy P (1997) Cyberculture. Rapport au Conseil de l'Europe. Editions Odile Jacob, Paris
Lioutas ED, Charatsari C (2020a) Big data in agriculture: does the new oil lead to sustainability? Geoforum 109:1–3. https://doi.org/10.1016/j.geoforum.2019.12.019
Lioutas ED, Charatsari C (2020b) Smart farming and short food supply chains: are they compatible? Land Use Policy 94:104541. https://doi.org/10.1016/j.landusepol.2020.104541
Lioutas ED, Charatsari C (2022) Innovating digitally: the new texture of practices in agriculture 4.0. Sociol Rural 62:250–278. https://doi.org/10.1111/soru.12356
Lioutas ED, Charatsari C, La Rocca G, De Rosa M (2019) Key questions on the use of big data in farming: an activity theory approach. NJASWageningen J Life Sci 90:100297. https://doi.org/10.1016/j.njas.2019.04.003
Lundvall BA, Dosi G, Freeman C (1988) Innovation as an interactive process: from user-producer interaction to the national system of innovation. In: Dosi G, Freeman C, Nelson RR, Silverberg G, Soete L (eds) Technical change and economic theory. Pinter Publishers, London, pp 349–369
Mannari C, Sportelli M, Meesala H, Okoye OF, Lepore F, Bacco M et al (2025) End-user requirements modelling: an experience report from digital agriculture. In: Hess A, Susi A (eds) International working conference on requirements engineering: foundation for software quality. Springer Nature Switzerland, Cham, pp 304–316

Mansoor S, Iqbal S, Popescu SM, Kim SL, Chung YS, Baek JH (2025) Integration of smart sensors and IOT in precision agriculture: trends, challenges and future prospectives. Front Plant Sci 16:1587869. https://doi.org/10.3389/fpls.2025.1587869
Markow J, Fieldsend AF, von Münchhausen S, Häring AM (2023) Building agricultural innovation capacity from the bottom up: using spillover effects from projects to strengthen agricultural innovation systems. Agric Syst 209:103670. https://doi.org/10.1016/j.agsy.2023.103670
McCampbell M, Schumann C, Klerkx L (2022) Good intentions in complex realities: challenges for designing responsibly in digital agriculture in low-income countries. Sociol Rural 62:279–304. https://doi.org/10.1111/soru.12359
McGrath K, Regan Á, Russell T (2024) A user-centred future for agricultural digital innovation: demonstrating the value of design thinking in an animal health context. J Agric Educ Ext 31:417–435. https://doi.org/10.1080/1389224X.2024.2397968
Mohr S, Höhler J (2023) Media coverage of digitalization in agriculture-an analysis of media content. Technol Forecast Soc Chang 187:122238. https://doi.org/10.1016/j.techfore.2022.122238
Montesclaros JML, Teng PS (2023) Digital technology adoption and potential in southeast Asian agriculture. Asian J Agric Dev 20:7–30. https://doi.org/10.22004/ag.econ.339167
Morozov E (2013) To save everything, click here: the folly of technological solutionism. Public Affairs
Murphie A, Potts J (2017) Culture and technology. Bloomsbury Publishing
National Agriculture and Food Research Organization (2024) Development of Japan's first generative AI for agriculture. Available at: https://www.naro.go.jp/english/topics/laboratory/01rcait/166200.html
Nelson RR, Nelson K (2002) Technology, institutions, and innovation systems. Res Policy 31:265–272. https://doi.org/10.1016/S0048-7333(01)00140-8
Nenonen S, Kjellberg H, Pels J, Cheung L, Lindeman S, Mele C et al (2014) A new perspective on market dynamics: market plasticity and the stability–fluidity dialectics. Mark Theory 14:269–289. https://doi.org/10.1177/1470593114534342
Nettle R, Ingram J, Ayre M (2025) Digiwork: how agriculture 4.0 is changing work for farm advisers. Front Sustainable Food Syst 9:1542007. https://doi.org/10.3389/fsufs.2025.1542007
Nooteboom B (2000) Institutions and forms of co-ordination in innovation systems. Organ Stud 21:915–939. https://doi.org/10.1177/0170840600215004
Oliver AL (2004) On the duality of competition and collaboration: network-based knowledge relations in the biotechnology industry. Scand J Manag 20:151–171. https://doi.org/10.1016/j.scaman.2004.06.002
Olmstead AL, Rhode PW (1994) The agricultural mechanization controversy of the interwar years. Agric Hist 68:35–53
Paulus M, Herrera B, Knierim A (2024) Who do German farmers trust when making decisions about digital technologies? An analysis of the trustworthiness of innovation actors. Stud Agric Econ 126:117–127. https://doi.org/10.7896/j.2825
Pauschinger D, Klauser FR (2022) The introduction of digital technologies into agriculture: space, materiality and the public–private interacting forms of authority and expertise. J Rural Stud 91:217–227. https://doi.org/10.1016/j.jrurstud.2021.06.015
Pfeiffer J, Gabriel A, Gandorfer M (2021) Understanding the public attitudinal acceptance of digital farming technologies: A nationwide survey in Germany. Agric Hum Values 38:107–128. https://doi.org/10.1007/s10460-020-10145-2
Preininger EM (2024) Is this the future? Image and imagination in visual discourses on digital farming in Austrian media. Geogr Ann Ser B, Hum Geogr 106:441–459. https://doi.org/10.1080/04353684.2023.2261132
Rajalahti R, Janssen W, Pehu E (2008) Agricultural innovation systems: from diagnostics toward operational practices. Agriculture & Rural Development Department, World Bank, Washington, DC
Regan Á (2019) 'Smart farming' in Ireland: A risk perception study with key governance actors. NJAS Wageningen J Life Sci 90:100292. https://doi.org/10.1016/j.njas.2019.02.003
Ridley M (2020) How innovation works: and why it flourishes in freedom. Harper, New York

Rotz S, Duncan E, Small M, Botschner J, Dara R, Mosby I et al (2019) The politics of digital agricultural technologies: a preliminary review. Sociol Rural 59:203–229. https://doi.org/10.1111/soru.12233

Ruder SL, Wittman H (2025) Agricultural data governance from the ground up: exploring data justice with agri-food movements. Big Data Soc 12:20539517251330182. https://doi.org/10.1177/20539517251330182

Rural Utilities Service (2024) Notice of funding opportunity for the rural eConnectivity Program for Fiscal Year 2024. Available at: https://www.federalregister.gov/documents/2024/02/21/2024-03484/notice-of-funding-opportunity-for-the-rural-econnectivity-program-for-fiscal-year-2024

Sannino A, Engeström Y (2018) Cultural-historical activity theory: founding insights and new challenges. Cult Hist Psychol 14:43–56. https://doi.org/10.17759/chp.2018140305

Schnack A, Bartsch F, Osburg VS, Errmann A (2024) Sustainable agricultural technologies of the future: determination of adoption readiness for different consumer groups. Technol Forecast Soc Chang 208:123697. https://doi.org/10.1016/j.techfore.2024.123697

Schnebelin É, Labarthe P, Touzard JM (2021) How digitalisation interacts with ecologisation? Perspectives from actors of the French Agricultural Innovation System. J Rural Stud 86:599–610. https://doi.org/10.1016/j.jrurstud.2021.07.023

Shilomboleni H, Schnurr MA (2025) Are disruptive agricultural technologies compatible with agroecology? npj Sustainable Agric 3:21. https://doi.org/10.1038/s44264-025-00064-2

Smits R, Kuhlmann S (2004) The rise of systemic instruments in innovation policy. Int J Foresight Innov Policy 1:4–32. https://doi.org/10.1504/IJFIP.2004.004621

Spielman DJ, Ekboir J, Davis K, Ochieng CM (2008) An innovation systems perspective on strengthening agricultural education and training in sub-Saharan Africa. Agric Syst 98:1–9. https://doi.org/10.1016/j.agsy.2008.03.004

Spielman DJ, Ekboir J, Davis K (2009) The art and science of innovation systems inquiry: applications to sub-Saharan African agriculture. Technol Soc 31:399–405. https://doi.org/10.1016/j.techsoc.2009.10.004

Stitzlein C, Fielke S, Fleming A, Jakku E, Mooij M (2020) Participatory design of digital agriculture technologies: bridging gaps between science and practice. Rural Extens Innov Syst J 16:14–23

Tereso MJA, Abrahão RF, Gemma SFB, Montedo UB, Menegon NL, Guarneti JE, Ribeiro IAV (2012) Work and technological innovation in organic agriculture. Work 41:4975–4978. https://doi.org/10.3233/WOR-2012-0041-4975

Toke D, Breukers S, Wolsink M (2008) Wind power deployment outcomes: how can we account for the differences? Renew Sust Energ Rev 12:1129–1147. https://doi.org/10.1016/j.rser.2006.10.021

USDA (2019) A case for rural broadband: insights on rural broadband infrastructure and next generation precision agriculture technologies. United States Department of Agriculture. American Broadband Initiative. Available at: https://www.usda.gov/sites/default/files/documents/case-for-rural-broadband.pdf

Utterback JM, Pistorius C, Yilmaz E (2018) The dynamics of competition and of the diffusion of innovations. MIT Sloan School Working Paper, pp 5519–5518

Wang D, Du X, Sun J, Guo X, Chen Y (2018) Synergy of national agricultural innovation systems. Sustainability 10:3385. https://doi.org/10.3390/su10103385

Weersink A, Fraser E, Pannell D, Duncan E, Rotz S (2018) Opportunities and challenges for big data in agricultural and environmental analysis. Ann Rev Resour 10:19

Digital Agricultural Ecosystems: Scientific Landscape and Competency Perspectives

Ivo Pierozzi Junior, Stanley Robson de Medeiros Oliveira, Márcia Izabel Fugisawa Souza, Tércia Zavaglia Torres, and Luiz Manoel Silva Cunha

1 Introduction

The integration of digital technologies into agricultural and agri-food systems has driven an epistemological transformation, significantly reshaping how knowledge is generated, validated, shared, and applied in the agri-food system. This transformation is not merely technical, but conceptual, and is expressed through a variety of designations—such as *digital agriculture*, *agriculture 4.0*, *smart agriculture*, and *digital farm*—that reflect both disciplinary distinctions (e.g., agronomy, engineering, economics, information science) and specific empirical applications.

These terminologies represent more than a lexical evolution: they indicate structural shifts in how humans interact with technology in production domains and across food value chains. A clearer distinction between the terms Information and Communication Technologies (ICT) and Digital Technologies helps contextualize this evolution. ICT formally emerged in the late 1990s, particularly within educational and policy settings in the United Kingdom, to describe the convergence of computing and telecommunications. Digital technologies, in contrast, became consolidated in the second half of the twentieth century as digitalization fundamentally transformed the processes of data generation, storage, management, and dissemination—reshaping how information is produced, organized, and shared globally (Castell 2018; Rocha et al. 2020; Chapagai and Dhakal 2023; Hasbie et al. 2023).

Building on these technological trajectories, digital agricultural ecosystems have emerged as complex and interconnected sociotechnical arrangements that integrate digital technologies, institutional actors, production processes, agricultural

I. P. Junior (✉) · S. R. de Medeiros Oliveira · M. I. F. Souza · T. Z. Torres · L. M. S. Cunha
Embrapa Digital Agriculture, Campinas, São Paulo, Brazil
e-mail: ivo.pierozzi@embrapa.br; tercia.torres@embrapa.br

F. da Silveira et al. (eds.), *Agriculture 4.0*,
https://doi.org/10.1007/978-3-032-24552-6_13

practices, and data flows. Their purpose is to drive innovation, foster sustainability, and enhance knowledge generation and application across agri-food systems.

To avoid recurring ambiguities in terminology, this chapter adopts a clear conceptual distinction. Agricultural systems are used to denote primary production processes—encompassing the practices, techniques, and management carried within rural production units (Hendrickson et al. 2008; Le Gal et al. 2011). The term agri-food systems refers to the broader level, which encompasses not only primary production but also processing, distribution, consumption, and the impacts associated with food value chains (High Level Panel of Experts on Food Security and Nutrition 2017; Mausch et al. 2020). Digital agricultural ecosystems, in turn, are understood as transversal socio-technical arrangements that integrate digital technologies, data flows, institutional actors, and human competencies, spanning and transforming both agricultural systems and agri-food systems (Wolfert et al. 2017; Klerkx et al. 2019). This harmonization clarifies the different levels of analysis and provides a more precise understanding of how digitalization is reconfiguring production practices, value chains, and institutional environments.

Box 1 summarizes the conceptual distinctions adopted in this chapter, clarifying the respective scopes of agricultural systems, agri-food systems, and digital agricultural ecosystems.

Box 1

Concept	Scope	Definition
Agricultural systems	Primary production	Farming, livestock, agroforestry, and other production practices are carried out within rural production units
Agri-food systems	Full value chain	Broader networks encompassing production, processing, distribution, and consumption, along with associated social, economic, and environmental impacts
Digital agricultural ecosystems	Transversal socio-technical arrangements	The integration of digital technologies, data flows, institutions, and human competencies that transform both agricultural systems and agri-food systems

Such digital agricultural ecosystems rely on the articulation of physical, logical, and cognitive infrastructures that create dynamic environments for producing, analyzing, and applying scientific and technical knowledge (Rose and Chilvers 2018; Pierozzi Junior et al. 2023). Within this framework, *innovation corridors* (Guo et al. 2025) play a pivotal role. These are territorial and institutional configurations that connect research centers, public agencies, private companies, startups, digital platforms, and competence networks. They enable continuous multidirectional flows of data, technologies, and knowledge across geographic regions, economic sectors, and value chains, thereby reinforcing both digital transformation and sustainability goals within agri-food systems.

A set of enabling technologies—such as the Internet of Things (IoT), Big Data Analytics, Artificial Intelligence (AI), Cloud Computing, Edge AI, and machine learning—is now central to this transformation. These advancements not only automate processes and support complex data analysis but also ensure seamless interoperability among systems and platforms. More than increasing efficiency and productivity in agricultural systems, they are fundamentally reshaping the very foundations of scientific knowledge production and application across agri-food systems (Wolfert et al. 2017; Klerkx et al. 2019; Benos et al. 2021).

The digital revolution has also profoundly impacted traditional agricultural knowledge systems—historically based on empirical and local knowledge—by not only enhancing them with data-driven approaches but also by providing tools such as sensors, drones, and satellite imagery allowing large-scale, real-time data collection on variables such as soil quality, climate conditions, plant health, and animal behavior. These data are processed through machine learning algorithms and other computational techniques to generate insights that inform decision-making across all stages of the production cycle (Singh et al. 2024; Wu 2024).

This epistemological shift represents a new paradigm in which digitalization not only integrates multiple knowledge sources but also enables real-time decisions and fosters the sustainability and resilience of agri-food systems. Nonetheless, for digital agricultural ecosystems to realize their full potential, critical challenges must be addressed—including digital literacy, equitable inclusion of technology, robust information governance, and the development of appropriate professional competencies.

In this context, competencies are understood as the integrated set of knowledge, skills, and attitudes required to effectively engage with digital agricultural ecosystems. Their development is essential for ensuring inclusive access to digital innovation, institutional capacity-building, and the sustainable transformation of agricultural and agri-food systems.

Accordingly, this chapter explores digital agricultural ecosystems as an emerging techno-scientific domain by analyzing both their scientific landscape and the competency frameworks required for their consolidation. The methodology combines a contextual literature review with a structured conceptual and bibliometric analysis to map key topics, actors, and trends. The subsequent sections first present the methodological procedures adopted, then discuss the results of the bibliometric and semantic analyses, and finally examine enabling technologies, systemic infrastructures, and human competencies as the three pillars of sustainable and inclusive digital transformation. The chapter concludes by outlining emerging trends, research frontiers, and policy implications, offering a comprehensive agenda for future work.

2 Methodology

This chapter adopts a mixed-methods approach that combines bibliometric techniques, semantic analysis, and conceptual modeling to map the scientific landscape of digital agricultural ecosystems and to identify the competency frameworks required for their consolidation. The approach builds directly on the conceptual framework outlined earlier, clarifying the roles of agricultural systems, agri-food systems, and digital agricultural ecosystems. The main objective was to identify the knowledge domains, enabling technologies, and core competencies essential for consolidating this emerging paradigm within the agri-food systems. To this end, the study was guided by two central questions: (i) What technical and scientific domains currently define the scientific landscape of digital agricultural ecosystems? and (ii) What competencies—understood as the combination of knowledge, skills, and attitudes—are essential for their sustainability and inclusive development?

The investigation was organized into four methodological stages:

1. *Contextual Literature Review*: A targeted review of the scientific literature on digital agriculture, competency development, and innovation in agri-food systems was conducted to identify key concepts and trends.
2. *Corpus Construction and Processing*: A textual corpus was created using computational tools to extract terminological and semantic patterns from selected metadata.
3. *Visualization and Network Analysis*: Conceptual networks were visualized and analyzed using the *VOSviewer*[1] software, enabling the identification of thematic clusters and knowledge structures.
4. *Taxonomy and Vocabulary Development*: Controlled vocabularies and taxonomic structures were developed to map conceptual categories, knowledge domains, and associated competencies, supported by artificial intelligence tools.

The study combined quantitative bibliometric techniques with qualitative, interpretive content analysis. A total of 1750 documents were retrieved from the Scopus database, covering the period from 2015 to 2024, using search terms such as "*digital agriculture*," "*precision agriculture*," and "*IoT in agriculture*." This timeframe was chosen because it corresponds to the intensification of research on digital agriculture following rapid advancements in IoT, machine learning, and big data (Barman et al. 2024, 2025; Bhardwaj et al. 2025; Paudel et al. 2025). The inclusion criteria encompassed peer-reviewed journal articles, book chapters, books, conference papers, and review articles published in English, Portuguese, or Spanish.

[1] www.vosviewer.com/

From the retrieved dataset, the fields “title,” “keywords,” and “abstract” were extracted to compose the corpus. Text preprocessing was performed using the Nocodefunctions[2] web platform, which supports multilingual Natural Language Processing (NLP) and generates semantic graphs for integration with the *VOSviewer* software (van Eck and Waltman 2014). These tools enabled the construction of bibliometric maps and conceptual networks, which were subsequently interpreted using artificial intelligence-based assistants, such as *Perplexity*,[3] to ensure thematic consistency and conceptual depth.

The qualitative and semantic interpretation employed axial coding, with manual classification of clusters based on a review of a sample of 120 documents. For instance, clusters such as “*rural 5G*” (semantic latency: 0.87) and “*geographic coverage*” (0.72) were logically grouped under the category Essential Infrastructures. Another cluster concerning Technological Regulation included terms such as “*GDPR-Agri*” (0.68) and “*governance*” (0.61), highlighting regulatory challenges specific to the digitalization of agri-food systems. In this context, semantic latency was utilized as a metric of association strength, indicating the degree of relevance of each term to its cluster.

The final analytical layer involved cross-referencing the identified clusters with the taxonomy proposed by the Agritech Observatory (2024), a digital platform that consolidates technological trends, key challenges, and strategic opportunities in agri-food systems. This overlay generated hybrid semantic maps that revealed points of convergence and gaps in competency development, thereby supporting strategic planning in training and policy-making.

This analysis enabled the creation of a controlled vocabulary comprising 230 terms and the development of a multi-level taxonomy of competencies, categorized as

- Technical (e.g., IoT operation, drone piloting).
- Strategic (e.g., data interoperability, infrastructure orchestration).
- Interpersonal and cognitive (e.g., problem-solving, ethical awareness, lifelong learning).

This competency framework provides a substantive basis for capacity-building initiatives, public policies, and institutional strategies aimed at promoting an inclusive, sustainable, and ethically grounded digital transformation of agri-food systems.

In summary, the adopted methodology was designed to elucidate the conceptual architecture of digital agricultural ecosystems and to identify the competencies required for their implementation and governance, with explicit attention to the epistemological, technological, and sociopolitical dimensions highlighted throughout this chapter.

---

[2] https://nocodefunctions.com/

[3] https://www.perplexity.ai/

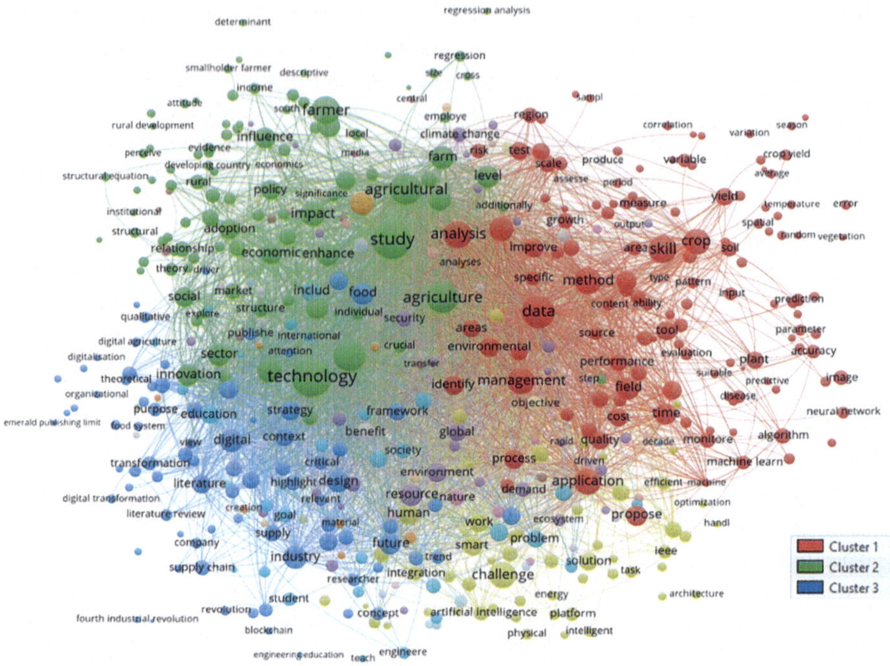

Fig. 1 Scientific landscape of digital agricultural ecosystems: conceptual co-occurrence map of concepts/terms based on the knowledge domain outlined in the textual corpus retrieved from the Scopus database

3 Analysis and Thematic Discussion

This section synthesizes the main evidence and insights drawn from the analyzed literature. It also presents the conceptual structure—referred to as the scientific landscape—of digital agricultural ecosystems and their associated competencies. Figure 1 illustrates this scientific landscape, generated using VOSviewer software, which provides a graphical representation of the analyzed textual corpus.

The analysis of this landscape reveals the complexity and multifaceted nature of the domain of digital agricultural ecosystems, organized around three major thematic cores identified through the conceptual segmentation of the network. The generated graph displays a network with 487 nodes—corresponding to terms or concepts extracted from the corpus—with sizes proportional to their relative frequency of occurrence, and 2134 edges representing normalized co-occurrences between terms. The analysis resulted in three main clusters, with a modularity score of $Q = 0.712$—a value indicating high internal cohesion of terms within each cluster and clear separation between groups, thereby demonstrating the consistency of the conceptual structure in the analyzed scientific landscape. Centrality and density metrics were applied to characterize the network; for instance, the term "machine learning" exhibited high centrality, connecting to 98 terms, while "interoperability" emerged as a bridging node, linking different thematic domains.

Thematic

Based on the conceptual structure of the network, the following section offers a detailed interpretation of the clusters, highlighting their thematic significance and implications for digital agricultural ecosystems.

3.1 Scientific Landscape and Thematic Domains of Digital Agricultural Ecosystems

The corpus analysis revealed a conceptual landscape structured into three clusters (conceptual groupings) characterized by high term co-occurrence (Fig. 1). These clusters constitute the first analytical axis of the study: the main thematic domains of digital agricultural ecosystems.

Cluster 1: Technological Core. Encompasses enabling technologies such as machine learning, IoT, Big Data, smart sensors, agricultural robotics, cloud and edge computing, and computer vision. Its technical density and degree of interconnectivity underscore its central role in the operation and development of digital agricultural ecosystems.

Cluster 2: Sustainability and Public Policies. Integrates terms such as circular economy, carbon footprint, Sustainable Development Goals (SDGs), and data governance. These themes associated within the water-energy-food nexus suggest integrated concerns with resource efficiency and environmental impact.

Cluster 3: Complex Systems Modeling. Covers multi-agent simulations, logistic curves of technology adoption, and systemic risks such as cascading failures. Its systemic approach is characterized by the interdependence of variables and the representation of agricultural and agri-food system dynamics in digital scenarios.

In addition to this conceptual segmentation—which constitutes the study's first analytical axis—the analysis also revealed a second axis, centered on the essential competencies required to operate, manage, and develop digital agricultural ecosystems. These competencies are organized into three complementary levels:

- *Technical competencies*: mastery of machine learning techniques; application of Internet of Things (IoT) technologies; and adherence to interoperability protocols.
- *Operational skills*: such as calibration of spectral sensors; implementation of multi-objective Decision Support Systems (DSS); and management of large volumes of data (Big Data).
- *Strategic behavioral attitudes*: adaptability to emerging technologies; cross-domain collaboration; and compliance with the General Data Protection Regulation (GDPR).

The analysis also highlighted the importance of essential infrastructure, particularly 5G communication networks in rural areas, which are critical for ensuring connectivity and low latency in agri-food systems operations (Tang et al. 2021).

From a practical standpoint, the findings reinforce the need for foundational public policies, notably the creation of regulatory testing environments (sandboxes), large-scale capacity-building initiatives, and incentives for the adoption of sustainable technological solutions. The use of tiny ML (embedded AI models) and open platforms emerges as a promising strategy to boost productivity and enhance digital sovereignty.

In summary, digital agriculture requires methodological approaches that integrate human competencies, technological infrastructure, and adaptive regulatory frameworks. This integration is essential to maximize the positive impacts of digital transformation, fostering more contextualized, resilient, and sustainable solutions.

3.2 Conceptual and Terminological Evolution

Although the analysis of the textual corpus revealed well-established concepts in the field of digital agricultural ecosystems, certain aspects of terminological evolution required a complementary interpretative approach. This included elements that predate the 2015–2024 time frame adopted in this chapter to construct the base corpus in the scientific landscape analysis of digital agricultural ecosystems. The conceptual trajectory of these ecosystems can be organized into three analytical phases, based on the recurrence of terms, resemanticization, and terminological hybridization.

Precursor Phase (1980–2010). This phase marks the initial adoption of digital technology in rural and agricultural systems. Concepts such as geoprocessing, Geographic Information Systems (GIS), and yield mapping emerged with the introduction of GPS in harvesters and spatial analysis tools. The transition from manual records to rudimentary digital systems led to the use of the term "electronic agriculture," while "precision agriculture" became established as the dominant concept (Franzen and Mulla 2015).

Convergence Phase (2010–2020). This phase is characterized by the diffusion of Agriculture 4.0, with a significant increase in terms such as "Big Data" (with an estimated relative growth of 320% after 2015), "IoT," "intelligent sensors," and "machine learning." During this period, expressions like "digital agriculture" were semantically integrated, often extended to include broader agri-food systems. Hybrid terms such as "AgroAPI" and "smart farmers" also began to emerge (França et al. 2020).

Complexification Phase (2020–present). This phase is marked by the emergence of neologisms and conceptual redefinitions. Notable examples include "tiny ML," "bioalgorithms" (the integration of biological data and artificial intelligence), and "quantum agriculture" (the experimental use of quantum computing in logistics and production chains). Previously well-established concepts such as "sustainability" and "interoperability" have incorporated new technical and regulatory dimensions. Meanwhile, blockchain has become established as essential infrastructure for ensuring traceability, security, and integrity of agri-food system data flows (Van Wassenaer et al. 2021; Han and Siebert 2022; Mo et al. 2024; Ryan et al. 2024; Souza et al. 2024).

Box 2 summarizes this evolution by highlighting emerging concepts, obsolete terms, and paradigmatic shifts, without requiring an extensive description of each.

Box 2: Conceptual Evolution of Digital Agricultural Ecosystems

Phases	Emerging terms	Outdated terms	Conceptual transformations
Precursor (1980–2010)	Electronic agriculture; GIS	Analog agriculture; manual recordkeeping	Initial digitalization; site-specific input management
Convergence (2010–2020)	Agriculture 4.0; Big Data; IoT	Traditional statistics; statistical learning	Network integration; process automation
Complexification (2020–)	Agriculture 5.0; tiny ML; bioalgorithms; agro-metaverse; quantum agriculture; blockchain	Classical sustainability	Technoscientific hybridization; digital immersion

Additional terms such as "cellular agriculture," "synthetic farm," and "agro-metaverse" illustrate the semantic expansion of the field and are included as examples due to their relevance at the frontiers of innovation (Stephens et al. 2018; Sargent et al. 2022; Nguyen et al. 2024).

The analysis also highlights key terminological challenges, including:

- *Critical polysemy*: approximately 23% of the terms analyzed exhibit multiple contextual meanings—for example, "Agriculture 4.0" and "Digital Agriculture," which are often used interchangeably, despite representing distinct approaches (Wolfert et al. 2017; Klerkx et al. 2019).

- *Lexical fragmentation*: the proliferation of competing neologisms such as "AgroIoT," "Smart Farming 2.0," and "Digital AgriFood Systems" hinders terminological standardization (Bronson and Knezevic 2016; Eastwood et al. 2019).
- *Regulatory anachronism*: on average, there is a gap of 3–5 years between the emergence of new technologies and their regulatory integration, which undermines governance and interoperability (Stilgoe 2020).

This evolution indicates the consolidation of a specialized metalexicon, in which each new term reflects transformations in the modes of knowledge production, management, and organization within agri-food systems (Klerkx et al. 2019). At the same time, there is a coexistence of zones of hyper-specialization—for example, in the application of algorithms for computer vision—and conceptual gaps, particularly regarding data ethics and digital governance (Wolfert et al. 2017). This tension reveals a misalignment between the pace of technological innovation and the terminological maturation of digital agricultural ecosystems.

3.3 Structuring Elements of Digital Agricultural Ecosystems: Human Competencies and Infrastructure

The sustainability and consolidation of digital agricultural ecosystems depend on two fundamental structuring elements: (i) human competencies and (ii) enabling infrastructures.

Human Competencies

Within digital agricultural ecosystems, human competencies play a central and multifaceted role that extends beyond conventional technical expertise. Farmers, technicians, and rural managers must cultivate skills ranging from searching for, interpreting, and applying digital information to analyzing data that supports strategic decision-making and fostering innovation in production processes (Dara et al. 2022). In addition to these technical skills, attitudes such as curiosity, openness to innovation, flexibility, creativity, and a commitment to lifelong learning are essential to succeed in dynamic, technology-driven environments. When combined with digital infrastructures—such as connectivity networks, smart sensors, data platforms, and decision support systems—these competencies become critical for the effective operation, adaptation, and continuous evolution of digital agricultural ecosystems, fostering a smarter, more sustainable, and competitive agriculture (Charatsari et al. 2023).

Enabling Infrastructures

Enabling infrastructures comprise connectivity (e.g., 5G networks, edge and cloud computing), smart sensor systems, interoperable data repositories, and digital platforms facilitating the integration and exchange of information. These infrastructures form the technological backbone of digital agricultural ecosystems, ensuring interoperability, scalability, and resilience across diverse contexts. Their availability is especially critical in rural areas, where robust connectivity determines the ability of farmers and institutions to participate fully in digital innovation (Rose and Chilvers 2018; Klerkx et al. 2019).

4 Results and Discussion

4.1 *Quantitative Structure and Metrics of the Scientific Landscape*

The mapping of competencies—understood as the combination of knowledge, technical skills, and behavioral attitudes—associated with digital agricultural ecosystems, based on the analyzed literature and the conceptual visualization generated by VOSviewer, enabled the identification of interconnection patterns among concepts and the grouping of key competencies into three major integrated dimensions: technical, sustainable, and strategic.

The technical dimension focuses on competencies aligned with the conceptual domain identified as the Technological Domain. These competencies involve the use and management of technologies applied to agricultural systems, such as IoT, smart sensors, and data analysis. Specific skills include the operation of smart sensors and the integration of edge-cloud systems (He et al. 2024), the application of machine learning algorithms, and the use of Big Data tools capable of processing over 1.2 terabytes per hour (Benos et al. 2021). This dimension also includes skills in agricultural automation, such as drone programming and the operation of Variable Rate Technology (VRT) machinery with geospatial precision (Sharma et al. 2025).

The sustainable dimension, associated with the conceptual domain of Sustainable Management and Public Policies, encompasses competencies that integrate technical knowledge, practical skills, and behavioral attitudes aimed at optimizing resource use and mitigating environmental impacts within agri-food systems. This includes, for example, the application of circular economy principles, which can reduce input usage by 15–30% (Panigrahi et al. 2024). It also requires competencies in material flow analysis and data-driven decision-making. Moreover, it involves the ability to monitor environmental impacts using Decision Support Systems (DSS) to calculate indicators such as water footprint and carbon footprint (Zhai et al. 2020; Lotfy et al. 2024), particularly in forecasting as water footprint of agricultural crops under different agroclimatic conditions.

The strategic dimension, associated with the conceptual domain of Complex Systems Modeling and Data Governance, encompasses competencies involving the integration of technical knowledge, systemic thinking, and decision-making abilities within complex organizational contexts. The conceptual mapping suggests that these competencies are essential for addressing the challenges of digital transformation in the agri-food systems, particularly in scenarios characterized by high uncertainty and continuous innovation (Klerkx et al. 2019).

Within this dimension, particular emphasis is placed on the ability to adapt to emerging protocols—updated biannually—and to manage interoperability among digital platforms, with up to 92% compliance with Open Geospatial Consortium (OGC) standards, which are designed to support data integration in agri-food systems (Calvet et al. 2022). Additionally, nodes with high betweenness centrality in the conceptual network represent competencies that act as bridges between technical knowledge and farmers' tacit knowledge and practices, as well as the ability to translate agronomic data into effective operational decisions in the field (Charatsari et al. 2023).

The development of these strategic competencies is strongly linked to the increasing technological reliance that directly impacts agricultural productivity. Technologies such as drones equipped with GIS and LiDAR, predictive analytics using Python and R, and IoT management through protocols like LoRaWAN and Sigfox contribute to more efficient input usage, reduced losses, and improved water resource management. However, the challenges of digital transformation within agri-food systems go beyond the technological and strategic dimensions—related to the operation and management of digital systems—as well as the sustainable dimension, which focuses on resource efficiency and environmental impact mitigation. Therefore, the consolidation of digital agricultural ecosystems requires professionals capable of integrating technical knowledge, strategic foresight, and sustainable practices to enhance the productivity, resilience, and sustainability of modern agricultural and agri-food systems (Klerkx et al. 2019; Rivera et al. 2023).

4.2 Essential Infrastructures and Systemic Competencies

Integrated infrastructures play a pivotal role in the design, implementation, and governance of digital agricultural ecosystems, enhancing the efficiency, sustainability, and competitiveness of agri-food systems. Key components critical to the effective functioning of these ecosystems include: IoT networks, 4G/5G communication technologies, integrated digital platforms, cloud storage, smart sensors with Edge AI, and blockchain-based traceability systems (Klerkx et al. 2019; Avila and Barbosa 2024).

IoT networks connect sensors, machinery, and devices, enabling process automation and real-time monitoring. Advanced communication networks ensure efficient and low-latency data transmission, which is crucial for operations in rural and remote areas. Integrated digital platforms and Agricultural Digital Twins aggregate and analyze environmental and operational data, allowing for simulations and holistic management of agricultural and agri-food systems (Wolfert et al. 2017; Klerkx et al. 2019).

Cloud computing enables the management of growing volumes of agri-food system data, supporting complex analyses using AI and Big Data techniques. Smart sensors equipped with Edge AI capabilities enable on-site data collection and processing, optimizing real-time responses. Finally, blockchain-based traceability systems ensure transparency, data integrity, and sustainability, particularly in the monitoring of environmental and carbon indicators (Kalyani and Collier 2021).

Beyond serving as technical assets, these critical infrastructures also function as nonhuman systemic competencies essential to digital agricultural systems. Just as human knowledge and behavior enable the practical application of technology, these structures fulfill three vital roles:

- Technical interoperability, ensured by networks such as rural 5G, with 87% coverage and latency below 50 ms, acting as the system's "communication pathways."
- Cognitive processing, supported by cloud computing platforms such as Azure FarmBeats, which serve as "operational memory," processing up to 1.2 terabytes per hour using machine learning algorithms.
- Adaptive resilience, provided by high-precision spectral sensors (with an error margin of $<\pm 2.5$ nm), functioning as "digital senses," capable of detecting water or nutrient stress before it becomes visible in the crops (Wolfert et al. 2017).

This perspective frames infrastructure as a form of biosystemic capacities—that is, nonhuman competencies functioning analogously to the vital functions of biological systems. These are technical capabilities embedded within systems, devices, and platforms, serving as essential components of digital agricultural ecosystems. Just as human competencies transform knowledge into practice through skills, expertise, and attitudes, infrastructures enable the execution of fundamental tasks in agro-digital system operations. They operate as the system's functional "organs": sensors act as digital senses, the cloud serves as operational memory, and communication networks function as neural pathways—demonstrating that the effective performance of the ecosystem relies on the synergy between human and techno-infrastructural capacities (Latour 2005).

The interaction between these infrastructures and human competencies—such as data analysis and decision-making—constitutes the connective tissue of digital agricultural ecosystems. The absence of any infrastructural layer, such as connectivity, creates a critical gap that compromises the performance, adaptability, and resilience of the production system (Eastwood et al. 2019).

4.3 Challenges to the Consolidation of Digital Agricultural Ecosystems

The qualitative analysis of the textual corpus revealed structural challenges that impede the development and consolidation of digital agricultural ecosystems within agri-food systems. One of the main obstacles is the lack of infrastructure and connectivity in rural areas, particularly on small farms and in remote regions. This deficiency makes it unfeasible to use sensors, digital platforms, and even basic operations such as issuing electronic invoices or accessing market information on a daily basis.

Connectivity is a technical prerequisite for the functioning of digital agricultural ecosystems. Its absence constitutes a structural barrier that can only be addressed through robust policies and targeted investments in digital inclusion in rural areas. This challenge is further intensified by the high costs associated with implementing and maintaining technologies, particularly for small- and medium-sized producers with limited access to credit and financing (Eastwood et al. 2019; Klerkx et al. 2019).

Another set of challenges is related to the human and organizational dimension. The lack of technical training and digital literacy undermines the effective adoption of available technologies within digital agricultural ecosystems. This scenario is further aggravated by the limited availability of technical assistance programs, continuing education, and digital inclusion initiatives. Compounding the issue is resistance to change, often driven by cultural factors such as attachment to traditional practices, distrust of technology, and lack of awareness regarding its potential benefits (Eastwood et al. 2019).

At the systemic level, additional barriers are also identified: limited access to public policies and credit lines, which constrain investment capacity; the lack of standardization and incompatibilities between systems and platforms, which hinder interoperability and compromise the efficient flow of data across agri-food systems; and uncertainty surrounding the sharing of sensitive data, which raises concerns about privacy and the potential misuse of information (Bronson and Knezevic 2016; Klerkx et al. 2019).

These challenges highlight the urgency of foundational public policies, sustained investments in digital infrastructure, large-scale technical training, and financial incentives to promote rural digital inclusion and enhance the sustainability of agricultural and agri-food systems in the face of ongoing technological transformation (Rose and Chilvers 2018).

5 Emerging Trends and Technologies

In recent years, the digital transformation of agriculture has advanced beyond foundational technologies such as precision farming and remote sensing. A new generation of tools and paradigms—ranging from intelligent automation to immersive simulations—is reshaping the operational, organizational, and epistemological dimensions of agricultural and agri-food systems. These innovations not only enhance productivity and efficiency but also open avenues for reimagining the relationship between humans, technology, and digital agricultural agroecosystems (Rose and Chilvers 2018; Klerkx et al. 2019). The following sections are organized to reflect this progression—from advanced paradigms and human–machine interfaces to autonomy, data collaboration, impact metrics, and resilience.

5.1 Agriculture 5.0 and Bio-algorithms

Agriculture 5.0 is emerging as an advanced paradigm that builds on the foundations of precision agriculture (4.0) by emphasizing self-optimization, contextual adaptation, and the synergistic integration of biological and computational systems. It represents a shift toward deeper cyber-physical convergence, characterized by decentralized intelligence, hyperconnectivity, and decision systems increasingly informed by biological data streams.

A core feature of this paradigm is the incorporation of bio-algorithms—computational techniques inspired by natural and biological systems. These include artificial neural networks (Kamilaris and Prenafeta-Boldú 2018), genetic algorithms (Katoch et al. 2021), artificial immune system models (Kidd 2019), and swarm intelligence (Nguyen 2024), all of which have been explored in agricultural and multi-robotic applications. These approaches are particularly well suited to addressing the nonlinear, multiscale, and dynamic complexities of agricultural systems—especially under conditions of uncertainty and resource constraint.

Agriculture 5.0 increasingly emphasizes the integration of genomics, phenomics, and environmental data, enabling enhanced precision phenotyping, adaptive crop management, and resilient agroecosystem design (Klerkx et al. 2019; Roth et al. 2025). This integration is not merely technical but also epistemological, as it transforms how knowledge is generated and operationalized in agricultural research and innovation.

Recent advances have demonstrated the effectiveness of evolutionary computation and swarm-based optimization in supporting climate-smart agriculture, optimizing nutrient uptake, and improving genotype selection under stress conditions such as drought and salinity (Maraveas et al. 2023). These bio-algorithms can be deployed in real-time decision-making environments, particularly when integrated with edge computing architectures and multi-omics platforms, thus enabling the fine-tuning of input applications, early disease detection, and predictive responses to environmental perturbations.

By drawing on principles of biomimicry and systems biology, Agriculture 5.0 fosters the design of adaptive, resilient, and efficient production systems within agricultural and agri-food systems, aligned with agroecological values and long-term sustainability goals.

5.2 Augmented and Virtual Reality (AR/VR)

Augmented Reality (AR) and Virtual Reality (VR) technologies are gaining momentum in agricultural innovation as tools that enhance visualization, simulation, and human–computer interaction in digital agricultural ecosystems. AR superimposes digital content onto real-world settings, while VR immerses users in fully virtual simulations. In agriculture, both technologies are being explored for applications such as training, crop management, equipment maintenance, and remote advisory services (Hurst et al. 2021; Anastasiou et al. 2023).

For instance, VR-based training platforms have shown significant potential for capacity building in farm machinery operation, as they allow users to practice in safe and controlled environments without risking physical resources (Cutini et al. 2023). In parallel, AR applications enhance real-time decision-making in the field by overlaying sensor data directly onto crops, thereby guiding users in tasks such as pest detection, irrigation optimization, and nutrient management (Huong et al. 2023).

Beyond technical applications, immersive technologies also offer pedagogical benefits, particularly in agricultural extension services and rural education, where they facilitate experiential learning and knowledge co-creation among diverse stakeholders. As noted by Naudé et al. (2024), these tools are particularly valuable in participatory planning and scenario modeling, enabling farmers and policymakers to jointly explore alternative land-use strategies or simulate the impacts of climate change.

Despite their potential, barriers remain, including high development costs, limited rural connectivity, and the need for interdisciplinary content design that is culturally and contextually appropriate. Continued research and investment are essential to fully integrate AR/VR into inclusive digital agricultural ecosystems.

5.3 Integrated Autonomous Systems

The convergence of robotics, drones, and the Internet of Things (IoT) has driven the emergence of integrated autonomous systems in agricultural and agri-food systems. These technologies collectively support automated sensing, data transmission, actuation, and real-time decision-making, reducing dependence on human labor and enhancing operational efficiency (Shamshiri et al. 2018).

Autonomous systems are being deployed in tasks such as precision seeding, targeted spraying, automated weeding, and harvesting, with significant implications for labor productivity, input efficiency, and environmental sustainability (Bechar and Vigneault 2016). For example, swarm robotics and drone fleets are now capable of coordinating actions across large farms, improving coverage and minimizing redundancy.

These systems are increasingly integrated into cloud-based platforms, enabling remote monitoring, predictive analytics, and adaptive feedback loops. When combined with edge computing, autonomous units can make localized decisions—such as altering irrigation patterns or adjusting harvesting routes—without relying solely on centralized control (Kalyani et al. 2023; He et al. 2024).

Nevertheless, the widespread adoption of these systems raises concerns about technical standardization, data governance, maintenance complexity, and the digital divide between large agribusiness and smallholder farmers. Ensuring accessibility, interoperability, and resilience across diverse agroecological contexts within agri-food systems remains a key challenge for future development.

5.4 Open Platforms and Collaborative Data Networks

Open platforms and collaborative data ecosystems constitute critical infrastructures for democratizing agricultural innovation, enabling interoperability, and fostering multi-actor collaboration within digital agricultural ecosystems. They serve as shared environments for the collection, analysis, and distribution of agri-food system data, typically relying on standardized APIs, metadata frameworks, and modular interfaces (Wolfert et al. 2017; van der Burg et al. 2021).

By promoting transparency, reusability, and co-ownership, these platforms create synergies among farmers, researchers, agri-tech developers, and public institutions. Initiatives such as GODAN (Global Open Data for Agriculture and Nutrition), AgGateway, and the EU Code of Conduct for Agricultural Data Sharing illustrate the growing momentum toward open innovation and interoperable digital infrastructures, reinforced by policy-oriented guidelines (Musker and Schaap 2018; van der Burg et al. 2021; Ryan et al. 2024).

In practice, collaborative networks support functions such as benchmarking, disease outbreak alerts, market intelligence, and crowdsourced problem-solving. At the policy level, they contribute to evidence-based decision-making, improve resource targeting, and foster more equitable value chains within agri-food systems (Ryan et al. 2024; Yeo and Keske 2024).

Nevertheless, persistent challenges remain regarding data privacy, security, licensing models, and the fair distribution of benefits. Addressing these issues requires robust governance mechanisms, participatory design processes, and international cooperation to ensure that open platforms serve the public good without reinforcing existing asymmetries (van der Burg et al. 2021; Ryan et al. 2024; Yeo and Keske 2024).

5.5 Social, Economic, and Environmental Impact Metrics

While digital technologies in agricultural ecosystems are often celebrated for their technical efficiency, there is a pressing need to evaluate their broader impacts—including social equity, economic viability, and environmental sustainability. Conventional performance indicators such as yield or input-use efficiency are insufficient to capture the complex transformations associated with digital transitions (Bronson and Knezevic 2016; Rotz et al. 2019).

Recent research highlights the importance of inclusive metrics frameworks that assess impacts across multiple domains: labor displacement or transformation, gender equity, data sovereignty, carbon emissions, biodiversity conservation, and food system resilience (FAO 2021; Rotz et al. 2019). Such metrics must be context-sensitive, capable of reflecting regional disparities, local knowledge systems, and historical inequities.

In response, international efforts such as the FAIR Digital Objects framework, the Digital Inclusion Index, and the One CGIAR Impact Assessment Platform are advancing tools to capture both quantitative and qualitative outcomes of digital agriculture within agri-food systems (CGIAR Independent Advisory and Evaluation Service 2022).

Nevertheless, the design and application of such metrics require interdisciplinary collaboration, methodological transparency, and capacity building among institutions and communities to ensure that results are interpreted effectively and translated into actionable strategies.

5.6 Resilience and Risk Management

Digital agricultural ecosystems hold significant potential for enhancing systemic resilience by enabling early detection, adaptive planning, and risk-informed interventions. In an era marked by climate variability, pest outbreaks, and market volatility, such capabilities are increasingly indispensable (Lyu et al. 2025; Intergovernmental Panel on Climate Change 2023). Technologies like remote sensing and AI-based forecasting models can contribute to early warning systems and contingency planning, reducing vulnerabilities and supporting recovery (Buchaillot et al. 2022). Blockchain-based traceability systems also play a role in strengthening transparency and resilience in agri-food supply chains (Demestichas et al. 2020). For example, mobile-based platforms are used in sub-Saharan Africa to alert farmers to fall armyworm outbreaks, enabling timely responses and reducing crop losses (Buchaillot et al. 2022). Moreover, resilience is not purely technical—it involves institutional robustness, social capital, and inclusive governance structures. Digital tools can enhance these dimensions when designed with attention to local participation, cultural contexts, and ethical frameworks (Béné et al. 2016; Folke et al. 2016).

However, excessive dependence on centralized systems, poorly designed AI models, or proprietary platforms can introduce new vulnerabilities. Therefore, future efforts should prioritize redundant, decentralized, and open systems capable of adapting to unforeseen disruptions and reinforcing community agency in risk management.

6 Future Research Perspective

Building on the technological trajectories outlined in the previous section, future research in digital agricultural ecosystems must adopt a critical, systemic, and transdisciplinary approach. Evaluating innovations solely through the lens of productivity or technological feasibility is no longer sufficient. Instead, research should address the complex interplay among sociotechnical systems, data governance regimes, agroecological contexts, and institutional dynamics. Accordingly, the subsections below outline a program that integrates conceptual issues, policy frameworks, collaborative infrastructure, human capacities, impact assessment, and technological foresight.

6.1 Integration of Traditional and Digital Agricultural Systems

As digital agricultural ecosystems advance, a key research frontier involves bridging algorithmic decision-making and ancestral agricultural knowledge. Smallholders, Indigenous peoples, and traditional communities often operate under socioecological logics that are poorly represented by conventional data-driven models. Empirical evidence indicates that digital technologies can be incorporated into agroecological and agricultural systems in ways that respect and reinforce local practices, as illustrated by a case study of a biodynamic vineyard in Italy (Meesala and Brunori 2025). Future studies should focus on co-design methodologies, participatory sensing, and ontologically pluralistic frameworks that enable the respectful integration of digital tools into local epistemologies.

This entails rethinking digital agriculture not as a one-size-fits-all innovation pipeline, but as a hybrid knowledge infrastructure, where semantic interoperability between worldviews—including agroecological, cosmological, and computational—becomes a methodological imperative (Khatoon and Ahmed 2021). Empirical research should assess how localized digital solutions (e.g., mobile apps, low-power IoT sensors, open geospatial platforms) can enhance resilience without reinforcing dependency, marginalization, or cultural homogenization, as argued in recent work exploring ethical, legal, and sociological trade-offs in digital agricultural ecosystems (Milian Gómez and Byttebier 2025). Here, comparative case studies and participatory evaluation will be critical to translate these principles into operational practices.

6.2 Public Policy, Incentives, and Regulatory Frameworks

Digital agricultural ecosystems are embedded in policy architectures that determine their use and accountability. Evidence from OECD[4]-related analyses indicates that while digital tools can improve the policy cycle, they also introduce governance

[4] Organisation for Economic Co-operation and Development. Available at: https://www.oecd.org/

risks that require safeguards for trust, privacy, and interoperability (McFadden et al. 2022).

Regulatory environments increasingly shape agricultural and agri-food data flows. Beyond general data-protection regimes such as the *General Data Protection Regulation* (GDPR) (European Union 2016) and Brazil's *Lei Geral de Proteção de Dados Pessoais* (LGPD) (Brasil 2018), the *EU Data Act* (applicable from 12 September 2025) establishes access and sharing rights for data generated by connected products and related services—rules that directly affect farm IoT equipment and platform ecosystems. Sector-specific instruments, such as the *EU Code of Conduct on Agricultural Data Sharing by Contractual Agreement*, complement these laws by clarifying contractual rights and responsibilities for agri-data (European Commission 2025).

Key topics for future research include:

- Comparative analysis of national strategies for digital agriculture and digital agricultural ecosystems, including how countries operationalize e-agriculture visions and action plans and align incentives across ministries and agencies.
- Interplay between data-protection laws (e.g., GDPR, LGPD) and agricultural data flows, focusing on consent, purpose limitation, portability, joint controllership in platforms, and the interface with sectoral sharing codes.
- Implications of platformization and algorithmic regulation for rural sovereignty, including lock-ins and power asymmetries in digital farming platforms and data-driven agri-environmental governance.
- Assessment of public–private partnerships (PPPs) in digital agricultural ecosystems, particularly for rural connectivity, digital infrastructure, and interoperable platforms that enable inclusive adoption.

Special attention should be given to the role of international organizations and multilateral forums (e.g., FAO,[5] ITU,[6] OECD, World Bank) in setting norms, funding mechanisms, and interoperability protocols—often amid representation gaps for small-scale food producers. Recent reviews underscore the need to strengthen trust (privacy, security, fair contracts), ensure meaningful participation, and monitor the actual uptake of voluntary data-sharing codes in the agri-food sector.

6.3 Collaborative Innovation and Open Data Ecosystems

Open innovation and data commons are central to the democratization of digital agricultural ecosystems. However, their operationalization raises questions about equity, consent, benefit sharing, and epistemic justice. Future research should examine governance models for collaborative innovation infrastructures, such as

[5] Food and Agriculture Organization of the United Nations. Available at: https://www.fao.org/

[6] International Telecommunication Union. Available at: https://www.itu.int

living labs, regional data spaces, cooperative cloud architectures, and decentralized platforms (Janssen et al. 2020).

Empirical studies are needed to:

- *Map institutional configurations* that foster trust and co-creation, including how governance arrangements, actor roles, and decision pathways (e.g., cooperatives, public–private partnerships, living labs) enable transparent rules, shared benefits, and durable collaboration.
- *Analyze incentive structures* for farmer participation, testing which mixes of monetary rewards, data services, technical support, recognition, and risk-sharing drive uptake and sustained engagement across farm sizes and contexts.
- *Evaluate open-licensing models* across regulatory environments, examining how licenses—e.g., Creative Commons (BY, CC0), Open Databases License (OdbL; attribution + share-alike), and bespoke data-sharing agreements—interact with data-protection, intellectual property, and competition laws to shape legal clarity, transaction costs, and reuse outcomes.
- *Examine data-stewardship roles and community data trusts*, including how stewards/trusts govern access, consent, and benefit sharing, and how their performance compares with firm-centric models in preventing extractive practices.

Moreover, research must address the political economy of data asymmetries, particularly when digital resources are controlled by large agritech corporations with opaque algorithms and closed infrastructures (Rotz et al. 2019; Fraser 2022). Attention to anticompetitive risks and vendor lock-in will be essential to align openness with equitable outcomes.

6.4 Human Skills and Capacity Development

The success of digital agricultural ecosystems depends on human agency, institutional readiness, and pedagogical innovation. Future research should identify and assess the digital competencies required across roles and contexts, including:

- *Data literacy and critical data interpretation*, including capabilities for data-quality assessment, provenance tracking, handling bias and uncertainty, and decision-oriented visualization.
- *Algorithmic accountability and explainability*, including transparent documentation (e.g., model cards/datasheets), auditable pipelines, impact assessments, and user-recourse mechanisms.
- *Ethical AI governance*, including privacy and consent management, fairness and nondiscrimination, human oversight, and benefit-sharing mechanisms.
- *Interdisciplinary collaboration* across agronomy, computer science, and social sciences, including co-design practices, cross-training curricula, shared vocabularies/ontologies, and participatory evaluation methods (Klerkx and Rose 2020).

Research should also prioritize gender-responsive, culturally appropriate, and inclusive training pathways, ensuring that marginalized groups are not excluded from digital transitions (Status of digital 2022). This requires the co-development of lifelong learning ecosystems that blend formal and informal education and leverage digital pedagogies—such as microlearning, gamification, and mobile learning hubs. Evaluative designs should track skills acquisition, transfer to practice, and organizational outcomes to close the loop between the capacity-building process and impact.

Finally, studies should examine the institutional barriers to capacity building—such as bureaucratic inertia, technocratic bias, or underinvestment in rural education infrastructure—and test organizational and policy interventions to overcome them.

6.5 Multidimensional Impact Assessments

There is an urgent need to move beyond reductionist performance metrics and adopt multidimensional frameworks that capture the systemic impacts of digitalization across agri-food systems. Research should develop and operationalize indicators spanning: labor dynamics (automation, deskilling/reskilling, new form of employment), environmental externalities (emissions, e-waste, energy use), data justice (ownership, access, commodification, surveillance), and institutional robustness and democratic control—an agenda consistent with foundational critiques of big data in food and agriculture and analyses of power asymmetries in smart farming (Bronson and Knezevic 2016).

Emerging initiatives provide useful starting points but require empirical validation, contextual adaptation, and stakeholder engagement: the FAIR Guiding Principles (for data stewardship and reusability), the FAIR Digital Objects concept (for machine-actionable, typed digital objects), digital inclusion indices (for access, skills, and readiness), and One CGIAR's results/impact dashboards (for outcome tracking and comparability) (Wilkinson et al. 2016; De Smedt et al. 2020; CGIAR Independent Advisory and Evaluation Service 2022; Status of digital 2022). Combining quantitative indicators with qualitative narratives will better capture unintended consequences and distributional effects.

Mixed-method designs—combining participatory indicator development, longitudinal studies, and ethnographic insights—are essential to ensure that transitions are not only measurable but also meaningful to those affected (Rotz et al. 2019; Klerkx and Rose 2020; Status of digital 2022).

6.6 Foresight on Emerging Technologies in Agriculture

A strategic research agenda must include technology assessment and anticipatory governance of cutting-edge innovations:

- *TinyML (Tiny Machine Learning):* enables ultra-low-power, on-device inference for local decision-making; standardized community benchmarks (MLPerf Tiny) now evaluate accuracy, latency, and energy efficiency (Banbury et al. 2021).
- *Agro-metaverse (XR/immersive environments)*: exploratory uses for collaborative scenario modeling, training, and extension are emerging, but the evidence base remains nascent and context-dependent (Naudé et al. 2024).
- *Quantum computing for agriculture*: early studies suggest potential for optimization and modeling (e.g., supply chains, remote-sensing analytics), yet costs, maturity, and energy implications remain open questions (Maraveas et al. 2024).
- *Synthetic biology and CRISPR-based decision support; neuro-symbolic AI for agroecological reasoning*: promising avenues that require domain-specific evaluation frameworks addressing risk, ethics, and context of use (FAO 2022; Tanny et al. 2023; DeLong et al. 2024).

These technologies carry significant ethical, energy, scalability, and equity concerns. Interdisciplinary foresight—linking socio-technical transition thinking with responsible innovation—can help anticipate path dependencies, unintended consequences, and regulatory gaps, and regulatory gaps, aligning trajectories with agroecological transitions and the Sustainable Development Goals (Rotz et al. 2019; Klerkx and Rose 2020). Scenario exercises and sandbox-style pilots can operationalize this foresight before large-scale deployment.

6.7 Final Considerations

The future research agenda outlined here underscores that digital agriculture should be explicitly reframed as digital agricultural ecosystems, understood as a socio-technical transformation rather than merely a technological upgrade. This reframing calls for a reflexive approach—continuous critical examination of assumptions, methods, data, and impacts—combined with interdisciplinarity and participatory governance, to foster equitable innovation pathways that are ecologically grounded, epistemologically diverse, and ethically robust. In practice, this means coupling infrastructure and policy advances with capacity building and inclusive evaluation, ensuring that benefits are equitably distributed across heterogeneous agri-food contexts.

7 Conclusions

This chapter offered an integrated analysis of digital agricultural ecosystems through the lenses of scientific-landscape mapping, conceptual modeling, and competency-oriented perspectives. Drawing on a corpus of 1750 scientific articles and a systematic bibliometric approach using VOSviewer, it illuminated core research domains, persistent knowledge gaps, and the evolving structure of innovation trajectories in digital agriculture.

The scientific mapping reveals a field anchored by high-centrality concepts—such as smart farming, the Internet of Things (IoT), and artificial intelligence—while simultaneously exposing underexplored areas, notably data governance, interoperability, and socio-environmental impacts. Thematic clusters around frontier technologies (e.g., blockchain and quantum computing) signal a turn toward more complex and at times speculative forms of digital transformation. These dynamics challenge traditional epistemologies and call for new modes of collaboration among scientific, technological, and institutional actors.

The chapter advances the view that digital infrastructure is not merely a technological enabler but a structuring capacity that co-evolves with human competencies. In this sense, connectivity, open-data platforms, edge computing, and immersive interfaces are integral to knowledge flows and decision-making—particularly when coupled with capacity building, participatory governance, and ethical design.

Identified trends—such as Agriculture 5.0, bio-algorithms, augmented and virtual reality, integrated autonomous systems, and open innovation platforms—underscore a dynamic, multifaceted research frontier. These trends reinforce the need to transcend purely technical assessments and to incorporate metrics that evaluate the social, environmental, and ethical dimensions of innovation. An explicit emphasis on risk management, resilience, and inclusive design emerges as foundational for advancing sustainable digital agricultural ecosystems.

Looking ahead, the chapter proposes a systemic, interdisciplinary, and inclusive research agenda across six priority domains: (1) integration of traditional and digital agricultural systems; (2) policy, incentives, and regulation; (3) collaborative innovation and open-data ecosystems; (4) human competencies and capacity development; (5) multidimensional impact assessment; and (6) anticipatory research on frontier technologies. Each domain underscores contextual relevance, sound governance, and equitable access to technological benefits—especially in regions marked by structural inequalities.

Finally, the chapter argues that digital agricultural ecosystems should be approached as a socio-technical transformation, rather than a mere technological upgrade, one that requires a critical, reflexive stance, transparency, and co-creation among diverse stakeholders. Only through such an integrated and critical approach can digital innovation effectively contribute to resilient, inclusive, and sustainable agri-food systems.

Acknowledgments The authors acknowledge the use of language editing services to improve clarity and consistency. The scientific content, conceptual framing, and interpretations remain the sole responsibility of the authors.

Competing Interests The author(s) has no competing interests to declare that are relevant to the content of this manuscript.

References

AgriTech Observatory (2024) Taxonomy and terminology. FAO. https://agritechobservatory.fao.org/taxonomy

Anastasiou E, Balafoutis AT, Fountas S (2023) Applications of extended reality (XR) in agriculture, livestock farming, and aquaculture: a review. Smart Agric Technol 3:100105. https://doi.org/10.1016/j.atech.2022.100105

Avila FR, Barbosa JLV (2024) Smart environments in digital agriculture: a systematic review and taxonomy. Revista Brasileira de Computação Aplicada 16(1):11–25. https://doi.org/10.5335/rbca.v16i1.15142

Banbury C, Reddi VJ, Torelli P, Holleman J, Jeffries N, Kiraly C, Montino P, Kanter D, Ahmed S, Pau D, Thakker U, Torrini A, Peter W, Cordaro J, Di Guglielmo G, Duarte J, Gibellini S, Parekh S, Tran H et al (2021) MLPerf Tiny Benchmark. arXiv. https://doi.org/10.48550/arXiv.2106.07597

Barman B, Singh R, Quader SW, Nath RK (2024) The rise of agriculture 4.0: a bibliometric analysis of research trends and future directions. Research Square. https://doi.org/10.21203/rs.3.rs-4948484/v1

Barman B, Singh R, Padaria RN, Quader SS, Das A, Ghoshi B, Munshi SA, Sultana S, Meghana N (2025) Mapping the evolution of agriculture 4.0: a bibliometric analysis of research trends. Research Square. https://doi.org/10.21203/rs.3.rs-4948484/v2

Bechar A, Vigneault C (2016) Agricultural robots for field operations: concept and components. Biosyst Eng 149:94–111. https://doi.org/10.1016/J.BIOSYSTEMSENG.2016.06.014

Béné C, Headey D, Haddad L, von Grebmer K (2016) Is resilience a useful concept in the context of food security and nutrition programme? Some conceptual and practical considerations. Food Security 8:123–138. https://doi.org/10.1007/s12571-015-0526-x

Benos L, Tagarakis AC, Dolias G, Berruto R, Kateris D, Bochtis D (2021) Machine leaning in agriculture: a comprehensive updated review. Sensors 21(11):3758. https://doi.org/10.3390/s21113758

Bhardwaj M, Kumar P, Singh A (2025) Bibliometric review of digital transformation in agriculture: innovations, trend and sustainable futures. J Agribusiness Dev Emerg Econ., Vol. ahead-of-print N° ahead-of-print. https://doi.org/10.1108/JADEE-02-2025-0083

Brasil (2018) Lei n° 13.709, de 14 de agosto de 2018. Dispõe sobre a proteção de dados pessoais e altera a Lei n° 12.965, de 23 de abril de 2014 (Marco Civil da Internet). Diário Oficial da União. https://www.planalto.gov.br/ccivil_03/_ato2015-2018/2018/lei/l13709.htm

Bronson K, Knezevic I (2016) Big Data in food and agriculture. Big Data Soc 3(1):1–5. https://doi.org/10.1177/2053951716648174

Buchaillot ML, Cairns J, Hamadziripi E, Wilson K, Hughes D, Chelal J, McCloskey P, Kehs A, Clinton N, Araus JL, Kefauver SC (2022) Regional monitoring of fall armyworm (FAW) using early warning systems. Remote Sens 14(19):1–19. https://doi.org/10.3390/rs14195003

Calvet E, Falcão R, Thom L (2022) Business process model for interoperability improvement in the agricultural domain using digital twins. arXiv. https://arxiv.org/abs/2206.08589

Castell M (2018) A sociedade em rede. Paz e Terra

CGIAR Independent Advisory and Evaluation Services (2022) Applying the CGIAR quality of research for development framework to process and performance evaluations: CGIAR evaluation guidelines. IAES. https://iaes.cgiar.org/sites/default/files/pdf/CGIAR-QoR4D-Evaluation-Guidelines.pdf

Chapagai A, Dhakal S (2023) Information and Communication Technology (ICT) in the modern world: prospects and challenges. Int J Innov Sci Res Technol 8(7):2080–2091. https://doi.org/10.5281/zenodo.8228674

Charatsari C, Michailidis A, Lioutas ED, Bournaris T, Loizou E, Paltaki A, Lazaridou D (2023) Competencies needed for guiding the digital transition of agriculture: are future advisors well-equipped? Sustainability 15(22):1–15. https://doi.org/10.3390/su152215815

Cutini M, Bisaglia C, Brambilla M, Bragaglio A, Pallotino F, Assirelli A, Romano E, Montaghi A, Leo E, Pezzola M, Maroni C, Menesatti P (2023) A co-simulation virtual reality machinery simulator for advanced precision agriculture applications. Agriculture 13(8):1–21. https://doi.org/10.3390/agriculture13081603

Dara R, Fard SMH, Kaur J (2022) Recommendations for ethical and responsible use of artificial intelligence in digital agriculture. Front Artif Intell 5:1–11. https://doi.org/10.3389/frai.2022.884192

De Smedt K, Koureas D, Wittenburg P (2020) FAIR digital objects for science—From data pieces to actionable knowledge units. Publications 8(2):1–17. https://doi.org/10.3390/publications8020021

DeLong LN, Fernández Mir R, Fleuriot JD (2024) Neurosymbolic AI for reasoning over knowledge graphs: a survey. arXiv. https://doi.org/10.48550/arXiv.2302.07200

Demestichas K, Peppes N, Alexakis T, Adamopoulou E (2020) Blockchain in agriculture traceability systems: a review. Appl Sci 10(12):1–22. https://doi.org/10.3390/app10124113

Eastwood C, Klerkx L, Ayre M, Della Rue B (2019) Managing socio-ethical challenges in the development of smart farming: from a fragmented to a comprehensive approach for responsible research and innovation. J Agric Environ Ethics 32:741–768. https://doi.org/10.1007/s10806-017-9704-5

European Commission (2025) Data Act explained. European Union. https://digital-strategy.ec.europa.eu/en/factpages/data-act-explained

European Union (2016) Regulation (EU) 2016/679 of the European Parliament and of the Council of 27 April 2016 on the protection of natural persons with regard to the processing of personal data and on the free movement of such data (General Data Protection Regulation). Official Journal of the European Union. http://data.europa.eu/eli/reg/2016/679/oj

FAO (2021) The state of food and agriculture 2021: making agrifood systems more resilient to shocks and stresses. FAO. https://doi.org/10.4060/cb4476en

FAO (2022) Gene editing and agrifood systems. FAO. https://doi.org/10.4060/cc3579en

FAO and ITU (2022) Status of digital agriculture in 47 sub-Sharan African countries. FAO, The International Telecommunication Union. https://doi.org/10.4060/cb7943en

Folke C, Biggs R, Norström AV, Reyers B, Rockström J (2016) Social-ecological resilience and biosphere-based sustainability science. Ecol Soc 21(3):4–16. https://doi.org/10.5751/ES-08748-210341

França RS, Ziviani F, Muylder CF (2020) Agricultural digitalization and digital transformation: The future of agricultural competitive excellence in the 4.0 environment. Braz J Dev 6(2):7240–7260. https://doi.org/10.34117/bjdv6n2-140

Franzen D, Mulla D (2015) A history of precision agriculture. In: Zhang Q (ed) Precision agriculture technology for crop farming. CRC Press, pp 1–19. https://www.taylorfrancis.com/chapters/oa-edit/10.1201/b19336-1/history-precision-agriculture-david-franzen-david-mulla

Fraser A (2022) You can't eat data?: Moving beyond the misconfigured innovations of smart farming. J Rural Stud 91:200–207. https://doi.org/10.1016/j.jrurstud.2021.06.010

Guo Y, Chen P, Wan Y, Zhu Y, Cao X, Zeng G (2025) Technology transfer in asymmetric corridors: theory and empirical evidence from China. Appl Geogr 178:103599. https://doi.org/10.1016/j.apgeog.2025.103599

Han H, Siebert J (2022) TinyML: a systematic review and synthesis of existing research. In: Proceedings of the 4th International Conference on Artificial Intelligence in Information and Communication. The Korean Institutes of Communications and Information Sciences. https://doi.org/10.1109/ICAIIC54071.2022.9722636

Hasbie SNR, Assim MISA, Taasim SI, Anuar FN (2023) Information and communication technology (ICT) adoption and community: a systematic literature review. Int J Acad Res Bus Soc Sci 13(15):153–168. https://doi.org/10.6007/IJARBSS/v13-i15/18799

He Q, Zhao H, Feng Y, Wang Z, Ning Z, Luo T (2024) Edge computing-oriented smart agricultural supply chain mechanism with auction and fuzzy neural networks. J Cloud Comput 13:1–18. https://doi.org/10.1186/s13677-024-00626-8

Hendrickson JR, Hanson JD, Tanaka DL, Sassenrath G (2008) Principles of integrated agricultural systems: Introduction to processes and definitio. Renew Agric Food Syst 23(4):265–271. https://doi.org/10.1017/S1742170507001718

High Level Panel of Experts on Food Security and Nutrition (2017) Nutrition and food systems: a report by the High Level Panel of Experts on Food Security and Nutrition of Committee on Word Food Security. HLPE. https://www.fao.org/3/i7846e/i7846e.pdf

Huong DT, Duy NTH, Tu PVM, Hanh HH, Yamada K (2023) Enhancing precision agriculture: An IoT-based smart monitoring system integrated with LoRaWAN, ML and AR. EAI Endorsed Trans Smart Cities 7(4):1–11. https://doi.org/10.4108/eetsc.7286

Hurst W, Mendoza FR, Tekinerdogan B (2021) Augmented reality in precision farming: concepts and applications. Smart Cities 4(4):1454–1468. https://doi.org/10.3390/smartcities4040077

Intergovernmental Panel on Climate Change (2023) Summary for policymakers. In: Lee H, Calvin K, Dasgupta D, Krinner G, Mukherji A, Thorne P, Trisos C, Romero J, Aldunce P, Barrett K, Blanco G, Cheung WWL, Connors SL, Denton F, Diogue-Niang A, Dodman D, Garschagen M, Geden O, Hayward B et al (eds) Climate change 2023: synthesis report. IPCC, pp 1–34. https://doi.org/10.59327/IPCC/AR6-9789291691647.001

Janssen M, Brous P, Estevez E, Barbosa LS, Janowski T (2020) Data governance: organizing data for trustworthy artificial intelligence. Gov Inf Q 37(3):1–8. https://doi.org/10.1016/j.giq.2020.101493

Kalyani Y, Collier R (2021) A systematic survey on the role of cloud, fog, and edge computing combination in smart agriculture. Sensors 21(17):1–27. https://doi.org/10.3390/s21175922

Kalyani Y, Bemeo NV, Collier R (2023) Digital twin deployment for smart agriculture in Cloud-Fog-Edge infrastructure. Int J Parallel Emergent Distrib Syst 38(6):461–476. https://doi.org/10.1080/17445760.2023.2235653

Kamilaris A, Prenafeta-Boldú FX (2018) Deep learning in agriculture: a survey. Comput Electron Agric 147(1):70–90. https://doi.org/10.1016/j.compag.2018.02.016

Katoch S, Chauhan SS, Kumar V (2021) A review on genetic algorithm: past, present, and future. Multimed Tools Appl 80:8091–8126. https://doi.org/10.1007/s11042-020-10139-6

Khatoon PS, Ahmed M (2021) Importance of semantic interoperability in smart agriculture systems. Trans Emerging Telecommun Technol 33(5):1–22. https://doi.org/10.1002/ett.4448

Kidd R (2019) Artificial immune systems: an overview for faulting actuators. Actuators 8(3):1–26. https://doi.org/10.3390/act8030053

Klerkx L, Rose D (2020) Dealing with the game-changing technologies of Agriculture 4.0: how do we manage diversity and responsibility in food system transtion pathways? Glob Food Sec 24:1–7. https://doi.org/10.1016/j.gfs.2019.100347

Klerkx L, Jakku E, Labarthe P (2019) A review of social science on digital agriculture, smart farming and agriculture 4.0: new contributions and a future research agenda. NJAS Wageningen J Life Sci 90–91:100315. https://doi.org/10.1016/j.njas.2019.100315

Latour B (2005) Reassembling the social: an introduction to actor-network-theory. Oxford University Press

Le Gal P-Y, Dugué P, Faure G, Novak S (2011) How does research address the design of innovative agricultural production systems at the farm level? A review. Agric Syst 104(9):714–728. https://doi.org/10.1016/j.agsy.2011.07.007

Lotfy AA, Abuarab MR, Farag E, Derardja B, Khadra R, Abdelmoneim AA, Mokhtar A (2024) Forecasting blue and green water footprint of wheat based on single hybrid, and stacking ensemble machine learning algoritms under diverse agro-climatic conditions in Nile Delta Egypt. Remote Sens 16(22):1–38. https://doi.org/10.3390/rs16224224

Lyu Z, Jing Z, Yang X (2025) Bridging the digital divide for sustainable agriculture: how digital adoption strengthens farmer livehood resilience. Front Sust Food Syst 9:1–13. https://doi.org/10.3389/fsufs.2025.1628588

Maraveas C, Asteris PG, Arvanitis KG, Bartzanas T, Loukatos D (2023) Application of bio and nature-inspired algorithms in agricultural engineering. Arch Comput Methods Eng 30:1979–2012. https://doi.org/10.1007/s11831-022-09857-x

Maraveas C, Konar D, Michopoulos DK, Arvanitis K, Peppas KP (2024) Harnessing quantum computing for smart agriculture: Empowering sustainable crop management and yield optimization. Comput Electron Agric 218:1–23. https://doi.org/10.1016/j.compag.2024.108680

Mausch K, Hall A, Hambloch C (2020) Colliding paradigms and trade-offs: agri-food systems and value chain interventions. Glob Food Sec 26:1–9. https://doi.org/10.1016/j.gfs.2020.100439

McFadden J, Casalini F, Antón J (2022) Policies to bolster trust in agricultural digitalisation: issues note. OECD. https://doi.org/10.1787/5a89a749-en

Meesala H, Brunori G (2025) Dynamics of using digital technologies in agroecological settings: a case study approach. Agriculture 15(15):1–22. https://doi.org/10.3390/agriculture15151636

Milian Gómez JF, Byttebier K (2025) Agroecological sustainability: exploring the intersection of digital agriculture, ethics and the right to food. Discov Agric 3(91):1–16. https://doi.org/10.1007/s44279-025-00228-1

Mo Y, Bier R, Li X, Daniels M, Smith A, Yu L, Kan J (2024) Agricultural practices influence soil microbiome assembly and interactions at different depths identified by machine learning. Commun Biol 7:1–16. https://doi.org/10.1038/s42003-024-07059-8

Musker R, Schaap B (2018) Global Open Data in Agriculture and Nutrition (GODAN) initiative partner network analysis. F1000Res 7:1–10. https://doi.org/10.12688/f1000research.13044.1

Naudé DH, Botha BS, Hugo L, Jordaan H, Lombard WA (2024) Extended reality in agricultural education: a framework for implementation. Educ Sci 14(12):1–22. https://doi.org/10.3390/educsci14121309

Nguyen LV (2024) Swarm intelligence-based multi-robotics: a comprehensive review. AppliedMath 4(4):1192–1210. https://doi.org/10.3390/appliedmath4040064

Nguyen A, Francis M, Windfeld E, Lhermie G, Kim K (2024) Developing an immersive virtual farm simulation for engaging and effective public education about the dairy industry. Comput Graph 118:173–183. https://doi.org/10.1016/j.cag.2023.12.011

Panigrahi KK, Mohanty A, Padhan SR, Shatpathy P, Bhoi P (2024) Optmitizing resource efficiency in agriculture: a comprehensive analysis of circular farming practices. Biotica Res Today 6(3):81–84. https://doi.org/10.13140/RG.2.2.13850.252

Paudel B, Riaz S, Teng SW, Kolluri RR, Sandhu H (2025) The digital future of farming: a bibliometric analysis of big data in smart farming research. Cleaner Circ Bioecon 10:1–17. https://doi.org/10.1016/j.clcb.2024.100132

Pierozzi Junior I, Visoli MC, Souza MIF, Cunha LMS, Vacari I, Torres TZ (2023) Information engineering: contributions to digital agriculture. In: Massruhá SMF, Leite MA d A, Oliveira SR d M, Meira CAA, Luchiari Junior A, Bolfe EL (eds) Digital agriculture: research, development and innovation in production chains. Embrapa, pp 145–163

Rivera G, Porras R, Florencia R, Sánchez-Solís JP (2023) LiDAR applications in precision agriculture for cultivating crops: a review of recent advances. Comput Electron Agric 207:1–23. https://doi.org/10.1016/j.compag.2023.107737

Rocha PCS, Jucá SCS, Silva SA (2020) The evolution of information and communication technologies from the perspective of Touraine, Bell and Castells. Res Soc Dev 8(5):1–12. https://doi.org/10.33448/rsd-v8i5.928

Rose DC, Chilvers J (2018) Agriculture 4.0: broadening responsible innovation in an era of smart farming. Front Sustainable Food Syst 2:1–7. https://doi.org/10.3389/fsufs.2018.00087

Roth L, Marzougui A, Walter A (2025) A review of the journey of field crop phenotyping: from trait stamp collections and fancy robots to phenomics-informed crop performance predictions. J Plant Physiol 311:1–12. https://doi.org/10.1016/j.jplph.2025.154542

Rotz S, Gravely E, Mosby I, Duncan E, Finnis E, Horgan M, LeBlanc J, Martin R, Neufeld HT, Nixon A, Pant L, Shalla V, Fraser E (2019) Automated pastures and the digital divide: how agricultural technologies are shapping labour and rural communities. J Rural Stud 68:112–122. https://doi.org/10.1016/j.jrurstud.2019.01.023

Ryan M, Atik C, Rijswijk K, Bogaardt M-J, Maes E, Deroo E (2024) The future of agricultural data-sharing policy in Europe: Stakeholder insights on the EU Code of Conduct. Humanit Soc Sci Commun 11:1–15. https://doi.org/10.1057/s41599-024-03710-1

Sargent D, Conaty WC, Tissue DT, Sharwood RE (2022) Synthetic biology and opportunities within agricultural crops. J Sustain Agric Environ 1(4):89–107. https://doi.org/10.1002/sae2.12014
Shamshiri RR, Weltzien C, Hameed IA, Yule IJ, Grift TE, Balasundram SK, Pitonakova L, Ahmad D, Chowdhary G (2018) Research and development in agricultural robotics: a perspective of digital farming. Int J Agric Biol Eng 11(4):1–14. https://doi.org/10.25165/j.ijabe.20181104.4278
Sharma V, Vaddevolu UBP, Bhambota S, Ampatzidis Y, Bayabil H, Singh A (2025) Variable rate technology and its application in precision agriculture: AE607, 1/2025. EDIS. https://doi.org/10.32473/edis-AE607-2025
Singh S, Suman, Kour P, Kumar S (2024) Digitalization and technology in agribusiness: trends, impacts, and future directions. Int J Biochem Res 8(11):27–29. https://doi.org/10.33545/26174693.2024.v8.i11a.2805
Souza KXS, Bolfe EL, Leite MAA, Bambini MD, Visoli MC, Silva FR, Esquerdo JCDM, Yassitepe JECT (2024) Quantum computing: current and potential applications in digital agriculture. Pesq Agrop Brasileira 59:1–14. https://doi.org/10.1590/S1678-3921.pab2024.v59.03753
Stephens N, Di Silvio L, Dunsford I, Ellis M, Glencross A, Sexton A (2018) Bringing cultured meat to market: technical, socio-political, and regulatory challenges in cellular agriculture. Trends Food Sci Technol 78:155–166. https://doi.org/10.1016/j.tifs.2018.04.010
Stilgoe J (2020) Who's driving innovation? New technologies and the collaborative state. Palgrave Macmillan. https://doi.org/10.1007/978-3-030-32320-2
Tang Y, Dananjayan S, Hou C, Guo Q, Luo S, He Y (2021) A survey on the 5G network and its impact on agriculture: challenges and opportunities. Comput Electron Agric 180:1–15. https://doi.org/10.1016/j.compag.2020.105895
Tanny T, Sallam M, Soda N, Nguyen N-T, Alam M, Shiddiky MJA (2023) CRISP/Cas-based diagnostics in agricultural applications. J Agric Food Chem 71(31):11765–11788. https://doi.org/10.1021/acs.jafc.3c00913
van der Burg S, Wiseman L, Krkeljas J (2021) Trust in farm data sharing: reflections on the EU code of conduct for agricultural data sharing. Ethics Inf Technol 23:185–198. https://doi.org/10.1007/s10676-020-09543-1
van Eck NJ, Waltman L (2014) Visualizing bibliometric networks. In: Ding Y, Rousseau R, Wolfram D (eds) Measuring scholarly impact. Springer. https://doi.org/10.1007/978-3-319-10377-8_13
Van Wassenaer L, van Hilten M, van Ingen E, van Asseldonk M (2021) Applying blockchain for climate action in agriculture: state of play and outlook. FAO, Wur. https://doi.org/10.4060/cb3495en
Wilkinson MD, Dumontier M, Aalbersberg IJ, Appleton G, Axton M, Baak A, Blomberg N, Boiten JW, da Silva Santos LB, Bourne PE, Bouwman J, Brookes AJ, Clark T, Crosas M, Dillo I, Dumon O, Edmunds S, Evelo CT, Finkers R et al (2016) The FAIR guiding principles for scientific data management and stewardship. Sci Data 3:1–9. https://doi.org/10.1038/sdata.2016.18
Wolfert S, Ge L, Verdouw C, Bogaardt M-J (2017) Big Data in smart farming - a review. Agric Syst 153:69–80. https://doi.org/10.1016/j.agsy.2017.01.023
Wu J (2024) Digitizing agriculture: strategic insights into traditional challenges and modern management solutions. Adv Econ Manag Polit Sci 121(1):215–219. https://doi.org/10.54254/2754-1169/121/20242631
Yeo ML, Keske CM (2024) From profitability to trust: factors shapping digital agriculture adoption. Front Sustainable Food Syst 8:1–15. https://doi.org/10.3389/fsufs.2024.1456991
Zhai Z, Fernán Martínez J, Beltran V, Martínez NL (2020) Decision support systems for agriculture 4.0: survey and challenges. Comput Electron Agric 170:1–16. https://doi.org/10.1016/j.compag.2020.105256

Knowledge and Learning in an Agri-Food Ecosystem Context

Andrea Knierim, Beatriz Herrera, and Angelika Thomas

We cannot teach another person directly; a person can only facilitate learning.

Carl Rogers

1 Introduction

Farmers operate in increasingly fast changing conditions, and uncertainty about, e.g. global market trends or climate change impacts, which nowadays matter almost everywhere. Pressures on farming result from a growing number of economic, environmental, and societal challenges such as digitalisation, resource depletion, civil and military conflicts that transgress national boundaries and affect the efficacy and efficiency of agricultural production in various ways (Rockström et al. 2020; Finger 2023; Richardson et al. 2023).

Also, new dietary regimes and changes in consumption behaviours, as well as raising, even contradicting expectations from consumers and citizens towards agricultural production and related services, as expressed in public policies and media exert a burden. Particularly, EU agricultural policies such as the European farm2fork strategy, the vision for agriculture, and the recently formulated agro-ecology paradigm reveal the manifold demands that exist with respect to farming practices (EC COM 2020; EC COM 2025).

Rapidly evolving digital technology progress is considered to come with an impressive array of opportunities for agriculture, e.g. fostering productivity and moderating negative externalities through precision farming, automated aerial and terrestrial devices, and data-driven decision support and management systems.

A. Knierim (✉) · B. Herrera
Department of Social Sciences in Agriculture, University of Hohenheim, Stuttgart, Germany
e-mail: andrea.knierim@uni-hohenheim.de

A. Thomas
Institute of Applied Agricultural Research, University of Nuertingen-Geislingen, Nürtingen, Germany

F. da Silveira et al. (eds.), *Agriculture 4.0*,
https://doi.org/10.1007/978-3-032-24552-6_14

However, there are also likely limitations and risks associated with the use of these technologies, such as unequal access for agricultural communities with structurally different characteristics, due e.g. to unfavourable farm characteristics and structures or a lack of skills and expertise among farm managers (Finger 2023). Risks are also likely to arise with respect to data privacy and data management security, and so-called 'unintended side effects' (Unseens) in the field of agro-ecology, food security, digitally mediated power concentration, and farmers' individual decision-making autonomy (Zscheischler et al. 2021).

Without doubt, digitalisation is considered an important driver and shaping force for societal change in general, and the agrifood sector in particular, and both its potentials but also its risks and dangers should not be underestimated (Wolfert et al. 2017). In their comprehensive review of the role of digitalisation for a global transformation towards a more sustainable future, the German Advisory Council for Global Change concludes that digitalisation should be weighed against whether and how it affects (i) human dignity, (ii) the maintenance of the planet's natural resources, (iii) the socio-economic and political inclusion and participation options for individuals within their respective societies, and (iv) the support offered for the expression and actualisation of one's personal traits and characteristics (WBGU 2019). Pursuing such a fundamental and comprehensive assessment of potential impacts requires differentiated knowledge, insight-driven engagement, and considerable learning endeavours and activities of individuals and collective entities within societies.

In this chapter, we discuss the challenges of 'Agriculture 4.0' digitalisation for enhancing knowledge exchange and learning for and with agricultural professionals in a user-oriented way and draw conclusions for future research. Hereby, we use the agri-food ecosystem concept to sketch the environment and illustrate the interdependencies in which farmers perceive their information needs and learning interests, express these to potential knowledge providers, and realise increased insights and improved understanding. We propose the AKIS (Agricultural Knowledge and Innovation System) concept to analyse how information access and knowledge exchange in agriculture can be understood at an institutionalised level, and enhanced through education, training, and advisory services organisations. Second, we present how learning, as the personal effort to acquire knowledge, competences, and skills, faces opportunities and challenges in this new context. We study the topic with a focus on the situation in the European Union, particularly drawing on examples from the German state Baden-Württemberg. Finally, we outline new possibilities on how the knowledge system approaches could be leveraged in order to face the new uncertainties and trends in the agrifood ecosystem.

2 Agrifood Ecosystem as a Learning Environment

Agriculture 4.0 is used as an abbreviation for the overall socio-technological progress induced by digital technologies, the composite of effects resulting from the application of digital innovations such as smart devices, data-driven operations, etc. for farming practices (Charatsari et al. 2022). The term emphasises the rapid, disruptive development of technological progress through digitalisation and at the same time, implies a focus on the technological side of agriculture. This focus is embedded in a broader picture with the term 'agri-food ecosystem', which evokes a comprehensive, albeit diversely interpretable image of the manifold factors influencing agricultural production, food processing, and consumption. With the aim to thoroughly explore farmers' knowledge and learning conditions and environments, and particularly how these factors change or are modified and impacted through digitalisation, we develop both the agri-food ecosystem concept as a contextual scheme and the AKIS concept as an analytical frame.

2.1 The Concept of the Agri-Food Ecosystem

The agri-food ecosystem concept comes with the ambition to comprehensively capture the multiple dimensions and spheres from which originate influencing factors on behavioural and institutional change towards a more sustainable agriculture. Digital technologies have a specific part therein and it is widely expected by politicians and scientists that they will allow for major leaps in efficiency and thus substantially decrease external environmental effects and may also have positive effects for farm organisation and easing of work (Walter et al. 2017; Wolfert et al. 2017).

What is an 'agri-food ecosystem'? A food system is described as to gather 'all the elements (environment, people, inputs, processes, infrastructures, institutions, etc.) and activities that relate to the production, processing, distribution, preparation and consumption of food, and the outputs of these activities, including socio-economic and environmental outcomes' (HLPE 2017:11). Ewert et al. (2023:358f) provide a short summary on agri-food system conceptions over time, highlighting both, the interplay and the implicit directionality of various elements such as drivers, actors, and outcomes. Hereby, the cross-cutting understanding of the term 'system' is that a certain number of elements are mutually related and interdependently connected in various ways, which may or may not be further specified. Second, they elaborate on agri-food systems with a teleological intention, which is here shaped by agroecology and characterised by sustainability features, the SDGs, the levels of agency and societal organisation, social actors, food value chain components, and codesign procedures and activities. Obviously, agri-food systems cannot conclusively be defined, but 'need to be determined in a context-specific way based on the specific problem to be addressed' (ibid*:360*). As another example for the term's fuzziness, we quote McGreevy et al. (2022:1011) who characterise global agri-food systems as 'contemporary knowledge, institutions, infrastructures, practices and crops that define the predominant patterns of food production and consumption', thus as a conglomerate of material and non-material entities and processes without further specifying relations and interactions among them.

Turning to the term 'ecosystem' in combination with the agri-food realm, how can, how should this be understood and defined? There is a big difference, whether we speak of an ecosystem in the context of natural sciences and particularly ecology, where it is understood as 'a system of interactions between living organisms, inanimate natural components and man-made components that interact with each other and with their environment in terms of energy, matter and information' (ANL 1991, *own translation*), or within an entrepreneurial, business management study context, where technology innovation is the core study object that is stimulated and directed by various influencing factors, summarised as ecosystems (Adomavicius et al. 2008). Wolfert et al. (2023:4) expand on the term and consider innovation ecosystems as embedded in business ecosystems; they specify 'a digital innovation ecosystem typically (....) concerns flows of technology and across people, organizations and institutions', (. . .) 'an integrated system of systems in which no actor, actant or system is greater than another' (ibid:4; Pigford et al. 2018:118). Sometimes, authors also use normative descriptions for innovation ecosystems, such as 'an ecosystem that is conducive to the innovation objectives' (Wolfert et al. 2023:2), or 'a loose network of actors with various backgrounds [is] involved in the creation and delivery of end products' (ibid:4). A more operational definition has been proposed by Pigford et al. (2018), illustrating the term as 'innovation platforms' or 'innovation community' (ibid:117) composed of a range of diverse actors with multiple interests, goals and influence levels who interact to foster an innovation. Even more nuanced, these authors stress the fact that actually, the term ecosystem presents an analogy of a socio-technical reality to a natural one, which may broaden the perspective and enrich the discussion, but does not necessarily indicate a similar kind of relationships or cause–effect links among the elements (ibid:119).

Thus, we conclude that the term 'agri-food ecosystem' and its derivatives fulfil the purpose of holistically integrating multiple components that matter for agricultural production from farm to fork topics and from local to international level actors and entities, with a particular focus on the roles and impacts of digital technologies. However, going beyond this general holism means defining elements, relations, and boundaries of the agri-food ecosystem into consideration. Particularly, the fuzziness of the term 'ecosystem' matters, as it obscures rather than reveals the effects and causalities around digitalisation processes and their impacts on knowledge and learning. We therefore propose in the next section the AKIS concept to focus and operationalise the mentioned interdependencies between agri-food actors' knowledge and learning interests and the use of digital technologies.

2.2 Importance of Knowledge in Agrifood Ecosystems

Given the speed of technological progress, it is not surprising that many voices stress the need for investing heavily in training and education to encourage and enable learning on how to handle and master digital technologies (WBGU 2019;

Zscheischler et al. 2021). Narrowing the focus to the agricultural sector, in order to make use of and to profit from digitalisation, there is a need for and pressure on farmers and other agri-food professionals to get informed, to stay updated, and to learn how to best keep track of the changes and challenges. How can, how should this happen at best, given the constantly increasing reservoir of information and knowledge? From a normative stance, one could argue that education, training, and advisory service providers are to engage with farmers' diverse needs, to adapt and improve their approaches and offers correspondingly. Policy makers, e.g. in the EU, aware of the eminent efforts needed, have started to set up corresponding, cross-cutting objectives and related measures, focusing the strengthening of agricultural knowledge and innovation systems (EU 2021). Alternatively, one may also argue that it is the individual farmers themselves who should be proactive and, in a targeted way, engage in learning and acquiring new skills and competences, both during the vocational education as well as through further training and lifelong learning measures.

In this ambiguity, the AKIS concept helps to clarify and analyse the multiple elements, both human actors and non-human infrastructures and, artefacts in their relations and interdependencies that matter for the generation, dissemination, acquisition and use of knowledge and innovations. The AKIS concept has proven its functionality to illustrate and translate complex configurations of knowledge infrastructures, and their interplay in terms of vertical and horizontal linkages and coordination mechanisms within and among subsystems (Blum 1991; EU SCAR 2012; Knierim et al. 2015). Applied to the situation in various EU member states, recent in-depth analyses showed for many of them deficits of the research and education subsystems to connect satisfyingly with farmers and farmer organisations so as to capture their knowledge needs and research interests and to respond to their problem-solving requests (Birke et al. 2022).

Other authors use the AKIS concept to structure and frame digitalisation-related research questions, spanning widely from macro-level topics such as ethical and political economy challenges, meso-scale organisation of knowledge provision mediated by digital infrastructures, and micro-level aspects of user orientation and service interactions among farmers and advisors (Klerkx et al. 2019). Ingram and Maye (2020) reflect on the implications of digitalisation for the agricultural knowledge body and examine the ways in which digital technologies likely impact farmers' knowledge acquisition and decision-making. They point to 'an increased learning load for farmers from using digital agricultural tools and the need to invest in human capital' (Ingram and Maye 2020:3). Beyond individual level, they stress the potential of digital technologies to support the creation of networks for knowledge exchange and collaboration or to replace interpersonal interaction with other AKIS actors such as advisors by formalised 'smart systems' tailored to provide expert knowledge and advanced analytical services directly to farmers (ibid:4). Finally, they point to two areas in which digitalisation is likely to have a clear effect on how an AKIS is functioning: on the one hand, the field of actors will expand to include software manufacturers and data-driven companies and projects etc. On the other hand, the availability and evaluation of very large data sets (big data) could create more opportunities for predicting and prescribing future developments with a new quality of exactness, which in turn may impact the importance of traditional, local, and practical farm-level knowledge (Zscheischler et al. 2021). However, the ability to transform this data into useful information and, ultimately, actionable wisdom relies on rules of access and use of this data, as well as on the integration of new services, e.g. providing interoperability, maintaining privacy, protecting legitimate data ownership interests, preventing the concentration of power, etc. (Brunori et al. 2025).

We conclude that the AKIS concept is appropriate to

- Specify knowledge infrastructures and corporate actors that play a role in digitalisation-related knowledge in the agri-food sector.
- Frame the appraisal of policies, programmes, and initiatives for advisory, education, and information services on digitalisation in agriculture.
- Address and analyse established and emerging organisations and institutions that mediate and support farmers' learning for digitalisation.

3 Knowledge Sharing and Learning Among Agricultural Professionals

This section presents how knowledge exchange and learning in agriculture can be understood and enhanced through education, training, and advisory services, making use of the AKIS concept and highlighting opportunities and challenges related to the rapid technological progress. We draw on facts and cases from Germany to illustrate our argumentation.

3.1 Knowledge Concepts

The literature distinguishes explicit, codified, and implicit, tacit types of knowledge (Blackler 1995; Polanyi 1966), frequently also associated with scientific and practical knowledge. Further differentiations are local, traditional, and indigenous knowledge, to name a few, commonly used in ethnography (Richards 1985). Equally relevant in our case is a distinction that originates from transdisciplinary research (Hirsch Hadorn et al. 2008), which structures and separates what we call 'facts' to describe realities (system knowledge), from the 'values' that are incorporated in set objectives and goals (target knowledge), and from the practice-related knowledge that makes the realisation of a change possible through respective behavioural changes (transformation knowledge).

Input to all types of knowledge is information, aggregated and ordered data that allows for the making of meaning (Ackoff 1989). Generally, information is stored in various forms, e.g. printed materials, audio and video materials, etc., but it can also come orally from family, friends, peers, experts or any other person. Also, information is contained in forms, colours, and configurations of the environment and, e.g. farmers can 'read' and understand the state of a field or cattle by information they gain from the appearance, shape, etc. of the studied entity.

While sources of information are manifold and may not be exhaustively enumerated, media that provide access to information are artifacts and clearly can be determined. Besides traditional and well-established media (print media, TV, Radio etc.), digital-configurated social media have increasingly gained momentum as carriers of information (platforms for storage and retrieval of printed, audio, and video-enhanced contents, devices to support communicative interaction and knowledge exchange, LLM-based tools to generate aggregated information, etc.). Huge expectations are nourished that digital tools may be prolific for fast and comprehensive access to relevant knowledge for farmers and support more sustainable and resilient farming practices (EC COM 2025/75). Critical voices point to the fact that the potential of digital tools is only open to those people who have acquired the respective competences ('digital literacy') so that they are well equipped to master and use them for their purposes, in other words, to participate in the digitalised societal context (Friesike and Sprondel 2022:31).

A white book on digitalisation in Germany recently highlighted the eminent increase of knowledge and learning requirements that farmers face with the ongoing diffusion of digital technologies in agriculture (Zscheischler et al. 2021:159ff). The authors argue that digitalisation may imply easement of working conditions and the generation of new knowledge for land management, but it definitely comes with new learning requirements for farmers to appropriately respond to digitally enhanced machinery, to correctly interpret and assess the quality and reliability of generated data, to master new ways of information exchange and technological integration ('system of systems') and to handle the risks and uncertainties of data protection and data privacy (Goller et al. 2021). Are German farmers fit for digitalisation-related knowledge acquisition, are they ready to face the learning efforts that are necessary to confidently use and profit from agriculture 4.0?

Empirically, there is no comprehensive evidence of farmers' state of information and knowledge on digital technologies and even less about their formal and further education activities in this field. In general, vocational education of farmers varies greatly within the EU (OECD 2024), and equally, within Germany (Destatis 2023). What are relevant sources, where can farmers turn to when seeking reliable information, training, and support? In the following, we make use of selected findings from empirical studies conducted as part of a nationally funded project on digitalisation in agriculture (www.diwenkla.de; duration 2020–2025). These include a standardised survey of farmers in the federal state of Baden-Württemberg and qualitative interviews with 40 experts (Paulus and Knierim 2023; Pfaff et al. 2023) as well as previously unpublished results of an AKIS analysis for Baden-Württemberg and a research report documenting publicly funded information, training and advisory measures on digitalisation in agriculture in Germany, Austria and Switzerland (Koroed et al. 2024). We complete these results with insights from other empirical studies from both the German national and selected states' level, with the overall aim to present the AKIS infrastructures and corporate actors concerned with digitalisation in agriculture and complement this with farmers' perspectives and appraisals of access to and use of information, training and further services for their learning interests and needs.

3.2 Digitalisation-Related Knowledge in the AKIS

Innovation and knowledge processes related to smart farming and agriculture 4.0 are driven by multiple actors (Kernecker et al. 2021). The use of the AKIS concept refers to the individual, collective and organisational generation, sharing, and utilisation of agriculture-related knowledge and technology and learning among the actors engaged in the agricultural sector and related spheres (Birke et al. 2022). Taking Germany as an example, there is a divide between the national and the federal states' level for the governance of the AKIS: public responsibility for agricultural policy as well as for education and training is allocated to the federal states (Birke et al. 2022; Knierim et al. 2022), while national-level public actors focus on lateral coordination and incentivising research and development programmes. At the national level, multi-level, farmer-based organizations and private entrepreneurs' associations have emerged as important players in linking and integrating information and data flows in the agricultural system. For instance, the chambers of agriculture, vocational education institutions, and selected state-level public advisory bodies established and maintain knowledge flows among AKIS actors. These organizations have created platforms for dialogue, collaboration, and knowledge sharing, which have helped to build trust and foster cooperation among stakeholders (Birke et al. 2022).

As for digitalisation in agriculture, there are no specific national infrastructures. However, funding of research projects targeting agricultural digitalisation is a national-level instrument. One major programme of the German Federal Ministry of Food and Agriculture was dedicated to promoting digitalisation in agriculture through 14 so-called 'trial fields', a set of highly interactive research projects, targeted to regional challenges, which spread all over the country. In frame of this programme, an AKIS study was conducted by one of the authors in the federal state of Baden-Württemberg centred around the identification of relevant actors and their connections and interactions to further the adoption and diffusion of digital technologies in crop farming.

Figure 1 displays an aggregated overview of actors involved in the development, dissemination, and application of digital technologies related to smart (arable) farming, following the AKIS visualisation method (Knierim and Birke 2023). Actors are grouped and situated according to their organisational characteristics (public, private, third sector), their professional orientation (farmer-led, research and education) and their predominant intervention level (local to regional, federal state, national level). While digitalisation is a predominant subject in many respects, only

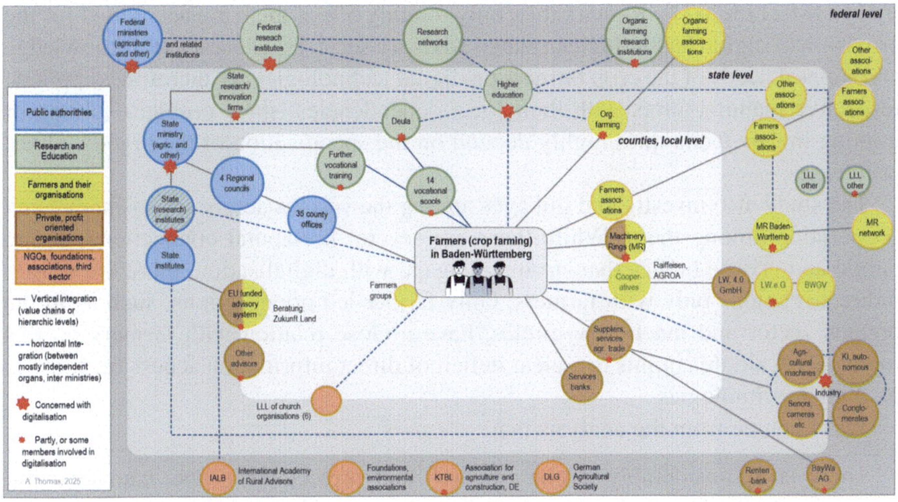

Fig. 1 AKIS overview for digitalisation in crop farming in Baden-Württemberg (2024)

those bodies are red-marked that engage in the development or spread of digital technologies and software for farms.

The overall picture presents a considerable actor diversity (translated by the distinct colours) and shows a numerical dominance of farmer-based (yellow) and private sector (brown) organisations, which reflects the pluralism of the AKIS. Looking at the red-marked organisations, we see digitalisation-related activities are unevenly spread among the groups, that public authorities (blue) and research institutions (green) are at least as active as the industry group and that most of these bodies are at a distance from the local farm level.

Our detailed data analyses reveal that federal and state-level ministries with subordinated institutes and research stations are promoting digitalisation and are part of the research landscape. In contrast, local agricultural state offices are assigned for public activities and information, and are only occasionally involved, despite being closest to farm-level realities. Regarding education and training, teaching on digital agricultural technologies is not fully integrated into the curricula. A few vocational schools, organisationally connected to state offices or to the ministry of education, are taking part in research projects to develop learning nuggets for digital agriculture. Another partner is the privately organised Deula, a publicly subsidised technology training unit that is part of a training network in Germany. Widely, higher education institutions are linked to many actors, whereby information and training for farmers, plus knowledge production with farmers, is limited by project-based funding. A considerable part of knowledge comes from private and profit-oriented organisations, including the agricultural machinery industry, software, and other service providers. A particular group of entrepreneurs is agricultural advisory organisations in Baden-Württemberg, private companies and freelancers, although their services are subsidised within the EU co-funding scheme

(EU COM 2115/2021). Remarkably, only roughly 5% of them are accredited for the provision of digital technologies-related advisory services. Closest to the knowledge and support needs of the crop farmers on digital technologies are therefore suppliers, respective manufacturers with their local dependencies, since problem solving or training in new techniques highly depend on the specification of the brands, types, or releases.

The study also investigated linkages among the various actors and exchange and cooperation among them. While there are several horizontal connections among public and private bodies that strongly engage with digitalisation, there are mostly indirect relations only with farmers. Only farmer-led organisations, such as in the organic sector and machinery circles, have a close relation with farmers on this topic. All in all, this points to a clear deficit of direct information access for farmers on digital technologies.

We conclude from the analysis that:

- There is a considerably strong research and project landscape, causing close network activities of private and public partners, including pilot farms.
- Closest to the present needs of farmers are private technique providers or pilot projects, but their outreach and capacities are limited; there is a lack of experts providing an overview and specific knowledge.
- Independent agricultural advisors (free-lance and companies) play a minor role as knowledge providers.
- Also, the integration of digital technologies topics into vocational education happens in an unsystematic, even sporadic manner and frequently depends on external initiatives and projects.

3.3 Farmers' Knowledge and Interests in Digitalisation

The fragmented and not yet digitalised AKIS situation can be complemented with data about farmers' perspectives from Baden-Württemberg. An online survey, conducted in 2021, targeted to explore farmers' information sources, support interests, and knowledge needs with respect to the adoption of digital technologies. Generally, agriculture in Baden-Württemberg is characterised by a relatively high number of small-scale farmers with an average farm size of 36 ha compared to the German average of 63 ha, and a high share of part-time farmers of 57% compared to the German average of 49% (Destatis 2021). The survey yielded a data set of 203 responses, stemming from a relatively high share of full-time farmers (58%) with relatively large farms (average size 73 ha).

In a self-assessment, these farmers were quite positive about their digital knowledge and competences. Two groups with 64% and 29% see themselves as professionals or advanced in competent use of basic IT equipment, and quite similar shares are confident in their handling of internet-based information and sources and their use for further training. Groups with 40% and 45% shares of all respondents consider themselves as professionals, respectively, advanced regarding the use of digital data sources as a decision-making aid. In contrast, for independent troubleshooting of technical problems, only 18% assess their level as a professional one, and another 55% assess themselves as advanced (Pfaff et al. 2023).

The reported knowledge sources influencing digitalisation-related decisions are magazines, journals, and other farmers in the first place and field days, exhibitions and technology and machinery providers coming second. Training and advisory services were not much considered: 18% of the responding farmers were not aware of education offers and another 23% declared not to be interested. Even larger was the number of farmers with no perceived need or no awareness of digital technology related advisory services (36% and 37%) (Paulus and Knierim 2023). While these figures from Baden-Württemberg reflect quite some reservation among farmers towards education and advisory support measures, figures from other federal states deviate from this picture, with, e.g. a 57% share of farmers ($n = 391$) in Rhineland-Palatine in favour of advisory services on digitalisation or a similar trend of preference for advisory services over education offers among Bavarian farmers (Gabriel and Gandorfer 2023). In summary, it should first be emphasised that no representative data is available on this topic for Germany or any of the federal states. Due to the online survey format, it seems likely that the responses to the survey are biased towards a positive perception of digitalisation. Secondly, there is a notable relation between an open or proactive approach to digital technologies of the responding farmers and their level of education, as well as the size and type of the farm business, which may be linked, for example, to an entrepreneurial mindset. Moreover, it is interesting to note and merits further investigation that how farmers differentiate their competences in the handling of digitalisation issues, and that there is a particular interest but not yet broad expertise in independent troubleshooting (Carolan 2017).

3.4 Learning as an Interactive Process

The acquisition of knowledge, which may also encompass practical knowledge such as skills or integrated knowledge such as competences, happens through the uptake and reorganisation of information by the individual person. This learning can take place by cognitive processes only, but it may also be combined with emotional and haptic experiences, which contribute to the embodiment of the information acquisition. Embodied information is what is finally considered personal knowledge. Thus, there are strong interdependencies between the processes of perception, learning and change of practice.

To characterise what learning means and how it can be operationalised for the field of agriculture, we build upon Sol et al. (2012) who summarised insights from educational thinkers and applied andragogues such as Piaget (1964), Festinger (1957), Vygotsky (1978) and Wenger (1998) and define 'learning as an interactive process that leads to some form of dissonance as a result to being exposed to alternative ways of seeing, knowing and understanding, coupled with a desire to overcome such dissonance by changing one's own thinking in sometimes subtle and sometimes more radical ways. As such, learning can result in a change in perception, knowledge and behaviour of individuals, organisations and/or groups' (Sol et al. 2012:37). The authors thus emphasise (i) the social nature of learning based on the evidence that learning is always linked to some kind of interaction with others, and (ii) the process of learning dissected in first, a change in perception, second, integration of new knowledge, and finally, change of behaviour. This change in perception may be a subtle, incremental one or a more radical, even disruptive one.

Learning pathways of agricultural professionals vary according to the national educational systems, including, for example, vocational training, apprenticeship, undergraduate, and postgraduate higher education. Eurostat data from 2013 demonstrates that around 70% of European farmers acquired their skills through practical experience alone, whilst less than 10% received full-time agricultural education (Augère-Granier 2017). Data from the same year shows that one-third of the sample underwent some agricultural training and that, in general, the educational level of the agricultural labour force is often below the national averages, and highly correlated with variables such as age and farm size (Augère-Granier 2017).

While professional rules and regulations of quality management and control are widely recognised for traditional education pathways, such institutions and measures are less common in new situations brought by digitalisation. The diversity of information sources and the variety of media to access information are impressive and still growing and accelerating. Risk of deskilling, and opportunities for reskilling or upskilling are mentioned in an increasing body of literature (Daum 2025; Glover and Sumberg 2025).

3.5 New Ways of Learning in Digital Environments

The potential of digital technologies can only be exploited if the corresponding skills are available, sometimes also referred to as 'digital participation' (Friesike and Sprondel 2022:31). Skills that can answer to the changes in the food system are described as '21st century skills and competencies', which allow the learners to benefit from the emerging new forms of socialisation and to contribute actively under a system where the main asset is knowledge (Ananiadou and Claro 2009). Interestingly, the twenty-first century skills bring together those skills considered necessary in the knowledge society, but where the use of ICT is not a necessary condition (Ananiadou and Claro 2009). That means that there is a growing importance of transversal complex abilities (critical thinking, creativity, collaboration, and communication) as well as digital and technological literacy (Köhler and Tobler 2025).

The discussion on skills needed by agricultural professionals involves multiple disciplines and sometimes approaches in a large spectrum of positivist versus constructivist approaches (Glover and Sumberg 2025). Related to how farmers acquire (or do not acquire) those skills, Stone (2025) identifies three modalities of learning: environmental learning (observing), social learning (emulating models based on social criteria), and didactic learning (influenced by off-farm entities). Agricultural learning can best occur when these modalities are balanced (Taylor and Bhasme 2025). Recognizing that skills in agriculture involve not only technical components but also social learning and complex knowledge production, there is a trend in recognizing the value of the integration of local and/or intrinsic knowledge with other different types of knowledge through participatory research methods (Hoffmann et al. 2007) and collective learning formats (Monteiro Moretti et al. 2023).

Thus, digitalisation has visible and unseen repercussions for the way agricultural professionals learn. From accessing information and acquiring knowledge to an increase in learning possibilities and the continuous update of innovations, there are several implications for farmers and other stakeholders in the agri-food sector. The range of knowledge and training sources increases, as does the degree of self-organisation, the singularisation of information, and the provision of tailor-made solutions. However, this comes with a lack of transparency in the market, which requires enormous efforts for checking and discerning the quality of sources.

3.6 Learning About Digitalisation in the Agri-Food Sector

Successful learning about digitalisation opportunities and the subsequent decision for and adoption of digital technologies relies on farmers' individual changes of behaviour, but similarly on an effective interplay among peers and with various other actors of the broader agri-food sector. In the following, we add further facts and cases from the German agri-food system to enrich the picture of its digitalisation progress and to illustrate where and how such interfaces and interplays are enacted and cooperation for knowledge exchange and learning takes place.

Recent data confirms that, principally, there is openness and interest among German farmers to learn more about the opportunities of digital technologies. Bitkom, a digital tech enterprises' association and European think tank, published survey results collected in 2020 from 500 farmers (with above 20 ha average farm size, randomly selected across Germany), according to which 78% see digitalisation as a chance, and 14% rather as a risk for their farm. However, this positive view is not yet fully translated into action, with only 17% of the respondents confirming a planned investment into digital tools and applications in the next 12 months, 43% being uncertain, and 39% not planning such an investment (Rohleder and Meinel 2022). Although the number of positive views has clearly increased between 2016 and 2020 (ibid), the moderate readiness to invest and adopt innovations in this field confirms other studies, which report on many farmers' cautious attitudes (Knierim et al. 2019; Kernecker et al. 2020). Asked about favourable political measures, there is almost unanimous agreement that more support for broadband expansion (96%), free access to public geo- and weather data (95%), and support of further education and training (90%) are required (Rohleder and Meinel 2022).

Again, there is no good overview of agricultural education and advisory measures in Germany, targeting digitalisation. Generally, the diversity of service providers in both fields is considerable, covering public, private, and civil society organisations, from which a good part are farmer-led, similarly (Knierim et al. 2022). In the frame of the above-mentioned national 'trial field' programme, a study was conducted to identify promising education and advisory projects and programmes that support the spread of digital tools and applications among farmers. A range of advisory service, further education and information offers were identified, namely in 8 federal states 21 advisory projects, 31 further education measures in 12 federal states and one as online format at national level and, 21 information offers in 10 federal states and two as online format. 13 out of the 21 advisory measures are conducted in 3 federal states, while the education and information activities are more evenly spread across the federal states with the exception of lower saxony (four advisory measures, seven education, and five information offers) (Koroed et al. 2024). Providers of the studied advisory measures comprise public and farmer-led corporate actors, while the further education and training offers are frequently provided in the frame of publicly funded programmes and by institutionalised public actors. Another important organisation in this field is the DEULA e.V., a not-for-profit association for education measures in the green sector, with satellites and training activities in almost all federal states.

The in-depth study of advisory offers in Northern and Central Germany revealed a considerable interest of farmers in specialised, mid-term technology introduction accompanying measures, e.g. the introduction of an automated milking system and similar farm-level technologies and management systems. In some states, there is explicit support for small and medium-scale farms, which is not only implemented through single-farm counselling, but also in the form of the creation and maintenance of farmer networks and through regular online events (Koroed et al. 2024).

Generally, the amount of information provision on digital tools in the agricultural sector is impressive, although difficult to structure and oversee. Therefore, the study focused on information provided by public authorities, notably competence centres and by national research and development programmes, such as the already mentioned programme 'field trials'. Overall, the provision of information plays a central role for these actors, particularly with regard to technological developments, market conditions, and legal requirements. In interviews, experts emphasised the challenge to channel the flood of information and to make it actually useful and relevant to farmers. This means, e.g., the integration of information with the offers of the education and advisory subsystem actors to ensure that the information is easily accessible, understandable, and immediately applicable (Koroed et al. 2024).

Summarising, the emerging picture of education and support options regarding digitalisation dedicated to farmers is diverse, and obviously, the offerings in the different federal states are uneven and not harmonised. Also, the heterogeneity of both topics and providers renders the creation of transparency a challenge and makes coordination and cooperation within the AKIS without further efforts and resources impossible. It is not surprising that intended users and beneficiaries of these offers may not easily find their way into the measures and programmes.

4 Conclusions

In this chapter, we aim at exploring the role of knowledge and learning for the transition of the agri-food sector and specifically for farmers towards a more integrated, systemic response to major challenges, which at the same time makes use of and profits from the advantages of digitalisation. We ground our conceptual approach in the agri-food ecosystem model and apply the AKIS concept as a framework for the analysis of knowledge provision for and learning among farmers and other actors. We make use of empirical data from Germany to report on the state of knowledge, and to illustrate and analyse how digitalisation becomes part of the body of knowledge of AKIS actors and matters for organisations and networks. In our conclusions, we highlight three aspects, namely the emerging collective of AKIS actors and infrastructures that interplay to support digitalisation, the degree of engagement from the farmers' side, and the roles and functions that advisory and education subsystems play in this respect.

4.1 No AKIS in Baden-Württemberg Yet that Suitably Supports Digitalisation

Looking at the example of the AKIS in Baden-Württemberg, focusing on the actors recognised for explicit engagement with agricultural digitalisation and on the linkages among them and with the ultimately envisaged beneficiaries, the farmers, it becomes evident that

- Knowledge, expertise, and interests to support digitalisation in agriculture are unevenly spread among and within the actor groups,
- More expertise and resources for digitalisation are located at national and state levels than at local and regional levels, while higher-level actors seem to have low to no direct linkages in this matter with farmers, only a certain number of farmers, notably those running pilot or demonstration farms and farmers with an entrepreneurial mindset, engage directly with private companies or publicly funded research projects and learn and profit from these close connections; and that impacts of such cooperation can not be observed in the wider community of the farmers or at the state level of AKIS.

Overall, we deduce that Baden-Württemberg is still at an early stage of a well-functioning 'digitalisation AKIS' with a need for (i) an improved creation of transparency of information and support options and opportunities, (ii) a raising awareness among all actors to be part of a knowledge system that profits from knowledge exchange and cooperation, and (iii) the adoption of mid- to long-term perspectives for knowledge integration even in the frame of short-term projects and initiatives.

Furthermore, the analysis shows and raises awareness on the group of private entrepreneurial actors that are quite numerously represented in the AKIS at local and regional levels, while little is known about the effectiveness of their contributions to information dissemination and knowledge exchange with farmers (cf. Kernecker et al. 2020).

4.2 How Fit Are Farmers for Digitalisation?

The picture of farmers' engagement with digital technologies gains more and more facets, although a widely observed attitude of reservation to adoption seems to decrease only slowly. The data presented for Germany and notably, Baden-Württemberg indicate a differentiated assessment of own knowledge and skills with respect to digital tools' use and adaptation, which reflects variations of farmers' interests, needs, and competences, but also structural effects of, e.g. farm size or farmers' level of formal education.

The data also shows that farmers proactively engage with a range of information sources to increase their knowledge and expertise in this field. However, they tend to use familiar and easily approachable sources and AKIS partners, which may not always be the best-informed ones.

Given the fact that there is no representative study, it is difficult to estimate how many farmers use further education and lifelong learning options to stay updated and increase their digital skills and knowledge. In the studied case, the self-assessment is quite positive and the perceived needs are moderate. However, this may be only one perspective of a well-informed, technology-savvy sample, while a national survey indicates that political support for further education measures is strongly welcomed.

Some of the facts are clearly ambiguous, if not contradictory: on the one hand, farmers consider themselves to be sufficiently or even comprehensively competent in dealing with and applying digital technologies, but a broader adoption in operational practice remains limited. On the other hand, farmers express the expectation that politicians should do more to promote education and training, while their own interest in taking advantage of such opportunities is significantly lower. Although these findings do not originate from the same group of farmers, there is likely a research gap regarding what content should be offered with what methods and media to effectively reach a broad array of practitioners and respond to their learning needs and interests.

4.3 *What Activities and Measures from Advisory and Education System Actors?*

Politically, there is a widely shared expectation that advisory services and education and training are effective means to support the knowledge exchange and learning that matters for the transition of the agri-food sector towards more sustainable practices. From the cases and data that were presented here, we can see that there are both relevant and effective initiatives of various actors from these two subsystems to engage with the challenge of supporting digitalisation in agriculture and eminent gaps and deficits in providing services and reaching farmers with their questions and needs.

Several remarkable initiatives were conducted by public sector actors and members of agricultural chambers. Both are organisations with a long-term perspective, considering not or not only for-profit interests, but also societal objectives and political expectations. In contrast, private sector advisory bodies have less incentives to embrace complex topics that may leverage not only advantages but also re-thinking and re-organisation issues with transaction costs, etc. Possibly, the recent trend towards privatisation of service provision may constitute a hindrance to a proactive spread of information and skills for digitalisation in the context of advisory services, an issue that would merit further studies. As for the training and education actors, here we constat a huge diversity of organisations which may make an effective support to the spread of digitalisation a challenge. The sketch of the AKIS situation in Baden-Württemberg revealed that there is little exchange and even less coordination among these bodies, so that the effectiveness of education and training offers remains very low. While some authors argue that technological changes come with the need for reskilling and upskilling (Daum 2025), there is still a challenge for teachers, trainers, and advisors as there is not (yet) a body of generic content, methods, and curricula to be shared and applied. Although services and structures supportive of farmers' knowledge acquisition are available, they are, however, fragmented, with gaps and deficits regarding coordination and quality management of content. Cooperation among AKIS intermediaries would respond to the increasingly specialised needs and interests of several actors. This, however, comes with structural and communication challenges, at least in pluralistic AKIS configurations in Germany (Knierim et al. 2015).

4.4 *Does the Agri-Food Ecosystem Concept Support Knowledge and Learning?*

Regarding the functionality of the agri-food system perspective and its progression as an agri-food ecosystem, we conclude that they are conducive to illustrate the bigger picture, especially for institutionalised actors in research and political decision-making spheres. They particularly allow for the recognition of the multiplicity of actors as well as their interactions and interplay that lead to changes and transitions. Making use of the AKIS concept in policies and research programmes gives a focus to actors and activities that matter for effective knowledge provision and learning, and interaction facilities.

Still, challenges related to digitalisation in agriculture are manifold, and the information and learning options and venues farmers and other actors have, are diverse and driven by practical needs, so that an integrative system perspective alone may not be sufficient to provide adequately concrete and detailed information. Those who offer information access and learning and problem-solving support must foresee and design individual, group- and network-based approaches. Very apparently, the understanding of human-tech spheres' interfaces and the question of how the respective actants mutually shape the degree of interaction and penetration matter and require multiple solutions, tailored to the concrete situations of technological, economic, and social conditions. Learning theory concepts and approaches from adult education and research on lifelong learning conditions should be considered. Pluralistic AKISs have the potential to offer a diversity of possible learning occasions, constellations, and formats. In this respect, not only directive government interventions, but equally AKIS governance efforts and initiatives are necessary from various sides. Local, regional and sectoral, small, steady and sometimes laborious steps will lead to the goal. Many actors have already begun this process in many places.

Competing Interests The author(s) has no competing interests to declare that are relevant to the content of this manuscript.

References

Ackoff RR (1989) From data to wisdom. J Appl Syst Anal 16:3–9

Adomavicius G, Bockstedt JC, Gupta A, Kaufmann RJ (2008) Making sense of technology trends in the information technology landscape. MIS Q 32(4):779–809. Available at: https://ink.library.smu.edu.sg/sis_research/2125

Ananiadou K, Claro M (2009) 21st century skills and competences for new millennium learners in OECD countries. OECD education working papers, no. 41. OECD publishing

ANL (Akademie für Naturschutz und Landschaftsökologie) (ed) (1991) Begriffe aus der Ökologie, Umweltschutz und Landnutzung, 2nd edn. ANL-Informationen Nr 4

Augère-Granier ML (2017) Agricultural education and lifelong training in the EU, Briefing. European Parliamentary Research Service EPRS. https://www.europarl.europa.eu/RegData/etudes/BRIE/2017/608788/EPRS_BRI(2017)608788_EN.pdf

Birke F, Bae S, Schober A, Wolf S, Gerster-Bentaya M, Knierim A (2022) AKIS in European countries: cross analysis of AKIS country reports from the i2connect project. Available under https://i2connect-h2020.eu/wp-content/uploads/2022/12/2022-12-02-AKIS-cross-analysis_updated.pdf

Blackler F (1995) Knowledge, knowledge work and organizations: an overview and interpretation. Organ Stud 16(6):1021–1010

Blum A (1991) What can be learned from a comparison of two agricultural knowledge systems? The case of The Netherlands and Israel. Agric Ecosyst Environ 33:325–339

Brunori G, Bacco M, Puerta-Piñero C, Borzacchiello MT, Stormer E (2025) Agri-food data spaces: highlighting the need for a farm-centered strategy. Data Brief 59:111388. https://doi.org/10.1016/j.dib.2025.111388

Carolan M (2017) Publicising food: big data, precision agriculture, and co-experimental techniques of addition. Sociol Rural 57(2):135–154. https://doi.org/10.1111/soru.12120

Charatsari C, Lioutas ED, Papadaki-Klavdianou A, Michailidis A, Partalidou M (2022) Farm advisors amid the transition to Agriculture 4.0: professional identity, conceptions of the future and future-specific competencies. Sociol Rural 62(2):335–362. https://doi.org/10.1111/soru.12364

Daum T (2025) Digitalization and skills in agriculture. Outlook Agric 54(2):171–181. https://doi.org/10.1177/00307270251336474

Destatis (Statistisches Bundesamt) (2021) Statistischer Bericht. Land- und Forstwirtschaft, Fischerei: Landwirtschaftliche Betriebe. URL: https://www.destatis.de/DE/Themen/Branchen-Unternehmen/Landwirtschaft-Forstwirtschaft-Fischerei/Landwirtschaftliche-Betriebe/_inhalt.html#238418 (29.12.2022)

Destatis (Statistisches Bundesamt) (2023) Statistischer Bericht. Land- und Forstwirtschaft, Fischerei: Landwirtschaftliche Betriebe - Arbeitskräfte und Berufsbildung der Betriebsleitung/Geschäftsführung. Published 03.05.2024. https://www.destatis.de/DE/Themen/Branchen-Unternehmen/Landwirtschaft-Forstwirtschaft-Fischerei/_inhalt.html (01.06.2025)

EC COM (2020): Communication from the commission to the european parliament, the council, the european economic and social committee and the committee of the regions. A farm to fork strategy. Communication 381 final. https://eur-lex.europa.eu/legal-content/EN/TXT/?uri=CELEX:52020DC0381

EC COM (2025): Communication from the commission to the european parliament, the council, the european economic and social committee and the committee of the regions. A vision for agriculture and food shaping together an attractive farming and agri-food sector for future generations. COM(2025) 75 final. https://eur-lex.europa.eu/legal-content/EN/TXT/PDF/?uri=CELEX:52025DC0075

EU (2021) Regulation (EU) 2021/2115 of The European Parliament and of the Council of 2 December 2021 establishing rules on support for strategic plans to be drawn up by Member States under the common agricultural policy (CAP Strategic Plans) and financed by the European Agricultural Guarantee Fund (EAGF) and by the European Agricultural Fund for Rural Development (EAFRD) and repealing Regulations (EU) No 1305/2013 and (EU) No 1307/2013. OJ L 435, 6.12.2021, p. 1–186

EU SCAR (2012) Agricultural knowledge and innovation systems in transition—a reflection paper. Standing committee on agricultural research (SCAR) collaborative working group AKIS, EC, Brussels. Available online at: https://scar-europe.org/akis-documents

Ewert F, Baatz R, Finger R (2023) Agroecology for a sustainable agriculture and food system: from local solutions to large-scale adoption. Ann Rev Resour Econ 15:351–381. https://doi.org/10.1146/annurev-resource-102422-090105

Finger R (2023) Digital innovations for sustainable and resilient agricultural systems. Eur Rev Agric Econ 50(4):1277–1309. https://doi.org/10.1093/erae/jbad021

Friesike S, Sprondel J (2022) Träge transformation. Welche Denkfehler den digitalen Wandel blockieren. Reclam

Gabriel A, Gandorfer M (2023) Adoption of digital technologies in agriculture—an inventory in a european small-scale farming region. Precis Agric 24(1):68–91. https://doi.org/10.1007/s11119-022-09931-1

Glover D, Sumberg J (2025) Skill and deskilling in an uncertain agricultural world: introduction to the special issue. Outlook Agric 54(2):123–132. https://doi.org/10.1177/00307270251337112

Goller M, Caruso C, Harteis C (2021) Digitalisation in agriculture: knowledge and learning requirements of German dairy farmers. Int J Res Vocat Educ Train 8(2):208–223. https://doi.org/10.13152/IJRVET.8.2.4

Hirsch Hadorn G, Hoffmann-Riem H, Biber-Klemm S, Grossenbacher-Mansuy W, Joye D, Pohl C, Wiesmann U, Zemp E (eds) (2008) Handbook of transdisciplinary research. Springer Netherlands

HLPE (2017) Nutrition and food systems. A report by the High Level Panel of Experts on Food Security and Nutrition of the Committee on World Food Security, Rome. https://icdasustainability.org/report/nutrition-and-food-systems/

Hoffmann V, Probst K, Christinck A (2007) Farmers and researchers: how can collaborative advantages be created in participatory research and technology development? Agric Hum Values 24(3):355–368

Ingram J, Maye D (2020) What are the implications of digitalisation for agricultural knowledge? Front Sustainable Food Syst 4:66. https://doi.org/10.3389/fsufs.2020.00066

Kernecker M, Knierim A, Wurbs A et al (2020) Experience versus expectation: farmers' perceptions of smart farming technologies for cropping systems across Europe. Precision Agric 21:34–50. https://doi.org/10.1007/s11119-019-09651-z

Kernecker M, Busse M, Knierim A (2021) Exploring actors, their constellations, and roles in digital agricultural innovations. Agric Syst 186:102952. https://doi.org/10.1016/j.agsy.2020.102952

Klerkx L, Jakku E, Labarthe P (2019) A review of social science on digital agriculture, smart farming and agriculture 4.0: new contributions and a future research agenda. NJAS Wageningen J Life Sci 90–91(1):1–16. https://doi.org/10.1016/j.njas.2019.100315

Knierim A, Birke FM (2023) Visualised AKIS diagnosis—an instrumental approach to support AKIS appraisal. EuroChoices 22(2):59–70. https://doi.org/10.1111/1746-692X.12397

Knierim A, Boenning K, Caggiano M, Cristóvão A, Dirimanova V, Koehnen T et al (2015) The AKIS concept and its relevance in selected EU member states. Outlook Agric 44(1):29–36. https://doi.org/10.5367/oa.2015.0194

Knierim A, Kernecker M, Erdle K, Kraus T, Borges F, Wurbs A (2019) Smart farming technology innovations—insights and reflections from the German Smart-AKIS hub. NJAS 90–91:100314. https://doi.org/10.1016/j.njas.2019.100314

Knierim A, Birke FM, Bae S, Schober A, Gerster-Bentaya M, Asensio P (2022) Der AKIS Ansatz—(wie) unterstützt er die Akteure im Sektor? Berichte über Landwirtschaft - Zeitschrift für Agrarpolitik und Landwirtschaft, Aktuelle Beiträge. https://doi.org/10.12767/BUEL.V100I1.378

Köhler K, Tobler S (2025) Yes, we can: empowering 21st century skills in a large introductory classroom through project-based learning. ETH Learn Teach J 5(1):2025. Department of Biology (D-BIOL), ETH Zurich

Koroed O, Gutekunst L, Ueltgesforth J, Knierim A (2024) Beratungsangebote zur Digitalisierung in der Landwirtschaft. Bericht im Rahmen von DiWenkLa, TP 3. https://hohpublica.uni-hohenheim.de/handle/123456789/17375 and https://doi.org/10.60848/12320

McGreevy SR, Rupprecht CDD, Niles D et al (2022) Sustainable agrifood systems for a post-growth world. Nat Sustainability 5:1011–1017. https://doi.org/10.1038/s41893-022-00933-5

Monteiro Moretti D, Baum CM, Ehlers M, Finger R, Bröring S (2023) Exploring actors' perceptions of the precision agriculture innovation system—a group concept mapping approach in Germany and Switzerland. Technol Forecast Soc Chang 189:122270. https://doi.org/10.1016/j.techfore.2022.122270

OECD (2024) Education at a glance 2024: OECD indicators. OECD Publishing, Paris. https://doi.org/10.1787/c00cad36-en

Paulus M, Knierim A (2023) Wissenstransfer zu digitalen Technologien in Baden-Württemberg: Ergebnisse einer Betriebsbefragung. Austrian J Agric Econ Rural Stud 32:97–104. https://doi.org/10.15203/OEGA_32.12

Pfaff SA, Paulus M, Schüle H, Thomas A (2023) Werden Landwirte zu IT-Spezialisten? Ein Mixed-Methods-Ansatz zur Erfassung der digitalen Kompetenzanforderungen und des Kompetenzstands der Betriebsleiter in Baden-Württemberg. Berichte über Landwirtschaft 101(2)

Pigford AAE, Hickey GM, Klerkx L (2018) Beyond agricultural innovation systems? Exploring an agricultural innovation ecosystems approach for niche design and development in sustainability transitions. Agric Syst 164:116–121

Polanyi M (1966) The tacit dimension. Chicago Press
Richards P (1985) Indigenous agricultural revolution. Ecology and food crops in West Africa. HarperCollins Publishers Ltd
Richardson K et al (2023) Earth beyond six of nine planetary boundaries. Sci Adv 9:eadh2458. https://doi.org/10.1126/sciadv.adh2458
Rockström J, Edenhofer O, Gaertner J, DeClerck F (2020) Planet-proofing the global food system. Nat Food 1:3–5. https://doi.org/10.1038/s43016-019-0010-4
Rohleder B, Meinel T (2022) Die Digitalisierung der Landwirtschaft. www.bitkom.org/
Sol J, Beers PJ, Wals AEJ (2012) Social learning in regional innovation networks: trust, commitment and reframing as emergent properties of interaction. J Clean Prod 49:35–43
Stone GD (2025) Skill talk: the struggle over agricultural decision-making. Outlook Agric 54(2):133–141. https://doi.org/10.1177/00307270251334169
Taylor M, Bhasme S (2025) Towards a theory of agrarian skilling (Or, why farmer knowledge does not stop at the edge of the field). Outlook Agric 54(2):142–149. https://doi.org/10.1177/00307270251335671. (Original work published 2025)
Walter A, Finger R, Huber R, Buchmann N (2017) Smart farming is key to developing sustainable agriculture. Proc Natl Acad Sci USA 114(24):6148–6150. https://doi.org/10.1073/pnas.1707462114
WBGU—German Advisory Council on Global Change (2019) Towards our common digital future. Flagship Report. WBGU, Berlin
Wolfert S, Ge L, Verdouw C, Bogaardt M-J (2017) Big data in smart farming–a review. Agric Syst 153:69–80. https://doi.org/10.1016/j.agsy.2017.01.023
Wolfert S, Verdouw C, van Wassenaer L, Dolfsma W, Klerkx L (2023) Digital innovation ecosystem in agri-food: design principles and organizational framework. Agric Syst 204:103558. https://doi.org/10.1016/j.agsy.2022.103558
Zscheischler J, Brunsch R, Griepentrog HW, Tölle-Nolting C, Rogga S, Berger G, Lehmann B, Strobel-Unbehaun T, Reichel C, Ober S, Scholz RW (2021) Landwirtschaft, Digitalisierung und digitale Daten. In: Scholz R, Beckedahl M, Noller S, Renn O (eds) DiDaT Weißbuch Verantwortungsvoller Umgang mit digitalen Daten—Orientierungen eines transdisziplinären Prozesses. https://doi.org/10.5771/9783748924111

Science Center for Development in Digital Agriculture

Luciana A. S. Romani, Jayme G. A. Barbedo, Alberto Paradisi, Durval Dourado Neto, and Silvia M. F. S. Massruhá

1 Introduction

Brazil is one of the leaders in tropical agriculture due to its soil and climate conditions, public policies in science, technology, and innovation (ST&I), and the promotion of rural entrepreneurship. The Ministry of Agriculture, Livestock and Supply (MAPA) (MAPA 2021) projects that grain production will reach 330–380 million tons in the 2030/31 harvest. For animal protein production, projections indicate a growth from the current 27–37 million tons. There are also new demands for bioeconomy products such as cotton, cellulose, milk, sugar, and fruit. As a result, the agricultural sector's share of GDP has reached 21.4% (CEPEA 2025).

Demographic, socio-economic, and environmental changes have combined to condition the growing demand for food and energy security. In this sense, there are two technological paths to consider: (i) improving organisms through genetic selection for productivity gains amid climatic and/or health extremes and (ii) good agricultural practices based on the circular economy, i.e., the sustainable use of renewable and fossil resources.

L. A. S. Romani (✉) · J. G. A. Barbedo
Embrapa Digital Agriculture, Campinas, Brazil
e-mail: luciana.romani@embrapa.br

A. Paradisi
CPQD, Campinas, Brazil

D. D. Neto
Esalq, USP, Piracicaba, Brazil

S. M. F. S. Massruhá
Embrapa, Brasília, Brazil

F. da Silveira et al. (eds.), *Agriculture 4.0*,
https://doi.org/10.1007/978-3-032-24552-6_15

The digital transformation of agriculture can contribute by promoting innovations in biotechnology, geotechnologies, digital markets, distribution, logistics, and certification, among others. There are some new technologies identified as critical to digital transformation: edge/cloud computing, Internet of Things (IoT), social media and networks, big data, data science, artificial intelligence, augmented reality, robotics, connectivity, machine learning, digital twins, automation, biotechnology, bioinformatics, and nanotechnology (Massruhá et al. 2023).

To secure and expand its leading position in sustainable markets for food, fiber, and bioenergy, Brazil must overcome inequalities through Research, Development, and Innovation (RD&I). National and international demands for food, nutritional, socio-environmental, and climate security have grown significantly in recent years. As a result, agribusiness researchers and professionals should improve the dissemination of technologies associated with digital agriculture, even among small and medium-sized producers. From this perspective, RD&I associated with the use of digital technologies offers opportunities to include small and medium-sized producers in the ongoing digital transformation process, taking into account the regional diversity of production and natural resources. Another opportunity relates to promoting increased connectivity in rural areas, allowing the deployment of digital technological devices at a lower cost. Connecting multiple actors in production chains can lead to a reduction in transaction costs. In addition, the integration of digital technologies into rural production processes effectively contributes to generating socio-economic impacts in the short and medium term. Technological training and the sharing of knowledge and practices can also accelerate the adoption of digital technological assets by the markets. Innovative business and operational models in agricultural digital transformation can enhance a company's ability to attract investment and scale its operations. Furthermore, it is essential to implement and strengthen sustainable intensification models through integrated production systems to enhance agricultural efficiency and productivity.

Therefore, progress in the adoption of digital technologies is crucial for Brazilian agriculture to maintain its productive capacity in an economically, environmentally, and socially sustainable manner. The participation of small and medium-sized rural producers in this transformation of the countryside is essential. To this end, Embrapa and the São Paulo State Research Foundation (FAPESP) have established the Science Center for the Development of Digital Agriculture (CCD-AD/Semear Digital).

The project aims to advance knowledge and generate solutions that meet the needs of small and medium-sized rural producers. These will help to reduce market imperfections and inequalities in the adoption of digital technologies, thereby promoting productivity and competitiveness gains, reducing costs, and increasing the efficiency of agricultural production. Planned actions include the establishment of Agrotechnological Districts (DATs) in different regions of the country to promote connectivity solutions in rural areas, as well as integrating digital technologies into agricultural production processes. These technologies involve applications such as artificial intelligence, remote sensing, automation, precision agriculture, traceability, and certification. A DAT is an integrated system comprising hardware, software, and connectivity installed in a specific region, usually limited to one municipality, which aims to address the practical needs of rural producers. By providing training and incentives to the local ecosystem, the DAT acts as a model or showcase for digital technologies and services that solve rural problems, which can be quickly and easily adopted and integrated into production processes.

2 Background

Advances in digital technologies such as big data (Wolfert et al. 2017), machine learning (Liakos et al. 2018), deep learning (Kamilaris and Prenafeta-Boldú 2018), the Internet of Things (Verdouw et al. 2019), remote sensing (Kayad et al. 2020), mobile applications (Michels et al. 2020), connectivity (Zervopoulos et al. 2020), blockchain (Chen et al. 2020), and artificial intelligence (Abraham et al. 2020) have contributed to the development of agricultural chains.

The search for Industry 4.0 digital services and technologies (Schwab 2018) has shaped development agendas. This quest is reflected in the United Nations' 17 Sustainable Development Goals (SDGs), and it is projected that digital transformation in agriculture will contribute to achieving these goals. According to Alves et al. (2019), 66% of rural establishments in Brazil are classified as very poor, while 22.6% are considered poor. This poverty is essentially due to market imperfections (Alves et al. 2019). Digital transformation could impact this situation in various ways, depending on the governance models and incentives adopted. RD&I efforts can generate high-cost innovations that serve mostly large producers, accentuating inequalities. In this sense, the Semear Digital Center aims to seek inclusive digital solutions that mitigate market imperfections and promote sustainability.

Another agenda gaining relevance in the corporate and financial markets is the adoption of Environmental, Social, and Governance (ESG) practices. Innovations and digital technologies have emerged to support organizations in implementing ESG policies and to facilitate hybrid working. New paradigms are also emerging. One example is the metaverse, which aims to integrate the physical and cyber worlds into a collective and shared space. This virtual space includes technologies such as virtual and augmented reality, involving concepts like social networking, digital twins, robots, and cyber bots.

Small producers face challenges due to a lack of resources and training, as well as geographical distance and poor internet connectivity. These factors restrict their ability to participate in non-digital sustainability initiatives (Kruk et al. 2021). Inclusive digital technologies offer alternatives to overcome such limitations, enabling the acquisition of data from the interaction between producers and natural resources and making this information accessible throughout the production and consumption chain.

In addition to mitigating market asymmetries and adjusting governance agendas, inclusive digital technologies have the potential to increase the competitiveness of Brazilian agricultural products and services by promoting gains in scale, quality, and production efficiency. From this perspective, Brazil is well placed to provide other countries with technological solutions for digital agriculture, as well as solutions for the sanitary control of agricultural commodities through tracking systems. For Brazil, the projected economic gain from digital technology innovations can reach US$ 21.1 billion in 2025 (Manyika et al. 2015).

In this context, the applications, challenges, and prospects of using digital technologies stand out (Bolfe et al. 2020). Among the advantages perceived by farmers were increased productivity, labor efficiency, production quality, and reduced environmental impact. The main challenges were high investment costs, limited Internet access, and the need for training in digital agriculture.

Small and medium-sized producers are demanding digital technologies "from the farm gate inwards." These technologies must be able to: (i) characterize the state of crops to enable appropriate interventions throughout the crop cycle (e.g., irrigation, fumigation, and fertilization), and (ii) increase productivity, saving resources and ensuring compliance. The main questions are: which sensors and data sources can be used; how this data relates to agronomic variables; and how this data can be obtained on the required scale and frequency, and at what cost. There is also a communication gap: how can this data be transmitted so that it can be integrated, analyzed, and returned to producers and autonomous systems to allow decision-making? From the "farm gate outwards," there is a need for traceability and certification, which adds value to the product from the farmer's perspective. For companies, it represents an opportunity in terms of the ESG agenda. These challenges are too substantial to be addressed by producers and the market alone within a short timeframe. Therefore, there is a sense of urgency for government action to optimize the articulation in the business environment.

A study by (Dibbern et al. 2025) analyzed the adoption of digital agriculture in Brazil, investigating the barriers and drivers of this process across different biomes. Central barriers included limited digital literacy, financial constraints, labor shortages, limited accessibility of service providers, and inadequate infrastructure. The drivers encompassed increased productivity, reduced costs, improved management, greater infrastructure availability and producer education. The main challenge lies in generating appropriate and affordable technologies that meet the needs of small and medium-sized producers. In this regard, establishing effective partnerships with the private sector that are committed to ESG principles is a priority. Furthermore, it is crucial to generate strategic policy interventions.

There is also a need to strengthen certification and traceability processes using digital technology, adding value to rural production chains (Gazolla and Aquino 2022). Food and health safety is driving new applications from input suppliers to rural producers, from marketing to transportation, and from distribution to end consumers (Dias et al. 2023).

3 Center of Science for Development in Digital Agriculture—Semear Digital

Semear Digital is a Science Center for the Development of Digital Agriculture funded by the São Paulo State Research Foundation (FAPESP). Its mission is to conduct Research, Development, and Innovation (RD&I) in emerging technologies to promote digital inclusion among small and medium-sized rural producers. The goal is to enhance productivity and competitiveness and to increase the economic impact by reducing costs and improving the efficiency of agricultural production at various stages of the production chain in a sustainable manner, taking into account economic, environmental, and social factors. To achieve its mission, the Center proposes solutions adapted to the various rural realities in the country, taking into account the regional diversity of production and natural resources, to reduce transaction costs among the multiple players in the production chains, improve socio-economic impacts in the short and medium term, promote training, increase the adoption of digital technological assets by producers and encourage new investments so that this digitization gains scalability.

Semear Digital is the product of a collaboration between the Brazilian Agricultural Research Corporation (Embrapa), the Telecommunications Research and Development Centre (CPQD), the Agronomic Institute of Campinas (IAC), the Institute of Agricultural Economics (IEA), the National Telecommunications Institute (Inatel), the Luiz de Queiroz College of Agriculture at the University of São Paulo (Esalq/USP), and the Federal University of Lavras (UFLA). These institutions collaborate and work together in an integrated way on different lines of research. In 2025, the project had over 120 people in its ranks.

The Semear Digital Center researches inclusive digital solutions for small and medium-sized producers, aiming to mitigate market imperfections and promote sustainability. For these producers, a lack of resources and training, as well as distance and disconnection, limits the effectiveness of non-digital sustainability initiatives. These initiatives include corporate social responsibility programs, certification, traceability, and extension services, which rely on costly sampling and documentation methods. Digital technologies are proving to be an effective way of overcoming these limitations by enabling the collection of data on interactions between producers and the use of natural resources, making this information accessible throughout the production and consumption chain. Digital technologies can also enhance scale, quality, and production efficiency.

The multidisciplinary and multi-institutional research, development, and innovation activities have been implemented in production environments (farms, ranches, etc.) in ten selected DATs: Alto Alegre (SP), Boa Vista do Tupim (BA), Breves (PA), Caconde (SP), Guia Lopes da Laguna (MS), Ingaí (MG), Jacupiranga (SP), Lagoinha (SP), São Miguel Arcanjo (SP), and Vacaria (RS) (Fig. 1).

The participating institutions have conducted the research activities collaboratively through six thematic axes. Axis 1 focuses on "Agrotechnological Districts and Socioeconomic Impact Assessment." Axis 2 focuses on "Connectivity and

Fig. 1 Maps with DATs in Brazil

Devices." Axis 3 focuses on "Artificial Intelligence and Remote Sensing." Axis 4 focuses on "Automation and Precision Agriculture." Axis 5 focuses on "Traceability and Digital Certification." Finally, Axis 6 focuses on communication and establishing partnerships.

3.1 *Axis 1: DAT and Socio-Economic Impact Assessment*

A methodology based on national indicators was used to define the DATs, such as Technical Assistance, Economy, Education, Land Structure, Infrastructure, Social Organizations, Population, and Social. This methodology also includes monitoring and evaluating the socio-economic impacts of the actions implemented by the other thematic axes within each DAT.

After selecting the municipalities of each DAT, the team prepared a socio-economic diagnosis for each location. These analyses determined the predominant agricultural activities, which were based either on cultivated area or economic representativeness. The aim was to align the needs of the production chains (CEPEA 2025) with the digital needs of producers and the available technological solutions. Each diagnosis was based on three themes: (a) socio-economic, (b) local infrastructure, and (c) support network. The first involves aspects such as the producer's level of education, production value, and agricultural and non-agricultural economic activities.

In terms of local infrastructure, the following aspects are decisive for the adoption of digital technologies by producers (Abraham et al. 2020): electricity supply, signal coverage from internet and telephone operators, as well as others that complement the diagnosis. These aspects consider the conditions of the DATs as laboratories for implementing the technologies to be developed in the project's thematic areas.

The support network addressed the level of social organization of rural producers in associations, and/or agro-industrial production and credit cooperatives, as these play a crucial role in identifying demands and finding technological solutions.

In the second stage, the team collected primary data and consolidated the results (T0). In the third stage, named T1, the authors will collect primary data from the producers two harvests after T0. This data will determine the additionality, causality, and impacts on the proposed dimensions.

3.2 Axis 2: Connectivity and Devices

Research teams are advancing connectivity solutions for rural and remote areas through the development of technologies for 5G and beyond. Key efforts include designing dual-band dielectric resonator antennas for Integrated Access and Backhaul (IAB) networks, deploying hybrid fiber/wireless architectures with Radio over Fiber (RoF) and Free Space Optics (FSO), and implementing cognitive radio communication over TV White Spaces. Innovations also extend to spectral sensing, with the development of a functional IoT prototype (SSIoT) and improvements in the mPRIDe detector, enabling efficient, low-latency spectrum monitoring. These initiatives are supported by fellowships focusing on algorithm development and hardware-software integration.

Hybrid network architectures combining terrestrial and non-terrestrial technologies are being modeled to expand rural coverage. These include scenarios using low Earth orbit satellites, Unmanned Aerial Vehicles (UAVs), and balloons alongside ground-based nodes. Several configurations explore regenerative and transparent satellite modes to optimize spectrum use and latency. These architectures are designed to support applications such as precision agriculture, autonomous vehicles, and large-scale IoT deployments. Practical experiments have shown high data throughput and energy efficiency, using technologies such as Power-over-Fiber and Visible Light Communication (VLC) integrated with RoF.

Future work includes refining antenna systems for full-spectrum 5G support, enhancing Open RAN frameworks, and developing simulation tools for network optimization in low-density regions. The integration of advanced radio resource management algorithms will address challenges related to spectral efficiency, latency, and network reliability. The aim is to provide scalable, cost-effective connectivity solutions for underserved areas, supporting digital inclusion and enabling emerging applications in smart farming and environmental monitoring.

3.3 Axis 3: Artificial Intelligence and Remote Sensing

The challenge of using artificial intelligence is to create algorithms able to cope with seasonal changes, the detection of new diseases, emerging pests, and new cultivars, among others. Most of the research carried out in the Center's AI axis is enabling, making it possible to create new technologies. In many cases, these technologies can be exploited in the form of services offered to producers, extension workers, or producer associations.

The AI research at Semear Digital involves the use of digital images and artificial intelligence for detecting and recognizing plant pests and diseases, as well as counting animals. Despite the potential of digital images combined with artificial intelligence to solve agricultural problems, there are often cases where images do not carry enough information to provide satisfactory answers. One possible solution is the fusion of different types of data using specific techniques. The biggest challenge involved in this type of approach is finding a way to effectively exploit the synergy and complementarity that exists between different types of data. Besides images, other types of data are being collected: soil, meteorological, and historical data related to the specific issues being addressed, such as the incidence of a particular disease.

Axis number three also carries out work in the field of Remote Sensing. Research in this area involves organizing geospatial databases and analyzing the spatio-temporal dynamics of land use and land cover at a regional level. It also focuses on improving mapping and monitoring methodologies through multi-sensor data harmonization/fusion techniques into crops, pastures, fruit growing, agroforestry systems, and crop-livestock-forest integration.

3.4 Axis 4: Automation and Precision Agriculture

Precision Agriculture (PA) is defined as a management strategy that gathers, processes, and analyzes temporal, individual, and spatial data, combining them with other information to support management decisions, considering estimated variability. The aim is to make the use of resources more efficient, increasing productivity, quality, profitability, and sustainability of agricultural production.

Since producers of small and medium-sized farms rarely collect data that can be used to consider PA concepts, the Center has invested in research in the area of automation and computer vision, developing prototypes for automatic data capture. Field scanning vehicles, capable of traveling through orchards between rows, have been used in precision fruit growing, and also vehicles equipped with sensors (cameras, GPS LIDARs, etc.) for large-scale terrain scanning and AI-based pattern detection.

3.5 *Axis 5: Traceability and Digital Certification*

Digital models for certifying the origin and traceability of products and services can be customized in a variety of ways, from simple systems based on auditing, inspection, and document scanning, to larger systems based on interconnected devices and sensors/actuators, which can provide a range of detailed information about a production system in rural establishments, associations, cooperatives, etc. To achieve this objective, two actions are carried out:

1. Considering the target audience of small and medium-sized producers, priority is given to the development and validation of intuitive, inexpensive, and independent digital technologies with persistent connectivity for auditable digital data through geolocation and the tokenization of smart contracts aimed at certifying geographical origin for traceability purposes.
2. Development of more elaborate models for generating auditable digital data with smart contracts to certify the traceability of processes for higher value-added product chains, and to intermediate and final links in production chains, such as small and medium-sized agro-industries and retailers

3.6 *Axis 6: Partnerships, Innovation, and Scientific Dissemination*

This axis aims to connect the DATs with ICTs and companies that generate technological solutions. Innovation actions aim to increase the generation of wealth in agribusiness through inclusive digital technologies, producing economic impacts in terms of the quantity, quality, and efficiency of rural production. Thus, new projects are incorporated to contribute to the Center's mission. These new projects bring new researchers, institutions, and assets generated.

Knowledge creation arises from the interaction between various areas of knowledge and different stakeholders in a complex, multidisciplinary, and networked manner. A technical-scientific mode of production can be employed to this end (Gibbons et al. 1994), with the aim of solving practical problems and advancing technology through applied research. The challenges related to communication and dissemination of research results, technologies, and innovations, as well as actions linked to the characteristics of the Semear Digital Center, are directly related to the target audiences: (i) small and medium-sized rural producers; (ii) extension workers; (iii) public managers; (iv) other actors, which could be agritechs, agro hubs, machinery and equipment companies, and input companies; and (v) society in general, to better understand the rural infrastructure for the production and marketing of agricultural products. To this end, the communication team has used various communication channels, like websites, newsletters, bulletins, and social networks.

4 Results and Discussion

The goals of the Center's first year focused on: determining the municipalities where the DATs would be installed, surveying the situation in the DAT areas before the Center began operating (T0), and starting research activities.

The definition of the DATs employed a methodology that took into account 33 indicators derived from national data, which were used to rank the municipalities in each of the regions considered. Based on this ranking, the ten municipalities closest to the median were selected. This criterion was used to avoid selecting municipalities with too favorable or precarious situation. The main results were a methodology for defining DATs, a database of indicators for all Brazilian municipalities, and a dashboard developed using Business Intelligence software. These actions have led to a considerable volume of publications.

Digital solutions are also being developed and validated:

- An application for managing and supporting tilapia production called "Aquicultura certa," which uses generative AI.
- Development of the "Manejatech" application for Açaí management.
- Improved applications for bee breeding (Zapbee/Infobee), which also use AI.
- Prototype of a solution for imaging and counting fruit, called SEEmear.

The Center has also generated two new projects that are currently underway: ArmaTrigo and CaféDigital. In addition, partnerships have been signed with companies and institutions (machinery/equipment, educational institutions, cooperatives, associations, rural unions, environmental companies, banks, etc.) that are supporting different actions at the Center.

As mentioned earlier, connectivity is essential for the use of digital technologies. For this reason, Semear Digital has sought support and resources to expand network coverage in rural areas where the project operates. One of the actions taken is the integration of national rural connectivity and innovation programs, such as the Wi-Fi Brasil program (GESAC[1]) of the Ministry of Communications and the ConectarAgro network.[2]

Another important aspect of the project is generating data to support the development of public agricultural policies. This information includes reports on the demands and socioeconomic data of producers operating in over 15 production chains across four national biomes. Most of these reports indicate the need for agricultural equipment that can only be purchased through government subsidies or special lines of credit.

In order to expand its role in public policy, one of the project's next steps is to review existing policies to determine their suitability for small and medium-sized producers. Additionally, the project intends to improve the dissemination of these policies and seek partners who can support producers in utilizing public policies.

Finally, Brazil has a long and significant history of Research, Development, and Innovation (RD&I) in tropical agriculture, and more recently in digital agriculture. The successful Semear Digital initiative demonstrates that it is possible to develop and promote the adoption of digital technologies among small and medium-sized producers. Some of these opportunities are:

(i) Integrate producers into the ongoing digital transformation process while considering regional production diversity and natural resources.
(ii) Promote connectivity of electronic devices in the field to reduce the implementation costs of digital technological assets.
(iii) Connect multiple actors in production chains to reduce transaction costs.
(iv) Integrate digital technologies into rural production processes effectively and simply, with the aim of generating socioeconomic impacts in the short and medium term.
(v) Promote technological training and the sharing of knowledge and practices to accelerate the adoption of digital technologies by the market.
(vi) Enable new business models and operations in the digital transformation process to attract investment for scalability.
(vii) Induce and strengthen models of sustainable intensification through integrated production systems to increase agricultural efficiency and productivity.

The Axis 6 team is also dedicated to scientific communication and dissemination actions to reach different audiences. Over the last 2 years, various media have published more than 200 news articles about Semear Digital.

Over the course of 3 years, Semear Digital has established more than 30 national partnerships and three international ones.

[1] The Electronic Government Citizen Service Program (GESAC), coordinated by the Ministry of Communications (MCOM) since 2002, offers free satellite broadband internet connection to citizens, with the aim of promoting digital inclusion throughout Brazil.

[2] https://www.conectaragro.com.br/

In 2025, the project launched the 3-year Procisur Semear Digital pilot in Latin America, implementing DATs in Argentina, Chile, Paraguay, and Uruguay. The objectives of this initiative are:

- To establish a viable and sustainable model for the digital transformation of agriculture in different countries.
- To assess the economic, social, and environmental impact of sustained use of digital solutions in each pilot location and at the regional level.
- To generate information that improves access to digital technologies for small and medium-sized rural producers.

As a result, Brazil is positioning itself as a key player in the global digital agricultural networks.

5 Conclusions

This chapter presented the ongoing activities within the Science Center for Development in Digital Agriculture, offering a brief overview of the initiatives and context as of 2025. All three pillars (connectivity, education, and digital technologies) are being actively addressed, with numerous initiatives and the introduction of new technologies. The ultimate goal is to improve the lives of rural populations in the selected areas. Arguably, the most relevant factor in achieving this objective in all ten locations is the project's ability to involve the local agricultural ecosystem as part of the efforts.

The establishment of DATs is a continuous process. Progress varies between regions, influenced by factors such as the level of local engagement, the specific production chains involved, and the pre-existing conditions before the project implementation. Although varying degrees of success are anticipated, current developments suggest that positive results will be achieved in all targeted areas.

Following the successful implementation of the Semear Digital project in ten Brazilian municipalities, there has been a demand to expand its activities to new municipalities in Brazil and abroad. To this end, the Semear Digital methodology is being refined to support these new initiatives.

As several identified demands cannot be met by the project's end in 2027, plans for continuity and expansion are being proposed. Semear Digital has the potential to become a programme supporting RD&I in digital agriculture for small and medium-sized agricultural producers, as well as supporting public policies for this sector.

Acknowledgments The authors would like to thank Fapesp (Proc. 2022/09319-9) for funding.

Competing Interests This study was funded by Fapesp (2022/09319-9). The authors have no conflicts of interest that are relevant to the content of this chapter.

References

Abraham ER, Mendes dos Reis JG, Vendrametto O, de Oliveira Costa Neto PL, Carlo Toloi R, de Souza AE, Oliveira Morais MD (2020) Time series prediction with artificial neural networks: an analysis using Brazilian soybean production. Agriculture 10(10):475

Alves E, da Silva G, Mello PF, Marra R (2019) Market imperfections and rural poverty. Revista de Política Agrícola 28(4):77–77

Bolfe ÉL, Jorge LADC, Sanches IDA, Luchiari Júnior A, da Costa CC, Victoria DDC, Ramirez AR (2020) Precision and digital agriculture: adoption of technologies and perception of Brazilian farmers. Agriculture 10(12):653

CEPEA (2025) Center for advanced studies on applied economics. Brazilian Agribusiness GDP

Chen Y, Li Y, Li C (2020) Electronic agriculture, blockchain and digital agricultural democratization: origin, theory and application. J Clean Prod 268:122071

Dias CN, Jardim F, Sakuda LO (eds) (2023) Radar Agtech Brazil 2023: map of the Brazilian startups of the agricultural sector. Embrapa, SP Ventures e Homo Ludens, Brasília

Dibbern T, Romani L, Massruhá S (2025) Drivers and barriers to digital agriculture adoption: a mixed-methods analysis of challenges and opportunities in Latin American. Sustainability 17(8):3676

Gazolla M, Aquino JRD (2022) Reinvention of family farming markets in Brazil: the novelty of digital marketing sites and platforms in times of COVID-19. Agroecol Sustain Food Syst 46(6):902–927

Gibbons M, Limoges C, Scott P, Schwartzman S, Nowotny H (1994) The new production of knowledge: the dynamics of science and research in contemporary societies. Sage, London

Kamilaris A, Prenafeta-Boldú FX (2018) Deep learning in agriculture: a survey. Comput Electron Agric 147:70–90

Kayad A, Paraforos DS, Marinello F, Fountas S (2020) Latest advances in sensor applications in agriculture. Agriculture 10(8):362

Kruk SR, Kloppenburg S, Toonen HM, Bush SR (2021) Digitalizing environmental governance for smallholder participation in food systems. Earth Syst Gov 10:100125

Liakos KG, Busato P, Moshou D, Pearson S, Bochtis D (2018) Machine learning in agriculture: a review. Sensors 18(8):2674

Manyika, J., Chui, M., Bisson, P., Woetzel, J., Dobbs, R., Bughin, J., Aharon, D. (2015). The internet of things: mapping the value beyond the hype

MAPA (2021) Ministry of agriculture, livestock and supply. In: Agribusiness projections: Brazil 2020/21 to 2030/31 long-term projections. MAPA/ACE, Brasília

Massruhá SMFS, de Leite MAA, de Oliveira SR, Meira CAA, Luchiari Júnior A, Bolfe EL (2023) Digital agriculture: research, development and innovation in production chains. Embrapa, Brasília, p 299

Michels M, Bonke V, Musshoff O (2020) Understanding the adoption of smartphone apps in crop protection. Precis Agric 21:1209–1226

Schwab K (2018) Shaping the fourth industrial revolution. World Economic Forum 2018, p 352

Verdouw C, Sundmaeker H, Tekinerdogan B, Conzon D, Montanaro T (2019) Architecture framework of IoT-based food and farm systems: a multiple case study. Comput Electron Agric 165:104939

Wolfert S, Ge L, Verdouw C, Bogaardt MJ (2017) Big data in smart farming–a review. Agric Syst 153:69–80

Zervopoulos A, Tsipis A, Alvanou AG, Bezas K, Papamichail A, Vergis S, Oikonomou K (2020) Wireless sensor network synchronization for precision agriculture applications. Agriculture 10(3):89

Innovation Portfolio Management for Agrifood Ecosystem Transformation in the Context of Agriculture 4.0: Lessons from Private Sector and a CGIAR Case Study

Marc Schut, Julien Colomer, Emma Proud, Iddo Dror, Enrico Bonaiuti, Edwin Kangethe, Lorena Esquivias, Donald Menzies, Christopher B. Barrett, Ismail Adeyemi, Benjamin Kumpf, Dagmawi Melaku, and Cees Leeuwis

1 Introduction

Global food systems face enormous challenges, including climate change, malnutrition, and biodiversity losses (Myers et al. 2017; von Braun et al. 2023). These challenges compromise the livelihoods of billions of farmers, consumers, and other food system actors worldwide, particularly in low- and middle-income countries. There is an overall belief that innovations can make a positive contribution to the food systems transformation required to deal with these challenges (Reardon et al.

M. Schut
CGIAR Office of Chief Scientist - Portfolio Performance Team, Leuven, Belgium

Knowledge, Technology and Innovation Group, Wageningen University, Wageningen, the Netherlands
e-mail: marc.schut@wur.nl; m.schut@cgiar.org

J. Colomer
Portfolio Performance Unit, CGIAR System Organization, Montpellier, France

E. Proud · L. Esquivias
Utrecht, The Netherlands

I. Dror
Portfolio Performance Unit, CGIAR System Organization, Montpellier, France

International Livestock Research Institute (ILRI), Nairobi, Kenya

E. Bonaiuti
International Center for Agricultural Research in the Dry Areas, Beirut, Lebanon

International Potato Center, Lima, Peru

E. Kangethe
Portfolio Performance Unit, CGIAR System Organization, Nairobi, Kenya

F. da Silveira et al. (eds.), *Agriculture 4.0*,
https://doi.org/10.1007/978-3-032-24552-6_16

2019; Klerkx and Begemann 2020; Blakeney 2022; Thornton et al. 2024). Over the past two decades, innovation has slowly, but steadily, gained presence in public, non-for-profit international (agricultural) research and development cooperation (henceforth referred to as public R&D organizations), influenced by the ambitious UN 2030 Sustainable Development Goals (SDG) agenda (Silva 2021).

Innovations are defined as new or improved products, processes and services, or policy and institutional arrangements that intend to produce value (Nagji and Tuff 2012). Innovations can be of a technological nature (a machine, new seed variety, digital tool or device) or a nontechnological nature (capacity development approach, policy, business model, gender norms) (Geldes et al. 2017). Innovation is characterized by processes of maturation, where ideas evolve into fully developed products or services. Once there is sufficient confidence that an innovation is proven to work, there is often the aspiration to make it available to more clients or in other geographies. This is also referred to as innovation scaling—an important precondition for achieving impact at scale (Woltering et al. 2019; Schut et al. 2020).

Most public R&D organizations manage not just one, but a wide variety of innovations, innovation projects, and interventions aimed at delivering (positive) societal impact (Proud et al. 2023). This is being referred to as an innovation portfolio. Depending on the size of a public R&D organization, a global portfolio could easily comprise of 1000s of innovations under development and in use. Innovation portfolio management is the practice of intentionally guiding an organization's innovation investments and decisions to advance its vision and strategy (OECD 2021). The closure of USAID, alongside broader global political trends, has significantly reduced the availability of public R&D funding. This has created momentum for approaches like innovation portfolio management, which aim to optimize (scarce) innovation and scaling resource use efficiencies for greater development impact.

D. Menzies
FCDO Energy, Climate and Environment Directorate, Glasgow, Scotland

C. B. Barrett
Cornell University, Ithaca, New York, USA

I. Adeyemi
Kwara State University, Malete, Nigeria

B. Kumpf
OECD, Innovation for Development Facility, Paris, France

D. Melaku
Knowledge, Technology and Innovation Group, Wageningen University, Wageningen, the Netherlands

International Livestock Research Institute (ILRI), Nairobi, Kenya

C. Leeuwis
Knowledge, Technology and Innovation Group, Wageningen University, Wageningen, the Netherlands

Agriculture 4.0, defined from an innovation portfolio management perspective, is the strategic and systematic alignment, management, and scaling of diverse digital, technological, and nontechnological agricultural innovations within portfolios. It emphasizes structured processes, transparent criteria, data-driven decision-making, and adaptive resource allocation to maximize efficiency, agility, and sustainable impact. From this viewpoint, Agriculture 4.0 is less about individual technologies and more about intentionally managing interrelated innovation bundles (Barrett et al. 2022) or packages (Sartas et al. 2020)—e.g., combining new technologies with finance, market, and policy solutions—within a portfolio framework to ensure that investments collectively achieve transformative and responsible agrifood system outcomes aligned with broader societal goals and sustainable development priorities.

Despite the growing momentum and need for innovation portfolio management approaches, most public R&D organizations have shown limited adoption of such strategies, with well-documented public sector case studies and experiences being even more scarce (Holden et al. 2018; Megersa 2019; Schut et al. 2024a). The objective of this chapter is twofold. First, we draw lessons from innovation portfolio management in the private sector and explore how these can support innovation portfolio management in the public sector, as discussed in Schut et al. (2024a). Second, we present and analyze a case study of how CGIAR, one of the largest global public agrifood research and innovation networks, is mainstreaming innovation portfolio management with the objective to produce best practices that can benefit other public R&D organizations, building on a case study published by Schut et al. (2024b).

2 Learning from Innovation Portfolio Management in the Private Sector[1]

2.1 Literature Review Methodology

The academic literature on innovation portfolio management in the private sector forms the main source of information for this section. Databases Google Scholar, Emerald, ScienceDirect, Springer, and Taylor and Francis were searched for "innovation portfolio management" in their title, abstracts, and other searchable fields. The initial results were screened against the following criteria: (i) whether the article was published in the English language, (ii) whether the publication is published in a peer-reviewed international journal, (iii) whether innovation portfolio management is well emphasized (mentioned >5 times), and (iv) whether the study was published

[1] This section draws heavily on material originally presented in Schut et al. (2024a).

after 2011 to ensure its relevance. Fifteen (15) publications were analyzed in more detail (Appendix) to answer two key questions: (1) What are the reasons why private sector organizations invest in innovation portfolio management? (2) What are the success factors for innovation portfolio management?

2.2 Main Reasons Why Private Sector Organizations Invest in Innovation Portfolio Management

Innovation portfolio management is generally defined as a systematic approach to optimizing or aligning internal processes, resources, capacities, partnerships, and performance to achieve organizational mission, vision, and goals (Spieth and Lerch 2014; Urhahn and Spieth 2014; Martinsuo and Vuorinen 2019; Röth et al. 2019). The literature reveals four main reasons why private sector organizations invest in innovation portfolio management:

1. *Enhance Transparent and Quality Decision-Making*: Adopting a systematized process with predefined innovation stages, decision-making points, and assessment criteria for innovation management (also referred to as stage-gating: Cooper 2008) will improve transparency and trustworthiness in strategic decision-making and prioritization (Spieth and Lerch 2014; Kock and Gemünden 2016; Röth et al. 2019; Pashley et al. 2020).
2. *Strengthen Organizational Agility and Mission-Orientedness*: Monitoring of innovation portfolios increases the adaptive capacity of an organization to respond to a changing, turbulent market or funding environment (Brook and Pagnanelli 2014; Kock and Gemünden 2016; Martinsuo and Vuorinen 2019), whilst using the organizational mission as its compass (Attar and Abdul-Kareem 2020).
3. *Facilitate Monitoring and Risk Management*: The tracking of innovations helps understanding which innovations should be assigned resources, put on hold, or be discarded (Holtzman 2014; Urhahn and Spieth 2014; Röth et al. 2019; Pashley et al. 2020). Risk management refers to ensuring a healthy and diverse portfolio that can deliver against organizational objectives in the short and long terms. Risk-Reward is a key metric of innovation portfolio management (Holden et al. 2018; Cooper and Sommer 2023).
4. *Improve Organizational Innovation Performance*: Innovation portfolio management provides important insights into the efficiencies and inefficiencies of organizational innovation processes (Holtzman 2014; Klingebiel and Rammer 2014; Sicotte et al. 2014; Spieth and Lerch 2014; Behrens 2016; Üçler 2018; Toh and Ahuja 2022). Such learning can be used to optimize organizational innovation processes and performance.

2.3 *Success Factors for Innovation Portfolio Management in the Private Sector*

The literature highlights several key success factors for innovation portfolio management in the private sector that can inform the use of such approaches in the public R&D sector.

1. *Systems to Ensure Timely Access to Quality Innovation Data and Analytics*: Timely access to quality information, data, and metrics is a key success factor for innovation portfolio management. Data can focus on tracking internal research, manufacturing, distributing processes, and cost information (Röth et al. 2019), as well as the broader innovation ecosystem data on competing or coopting organizations, supply chains, customer needs, and uptake in a holistic way (Holtzman 2014; Kock and Gemünden 2016; Huvaj and Johnson 2019). Here, literature acknowledges the difficulties in designing methodology for measuring specific innovation metrics and tracking the dynamic nature of innovation processes (Brook and Pagnanelli 2014; Spieth and Lerch 2014). To ensure that the information is of sufficient quality, timely, and relevant, innovation management systems that provide procedures and protocols for data collection, curation, and reporting systems are essential (Urhahn and Spieth 2014; Üçler 2018). Data and systems should be able to inform decision-making and resource allocation based on harder criteria (e.g., return on investment, technological uncertainty, consumer demand) as well as softer criteria (e.g., opinions, company identity and reputation, management support, political sense-making) (Behrens 2016).
2. *Clarity on Assessment Criteria, Resource Allocation Strategies, and Governance Mechanisms*: Establishing explicit innovation assessment criteria aligned with organizational vision, such as balancing financial metrics with strategic and social objectives and managers' tacit business intelligence, prevents excessive risk-aversion and promotes a diversified innovation portfolio (Behrens 2016; Röth et al. 2019; Pashley et al. 2020; Cooper and Sommer 2023). Managing a portfolio of innovations by—for example—only assessing financial criteria (e.g., return on investment, profitability) may lead to favoring low-risk innovations (Cooper 2013) which may limit the organization's growth into new, niche markets (Üçler 2018; Huvaj and Johnson 2019; Pashley et al. 2020). A key prerequisite for resource allocation and prioritization is having clarity on an organization's innovation portfolio ambition (Klingebiel and Rammer 2014). Clarity on the specific portfolio objectives and outcomes (what to achieve) needs to go hand in hand with clarity on the path to reach those objectives and outcomes (how to achieve). Simply setting the goal is not enough; goals must be supported by transparent resource allocation strategies (Cooper 2013). Such resource allocation strategies may be calibrated differently at early and advanced innovation stages, ensuring appropriate investment breadth and depth, balancing the need for experimentation and selectivity (Klingebiel and Rammer

2014). Roles and responsibilities of the various actors involved in innovation portfolio management need to be clearly defined. Such governance processes and mechanisms are essential to manage the inherent tensions and trade-offs in innovation portfolio management (Röth et al. 2019).

3. *Invest in Organizational Innovation Capability and Culture Growth*: Effective innovation portfolio management requires strategic investment in organizational innovation culture and capacity. It is as important as having the right metrics and systems (Cooper 2013). Essential elements include fostering an innovation mindset, supporting creativity through experimentation and learning from failure, clearly defining success, using tailored metrics, and balancing risk to encourage diverse innovations (Holtzman 2014). Such innovation culture change should be mandated and nurtured by senior leadership in word and action. It is important to acknowledge that for employees, the systematized management of the creative innovation processes feels counterintuitive. This may result in resistance to innovation portfolio management. Sensitivity to employee concerns and employees' active involvement in innovation portfolio management design can enhance trust, capability development, and commitment (Nagji and Tuff 2012; Sicotte et al. 2014; Üçler 2018).

2.4 Benefit of Innovation Portfolio Management in the Public Sector

Especially for larger public nonprofit R&D organizations, the benefits of innovation portfolio management approaches seem obvious (Holden et al. 2018). Similar to the private sector, public nonprofit R&D operates in an unpredictable environment where resources are scarce, and making well-informed and transparent resource allocation decisions is important. Public nonprofit R&D often has a very clear mission, and innovation portfolio management can improve how organizations monitor and learn, manage risks, and prioritize investments against that mission (Urhahn and Spieth 2014; Üçler 2018). Having a systematic approach to innovation management can improve credibility and trust for both internal stakeholders (the researchers and innovators) as well as for external stakeholders (funders and partners).

There are several key differences between innovation portfolio management in the private and public nonprofit R&D sectors. Some of the typical differences and adaptation needs are summarized in Table 1.

Some of the more critical differences between private and public sector innovation portfolio management are discussed in a bit more detail below.

1. *Collaboration, Integration, and Partnerships for Innovation*: Different views exist on whether and how to manage interdependencies across innovations in a portfolio. Röth et al. (2019) emphasize that in the private sector, interdependencies provide a risk as they reduce the chance of terminating individual

Table 1 Comparison of typical private and public sector innovation portfolio management dimensions and adaptation needs

Innovation portfolio management dimension	Private sector (typical)	Public R&D/nonprofit (typical)	Adaptation from private to public innovation portfolio management
Mission & objective function	Shareholder value; growth; competitiveness	Mission impact; public value; SDGs; equity & inclusion	Make the portfolio intent explicit and assess the portfolio along its intent/ mission
Success metrics	Return on Investment, Net Present Value, payback, market share	Progress toward outcomes/impact, use/adoption, cost-effectiveness	Use dual scorecards: (i) Impact (use/readiness, beneficiary reach, equity) and (ii) Performance (unit cost, lead time, quality).
Time horizons	Short–medium (quarters–3 years) + selected long bets	Medium–long (3–10+ years), policy cycles	Diversify portfolio diversity in terms of disruptive, radical, and incremental innovation and required time horizons
Risk–return logic	Financial risk–return frontier; kill fast	Mission risk–return (political, social, ethical) with higher tolerance for uncertainty	Define mission risk–return explicitly (e.g., tech, delivery, political, ethical risks vs. impact potential)
Interdependencies	Often minimized to ease innovation termination	Often essential for impact (bundles, packages, policy alignment)	Map bundle/package interdependencies and invest with partners as coherent missions, not silos.
Partnerships	Seen as a potential liability	Essential for success	Ensure co-investment and productive/ mutually beneficial partnership with public/ private partners
Data systems	Product/market analytics, unit economics	Results frameworks, monitoring, evaluation, and learning	Couple results and performance monitoring with portfolio analytics dashboards and management systems
Resource allocation	Stage-gate budgets; real-options logic	Earmarking by funders; annual public cycles	Create ring-fenced pools by pipeline/readiness; precommit "option" funding for promising cases.
Governance	Product councils; empowered portfolio managers; fast kill	Committee- and consensus-based; slower and political	Establish portfolio boards with delegated authority; time-boxed decisions; transparent rationales.

(continued)

Table 1 (continued)

Innovation portfolio management dimension	Private sector (typical)	Public R&D/nonprofit (typical)	Adaptation from private to public innovation portfolio management
Incentives	Bonuses, equity, and team recognition for outcomes	Compliance-oriented, visibility, recognition for progress toward impact	Incentives that reward team progress to impact, rather than individual innovations/ outputs
Organizational culture	Hierarchical, "kill to win" culture	Less hierarchical, more path dependency, sunk-cost bias	Track discontinuation rates & reasons; reward high-quality termination decisions.
Agility vs. strategy	High tactical agility plus strategic bets	Strategic mandates; risk of "stop-start" due to budget and need for novelty to satisfy funders	Protect long-term missions with multiyear envelopes; use shorter-term reports–reflect–plan to stay agile.
Accountability & transparency	To investors/boards: commercial sensitivity	To public/funders: high transparency expectations	Publish portfolio health reports (mix, readiness, adoption, equity) and decisions against criteria.

innovation projects. In the public R&D sector, however, the dominant majority of global development challenges (such as climate change or biodiversity losses) are multidimensional and -disciplinary, cut across multiple organizational mandates and sectors, and can therefore not be neatly boxed (Hall and Clark 2010). As a result, innovation to address such global challenges often emerges from an organization's ability to bundle different types of innovations, knowledge, or solutions (Barrett et al. 2022), and to embed these in context-specific market, finance, and policy packages (Sartas et al. 2020).

2. *Agility Versus Strategic Long-Term R&D*: Ongoing responses to changes and innovation *start-stop* culture increase transaction costs and compromise strategic visioning and organizational planning (Spieth and Lerch 2014; Kock and Gemünden 2016; Üçler 2018). Though agility is often presented as good practice, it may be perceived as focusing on *quick wins* and addressing immediate needs, over investment in strategic long-term transformation agendas and pathways, which is a core mandate of many public nonprofit R&D organizations.
3. *Organizational Innovation Cultures*: Innovation portfolio management goes hand in hand with organizational culture change and growth. Innovation portfolio management seeks to bring greater balance between the support of new ideas and the scaling of existing innovations for broad societal benefit (Kumpf and Proud 2022). Historically, public R&D employees might focus more on novelty and testing new ideas, and be less equipped or interested in supporting the scaling of these innovations with partners. It is important to demystify some innovation portfolio management misconceptions (e.g., that it increases top-down decision-

making, innovation cannot be measured), and emphasize its advantages (e.g., that it provides a more transparent framework for risk management, that it can increase space for experimentation and learning).

2.5 An Operational Framework for Innovation Portfolio Management in the Public R&D Sector

For innovation portfolio management to succeed in the public nonprofit R&D sector, it is important to acknowledge the key similarities and differences compared to the private sector. Similarities are related to the importance of having clarity on the organization's innovation portfolio intent and having transparent hard and soft criteria and resource allocation strategies to ensure a healthy and diverse innovation portfolio that delivers against the short- and long-term organizational mission. Differences can be found in the importance of internal collaboration and external partnerships for innovation to address complex or wicked global challenges; the organizational innovation cultures and mindsets, and how these are being reinforced by funding and incentive mechanisms; and finally, the tension that may result from a focus on immediate needs and emergent opportunities, over investment in strategic long-term transformation agendas and pathways.

There are no *plug-and-play* solutions for innovation portfolio management in the public nonprofit R&D sector. Figure 1 illustrates how innovation portfolio management is as much about *Methods* (systems to have timely access to quality data on innovation and transparent criteria to assess the portfolio) and *Mechanisms* (incentives, resource allocation strategies, and governance approaches) as it is about growing the organizational *Mindsets* (the underlying cultural and capacity growth dimension, and organizational leadership). This framework is inspired by Brink's theory of Behavioral Innovation, which is grounded in behavioral science, psychology, and innovation practice (Brink 2022). Brink uses this framework to inform innovation practice at multiple scales. Table 1 provides more detail on the specific design elements of an innovation portfolio management approach. What constitutes the right set of Methods, Mechanisms, and Mindsets will vary across organizations, and has to be tailored to their mission, portfolio intent, and organizational culture.

3 Case Study on Innovation Portfolio Management in CGIAR[2]

The framework presented in Fig. 1 and Table 2 provides a basis for presenting a case study on innovation portfolio management in CGIAR: a global partnership

[2] This section draws heavily on material originally presented in Schut et al. (2024b).

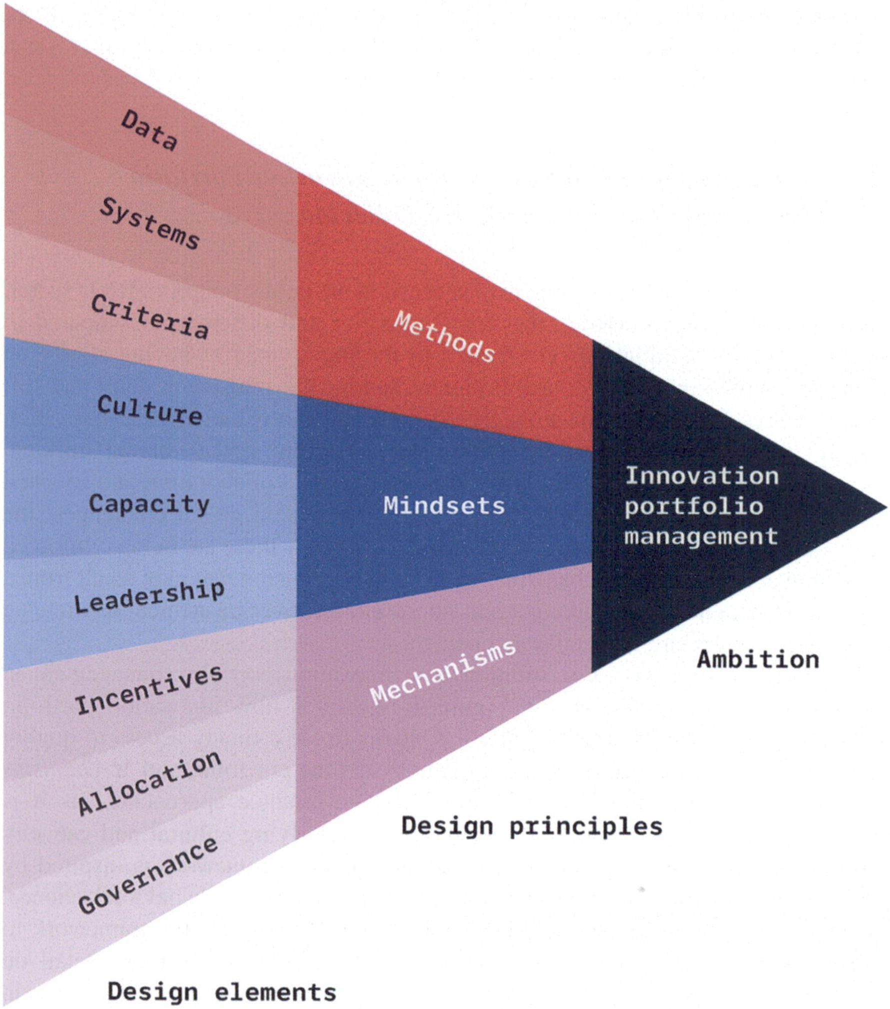

Fig. 1 Critical design principles and elements that need to be considered for innovation portfolio management. (Proud et al. 2023)

of research organizations dedicated to addressing complex agricultural challenges in collaboration with governments and the private and public sectors. The CGIAR is a consortium of centers,[3] the first of which were established in the late 1960s. Formed in 1971 on the heels of the first celebrations of the Green Revolution in 1968, CGIAR was tasked with extending and translating that revolution in agricultural production to new countries and new crops. Over more than five

[3] https://www.cgiar.org/research/research-centers/

Table 2 Basic description of innovation portfolio management design principles and elements

Design principles	Design elements	Basic description
Methods	Data	Information and metrics that are needed to support strategic decision-making
	Systems	Timely access to quality and evidence-based data and information
	Criteria	Hard and soft criteria to analyze and compare across the portfolio
Mindsets	Culture	Organizational culture growth that supports the portfolio management approach
	Capacity	Knowledge, skills, and attitudes required for innovation portfolio management
	Leadership	Strategic engagement of leaders/champions to support organizational culture growth
Mechanisms	Incentives	Policies, planning, and reward structures to enable innovation portfolio management
	Allocation	Clarity on how financial, human, and other resources are allocated across the portfolio
	Governance	Procedures, roles, and responsibilities related to strategic sense- and decision-making

decades, the CGIAR has undergone several reforms as donor funding models and stakeholder priorities shifted, as well as accountability and reporting mechanisms for accessing funds have evolved (McCalla 2014; Leeuwis et al. 2018). Similar to other organizations in the public R&D sector, there is increased pressure on CGIAR to demonstrate how investments in science and innovation contribute to objectives such as those formulated under the United Nations' SDGs (Penfield et al. 2014; Glover et al. 2016; Leeuwis et al. 2018). An important element of this is ensuring a more coherent CGIAR innovation portfolio (CGIAR System Organization 2021).

3.1 *Case Study Methodology*

Given the limited number of public sector organizations applying innovation portfolio management, an in-depth case study approach is used. A case study approach is a research method that involves deep investigation and analysis of a particular organization or phenomenon. It focuses on studying a specific case within its real-life context, collecting qualitative and/or quantitative data through various sources such as interviews, observations, documents, and archival records (Yin 2009). Data for this section was collected through autoethnography (Adams et al. 2017); the authors' active participation in the design, operationalization, and implementation of CGIAR's innovation portfolio management approach. As much as possible, events, activities, and decisions will be supported by written reports and other types of documentation and records. In addition, this section presents some basic 2022-2024 innovation data extracted from the CGIAR Performance and Results Management System (PRMS). 2025 Data was not yet available during the development of this book chapter, but can be accessed through the CGIAR Results Dashboard (see Sect. 3.2.1).

3.2 CGIAR's Innovation Portfolio Management Approach

In 2019, a CGIAR reform was initiated that was generally referred to as the transition toward "One CGIAR" (CGIAR System Organization 2019). An important element of the transition was to develop a 2030 Research and Innovation Strategy anchored in a unifying mission of "Ending hunger by 2030—through science to transform food, land and water systems in a climate crisis," (CGIAR System Organization 2021). The 2030 Strategy puts innovation and scaling at the heart of the CGIAR mission. CGIAR's Performance and Results Management Framework (PRMF) is the companion document to the overall CGIAR 2030 Research and Innovation Strategy (CGIAR System Organization 2021). It describes the key components of CGIAR's processes, tools, and performance management measures that will support portfolio delivery. A key novelty in the PRMF is the Scaling Readiness concept (Sartas et al. 2020). Scaling Readiness is an evidence-based approach to support the design, implementation, and monitoring of strategies to support innovation and scaling along an impact pathway. This approach embodies a new way of defining innovations and managing them to scale through innovation packages. Its standardized methods and metrics allow for aggregation of data, aimed at informing the management of an innovation portfolio (Schut et al. 2024b).

The PRMF is operationalized through CGIAR's Technical Reporting Arrangement (CGIAR System Organization 2022) and the Performance and Results Management System (PRMS). The CGIAR Technical Reporting Arrangement fulfills the system-level programmatic reporting requirements set out in the Standard Provisions annexed to the Funding Agreement or Arrangement signed between each funder and the CGIAR System Organization. It provides operational detail on the content, timing, format, standards, and scope of technical reporting applicable to all core funded investments, representing approximately 40% (370 million USD in 2024) of the overall CGIAR portfolio funding.

3.2.1 Methods for CGIAR Innovation Portfolio Management

Data CGIAR has developed a unified framework for reporting innovation development and use across its portfolio. Standard Indicator Description Sheets (SIDS)—that outline key parameters against which innovation data is collected—were

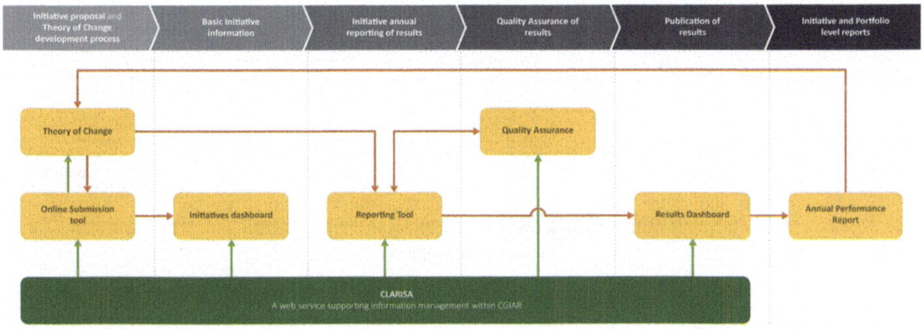

Fig. 2 Illustration of the PRMS ecosystem and data flows. (Author M. Almanzar, CGIAR)

developed in 2022 and updated annually.[4] From 2023 onwards, mandatory updating of previously reported innovation results has resulted in dynamic trend data on—for example—how innovations mature over time. This will also allow CGIAR to demonstrate which innovation development or scaling processes were discontinued or put on hold, and for which reasons. For each reporting period, two key metrics are captured: innovation readiness and innovation use. Both metrics emerge from the Scaling Readiness approach (www.scalingreadiness.org) that CGIAR has embraced as the main framework for innovation portfolio management (Kohl 2024).

Systems To ensure innovation data is collected in a standardized and timely way, CGIAR developed a Performance and Results Management System (PRMS). PRMS is a web-based Management Information Ecosystem designed to collect, curate, and report data on research results achieved by multiple projects and programs structured in a portfolio (Fig. 2).

A key element is a rigorous Quality Assurance (QA) process ensuring that submitted innovation data (as well as all other results) is evidence-based. For example, innovation readiness and use claims need to be accompanied by supporting evidence that is examined by an AI helper and two human reviewers. All QA-ed data are presented and accessed by CGIAR funders and stakeholders from the CGIAR Results Dashboard (Fig. 3).

Criteria Developing a set of innovation portfolio management criteria can guide innovation portfolio decision-making and prioritization. Criteria may be of a strategic, financial, or risk-related nature, so investments can be re-prioritized or stopped in case minimum requirements are not met. CGIAR does not have an agreed-upon set of criteria to manage its innovation portfolio. Nevertheless, the data that is currently being collected could inform key decision-making criteria outlined in Table 3.

---

[4] The SIDS are available through the CGIAR Performance and Results Hub: https://sites.google.com/cgxchange.org/cgiarprhub/home

Fig. 3 The CGIAR results dashboard with 2024 data, with dedicated pages for innovation development and innovation use

3.2.2 Mindsets for CGIAR Innovation Portfolio Management

Culture Initial engagements with various groups of CGIAR innovation portfolio managers revealed some pushback and concerns. A change-management expert was engaged to better understand portfolio managers' needs, objectives, and fears. "Start small and work with early adopters" became an important guiding principle. CGIAR projects could indicate whether they wanted to be part of an early adopter group that would receive dedicated support. To accommodate fears, the Scaling Readiness approach underwent a redesign process that significantly reduced the time and effort needed to report innovations. The engagement with the early adopter groups fostered a degree of trust, and peer-to-peer exchange helped to overcome some of the initial resistance. Furthermore, positive feedback from Executive Management, CGIAR's System Board and System Council, as well as funders, also contributed to increased acceptance across the organization. During the past 3 years, CGIAR has organized a "Week of Scaling" to bring together CGIAR experts, partner organizations, and funders around the theme of innovation scaling (Schut and Pircher 2022). Innovation and scaling culture and community growth are key outcomes of these events and sessions on innovation portfolio management, which allowed colleagues and partners to familiarize themselves with this topic (Photo 1). With support of the CGIAR Scaling for Impact Program, regional Scaling Hubs are being established, which form another mechanism to grow CGIAR and partner innovation and scaling culture and community.

Capacity Innovation portfolio management capacity was strengthened during the numerous information, training, and drop-in sessions that were organized. During such sessions, the key innovation management concepts and operating procedures were explained, and questions were answered. Second, CGIAR developed an e-learning course on innovation and scaling (https://innovationandscaling.thinkific.com) that was promoted throughout the system. To date, more than 3000 people have enrolled since its launch in April 2022. This free, 3-h course provided background on the key concepts underlying the CGIAR innovation portfolio management approach. Third, approximately 100 CGIAR scaling experts/facilitators were trained on the Scaling Readiness approach. Graduates were certified and added to a CGIAR Scaling Directory (https://www.cgiar.org/scalingdirectory/). Last, the internal capacity grew from 1 to 10 people between 2021 and 2024, with dedicated expertise on data analytics, artificial intelligence, project management, scaling community growth, scaling strategies, and responsible scaling. In 2023, CGIAR initiated the recruitment of a senior innovation portfolio management advisor.

Leadership In 2022, a needs assessment was carried out amongst different groups of innovation portfolio managers in CGIAR. Nine groups of portfolio managers were identified, with a total of 17 representatives from these groups taking part in a 30-min interview. It identified the needs of the managers and the overlaps and gaps between the approach that was under development, and it offered an opportunity to raise awareness and engage CGIAR's leadership. Leadership support and engagement have varied across groups of leaders. There is overall support for the innovation portfolio management concept, and agreement that systematic innovation portfolio management can support CGIAR in achieving its 2030 Research and Innovation Strategy. Also, the support from some of the main CGIAR funders has increased the internal validity of the innovation portfolio management work. Leadership engagement in developing innovation portfolio management criteria,

Table 3 Examples of criteria that inform CGIAR innovation portfolio management

Criteria	Explanation	CGIAR innovation portfolio example
Innovation risk-reward diversity and optimization	Data collected on the innovation nature (incremental, radical, disruptive) can inform risk-return discussion, with incremental innovations often having lower risks and lower returns, and disruptive innovations often having higher risks, but also higher potential returns.	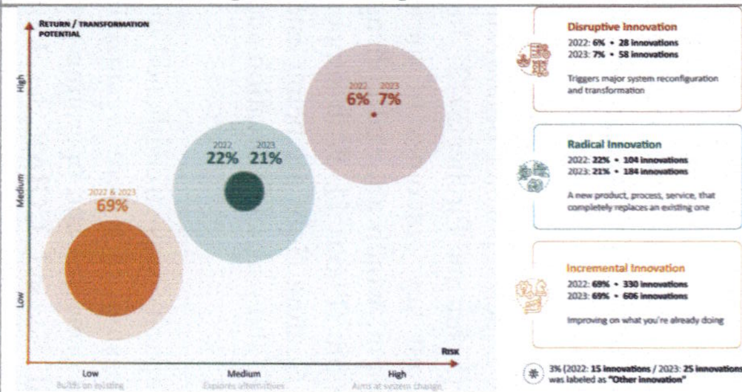 In 2022 and 2023, CGIAR's innovation portfolio included 69% incremental, 22/21% radical innovations, and 6/7% disruptive innovations[a].
Investment and its returns	Estimated USD investment in innovations offers a starting point for: (i) tracking innovation investments over time, (ii) different types of return[b] on investment, (iii) relative investment requirement for innovations of a specific type, nature, or readiness level, and (iv) changes in CGIAR and partner co-investment in innovation and scaling.	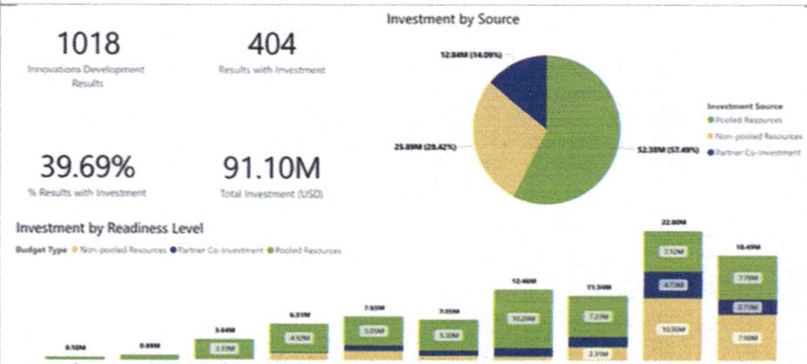 2024 USD data on ~40% of innovations under development shows partner and bilateral project co-investment increases as innovations get more scaling-ready.

Innovation portfolio diversity	Portfolio diversity in terms of type/ nature of innovations, types of SDGs targeted, geofocus of activities, private vs. public sector scaling trajectories, readiness levels, and partnerships.	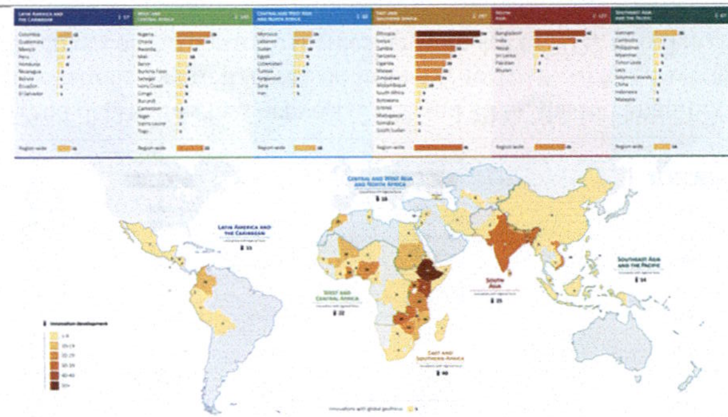 2022 results show innovation development in 64 countries, with the top countries being Ethiopia (54), Kenya (48), and Bangladesh (42) ($n = 477$).

(continued)

Table 3 (continued)

Criteria	Explanation	CGIAR innovation portfolio example
Portfolio progress	A key innovation and scaling performance metric is evolution in readiness and use. The expectation is that innovations 'mature' over time. As part of investment, ideas are developed into proof of concept that then become prototypes which are tested under controlled and less controlled conditions. At some point, innovations transition into actual use. Often first with partners as part of pilot project, but ultimately by target beneficiaries under real-world conditions and without external support. Tracking that evolution over time provides essential input into resource prioritization/stage-gating, accountability to investors, and organizational learning.	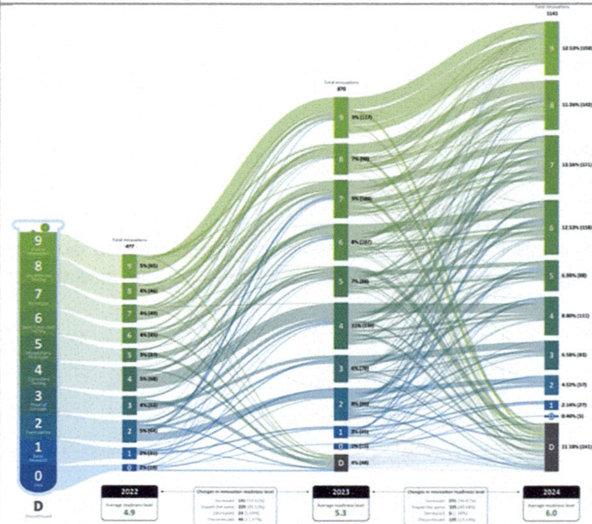 Trend data shows that between 2022 and 2024, the total number of innovations reported increased from 477 to 1018, with the average readiness of the portfolio increasing from 4.9 in 2022 to 6.0 in 2024.

Acceptance/ kill rates	As innovations evolve from idea toward a proven, scalable product, service, or institutional arrangement, decisions need to be made on which innovation development trajectories to advance, and which ones to put on hold or stop. Having insights into the reasons for discontinuation can provide important lessons for improving CGIAR's innovation and scaling performance.	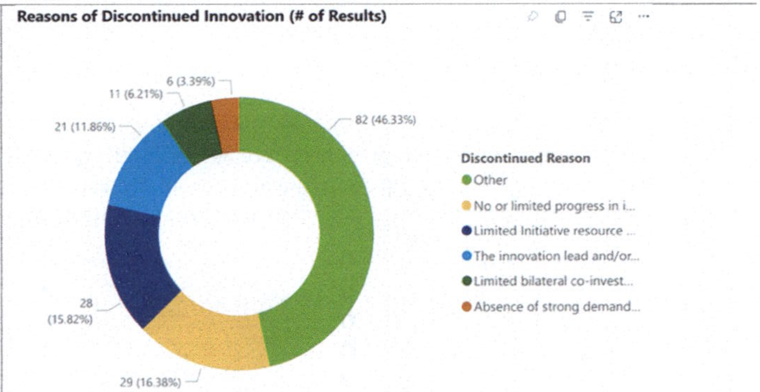 2024 data on discontinued innovations (~16% of the total portfolio).
Responsible innovation/ scaling	Responsible innovation and scaling captures (i) innovation use by vulnerable groups, (ii) the extent to which innovations contribute to promoting Gender Equality and Social Inclusion (GESI), and (iii) responsible innovation to limit unintended negative consequences/ impacts	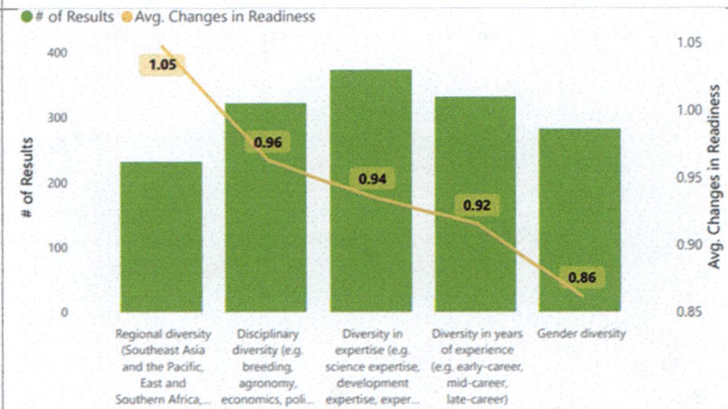 2024 data showing the relation between innovation team composition and average changes in innovation readiness.

(continued)

Table 3 (continued)

Criteria	Explanation	CGIAR innovation portfolio example
Innovation bundles and packages	Demonstrates the degree to which innovations are interconnected with (i) other complementary innovations or solutions provided by the CGIAR (innovation bundles), and (ii) broader context-specific sociocultural, finance, market, policy enablers provided by other organizations that can be enabling or constraining for broad accessibility and use of innovations (innovation packages)	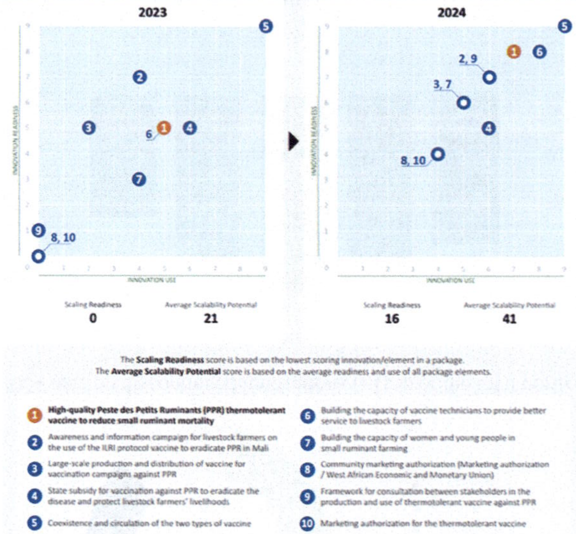 The 2023–2024 evolution of the innovation package and scaling readiness assessment for the high-quality Peste des Petits Ruminants (PPR) thermotolerant vaccine to reduce small ruminant mortality in Mali.

[a]The remaining innovations were labeled as "other."

[b]Return can be defined as changes in readiness/use levels, as well as the ultimate returns related to innovation usage (e.g., changes in farm household productivity, income, health, etc.)

governance mechanisms, and resource allocation strategies is less strong and needs more deliberate efforts for the approach to deliver value.

3.2.3 Mechanisms for CGIAR Innovation Portfolio Management

Incentives A key incentive mechanism in CGIAR is its Technical Reporting Arrangement—endorsed by the CGIAR System Council in June 2022 (CGIAR System Organization 2022)—which is applicable to all CGIAR Initiatives and Programs that receive core funding. The arrangement describes the key parameters of CGIAR's technical reporting as well as details on content, timing, format, standards, and scope of reporting. Furthermore, the reported innovation data features prominently on the CGIAR Results Dashboard, as well as informs various other reports, including the CGIAR Portfolio Narrative, which has a dedicated section on Innovation Portfolio Management (CGIAR System Organization 2023).

Another incentive mechanism is the CGIAR Scaling Fund. Only formally reported and QA-ed innovations with readiness 7 and higher are eligible to enter a competitive process that offers expert and financial support for scaling optimization and the design of scaling strategies. In 2023, three innovation teams[5] were awarded USD125,000 and technical scaling support. They were selected from an initial pool of 28 applicants. In 2025, a new Scaling Challenge is being launched in CGIAR.

Allocation Resource allocation strategies differ from criteria in the sense that allocation strategies focus on how to distribute resources (such as budget and human capacity) efficiently among various innovation projects. Meanwhile, criteria refer to the specific evaluation and prioritization factors used to assess the potential value and alignment of these projects with strategic goals. CGIAR has not clearly defined its resource allocation strategy against specific criteria such as innovation risk profiles, innovation readiness, or innovation pipelines. Nevertheless, the innovation portfolio management approach has influenced the design of the 2025–2030 CGIAR business cycle, with eight Science Programs focusing on developing and improving the readiness of (new) innovation, and a dedicated Scaling for Impact Program focusing on improving the effective use of innovations (Fig. 4). The Scaling for Impact Program has an annual budget of ~USD27M and brings together CGIAR and partner capacity and tools on innovation scaling. The Scaling Readiness approach is at the heart of the Program, which has a dedicated Activity Cluster on innovation portfolio management (CGIAR 2024).

Governance The governance mechanism for innovation management is currently embedded in a so-called "report-reflect-plan" annual adaptive management cycle (CGIAR System Organization 2023). The Reflect process within this cycle enables

[5] https://www.cgiar.org/news-events/news/announcing-the-winners-of-a-first-round-of-funding-to-take-agricultural-innovations-to-scale-in-africa/ and https://www.cgiar.org/news-events/news/meet-this-years-winners-of-cgiars-scaling-fund/

Photo 1 Scaling week participants exchange on the beautiful grounds of CGIAR-ILRI campus in Nairobi, in a festival-like ambiance. Visible from left to right: Marie Michele Codja (Technologies for African Agricultural Transformation (TAAT) program), Ravinder Grover (HarvestPlus Solutions), Hezekiah Agwara (CGIAR Independent Advisory and Evaluation Service), Kristen Becker (Feed the Future Innovation Lab for Horticulture), Samuel Mugambi (Wageningen University), Nekesah T. Wafullah (Alliance of Bioversity International and CIAT). (Photo credit: CGIAR—all people in the photo have granted written permission for their image to appear in this book chapter)

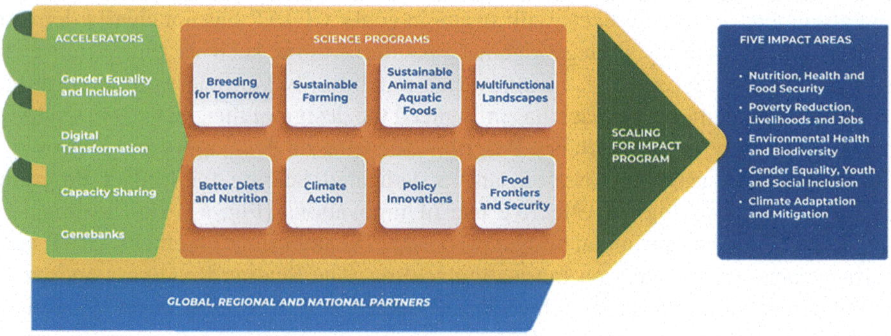

Fig. 4 Schematic representation of the CGIAR 2025–2030 portfolio with innovation portfolio management as an essential instrument to ensure portfolio cohesion, collaboration, and delivery

Programs to assess their results and make recommendations for how to adapt and evolve in the context of changing capacities, budgets, and circumstances. A summary of the results of this process is included in the Annual Technical Reports to demonstrate how the Reflect process facilitated strategic decisions to prioritize, and in some cases de-prioritize investment in innovation within the Initiatives and reallocate funds to respond to lessons learnt. Key areas of adaptation include: (1) Revision of scope, targets, and theory of change due to budget fluctuations which requires constant prioritization of innovations, (2) better internal and external coordination with specific reference was made to the design of innovation packages as a mechanism to increase complementarity, collaboration, and coordination, and (3) meeting demand: better linking the "supply" of innovations to concrete stakeholder "demand." Different types of projects, countries, Science Group, and CGIAR-center level portfolio analytics were created and published. An example can be found here: https://cgspace.cgiar.org/handle/10568/130462.

3.3 Current State and Next Steps for Innovation Portfolio Management in CGIAR

A reflection on the current practice of innovation portfolio management in CGIAR shows that the various principles and elements are at different stages of development and implementation. Table 4 offers an analysis that distinguishes between the following statuses:

- Optimal implementation (++): Current state is desired and meets expectations.
- Functional with enhancements (+): Presently operational, but with room for improvement.
- In progress (+/−): Undergoing development and refinement, not fully operational yet.
- Monitored but incomplete (−): Not yet implemented, but actively explored.
- Out of active consideration (—): Not in use and not under active consideration.

Based on the analysis presented in Table 4, the following actions for advancing innovation portfolio management in CGIAR are critical:

Clarity on Innovation Portfolio Management Ambition The starting point for considering the "right" mix or configuration of methods, mindsets, and mechanisms is a clear innovation portfolio management ambition. An important next step in CGIAR's journey would be for leadership to define this ambition. In 2020, CGIAR defined four key principles for its Performance and Results Management Framework 2022–2030 (CGIAR System Organization 2020):

1. Enable transparent, evidence-based allocation of resources.
2. Support reflection, learning, and adaptive management of CGIAR Initiatives.
3. Facilitate performance management using specific indicators and metrics.
4. Encourage innovation, creativity, and action.

Table 4 Assessment of the current state of different design principles and elements of the innovation portfolio management approach in CGIAR

Design principles	Design elements	Current status	Details
Methods	Data	+	Key metrics, indicators, and parameters are reviewed and updated annually
	Systems	+	Functional PRMS system and related QA module and dashboard are improved and updated annually
	Criteria	—	Key criteria for data analysis and interpretation have not been defined/agreed upon
Mindsets	Culture	+/−	Ongoing culture growth process focused on improving acceptance and uptake of the approach
	Capacity	+/−	Focus on training experts to lead innovation portfolio management in various initiatives
	Leadership	+	Expressed need for and growing enthusiasm about what innovation portfolio management offers
Mechanisms	Incentives	+	Innovation data collection embedded in project lifecycles and technical reporting
	Allocation	—	No resource allocation strategy has been defined and agreed upon—Needs to be initiated
	Governance	−	Part of the report-reflect-plan cycle, but the rigor in decision-making and reporting can be improved

These principles can be further operationalized to guide transparent and evidence-based innovation portfolio decision-making and resource allocation along the CGIAR 2030 Research and Innovation Strategy.

Move Toward an Innovation Portfolio Management Approach that Captures All CGIAR Innovations At the moment, CGIAR's innovation portfolio management approach covers approximately 40% of the CGIAR budget invested in innovation and scaling. Inclusion of the most important noncore funded innovation and scaling work is needed to empower true and holistic management of CGIAR's portfolio of innovation and scaling investments and better supply-demand alignment.

Better Linkages Between Supply and Demand There is an overall need to better connect CGIAR innovation with the needs and demands of innovation users and/or national or regional priorities, as was mentioned by CGIAR innovation portfolio managers. Until today, the CGIAR innovation portfolio management approach has mainly focused on mapping innovations that CGIAR and partners have on offer. How that "supply" is informed by and linked to innovation user "demand" is deemed important, but currently lacks well-defined mechanisms. In

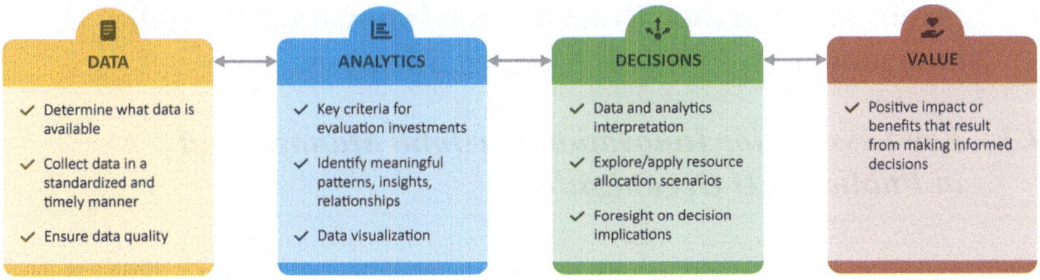

Fig. 5 Relationship between data, analytics, decisions, and value as critical stages in innovation portfolio management

defining its innovation portfolio ambition, CGIAR has a unique opportunity to clarify what constitutes a "healthy" innovation portfolio. This involves articulating a clear ambition for responsible (and demand-driven) innovation and scaling for food system transformation. The CGIAR Scaling for Impact Program has a dedicated area of work on demand intelligence (CGIAR 2024).

Criteria, Allocation Strategies, and Governance for Decision-Making Although CGIAR collects high-quality innovation data, supported by evidence and a quality assurance process, there is no agreement yet on which metrics and analytics will inform decision-making (criteria), how that will trigger resource distribution (allocation), and what are the procedures and responsibilities through which such decisions are made (governance). Agreement on such criteria, allocation strategies, and governance mechanisms for decision-making will allow CGIAR to derive real value from mainstreaming its innovation portfolio management approach (Fig. 5).

Grow Organizational Culture and Capacity for Portfolio Management Though CGIAR has invested in organizational culture and capacity growth, this will require continued attention over a longer period of time. Historically, CGIAR has focused on conducting scientific research to advance technological innovation (McIntire and Dobermann 2023). The systematized management of the creative process of science-based innovation development may feel counterintuitive to scientists and is likely to be met with resistance that needs to be acknowledged and handled sensibly. Yet, having a level of standardization and a common set of indicators and criteria across an organization is essential to ensure the innovation portfolio is well-balanced and responsible, as well as aligned with the short- and long-term organizational mission and strategy. Leadership and portfolio managers—who increasingly request and use the innovation portfolio data and analytics for their strategic decision-making, communication, and fundraising—should play a more prominent role in the change management process and foster broader acceptance of a portfolio management approach.

4 Lessons and Further Research for Innovation Portfolio Management in Public R&D Organizations

4.1 Key Lessons for Innovation Portfolio Management in Public R&D Organizations

With studies on innovation portfolio management in the public sector being scarce (Holden et al. 2018; Schut et al. 2024a), the CGIAR case offers unique insights and learning opportunities for other public R&D organizations that consider an innovation portfolio management approach. Key lessons learned are as follows:

Lesson 1: Take a Holistic View of Innovation Portfolio Management.

It is essential to take a broad perspective on innovation portfolio management that considers methods, mindsets, and mechanisms. There is a general tendency to focus on methods and quality data-collection systems, but without the appropriate investments in incentives and governance mechanisms, capacity, behavioral, and cultural growth, the process is unlikely to be successful. It helps to view innovation portfolio management as an organizational change process that needs to be adequately managed and supported through culture, capacity growth, and incentive mechanisms.

Lesson 2: Ensure the Innovation Portfolio Management Approach Is Fit-for-Purpose.

It is important that the innovation portfolio management approach is fit-for-organizational-purpose. Acknowledging the organizational mandate, strengths, and weaknesses, and interdependencies for achieving innovation impact form important inputs for designing and implementing an innovation portfolio management approach that truly delivers value. For CGIAR, for example, it is important to acknowledge the interconnectedness with other innovation and scaling partners. CGIAR is an important player in the agrifood innovation ecosystem, but cannot achieve impact by actively partnering with government, private, and public sector organizations. Hence, having an innovation management approach that stimulates innovation bundling and packaging with other organizations is important.

Lesson 3: Involve Leadership in Defining Portfolio Goals, Criteria, Strategies, and Governance Mechanisms.

The CGIAR innovation portfolio management journey has—so far—focused on developing systems for data collection, quality assurance, and reporting, and on incentive mechanisms to ensure timely data reporting. To ensure that data and analytics influence decision-making and create organizational value, it is important that organizations consider defining a clear innovation portfolio ambition/intent and criteria, allocation strategies, and governance mechanisms to actively manage the innovation portfolio. This is preferably done during the early stages of designing the innovation portfolio management approach.

Lessons 4: Start Small, Focus on Minimal Viable Product and Pockets of Support.

For many, the systematized management of the creative process of science-based innovation may feel counterintuitive. Yet, having a well-balanced portfolio is important to achieve an organizational mission and impact (Nagji and Tuff 2012). It is good to realize that innovation portfolio management is not necessarily embraced by the entire organization. CGIAR started with early-adopter initiative teams, developed simple Minimum Viable Product (MVP) tools, and showed early results, which created modest enthusiasm and a starting point for mainstreaming, growth, and more advanced innovation portfolio management. In many other public sector organizations, a similar approach may be useful.

Lesson 5: Ensure Strong Incentive Mechanisms.

Incentive mechanisms relate to both intrinsic and external motivation. Intrinsic motivation refers to the motivation that comes from within the individual. External motivation comes from external factors or rewards. In the CGIAR case, both are being supported. Intrinsic motivation is supported through having access to state-of-the-art innovation data and communication materials, such as the innovation profile and innovation portfolio management reports, and being part of the CGIAR innovation and scaling community (e.g., the Scaling Week). External motivation and incentive mechanisms include the Technical Reporting Arrangement that embeds portfolio management data collection and decision-making in project design, implementation, and reporting cycles, as well as making it part of a process to access (scaling) funding or other types of support.

Lesson 6: Demonstrate How Innovation Portfolio Management Adds Value.

Innovation portfolio management can produce powerful data analytics and visualizations that can support engagement with partners, funders, and the general public. Based on 3 years of innovation data collection, CGIAR received various requests from partners and funders to share the innovation portfolio management approach. Analytics were also requested internally by CGIAR-centers and country and thematic leadership to showcase results as part of strategic communication and events. An independent evaluation by the Global Community of Practice on Scaling concluded that: "no other official donor or research/organization that has done anywhere near as much [than CGIAR] to integrate scaling into the innovation process" (Kohl 2024), referring to the CGIAR innovation portfolio management approach.

4.2 Research Frontiers: Gaps and Opportunities for Advancing Innovation Portfolio Management in the Public Sector

Despite growing interest in innovation portfolio management within public R&D organizations, systematic evidence on its design, implementation, and effectiveness remains limited. Building on the CGIAR experience and lessons from private-sector practice, several research directions emerge to deepen understanding and guide the evolution of managing innovation portfolios in mission-oriented, multistakeholder environments. We distinguish between three types of research frontiers.

A first area of future research should assess the effectiveness of prioritization criteria and performance metrics used to guide resource allocation and strategic decision-making in public R&D portfolios. This includes a better understanding of:

- Metrics and composite indicators to best capture mission-oriented "risk–return" dynamics—balancing short-term deliverables with long-term transformative goals.
- Advantages and disadvantages of the different types of rules, processes, and governance mechanisms to guide portfolio decision-making.
- The predictive validity of indicators such as (changes in) innovation readiness and innovation use levels, partner engagement and co-investment, and demand in real-world impact pathways.

A second area of frontier research should examine how organizational mindsets, culture, incentives, and leadership styles influence portfolio management adoption and performance. Research themes include:

- Scientists and managers perceptions on systematized innovation management, and insights on resistance or adaptation to inform change-management strategies.
- Leadership behaviors and incentive mechanisms that foster experimentation, learning, and failing forward, while ensuring accountability and resource use efficiency.
- The role of training, peer learning, and community-building to accelerate the internalization of innovation portfolio management principles.

A third of frontier research focuses on demand, partnerships, and interconnectivity. Innovation portfolios in the public sector are embedded within broader ecosystems of funders, implementers, and users. Future research should provide insights into how portfolio management interacts with sectoral demand and partnership dynamics, particularly in resource-constrained settings. Relevant avenues of research include:

- Mechanisms for demand intelligence—how (changing) user needs, policy priorities, and market signals are captured and fed into portfolio decisions.
- Investigating how portfolio data can support alignment between innovation pipelines and investment pipelines of governments, International Financial Institutions, and private partners.
- Mechanisms that support sector-level management of innovation portfolios across public and private funders, innovator, implementer, and user organizations through data and system interoperability.

5 Conclusion and Broader Reflections on Innovation Portfolio Management for Agriculture 4.0

Innovation portfolio management is critical for realizing the transformative potential of Agriculture 4.0, especially within public R&D organizations working on complex problems in dynamic agrifood innovation ecosystems. Agriculture 4.0 requires strategic alignment and integrated management of innovations that support, underpinned by structured processes, transparent criteria, data-driven decision-making, and adaptive management of resources. Beyond individual innovation advancements, Agriculture 4.0 emphasizes systematic integration and scaling of innovation bundles and packages, recognizing that many public R&D organizations, including CGIAR, are part of a broader, interconnected system.

Given the inherent complexity and multidimensional nature of global agrifood system challenges, no single organization can achieve transformation independently. Effective partnerships and collective action by public R&D organizations, government, development organizations, and private sector organizations around development goals are essential.

For public R&D organizations, the implications include:

- Adoption of holistic innovation portfolio management frameworks that integrate robust methods (data, systems, and criteria), supportive mindsets (organizational culture, leadership, and capacity), and effective mechanisms (incentives, resource allocation, and governance).
- Prioritization of adaptive, transparent, and evidence-based decision-making processes to align innovation investments with strategic priorities and market demands.
- Strengthening and leveraging partnerships to collectively develop and scale innovations effectively within agrifood innovation ecosystems.

To move forward, the public R&D sector needs to become more aware of the costs and benefits of innovation portfolio management. It is important to recognize that innovation portfolio management is not a mere administrative task but a catalyst for organizational change. It requires strong leadership, phased implementation, robust incentive structures, and demonstration of tangible value addition. Moreover, integrating innovation portfolio management into broader organizational culture growth processes is paramount, aligning human resources, partnerships, finances, legal frameworks, and risk management to ensure that innovation becomes ingrained in the organization's mission.

Acknowledgements The authors would like to thank the CGIAR Initiative on Diversification in East and Southern Africa (Ukama Ustawi), the CGIAR Scaling for Impact Program, and all funders who support this work through their contributions to the CGIAR Trust Fund (https://www.cgiar.org/funders/). This publication was funded/co-funded by the European Union. Its contents are the sole responsibility of the authors and do not necessarily reflect the views of the European Union. The authors are grateful to all CGIAR advisory bodies and colleagues for their continued support of the innovation portfolio management journey.

Competing Interests The author(s) has no competing interests to declare that are relevant to the content of this manuscript.

Note by the Authors and Declaration of AI-Assistance in the Writing Process The authors declare that this book chapter is a merger and expansion of two published scientific publications in Taylor & Francis' Innovation and Development journal (https://doi.org/10.1080/2157930X.2024.2400779) and Elsevier's Agricultural Systems journal (https://doi.org/10.1016/j.agsy.2024.103907). The author(s) used ChatGPT-4.0 to improve readability and language, and as a search engine. The authors carefully reviewed and edited the content and take full responsibility for the content of the publication.

Appendix: Overview of the 15 Publications Used in the Literature Review

Citation (in alphabetic order)	Online access (URL)	Open access
Behrens, J., 2016. A lack of insight: an experimental analysis of R&D managers' decision making in innovation portfolio management. *Creativity and Innovation Management* 25, no. 2: 239–250.	https://onlinelibrary.wiley.com/doi/abs/10.1111/caim.12157	N
Brook, J. W., and F. Pagnanelli, 2014. Integrating sustainability into innovation project portfolio management–A strategic perspective. *Journal of Engineering and Technology Management* 34: 46–62.	https://www.sciencedirect.com/science/article/pii/S0923474813000830	N
Holtzman, Y., 2014. A strategy of innovation through the development of a portfolio of innovation capabilities. *Journal of Management Development* 33, no. 1: 24–31.	https://www.emerald.com/insight/content/doi/10.1108/JMD-11-2013-0138/full/html	N
Huvaj, M. N., and W.C. Johnson, 2019. Organizational complexity and innovation portfolio decisions: Evidence from a quasi-natural experiment. *Journal of Business Research* 98: 153–165.	https://www.sciencedirect.com/science/article/abs/pii/S0148296318306660	N
Klingebiel, R. and C. Rammer, 2014. Resource allocation strategy for innovation portfolio management. *Strategic Management Journal* 35, no. 2: 246–268.	https://onlinelibrary.wiley.com/doi/abs/10.1002/smj.2107	N
Kock, A., and H. G. Gemünden, 2016a. Antecedents to decision-making quality and agility in innovation portfolio management. *Journal of Product Innovation Management* 33, no. 6: 670–686.	https://onlinelibrary.wiley.com/doi/abs/10.1111/jpim.12336	N

(continued)

Citation (in alphabetic order)	Online access (URL)	Open access
Kock and Gemünden (2016). How strategic orientation moderates the relationship between innovation portfolio management and success. In *R&D Management Conference 2016: From Science to Society: Innovation and Value Creation* Cambridge, UK: 1–14.	Retrieved from the authors' private ResearchGate repository	Y
Martinsuo, M., and L. Vuorinen, 2019. Uncertainties and dynamics in the practice of innovation project portfolio management. In *Proceedings of the 26th Innovation and Product Development Management Conference*. European Institute for Advanced Studies in Management, EIASM, 2019.	https://trepo.tuni.fi/handle/10024/135967	Y
Pashley, D., T. Tryfonas, A. Crossley, C. Setchell, and S. Karatzas, 2020. Innovation Portfolio Management for Small-medium Enterprises. *Journal of Systems Science and Systems Engineering* 29, no. 5: 507–524.	https://link.springer.com/article/10.1007/s11518-020-5467-z	N
Roeth, T., P. Spieth, and D. Lange, 2019. Managerial political behavior in innovation portfolio management: A sensegiving and sensebreaking process. *Journal of Product Innovation Management* 36, no. 5: 534–559.	https://onlinelibrary.wiley.com/doi/abs/10.1111/jpim.12501	N
Sicotte, H., N. Drouin, and H. Delerue, 2014. Innovation portfolio management as a subset of dynamic capabilities: Measurement and impact on innovative performance. *Project Management Journal* 45, no. 6: 58–72.	https://journals.sagepub.com/doi/abs/10.1002/pmj.21456	N
Spieth, P., and M. Lerch, 2014. Augmenting innovation project portfolio management performance: the mediating effect of management perception and satisfaction. *R&D Management* 44, no. 5: 498–515.	https://onlinelibrary.wiley.com/doi/abs/10.1111/radm.12092	N
Toh, P. K., and G. Ahuja, 2022. Integration and appropriability: A study of process and product components within a firm's innovation portfolio. *Strategic Management Journal* 43, no. 6: 1075–1109.	https://onlinelibrary.wiley.com/doi/abs/10.1002/smj.3351	N
Urhahn, C., and P. Spieth, 2014. Governing the portfolio management process for product innovation—A quantitative analysis on the relationship between portfolio management governance, portfolio innovativeness, and firm performance. *IEEE Transactions on Engineering Management* 61, no. 3: 522–533.	https://ieeexplore.ieee.org/abstract/document/6849434/	N
Üçler, Ç., 2018. Portfolio management for innovation ideas: weighing antecedents with AHP. *Journal of Entrepreneurship and Innovation Management* 7, no. 1: 25–52.	https://www.betadergi.com/jeim/yonetim/icerik/makaleler/93-published.pdf	Y

References

Adams TE, Ellis C, Jones SH (2017) Autoethnography. In: The international encyclopedia of communication research methods, pp 1–11

Attar M, Abdul-Kareem A (2020) The role of agile leadership in organisational agility. In: Akkaya B (ed) Agile business leadership methods for Industry 4.0. Emerald Publishing Limited, pp 171–191

Barrett CB, Benton T, Fanzo J, Herrero M, Nelson RJ, Bageant E, Buckler E, Cooper K, Culotta I, Fan S, Gandhi R, James S, Kahn M, Lawson-Lartego L, Liu J, Marshall Q, Mason-D'Croz D, Mathys A, Mathys C, Mazariegos-Anastassiou V, Miller A, Misra K, Mude A, Shen J, Sibanda LM, Song C, Steiner R, Thornton P, Wood S (2022) Socio-technical innovation bundles for agri-food systems transformation. Palgrave Macmillan, Cham

Behrens J (2016) A lack of insight: an experimental analysis of R&D managers' decision making in innovation portfolio management. Creat Innov Manag 25:239–250. https://doi.org/10.1111/caim.12157

Blakeney M (2022) Agricultural innovation and sustainable development. Sustainability 14:2698

Brink (2022) What we mean by "behavioural innovation", London. Accessed 31 Jan 2024 from: https://hellobrink.co/what-we-mean-by-behavioural-innovation/

Brook JW, Pagnanelli F (2014) Integrating sustainability into innovation project portfolio management—a strategic perspective. J Eng Technol Manag 34:46–62. https://doi.org/10.1016/j.jengtecman.2013.11.004

CGIAR (2024) Scaling for impact program - full design document. CGIAR, Montpellier, p 108. Available at: https://hdl.handle.net/10568/162816

CGIAR System Organization (2019) One CGIAR: a bold set of recommendations to the System Council. CGIAR System Organization, Chengdu. Available at: chrome-extension://efaidnbmnnnibpcajpcglclefindmkaj/https://storage.googleapis.com/cgiarorg/2019/11/SC9-02_SRG-Recommendations-OneCGIAR.pdf

CGIAR System Organization (2020) CGIAR performance and results management framework 2022-2030. Companion document to the CGIAR 2030 research & innovation strategy. CGIAR system organization, Montpellier

CGIAR System Organization (2021) CGIAR 2030 research and innovation strategy: transforming food, land, and water systems in a climate crisis. CGIAR System Organization, Montpellier. Available at: https://cgspace.cgiar.org/handle/10568/110918

CGIAR System Organization (2022) CGIAR technical reporting arrangement in CGIAR technical reporting arrangement. CGIAR System Organization, Montpellier. Availble at: https://storage.googleapis.com/cgiarorg/2022/06/CGIAR-Technical-Reporting-Arrangement-June2022.pdf

CGIAR System Organization (2023) CGIAR portfolio narrative 2022. CGIAR System Organization, Montpellier. Available at https://www.cgiar.org/portfolio-narrative

Cooper RG (2008) Perspective: the stage-gate® idea-to-launch process—update, what's new, and NexGen systems. J Prod Innov Manag 25:213–232. https://doi.org/10.1111/j.1540-5885.2008.00296.x

Cooper RG (2013) Where are all the breakthrough new products?: using portfolio management to boost innovation. Res Technol Manag 56:25–33. https://doi.org/10.5437/08956308X5605123

Cooper RG, Sommer AF (2023) Dynamic portfolio management for new product development. Res Technol Manag 66:19–31. https://doi.org/10.1080/08956308.2023.2183004

Geldes C, Felzensztein C, Palacios-Fenech J (2017) Technological and non-technological innovations, performance and propensity to innovate across industries: the case of an emerging economy. Ind Mark Manag 61:55–66. https://doi.org/10.1016/j.indmarman.2016.10.010

Glover D, Sumberg J, Andersson JA (2016) The adoption problem; or why we still understand so little about technological change in African agriculture. Outlook Agric 45:3–6. https://doi.org/10.5367/oa.2016.0235

Hall A, Clark N (2010) What do complex adaptive systems look like and what are the implications for innovation policy? J Int Dev 22:308–324. https://doi.org/10.1002/jid.1690
Holden A, Cassidy J, Hallberg K, Marsh W (2018) Developing innovation portfolios for the public sector. Deloitte Insights
Holtzman Y (2014) A strategy of innovation through the development of a portfolio of innovation capabilities. J Manag Dev 33:24–31. https://www.emerald.com/insight/content/doi/10.1108/JMD-11-2013-0138/full/html
Huvaj MN, Johnson WC (2019) Organizational complexity and innovation portfolio decisions: evidence from a quasi-natural experiment. J Bus Res 98:153–165
Klerkx L, Begemann S (2020) Supporting food systems transformation: the what, why, who, where and how of mission-oriented agricultural innovation systems. Agric Syst 184:102901. https://doi.org/10.1016/j.agsy.2020.102901
Klingebiel R, Rammer C (2014) Resource allocation strategy for innovation portfolio management. Strateg Manag J 35:246–268. https://doi.org/10.1002/smj.2107
Kock A, Gemünden HG (2016) Antecedents to decision-making quality and agility in innovation portfolio management. J Prod Innov Manag 33:670–686. https://doi.org/10.1111/jpim.12336
Kohl R (2024) Mainstreaming scaling: a case study of the CGIAR of international agricultural research centers, mainstreaming scaling initiative case studies. Available at: https://scalingcommunityofpractice.com/mainstreaming-scaling-a-case-study-of-cgiar/: Global Scaling Community of Practice
Kumpf B, Proud E (2022) The adoption of innovation. In: Stanford social innovation review. https://doi.org/10.48558/TC91-MA61
Leeuwis C, Klerkx L, Schut M (2018) Reforming the research policy and impact culture in the CGIAR: integrating science and systemic capacity development. Glob Food Sec 16:17–21. https://doi.org/10.1016/j.gfs.2017.06.002
Martinsuo M, Vuorinen L (2019) Uncertainties and dynamics in the practice of innovation project portfolio management, innovation and product development management conference. European Institute for Advanced Studies in Management, EIASM
McCalla AF (2014) CGIAR reform - why so difficult? In: Review, reform, renewal, restructuring, reform again and then "the new CGIAR" - so much talk and so little basic structural change - why? Department of Agricultural and Resource Economics, UC Davis. Retrieved from https://escholarship.org/uc/item/7h04960c
McIntire J, Dobermann A (2023) The CGIAR needs a revolution. Glob Food Sec 38:100712. https://doi.org/10.1016/j.gfs.2023.100712
Megersa K (2019) Designing and managing innovation portfolios, K4D helpdesk report. Institute of Development Studies, Brighton
Myers SS, Smith MR, Guth S, Golden CD, Vaitla B, Mueller ND, Dangour AD, Huybers P (2017) Climate change and global food systems: potential impacts on food security and undernutrition. Annu Rev Public Health 38:259–277. https://doi.org/10.1146/annurev-publhealth-031816-044356
Nagji B, Tuff G (2012) Managing your innovation portfolio. Harvard business review May 2012 issue
OECD (2021) Public sector innovation facets: innovation portfolios. Observatory of Public Sector Innovation (OPSI). Organisation for Economic Co-operation and Development (OECD)
Pashley D, Tryfonas T, Crossley A, Setchell C, Karatzas S (2020) Innovation portfolio management for small-medium enterprises. J Syst Sci Syst Eng 29:507–524. https://doi.org/10.1007/s11518-020-5467-z
Penfield T, Baker MJ, Scoble R, Wykes MC (2014) Assessment, evaluations, and definitions of research impact: a review. Res Eval 23:21–32. https://doi.org/10.1093/reseval/rvt021
Proud E, Schut M, Kumpf B (2023) Innovation portfolio management in the public sector: a matter of methods, mindsets and mechanisms. Observatory of Public Sector Innovation (OPSI). Organisation for Economic Co-operation and Development (OECD). Available at: https://oecd-opsi.org/blog/innovation-portfolio-management-public-non-profit-sector/

Reardon T, Echeverria R, Berdegué J, Minten B, Liverpool-Tasie S, Tschirley D, Zilberman D (2019) Rapid transformation of food systems in developing regions: highlighting the role of agricultural research & innovations. Agric Syst 172:47–59. https://doi.org/10.1016/j.agsy.2018.01.022

Röth T, Spieth P, Lange D (2019) Managerial political behavior in innovation portfolio management: a sensegiving and sensebreaking process. J Prod Innov Manag 36:534–559

Sartas M, Schut M, Proietti C, Thiele G, Leeuwis C (2020) Scaling readiness: science and practice of an approach to enhance the impact of research for development. Agric Syst 183:102874. https://doi.org/10.1016/j.agsy.2020.102874

Schut M, Pircher T (2022) Time to scale is now-now: CGIAR-GIZ scaling networking and learning event. In: In time to scale is now-now: CGIAR-GIZ scaling networking and learning event. CGIAR week of scaling. CGIAR System Organization, Nairobi

Schut M, Leeuwis C, Thiele G (2020) Science of scaling: understanding and guiding the scaling of innovation for societal outcomes. Agric Syst 184:102908. https://doi.org/10.1016/j.agsy.2020.102908

Schut M, Adeyemi I, Kumpf B, Proud E, Dror I, Barrett CB, Menzies D, Colomer J, Leeuwis C (2024a) Innovation portfolio management for the public non-profit research and development sector: what can we learn from the private sector? Innov Dev 15(3):689–707. https://doi.org/10.1080/2157930X.2024.2400779

Schut M, Colomer J, Proud E, Bonaiuti E, Dror I, Kang'ethe E, Esquivias L, Leeuwis C (2024b) Innovation portfolio management for responsible food systems transformation in the public sector: lessons, results and recommendations from CGIAR. Agric Syst 216:103907. https://doi.org/10.1016/j.agsy.2024.103907

Sicotte H, Drouin N, Delerue H (2014) Innovation portfolio management as a subset of dynamic capabilities: measurement and impact on innovative performance. Proj Manag J 45:58–72. https://doi.org/10.1002/pmj.21456

Silva AL (2021) Innovation in development cooperation: emerging trajectories and implications for inclusive sustainable development in the 21st century. Innov Dev 11:151–171. https://doi.org/10.1080/2157930X.2020.1807100

Spieth P, Lerch M (2014) Augmenting innovation project portfolio management performance: the mediating effect of management perception and satisfaction. R&D Manag 44:498–515

Thornton P, Mason D'Croz D, Kugler C, Remans R, Zornetzer H, Herrero M (2024) Enabling food system innovation: accelerators for change. Glob Food Sec 40:100738. https://doi.org/10.1016/j.gfs.2023.100738

Toh PK, Ahuja G (2022) Integration and appropriability: a study of process and product components within a firm's innovation portfolio. Strateg Manag J 43:1075–1109. https://doi.org/10.1002/smj.3351

Üçler Ç (2018) Portfolio management for innovation ideas: Weigthining antecedents with AHP. J Entrep Innov Manag 7:25–52

Urhahn C, Spieth P (2014) Governing the portfolio management process for product innovation—a quantitative analysis on the relationship between portfolio management governance, portfolio innovativeness, and firm performance. IEEE Trans Eng Manag 61:522–533

von Braun J, Afsana K, Fresco LO, Hassan MHA (2023) Science and innovations for food systems transformation. Springer, Cham

Woltering L, Fehlenberg K, Gerard B, Ubels J, Cooley L (2019) Scaling—from "reaching many" to sustainable systems change at scale: a critical shift in mindset. Agric Syst 176:102652. https://doi.org/10.1016/j.agsy.2019.102652

Yin RK (2009) Case study research: design and methods, 4th edn. SAGE, Los Angeles/London/New Dehli/Singapore/Washington, DC

How International Exchange Is Supporting Agriculture and Food 4.0: Global Intermediary Organizations and Spaces in AgriFoodTech Start-Up Ecosystems

Laurens Klerkx and Federico Bert

1 Introduction

Agriculture 4.0 and Food 4.0 have emerged strongly as drivers of the development and transformation of the agrifood sector worldwide (da Silveira et al. 2021; Rosenthal et al. 2021; Hassoun et al. 2024). Though building on earlier technological developments in fields such as digitalization and biotechnology, they present more pervasive transformations (e.g., digital transformation, cellular food production). Furthermore, a distinctive feature is that they embrace a different form of innovation and entrepreneurship, inspired by technology companies that have emerged in Silicon Valley, such as Google, Apple, and Meta (Fairbairn et al. 2022; Klerkx and Villalobos 2024). This tech entrepreneurship model is characterized by seeing technology start-ups as the drivers of innovation, as opposed to public agricultural research organizations or larger incumbent agricultural technology firms such as Bayer, Syngenta, and John Deere, and started to take off in agriculture and food in the mid-2010s (Waltz 2017). Typically, tech start-ups follow a pathway of agile development, launching their product quickly, and then improving it through user feedback, backed by generous amounts of seed and venture capital which they acquire by pitching their ideas to investors (Cockayne 2019; Birch 2023; Sippel and

L. Klerkx (✉)
Departamento de Economía Agraria, Facultad de Ciencias Agrarias, Universidad de Talca, Talca, Chile

Knowledge, Technology and Innovation Group, Wageningen University, Wageningen, The Netherlands

Center for Advancing Agri-Food System Transformation (CA2FST), Santiago, Chile
e-mail: Laurens.Klerkx@wur.nl; laurens.klerkx@utalca.cl

F. Bert
Interamerican Institute for Cooperation on Agriculture, San José, Costa Rica

F. da Silveira et al. (eds.), *Agriculture 4.0*,
https://doi.org/10.1007/978-3-032-24552-6_17

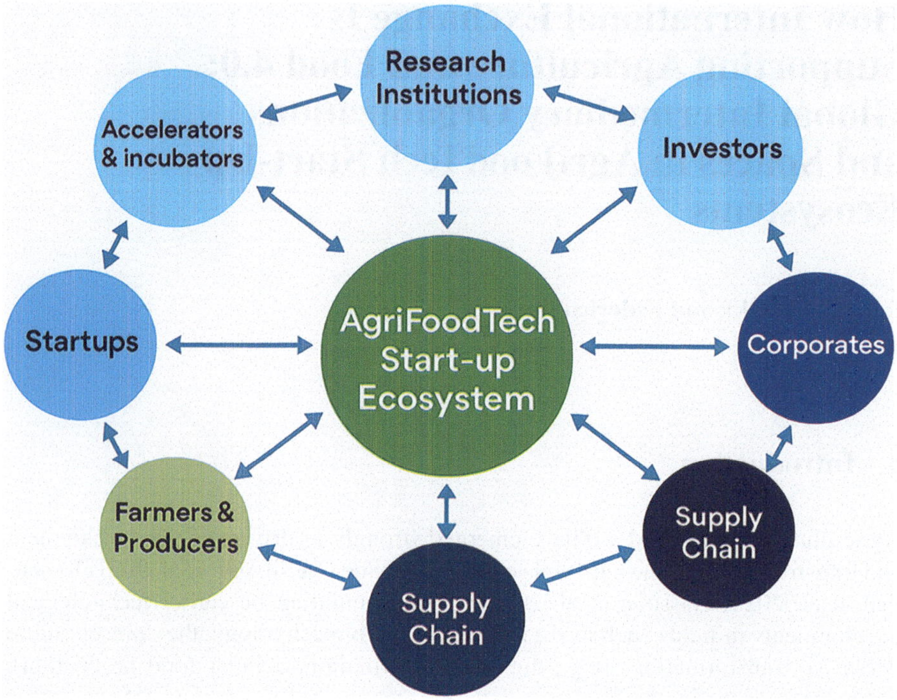

Fig. 1 Typical actors in an AgriFoodTech start-up ecosystem. (Source: own elaboration)

Dolinga 2023). The aim is then to grow quickly and become so-called scale-ups, which increase in market valuation and generate a market share that then enables them to create return on investment (Moro-Visconti 2021; Pfotenhauer et al. 2021). Though this ideal type Silicon Valley model for tech start-up has been criticized not to be a perfect fit for the agrifood sector, for example in terms of development times needed, the tension between broad scaling and locally specific farming conditions, and the need to engage with large incumbent agribusiness to make use of their infrastructures and platforms (De Bernardi and Azucar 2020c; Fairbairn and Reisman 2024; Klerkx and Villalobos 2024; Mac Clay et al. 2024; Fernandez-Vidal and Alarcon 2025; Gaitán-Cremaschi and Klerkx 2026), the AgriFoodTech start-up approach to innovation and entrepreneurship continues to develop.

AgriFoodTech start-ups are embedded in, and supported by, and ecosystem of actors that provides support in the form of coaching and mentoring of nascent entrepreneurs and resources such as knowledge and funding, which include science parks, incubators, accelerators, venture capital investors, angel investors, science organizations (Mungila Hillemane et al. 2019; Stam and van de Ven 2021) (see Fig. 1 for a schematization). AgriFoodTech start-up systems can thus be characterized as both entrepreneurial and innovation ecosystems. Specialized AgriFoodTech start-up ecosystems have emerged in many countries around the world (De Bernardi and Azucar 2020a; Mikhailov et al. 2021; Newell et al. 2021; Peña 2021; Lachman and López 2022; Norouzi et al. 2023; Singh and Kapoor 2025), with the aim of strengthening national capacity in creating AgriFoodTech start-ups and facilitate their growth and sustainable establishment, either as an independent company or by being absorbed by an incumbent agrifood company. An important feature of these start-up ecosystems is that they incorporate many intermediary functions and spaces, meaning spaces where connections are made, and interactions are facilitated between ecosystem actors (Klerkx and Leeuwis 2009; Stephens and Wolf 2023; Colovic et al. 2025; Gaitán-Cremaschi and Klerkx 2026). For example, a start-up incubator that helps start-ups find the right investor or scientist by brokering connections personally or by organizing spaces where these actors can meet, such as agtech and food tech conferences and fairs. While such intermediary work can take place within a country's AgriFoodTech ecosystem (Nettle et al. 2025; Gaitán-Cremaschi and Klerkx 2026), given the global character of start-ups in terms of their ambitions, knowledge and technology flows, and venture capital flows (Cobben et al. 2022; Burström et al. 2023), there are also intermediary actors and spaces that operate between national AgriFoodTech start-up ecosystems, explicitly and implicitly. There is however not much known about the kinds of global intermediary AgriFoodTech intermediaries and spaces, and this chapter aims to present an overview and a characterization, illustrated with examples.

2 Typical Roles of Intermediaries in Innovation and Entrepreneurship

The roles and contributions of intermediaries and intermediary spaces in innovation and entrepreneurship have been well documented, in general but also specifically for the agrifood sector (Howells 2006; Klerkx and Leeuwis 2009; Batterink et al. 2010; Kivimaa et al. 2019a), and also intermediaries that operate internationally have been described (Stadtler and Probst 2012; Ma et al. 2014; Manning and Roessler 2014). These intermediaries are important in innovation and entrepreneurship because they help resolve different disconnects and gaps that may exist between actors in the ecosystem (such as difficulty of finding each other, difficulties of understanding each other, difficulties in interpreting and incorporating new knowledge). Intermediaries are actors that act as "in-betweens" and "network orchestrators," making connections between entrepreneurial and innovation ecosystems actors, by articulating demand and supply and matching it, transferring knowledge, and enhancing understanding, learning, and resource exchange between actors who come from different public and private domains and have different organizational cultures, incentives, and knowledge systems. Typically, these intermediaries also create intermediary spaces, both online and offline, which enable learning and exchange to happen, such as co-working spaces, workshops, conferences, and discussion platforms (Cabral and Winden 2024; Crispim et al. 2024).

Based on a review of literature and experiences as regards intermediation in different fields (agriculture, health, international development), Harvey et al. (2012) propose a classification that characterizes intermediaries on the basis of whether

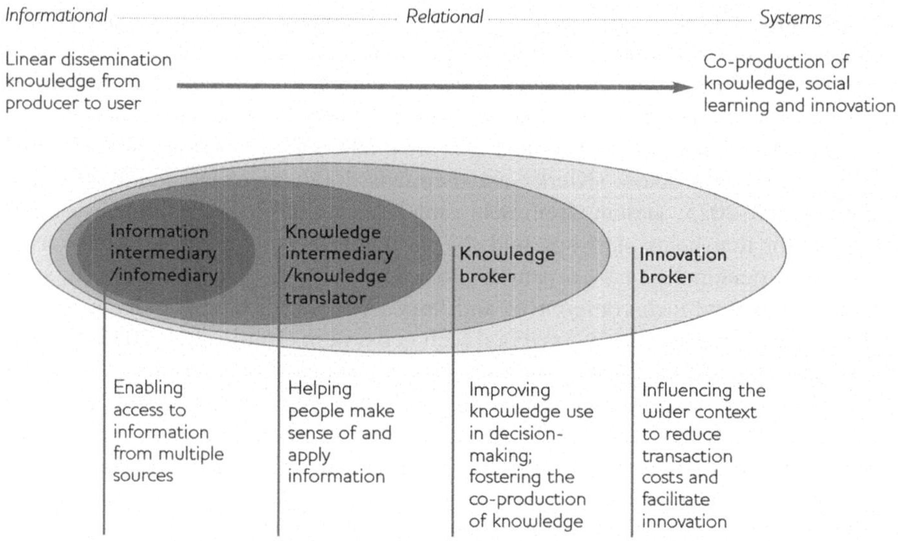

Fig. 2 A typology of intermediaries. (Source: Harvey et al. 2012)

they have informational, relational, and systems functions (see Fig. 2). This connects to a scale which moves from linear dissemination of information, to more tailored application of knowledge, to being broader facilitators of entrepreneurial and innovation ecosystems. The types of intermediaries are not mutually exclusive; an individual or organization doing intermediary work may represent different types of intermediaries.

In what follows, we will use is typology of intermediaries to characterize intermediary actors and related spaces in enabling international exchange in the global AgriFoodTech system, and reflect on their contributions to AgriFoodTech ecosystems and start-up development. We will illustrate these with some brief and more extensive examples.

3 A Characterization of Global Intermediary Organizations and Spaces in AgriFoodTech Start-Up Ecosystems

3.1 Global AgriFoodTech Information Platforms

In the global AgriFoodTech ecosystem, there are several organizations that provide news on developments in AgriFoodTech, such as investment news (total investment trends, investment rounds closed), AgriFoodTech segment development news (on for example, vertical farming, alternative protein, biological inputs), entry and exit of start-ups (the latter can concern both acquisition or exit, or bankruptcy and ceasing of operation), technological and consumer trends and how they may influence a particular AgriFoodTech segment, and interviews with founders and investors sharing experiences and views.

The contribution of these platforms thus lies in signaling investment opportunities or reversely, showing what investment opportunities may be hyped, sharing experiences of what works and what does not work in AgriFoodTech development (also by showing stories of failed start-ups and declining AgriFoodTech categories), and providing critical perspectives that challenge the modes of working of start-up founders, ventural capital actors, and other AgriFoodTech ecosystem players alike.

A pioneer organization in this field has been AgFunder News (https://agfundernews.com/), and there are multiple others, such as Foodtech Weekly (https://www.foodtechweekly.com/), Green Queen Media (https://www.greenqueen.com.hk/), Agritecture (https://www.agritecture.com/blog), AgTech So what? (https://www.agtechsowhat.com/), IGrow News (https://igrownews.com/), Tribu (https://tribu.la/), and Agritech Digest (https://agritechdigest.com/). Beyond their websites, these organizations often also have an active social media presence and act as influencers on these virtual intermediary spaces (such as LinkedIn and Instagram). Some also take an explicitly critical stand in these online communities, such as Ag Uncensored (https://www.linkedin.com/company/aguncensored/). Many of the influencers connected to news media and active in virtual intermediary spaces are also very present in physical intermediary spaces (which will be discussed further below). While many of the aforementioned organizations have a private character, there are also information intermediaries that are more public in nature, such as the Smart AgriFood Observatory of the Politecnico di Milano and the University of Brescia (https://eng.osservatori.net/smart-agrifood-eng/).

Though most are global in scope, some also have an additional regional focus, such as Tribu which is connected to Latin America, and Agritech Digest, which features a lot of news on Africa. Some, such as Agfunder, are broad in terms of coverage (both agtech and foodtech), whilst others zoom in on Foodtech, such as Foodtech Weekly, or are even more specialized, such as Architecture, which focuses on vertical farming and other sorts of controlled environment agriculture. Global AgriFoodTech news platforms can mostly be seen as information intermediaries that facilitate the global flow of information on AgriFoodTech developments. However, some of them also fulfill knowledge translator and knowledge roles, which leads us to the next category.

3.2 Internationally Operating AgriFoodTech Consultants

In the categories knowledge intermediary/knowledge translator and knowledge broker, with some also taking innovation broker roles as they connect to sources beyond knowledge (e.g., to investments), it can be witnessed that agriculture 4.0 and food 4.0 have given rise to dedicated advisory and consultancy services for AgriFoodTech start-ups, farmers, and other agrifood value chain stakeholders. This advice is focused, for example, on building digital literacy of farmers, assisting potential investors to better understand start-ups, and helping tech start-ups better understand farmers and the agrifood sector as a whole (Eastwood et al. 2019; Klerkx 2020; Nettle et al. 2025). Some of them also provide analytical services and applied market research. Whilst such advice and consultancy can be targeted at national or regional AgriFoodTech ecosystem actors (e.g., Agceleration Advisory service, https://www.agceleration.com/, Bitwise Agronomy https://bitwiseag.com/), often these advisors operate internationally, and some of them advertise this explicitly.

The contribution of such AgriFoodTech consultants is that they can help in better design of AgriFoodTech products and services by connecting the realities of prospective adopters and users with start-up founders and their teams of developers of AgriFoodTech, they provide a "reality check," and can enhance start-up development in this way, by helping start-ups to avoid fallacies and self-fulfilling prophecies. They bring in knowledge on how the process of innovation adoption works, and what the typical barriers are in this process. Also, since these advisors see many realities and experiences on the ground, they bring these from one AgriFoodTech ecosystem to the other, stimulating global knowledge flows. In addition, some of them specifically act as "intercultural bridge builders" as they specialize in operating between two regions, making connections between and enhancing understanding between different start-up ecosystems and their respective entrepreneurial and business cultures.

Examples include the aforementioned Agritecture and AgFunder, AgThentic (https://agthentic.com/), Digital Food Lab (https://digitalfoodlab.com/), Bright Green Partners (https://brightgreenpartners.com/) and Agritierra (https://agritierra.com) which have existed for over 5 years, but also more recent players such as Agtech Pro (https://www.agtech-pro.com/), Empower Agri (https://empoweragri.com/), Giga (https://gigafutures.com/), and G3 partners (https://g3partners.co/). These consultants often explicitly indicate that they can help make connections with other countries and markets, and operate cross-culturally, sometimes specializing (e.g., AgriTierra focuses on connections between the UK and Latin American AgriFoodTech ecosystems, G3 partners focuses on the Middle East and Near Asia (MENA) region, Empower Agri helps agtech start-ups enter the European Union, and Agtech Pro focuses on connections between Europe and North America).

Within this category of AgriFoodTech consultants, some also have an analysis branch that does trend analysis, market scans, and investment overviews (e.g., AgFunder, Digital Food lab, but also organizations such as The Good Food Institute, which calls itself a "nonprofit think tank"—https://gfi.org/), which also feeds into the AgriFoodTech news platforms discussed earlier. In terms of global exchanges, interesting in this regard is also the emergence of global AgriFoodTech consultant networks, such as the recently established (in 2024) AgTech Advisory Collective (https://www.agtech-collective.com), whose value proposition is to combine expertise from different parts of the world to enable global reach of AgriFoodTech actors such as start-ups, scale-ups, corporate incumbents, and investors.

While many of these AgriFoodTech consultants are private, some public organizations, such as CGIAR and the Interamerican Development Bank, provide applied research and manuals in this space. Examples of their work include the "Playbook for the Design and Implementation of Food Systems-Focused Accelerator Programs in Africa" and comparative country studies such as the report "Desarrollo Agtech en la Región Andina: casos de éxito y lecciones para el futuro" (Agtech development in the Andean Region: successful cases and future lessons) (Bert et al. 2023).

3.3 Globally Oriented Venture Capital Investors

Having similar roles as AgriFoodTech consultants, mainly as knowledge and innovation brokers, there is also a category of explicitly internationally operating venture capital investors. Some combine their roles as AgriFoodTech news platforms and AgriFoodTech consultancy with investor roles, such as AgFunder, which operates under the same name for all three roles, or AgThentic, which shares its leadership team with Tenacious Ventures. Also, others, such as The Yield Lab, have multiple roles. It is a globally active venture capital fund, which uses its presence in North America, Latin America, Europe, and Asia to "maintain close working relationships across the funds pooling together our knowledge to leverage our global insights, deep sectoral expertise, and strategic relationships for the benefit of our 90+ portfolio companies and investors" (from https://www.theyieldlab.com/). In addition, the YieldLab has its "Yield Lab Institute" which does applied research that feeds into information platforms, similar to players such as AgFunder. Other VCs are following suit, such as Rockstart, which is based in both Europe and Latin America (https://rockstart.com/), Eatable Adventures, active in Europe, North and South America (https://eatableadventures.com/), and 2811 ventures and its venture studio Reciprocal (https://2811ventures.com/, https://reciprocal.studio/). The latter two also collaborate with a lot of public AgriFoodTech support programs (see next section). Hence, particularly in roles as innovation brokers, these VCs foster local to global connections between different start-up ecosystem actors.

Their contribution lies in that they look at AgriFoodTech through an investor perspective, having experience with filtering and selecting entrepreneurial ideas, and see what works and what does not work in different regions. This gives insight into where differences lie in markets for AgriFoodTech around the globe, particularly when this is coupled with a strong analytical branch, enhancing more evidence-based decision-making on investment in start-ups. This evidence again feeds into the trend analysis discussed in Sect. 3.1. Since they are investors, they also have "skin in the game" and an interest in the advancement of AgriFoodTech, and can steer toward broader sustainability contributions of AgriFoodTech by working with particular sorts of start-ups. However, this interest can also influence the way they communicate and make information available, inducing certain bias, since they often wear multiple hats of investor, information platform, and innovation consultant. This risk of mixed roles is known to influence the credibility of intermediaries (Klerkx and Leeuwis 2009). However, they do disclose what different roles they have in start-up ecosystems and what the context and purpose of the information and advice provided.

3.4 Multicountry and Cross-Border Public and Public-Private AgriFoodTech Support Programs and Organizations

The aforementioned actors playing intermediary roles in the global AgriFoodTech start-up (and scale-up) ecosystem are largely private. However, there are also several public and public-private AgriFoodTech support programs and organizations acting as knowledge and innovation brokers. For example, the European Union has made considerable investments in projects focused on AgriFoodTech, which have helped build the digital ecosystem. For example, Wolfert et al. (2023) report on a series of projects which gradually built an ecosystem for digital agriculture development in the European Union across all technology readiness levels (see Fig. 3), which continues to be expanded with facilities such as joint data spaces and joint facilities for AI and robotics piloting (The European Testing and Experimentation Facilities for Agrifood Innovation—TEF Agrifood, https://agrifoodtef.eu/). Also, the European Union co-funds EIT Food (https://www.eitfood.eu/), a pan-European network of incubators and accelerators (De Bernardi and Azucar 2020b) supporting open innovation, and acting as an investor. The United Nations supports AgriFoodTech ecosystems aimed at development in lower-income countries, through the WFP Innovation Accelerator, which has, besides more locally focused incubators and accelerators, a Global Innovation Community. This is a network of regional and country-based Innovation Hubs to identify, nurture, pilot, and scale innovative ideas (https://innovation.wfp.org/wfp-innovation-network). An example of a worldwide public-private organizations is the Food Innovation Hubs from the World Economic Forum (https://initiatives.weforum.org/food-innovation-hubs/home), which connects incumbents and start-ups, and the Thought for Food community (https://thoughtforfood.org/), which fosters communities of AgriFoodTech innovators. Both operate globally and hence foster international information and knowledge exchange. At a country level, examples are Chile Global Ventures (https://chileglobalventures.cl/), which, though not only focused on AgriFoodTech, is aimed at connecting start-ups with international investors and corporates, and organizations such as ProChile (https://www.prochile.gob.cl/en) and ChileMass (https://chilemass.org/2024/), which organize exchanges between Chilean start-ups and the California and Massachusetts-based tech ecosystems.

Multilateral, regional, and global technical cooperation and financing organizations (IICA, FAO, IDB, CAF, IFAD, World Bank, etc.) have also shown growing interest in Agriculture 4.0 and Food 4.0, particularly in the context of the role that new technologies—especially digital technologies—play in the transition toward more productive, sustainable, resilient, and inclusive agrifood systems (Lajoie-O'Malley et al. 2020; WorldBank 2021). Consequently, these organizations—regardless of their differing mandates—have increasingly developed actions and programs aimed at promoting the development and adoption of technologies, with particular attention to avoiding negative impacts associated with digital transformation, especially in scenarios characterized by wide gaps among actors (in terms of connectivity, skills, etc.).

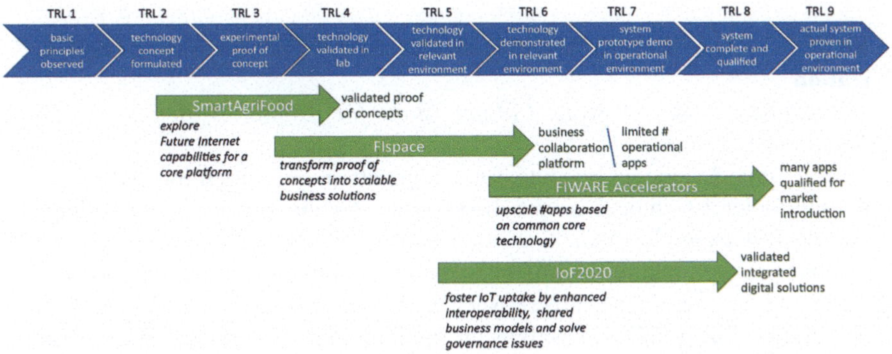

Fig. 3 Sequential EU-funded projects which helped create a digital innovation ecosystem. (Source: Wolfert et al. 2023)

The initiatives carried out by these organizations (e.g., open innovation programs, agtech financing initiatives, diagnostic studies, convening spaces for stakeholders, etc.) place special emphasis on fostering connections among the various actors that are shaping the emerging innovation ecosystems, often acting as intermediaries—and in some cases, as intermediaries of intermediaries. These institutions also play a critical role in raising awareness and placing the issue of agricultural digital transformation on the agendas of governments and policymakers, supporting the transformation processes of member countries through a range of instruments, from policy recommendations to financing. Their contribution hence lies in gathering best practices, taking a holistic perspective, taking a big picture angle at AgriFoodTech and start-up ecosystem development, and bringing this to the world of public policymaking, informing how public policies can foster optimal AgriFoodTech start-ups and ecosystem development support programs, complementing private sector roles of, for example, venture capital actors and multinational companies. Box 1 goes deeper into the experience of one of these organizations, IICA.

Furthermore, some countries have explicitly set up bilateral international exchange between AgriFoodTech ecosystems, such as the German-Argentinean and German-Uruguayan Dialogo Agropecuario (https://dialogoagro.uy/, and https://agrinnova.tech/), as well as exchange between the UK, the USA, and Latin America supported by the Agritierra consultancy mentioned above, for example through the recent ARISE program (https://www.agri-tech-e.co.uk/civicrm/event/info/?reset=1&id=27316#About-ARISE) which organized knowledge exchanges in Sao Paulo (Brazil), Saint Louis (United States) and Cambridge (United Kingdom). These are both instances of active knowledge and innovation brokering between two or more countries, which contribute to international collaboration, enhancing bilateral and multilateral flows of knowledge and enhancing investment opportunities and collaborations between start-ups, research institutes, and other ecosystem actors in the participating countries.

Box 1: The Experience of IICA in Stimulating Agrifood System Digitalization

International Public-Private Collaboration for Digital Agriculture: The Experience of IICA

The Inter-American Institute for Cooperation on Agriculture (IICA)—the multilateral technical cooperation agency for the Americas—is perhaps the clearest example of the importance that multilateral organizations are assigning to Agriculture 4.0 and the role they are playing in driving it forward. Through several early initiatives—such as studies on rural connectivity (2020) and digital skills (2021), as well as the creation of the Interpretive Center for the Agriculture of the Future (CIMAG: https://www.cimag.iica.int/inicio) and the FABLab (Digital Fabrication Laboratory)—IICA progressively developed a strategic approach that culminated in 2022 with the formal establishment of its Agrifood Digitalization Program (i.e., the topic was formally incorporated into the Institute's organizational structure).

This Program consolidated a dedicated team of experts that launched various activities aimed at: raising awareness among member countries of the importance of digital transformation and digital innovation ecosystems; identifying and disseminating existing digital technologies and experiences; and producing high-quality information (e.g., studies) with actionable recommendations for the design of both public and private policies. Among its most prominent actions is the annual organization of the Digital Agriculture Week, which, since 2022, has become a key meeting space for agtech startups, investment funds, public innovation institutions, producer organizations, and experts.

Building on this, IICA also spearheaded the creation of the Agtech Network of the Americas and the Agtech Accelerator Program in partnership with Endeavor, further strengthening its role as a facilitator of public-private partnerships and an active promoter of a dynamic and inclusive agrifood digital ecosystem in the region. Equally important have been its mappings and analyses of public policies that countries are implementing to promote digitalization, as well as studies on the state of agtech innovation ecosystems across different regions of the continent, which have collectively generated multiple recommendations for policymakers.

3.5 *Multinational Companies' Incubators and Accelerators*

In order to capture cutting-edge innovations, following principles of open innovation, many incumbent multinational agrifood firms within different segments and verticals of the food value chain (inputs, machinery, finance, food processing) have become engaged in AgriFoodTech and launched their own incubator and accelerator programs. These are inherently international, given the multi-country operations of these multinationals, and fulfill knowledge and innovation brokering roles. An example is FoodBytes by Rabobank, which states that it "is the entry point into Rabobank's global network for start-ups of all stages." The hub offers exposure to potential partners and funders—all focused on food and ag—as well as connectivity with relevant Rabo product groups, including strategic advisory, direct investment, and equipment financing (https://www.foodbytesworld.com/about-us/network/). Examples of agricultural input multinationals with inhouse incubators are Leaps by Bayer (https://leaps.bayer.com/), and Shoots by Syngenta (https://shootsbysyngenta.com/accelerator). Within Foodtech, large multinationals such as Danone and Unilever have hubs such as the Unilever Foundry (https://www.theunileverfoundry.com/home.html) and Danone ventures (https://www.danoneventures.com/). Besides providing support to start-ups in the form of mentoring and enhancing access to industry knowledge, the international scope of these companies can both enhance access to knowledge about worldwide markets and assist start-ups in developing a scaling strategy, which can also be achieved by being incorporated into the multinationals' operations by being acquired.

3.6 International Gatherings as Physical Global Intermediary Spaces

While many of the abovementioned actors are connected to permanent intermediary spaces, such as co-work locations and start-up hubs which are designed to stimulate interactions between founders (which often have international backgrounds), such as Startlife at Wageningen University (https://startlife.nl/), there are also temporary global intermediary spaces in the form of international events. Some of these take place in multiple countries under the same name and concept, such as the World Agritech Innovation Summits held in San Francisco, Dubai, London, Mexico City, and Sao Paul, and Future Food Tech held in London, Chicago, and San Francisco, which are both organized by a specialized events company, Rethink (https://rethinkevents.com/events/foodtech/, https://rethinkevents.com/events/agritech/). There is also Greentech Global, organizing events in Amsterdam (https://www.greentech.nl/amsterdam) and Querétaro in Mexico (https://www.greentech.nl/americas). However, there are also several events only taking place in one country, such as F&A Next in the Netherlands, which has connections with StartLife (https://www.fanext.com/), and the Week of Digital Agriculture (Semana de la Agricultura) from IICA (semanaad.iica.int)—see Box 1—but also these attract foreign visitors and have speakers from other countries. The contribution of these spaces is in the unexpected and serendipitous encounters that spark new ideas and opportunities, the personal connections made that enable to see whether there is a good fit for collaboration, and the informal exchanges that lead to more formal follow-up.

In summary, there are many intermediaries of different types which operate between countries, with different organizational set-ups, roles, and foci. Table 1 provides a summarizing overview.

Table 1 overview of global agrifoodtech intermediary organizations and spaces

Category of global AgriFoodTech intermediary	Type of intermediary	Examples
Global AgriFoodTech information platforms	Information intermediary	AgFunder News—https://agfundernews.com Foodtech Weekly—https://www.foodtechweekly.com Green Queen Media—https://www.greenqueen.com.hk Agritecture—https://www.agritecture.com/blog AgTech . . . So What?—https://www.agtechsowhat.com IGrow News—https://igrownews.com Tribu—https://tribu.la Agritech Digest—https://agritechdigest.com Ag Uncensored—https://www.linkedin.com/company/aguncensored Smart AgriFood Observatory (Politecnico di Milano & University of Brescia)—https://eng.osservatori.net/smart-agrifood-eng
Internationally operating AgriFoodTech consultants	Knowledge translator Knowledge broker Innovation broker	*AgriFoodTech Advisory & Consultancy Services* Agceleration Advisory Service—https://www.agceleration.com Bitwise Agronomy—https://bitwiseag.com Agritecture—https://www.agritecture.com AgFunder—https://agfundernews.com AgThentic—https://agthentic.com Digital Food Lab—https://digitalfoodlab.com Agtech Pro—https://www.agtech-pro.com Empower Agri—https://empoweragri.com/ Giga Futures—https://gigafutures.com G3 Partners—https://g3partners.co Agritierra—https://agritierra.com Bright Green Partners—https://brightgreenpartners.com/ *Trend & Market Analysis* The Good Food Institute—https://gfi.org *Global Consultant Networks* AgTech Advisory Collective—https://www.agtech-collective.com *Public Organizations & Research Initiatives* CGIAR—https://www.cgiar.org Inter-American Development Bank—https://www.iadb.org

(continued)

Table 1 (continued)

Category of global AgriFoodTech intermediary	Type of intermediary	Examples
Globally oriented venture capital investors	Knowledge broker Innovation broker	AgFunder—https://agfundernews.com*(Operates as a news platform, consultancy, and investor)* AgThentic—https://agthentic.com*(Shares leadership with Tenacious Ventures)* Tenacious Ventures—https://www.tenacious.ventures The Yield Lab—https://www.theyieldlab.com*(Has multiple regional funds + a research arm: Yield Lab Institute)* Rockstart—https://rockstart.com Eatable Adventures—https://eatableadventures.com 2811 Ventures—https://2811ventures.com Reciprocal (Venture Studio of 2811 Ventures)—https://reciprocal.studio
Multicountry and cross-border public and public-private AgriFoodTech support programs and organizations	Knowledge broker Innovation broker	*Public & Public-Private AgriFoodTech Support Programs* TEF Agrifood (European Testing and Experimentation Facilities for Agrifood Innovation)—https://agrifoodtef.eu EIT Food (European Institute of Innovation and Technology)—https://www.eitfood.eu WFP Innovation Accelerator (World Food Programme)—https://innovation.wfp.org/wfp-innovation-network Food Innovation Hubs (World Economic Forum)—https://initiatives.weforum.org/food-innovation-hubs/home Thought For Food—https://thoughtforfood.org Chile Global Ventures—https://chileglobalventures.cl/ ProChile—https://www.prochile.gob.cl/en ChileMass—https://chilemass.org/2024/ *Multilateral, Regional, & Global Technical Cooperation Organizations* IICA (Inter-American Institute for Cooperation on Agriculture)—https://iica.int FAO (Food and Agriculture Organization of the United Nations)—https://www.fao.org IDB (Inter-American Development Bank)—https://www.iadb.org CAF (Development Bank of Latin America and the Caribbean)—https://www.caf.com IFAD (International Fund for Agricultural Development)—https://www.ifad.org World Bank—https://www.worldbank.org *Bilateral International Exchange Programs* German–Argentine Dialogo Agropecuario—https://dialogoagro.uy German–Uruguayan Dialogo Agropecuario/Agrinnova—https://agrinnova.tech

(continued)

Table 1 (continued)

Category of global AgriFoodTech intermediary	Type of intermediary	Examples
Multinational companies incubators and accelerators	Knowledge broker Innovation broker	FoodBytes by Rabobank—https://www.foodbytesworld.com/about-us/network Leaps by Bayer—https://leaps.bayer.com Shoots by Syngenta—https://shootsbysyngenta.com/accelerator Unilever Foundry—https://www.theunileverfoundry.com/home.html Danone Ventures—https://www.danoneventures.com
International gatherings as physical global intermediary spaces	Information intermediary Innovation broker	*Permanent Intermediary Spaces* StartLife (Wageningen University Start-up Hub)—https://startlife.nl *International Event Series (Temporary Global Intermediary Spaces)* World Agritech Innovation Summits *(San Francisco, Dubai, London, Mexico City, São Paulo)*—https://rethinkevents.com/events/agritech Future Food-Tech Summits *(London, Chicago, San Francisco)*—https://rethinkevents.com/events/foodtech Greentech Global (Amsterdam & Querétaro) Amsterdam: https://www.greentech.nl/amsterdam Americas (Querétaro): https://www.greentech.nl/americas *Single-Country Events with Global Reach* F&A Next (Netherlands)—https://www.fanext.com

4 Conclusion

This chapter aimed to show the diversity of intermediaries that build bridges and make connections and facilitate collaboration between different countries' AgriFoodTech start-up ecosystems, and provide an initial overview and characterization. Given that this only enables a glimpse into the multiple global intermediary organizations and spaces in AgriFoodTech start-up ecosystems, deeper research on how they foster meaningful collaborations is needed to validate whether and how they add value to AgriFoodTech innovation and entrepreneurship. This implies going beyond the descriptive and tentative analysis presented in this chapter and having a critical, more empirical and in-depth look at the different sorts of intermediaries listed in this chapter. Some key areas and questions for further analysis include:

1. Looking at the drivers of intermediaries and intermediary spaces in AgriFoodTech, are these merely oriented toward growth of start-ups, market share, and return on investment, or (following Stephens and Wolf 2023; Gaitán-Cremaschi and Klerkx 2026) transformative mission and impact oriented, or a combination of both? So to what extent are they driven by purpose, or by profit? This connects with ideas that intermediaries can play important roles in

the transition process toward more sustainable systems (Kivimaa et al. 2019a), but sometimes also just benefit from creating noise and hype (Fairbairn et al. 2022).

2. The AgriFoodTech space is a fast-developing space, now also moving from the paradigms of Agriculture and Food 4.0 to the paradigms of Agriculture 5.0 and Food 5.0 (Fraser and Campbell 2019; Hassoun et al. 2024). How do existing intermediaries adapt to future trends and potentially disruptive new innovations in the AgriFoodTech sector, such as Agriculture 5.0, bio-algorithms, and digital twins, or do additional specialized intermediaries develop in these fields?
3. To what extent do different sorts of intermediaries complement each other in different stages of start-up development and growth, as it has been noted that different intermediaries have different roles and contributions over time (Kivimaa et al. 2019b)? Or rather, does it create more complexity in start-up ecosystems, having so many different sorts of actors fulfilling similar intermediary roles?
4. Credibility and impartiality have been noted as key characteristics of intermediaries (Klerkx and Leeuwis 2009), but several of the intermediaries hold dual roles (e.g., knowledge broker and investor), which may induce certain biases despite clear disclaimers of these intermediaries. It is still unknown how such dual roles play out, so a main question is how credibility and impartiality are assessed by those that work with AgriFoodTech intermediaries?
5. What global flows of knowledge, capital, mentoring, and other resources do intermediaries actually facilitate (following Klerkx and Villalobos 2024), and how does this create collaboration, exchange, and integration between particular national AgriFoodTech ecosystems in global ecosystems?
6. How do intermediaries develop different approaches, ways of working, and goals, depending on whether they are public or private, and how does public policy play a role in the functioning of global intermediary organizations and spaces?
7. To what extent do global intermediaries in AgriFoodTech ecosystems really enhance success and scaling and ultimately impact of start-ups, given that the use and market share of several AgriFoodTech solutions remain low?
8. As many of the contributions of intermediaries are rather intangible and sometimes invisible (Klerkx and Leeuwis 2009), what are suitable indicators for measuring the economic, environmental, and social impact of these globally oriented AgriFoodTech ecosystem intermediaries?

Given that AgriFoodTech is a dynamic field, in which different product and service categories continuously evolve in terms of growth and decline in numbers of start-ups and investment, it is likely that the intermediaries that facilitate start-up ecosystems are also dynamic. Hence, it is important to continue tracking and assessing global intermediary organizations and spaces in AgriFoodTech start-up ecosystems, inspired by the lines of enquiry outlined above and by raising other critical questions.

Competing Interests The author(s) has no competing interests to declare that are relevant to the content of this manuscript.

Funding The work underpinning this chapter was funded by the Chilean National Agency for Research and Development (ANID) through FONDECYT Regular Grant No. 1250482 (2025) as well as ANID CA2FST CIN250015.

References

Batterink MH, Wubben EFM, Klerkx L, Omta SWF (2010) Orchestrating innovation networks: the case of innovation brokers in the agri-food sector. Entrep Reg Dev 22:47–76

Bert FE, Lachman J, Del Río JA (2023) Desarrollo Agtech en la Región Andina: casos de éxito y lecciones para el futuro. Inter-American Development Bank, Washington, DC

Birch K (2023) Reflexive expectations in innovation financing: an analysis of venture capital as a mode of valuation. Soc Stud Sci 53:29–48

Burström T, Lahti T, Parida V, Wartiovaara M, Wincent J (2023) A definition, review, and extension of global ecosystems theory: trends, architecture and orchestration of global VCs and mechanisms behind unicorns. J Bus Res 157:113605

Cabral V, Winden W (2024) Exploring the coworking space as an innovation intermediary: a case study in Amsterdam. Int J Knowl Based Dev 14:87–111

Cobben D, Ooms W, Roijakkers N, Radziwon A (2022) Ecosystem types: a systematic review on boundaries and goals. J Bus Res 142:138–164

Cockayne D (2019) What is a startup firm? A methodological and epistemological investigation into research objects in economic geography. Geoforum 107:77–87

Colovic A, Caloffi A, Rossi F, Russo M (2025) Institutionalising the digital transition: the role of digital innovation intermediaries. Res Policy 54:105146

Crispim RT, Camboim GF, Zawislak PA (2024) Collaborative spaces as intermediaries for small business innovation. Int J Innov Technol Manag 21:2450036

da Silveira F, Lermen FH, Amaral FG (2021) An overview of agriculture 4.0 development: systematic review of descriptions, technologies, barriers, advantages, and disadvantages. Comput Electron Agric 189:106405

De Bernardi P, Azucar D (2020a) Entrepreneurial food ecosystem: strategic driver to boost resilience and sustainability. In: Innovation in food ecosystems. Contributions to management science. Cham, Springer, pp 135–159

De Bernardi P, Azucar D (2020b) A European food ecosystem: the EIT food case study. In: De Bernardi P, Azucar D (eds) Innovation in food ecosystems: entrepreneurship for a sustainable future. Springer, Cham, pp 245–280

De Bernardi P, Azucar D (2020c) Startups and knowledge sharing in ecosystems: incumbents and new ventures. In: De Bernardi P, Azucar D (eds) Innovation in food ecosystems: entrepreneurship for a sustainable future. Springer, Cham, pp 161–188

Eastwood C, Ayre M, Nettle R, Dela Rue B (2019) Making sense in the cloud: farm advisory services in a smart farming future. NJAS Wageningen J Life Sci 90–91:100298

Fairbairn M, Reisman E (2024) The incumbent advantage: corporate power in agri-food tech. J Peasant Stud 51:1331–1354

Fairbairn M, Kish Z, Guthman J (2022) Pitching agri-food tech: performativity and non-disruptive disruption in Silicon Valley. J Cult Econ 15:652–670

Fernandez-Vidal J, Alarcon S (2025) Financing agricultural innovation: challenges and alternatives to venture capital in the AgTech sector. Food Policy 136:102967

Fraser EDG, Campbell M (2019) Agriculture 5.0: reconciling production with planetary health. One Earth 1:278–280

Gaitán-Cremaschi D, Klerkx L (2026) The roles of agri-food tech start-up ecosystems in strategic niche management for food system transitions. Environ Innov Soc Transit 60:101140

Harvey B, Lewin T, Fisher C (2012) Introduction: is development research communication coming of age? IDS Bull 43:1–8

Hassoun A, Jagtap S, Trollman H, Garcia-Garcia G, Duong LNK, Saxena P, Bouzembrak Y, Treiblmaier H, Para-López C, Carmona-Torres C, Dev K, Mhlanga D, Aït-Kaddour A (2024) From Food Industry 4.0 to Food Industry 5.0: identifying technological enablers and potential future applications in the food sector. Compr Rev Food Sci Food Saf 23:e370040

Howells J (2006) Intermediation and the role of intermediaries in innovation. Res Policy 35:715–728

Kivimaa P, Boon W, Hyysalo S, Klerkx L (2019a) Towards a typology of intermediaries in sustainability transitions: a systematic review and a research agenda. Res Policy 48:1062–1075

Kivimaa P, Hyysalo S, Boon W, Klerkx L, Martiskainen M, Schot J (2019b) Passing the baton: how intermediaries advance sustainability transitions in different phases. Environ Innov Soc Trans 31:110–125

Klerkx L (2020) Advisory services and transformation, plurality and disruption of agriculture and food systems: towards a new research agenda for agricultural education and extension studies. J Agric Educ Ext 26:131–140

Klerkx L, Leeuwis C (2009) The emergence and embedding of innovation brokers at different innovation system levels: insights from the Dutch agricultural sector. Technol Forecast Soc Chang 76:849–860

Klerkx L, Villalobos P (2024) Are AgriFoodTech start-ups the new drivers of food systems transformation? An overview of the state of the art and a research agenda. Glob Food Secur 40:100726

Lachman J, López A (2022) The nurturing role of the local support ecosystem in the development of the Agtech sector in Argentina. J Agribusiness Dev Emerg Econ 12:714–729

Lajoie-O'Malley A, Bronson K, van der Burg S, Klerkx L (2020) The future(s) of digital agriculture and sustainable food systems: an analysis of high-level policy documents. Ecosyst Serv 45:101183

Ma X-F, Kaldenbach M, Katzy B (2014) Cross-border innovation intermediaries – matchmaking across institutional contexts. Tech Anal Strat Manag 26:703–716

Mac Clay P, Feeney R, Sellare J (2024) Technology-driven transformations in agri-food global value chains: the role of incumbent firms from a corporate venture capital perspective. Food Policy 127:102684

Manning S, Roessler D (2014) The formation of cross-sector development partnerships: how bridging agents shape project agendas and longer-term alliances. J Bus Ethics 123:527–547

Mikhailov A, Oliveira C, Padula AD, Reichert FM (2021) Californian innovation ecosystem: emergence of agtechs and the new wave of agriculture. Innov Manag Rev 18:292–307

Moro-Visconti R (2021) FoodTech and AgriTech startup valuation. In: Moro-Visconti R (ed) Startup valuation: from strategic business planning to digital networking. Springer, Cham, pp 363–390

Mungila Hillemane BS, Satyanarayana K, Chandrashekar D (2019) Technology business incubation for start-up generation: a literature review toward a conceptual framework. Int J Entrep Behav Res 25:1471–1493

Nettle R, Ingram J, Ayre M (2025) Digiwork: how agriculture 4.0 is changing work for farm advisers. Front Sustainable Food Syst 9:1542007

Newell R, Newman L, Mendly-Zambo Z (2021) The role of incubators and accelerators in the fourth agricultural revolution: a case study of Canada. Agriculture 11:1066

Norouzi A, Sadighi H, Abbasi E, Fami HS, Aski HM (2023) Strategic analysis of startup ecosystem in Iran's agricultural sector. J Glob Entrep Res 13:1

Peña I (2021) Tecnolatinas 2021: the LAC startup ecosystem comes of age. Inter-American Development Bank, Washington, DC

Pfotenhauer S, Laurent B, Papageorgiou K, Stilgoe J (2021) The politics of scaling. Soc Stud Sci 52:3–34

Rosenthal A, Maciel Guedes AM, dos Santos KMO, Deliza R (2021) Healthy food innovation in sustainable food system 4.0: integration of entrepreneurship, research, and education. Curr Opin Food Sci 42:215–223

Singh N, Kapoor S (2025) Agtech platforms: complementors and value propositions. Tech Anal Strat Manag 37(11):1294–1308
Sippel SR, Dolinga M (2023) Constructing agri-food for finance: startups, venture capital and food future imaginaries. Agric Hum Values 40:475–488
Stadtler L, Probst G (2012) How broker organizations can facilitate public–private partnerships for development. Eur Manag J 30:32–46
Stam E, van de Ven A (2021) Entrepreneurial ecosystem elements. Small Bus Econ 56:809–832
Stephens P, Wolf SA (2023) Agritech entrepreneurship, innovation intermediaries, and sustainability transitions: a critical analysis. J Innov Econ Manag 42(3):43–72
Waltz E (2017) Digital farming attracts cash to agtech startups. Nat Biotechnol 35:397–398
Wolfert S, Verdouw C, van Wassenaer L, Dolfsma W, Klerkx L (2023) Digital innovation ecosystems in agri-food: design principles and organizational framework. Agric Syst 204:103558
WorldBank (2021) What's cooking: digital transformation of the agrifood system. World Bank, Washington, DC

AgriFoodTech Start-up Ecosystems: The Case of Chile

Daniel Gaitán-Cremaschi, Laurens Klerkx, and Pablo Villalobos

1 Introduction

Food systems worldwide are experiencing significant pressures from climate change, resource scarcity, population growth, and changing consumer preferences. These challenges are driving increased interest in innovation to make food systems more efficient, sustainable, and resilient (Herrero et al. 2020).

In response, technology-focused and science-based ventures known as Agri-FoodTech start-ups are becoming important players in transforming how food is produced, processed, distributed, and consumed (Klerkx and Villalobos 2024). These are typically young, innovation-driven companies with scalable business models that operate in highly uncertain environments (Cockayne 2019; Rodriguez Cárdenas 2024). They use digital technologies, biotechnology, artificial intelligence,

D. Gaitán-Cremaschi
Departamento de Economía Agraria, Facultad de Ciencias Agrarias, Universidad de Talca, Talca, Chile

L. Klerkx
Departamento de Economía Agraria, Facultad de Ciencias Agrarias, Universidad de Talca, Talca, Chile

Center for Advancing Agri-Food System Transformation (CA2FST), Santiago, Chile

Knowledge, Technology and Innovation Group, Wageningen University, Wageningen, The Netherlands

P. Villalobos (✉)
Departamento de Economía Agraria, Facultad de Ciencias Agrarias, Universidad de Talca, Talca, Chile

Center for Advancing Agri-Food System Transformation (CA2FST), Santiago, Chile
e-mail: pvillal@utalca.cl

F. da Silveira et al. (eds.), *Agriculture 4.0*,
https://doi.org/10.1007/978-3-032-24552-6_18

and data analytics, among others, to solve problems across the food system, from farm production to consumer delivery and waste reduction.

These start-ups are part of and shape what experts refer to as Agriculture 4.0 and Food 4.0. These terms reflect the increasing use of technologies such as sensors, robotics, digital platforms, and data systems to modernize agriculture and food systems. The goal is to make these systems more productive, efficient, transparent, sustainable, and better aligned with evolving societal demands (Klerkx and Villalobos 2024).

Most research and investment in AgriFoodTech start-ups have focused on North America, Europe, and parts of Asia, where mature financial markets, strong research institutions, and consolidated agri-food industries create favourable conditions. In contrast, Latin America has received comparatively less attention, despite being a major food-producing and exporting region with growing entrepreneurial activity (IDB Invest and Accenture 2024; Vergara and Barrett 2025). Nonetheless, there are emerging academic and policy-oriented studies, on countries like Brazil, Argentina, the Andean Region, and Central America (Embrapa et al. 2019; Vitón 2020; CEPAL and FAO 2021; Lachman et al. 2022; Lachman and López 2022; Bert et al. 2023).

Within Latin America, Chile stands out as a particularly relevant case, which so far has not been documented. The country has built a strong agri-food sector oriented towards exports, supported by its unique geography, favourable agroecological conditions, skilled human capital, open trade policies, and robust phytosanitary systems. Its counter-seasonal production gives it an additional advantage in global markets. These conditions have enabled the development of competitive value chains in products such as fresh fruit, wine, seafood, and processed foods (HNRG 2021; Endeavor 2022; Rodriguez Cárdenas 2024).

Building on this foundation, a new generation of AgriFoodTech start-ups has begun to emerge in Chile over the past decade, responding to both domestic and global challenges and opportunities. In 2024, the sector attracted 58 million dollars in investment (AgFunder 2025), placing Chile among the top three countries in Latin America for start-up density and early-stage activity (Brixton Ventures Lab 2024). This growth reflects not only entrepreneurial momentum but also strong public support, and broader institutional reforms that have strengthened Chile's innovation ecosystem and enhanced national competitiveness (ICEX 2024).

These start-ups do not operate in isolation. They are part of a broader ecosystem that includes investors, public innovation agencies, business incubators and accelerators, corporate partners, research centres, and industry associations, among others (Klerkx and Villalobos 2024). Together, these actors perform key functions that support the development and scaling of start-up-led innovations, including funding, technical assistance, infrastructure, knowledge exchange, legitimization, and access to markets.

This chapter provides an overview of Chile's emerging AgriFoodTech start-up ecosystem. Rather than focusing on individual companies, it maps the ecosystem's structure, outlines the roles of key actors, and examines the main factors that enable or constrain start-up-driven innovation. The goal is to offer a practical case that can inform discussions on how AgriFoodTech ecosystems develop in emerging market contexts. It is based on a review of reports and sector studies (e.g. Endeavor 2022), complemented by expert interviews conducted between 2023 and 2024 as part of ongoing academic research (Gaitán-Cremaschi and Klerkx 2026) and a master's thesis project (Rodriguez Cárdenas 2024).

2 Mapping the AgriFoodTech Ecosystem in Chile

2.1 What Is an AgriFoodTech Ecosystem?

An *AgriFoodTech ecosystem* refers to the evolving network of actors, institutions, resources, and interactions that collectively drive innovation in food and agriculture through the use of technology (Mikhailov et al. 2021; Newell et al. 2021). It includes far more than start-ups alone. It encompasses the broader enabling environment in which these ventures operate. Start-ups do not innovate in isolation; they are embedded in systems that shape the type, direction, and outcomes of their innovation efforts.

The concept draws from broader theories of innovation and entrepreneurial ecosystems (Freeman 1988; Lundvall 1992; Moore 1993; Adner 2017; Granstrand and Holgersson 2020; Kansheba and Wald 2020; Stam and Van de Ven 2021), which emphasize interdependence among actors combining complementary capabilities for innovation and venture development.

In the AgriFoodTech domain, a typical ecosystem includes (Mikhailov et al. 2021; Newell et al. 2021; Klerkx and Villalobos 2024):

- *AgriFoodTech start-ups*: Early-stage ventures developing technological and/or science-based solutions for agri-food system challenges.
- *Public innovation agencies*: Government bodies providing grants, subsidies, policy support, and innovation programs.
- *Research institutions*: Universities, public research centres, and technology centres generating scientific, technical, and applied knowledge.
- *Financial investors*: Angel investors, venture capital funds, impact investors, and other actors providing capital for venture development and scaling.
- *Business support organisations*: Incubators, accelerators, and innovation hubs providing mentoring, infrastructure, networking, and business development support.
- *Corporate partners*: Established agri-food firms serving as customers, distributors, investors, or strategic partners.
- *Industry intermediaries*: Trade associations, cooperatives, sector organisations, and platforms connecting ecosystem actors and supporting coordination.
- *Clients and users*: Farmers, distributors, processors, retailers, supply chain operators, and consumers who validate, use, and create demand for new technologies.

Following ideas from Hekkert et al. (2007), these AgriFoodTech start-up ecosystems are expected to perform several interrelated functions for innovation and venture creation and establishment to occur:

- *Entrepreneurial activity*: The creation and scaling of start-ups that develop new technologies and business models to address challenges in food and agriculture.
- *Knowledge development and validation*: Research, experimentation, and testing that generate new knowledge and demonstrate technological feasibility across farming, food processing, and consumer applications.
- *Knowledge diffusion*: The exchange of ideas, methods, and practices through networks, business support organizations, demonstration platforms, and extension services.
- *Guidance of the search*: The alignment of innovation efforts with strategic goals such as export competitiveness, food security, nutritional quality, climate resilience, food safety, and sustainable production and consumption.
- *Market formation*: The creation of adoption opportunities through early user engagement, value chain partnerships, supportive regulation, certification schemes, and public procurement.
- *Resource mobilization*: The provision of financial capital, skilled human resources, infrastructure, and other assets needed to sustain innovation and enable scale-up.
- *Legitimation*: The development of institutional and societal trust in new technologies through certifications, endorsements, public visibility, and stakeholder support.

Beyond actors and functions, innovation and entrepreneurial ecosystems require enabling conditions: the infrastructure and institutional frameworks that support innovation activities. These include digital connectivity, testing facilities, pilot centres, specialized laboratories, and regulatory frameworks governing data privacy, quality standards, food safety protocols, biotechnology approvals, and intellectual property protection. The design and implementation of these conditions can either facilitate or hinder innovation, particularly during early venture stages when start-ups face higher uncertainty and resource constraints (Lachman 2023).

While many AgriFoodTech ecosystems have adopted elements of the Silicon Valley model, such as venture capital funding, accelerator programs, and university-industry partnerships, many adapt these approaches to local circumstances. In contexts characterized by limited infrastructure, fragmented markets, or different institutional arrangements, direct replication of the Silicon Valley model may be neither feasible nor optimal (Fairbairn and Guthman 2020; Klerkx and Rose 2020). Instead, ecosystems are likely to evolve along different trajectories, shaped by local regulatory environments, market dynamics, and innovation cultures (Klerkx and Villalobos 2024).

The Chilean ecosystem illustrates how global innovation models are selectively adapted to local contexts. For example, while public initiatives like *Start-Up Chile* reflect elements associated with the Silicon Valley model, such as accelerator programs and an international outlook, the structure and evolution of the ecosystem

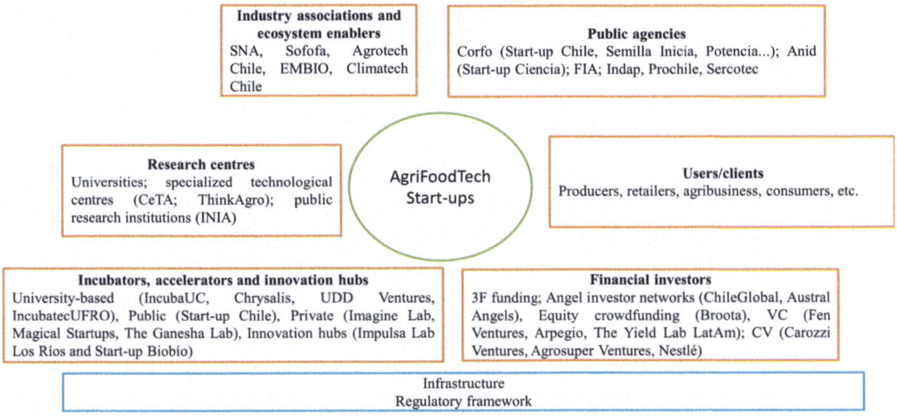

Fig. 1 A typical AgriFoodTech ecosystem structure, with illustration of Chilean organizations, highlighting actor categories and their relationships. It is important to note that a single organization may play multiple roles. For example, a university may be both a research centre and the host of an incubator. (Figure adapted from Lachman 2024)

appear shaped by national priorities and institutional conditions. These include a prominent role for public-sector support, a focus on export-oriented agri-food sectors, and constraints related to venture capital availability and actor coordination. Figure 1 shows a typical AgriFoodTech ecosystem structure, with illustration of Chilean organizations, and these dynamics are explored further in the following sections.

2.2 *Actor Categories and Roles in the Chilean Ecosystem*

2.2.1 Start-ups

Start-ups are central to Chile's AgriFoodTech ecosystem and contribute to multiple core innovation system functions. They drive entrepreneurial activity and experimentation, develop technologies across the value chain, and introduce agricultural and food innovations that support market formation. Many also help guide the direction of the transition through sustainability-oriented models. Through collaborations, technology transfer, and public recognition, they also enable knowledge diffusion and legitimation.

In 2024, Chile accounted for approximately 12% of the region's 618 identified AgriFoodTech start-ups ranking as the third most significant ecosystem after Argentina (28%) and Mexico (24%), not including Brazil (Brixton Ventures Lab 2024). This relative concentration reflects strong entrepreneurial activity, the foundational function of any innovation ecosystem. The emergence of globally oriented ventures and unicorns like NotCo also points to the system's capacity to nurture start-ups with international potential.

Consistent with this global orientation, many Chilean AgriFoodTech ventures aim to internationalize early due to small domestic demand and thin late-stage capital (see Sect. 3 "*Challenges in Chile's AgriFoodTech ecosystem*"). To do so, start-ups leverage international accelerators and soft-landing bridges, build visibility in foreign ecosystems, and enter strategic alliances with international incumbents, for example, NotCo's joint venture with Kraft Heinz (The Kraft Heinz Not Company LLC) supporting co-branded plant-based launches in the United States[1] (see Sect. 2.2.6.*"Ecosystem enablers -International integration & foreign hubs"* for mechanisms and programs).

These ventures address different segments of the agri-food value chain (see Table 1 for examples). On farms, they improve how crops are grown and monitored. These include precision agriculture platforms that use sensors and data analysis to help farmers optimize water and fertilizer use. Biotechnology companies develop biological alternatives to chemical fertilizers and pesticides, often using microorganisms found in Chilean soils. Other start-ups work on alternative growing methods like hydroponic systems and vertical farms that allow production in controlled environments with less land and water. Between farm and consumer, start-ups improve processing and distribution. These companies use artificial intelligence to optimize food recipes and quality control. Others implement blockchain technology to track products from farm to table, which is particularly important for Chilean agricultural exports that must meet strict international standards. Digital platforms connect farmers directly with buyers, reducing intermediaries and improving market access for small producers. Consumer-focused innovations represent the most active area of development. Companies developing alternative proteins use Chile's marine resources like algae and its diverse fungi to create new food products. These ventures employ advanced fermentation techniques and emerging cellular agriculture methods. Many start-ups also focus on circular economy approaches, converting agricultural waste from breweries and food processing into valuable inputs for farming or functional food ingredients (Endeavor 2022; Gaitán-Cremaschi and Klerkx 2026).

AgriFoodTech start-ups use different strategies to enter markets and grow their businesses. Some work closely with established agricultural companies providing technologies that improve existing operations (e.g. *Instacrops*). Others develop proprietary platforms or data systems that they license to larger companies. Some create entirely new products or services for specific market segments, such as plant-based foods (e.g. *NotCo*) or direct-to-consumer platforms (e.g. Savia) that bypass traditional distribution channels. A smaller group pursues more transformative innovations in areas like cellular agriculture (e.g. *Luyef*) or regenerative agriculture (e.g. *Grassland Analysis*[2]) that could potentially change how food systems operate.

[1] https://notco.com/mx/partnerships

[2] http://grasslandanalysis.cl/

Table 1 Examples of Chilean agri-food tech start-ups per upstream and middle stream category

Category	Description	Example of Chilean start-ups
Ag Biotechnology	Inputs in crop and livestock farming such as genetics, microbiome, reproduction, animal health, genetically modified organisms, biofertilizers, and biopesticides to improve productivity and sustainability	*Pewman innovation*[a]: Microbial bio stimulants combating frost and drought using extremophile microbes. *NeoCrop technologies*[b]: A precision breeding platform that develops stress-tolerant seeds using genome editing (CRISPR), speed breeding, and gene selection software.
Farm management software, sensing, & IoT	Solutions that integrate big data analytics and decision-support software from a variety of sources, including sensors, weather forecasts, and crop health assessments to provide comprehensive insights	*Instacrops*: IoT sensors and AI for precision farming decisions. *Neltume Agro*[c]: Solar-powered stations using cameras and sensors to track pest pressure and crop phenology for precision input use.
Novel farming systems	Innovative methods of producing food beyond traditional open-field agriculture—e.g. vertical farms, controlled environment agriculture, insect farming, aquaculture tech, etc.	*AgroUrbana*[d]: hydroponics, LED spectrum lighting, and advanced climate control to produce premium, pesticide-free greens year-round. *Farmtastica*[e]: Modular controlled-environment farms for commercial vegetable production in urban and extreme zones.
Farm robotics, mechanization, & equipment	On-farm machinery, automation, drone manufacturers, and crop equipment	*Agrotrac*[f]: Provides GPS-based tracking and management systems for farm machinery, enabling real-time monitoring and precision fleet operations. *HardDrones*[g]: Develops drone-based solutions for precision agriculture, offering aerial monitoring and crop health diagnostics using sensor and imaging technologies.
Agribusiness marketplaces	Connect players in the agricultural value chain via digital platforms, whether it is buying/selling products, equipment sharing, financing, or inputs.	*AgroMatch*[h]: Airbnb of agricultural machinery. Equipment rental marketplace for farmers. *Finka*[i]: First agricultural crowdfunding platform, enabling individuals to invest in farming projects and farmland through a digital agri-food fintech marketplace. *Savia*[j]: digital platform that shortens the food value chain by directly connecting urban buyers (e.g. restaurants) with producers through fair trade principles, reducing food loss, and promoting access to fresh, healthy, and high-quality products.

(continued)

Table 1 (continued)

Category	Description	Example of Chilean start-ups
Innovative food	Novel food & ingredient such as alternative proteins (plant-based, cultured meat), novel ingredients, and food tech R&D	*NotCo*: AI-formulated plant-based food products like NotMilk. *Luyef*[k]: Lab-grown meat and myoglobin ingredients via fermentation.
Bioenergy & biomaterials	Non-food ag outputs—turning biomass into energy, bio-based materials, innovative ag inputs (e.g. biofuels, bioplastics)	*F4F*[l]: Black soldier to transform food waste into insect-based protein and fertilizer. *ByBug*[m]: Modified insects to produce pharmaceutical proteins.
Midstream technologies	Technologies in food safety and traceability tech, logistics and transport solutions, and processing technology	*PolyNatural*[n]: Natural coating (shelf-life) replacing wax on fruit. *Ecoterra*[o]: Traceability for certified humane and sustainable dairy/egg supply chains.

The focus is on upstream and middle stream, reflecting the more developed nature of these segments in Chile compared to solutions downstream. Categories are taken from *AgFunder* (https://agfunder.com/), which is a global AgriFoodTech investment platform that also publishes widely used industry classifications and annual investment trend reports

[a]https://pewmaninnovation.net/
[b]https://neocroptech.com/
[c]https://neltu.me/
[d]https://www.agrourbana.ag/
[e]https://www.farmtastica.com/
[f]https://agrotrac.cl/
[g]https://harddrones.com/
[h]https://agromatch.cl/
[i]https://finka.cl/
[j]https://saviamercado.cl/
[k]https://luyef.com/
[l]https://f4f.cl/
[m]https://www.bybug.io/
[n]https://polynatural.com/
[o]https://ecoterra.cl/

These approaches reflect the strategic framework developed by Gans et al. (2018), which emphasizes how technology companies must carefully choose their market entry strategies when introducing innovations into established industries.

In parallel, many Chilean AgriFoodTech start-ups position themselves as contributing to at least one Sustainable Development Goal (SDG), particularly in areas like responsible production, climate action, or food security. This sustainability orientation is partly self-motivated but also increasingly encouraged by key ecosystem actors. For example, some public programs such as *Crea y Valida Foco en Sostenibilidad* and *Start-Up Chile*, as well as some venture capital funds and accelerators, now include sustainability or impact-related language in their calls and selection criteria.

2.2.2 Public Agencies for the Promotion of Science, Innovation, and Entrepreneurship

Chile's AgriFoodTech ecosystem is anchored in a robust public policy framework in which public agencies play a central role in several functions. They contribute to resource mobilization, supporting knowledge development and diffusion, fostering market formation, and guiding the direction of innovation through strategic programs. Some also contribute to legitimation by endorsing ventures and amplifying their visibility, and they foster the decentralization of innovation efforts.

At the core is the Chilean Economic Development Agency *(CORFO)*, which offers a comprehensive portfolio of instruments covering the entire start-up lifecycle. These range from early-stage co-financing (e.g. *Semilla Inicia, Semilla Expande*) to support for venture capital markets, angel networks, crowdfunding platforms, and investment funds. *CORFO* also funds incubators, accelerators, and applied research centres (Seedstars 2019; ICEX 2024; Rodriguez Cárdenas 2024) and supports infrastructure for pilot/prototyping (see Table 2 for ecosystem instruments). Many instruments operate through national and regional calls, helping to reduce reliance on the Metropolitan Region. For example, in the Maule region of Chile, there is ThinkAgro,[3] which focuses on supporting digital transformation, which started with a grant from CORFO.

CORFO's flagship *Start-Up Chile* initiative has supported over 3000 ventures since 2010, including many in AgriFoodTech, through equity-free funding, global exposure, and integration into entrepreneurial networks (Startup Chile 2025). In parallel, CORFO has recently launched *StartupLabs,*[4] a national network of deep-tech hubs that combine shared laboratories, prototyping facilities, and venture support to reduce territorial and infrastructure gaps. The first hub, *StartupLab.01*[5] (formerly Startup Campus), is in Santiago and jointly managed with Fundación Chile. *CORFO* also leads thematic and strategic platforms such as *Transforma Alimentos,*[6] which promotes public–private coordination for agri-food upgrading through regional nodes and challenge-based projects and supports specialized centres like the Food Innovation Technological Centre (*CeTA*[7]) and *ThinkAgro,*[8] which provide infrastructure and technical expertise for product development and scale-up (CORFO 2025a; Transforma Alimentos 2025).

Other public institutions complement *CORFO*'s role. The National Agency for Research and Development *(ANID)* supports science-based entrepreneurship

[3] https://www.thinkagro.cl/

[4] https://www.latercera.com/pulso/noticia/startuplabs-la-apuesta-del-gobierno-para-crear-una-red-nacional-de-hubs-de-emprendimientos-de-base-cientifico-tecnologica/CNIOKVHZR5EKVOX54LVEB6OL4M/

[5] https://startuplab01.com/

[6] https://transformaalimentos.cl/

[7] https://cetalimentos.cl/

[8] https://www.thinkagro.cl/

Table 2 Support mechanisms for AgriFoodTech start-ups in the Chilean ecosystem at the national and regional levels

Actor	Instrument type	Description	Stage focus	Ecosystem functions this actor and/ or instrument contributes to[a]
AgriFoodTech start-ups				EA, MF, GS, KD, KDi, L
Public agencies				RM, KD, KDi, MF, GS, L
CORFO—Start-up Chile	Equity-free grants, mentoring, global networks	Three-stage accelerator: Build (problem–solution fit), Ignite (product–market fit), Growth (early scaling). National reach and global networks	Ideation to MVP (Minimum viable product) to scaling	
CORFO—Semilla Inicia/ Semilla Expande/ Crea y Valida/ Consolida y Expande Innovación	Non-reimbursable grants	Cover ideation/ prototyping to validation and scaling. National and regional calls	Seed to grow and scale-up	
CORFO—Potencia program	Co-funding for incubators, accelerators, and ecosystem builders	Development and strengthening of incubators and accelerators, with regional capacity-building	Enhancing ecosystem infrastructure	
CORFO—StartupLabs	Deep-tech hubs with shared laboratories and venture support	Network of public–private hubs (e.g. StartupLab.01 with Fundación Chile) offering shared labs, pilot/ prototyping support; 2025 regional expansion calls	Idea to validation and early scaling	
CORFO—Transforma Alimentos	Sectoral coordination program	National platform aligning agri-food upgrading with regional nodes and partners, challenge-based projects and market formation	All stages	

(continued)

Table 2 (continued)

Actor	Instrument type	Description	Stage focus	Ecosystem functions this actor and/ or instrument contributes to[a]
CORFO—VC co-investment & Invierte, ScaleX	Public–private co-investment (Invierte), venture funding platforms, and venture exchange/ secondary capital markets (ScaleX). CORFO mobilizes private capital indirectly through funds and platforms rather than direct start-up investment.	Mobilizes private early-stage capital via Invierte (matching VC/angel funds and supporting equity-crowdfunding) and deepens markets through ScaleX, the Santiago Stock Exchange venture segment enabling regulated equity raises, liquidity, and visibility. Nationwide	Seed to scale-up	
ANID—Start-Up Ciencia	Conditional, performance-based grants	Supports early-stage, science-based start-ups to validate technologies, develop business models, and prepare for market entry. Focused on high-impact, research-intensive ventures with TRL (Technology Readiness Level) 3–6	Early-stage, research-to-market (science-based MVP)	
FIA—Innovation projects, pilot programs, and thematic calls	Non-reimbursable public grants	Supports agri-food innovations, with a focus on small- and medium-scale producers, SMEs, cooperatives, and gender-inclusive approaches. Variable regional implementation	Co-development, validation, and early adoption	
INDAP—Tech transfer and investment support programs	Subsidies, adoption incentives, technical assistance for smallholder farmers	Facilitates access to agri-tech solutions for smallholder farmers through co-financed equipment, training, and extension services. Regional execution varies	Adoption and diffusion (post-validation)	

(continued)

Table 2 (continued)

Actor	Instrument type	Description	Stage focus	Ecosystem functions this actor and/ or instrument contributes to[a]
ProChile (GoGlobal, export promotion services, Global X/Global X Plus)	Soft-landing, trade facilitation, co-financed missions	Supports Chilean start-ups and SMEs enter and scale in international markets through export-readiness training, soft-landing programs (GoGlobal; Global X/Global X Plus for Europe), regulatory and market-validation support, and co-financed participation in trade fairs; includes investor/corporate matchmaking and in-market mentoring	Growth and export expansion	
Research institutions and universities				KD, KDi, RM, GS, L
Universities & university-based incubators	Pre-incubation and incubation services; often CORFO-funded. Examples: IncubaUC (PUC), OpenBeauchef (U. de Chile), Chrysalis (PUCV), UDD Ventures, IncubatecUFRO, IncubaUdeC (U. de Concepción), Austral Incuba (U. Austral de Chile)	Provide early-stage support including mentoring, business model validation, IP and regulatory guidance, and investor readiness. Often linked to university R&D outputs or entrepreneurial programs. Regional universities promote decentralization	Core focus idea to seed stage	
INIA—Partnerships & pilot programs	In-kind support, co-developed pilots, applied R&D collaboration	Access to experimental stations, labs, and experts for real-world trials and validation. Regional stations enable place-based pilots	Prototype to field validation	
Specialized research & technology tech transfer centres	Pilot infrastructure, R&D platforms, co-development services. Examples: Thinkagro, CeTA	Technical infrastructure, applied R&D capabilities, and field validation services for start-ups and SMEs. Multi-regional facilities and start-up–industry linkage	Product development to scale-up and adoption	

(continued)

Table 2 (continued)

Actor	Instrument type	Description	Stage focus	Ecosystem functions this actor and/ or instrument contributes to[a]
Incubators, accelerators, and innovation hubs				EA, KD, KDi, RM, MF, L
Incubators and accelerators (public, private, hybrid)	Mentoring, infrastructure, networking, funding (equity and non-equity). Examples: incubators often university-based or regionally anchored (see Sect. 2.2.3 *"Research institutions and universities"*). Accelerators often public (Start-up Chile) and private (e.g. Imagine Lab, Magical Startups, The Ganesha Lab).	Incubators focus on ideation, prototyping, and business model validation. Accelerators emphasize investment readiness, scaling, and market access, often linking start-ups with corporate or international partners. Many programs operate hybrid models that combine incubation and acceleration functions. Many regional, some national and international.	From ideation and validation (incubation) to pre-seed and early scale-up (acceleration)	
Innovation hubs (regional, sectoral)	Networking, technical assistance, training, local pilot facilitation. Examples: Impulsa Lab Los Ríos and Start-up Biobío, Polo Maule Innova Agro 4.0, ThinkAgro.	Offer mentoring, capacity-building programs, and access to regional partners and infrastructure. Often facilitate coordination with public programs and universities. Many are regionally based.	From ideation to early validation and local pilot testing	

(continued)

Table 2 (continued)

Actor	Instrument type	Description	Stage focus	Ecosystem functions this actor and/ or instrument contributes to[a]
Financial investors				RM, MF, L, GS
Angel investor networks	Seed equity, investor mentoring, strategic connections. Some are nationally oriented (ChileGlobal), others regional (Austral Angels)	Early-stage investment groups providing capital, business mentoring, and access to networks	Seed to early-stage	
Equity crowdfunding platforms	Equity-based crowdfunding. Examples: Broota	Online platforms that enable start-ups to raise public equity investment. Facilitates early-stage financing and community engagement	Pre-seed to seed stage	
Venture Capital (VC) funds	Equity investment; some operate hybrid accelerator—VC models. Examples: Fen Ventures (private—national/ regional); Arpegio (private—national); Genesis Ventures (private—national; hybrid); Araucaria (private—national/ regional); The Yield Lab LatAm (private—regional/ international; hybrid)	Private investment funds that provide capital and strategic support to start-ups. Some, like The Yield Lab, also offer mentoring, technical assistance, or broader ecosystem support, operating at the intersection of VC and acceleration. Often co-invest with CORFO and/or international funds	Early growth to scale-up (Series A/B)	

(continued)

Table 2 (continued)

Actor	Instrument type	Description	Stage focus	Ecosystem functions this actor and/ or instrument contributes to[a]
Corporate Venture (CV) funds	Corporate equity investment, strategic co-investment. Examples: Carozzi Ventures, Agrosuper Ventures, The Yield Lab LatAm (with Nestlé investment)	Investment vehicles established by incumbent food companies to fund and collaborate with start-ups aligned with their innovation and sustainability strategies. These funds combine capital with strategic value, such as scale-up support, R&D collaboration, and distribution access	Seed to growth-stage	
Industry associations and ecosystem enablers				KDi, L
AgriFoodTech industry associations	Advocacy, community building, industry promotion. Examples: Agrotech Chile, EMBIO (Association of Biotech companies in Chile), ClimateTech Chile	Sector-specific gremial organizations that represent start-ups and scale-ups working at the intersection of food, biotech, and climate innovation. These associations promote sector visibility, facilitate collaboration, organize events, and engage with policymakers	All stages (especially visibility, legitimacy, and ecosystem coordination)	
Industry innovation hubs	Multi-stakeholder coordination platforms. Example: Sofofa Hub	Cross-sector innovation hubs that connect start-ups with large firms to validate technologies, facilitate real-world piloting, and accelerate integration into national or export-oriented value chains. Emphasis is placed on shared infrastructure, regulatory articulation, and pre-competitive collaboration	Growth to scale-up (pilot to adoption)	
Users/clients				MF, KD, KDi, RM, GS

[a]Ecosystem functions: *EA* Entrepreneurial activity, *KD* Knowledge development, *KDi* Knowledge diffusion, *RM* Resource mobilization, *MF* Market formation, *GS* Guidance of the search, *L* Legitimation

through *Start-Up Ciencia,* combining grants with access to research infrastructure and expertise, particularly for science-intensive ventures with long development cycles (biotech, agri-bio, new materials, etc.) (ANID 2025). The Foundation for Agricultural Innovation *(FIA)* and the Agricultural Development Institute *(INDAP)* promote innovation adoption among small and medium-scale producers. Both are regionally implemented through their local directorates and programs. *INIA* complements these efforts and offers technical collaboration and access to experimental stations, such as Smart Field, which is focused on digital agriculture.[9] Also, regional administrations fund agrifoodtech hubs, such as Polo Maule Innova Agro 4.0 in the Maule region, which is focused on supporting agtech start-ups.[10] *ProChile*[11] facilitates internationalization through market access, and soft-landing services (see Sect. 2.2.6.*"Ecosystem enablers – International integration & foreign hubs"*) and the Technical Cooperation Service *(SERCOTEC)* provides regionally targeted seed funding with a focus on inclusivity (ICEX 2024).

2.2.3 Research Institutions and Universities

Research institutions and universities are key actors in Chile's AgriFoodTech ecosystem. They mainly contribute to knowledge development and diffusion, support early-stage resource mobilization, and play an important role in legitimation. Some also participate in the guidance of the search by aligning research agendas with national and regional priorities.

Universities conduct fundamental and applied research across agri-food relevant disciplines, and train specialized professionals. They provide laboratories, pilot plants, and experimental fields distributed across regions, which are resources that are costly or inaccessible for early ventures. Key institutions include Pontificia Universidad Católica de Chile, Universidad de Chile, Universidad de Concepción, Universidad de Talca, and Universidad Austral de Chile (Gaitán-Cremaschi and Klerkx 2026). Universities also act as intermediaries between research and entrepreneurship. Through university-affiliated incubators (often supported by *CORFO*) and technology transfer offices, they offer mentoring, technical and business advice, IP/regulatory guidance, and connections to seed instruments and early-stage investors. Many incubators are regionally embedded, linking start-ups to local contexts (see Table 2) (Gaitán-Cremaschi and Klerkx 2026).

Specialized research and technology centres, including *INIA*, *ThinkAgro*, and *CeTA*, complement universities. They provide applied R&D, experimental sites, pilot-scale and prototyping infrastructure, and university-anchored hubs that link firms and start-ups to validation and adoption. Many focus on national priority areas such as food safety, sustainable production, and value-added innovation, and some

[9] https://www.smartfieldinia.cl/

[10] https://polomauleinnova.cl/

[11] https://www.prochile.gob.cl/

provide regional prototyping and scale-up infrastructure in key value chains (CeTA 2025; INIA 2025; ThinkAgro 2025).

In addition to these roles, universities and research centres also lend credibility to new ventures. Their scientific reputation, formal validation processes, and links to public institutions and industries help de-risk technologies for regulators, farmers, agri-food companies and investors, supporting broader adoption.

2.2.4 Incubators, Accelerators, and Innovation Hubs

Incubators, accelerators, and innovation hubs are core intermediaries in Chile's AgriFoodTech ecosystem. They contribute primarily to entrepreneurial activity, knowledge development, knowledge diffusion, resource mobilization, market formation, and legitimation by supporting early-stage ventures with tailored services and institutional visibility.

Chile has built a dynamic network of such organizations over the past decade, including public programs, university-affiliated incubators (see Sect. 2.2.3. *"Research institutions and universities"*), and private accelerators with both regional and national focus (see Table 2). While originally generalist, few have progressively adapted to the needs of AgriFoodTech start-ups, offering business model support, technical and commercial validation, legal advisory, and access to co-working and prototyping infrastructure (Von Zedtwitz and Grimaldi 2006; Bruneel et al. 2012; ICEX 2024). These programs also confer visibility and credibility. Admission into a reputable incubator or accelerator enhances legitimacy, facilitates access to investors and partners, and supports iterative product refinement through structured exposure to market feedback and mentoring networks (Mungila Hillemane et al. 2019; Rodriguez Cárdenas 2024).

More recently, corporate-led programs have emerged (see Table 2), reflecting a shift toward demand-driven models. Unlike traditional accelerators, these initiatives are structured around concrete operational challenges and provide not only funding and mentorship but also access to distribution channels, regulatory expertise, and sector-specific know-how.

Regional innovation platforms complement these efforts by integrating incubation, acceleration, and ecosystem-building roles. Initiatives like *Impulsa Lab Los Ríos*[12] and *Startup Biobío*[13] serve as territorially anchored hubs that promote local entrepreneurship and strengthen regional innovation capacities. By combining sectoral specialization with public–private coordination, they help mitigate geographic disparities and expand support in regions with natural resource advantages relevant to AgriFoodTech (CORFO 2025b; Startup Biobío 2025).

[12] https://www.fomentolosrios.cl/

[13] https://www.startupbiobio.com/

2.2.5 Financial Investors in AgriFoodTech

Financial investors in AgriFoodTech contribute primarily to resource mobilization, and also play key roles in market formation, legitimation, and guidance of the search by shaping demand, validating technologies, and signalling strategic priorities.

Access to financial capital is essential for developing and scaling AgriFoodTech ventures, contributing directly to resource mobilization (Fig. 2) (SEGIB 2023). The financing system follows a clear development pattern reflecting Chile's evolving innovation funding. According to Endeavor (2022), 80% of Chilean start-ups initially depend on informal financing from personal savings and family, friends, and networks (*3F funding*). While crucial for early testing, this capital becomes insufficient for technology development, regulatory compliance, and market entry.

The funding gap between early testing and commercial success, the "valley of death," particularly challenges AgriFoodTech ventures, which face long development cycles, substantial infrastructure needs, and complex regulatory processes. Angel investor networks have emerged as intermediaries, providing capital access and industry expertise through national organizations like *ChileGlobal Angels*[14] (Fundación Chile's investment arm), regional networks such as *Austral Angels*[15] affiliated with Dadneo, and crowdfunding platforms like *Broota* that enable small investments from multiple investors (Seedstars 2019; ACVC 2024; Rodriguez Cárdenas 2024; Austral Angels 2025; Broota 2025; Gaitán-Cremaschi and Klerkx 2026).

As start-ups mature, they access venture capital funds. The Chilean Venture Capital Association reports nearly USD 900 million invested in local or foreign start-ups in 2023, positioning Chile among Latin America's most active markets (ACVC 2024). Venture capital typically involves significant equity investments in exchange for larger funding rounds and board representation. While most funds are generalist, some actively pursue agri-food innovation. Several operate hybrid models combining acceleration services with investment, creating support systems beyond capital provision. Some align with sustainability goals, reflecting investor interest in environmental and social outcomes alongside financial returns. Funds present in Chile span national, regional, and international levels (see Table 2).

Family offices are an emerging investment source, often linked to Chile's agricultural sector or regional development interests. Corporate investors, such as Carozzi and Agrosuper, also play a growing role through dedicated innovation arms. They typically engage through strategic partnerships, licensing agreements, or minority equity investments, seeking technologies that enhance operations or enable access to new markets. These actors not only provide financial resources but also open commercial pathways and lend credibility to early-stage ventures through their endorsement (see Table 2).

---

[14] https://chileglobalangels.cl/

[15] https://australangels.com/

a

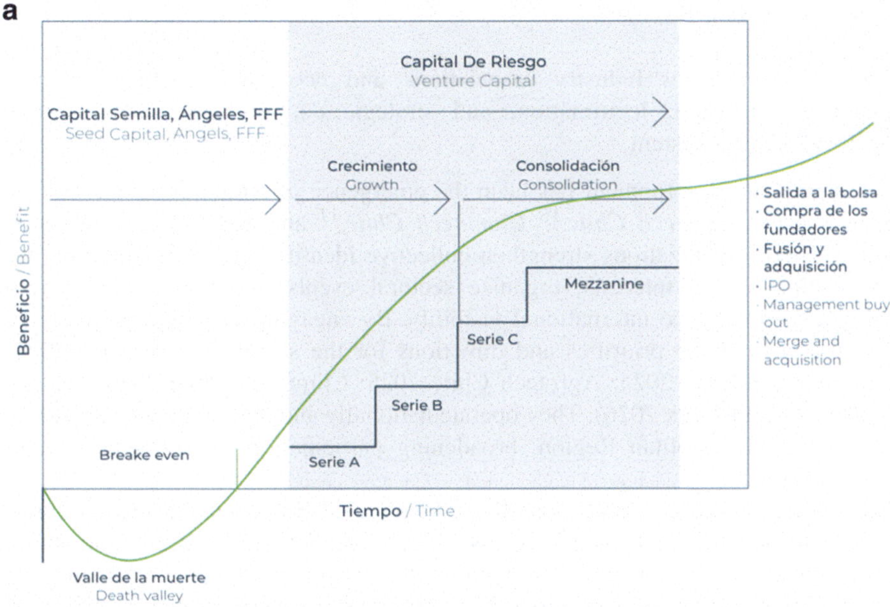

b

Etapa inicial	La mayoría de los emprendimientos comienzan con fondos que normalmente son aportados por los propios **fundadores/emprendedores**.
Ronda pre-seed	Con los primeros pasos en el desarrollo tecnológico y del modelo de negocio, comúnmente los emprendedores acuden a la **familia y amigos**, así como a incubadoras públicas o privadas, para obtener financiamiento.
Ronda Seed o semilla	Esta es una etapa fundamental en el desarrollo emprendedor, ya que permite alcanzar un producto mínimo viable y dar los primeros pasos comerciales. En ella, pueden participar del financiamiento **aceleradoras de negocios e inversores ángeles, además de instituciones públicas**.
Serie A	Una vez que el producto o servicio ha sido validado en el mercado y con la actividad comercial en marcha y expansión, la siguiente ronda de financiamiento suele ser la ronda denominada "Serie A", que comúnmente coincide con la primera inversión que realizan muchos **fondos de capital de riesgo**.
Serie B	Los emprendimientos que logran establecerse comercialmente en el mercado y logran un crecimiento sostenido de su negocio, recurren a la "Serie B" que les permita expandir y escalar su negocio, tanto a nivel de desarrollo tecnológico como a nivel de expansión geográfica y sectorial. En esta instancia, los **fondos de capital de riesgo, así como inversores privados y empresas son los principales inversores**.
Serie C/D y posterior	A esta altura del proceso de desarrollo de los emprendimientos, es posible que los mismos ya cuenten con recursos propios para financiar su crecimiento. Sin perjuicio de ello, es común ver nuevas rondas en las cuales participan no sólo **fondos de capital de riesgo y empresas**, sino también **fondos de capital privado**, que buscan inversiones en empresas ya establecidas y con un perfil de crecimiento atractivo.

Fig. 2 Financing stages and investor types across the start-up lifecycle in Chile. This figure illustrates the typical financing trajectory of start-ups in Chile, highlighting the different investor types involved at each stage, from initial informal funding (3F) through angel investment, venture capital, and later-stage funding. (Source: a. ACVC 2024; b. SEGIB 2023)

2.2.6 Ecosystem Enablers

Industry Associations Industry associations and ecosystem enablers support knowledge diffusion, legitimation, and strategic coordination across Chile's AgriFoodTech ecosystem.

A significant development has been the emergence of entrepreneur-led associations such as *Agrotech Chile,*[16] *Climatech Chile,*[17] and *EMBIO A.G.* (Biotech Chile). These organizations strengthen collective identity support legitimation by representing shared interests, organize sectoral events, promote public–private dialogue, and enhance international visibility. By engaging in policy discussions, they also help shape priorities and directions for the sector (El Mercurio 2024; Rodriguez Cárdenas 2024; Agrotech Chile 2025; Climatech Chile 2025; Gaitán-Cremaschi and Klerkx 2026). They operate nationally and regionally with activities outside the Metropolitan Region, broadening participation for start-ups located outside Santiago.

Traditional industry associations like *SNA* (Sociedad Nacional de Agricultura), *ChileAlimentos* (the food processors' association) and *Sofofa* have begun interacting more actively with start-ups. Responding to sustainability challenges and the growing role of innovation in productivity and competitiveness, they co-organize innovation forums, sponsor award programs, and help articulate industry needs to policymakers. Platforms like *Sofofa Hub*[18] foster structured engagement between start-ups and established firms, enabling early-stage validation, facilitating adoption, and enhancing visibility through institutional backing (Rodriguez Cárdenas 2024; SNA 2024; Sofofa Hub 2025; Gaitán-Cremaschi and Klerkx 2026).

International Integration and Foreign Hubs Having a highly export oriented agri-food sector, Chile has always actively worked on enhancing international connections (Nelson 2007; Klerkx and Guimón 2017), and this is also manifesting in Agri-FoodTech. Public connectors, especially *ProChile*, operate export-readiness and soft-landing/immersion programs (e.g. Go Global—Miami/Peru/Colombia/Mexico, Global X/Global X Plus—EU)[19] and organize multi-city hub foreign missions like Chile Tech Tour 2025. They often collaborate with associations such as *Agrotech Chile* on sector specific foreign missions.[20] Hybrid/private platforms, accelerators and soft-landing bridges extend these international linkages. *Chile Global Ventures*[21] mobilizes angel/VC capital via *ChileGlobal Angels* and fosters

[16] https://agrotech-chile.cl/

[17] https://www.climatech-chile.com/

[18] https://sofofahub.cl/

[19] https://www.prochile.gob.cl

[20] https://www.prochile.gob.cl/noticias/detalle-noticia/2025/03/13/chile-impulsa-la-innovaci%C3%B3n-agr%C3%ADcola-en-ee.uu.%2D%2Ddelegaci%C3%B3n-de-empresas-agtech-viaja-al-world-agritech-innovation-summit-con-apoyo-de-prochile

[21] https://chileglobalventures.cl/

corporate-venturing links; *Start-Up Chile* and university incubators build investor-readiness and soft-landing capacity; and initiatives such as *ChileMass*[22] provide a structured bridge to the Boston ecosystem through digital and immersion phases. In Europe, ventures leverage Global X immersions and ad hoc links such as to the Swiss innovation ecosystem (e.g. Pewman Innovation's immersion after being recognized "Start-up of the year"[23]) for mentoring, standards navigation, and visibility. Asia-facing linkages are nascent and selective, channelled mainly through *ProChile* trade missions and exploratory contacts. Complementary infrastructure such as ScaleX seeks to improve growth-stage capital visibility to foreign investors (see Table 2). Hemispheric conveners such as IICA[24] Digital Agriculture Week convene and coordinate cross-American cohorts and investment-bridging activities, linking Chilean ventures with regional investors and peers. Collectively, these connectors foster transnational innovation linkages, facilitating capital access, market entry, knowledge exchange, and contributing to resource mobilization, market formation, legitimation, and guidance of search.

2.2.7 Users and Clients as Innovation Partners

Users and clients, including farmers, agro-industrial companies, exporters, supermarkets, and consumers contribute primarily to market formation, knowledge development, resource mobilization, and guidance of the search, though the depth of their engagement varies widely across actor types.

These actors increasingly participate in the design, validation, and refinement of technologies and business models through open innovation challenges, early procurement, and pilot collaborations, often supported by public instruments and industry platforms (Lachman 2023; Rodriguez Cárdenas 2024; Gaitán-Cremaschi and Klerkx 2026). Their involvement goes beyond conventional client-supplier relations, providing feedback, operational insights, distribution networks and occasionally investment or strategic alliances.

For established firms, these collaborations enable exploration without disrupting core operations, while for start-ups, they provide essential partners for piloting and early market traction (Mungila Hillemane et al. 2019; Kansheba and Wald 2020; Fairbairn and Reisman 2024; IDB Invest and Accenture 2024; Mac Clay et al. 2025). Agro-industrial firms and exporters are key allies due to their role in (global) value chains. Through challenge-driven partnerships, open innovation programs and early procurement agreements, they facilitate market formation, commercial validation and scaling (Rodriguez Cárdenas 2024; Gaitán-Cremaschi and Klerkx 2026).

In the retail sector, large retailers play central roles beyond distribution. They shape packaging, labelling, certification, and consumer expectations. Landmark

[22] https://chilemass.org/2024/

[23] https://www.litoralpress.cl/SimbiuPDF/2025/03/27/5802508.pdf

[24] https://iica.int/es/

collaborations such as NotCo–Walmart's launch of NotMayo demonstrate how distribution partnerships can be decisive for scaling (LiveKindly 2020).

Consumers, particularly those seeking functional and novel foods and plant-based proteins, shape innovation trajectories through their preferences. Their purchasing decisions and sustainability preferences help start-ups refine their products and signal changing demand patterns, which in turn can influence the strategies of incumbent firms and the overall direction of the AgriFoodTech ecosystem (Rodriguez Cárdenas 2024; Gaitán-Cremaschi and Klerkx 2026).

By contrast, smallholders and rural SMEs are usually reached indirectly, through corporate buyers, distributors, or cooperatives, rather than via direct partnerships/sells with start-ups (Gaitán-Cremaschi and Klerkx 2026). Their adoption of AgriFoodTech solutions remains limited due to infrastructural, economic, human-capital, geographic, and cultural barriers. Connectivity gaps persist in many rural areas. High relative costs discourage investment, while low digital skills hinder effective use. Geographic dispersion raises transaction and logistics costs for start-ups, and conservative production cultures and risk aversion slow experimentation (Subtel 2024; Gaitán-Cremaschi and Klerkx 2026).

3 Challenges in Chile's AgriFoodTech Ecosystem

Chile's AgriFoodTech ecosystem has benefitted from growing public and private support to promote innovation and entrepreneurship. However, critical challenges remain that limit its ability to perform key innovation system functions in a balanced and sustained way, particularly when compared to more advanced ecosystems in FinTech or countries like Brazil, Colombia, and Argentina.

3.1 Institutional Fragmentation and Policy Coordination Gaps

Although *CORFO* and *ANID* lead a relatively robust support structure, institutional coordination across the public sector remains weak. Sector-specific agencies such as *FIA* and *INDAP* are poorly integrated into the AgriFoodTech start-up ecosystem, with only 3.4% of start-ups reporting support from them in 2022 (Endeavor 2022). This fragmentation constrains critical ecosystem functions such as entrepreneurial activity, resource mobilization and knowledge diffusion, particularly in areas where public agricultural expertise is essential. These gaps also reflect broader structural limitations, including institutional rigidity and limited adaptability to emerging innovation models (Rodriguez Cárdenas 2024).

Moreover, public support instruments frequently operate in silos and lack a shared long-term vision or coordinated roadmap for food system transformation. This weakens the guidance of the search function and hampers the emergence of system-wide learning and alignment. Although recent efforts, such as sustainability-oriented programs and strategic platforms like *Transforma Alimentos*, have sought to bridge public–private agendas, their contribution to building coherence and long-term orientation remains modest when compared to more consolidated sectors like renewable energy (Rodriguez Cárdenas 2024; Kok et al. 2025; Gaitán-Cremaschi and Klerkx 2026).

3.2 Sustainability Focus

Although sustainability is increasingly part of the ecosystem's discourse, it is not yet institutionalized as a core driver of innovation. There is still no robust mechanism that aligns public support, private investment, and entrepreneurial activity with long-term environmental or social goals. As a result, guidance of the search remains underdeveloped, and sustainability concerns are only weakly embedded in resource mobilization and legitimation processes (Gaitán-Cremaschi and Klerkx 2026).

3.3 Geographic Concentration and Infrastructure Disparities

Chile's start-up activity remains highly centralized in Santiago, which limits the development of core functions such as entrepreneurial activity, knowledge creation, and resource mobilization in rural and regional areas. Although regional assets have expanded, access to specialized laboratories, pilot plants and reliable high-speed broadband is more uneven outside the capital, constraining prototyping, validation and scaling of innovations tailored to location-specific agri-food systems (Seedstars 2019; Rodriguez Cárdenas 2024; Subtel 2024; Gaitán-Cremaschi and Klerkx 2026).

In response, Chile is decentralizing entrepreneurial support through: (i) *CORFO*'s StartupLabs (startuplab.01), and programs (e.g. Potencia); (ii) university-affiliated incubators outside the Metropolitan Region such as *Austral Incuba* (Valdivia), *Incubatec-UFRO* (Temuco) and *IncubaUdeC* (Concepción) that mobilize local labs and experimental stations; (iii) sectoral centres such as *CeTA Alimentos* and *ThinkAgro*; and (iv) regional initiatives including *Impulsa Lab Los Ríos* and *Startup Biobío*. Collectively, these efforts expand access to mentoring, infrastructure and pilot partners beyond the Metropolitan Region (see Table 2 for actors and instruments with regional reach). Despite these advances, thin early-stage and regional investor presence, fragmented coordination, and limited specialized talent, still push many ventures to relocate or commute to Santiago for funding, visibility and advanced facilities (Rodriguez Cárdenas 2024; Gaitán-Cremaschi and Klerkx 2026).

3.4 Financial Ecosystem Constraints

The sector lacks dedicated investment vehicles tailored to the specific needs of AgriFoodTech ventures. Start-ups in this space face distinct structural barriers, such as long validation cycles, high infrastructure demands, exposure to climatic and regulatory risks, and extended return horizons, that are often misaligned with the expectations of generalist venture capital. These conditions hinder the development of investable propositions and delay the demonstration of market fit, thereby deterring investors who typically prioritize sectors with faster, more scalable returns, such as fintech, e-commerce, or SaaS models (Rodriguez Cárdenas 2024; Gaitán-Cremaschi and Klerkx 2026). Consequently, Chilean AgriFoodTech start-ups, like their regional counterparts, capture only a marginal share of total venture capital, with funding concentrated in early-stage rounds (pre-seed to Series A) (IDB Invest and Accenture 2024).

In the absence of adequate follow-on capital (ACVC 2024), many start-ups resort to personal savings, angel investors, and non-dilutive public support, particularly *CORFO* grants, as primary funding mechanisms to sustain operations and validate technologies (IDB Invest and Accenture 2024; Rodriguez Cárdenas 2024; Gaitán-Cremaschi and Klerkx 2026). While these sources play a critical role in enabling early experimentation and lowering barriers to entry, they are not designed to support long-term scaling. As a result, a growing number of early-stage ventures become structurally reliant on public grants. In some cases, the lack of commercial pressure to scale, combined with opportunities to repeatedly access grant funding by pivoting among early-stage concepts, fosters a cycle of low-risk experimentation. This dynamic can inadvertently stall progress toward technological maturity, market validation, and broader resource mobilization (Gaitán-Cremaschi and Klerkx 2026).

3.5 Regulatory Framework

The regulatory framework in Chile presents a mix of enabling and constraining features for the development of the AgriFoodTech ecosystem. On the enabling side, supportive instruments such as the venture capital law (Ley Única de Fondos N° 20.712) and R&D tax incentives (Ley de Incentivo Tributario a la Inversión en I+D) aim to stimulate private investment and innovation activity. These tools contribute to the broader functions of resource mobilization and entrepreneurial activity. However, these incentives coexist with regulatory bottlenecks that affect start-ups, especially those operating at the technological frontier (ICEX 2024; Rodriguez Cárdenas 2024). Approval processes are often slow, fragmented, or ill-adapted to novel technologies, which delays time-to-market and undermines investor confidence (Rodriguez Cárdenas 2024). As a result, the functions of market formation, guidance of the search, and legitimation are negatively affected.

In food tech, for example, rigid labelling regulations and recent proposals like the so-called "Foodtech Law," which would limit the use of traditional food names for plant-based products, risk restricting product innovation and consumer acceptance (Rodriguez Cárdenas 2024). Likewise, the absence of a dedicated legal framework for data governance in agri-digital systems leaves key issues such as ownership, interoperability, and usage rights unresolved. This legal ambiguity introduces uncertainty, limits investor engagement, and constrains the development of scalable, data-intensive business models in the sector.

3.6 Market Access and End-User Engagement

AgriFoodTech start-ups in Chile face considerable challenges in scaling their operations cost-effectively, both in the domestic market and across Latin America. A key limitation lies in the difficulty of transferring standardized technological solutions across the region's highly heterogeneous agri-food sector. Differences in farm size, geographic dispersion, topography, and producer profiles complicate market formation and reduce the applicability of one-size-fits-all innovations (IDB Invest and Accenture 2024; Rodriguez Cárdenas 2024; Gaitán-Cremaschi and Klerkx 2026).

Nevertheless, the ecosystem shows relative strength in user integration, with some start-ups actively involving farmers (particularly large scale), food producers and other users/clients through co-design processes and field-based pilot projects. However, such engagement remains uneven. Cultural conservatism, low trust, and power asymmetries, particularly when dealing with large agro-industrial firms or retail chains, can reduce collaboration opportunities, delay adoption, or force start-ups to adjust solutions to incumbent needs rather than those of end users (Gaitán-Cremaschi and Klerkx 2026).

Support for legitimation is gradually improving. Sector-specific associations such as *Agrotech Chile* and *EMBIO A.G.* are helping to close knowledge gaps between start-ups and traditional stakeholders, while also enhancing the visibility of AgriFoodTech ventures. More established business associations are beginning to engage with innovation-oriented firms, suggesting a slow but growing acceptance of new actors within the agri-food sector (Gaitán-Cremaschi and Klerkx 2026).

In the FoodTech segment, additional barriers to market formation persist. Higher price points relative to conventional alternatives, entrenched consumer preferences, and scepticism toward unfamiliar ingredients or technologies continue to limit adoption (IDB Invest and Accenture 2024).

3.7 Inclusion of Small and Medium Producers and Rural Communities

Although Chile's AgriFoodTech ecosystem is dynamic, integrating smallholders, medium producers, and rural SMEs remains difficult. Extension and advisory programs target these groups, yet adoption gaps persist (Gaitán-Cremaschi and Klerkx 2026). Start-ups frequently pilot and commercialize in export-oriented chains, with smaller actors reached indirectly via buyer linkages and cooperatives. A few initiatives directly bridge this gap, most notably *ThinkAgro*, which supports SMEs in Maule/Ñuble to adopt Agriculture 4.0 solutions, and *INIA*'s *SmartField,*[25] which co-validates digital tools with growers under real conditions. Others provide indirect entry points. *INDAP programs*[26] strengthen organizations, co-finance on-farm investments, and deliver specialized technical assistance, and regional platforms such as *Maule Alimenta*[27] and the *Polo Maule Innova Agro 4.0* convene producers and technology providers for training, networking, and early validation. However, whether these initiatives translate into sustained adoption or scaling of start-up developed technologies and services among smallholders, medium producers, and rural SMEs remains uncertain, as systematic evidence is scarce.

4 Future Perspectives

Looking ahead (2025–2035), Chile's agri-food technology ecosystem is likely to experience selective, rather than generalized, growth, shaped by the continuity and implementation of public innovation policies, the availability of later-stage private investment, and export-market signals. Persistent water scarcity and climate variability (Garreaud et al. 2025) will continue to pose major production and adaptation challenges, especially for fruit and wine value chains. In these sectors, competitiveness increasingly depends on efficient resource use, quality management, and compliance with international traceability and sustainability standards. Furthermore, AgriFoodTech start-ups are increasingly geared towards and centred on supporting broader food systems transformation.

On the policy supply side, public innovation agencies (e.g. *CORFO, FIA, INIA*) have increasingly promoted initiatives aligned with Agriculture 4.0/Agro 4.0 (e.g. FIA 2022—National Promotion Program for Agriculture 4.0). These agendas are expected to reinforce the adoption of digital decision-support tools and IoT/AI applications, advanced water management technologies, traceability

[25] https://www.smartfieldinia.cl/

[26] https://www.indap.gob.cl/plataforma-de-servicios

[27] https://maulealimenta.cl/

systems, and biotechnological innovation. Furthermore, increasingly public support is "mission-oriented" and requires clear articulation of contributions to the Sustainable Development Goals (Kok et al. 2025), as well as stronger reflection on what transformation pathways AgriFoodTech aims to support, with concepts such as regenerative and agroecological farming, circular food systems, and bioinputs and bioeconomy.

Particular technology fields and verticals may become key Chilean contributions to the broader AgriFoodTech field, given particular strengths in these. Within biotechnology, Chile's status as a counter-season seed R&D and export multiplication hub is expected to remain a strategic niche, supported by stringent phytosanitary controls and geographic isolation advantages (Sánchez 2020; Seeworld 2023). In alternative proteins (precision-fermentation and cultivated products), developments are likely to remain nascent domestically due to high capital intensity, limited local market scale, and the still-emerging regulatory framework for novel foods (MINSAL 2025). The marine bioeconomy stands out as an emergent frontier. Seaweed and macroalgae applications, spanning biostimulants, biomaterials, and functional ingredients, are advancing supported by a seaweed aquaculture roadmap (TNC 2025) and complementary regional measures, signalling increasing policy attention to this domain. Finally, ecosystem dynamics are consolidating around public–private networks. *Start-Up Chile* and related accelerator programs are expected to continue seeding early-stage agri-food tech ventures. At the same time, sectoral associations such as *AgroTech Chile*, together with formal cooperation agreements with agencies like *FIA*, are tightening links among start-ups, research actors, producers, and public institutions, gradually integrating the agri-food tech ecosystem into Chile's broader agricultural innovation system.

5 Conclusion

Chile's AgriFoodTech ecosystem has evolved significantly in recent years, driven by a dynamic early-stage start-up community, targeted public programs, and an expanding network of support institutions. This chapter has highlighted both notable achievements and persistent structural challenges. Functions such as entrepreneurial activity, knowledge development, and early-stage resource mobilization are relatively well developed. The presence of high-quality research institutions and public–private centres further strengthens the knowledge base that underpins innovation.

However, several ecosystem functions remain underdeveloped. Resource mobilization beyond early stages is hindered by the scarcity of sector-specific investment vehicles, limited later-stage capital, and uneven access to infrastructure for piloting and scaling. Market formation is constrained by fragmented demand, the small domestic market, regulatory uncertainty, and conservative attitudes among agro-industrial incumbents, which together slow the adoption of emerging solutions. Directionality also remains weak. Despite strategic initiatives such as *Transforma Alimentos*, the ecosystem lacks a shared long-term vision and effective coordination mechanisms to align public and private actors around structural transformation goals. That said, it is important to recognize that start-up ecosystems often thrive on decentralized innovation, entrepreneurial autonomy, and market experimentation. These characteristics can foster creativity and responsiveness but may also hinder coordinated, long-term efforts which are typical for mission-oriented innovation for food systems transformation (Klerkx and Begemann 2020). Legitimation, meanwhile, is advancing through public endorsement and the rise of sectoral associations, yet remains fragile among traditional actors, where cultural and institutional barriers to agri-food innovation persist.

Looking ahead, strengthening Chile's AgriFoodTech ecosystem will require:

- Stronger coordination across public institutions to reduce fragmentation and guide long-term strategy.
- Continued focus on connecting the Chilean AgriFoodTech start-up ecosystems to ecosystems elsewhere, to enable international exchange and collaboration, and also position particular AgriFoodTech categories in which Chile excels internationally.
- Increased investment in sector-specific infrastructure and financing mechanisms tailored to the needs of AgriFoodTech ventures, including more diversified capital sources and blended financing models that combine public, private, and impact-driven funds to de-risk investments and support scaling.
- Build regional innovation capacities and ecosystems to integrate diverse agricultural profiles to reduce territorial disparities, foster small and medium producer participation, and strengthen linkages between start-ups, research institutions, and territorial innovation platforms.
- Greater regulatory flexibility to accommodate emerging technologies, streamline approval processes, and enable timely adaptation to innovation trends without compromising safety or accountability.
- A clearer national roadmap that balances market opportunities with sustainability and food system transformation objectives.
- Enhanced monitoring and feedback mechanisms to track the ecosystem's evolution, identify bottlenecks, and adjust policies over time. This includes regular mapping of start-ups, investment flows, and regional disparities to inform support strategies.

Acknowledgements The work underpinning this chapter was supported by the Chilean National Agency for Research and Development (ANID) through FONDECYT Regular Grant No. 1250482 (2025) and through the Center for Advancing Agri-Food System Transformation (CA2FST), ANID CA2FST CIN250015. Additional financial support was provided by the Dirección de Investigación of the Vicerrectoría Académica, Universidad de Talca.

Competing Interests The author(s) has no competing interests to declare that are relevant to the content of this manuscript.

References

ACVC—Asociación Chilena de Venture Capital (2024) Annual Report 2024. Retrieved from https://www.acvc.cl

Adner R (2017) Ecosystem as structure: an actionable construct for strategy. J Manag 43(1):39–58

AgFunder (2025) Developing Markets Agri-FoodTech Investment Report 2025

Agrotech Chile (2025) Retrieved from https://agrotech-chile.cl

ANID—Agencia Nacional de Investigación y Desarrollo (2025) Start-Up Ciencia 2025. Retrieved from https://anid.cl/concursos/startup-ciencia-2025/

Austral Angels (2025) Retrieved from https://australangels.com/

Bert FE, Lachman J, Del Río JA (2023) Desarrollo Agtech en la Región Andina: casos de éxito y lecciones para el futuro. Giles Álvarez L (ed)

Brixton Ventures Lab (2024) Mapping the Agtech and Foodtech ecosystem in Latin America. Retrieved from https://www.brixtonventures.com/post/ecosistema-agtech-latam-2024

Broota (2025) Retrieved from https://inversion.broota.com/

Bruneel J, Ratinho T, Clarysse B, Groen A (2012) The evolution of business incubators: comparing demand and supply of business incubation services across different incubator generations. Technovation 32(2):110–121

CEPAL & FAO (2021) Digitalización y cambio tecnológico en las mipymes agrícolas y agroindustriales en América Latina. Comisión Económica para América Latina y el Caribe (CEPAL)/Organización de las Naciones Unidas para la Alimentación y la Agricultura (FAO). Retrieved from https://bibliotecadigital.ciren.cl/handle/20.500.13082/147699

CeTA—Centro Tecnológico para la Innovación Alimentaria (2025) Retrieved from https://cetalimentos.cl/

Climatech Chile (2025) Retrieved from https://www.climatech-chile.com

Cockayne D (2019) What is a start-up firm? A methodological and epistemological investigation into research objects in economic geography. Geoforum 107:77–87

CORFO (2025a) Programas Tecnológicos Estratégicos. Retrieved from https://www.corfo.cl/sites/cpp/pte-programas-tecnologicos-estrategicos

CORFO (2025b) Impulsa lab regional news. Retrieved from https://corfo.cl/sites/cpp/sala_de_prensa/regional/movil/25_03_2025_impulsa_lab

El Mercurio (2024) Un tremendo avance para la industria biotech. Retrieved from https://www.linkedin.com/posts/cbt-chile_un-tremendo-avance-para-la-industria-biotech-activity-7306015367905136641%2D%2DdmV/

Embrapa, SP Ventures & Homo Ludens (2019) Radar Agtech Brasil. Retrieved from https://radaragtech.com.br/

Endeavor Chile (2022) Radiografía Agtech. Retrieved from https://www.endeavor.cl/publicaciones/radiografia-agtech/

Fairbairn M, Guthman J (2020) Agri-food tech discovers silver linings in the pandemic. Agric Hum Values 37:587–588

Fairbairn M, Reisman E (2024) The incumbent advantage: corporate power in agri-food tech. J Peasant Stud 51(6):1331–1354

FIA - Fundación para la Innovación Agraria (2022) Plan nacional para el fomento de la Agricultura 4.0. Santiago de Chile.

Freeman C (1988) Japan: a new national system of innovation? In: Dosi G (ed) Technical change and economic theory. Pinter, London, pp 330–348

Gaitán-Cremaschi D, Klerkx L (2026) The roles of agri-food tech start-up ecosystems in strategic niche management for food system transitions. Environ Innov Soc Transit 60:101140

Gans JS, Scott E, Stern S (2018) Strategy for start-ups. Harv Bus Rev 96(3):44–51

Garreaud R, Boisier JP, Álvarez-Garreton C, Christie D, Carrasco-Escaff T, Vergara I et al (2025) Hyperdroughts in Central Chile: drivers, impacts and projections. EGUsphere, pp 1–31

Granstrand O, Holgersson M (2020) Innovation ecosystems: a conceptual review and a new definition. Technovation 90:102098

Hekkert MP, Suurs RA, Negro SO, Kuhlmann S, Smits RE (2007) Functions of innovation systems: a new approach for analysing technological change. Technol Forecast Soc Chang 74(4):413–432

Herrero M, Thornton PK et al (2020) Innovation can accelerate the transition towards a sustainable food system. Nat Food 1:266–272

HNRG (2021) Growing agricultural investment opportunities in Chile. Retrieved from https://www.globalaginvesting.com/wp-content/uploads/2021/07/Ag-In-Chile_LR.pdf

ICEX España Exportación e Inversiones (2024) Ecosistema del Emprendimiento en Chile. Retrieved from https://www.icex.es/es/quienes-somos/donde-estamos/red-exterior-de-comercio/CL/documentos-y-estadisticas/estudios-e-informes/visor-de-documentos.o-d%2D%2Decosistema-del-emprendimiento-en-chile%2C-2024.doc105202409

IDB Invest & Accenture (2024) Sustainable digital impact: AgriFoodTech start-up ecosystem in Latin America & the Caribbean. Retrieved from https://idbinvest.org/en/publications/sustainable-digital-impact-agrifoodtech-startup-ecosystem-latin-america-and-caribbean

INIA—Instituto de Investigaciones Agropecuarias (2025) Sitio oficial del INIA Chile. Retrieved from https://www.inia.cl

Kansheba JMP, Wald AE (2020) Entrepreneurial ecosystems: a systematic literature review and research agenda. J Small Bus Enterp Dev 27(6):943–964

Klerkx L, Begemann S (2020) Supporting food systems transformation: the what, why, who, where and how of mission-oriented agricultural innovation systems. Agric Syst 184:102901

Klerkx L, Guimón J (2017) Attracting foreign R&D through international centres of excellence: early experiences from Chile. Sci Public Policy 44:763–774

Klerkx L, Rose D (2020) Dealing with the game-changing technologies of Agriculture 4.0: how do we manage diversity and responsibility in food system transition pathways? Glob Food Secur 24:100347

Klerkx L, Villalobos P (2024) Are Agri-foodTech start-ups the new drivers of food systems transformation? An overview of the state of the art and a research agenda. Glob Food Sec 40:100726

Kok KP, Gaitán-Cremaschi D, Klerkx L, Villalobos P (2025) Context and proto-institutions in the emergence of transformative innovation policy: insights from Chile. Sci Public Policy 52(4):597–612

Lachman J (2023) Metodología consensuada sobre los alcances del concepto de "ecosistema agtech" para su estudio por parte del PROCISUR. Documento de trabajo, FONTAGRO. Retrieved from https://www.fontagro.org/new/uploads/productos/19885_Fortalecimiento-_Producto_2_(Apr).pdf

Lachman (2024) Nuevos patrones de innovación y los desafíos emergentes para los sistemas de CyT en agro. Foro Estratégico de Quito. Retrieved from https://digital.fontagro.org/wp-content/uploads/2024/07/Jeremias-Lachman_Panel_1.pdf

Lachman J, López A (2022) The nurturing role of the local support ecosystem in the development of the Agtech sector in Argentina. J Agribusiness Dev Emerg Econ 12(4):714–729

Lachman J, Pereyra M, Tacsir E (2022) Agtech: startups y nuevas tecnologías digitales para el sector agropecuario: casos de estudio de Uruguay. Cuyonomics Investigaciones en Economía Regional 6(10):13–34

LiveKindly (2020) Retrieved from https://www.livekindly.com/walmart-chile-notcos-vegan-mayo-artificial-intelligence/

Lundvall BA (1992) National systems of innovation: towards a theory of innovation and interactive learning, vol 242. Pinter, London

Mac Clay P, Bort G, Feeney R, Sellare J (2025) Industrial dynamics and business strategies in the emergence of agricultural biological inputs. Bus Strateg Environ 34:6472

Mikhailov A, Oliveira C, Padula AD, Reichert FM (2021) Californian innovation ecosystem: emergence of agtechs and the new wave of agriculture. Innov Manag Rev 18(3):292–307

MINSAL (2025) Retrieved from https://www.easglobal.com/post/chile-opens-consultation-on-first-ever-regulatory-framework-for-novel-foods-and-ingredients

Moore JF (1993) Predators and prey: a new ecology of competition. Harv Bus Rev 71(3):75–86

Mungila Hillemane BS, Satyanarayana K, Chandrashekar D (2019) Technology business incubation for start-up generation: a literature review toward a conceptual framework. Int J Entrep Behav Res 25(7):1471–1493
Nelson RC (2007) Transnational strategic networks and policymaking in Chile: CORFO's high technology investment promotion program. Lat Am Polit Soc 49:149–181
Newell R, Newman L, Mendly-Zambo Z (2021) The role of incubators and accelerators in the fourth agricultural revolution: a case study of Canada. Agriculture 11(11):1066
Rodriguez Cárdenas N (2024) Harvesting innovation: navigating enablers and challenges for Chilean startups on the path to sustainable food ecosystems (master's thesis). Utrecht University, The Netherlands
Sánchez MA (2020) Chile as a key enabler country for global plant breeding, agricultural innovation, and biotechnology. GM Crops Food 11(3):130–139
Seedstars (2019) Chile: the leading LatAm start-up ecosystem. Retrieved from https://www.seedstars.com/content-hub/insights/chile-the-leading-latam-startup-ecosystem
SeedWorld (2023) Retrieved https://www.seedworld.com/latam/2023/08/21/chiles-seed-sector-reaping-globalizations-rewards
SEGIB (2023) El Ecosistema AgTech-FoodTech en Iberoamérica: Oportunidad Local y Liderazgo Global. Secretaría General Iberoamericana. Retrieved from https://www.segib.org/wp-content/uploads/Informe-AGTECH_ES.pdf
SNA (2024) Retrieved from https://www.sna.cl/2024/03/27/agrotech-se-une-a-la-sna-y-promete-transformar-la-agricultura-chilena/
Sofofa Hub (2025) Innovation platform for industrial innovation. Retrieved from https://sofofahub.cl
Stam E, Van de Ven A (2021) Entrepreneurial ecosystem elements. Small Bus Econ 56(2):809–832
Startup BioBio (2025) Regional innovation and entrepreneurship platform. Retrieved from https://www.startupbiobio.com
Start-Up Chile (2025) National start-up accelerator program by CORFO. Retrieved from https://startupchile.org
Subtel (2024) Retrieved from https://www.subtel.gob.cl/wp-content/uploads/2025/02/Informe-Final-Subtel-Acceso-y-Uso-Internet-2024.pdf?utm
The Nature Conservancy (TNC) (2025) Roadmap for the sustainable development of seaweed farming in Chile 2025
ThinkAgro (2025) Centro de Extensionismo Tecnológico para la Agroindustria. Retrieved from https://www.thinkagro.cl
Transforma Alimentos (2025) National strategic program for agri-food innovation. Retrieved from https://transformaalimentos.cl
Vergara JM, Barrett CB (2025) Private capital investments in Agrifood-Tech startups in South America, 2007–2022. Agribusiness
Vitón R (2020) Innovación Agrotech en América Central y el Caribe: Oportunidades y desafíos frente al cambio climático. Saini E (ed)
Von Zedtwitz M, Grimaldi R (2006) Are service profiles incubator-specific? Results from an empirical investigation in Italy. J Technol Transf 31(4):459–468

Startup Ecosystems and Business Models of Agtech Startups

Débora Monteiro Moretti and Stefanie Bröring

1 Introduction

The pursuit for efficiency and sustainability has driven innovation and change throughout the industrial landscape. Agriculture is a particularly challenging industry due to several reasons, including its fundamental role in sustaining human life, its exposure to environmental unpredictability, and the complex network of stakeholders involved (Rose et al. 2018; Gerhardt et al. 2022). Considering the complexity of this evolutionary path of agriculture, that encompasses not only simply new technologies but a profound change in business models (both at firm and farm levels) and supply chain relationships, this book chapter looks to startup ecosystems, strategies pursued by its players, and real-world business models at play.

On the one side, agriculture has been pushed by new technologies such as Artificial Intelligence, robotics, Internet of Things, and gene editing with the promise of enhanced efficiency (higher production with less input and sometimes, work labor) (Araújo et al. 2021; Pandeya et al. 2022). On the other hand, the sector faces an "environmental pull," mounting pressures from climate change, resource scarcity, and shifting consumer demands (Rose et al. 2020). These challenges are thought to reinforce the need for transformation and represent not only threats to current, established businesses—but also a fertile ground for innovation.

This transformation is not occurring in isolation inside organizations. Looking to the myriad of firms offering innovative solutions to agriculture, three players

D. M. Moretti (✉)
Professorship of Electrobiotechnology, Technical University of Munich, Straubing, Germany

S. Bröring
Chair of Entrepreneurship and Innovative Business Models, Ruhr University Bochum, Bochum, Germany

F. da Silveira et al. (eds.), *Agriculture 4.0*,
https://doi.org/10.1007/978-3-032-24552-6_19

stand out: First, incumbent firms such as John Deere, Bayer, and BASF (large input suppliers) that are actively investing in or acquiring new technologies to remain competitive. Proactively or reactively, such established companies may undergo thorough strategic renewals, which impact their routines, structure, products and services or even business models (Agarwal and Helfat 2009; Dattée et al. 2017; Dyck et al. 2023) . Second, at the same time, companies from adjacent industries—such as electronics, software, and telecommunications—are entering the agricultural domain, bringing with them new capabilities and business models (Sarto et al. 2021; Gerhardt et al. 2022). These companies, also called *diversifying entrants*, normally bring technological infrastructure and knowledge of how to deal and create value with these new, digital technologies (Chen et al. 2017; Bohnsack et al. 2024). Here Birner et al. (2021) separate multinational software and big data firms, such as IBM and Microsoft, from nonagriculture "hardware" companies, such as Bosch. And third, most notably, startups are playing a catalytic role in this shift, introducing not only novel but unique solutions that challenge established practices and open new markets (Birner et al. 2021). These developments are not only technological—they are deeply strategic. As new entrants bring fresh perspectives and digital-native approaches, they also introduce new business models that contrast sharply with those of traditional players (Wolfert et al. 2022; Cavazza et al. 2023).

Among these players, the rise of entrepreneurial ventures drives new value creation and economic development (Grande et al. 2025). Understanding the dynamics that support and sustain these ventures—often through what are termed *startup* or *entrepreneurial* ecosystems—in the agricultural sector therefore attracts significant attention and effort from both private and public sectors (Gerhardt et al. 2022).

In this chapter, we explore how startup ecosystems and business models are evolving in the context of Agriculture 4.0. We examine the origins and support structures of agri-tech startups, the phases of their development, and the critical—but often overlooked—role of the farmer as both customer and ecosystem participant. We also analyze how deep tech ventures, often emerging from academic research, navigate the path from scientific insight to market impact. Finally, we investigate how startups position themselves within complex agricultural ecosystems, using business model innovation and strategic control points to scale and sustain their operations. By unpacking these dynamics, this chapter aims to provide a nuanced understanding of how innovation in agriculture is being shaped—not just by technology, but by the interplay of actors, institutions, and evolving business logics.

2 Startups as the Avant-Garde

Startup ecosystems represent intricate networks of interconnected entities, including entrepreneurs, established corporations, investors, research institutions, and government agencies, that collectively contribute to the creation and growth of new ventures (Guindalini et al. 2021). They foster a dynamic environment where novel ideas and technologies can flourish (Dutia 2014). Their success can be strongly influenced by the surrounding ecosystems and, the other way around, ecosystems can actively facilitate their entrepreneurial path (Klerkx and Villalobos 2023).

The concept of ecosystem itself is fluid and evolves as technology and society roles also do. They can also be understood as entrepreneurial ecosystems, but, in particular, startups are companies where innovation is in its core, offering new value propositions or solutions to demands still to mature. And it is due to its very unknown path that they rely on the ecosystem to navigate the market.

As put by Autio et al. (2017) "*Entrepreneurial ecosystems are the only cluster type where cluster externalities and cluster-specific structural elements are explicitly organized around the entrepreneurial process of opportunity discovery, pursuit, and scale-up of new ventures*" (Autio et al. 2017). The dynamic nature of ecosystems is due to the focus on the relationships between the different actors that not only change over time but can evolve into interdependencies that can be influenced and orchestrated. Thus, the success of startups is deeply intertwined with the broader ecosystem in which they operate (Eslamloo et al. 2025), during their whole life cycle, which can be divided into three main phases: Standup, Startup, and Scaleup.

2.1 *The Startup Lifecycle: Standup, Startup, and Scaleup*

The development of startups in general, but also in agriculture, follows a dynamic and iterative process that can be broadly divided into three interrelated phases: standup, startup, and scaleup (Rossetti et al. 2018). These phases are not merely chronological—they reflect distinct sets of activities, knowledge flows, and ecosystem interactions that shape the trajectory of entrepreneurial ventures. Drawing on the framework proposed by (Autio et al. 2017), we can understand these phases as embedded within entrepreneurial ecosystems that combine digital and spatial affordances to support business model innovation and opportunity pursuit.

2.1.1 Standup: From Intention to Entrepreneurial Entry

The standup phase encompasses the earliest stage of the entrepreneurial journey, where individuals or teams self-select into entrepreneurship. In agriculture, this phase is often catalyzed by exposure to pressing challenges—such as climate variability, soil degradation, or labor shortages—that inspire problem-solving behavior. In the standup phase, innovative ideas take shape, where nascent startups are nurtured by universities, incubators, and research institutions, providing them with essential resources, mentorship, and early-stage funding (Eslamloo et al. 2025). Entrepreneurial ecosystems play a critical role here by offering mechanisms that attract and motivate talent, such as:

- Hackathons and innovation challenges focused on agtech themes,
- University-based entrepreneurship programs,
- Networking events and ideation workshops,
- Talent scouting by accelerators and incubators.

These mechanisms help cultivate what Autio et al. (2017) call "architectural knowledge"—shared heuristics and practices about how to organize for business model innovation. In agriculture, this might include understanding how to validate a solution with farmers, navigate regulatory hurdles, or integrate with existing farm management systems.

2.1.2 Startup: Experimentation and Business Model Discovery

The startup phase begins once a team forms around a venture idea and starts developing a solution. This phase is characterized by intense experimentation, particularly around the business model. In agtech, startups must often test not only the technical feasibility of their innovations (e.g., sensors, AI models, biological inputs) but also their fit with the complex and heterogeneous needs of farmers.

Entrepreneurial ecosystems support this phase through:

- Accelerators and incubators offering coaching and seed funding.
- Venture builders that hire co-founders to work on a new, many times provided solution.
- Coworking spaces and makerspaces for prototyping.
- Access to early adopters and pilot farms.
- Peer learning and horizontal knowledge spillovers.

Autio et al. (2017) emphasize that the startup phase is where cluster-level architectural knowledge—i.e., deep knowledge of how business would work in that specific context—is most actively cultivated and disseminated. In agriculture, this includes learning how to navigate long sales and seasoned cycles, demonstrate ROI to skeptical farmers, and build trust in data-driven tools. Such knowledge can be organized and disseminated by agtech-focused business incubators and accelerators, especially if their selection criteria focus on specific value propositions to farmers, such as technologies that enhance economic viability of farming by increasing crop yield and predictability (Grande et al. 2025). Importantly, the startup phase extends beyond legal incorporation—it continues until a robust and scalable business model is discovered.

2.1.3 Scaleup: Growth, Integration, and Ecosystem Positioning

Once a startup has validated its business model, it enters the scaleup phase. This stage involves expanding operations, entering new markets, and securing larger investments. In agriculture, scaleup is particularly challenging due to the sector's seasonality, regional diversity, and sometimes entrenched incumbent relationships.

Ecosystem support in this phase includes:

- Access to venture capital and growth-stage investors.
- Mentorship from experienced entrepreneurs and industry veterans.
- Cross-cluster linkages (e.g., partnerships with incumbents, entry into global markets).
- Infrastructure for scaling (e.g., cloud platforms, logistics networks).

  Successful scaleups often feed back into the ecosystem by reinvesting capital, mentoring new founders, and contributing to the collective knowledge base (Autio et al. 2017). Far from being altruistic, the reinvestment of capital in the ecosystem can be represented by the example of CropX that started aggressive acquisitions to position itself as one stop solution for farming resource optimization. On the other end, incumbent firms such as John Deere offer different possibilities: (1) A startup collaboration program for testing new solutions from startups; investments/acquisitions to fuel venture development; and (2) platforms, where new solutions can be integrated and offered to their customers.

  As we will discuss in further sections, incumbents have been pivotal for many agtech startups to scale their solutions, either through partnerships or being acquired. Interestingly, despite increasing power concentration, new value propositions can impose architectural innovations and shake established ecosystems (Bohnsack et al. 2013).

2.1.4 Visualizing the Lifecycle

Figure 1 below illustrates the three phases of the startup lifecycle—standup, startup, and scaleup—along with their associated processes, needs, ecosystem actors, and structural supports. It highlights how each phase is embedded within a broader entrepreneurial ecosystem that provides the necessary scaffolding for venture creation, validation, and growth. The figure also emphasizes the evolving nature of support mechanisms and the increasing complexity of needs as startups progress toward scale.

2.2 The Missing Link: Farmers and Advisors as Ecosystem Actors

In the discourse on startup ecosystems, much attention is given to capital, talent, and technology. Yet, in the context of Agriculture 4.0, one critical group of actors is often overlooked: the users of agricultural innovation—farmers—and their closest knowledge intermediaries—advisors.

	STANDUP PHASE	STARTUP PHASE	SCALEUP PHASE
Key Processes and mechanisms	Motivation, talent and concept development, opportunity identification	Product development, problem-solution validation, Architectural knowledge cultivation, business model experimentation, team building	Resourcing, cross-cluster linkages, infrastructure provisioning, team building, Internationalization, Scaling operations
Needs	Ideation Inspiration and role model Entrepreneurial education and training Networking opportunities	Customer access for validation; Benchmarking with other startups and founders; Initial capital; Set and execute go-to-market strategy;	Supply of specialized human capital, Grow while building internal culture; Leadership training; Financial control Growth capital
Structural elements from ecosystem	Hackathons, innovation challenges, entrepreneurship programs, networking events	Coworking & makerspaces, networking events, start-up academies, crowdfunding, acceleration programs	Business angels, crowdfunding, venture capital, mentoring initiatives, business events.
Ecosystem actors	Universities, community events, Venture-Builders, startup programs	Universities, Venture-Builders, Angel investors, Large corporates, Government funding, other intermediary agents	Government, Investors and specialized banks, Suppliers, Complementors, customers, Large corporates (incumbents)
Outcomes	Entrepreneur network, new venture concepts	New ventures, architectural knowledge on business model innovation and entrepreneurial opportunity pursuit and scale-up	Scaled-up ventures, M&As, IPOs, trade sales;

Fig. 1 Startup lifecycle

2.2.1 Farmers: Embodied Knowledge and Situated Expertise

Farmers are not passive adopters—also not simply "users" for that matter—of technology and to neglect them as fundamental stakeholders in the ecosystem can reinforce the misalignment of perceptions toward new technologies (Moretti et al. 2023). However, an active startup ecosystem, that wants to support new ventures, must find a way to include farmers into innovation activities, perhaps starting by changing the view that farmers are not innovators. As van der Velden et al. (2023) argue, in the digital world, they can be "*cyborg farmers*"—hybrid actors who integrate digital tools with embodied, sensory knowledge of their land. Precision agriculture technologies, such as sensors and drones, do not replace farmers' intuition; rather, they extend it. Farmers interpret data through the lens of years of lived experience—walking fields, feeling soil, and observing crops. This embodied knowledge is essential for making sense of digital outputs, which often lack context without human interpretation.

Startups that fail to recognize this embodied dimension risk designing tools that are technically sophisticated but practically irrelevant (Fairbairn et al. 2022). The assumption that data alone can drive decisions ignores accumulated expertise that farmers bring to the table. As van der Velden et al. (2023) show, even the most advanced decision-support systems must be validated through field visits, soil digging, and sensory engagement. Farmers want tools that augment—not replace—their judgment.

Even if new technologies demonstrate clear benefits, their adoption is also heavily influenced by (1) how well they integrate with farming practices and other already adopted solutions and (2) what is done with the data that are collected in many offered solutions. Data ownership concerns are paramount for farmers, often outweighing perceived technological advantages (Kernecker et al. 2020). The question here is not only about privacy, but sovereignty, as farmers seek to control their own data and leverage them for their own benefit, rather than having them exploited by third parties (Bronson 2019).

Indeed, farmers could be considered the group facing the most significant pressure in the food value chain, with prices controlled by downstream actors and input costs dictated by upstream suppliers. This precarious position underscores the imperative for digital agriculture innovations to demonstrably improve farmers' economic agency, rebalance power asymmetries, and support farm resilience—including for smallholders (Rose et al. 2020).

2.2.2 Advisors: Translators, Mediators, and Ethical Anchors

Farm advisors play a pivotal but underappreciated role in the agri-tech ecosystem and yet are undergoing a profound identity shift in the face of Agriculture 4.0 (Charatsari et al. 2022). They are no longer just technical experts—they are becoming data interpreters, technology translators, and ethical mediators. Advisors help farmers make sense of complex digital tools, assess their relevance, and integrate them into existing practices. They should be integrated in the development process of new technologies and mediate the translation of R&D activities to farms.

However, this transition is not without tension. Advisors often feel caught between the promises of agri-tech providers and the realities of farm life. They face new responsibility gaps, as they are expected to vouch for technologies they may not fully understand or control. Moreover, their traditional role as trusted, experience-based guides is being challenged by a growing reliance on algorithmic decision-making.

Charatsari et al. (2022) also highlight that advisors' engagement with digital tools is shaped by their own "digi-grasping"—their ability to navigate hybrid digital–physical workspaces. Their professional identity is in flux, and their capacity to act responsibly depends on how well they can anticipate the social and ethical implications of technology adoption.

2.2.3 Ecosystems that Are more Sustainable and Inclusive

Smart farming is not just about technology—it is about networks, governance, and value creation across the entire agri-food system (Wolfert et al. 2017). In this context, both farmers and advisors are essential nodes in the data value chain. They are not just end-users; they are co-creators of meaning, validators of utility, and gatekeepers of trust. After a boom of investments in precision agriculture, investors learned that fancy technology and good promises do not convince farmers to pay for new technologies. Startups that want to scale must therefore engage both farmers and advisors in a meaningful way.

Without farmer input, technologies may lack relevance. Without advisor support, they may lack credibility. Moreover, the success of agtech depends not only on technical performance but also on social embedding—on whether tools align with local practices, values, and knowledge systems (Eastwood et al. 2017; Fleming et al. 2018).

Therefore, resilient and impactful agtech ecosystems facilitate that startups:

- *Co-design with farmers*, recognizing their embodied knowledge and decision-making styles.
- *Collaborate with advisors*, treating them as strategic partners in technology translation and adoption.
- *Acknowledge the ethical dimensions* of digital agriculture, including issues of data ownership, responsibility, and trust.

In this regard, Baarbé et al. (2019) compare data governance models of different platforms. In specific, the authors analyze the examples of a large US-based agribusiness (John Deere); an UK NGO that works with smallholder farmers (Plantwise); and a social enterprise developing catch solutions for fishers in South Africa (Abalobi). Among the three cases, Abalobi is highlighted as a prime example of a platform prioritizing stakeholder co-ownership of data, where fishers actively participate in shaping the data infrastructure and its governance. They decide, in multiple steps, whether to share their data with third parties and are able to benefit from the pool when doing it.

In the case of John Deere and several other platforms, the companies recognize the importance of data ownership to farmers and inform them that they are the sole owners of their data. Yet, the terms of use of their platforms may create what Ruder (2024) calls the "*illusion of data ownership*." In such cases, farmers own the data they entered into the platform, but have no control over data and analyses performed by the platform, nor do they have the chance to benefit directly from the aggregated pool.

Hackfort (2021) highlight the five main patterns of inequalities in Agriculture 4.0: (1) in control over technology development; (2) unequal distribution of benefits from the use of digital technologies; (3) uneven sovereignty over data and hardware; (4) in skills and knowledge ("digital literacy"); and (5) in problem definition and problem-solving capacities. Most of the times, such inequalities either negatively affect farmers or only benefit the large-scale farms. This exacerbates existing power asymmetries within the agricultural value chain, further marginalizing small-scale producers who may struggle to access or afford the necessary technological tools and certifications (Prause et al. 2020). Therefore, with which farmer a startup wants to interact, or to which farmer a new company is focusing to tailor its offerings will significantly influence its strategic approach and potential for sustainable market penetration.

Finally, most of the new value propositions in agriculture are sustainability-oriented. Startup ecosystems should not only support the development of new technologies but also nurture sustainable innovation. In this regard, incubators, accelerators, and venture capitalists already try to pose themselves as drivers of sustainability (Negra et al. 2019). However, such objectives have been criticized, as the only sustainability those ecosystems support is the economical sustainability of the startup and their own ROI (Rose et al. 2020). The growth of sustainable startup ecosystems is multi-faceted approach, that can begin with the sustainability-oriented entrepreneur, but must reflect in the action of all actors involved (Chaudhary et al. 2023). Interesting examples of funded projects that supported startups including responsible innovation principles are presented by Wolfert et al. (2022), indicating that, at least, government-funded projects are looking in that direction.

In the next section, we present different origins of startups and another framework for the support of new venture development, dedicated to the case of university spin-offs.

2.3 Deep Tech and Academic Entrepreneurship in Agtech

2.3.1 The Relevance of Deep Tech in Agtech Innovation

Deep tech ventures—characterized by high R&D intensity, long development cycles, and transformative potential—are increasingly central to agtech innovation (Grande et al. 2025). These startups often emerge from academic or research institutions and rely on cutting-edge scientific discoveries to address complex agricultural challenges such as climate resilience, resource efficiency, and food security (Klerkx and Villalobos 2023).

The European Commission's JRC (2025) report highlights that deep tech startups, particularly those emerging from Research and Technology Organizations (RTOs), are uniquely positioned to drive industrial transformation (Grande et al. 2025). In USA, startups that filed for patents were also more likely to be acquired (Graff et al. 2024), reinforcing the importance of good quality and continuous research & development. Their capacity to leverage breakthrough technologies and create new markets makes them essential actors in the agtech ecosystem.

However, deep tech startups face unique challenges related to market uncertainty, technology development, and the need for strong linkages to knowledge producers (Grande et al. 2025). These ventures often require specialized support mechanisms to bridge the gap between scientific discovery and commercial application (Bobier et al. 2022; Căpățînă et al. 2024).

Table 1 Different deep tech companies and their origins

Startup	Origin	Focus area	Funding source
Agreena (Denmark)	University research	Carbon farming, regenerative agriculture	VC, EU grants
Ecorobotix (Switzerland)	Research institute spin-off	Precision weeding robots	Series B funding
FarmInsect (Germany)	University accelerator	Insect farming for animal feed	Series A, EU grants
Nunatak (Argentina)	University accelerator	Ag-biotech, supported by Eatable Adventures	Accelerator, corporate (Grupo Bimbo)
Kilimo (Argentina)	Independent entrepreneurs with university ties	AI-driven irrigation optimization	VC (SP Ventures), partnerships
Bosch Deepfield (Germany)	Corporate spin-off (Bosch)	Smart agriculture sensors	Internal Bosch funding
365FarmNet (Germany)	Corporate spin-off (CLAAS)	Farm management software and data integration	Funded by CLAAS
Symbiomics (Brazil)	Corporate spin-off (Corteva)	AI development of agricultural biologicals	Corporate investment
DigiFarm (Kenya)	Corporate spin-off (Safaricom)	Multi-sided platform including advisory, provision of inputs, financing, off-take services	Corporate funding
Formo (Germany)	Venture builder (Atlantic Food Labs)	Precision fermentation for animal-free dairy	Seed and Series A funding.
Solinftec (Brazil)	Independent entrepreneurs	AI & robotics for precision agriculture	VC, Series D ($60 M)
Verqor (Mexico)	Independent entrepreneurs, local accelerators	Embedded finance and AI for ag-credit	VC, local ag-fintech investors
SwiftVEE (South Africa)	Independent entrepreneurs, local incubator	AI livestock trading marketplace	Grants, incubator, angel/VC

2.3.2 University Spin-Offs

Deep Tech startups have different origins. They can stem from university spin-offs, corporate research labs, venture builders, or independent entrepreneurial endeavors. Table 1 shows examples of startups steaming from these different sources. Although all of them can be connected to the development of high technologies and can have strong R&D, university spin-offs are companies that result from research carried out in a university or public research institution (Frietsch et al. 2021).

University spin-offs (USOs) and innovative start-ups (ISs) undergo distinctive paths in agtech ecosystems. Modina et al. (2023) provide empirical evidence from

Italy showing that ISs tend to outperform USOs in terms of sales growth, especially in high R&D sectors. USOs often act as extended scientific laboratories, requiring targeted support to transition into market-ready ventures (Coad 2025). This suggests that while academic entrepreneurship contributes valuable scientific knowledge, it often lacks the commercial orientation and market connectivity needed for rapid scaling and struggle to translate scientific breakthroughs into commercially viable products (Graff et al. 2024; Coad 2025). Therefore, spin-offs deserve specific nurturing within startup ecosystems (Reutimann 2014).

In that matter, a recent report from the European Commission emphasizes the pivotal role of Research and Technology Organizations (RTOs) in bridging the gap between research and industrial application (Grande et al. 2025). Different from Universities, RTOs are dedicated to bridge the gap between research and industry. Simple and fast technology transfer is their mission. Therefore, their Technology Transfer Offices (TTOs) support spin-off creation through structured programs, venture builders, and equity-for-services models, where the RTO partner with the startup to solve technological problems.

But more than offering strong IP portfolios, due to close and frequent contact with industrial partners, these organizations are able to identify real-world opportunities. Thus, they may provide a truthful problem to be worked on and provide directly a potential customer for a specific solution. However, one may pose the question if such alignment rather incentivizes incremental technologies instead of disruptive innovations.

To support the growth of spin-offs, Guindalini et al. (2021) propose a framework that is not based on three, but four phases, emphasizing the need of a strong value proposition and scalable business model, for the success of the company.

2.4 A Process Framework for Academic Entrepreneurship: From Scientific Discovery to Market Impact

Drawing on a systematic review and network analysis of the academic entrepreneurship literature, Guindalini et al. (2021) identified a four-phase model that captures the dynamic, iterative, and multi-level nature of research commercialization. The framework comprises the following stages: *Research Insight, Unlock Potential, Commercialize,* and *Scale Company*.

2.4.1 Research Insight (Discovery Phase)

This initial phase centers on the generation of novel scientific knowledge and the personal motivations that drive academic engagement in entrepreneurial activities. The authors emphasize the role of individual-level factors—such as entrepreneurial self-efficacy, intentions, and prior exposure to business environments—in shaping researchers' propensity to pursue commercialization. Structural barriers, including gender disparities, lack of business acumen, and institutional disincentives, are also highlighted as critical impediments. Entrepreneurship education is identified as a key enabler, fostering the competencies and confidence necessary for venture creation.

2.4.2 Unlock Potential (Validation Phase)

The second phase addresses the crucial, yet often overlooked, process of validating the commercial viability of scientific discoveries. This involves identifying how a research output creates value for specific stakeholders and developing a sustainable business model. Guindalini et al. (2021) draw heavily on the *Lean Startup methodology* (Blank and Dorf 2012; Blank 2013), particularly the iterative processes of *customer discovery* and *customer validation*, which are designed to ensure a strong fit between the problem being addressed and the proposed solution. These steps are essential to avoid *premature scaling*, a common cause of failure in both commercial and academic ventures. Programs such as the *NSF I-Corps* in the United States and the *ON Program* in Australia are cited as successful implementations of this approach, demonstrating significant increases in spin-out rates and venture funding.

In Germany, universities like the *Technical University of Munich* and its multiple agencies as the *UnternehmerTUM* and the *TUM Venture Labs* have programs that combine research with entrepreneurial training, divided into phases. For example, the Xplore program focuses on initial ideas and ensures that the solution fits the problem by using customer interviews and design thinking. The Xpreneurs program, led by the incubator UnternehmerTUM, then refines the startup's business model to prove it can grow (Brigl et al. 2024). Similarly, in Latin America, a major reference is the government-funded program called StartUp Chile (Villegas-Mateos 2020), that inspired other countries, such as Brazil. StartUp Chile also offers a three-stage program to help companies go through all phases and grow in a more robust way.

2.4.3 Commercialize (Customer Creation Phase)

In this phase, the focus shifts to building a viable enterprise around the validated innovation. The institutional environment plays a pivotal role, with universities and research organizations providing critical infrastructure such as TTOs, incubators, accelerators, and entrepreneurship training (Gubitta et al. 2015). These support mechanisms help bridge the gap between academic research and market application by facilitating access to funding, mentorship, and industry networks. In contrast, the authors underscore the importance of fostering an entrepreneurial culture within academia to encourage sustained engagement in commercialization activities. On a specific note, diversity of knowledge and skills within an entrepreneurial team is crucial for success (Marullo et al. 2018). Moreover, leveraging external knowledge enhances a startup's prospects. This is especially important for university spin-offs, which often have teams solely from the laboratory.

2.4.4 Scale Company (Impact Creation Phase)

The final phase involves scaling the venture for broader impact, aligning with external ecosystem factors like government policy and access to capital and talent. Guindalini et al. (2021) use the Triple Helix model—collaboration among academia, industry, and government—to show how systemic support can help academic spin-offs succeed. While they suggest that a well-functioning innovation ecosystem should enable the sustainable growth of research-based enterprises, this perspective might be too optimistic given the (or the lack of) intense competition in sectors like agtech or simply how the institutional arrangement is designed (Fini et al. 2016).

Not only that, scaling a company is no easy task. It requires building a functional organization with complementary and diverse skill sets, while being agile enough to compete against established companies (Gans 2016; Murmann and Vogt 2022). Whereas many startups begin their journey already with an exit in mind, i.e., "an end" for the company by being integrated into larger corporations (Brueller and Capron 2021), others envision being successful on their own and becoming leaders.

For academic entrepreneurship to thrive, companies must be ready for fierce competition and adapt their business models as the ecosystem changes. In the case of agtech, the oligopolistic scenario offers the opposite of fierce competition, or the lack of possibility to compete (Dörr 2018; Sauvagerd et al. 2024). Often, venture capital is used not only to finance companies but also to strengthen their know-how about the industry and facilitate interaction with potential customers, although the success of this interaction can also be mediated by ecosystem actors (Gubitta et al. 2015; Rodríguez-Gulías et al. 2017). On the one hand, bold venture capitalists could finance a company that is willing to disrupt such oligopolistic behavior of the agtech industry. On the other, many are not resilient enough and succumb to the pressures of industry incumbents, often leading to acquisitions that consolidate market power rather than foster true disruption.

Considering the competition landscape of agriculture, we further analyze the endeavors of agribusinesses to cope with digital transformation.

3 Incumbents and Diversifying Entrants in Agriculture: Ecosystem in Motion

The agricultural sector has long been dominated by a handful of powerful incumbents—century-old firms like John Deere, Bayer, and BASF—that have shaped the industry through scale, infrastructure, and deep-rooted relationships with farmers (Fairbairn and Reisman 2024). Incumbents in general, due to market experience, have what is called as "traditional industry knowledge" and an important advantage—a deep customer relationship and thus, an established distribution network (Tripsas 1997; Bohnsack et al. 2024). However, the emergence of Agriculture 4.0 has opened the door for cross-industry entrants from sectors such as IT and telecommunications, bringing expertise in areas like data analytics, sensor technology, and connectivity (Gerhardt et al. 2022).

These new entrants do not compete head-to-head with incumbents, but rather bring complementary capabilities and new business models into the agricultural landscape. Therefore, before discussing the role of startups, it is important to understand the strategic moves of incumbents and diversifying entrants.

3.1 Incumbents Come into Play

Established firms typically possess extensive resources, distribution networks, and brand recognition, giving them a substantial competitive advantage (Gerhardt et al. 2022). Nevertheless, they encounter challenges in terms of agility, flexibility, and the capacity to embrace disruptive technologies, due to deeply ingrained organizational structures and operational methods (Remané et al. 2022). These limitations often stem from a resistance to deviate from well-established and dependable revenue streams (Bohnsack et al. 2024).

In order to understand the strategies pursued by such incumbents to renew their business, we look to corporate venturing initiatives (company acquisitions and corporate investments) and use the comprehensive Pitchbook database to map the business relationships between companies from both sectors (agriculture on one side and ICT, sensors, robotics, etc., on the other side). Such initiatives provide a rich ground to acquire not only market and ready-to-use technologies but also new knowledge and capabilities that are missing to the acquiring firm, indicating a path to navigate in the new ecosystem. It is important to note that we do not discern between investments and acquisitions—only when explicitly mentioned—using "investment" or "investor" as terms for all kinds of deals. Furthermore, we focus on corporate investments, therefore excluding, whenever possible, investments coming from capital venture firms. Pitchbook has over 1.5 million investment deals and roughly four million companies mapped in their database providing insightful data of different corporate venturing activities (Cumming and Zhang 2018).

To first map the evolution of Agtech ecosystem, we found that 5482 deals were closed between 2000 and 2020. Deals were tagged within Agtech as "vertical," but the nature of the purchased/invested firm is diversified. Narrowing down and focusing on deals between 2015 and 2020 from the top 10 industries ("Primary industry code" of the purchased/invested company), we see an interesting growth of investments on productivity software (Fig. 2).

To further investigate this phenomenon, Fig. 3 focuses only on those deals and presents further evidence that may support such trend: in the graph, the black line represents the deals filtered by sector, i.e., by the IT sector. The classification of the

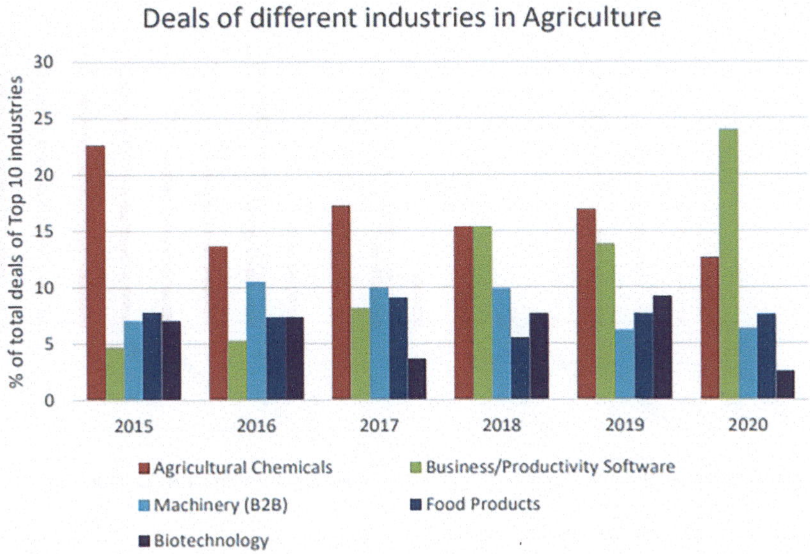

Fig. 2 Deals made by agtech companies

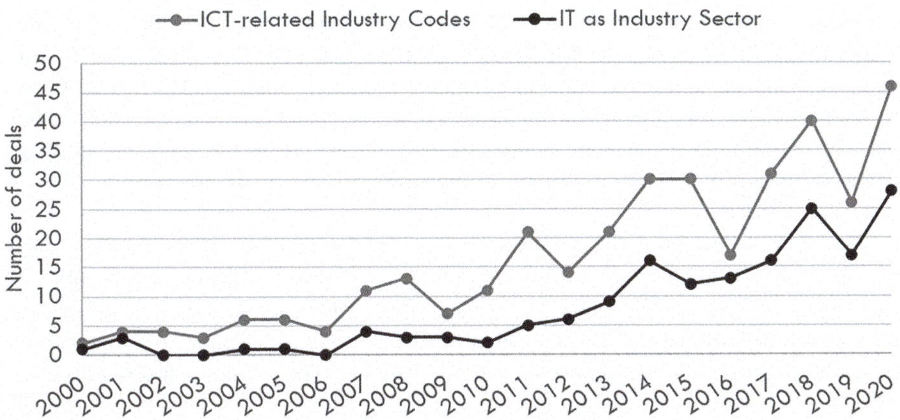

Fig. 3 Deals made by agtech companies only in IT/ICT companies

Pitchbook database, however, excludes from IT some industries that are important to the context of precision agriculture, such as "Aerospace and Defense," "Media and Information Services" or "Machinery." Therefore, the gray line represents the number of deals filtered by "primary industry code," hence including industries beyond the so-called IT sector, comprising what we call here the ICT sector.

As our motivation is to investigate whether incumbents reach for new technologies that can bring new value propositions to agriculture, we examined the presence and frequency of cross-industry (or cross-sector) investments investigating

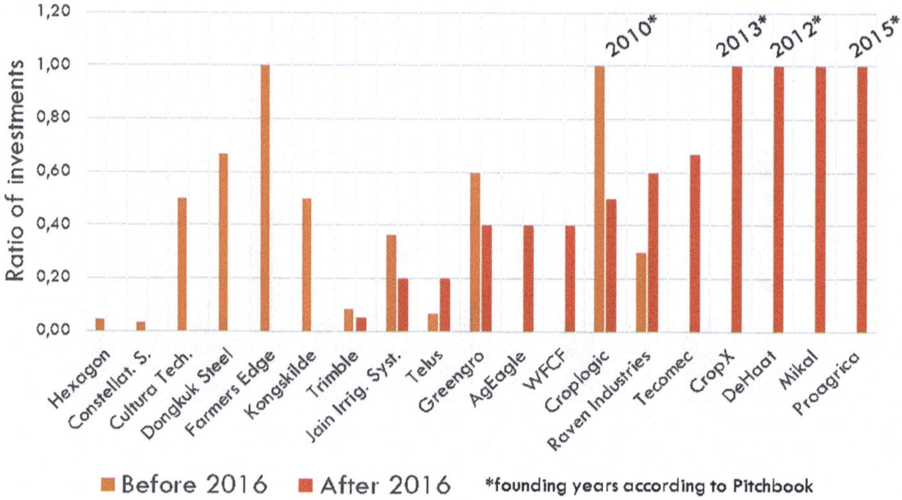

Fig. 4 Deals in agriculture made by IT companies

the nature of the investors. The total number of deals in agriculture was this time filtered by the investors belonging to ICT-related primary industry codes. This filter encountered 90 investors, of which only 19 have made two or more investments (Fig. 4). This is somewhat surprising that companies such as Microsoft, Cisco or Alphabet do not appear among this list. It is, however, possible that either those investments are classified in industries not covered by the research, they are investing in companies through capital investors, or the deals are simply missing from the platform.

The figure shows not the absolute number of investments performed by each company but the ratio between investments in agriculture and the total number of investments from each company. We chose to present results this way as we seek to demonstrate the interest of those companies in the sector. Trimble, for instance, is the leader in terms of absolute number of investments. The company has made, since 1990, 10 investments (mostly acquisitions) in agriculture. This number, however, represents only 10% of all deals closed by the company, which provides solutions for several different industries.

Following this rationale, in order to understand if such interest is growing or decreasing, we further separate the investments by date, highlighting those closed before and after 2016 (until 2021). Figure 4 not only indicates a higher interest in the sector (100% of investments in agriculture) but also highlights that young companies tend to lead the investments in the recent years. CropX is a remarkable example, as the company started doing acquisitions just a couple of years after foundation.

A similar approach was performed to see the interest of agriculture companies in the ICT sector. We first searched for all deals tagged within the "IT industry" or within the "Technology Media and Telecom (TMT)" and then filtered those deals

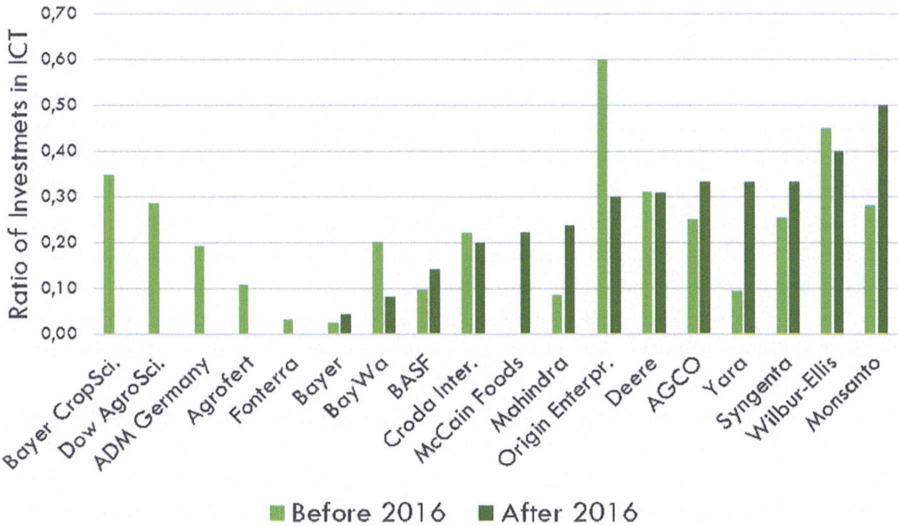

Fig. 5 Deals in IT made by agtech companies

by the nature of the investor. The 20 investors with the higher (absolute) number of investments in ICT tagged within the "Agtech" vertical were analyzed (Fig. 5).

Here although not necessarily growing, it is clear the interest from the incumbent firms in the sector. In reaction to the new entrants and to the change in technology, established companies seem to engage and lead investments in a sector previously not related to agriculture. As could be expected, most of the companies performing these deals in ICT are machinery producers, i.e., they look for opportunities that bring enhancements to their own technology, by searching for technological complementary partners.

In conclusion, current trends indicate that the agriculture sector is not only attractive to the Information and Communication Technology sector, but also that agricultural incumbents are actively seeking partnerships with ICT companies (Mikhaïlov et al. 2021). These findings align with discussions by other scholars (Birner et al. 2021; Dyck et al. 2023), notably Sauvagerd et al. (2024), who introduces the concept of "*oligopolist platformisation*," where leaders in agriculture and ICT establish strategic partnerships that reinforce their leadership positions. The abundant offer of services within a single platform and the technical barriers posed by Ag machinery manufactures create lock-in effects to hold farmers into their ecosystem, possibly impeding them to look for better and cheaper solutions and, more importantly, concentrating data in the hands of a few companies (Dörr 2018; Prause et al. 2020; Sauvagerd et al. 2024).

4 A Closer Look to Business Models

In the context of Agriculture 4.0, business models are not static blueprints but dynamic systems that evolve in tandem with technological capabilities and ecosystem dynamics. By definition, business models are a system of interrelated activities and resources from a focal firm and key partners that allow the *creation, delivery* and *monetization of value* (Zott and Amit 2009; Brea 2023). Following the current technological trends, especially digitalization, all three dimensions are affected.

Take big data: the amount of data generated in modern agriculture is massive, however, big data itself does not represent a competitive advantage (Mendes et al. 2021), which requires that business models are designed to effectively create value. In this regard, data-driven business models have the potential to generate value by converting spatially explicit big data into information and advice not only for farmers but also for regulatory authorities who may use the data for surveillance and control (Walter et al. 2017). On a specific note, Birner et al. (2021) comments that input firms are changing from product-based business models to service-based business models. For instance, a fertilizer company instead of simply selling fertilizer, offers a "weed-free" field. Although the data presented in the section before support the indication of this shift, the real implementation of the transformation seems to be only in the beginning (Rijswijk et al. 2019).

For IoT solutions, Sarto et al. (2021) argues that business models structuration varies depending on the ecosystem role adopted by the company. The authors make differentiation between "*Application*" businesses where companies are focused on the analysis of raw data and the provision of services based on such information; "*Network*" businesses are responsible for transmitting information and to whom privacy and IT infrastructure are crucial resources; "*Perception*" businesses are focused on the provision of sensors, actuators, and smart components and normally work with a "sensing-and-actuation-as-a-service" business model.

The change impacts not only the private sector. As transparency and trust are major issues, especially in agriculture where data access and ownership could shift power to the hands of corporations and potentially harming farmers (Wolfert et al. 2017), neutral initiatives that oversee data handling and promote interoperability may become a key element of future governance of data-driven business models (Fleming et al. 2018). Here we can see examples stemming from both private and public spheres. Agrirouter, backed by the agtech industry, is an example in Germany, focused on bringing together and smoothing interoperability between small and medium-size enterprises. On the other hand, national-level policies such as X-Gaia in Europe (Gaia-X Community 2021), which generated transparent agricultural platforms in several countries, such as Finland, Belgium, and Germany. Such platforms offer the possibility for different services to connect their data and be part of the ecosystem. Suppliers thus adhere to the same guidelines and farmers or consumers, depending on the platform, have full control of who has access or power to use generated data.

If not directly created or incentivized by the government, some initiatives are purely backed by public funding or donors, not only to maintain data non-privatized, but also because users can lack the resources to pay for such services. The Namibian Livestock Identification and Traceability System (NamLITS) is one such case, providing livestock identification and traceability in order to ensure consumer and animal health (Birner et al. 2021).

These last examples, however, indicate a need for sustainable business models, especially when data sharing initiatives are emerging in the AgriFood system (Gaia-X Community 2021). They are examples of firms or initiatives that explore a piece of the ecosystem, but not necessarily engage in verticalization or integration of processes. Brea (2023) brings some examples of orchestrated services in what he calls the "ecosystem business model." In this case, a company centralizes all dimensions of the business model (Value creation, delivery and capture—or monetization), whereas partner and connect with several other companies that have specific roles (value creator or carrier or capturer) in the ecosystem.

The author explains how the incumbent Phillips start engaging with different companies to complement capabilities. To leverage its service through already installed healthcare equipment in facilities all over the world, Phillips initially partnered with Salesforce to manage the clinical trial data. As a result, a cloud-based platform was created, where nowadays, thousands of service providers, consumers, and healthcare businesses interact. In this case, Phillips acts as the orchestrator, whereas important partners have specific roles: For instance, Salesforce and AWS Amazon deliver value whereas Samsung and Qualcomm capture an important share of the value through connected physical products (Brea 2023).

It is important to note that an established firm like Phillips, despite its numerous business arms, has not completely transformed its business model to solely adhere to the ecosystem model. Nevertheless, this strategic shift highlights a global trend in market evolution. For instance, von Bismarck-Osten (2021) highlights the example of six sub-Saharan African platforms, among which is Digifarm, a spin-off of Safaricom, a Kenyan telecommunications company. Digifarm successfully attracted strategic partners to deliver integrated digital services to smallholder farmers, ensuring simplicity and profitability for users, which significantly accelerated its adoption.

4.1 Orchestration in Agtech

Agriculture is not an exception to this kind of change within incumbents. Dyck et al. (2023) analyses two incumbent firms (AGCO and CompCo) and their pursuit to enter the ecosystem business model. The authors compare, among several characteristics, how open or controlling the engagement with third part developers was and how value capture differed. In the case of AGCO, the company was very strict to which services would be offered on its platform and got a share of the revenues from complementary firms. In contrast, COMPCO adopted an open model where the firm did not receive direct payments but indirectly benefited from the expansion of services available. In both cases, again, ownership of data and transparency were concerns presented by potential users (Dyck et al. 2023).

What we thus see is an increasing complexity in the configuration of business models for agricultural startups, moving away from linear models where companies are focused on offering technology or services to the farmers, to more sophisticated business models where companies can act as orchestrators between all agriculture stakeholders.

In this regard, Phillips et al. (2019) propose a 2 × 2 typology based on two strategic attributes to map innovation spaces:

- *Organization of Innovation*: Top-down (formal, hierarchical) versus Bottom-up (ad hoc, producer-driven).
- *Interoperability*: Closed platforms (proprietary, formal use rules) versus Open platforms (full interoperability).

These attributes are used to map four business models' architypes, described in Fig. 6.

- *Type I (Top-down, Closed): Corporate Model*: Characterized by vertical integration and proprietary systems, often by large incumbents like John Deere.
- *Type II (Top-down, Open): Strategic Networks*: Involve purposeful relationships among actors aiming for open operability, often industry collectives or large companies seeking to curate open-source platforms (e.g., Farmobile, Open Ag Data Alliance).
- *Type III (Bottom-up, Open): Perfect Competition/Hacking Universe*: Characterized by a multitude of entrepreneurs offering open and interoperable solutions, with low barriers to entry (e.g., AgSky with drones).
- *Type IV (Bottom-up, Closed): Primordial Systems*: Informally organized with little external interoperability, often disruptive new entrants creating new market mechanisms (e.g., FarmLead for grain marketing).

Data-rich platforms, especially those of Types II and III, exhibit significant network effects, meaning that benefits increase as the number of users grows. This implies that once a company gains a certain lead in terms of members, it can become the de facto platform, driving many others out of the market. Phillips et al. (2019) explicitly state that "*once one of these competing firms/models gains a bit of a lead in membership, it could become the de facto platform, driving many of the others out of the market.*" This suggests that while there is a "multitude of entrepreneurs" in Type III and strategic networks in Type II, the long-term competitive landscape may consolidate around a few dominant platforms, potentially even a Type I player like John Deere expanding into Type II.

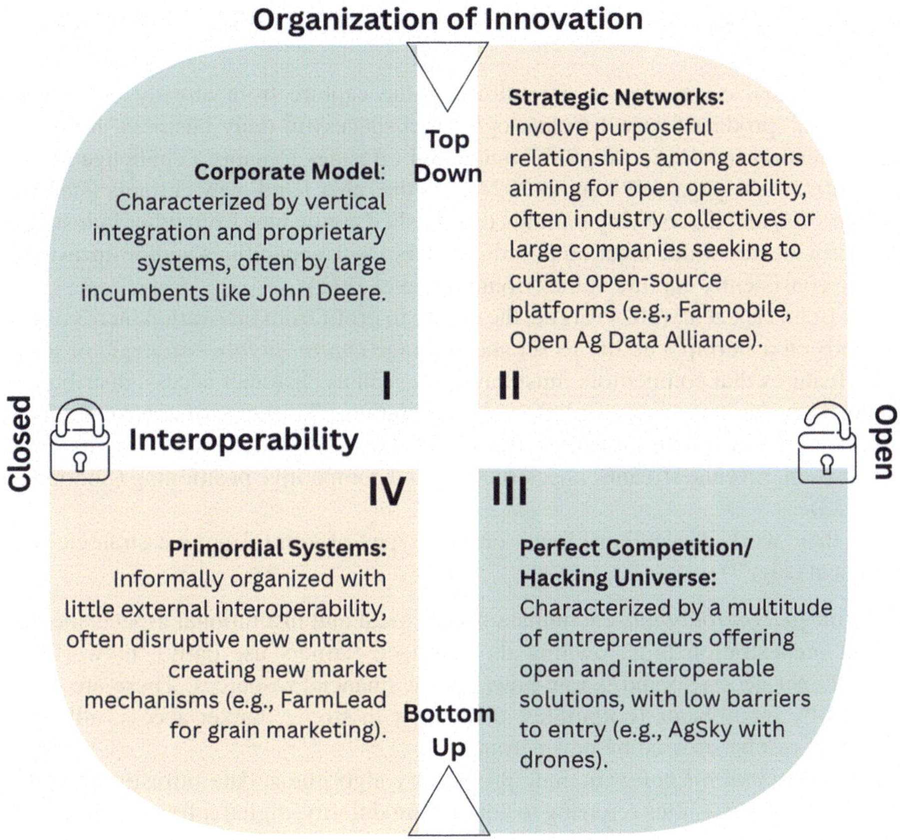

Fig. 6 Four archetypes of business models in agriculture according to Phillips et al. 2019

This is a critical consideration for investors and startups, although such an effect can be hard to track, when it is still evolving. Even if companies are born, sold, or die in 5-year cycles, markets do not change that fast. An observation in power dynamics can take multiple generations or be only visible after a complete technological transformation wave, as the one being pushed by artificial intelligence currently. Moreover, as well noted by Sauvagerd et al. (2024), the agricultural industry is per se more complex than other industries. Even by focusing on the farmer as central point, there is no single platform that controls all the relevant data flows, which creates barriers to data integration and leave farmers and stakeholders to deal with all restrictions and countless user interfaces.

5 The Perspective of Control Points and Strategic Positioning

For traditional companies in agriculture, value capture from already developed, "physical," product-focused technologies is a successful daily business; however, value creation from new, "digital," integrated offerings remains a challenge (Sarto et al. 2021; Alcayaga and Hansen 2024). On the other hand, new entrants (startups or spin-offs) and diversifying entrants (established companies from other industries) may provide new value creation possibilities, but lack capabilities and/or infrastructure to consistently capture value (Bohnsack et al. 2024).

As Bohnsack et al. (2024) argue, the ability to profit from innovation increasingly depends on a startup's ability to set and leverage *control points*—strategic or technical features that competitors must navigate, such as customer access, distribution channels, proprietary data, or platform orchestration. The choice of control points is deeply intertwined with a startup's overarching business model, influencing its value proposition, revenue streams, cost structure, and competitive positioning (Sharma et al. 2023).

In their work, the authors focus on two types of control points: strategic and technical (Fig. 7).

- *Strategic control points* encompass sociological and institutional aspects including orchestration (e.g., owning the interface farmers use daily), networking, customer access, brand, know-how, agility, financial resources. These are often held by incumbents (corporates, dealers) with strong market access, although agility can be stressed by new entrants.
- *Technical control points* include proprietary algorithms, data infrastructures, or unique devices—thus referring to content, modularity, digital infrastructure, data, scalability, and (unique) solution. These are typically developed by deep-tech startups or research spin-offs. In the case of agriculture, we have seen cross-industry endeavors, where companies born in other fields, such as Information Technology, are also providing agtech solutions. Such companies are called "diversifying agents" and also tend to dominate technical control points.

  Bohnsack et al. (2024) further distinguish *generic control points*—such as modularity and scalability—that are prerequisites for participating in digital ecosystems, and *institutional boundaries* (e.g., regulation, subsidies) that shape the feasibility and value of control points.

  Startups that succeed in Agriculture 4.0 often combine both types, although this is rather challenging. For example, *ZERO*, a vertical farming company studied by Cavazza et al. 2023, developed a proprietary AI system ("ROOT") and modular architecture, giving it technical control. At the same time, it built strategic partnerships and educational initiatives to shape the ecosystem and influence policy—thus gaining strategic control. On the other hand, corporate startups are born already not only with technical control points, focusing on unique solutions for substantial problems, but also with strategic control points, such as financial resources, customer access and know-how (Bohnsack et al. 2024).

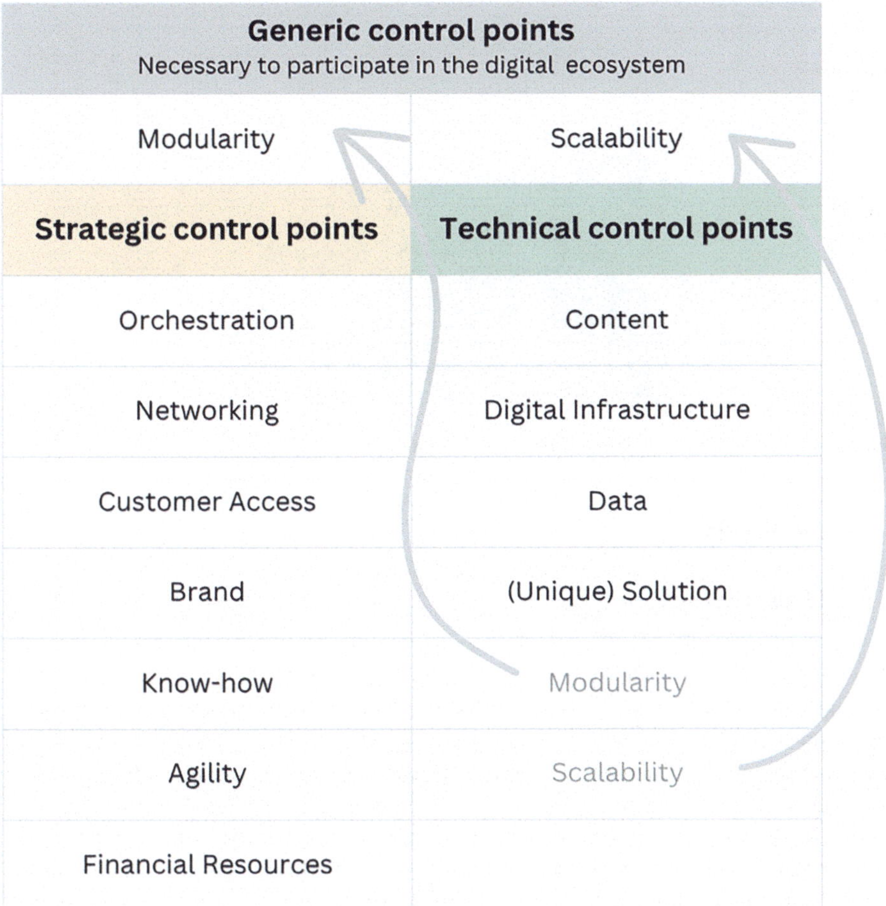

Fig. 7 Generic, strategic, and technical control points, according to Bohnsack et al. 2024

Obtaining or leveraging these control points is influenced by digitalization, which is transforming the industrial architecture into a layered architecture, connecting pure physical products to pure digital ones. This transformation influences how value is created and captured in emerging digital business ecosystems (Bohnsack et al. 2024).

From pure physical to pure digital, the authors define the following five layers (Fig. 8).

Startups that move "up the stack" (from physical to digital services) tend to increase their value capture potential. However, this also requires acquiring new knowledge—both technical and industry-specific. Tell et al. (2016) emphasized that this learning process is often iterative and collective, involving experimentation, storytelling, and ecosystem engagement.

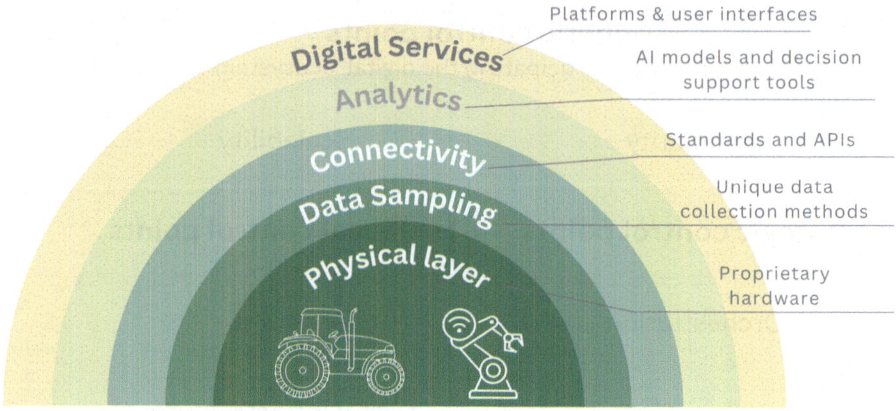

Fig. 8 The five layers of modern industrial architecture

6 Case Study

6.1 Methodology

This chapter adopts an analytical case study approach (George and Bennett 2007; Yin and Campbell 2018), applying established theoretical frameworks to interpret the business model strategies of agtech startups. The structured cross-comparison of cases and the application of theory through real-world cases offer a rich illustration of strategic positioning in Agriculture 4.0. Specifically, we analyze the last two cohorts of the Agtech Vertical from the Plug & Play accelerator, with a deeper focus on NutriSen, one of the participating startups. Our methodology is grounded in the use of established theoretical frameworks to interpret empirical data from real-world startup cases.

6.2 Case Selection and Data Sources

The Plug & Play Agtech Vertical was selected as a relevant and contemporary accelerator program that brings together early-stage startups, corporate partners, and investors in the agricultural technology domain (Ferrigno et al. 2024). The startups analyzed in this chapter were selected from the two most recent Plug & Play Agtech batches (2023–2024), ensuring the relevance and timeliness of the data.

NutriSen was chosen for in-depth analysis due to its strong research foundation, active participation in Plug & Play, and its alignment with the themes of Agriculture 4.0, including precision fertilization, sensor technology, and data-driven decision-making. Data sources include:

- Publicly available company descriptions and value propositions.
- Internal Plug & Play documentation and startup pitch materials.
- NutriSen's business plan and ethnographic insights gathered during the acceleration process of Plug & Play.

6.3 Data Analysis

We conducted a structured content analysis of each startup's business model, coding for the presence of specific layers and inferring control points based on value creation logic and ecosystem roles. A heatmap visualization was used to compare the technological scope of each company. For NutriSen, we applied the full analytical framework to reconstruct its business model evolution, ecosystem positioning, and strategic control points.

This methodology allows us to bridge theory and practice, offering insights into how startups navigate the complex and evolving landscape of Agriculture 4.0.

7 Results

7.1 Startup Business Models According to the Layered Modular Architecture

To understand real world applications of the layered modular architecture and the different control points present in each of them, we use the two last cohorts of agtech companies from the Plug & Play Acceleration Program. By mapping the different solutions proposed by the startups across the layers, we can understand what the dominant strategy is used by the startups, if they focus on one layer, as was more common in a non-digital logic, or, if they have a more comprehensive solution, representing the evolution of Agriculture.

In total, 26 companies were accelerated by Plug & Play in the Agtech Vertical, in the batches of 2023 and 2024. Figure 9 presents all the 26 companies, categorized by their offering to the agricultural supply chain and Fig. 10 presents the verticals on which they are inserted beyond Agtech. The category "agtech" refers to the companies that are categorized only in that vertical.

For the layered, control points and innovation archetypes analysis, we excluded the companies that had biotech solutions as core businesses. Many of the papers that analyze the new technologies in the Agriculture 4.0 do not include biotech solutions into the matrix, and focus on digitalization processes (Phillips et al. 2019; Vlachopoulou et al. 2021; Bohnsack et al. 2024).

Next we assess the business model of the Agtech companies through the lens of the layered modular architecture, adapted to the agricultural context. We categorize each company's primary value proposition and offerings into one or

Fig. 9 Plug & Play Agtech startups across the agricultural supply chain

more layers. As expected, most of the companies focus on the more digital layers—i.e., Analytics and Digital Services. Some of them, such as Blomso or Ground Truth, offer solutions across multiple layers, including their own hardware (in the case of Ground Truth), proprietary data sampling methods, connectivity solutions, advanced analytics, and user-friendly digital services (Bohnsack et al. 2024). Others such as RawData and Gripp are very focused on solving one specific although acute

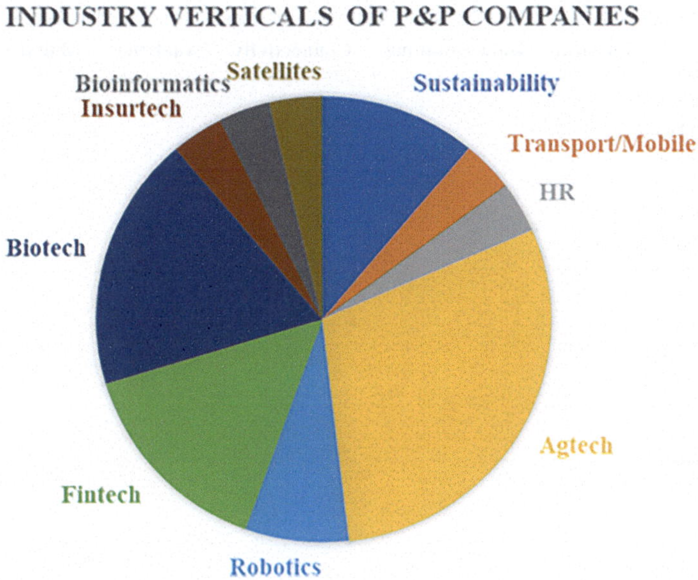

Fig. 10 Verticals where Plug & Play companies are included, in addition to agtech

and widespread customer problem and build its core competence around only one layer.

A relevant observation is that, according to the Plug & Play database, many companies initially position themselves as offering solutions primarily in the Data and Analytics layer, Table 2, but do not necessarily provide connectivity tools, which can limit their strategy.

Once the solutions of the Agtech startups are mapped into the layers, we can understand what the main control points are disputed by the startups (Table 3).

7.2 *Analysis of Layer Coverage*

Most Plug & Play companies operate in the "Digital Service" and "Data Analytics" layers. As expected, this indicates strong market demand and an entrepreneurial focus on software-based solutions, data interpretation, and decision-support tools in agtech. This concentration in the upper layers of the LMA reflects the migration of value from hardware to software and data, as predicted by theory. However, it also suggests potential challenges in integrating these digital services with the physical layers of farming.

Several companies span multiple layers, often combining physical devices/data sampling with connectivity and data analytics (e.g., Mapana, Atolla Tech, Biome Makers, Blomso, Ground Truth Agriculture, Upside Robotics). This integration

Table 2 Heatmap of the layers that describe the product offerings of Plug & Play Agtech companies

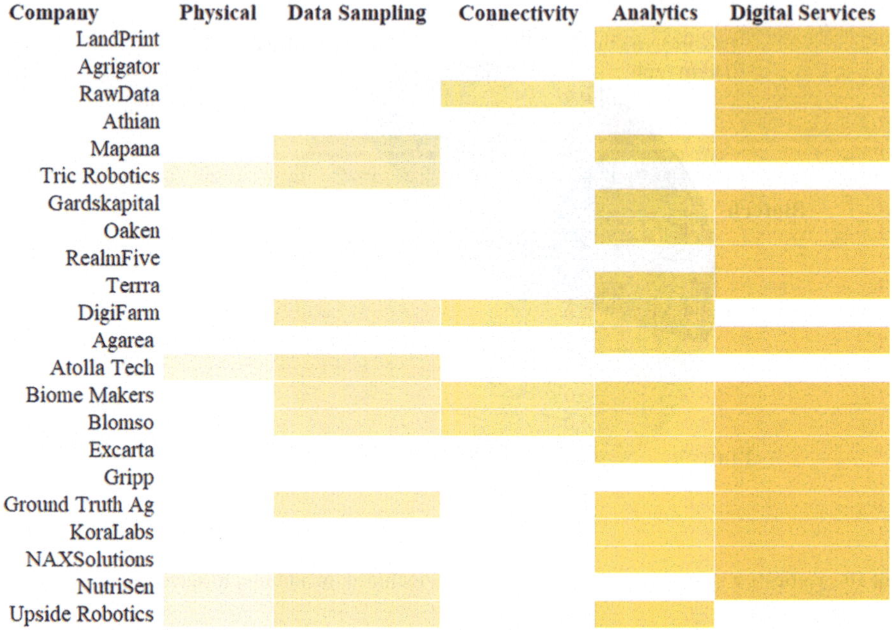

Intensity of colors illustrates how digital the companies' offerings are—the darker the color, the more digital the offering is

suggests a trend toward holistic solutions rather than isolated components, which are crucial for creating ecosystem value propositions.

The single analysis of business models based on public information available of new companies is limited and can oversee business models that integrate different layers through partnerships. Such companies should be followed throughout time so as to better understand in which types of business models they engage—for instance, more controlling and closed, keeping in-house the knowledge to act in all layers or less controlling and more open, focusing on one or two layers and relying on partnerships for other layers (Phillips et al. 2019).

7.3 *Analysis of Control Point Strategies*

Many companies explicitly utilize technical control points such as "Data" (Agrigator, Athian, Mapana, Biome Makers, Blomso, Excarta, Ground Truth Agriculture, NAXSolutions, NutriSen, Upside Robotics), "AI/Machine Learning" (Agrigator, Mapana, Biome Makers, Blomso, Excarta, NAXSolutions, Upside Robotics), and "Unique Solution" (Tric Robotics, DigiFarm, M4Life). The overwhelming emphasis on "Data Analytics" and "Digital Service" layers, often explicitly mentioning

Table 3 Heatmap of the technical control points potentially owned by Plug & Play Agtech companies

	Primary Inferred Technical Control Points			
Company	**Content**	**Digital Infrastructure**	**Data**	**Unique Solution**
LandPrint				Platform
Agrigator				AI/ML
RawData				Platform
Athian				Platform
Mapana				AI/ML
Tric Robotics				Robotics
Gardskapital				Fintech model
Oaken				SaaS Platform
RealmFive				Modeular SmartSites
Terra				Carbon Credit Technology
DigiFarm				Field Boundary Detection
Agarea				Fintech solution
Atolla Tech				Pest Detection
Biome Makers				AI/ML
Blomso				AI/ML
Excarta				AI/ML
Gripp				Tracking Platform
Ground Truth Agriculture				Automated Grading
KoraLabs				CRM Platform
NAXSolutions				AI/ML
NutriSen				Nutrient Sensing
Upside Robotics				AI-Drone Robots

"AI," "machine learning," and "data," confirms that data are seen as a primary technical control point in agritech.

Companies are building models around collecting, processing, and monetizing agricultural data. Bohnsack et al. (2024) mentioned "owning customer data" as a technical control point. The Plug & Play dataset demonstrates that numerous companies (e.g., Agrigator, Biome Makers, Blomso, NAXSolutions) have AI-driven data analysis for predictions, recommendations, and optimization as a core value proposition. This means that access to and sophisticated processing of agricultural data are becoming a possible critical bottleneck and control point, allowing these companies to capture significant value by offering "intelligence services," if they are able to demonstrate actionable results.

8 Key Observations and Discussion

8.1 Dominant Business Model Archetypes and Emerging Trends

The Plug & Play data strongly indicates a shift from purely physical product sales to digital services, data analytics, and platform-based models. This aligns with the concept of "value migration."

The rise of "as-a-Service" (XaaS) models is evident. Many companies offer "as-a-service" models implicitly or explicitly (e.g., Farmers Edge in Vlachopoulou et al. (2021), Upside Robotics in Plug & Play). This reduces acquisition costs for farmers and enables recurring revenue streams.

A significant and unique trend observed in the Plug & Play data is the strong convergence of agritech with fintech and sustainability. Companies like Gardskapital, Oaken, Terrra, Agarea, LandPrint, and Athian demonstrate this integration. This suggests that financial mechanisms and environmental impact are becoming central to new value propositions in agriculture. The emergence of companies like LandPrint and Terrra, which focus on measuring and monetizing environmental quality or carbon removal, indicates a shift where sustainability is no longer just a regulatory burden or a cost to be minimized, but a direct source of revenue and a core business model.

Traditionally, sustainability initiatives were often viewed as cost centers or compliance requirements. However, companies like Terrra explicitly create value by "*democratizing wetlands carbon removal through a technology platform to measure, manage, and issue carbon credits*." This transforms an environmental benefit into a tradable asset, creating new revenue streams for farmers and for the agritech companies that facilitate it. This is a significant shift in how economic value is perceived and captured in the sector.

The relevance for the current analysis is to understand if such startups or new entrants acting in fintech and sustainability will keep focusing on the digital layer or will manage to insert themselves into the layered ecosystem. For instance, in the study by Zaher et al. (2013), the change in tillage to no-tillage systems is analyzed and potential revenue from carbon credits is quantified. Similarly, companies like Farmers Edge that aim to quantify carbon removal are already connected to platforms such as the John Deere Data Center (Sauvagerd et al. 2024). While they can potentially balance power asymmetries enabling farmers to capture value in other ways, their connection with established platforms can again create lock-in effects, attaching farmers to a specific platform.

8.2 The Critical Role of Data, AI, and Interoperability

The widespread emphasis on data collection, analysis, and AI/ML among Plug & Play companies (Agrigator, Mapana, Biome Makers, Blomso, Excarta, NAXSolutions, Upside Robotics) confirms that data are the primary asset and a primary technical control point in digital agriculture.

AI is being utilized to go beyond descriptive data, providing predictive and prescriptive insights (e.g., plant health prediction, fertilizer recommendations, yield optimization, risk reduction). This enhances decision-making for farmers and other stakeholders.

While some companies may initially pursue proprietary solutions, the need for seamless data flow across the layered modular architecture and the desire for holistic solutions will drive greater interoperability. Companies like RawData and DigiFarm explicitly offer API-based solutions, indicating some commitment to open connectivity, but however not so frequently or at least, not openly.

Although open interoperability fosters ecosystem growth and network effects, it can dilute a company's control points (Phillips et al. 2019). Conversely, closed systems offer control but risk limiting adoption and value co-creation (Dyck et al. 2023). The Plug & Play data suggest a leaning toward more open and integrated solutions, implying that the benefits of ecosystem participation (access to more data, broader customer base, complementary services) often outweigh the desire for strict proprietary control for startups—or is simply seen as the only option to be able to grow.

Phillips et al. (2019) highlight John Deere's initially closed system and its later shift to the ADAPT open source. The Plug & Play data show many companies operating in "Connectivity" (RawData, DigiFarm, Biome Makers, Blomso, Ground Truth Agriculture, NutriSen, Upside Robotics) and offering services "via API" (DigiFarm, RawData). This suggests that while large incumbents may try to maintain closed systems, startups often adopt open approaches to gain traction and integrate into existing or emerging ecosystems, betting on network effects for long-term success, even if it means less initial control.

8.3 Strategic Control Points from the Start?

As new entrants, startups typically start with strong technical control points (unique solutions, data, AI), but often lack strategic control points such as established brands, broad customer access, or deep industry knowledge. Their challenge lies in learning about the industry and building these strategic assets, often through partnerships or by demonstrating agility.

On one hand, startups compensate for limited financial resources and industry knowledge with agility, rapidly addressing niche problems and providing a superior user experience. This means that *agility* is one strategic control point that they may own since the beginning of their journey. On the other hand, several startups, especially the ones that attract venture capital money, have the chance to acquire *know-how* about industry archetype through hiring C-Level employees that had previous industry in the Agtech industry. Moreover, several of the original founders already have industry knowledge from previous endeavors or industry experience. This is the case of companies such as Ground Truth, Blomso, DigiFarm, RealmFive, and Oaken.

While Plug & Play provides acceleration, the long-term challenge for startups remains securing sufficient financial resources to scale and compete with larger players or to achieve "*integrated system of systems*" status. Many Plug & Play companies start by addressing highly specific technical niches (e.g., pest management via UV light, soil nutrient mapping). Their opportunity lies in expanding their technical control points across more layers or leveraging them to build strategic control points, potentially becoming orchestrators or key partners in larger ecosystems. The Plug & Play data show companies like Tric Robotics and Blomso focusing on very specific problems (pest control, soil sensing). This aligns with the new entrant strategy of occupying technical control points in niches. However, for sustained growth, they must deepen their technical control (e.g., expand their AI capabilities) or build strategic control points (e.g., partnerships with large equipment manufacturers or chemical companies) to integrate their solutions into broader agricultural operations, moving toward the "integrated system of systems player" (Bohnsack et al. 2024).

Remains the question of whether startup ecosystem players are aware of such possibilities. Accelerators and Incubators focus on teaching startup founders to refine their value proposition and find the best target market (Guindalini et al. 2021). Long-term strategies such as those based on layered architecture and control points are too abstract and normally not mentioned to such an early-phase company. The way to develop strategic control points is implicitly left to partnering with venture capital or to the Exit strategy—which normally reinforce the power of established firms (Fairbairn and Reisman 2024).

8.4 Implications for Incumbents and Diversifying Entrants

Traditional agricultural incumbents (e.g., machinery manufacturers, chemical companies) face the challenge of acquiring component knowledge and establishing technical control points in digital layers, while leveraging their existing strategic control points (brand, customer access). They may partner with or acquire startups from accelerators like Plug & Play to bridge this knowledge gap.

Technology companies from adjacent industries (e.g., software providers, general technology providers) can enter agtech by leveraging their technical control points (e.g., cloud infrastructure, AI expertise) and then learning industry-specific knowledge through collaborations. The dynamic nature of digital business ecosystems necessitates "co-opetition" (collaboration among competitors). Incumbents, diversifying entrants, and new entrants (like Plug & Play companies) often need to partner to define standards, ensure interoperability, and co-create value for the overall ecosystem value proposition. Bohnsack et al. (2024) mentioned "co-opetition" in the context of openness and discussed how incumbents and diversifying entrants can "team up in a coalition." The Plug & Play data, while focused on startups, imply this dynamic. For example, a company like Biome Makers (soil data intelligence) would naturally seek partnerships with input suppliers or large farm management platforms. This suggests that the future of agtech is not just about competition between individual companies, but about strategic alliances and ecosystem orchestration, where control points are negotiated and shared.

8.5 Inner Case Study: NutriSen

The NutriSen GmbH, one of Plug & Play alumni, indicates how deep-tech academic research can evolve into a scalable Agtech venture through strategic ecosystem engagement, iterative business model development, and control point positioning. Founded in 2023 in Straubing, Germany, NutriSen emerged from the EIC Transition project LiveSen-MAP, with the mission to commercialize a novel nitrate biosensor for precision fertilization.

The company leverages its academic origins in enzyme-based biosensors to provide real-time soil nutrient data, initially targeting nitrate levels in wheat cultivation, and subsequently expanding to other crops.

8.5.1 Applying the Four-Phase Framework

Using Guindalini et al. (2021) four-phase framework, NutriSen's journey can be mapped as follows:

Research Insight

The company's origins trace back to Prof. Nicolas Plumeré's early academic work in chemistry and biology, where he observed firsthand the challenges farmers face in managing plant nutrition. This experience catalyzed a decade-long research trajectory, resulting in multiple patents and a robust biosensor platform capable of measuring nitrate levels in-field with laboratory precision.

Unlock Potential

The team began exploring commercialization in 2020, supported by German government programs and accelerator initiatives. Interviews with farmers and advisors revealed a strong market pull for a reliable, easy-to-use nitrate sensing solution. The team validated the technology's relevance through field trials and began building a network of partners, including agricultural chambers, machinery cooperatives, and research institutions.

Commercialize

NutriSen was formally incorporated in 2023 and launched, under the EIC Transition LiveSen-MAP (Technical University of Munich), its first large-scale field trials across Germany and Southern Europe in 2025. The company is planning to adopt a subscription-based model, bundling test strips, a measuring device, and access to a fertilization recommendation platform. This model was designed to reduce investment barriers, avoid lock-in effects, and build trust through value creation. Early adopters were identified through ethnographic interviews, with two key personas—curious, data-driven farmers and experienced agricultural advisors—emerging as primary diffusion agents.

Scale Company

NutriSen's roadmap includes transitioning from a hardware-based to a data-driven business model. By leveraging machine learning, the company aims to generate regional fertilization forecasts for entire regions based on measurements from few fields. This shift will enable scalability and reduce marginal costs. Such shift is also farmer-centric, as it brings similar value as the actual measurements, without the associated work in the field.

8.5.2 Nutrisen's Business Model

NutriSen's business model reflects a layered approach:

- *Initial Phase*: Hardware + software bundle (test strips, device, platform access).
- *Mid-Term*: Subscription-based with tailored packages per hectare.
- *Long-Term*: Data-driven platform offering predictive fertilization maps, decoupled from hardware.

This evolution mirrors the transition from technical to strategic control points, incorporating multiple layers from physical to digital products. Initially, NutriSen's value lay in its proprietary biosensor and reader (technical control; physical product). Over time, by building a data-driven platform and developing APIs to connect with farm management systems, the company advance in incorporating multiple layers of product architecture.

In this path, there is also opportunity to build strategic control through customer relationships, ecosystem integration, and data ownership. As it is part of its mission to "*democratize nutrient measurement,*" the company seems to seek strategic control points such as *Brand* and *Networking*. The founders claim that its technology can help advancing the digitalization of agriculture, which reinforces the necessity to be able to connect to different platforms and products.

It is notable that farmers play a crucial role for NutriSen, not only as potential customer/users, but truly involved in technology development. For instance, expert farmers supported the company during the field trials, which are essential for the company to reach a critical number of users/measurements to build trust and make the forecasting platform useful. Customer access is thus crucial, not only from the financial point of view, but to ultimately fulfill the company's long-term goal.

While the NutriSen case illustrates control point dynamics within an emerging agtech venture, it primarily serves to inform strategic planning for future developments. However, the effectiveness of such a strategy is inherently contingent upon the technology's successful market deployment and widespread adoption by farmers.

9 Conclusions and Future Outlook

This chapter examined the evolving role of agtech startups in shaping the future of Agriculture 4.0. Far from being mere technology providers, startups are emerging as ecosystem architects, business model innovators, and agents of systemic change. Their ability to scale and sustain impact, however, is deeply intertwined with the broader institutional and competitive landscape in which they operate.

A central insight from our analysis is the importance of *control points*—both technical (e.g., proprietary algorithms, data infrastructures) and strategic (e.g., customer access, ecosystem orchestration, brand)—in determining value capture and influence within agtech ecosystems. Startups must navigate these control points carefully, often contending with powerful incumbents who possess entrenched advantages in market access, infrastructure, and institutional legitimacy.

Institutional forces—including regulatory frameworks, public funding schemes, data governance policies, and rural infrastructure investments—play a pivotal role in shaping the opportunities and constraints for agtech startups. These forces can either reinforce existing power asymmetries or serve as levers for democratizing innovation. For instance, policies that promote *interoperability, data sovereignty*, and *inclusive innovation*, as in the Gaia-X example, can help level the playing field, enabling startups to challenge oligopolistic structures and co-create value with farmers. Conversely, weak or ambiguous regulation around data ownership and platform governance may entrench incumbent dominance and limit the transformative potential of new entrants.

9.1 Limitations

This study is exploratory and primarily conceptual, drawing on illustrative examples and secondary sources. While it offers a framework for understanding startup dynamics in agtech ecosystems, it does not provide a comprehensive empirical mapping across geographies or subsectors. Although we tried to cover examples coming from different continents, the analysis is skewed toward high-income contexts, where digital infrastructure and venture capital are more developed.

9.2 Future Research

To build on these insights, future research should:

Map institutional influence: Agriculture is an extremely regulated industry, but there is a gap regarding how national and regional policies, funding mechanisms, and regulatory environments can really impact power imbalances in the ecosystem.

Explore farmer inclusion: Study how startups engage with diverse farmer personas and how institutional support (e.g., extension services, digital literacy programs) affects adoption and co-creation.

Assess the impact of acquisitions in middle-income countries: There is few or no data about the impact of the mentioned "*oligopolistic platformisation*" process in countries from the global south. Nor is there any evidence connecting deep tech startups / spin-offs and their relationship with incumbents. A more regional approach is lacking.

Develop inclusive innovation metrics: Create frameworks to assess whether agtech ecosystems are delivering equitable outcomes in terms of access, participation, and benefit-sharing.

Ultimately, the promise of Agriculture 4.0 hinges not only on technological breakthroughs but also on the institutional scaffolding that supports or constrains innovation. A power-aware, policy-sensitive approach is essential to ensure that agtech ecosystems foster not just economic growth, but also environmental sustainability and social equity.

Competing Interests The authors have no conflicts of interest that are relevant to the content of this chapter.

References

Agarwal R, Helfat CE (2009) Strategic renewal of organizations. Organ Sci 20(2):281. https://doi.org/10.1287/orsc.1090.0423

Alcayaga A, Hansen EG (2024) Smart circular economy as a service business model: an activity system framework and research agenda. R D Manag [Preprint] 55:508–530. https://doi.org/10.1111/radm.12707

Araújo SO et al (2021) Characterising the agriculture 4.0 landscape—emerging trends, challenges and opportunities. Agronomy 11:667

Autio E et al (2017) Digital affordances, spatial affordances, and the genesis of entrepreneurial ecosystems. Strateg Entrep J 12(1):72. https://doi.org/10.1002/sej.1266

Baarbé J, Blom M, de Beer J (2019) A proposed 'agricultural data commons' in support of food security. Afr J Inf Commun [Preprint] 23:1–33. https://doi.org/10.23962/10539/27534

Birner R, Daum T, Pray CE (2021) Who drives the digital revolution in agriculture? A review of supply-side trends, players and challenges. Appl Econ Perspect Policy. Wiley 43:1260. https://doi.org/10.1002/aepp.13145

Blank S (2013) Why the lean start-up changes everything. Harv Bus Rev 91(5):64–72.

Blank S, Dorf B (2012) The startup owner's manual: The step-by-step Guide for building a great company. Pescadero, CA: K&S Ranch, Inc.

Bobier J-F et al (2022) Can Europe create its own deep-tech giants?
Bohnsack R, Pinkse J, Kolk A (2013) Business models for sustainable technologies: exploring business model evolution in the case of electric vehicles. Res Policy 43(2):284. https://doi.org/10.1016/j.respol.2013.10.014
Bohnsack R et al (2024) Profiting from innovation when digital business ecosystems emerge: a control point perspective. Res Policy 53(3):104961. https://doi.org/10.1016/j.respol.2024.104961
Brea E (2023) A framework for mapping actor roles and their innovation potential in digital ecosystems. Technovation 125:102783. https://doi.org/10.1016/j.technovation.2023.102783
Brigl M et al (2024) From lab to leader unlocking Europe's €8 trillion deep tech opportunity
von Bismarck-Osten M (2021) Understanding strategic decisions of digital agricultural platform companies: six case studies of Sub-Saharan African platforms. SSRN Electron J [Preprint]. https://doi.org/10.2139/ssrn.3849193
Bronson K (2019) Looking through a responsible innovation lens at uneven engagements with digital farming. NJAS Wageningen J Life Sci 1:1. https://doi.org/10.1016/j.njas.2019.03.001
Brueller NN, Capron L (2021) Acquisitions of startups by incumbents: the 3 Cs of co-specialization from startup inception to post-merger integration. Calif Manag Rev 63(3):70. https://doi.org/10.1177/0008125621996502
Căpățînă A, Bleoju G, Kalisz D (2024) Falling in love with strategic foresight, not only with technology: European deep-tech startups' roadmap to success. J Innov Knowl 9(3):100515. https://doi.org/10.1016/j.jik.2024.100515
Cavazza A et al (2023) Artificial intelligence and new business models in agriculture: the 'ZERO' case study. Manag Decis [Preprint] 63(10):3244–3261. https://doi.org/10.1108/md-06-2023-0980
Charatsari C et al (2022) Farm advisors amid the transition to Agriculture 4.0: professional identity, conceptions of the future and future-specific competencies. Sociologia Ruralis 62:335
Chaudhary S et al (2023) Connecting the dots? Entrepreneurial ecosystems and sustainable entrepreneurship as pathways to sustainability. Bus Strateg Environ 32(8):5935. https://doi.org/10.1002/bse.3466
Chen T, Qian L, Narayanan VK (2017) Battle on the wrong field? Entrant type, dominant designs, and technology exit. Strateg Manag J 38(13):2579. https://doi.org/10.1002/smj.2669
Coad A (2025) 'The company I keep is not corporate enough': exploring the specificities of university startups. J Technol Transf [Preprint]:1–31. https://doi.org/10.1007/s10961-025-10228-4
Cumming DJ, Zhang M (2018) Angel investors around the world. J Int Bus Stud 50(5):692. https://doi.org/10.1057/s41267-018-0178-0
Dattée B, Alexy O, Autio E (2017) Maneuvering in poor visibility: how firms play the ecosystem game when uncertainty is high. Acad Manag J 61(2):466. https://doi.org/10.5465/amj.2015.0869
Dörr F (2018) Food regimes, corporate concentration and its implications for decent work. In: Decent work deficits in southern agriculture measurements, drivers and strategies. Rainer Hampp Verlag
Dutia S (2014) AgTech: challenges and opportunities for sustainable growth. Innovations Technol Governance Glob 9:161. https://doi.org/10.1162/inov_a_00208
Dyck MV et al (2023) From product to platform: how incumbents' assumptions and choices shape their platform strategy. Res Policy 53(1):104904. https://doi.org/10.1016/j.respol.2023.104904
Eastwood C et al (2017) Managing socio-ethical challenges in the development of smart farming: from a fragmented to a comprehensive approach for responsible research and innovation. J Agric Environ Ethics 32:741. https://doi.org/10.1007/s10806-017-9704-5
Eslamloo F, Brown K, Biyouki ZA (2025) Successful commercialisation in startup accelerator programmes: how different startup ecosystems matter. J Innov Knowl 10(3):100688. https://doi.org/10.1016/j.jik.2025.100688
Fairbairn M, Kish Z, Guthman J (2022) Pitching agri-food tech: performativity and non-disruptive disruption in Silicon Valley. J Cult Econ 15(5):652. https://doi.org/10.1080/17530350.2022.2085142

Fairbairn M, Reisman E (2024) The incumbent advantage: corporate power in agri-food tech. J Peasant Stud 51(6):1331. https://doi.org/10.1080/03066150.2024.2310146
Ferrigno G, Sarto ND, Minin AD (2024) How do large corporations engage with startups? The rise of corporate service providers. Tech Anal Strat Manag 37:1. https://doi.org/10.1080/09537325.2024.2398058
Fini R et al (2016) Institutional determinants of university spin-off quantity and quality: a longitudinal, multilevel, cross-country study. Small Bus Econ 48(2):361. https://doi.org/10.1007/s11187-016-9779-9
Fleming A et al (2018) Is big data for big farming or for everyone? Perceptions in the Australian grains industry. Agron Sustain Dev 38(3):1–10. https://doi.org/10.1007/s13593-018-0501-y
Frietsch R et al (2021) Spin-offs from public research organisations in Germany: a comprehensive analysis based on bibliometric, Patent, Website and Company Register Data. Fraunhofer ISI
Gaia-X Community G-X (2021) Gaia-X Domain Agriculture. Position Paper. Federal Ministry of Economic Affairs and Energy. Available at: https://www.bundeswirtschaftsministerium.de/Redaktion/EN/Publikationen/Digitale-Welt/211116-pp-agriculture.pdf?__blob=publicationFile&v=3
Gans J (2016) The disruption dilemma, 5th edn. The MIT Press
George AL, Bennett A (2007) Case studies and theory development in the social sciences. MIT Press. Available at: https://ci.nii.ac.jp/ncid/BA71806379
Gerhardt C et al (2022) Framework for the digital transformation of the agricultural ecosystem. In: Springer eBooks. Springer Nature, p 59. https://doi.org/10.1007/978-3-662-64378-5_2
Graff G, de Grazia C, Rada N (2024) The contribution of startups, venture finance, and patenting to innovation in U.S. agriculture. SSRN Electron J. https://doi.org/10.2139/ssrn.4892791
Grande S et al (2025) Deep tech entrepreneurship in Europe and the crucial role of RTOs fostering impactful industrial spin-offs
Gubitta P, Tognazzo A, Destro F (2015) Signaling in academic ventures: the role of technology transfer offices and university funds. J Technol Transf 41(2):368. https://doi.org/10.1007/s10961-015-9398-7
Guindalini C, Verreynne M, Kastelle T (2021) Taking scientific inventions to market: mapping the academic entrepreneurship ecosystem. Technol Forecast Soc Chang 173:121144. https://doi.org/10.1016/j.techfore.2021.121144
Hackfort S (2021) Patterns of inequalities in digital agriculture: a systematic literature review. Sustainability 13(22):12345. https://doi.org/10.3390/su132212345
Kernecker M, Busse M, Knierim A (2020) Exploring actors, their constellations, and roles in digital agricultural innovations. Agric Syst 186:102952. https://doi.org/10.1016/j.agsy.2020.102952
Klerkx L, Villalobos P (2023) Are AgriFoodTech start-ups the new drivers of food systems transformation? An overview of the state of the art and a research agenda. Glob Food Sec 40:100726. https://doi.org/10.1016/j.gfs.2023.100726
Marullo C et al (2018) 'Ready for take-off': how open innovation influences startup success. Creat Innov Manag 27(4):476. https://doi.org/10.1111/caim.12272
Mendes JAJ et al (2021) In search of maturity models in agritechs. IOP Conf Ser Earth Environ Sci. IOP Publishing 893:22083. https://doi.org/10.1088/1755-1315/839/2/022083
Mikhaïlov AA et al (2021) Californian innovation ecosystem: emergence of agtechs and the new wave of agriculture. Innov Manag Rev 18(3):292. https://doi.org/10.1108/inmr-12-2018-0098
Modina M et al (2023) Innovation ecosystems: a comparison between university spin-off firms and innovative start-ups. Evidence from Italy. Int Entrep Manag J 20(2):575. https://doi.org/10.1007/s11365-023-00857-2
Moretti D et al (2023) Exploring actors' perceptions of the precision agriculture innovation system – a group concept mapping approach in Germany and Switzerland. Technol Forecast Soc Chang 189:122270. https://doi.org/10.1016/j.techfore.2022.122270
Murmann JP, Vogt F (2022) A capabilities framework for dynamic competition: assessing the relative chances of incumbents, start-ups, and diversifying entrants. Manag Organ Rev 19(1):141. https://doi.org/10.1017/mor.2022.37

Negra C et al (2019) Sustainable agri-food investments require multi-sector co-development of decision tools. Ecol Indic 110:105851. https://doi.org/10.1016/j.ecolind.2019.105851
Pandeya C et al (2022) Smart agriculture: technological advancements on agriculture—a systematical review. In: Deep learning for sustainable agriculture. Academic Press, pp 1–56
Phillips PWB et al (2019) Configuring the new digital landscape in western Canadian agriculture. NJAS Wageningen J Life Sci 90–91(1):1. https://doi.org/10.1016/j.njas.2019.04.001
Prause L, Hackfort S, Lindgren M (2020) Digitalization and the third food regime. Agric Hum Values 38(3):641. https://doi.org/10.1007/s10460-020-10161-2
Remané G, Schneider S, Hanelt A (2022) Digital business model types: understanding their mechanisms as recipes to commercialise digital technologies. Int J Innov Manag 26(3):2240019. https://doi.org/10.1142/s1363919622400199
Reutimann H (2014) Commercializing academic research results and the role of spin-off companies. CHIMIA Int J Chem 68(12):874. https://doi.org/10.2533/chimia.2014.874
Rijswijk K, Klerkx L, Turner J (2019) Digitalisation in the New Zealand agricultural knowledge and innovation system: initial understandings and emerging organisational responses to digital agriculture. NJAS Wageningen J Life Sci 1:1. https://doi.org/10.1016/j.njas.2019.100313
Rodríguez-Gulías MJ et al (2017) The role of venture capitalist to enhance the growth of Spanish and Italian university spin-offs. Int Entrep Manag J 14(4):1111. https://doi.org/10.1007/s11365-017-0489-9
Rose DC et al (2018) Agriculture 4.0: broadening responsible innovation in an era of smart farming. Front Sustainable Food Syst 2:87
Rose DC et al (2020) Agriculture 4.0: making it work for people, production, and the planet. Land Use Policy 100:104933
Rossetti F, Daniel N, Melisande C (2018) The Startup Europe ecosystem. Analysis of the Startup Europe projects and of their beneficiaries. RePEc Res Pap Econ [Preprint]. https://doi.org/10.2760/78946
Ruder S-L (2024) The 'terms and conditions' of surveillance capitalism: theorizing agricultural data policy and governance. J Peasant Stud 52:1. https://doi.org/10.1080/03066150.2024.2429480
Sarto ND et al (2021) One size does not fit all. Business models heterogeneity among Internet of Things architecture layers. Tech Anal Strat Manag 34(7):787. https://doi.org/10.1080/09537325.2021.1921138
Sauvagerd M, Mayer M, Hartmann M (2024) Digital platforms in the agricultural sector: dynamics of oligopolistic platformisation. Big Data Soc 11(4). https://doi.org/10.1177/20539517241306365
Sharma R et al (2023) Interlinkages between digital-social entrepreneurship and technological capabilities for sustainable value creation. J Glob Inf Manag 31(1):1. https://doi.org/10.4018/jgim.328515
Tell J et al (2016) Business model innovation in the agri-food sector: a literature review. Br Food J. Emerald Publishing Limited 118:1462. https://doi.org/10.1108/bfj-08-2015-0293
Tripsas M (1997) Unraveling the process of creative destruction: complementary assets and incumbent survival in the typesetter industry. Strateg Manag J 18:119. https://doi.org/10.1002/(sici)1097-0266(199707)18:1+<119::aid-smj921>3.3.co;2-s
van der Velden D et al (2023) Cyborg farmers: embodied understandings of precision agriculture. Sociol Rural [Preprint] 64:3. https://doi.org/10.1111/soru.12456
Villegas-Mateos A (2020) Regional entrepreneurial ecosystems in Chile: comparative lessons. J Entrep Emerg Econ 13(1):39. https://doi.org/10.1108/jeee-11-2019-0168
Vlachopoulou M et al (2021) Analyzing AgriFood-Tech e-business models. Sustainability 13(10):5516. https://doi.org/10.3390/su13105516
Walter A et al (2017) Smart farming is key to developing sustainable agriculture. Proc Natl Acad Sci 114(24):6148. https://doi.org/10.1073/pnas.1707462114
Wolfert J et al (2017) Big data in smart farming – a review. Agric Sys. Elsevier BV 153:69. https://doi.org/10.1016/j.agsy.2017.01.023

Wolfert J et al (2022) Digital innovation ecosystems in agri-food: design principles and organizational framework. Agric Syst 204:103558. https://doi.org/10.1016/j.agsy.2022.103558
Yin RK, Campbell DT (2018) Case study research and applications: design and methods. SAGE Publications, Inc. Available at: http://cds.cern.ch/record/2634179
Zaher U et al (2013) Life cycle assessment of the potential carbon credit from no- and reduced-tillage winter wheat-based cropping systems in Eastern Washington State. Agric Syst 122:73. https://doi.org/10.1016/j.agsy.2013.08.004
Zott C, Amit R (2009) Business model design: an activity system perspective. Long Range Plan 43:216. https://doi.org/10.1016/j.lrp.2009.07.004

Sowing Innovation Through Relational Architectures: Entrepreneurial Routes to Twin Transition in Remote Rural Contexts

Marcello De Rosa, Luca Bartoli, Concetta Cardillo, and Tiziana Pagnani

1 Introduction

Digital solutions are emerging as potentially disruptive technologies that can reshape agrifood systems and innovation paths in rural areas (Klerkx 2020). Disruption mechanisms activated through digitalization are boosted by digital affordances, which make it possible to perform existing functions more effectively or perform entirely new ones (Autio 2017). However, the outcomes of this transformative process are not linear; they are influenced significantly by the entrepreneurial profiles of individual farmers, that is, by the *farmers' ability to change or abandon old models and enter new phases of agriculture characterised by new challenges* (Cheriet et al. 2020, p. 9). Consequently, digital transitions boost new entrepreneurial models characterized by the changing locus of entrepreneurial opportunities and the change of entrepreneurial practices (Autio 2017). The process is particularly complex when the digital transition is joined to adopting sustainable farming practices, depicting cases of twin transition (Brunori 2022). The twin transitions occur within a "constrained institutional context" that can prove to be "(un)conducive" (Gittins et al. 2022; Gittins and McElwee 2024) and, as underlined in the literature, depending on effective mechanisms of knowledge transfer and innovation adoption. This chapter posits that adopting digital technologies is realized within an entrepreneurial process of knowledge anchoring shaped by different entrepreneurial traits. Set against this background, this chapter aims at:

M. De Rosa (✉) · L. Bartoli · T. Pagnani
Department of Economics and Law, University of Cassino and Southern Lazio, Cassino, Italy
e-mail: mderosa@unicas.it

C. Cardillo
Council for Agricultural Research and Economics (CREA), Research center for Policies and Bioeconomy, Rome, Italy

F. da Silveira et al. (eds.), *Agriculture 4.0*,
https://doi.org/10.1007/978-3-032-24552-6_20

(a) Investigating how farmers-entrepreneurs reshape their relational architecture as a consequence of the introduction of digital tools;
(b) Exploring the entrepreneurial identity of farmers adopting digital tools.

2 Literature Review

The twin transition (Brunori 2022) is a sociotechnical process (Fielke et al. 2020; Klerkx and Rose 2020; Rijswijk et al. 2021; Schnebelin et al. 2021; Metta et al. 2022) that is profoundly shaped by its contextual conditions. As pointed out by Benedetto and Forleo (2024, p. 7), "The integration of digitalization and sustainability requires not only changes in functions and the inclusion of new actors but also improved coordination at all system levels, among networks of actors, and through their interaction channels." Therefore, understanding the compatibility or incompatibility of this process with specific "reference contexts" (Baker and Welter 2020; Lioutas and Charatsari 2020) is essential for its successful implementation. In their study, Aubry et al. (2022) challenge the technocentric vision prevalent in digitalization research by advocating for a broader perspective that endogenizes social and institutional aspects. This more expansive view, aligned with Responsible Research and Innovation (RRI) approaches, takes into account the social and ethical aspects digital solutions may produce in the agricultural sector (Bronson 2019; Gremmen et al. 2019). Furthermore, within the new perspective, recent literature has emphasized, on the one side, directionality issues concerning the choice of the most suitable digital solution within specific "contexts" and, on the other side, risks of non-neutrality regarding the exclusion of potential beneficiaries from the adoption process (Fielke et al. 2020; Klerkx and Rose 2020; Schnebelin et al. 2021).

A coherent analysis of these mechanisms should consider the multidimensional nature of the "context," such as business, social, institutional, and spatial context (Welter 2011). Profound differences characterize rural areas (where context), the heterogeneity of the actors involved (who context), the strictly technical dimension of the innovation paths (what context), the path of transformation, and the identified objectives (what missions) (Klerkx 2020). While "context" undoubtedly influences adoption mechanisms, it is also crucial to analyze how digitalization acts as a "context enabler" (Baker and Welter 2020), influencing the trajectories of "smart rural development" (Torre et al. 2021). Consequently, the stock of "pertinent" knowledge required to adopt a digital innovation has not been considered homogeneous or a substitute for previously acquired knowledge. Irwin (1995) emphasized that "the point is not to sweep aside one knowledge form and to replace it with another. Instead, we need to recognize the contextual and partial nature of all forms of understanding." Thus, new combinatorial knowledge (Crevoisier and Jeannerat 2009) and new "knowledge cultures" (Tsouvalis et al. 2000) emerge. How these new forms of knowledge are created and combined, and how (and if) this can be considered as an entrepreneurial process deserves more attention in the literature. This chapter tries to fill this gap in the literature by providing an entrepreneurial perspective in analyzing new dynamics of knowledge creation and anchoring. More precisely (Crevoisier 2016), we explore this issue through the lens of the anchoring milieu by positing that each rural territory can be characterized by different anchoring milieus, in which anchoring is not synonymous with imitation but also with differentiation based on the specificity of the context. Crevoisier (2016) clearly defines the anchoring milieu: "The anchoring milieu can be defined as a set of local players who interact locally and with distant and/or mobile players on the basis of an institutional mix of authorship and ownership in order to generate ever more advanced (more efficient, exclusive and/or meaningful) knowledge and higher revenues." From the perspective of rural entrepreneurship (Korsgaard et al. 2015; Gittins and McElwee 2023), it suggests that an entrepreneurial anchoring of combinatorial knowledge preserves the territorial embeddedness and the territorially anchored mechanisms driving the adoption of new knowledge, specifically digital solutions. Therefore, redesigning combinatorial knowledge necessary for adopting digital solutions involves synthesizing newly created knowledge with indigenous local knowledge. This indigenous local knowledge is the system of cultures, customs, and traditions that characterize a particular local context (Peddi et al. 2023). Against this backdrop, a critical question arises: to what extent does indigenous local knowledge influence the adoption of digital solutions? In this chapter, the combination of existing and the anchoring process of newly created (and distant) knowledge is analyzed through the lens of the entrepreneurial process of creation–recreation of social capital by splitting sources of knowledge according to the "classic" categories of social capital: bonding, bridging, and linking ties (Woolcock 1998; Tisenkopfs et al. 2008). Consequently, the knowledge anchoring and creation of combinatorial knowledge is understood as a two-way process of entrepreneurial creation of knowledge that is affected by the context of reference and by institutional entrepreneurship, which captures the agency of entrepreneurs toward their institutional contexts (e.g., Battilana et al. 2009) and is realized through the strategic reconfiguration of the relational architecture mixing local indigenous and external knowledge.

3 Methodology

The methodology adopted for this research is based on a case-study approach (Yin 2018), focusing on digital innovation adoption in the "Comino Valley" biodistrict, located in Southern Lazio (Italy). Therefore, our case study fits with the analysis of a twin transition, where digital innovation is located within an area specializing in organic products. More precisely, a pluralistic view, trying to synthesize either exploratory ("what" question) or explanatory ("why" and "how" questions) perspective on the case study, will be adopted (Yin 2018).

This food biodistrict falls within the broader framework of Italian Food Districts, regulated by the Law n. 205/2017, art.1. Its objective is to boost place-based strategies for agrifood and rural development, reflecting the renewed interest in

Table 1 Types of social capital

Types of relations	Proxy for
Family members/friends	*Bonding ties*
Other farms, processing industries, or trade companies Private or public farm advisors Local associations Other agricultural producers Producer organizations	*Bridging ties*
Banks Universities Regional organisms Training courses	*Linking ties*

localized agrifood development strategies (Pecqueur et al. 2024). As highlighted on the website of the Italian Ministry of Agriculture,[1] the *Food district is a strategic tool aimed at fostering territorial development, cohesion, and social inclusion by promoting the integration of activities characterized by territorial proximity. Food Districts also aim to ensure food security, reduce the environmental impact of production, and minimize food waste. Another fundamental goal is to safeguard the territory and rural landscape through agricultural and agrifood activities. The Food Districts model is also intended to revitalize the experiences of rural districts already present in the national territory, as well as to encourage the creation of new realities through access to dedicated funding*. To boost a territorial approach to the agrifood system, a common financial framework is provided to support district projects through a national tool, the District Contract Agreement (Toccaceli and Pacciani 2023).

The specific biodistrict analyzed is located in the Comino Valley, which is identified as rural area with complex problems of development within the National Strategic Plan. As emphasized in the Long-Term Vision for Rural Areas 2040 (European Commission 2021), digitalization is considered as a fundamental goal to relaunch sustainable rural development. Moreover, the chosen Biodistrict could represent an interesting example of "twin transition," including ecological and digital dimensions (Brunori 2022).

An online questionnaire has been submitted to 50 farms within the Biodistrict to explore various dimensions of farmers' networks, attitudes, and characteristics. The questionnaire is articulated into four sections:

- The first section focused on assessing the knowledge-sharing networks of farmers either before or after the adoption of digital solutions. These networks were classified according to the three types of social capital: bonding, bridging, and linking ties (see Table 1).

[1] https://www.politicheagricole.it/flex/cm/pages/ServeBLOB.php/L/IT/IDPagina/14159

- The second section aimed to evaluate farmers' awareness and understanding of digital solutions alongside their adoption and implementation (which means maintenance) of these technologies.
- The third section comprised multiple subsections designed to classify farms based on key characteristics. A dedicated subsection evaluated farmers' trust in various institutions integral to their activities, including political bodies, local authorities, civil society organizations, research institutions, producer associations, and agricultural consultants. Respondents rated their trust on a Likert scale ranging from 1 (no trust) to 5 (maximum trust). Another subsection, inspired by Vesala et al. (2007), assessed the entrepreneurial identity of farmers by taking into account either economic (proactiveness, risk-taking, innovativeness) or individual (self-efficacy, personal control, optimism) values. Afterward, a three-item scale based on Teece's (2007) dynamic capability framework (considering sensing, seizing, and transforming capabilities) was employed to investigate farmers' adaptive and strategic capabilities.
- The final section of the questionnaire aimed to capture respondents' personal and farm-level characteristics (McElwee and Smith 2012). Variables in this section included age, education level, family composition, Standard Output,[2] and marketing channels.

The level of digital innovation was determined based on the tools currently adopted by the farmer. A high level of innovation is attributed to farmers utilizing advanced technologies such as drones, sensors, traceability software, and highly automated systems. A medium level corresponds to adopting tools like weather stations, radio collars, management software, or semi-automated equipment and machinery. Farmers who have not implemented any of these innovations are classified as having a basic level of technological innovation.

Cronbach's alpha was calculated to confirm the internal reliability of all scales used in the analysis, while the Kaiser-Meyer-Olkin (KMO) index and Bartlett's sphericity test were applied to ensure the sample's suitability for factor analysis (Verbeke and Viaene 1999).

A multistep methodology was implemented in this study to address the research questions. In the first step, a Principal Component Analysis (PCA) was conducted using the eigenvalue criterion (≥ 1) on variables related to farmers' knowledge-sharing networks, both before and after the adoption of digital solutions, as well as their trust in institutions relevant to their activities. This procedure reduced

[2] The Standard Output (SO) represents the monetary value of a given agricultural production, which includes sales at the farm gate price. Based on the EU regulations (EC Reg. 1242/2008 and following), each Member State must produce the SO coefficients every 3 years for a list of agricultural activities defined in the calculation handbook harmonized at the European level. The SO coefficients are determined at the regional level based on a 5-year average of the yields and prices of the products concerned and are applied to the hectares cultivated and the heads of animals raised by each farm. These coefficients are used for the typological classification of agricultural holdings collected in the farms statistical surveys and for the definition of the economic potential of agricultural holdings in the context of RDP tenders.

the original information into eight principal components that synthesize the most significant patterns in farmers' relational and institutional contexts. In the second step, data were further explored through a hierarchical cluster analysis (HCA) performed by applying Ward's linkage method with squared Euclidean distance. The clustering was based on the component scores obtained from the PCA, complemented by a set of farm- and farmer-specific variables, as detailed below:

- The eight component scores derived from the PCA.
- Farmers' knowledge of digital innovations (rated on a scale from 1 to 5 for each innovation considered).
- Their level of digital innovation (classified as high, medium, or basic).
- The 6 items from Vesala et al. (2007) assessing the entrepreneurial identity of farmers.
- The 3-item scale from Teece's (2007) capturing dynamic capabilities.
- Farmers' age.
- Education level.
- The farm's Utilized Agricultural Area.
- The farm's Standard Output.

The Duda-Hart goodness-of-fit index was applied to determine the optimal number of clusters identified as five, providing a robust solution (Halpin 2016). To characterize the clusters, analysis of variance (ANOVA) and Pearson's chi-square tests were employed to examine differences across the clusters based on the collected variables. All statistical analyses were performed using STATA software, version 18.

4 Results

4.1 Descriptive Results

Respondents to the questionnaire are predominantly male (64%). Pointing on the age, the 60% of the respondents are young under 40, 34% belong to the age class 40–65 years, whereas only 6% are over 65 years. Nearly 50% of the sample holds a high school diploma, whereas merely 10% have completed postgraduate studies. Regarding family composition, results reveal that 32% of the respondents are couples with children, 28% single individuals, 16% are families with two or more nuclei, 16% are couples without children, and 8% are single-parent families. Concerning farm size, 28% of the farms are small (span 1–5 ha), another 28% are 5–10 ha, and 24% are 10–20 ha. Standard Output (SO) distribution indicates that 42% of farms generate between €8000 and €25,000, while 18% fall into the €25,000–€50,000 range, and another 18% into €50,000–€100,000. Regarding agricultural size, 28% of farms have 1–5 ha, another 28% have 5–10 ha, and 24% have 10–20 ha.

Table 2 Entrepreneurial identity of farmers

	Sample	
	Mean	Std. dev.
Self-categorization	3.46	1.09
Self-efficacy	3.22	1.07
Optimism	3.48	1.09
Personal control	3.84	1.02
Risk-taking	4.26	0.85
Growth-orientation	3.98	0.94

The Standard Output (SO) is most commonly between €8000 and €25,000 (42%), followed by €25,000–€50,000 (18%) and €50,000–€100,000 (18%).

Farmers rely significantly on bonding ties for ordinary activities, with relatives being the most influential source. Bridging ties, particularly private consultants and other processing companies, play a moderate role, while linking ties, such as training courses and banks, also contributes substantially.

For digital solutions, relatives remain an essential source of knowledge (mean: 3.28), while friends are less significant (mean: 2.70). Among bridging ties, private consultants (mean: 3.52) are highly significant, while public consultants (mean: 2.74) and local associations (mean: 2.18) are less utilized. Other processing companies (mean: 2.64) and producer organizations (mean: 2.82) are moderately relevant. Training courses (mean: 3.54) and university education (mean: 3.00) are the most relevant linking ties for digital solutions. Banks (mean: 2.92), research centers (mean: 2.96), and suppliers (mean: 2.38) are less utilized.

Trust levels vary significantly across different institutions and groups. Farmers exhibit the highest trust in consultants (mean: 3.72), agricultural producers (mean: 3.46), and research institutions or universities (mean: 3.34). In contrast, they display lower trust toward political institutions (mean: 2.20) and local authorities (mean: 2.46).

As far as entrepreneurial traits are concerned (Table 2), farmers scored the highest on risk-taking (mean: 4.26) and growth orientation (mean: 3.98), reflecting a strong entrepreneurial mindset. Other traits, such as personal control (mean: 3.84) and optimism (mean: 3.48), are also prominent. However, self-efficacy is comparatively lower (mean: 3.22).

Farmers demonstrate a relatively high ability to implement significant changes to adapt to digital innovations (mean: 3.72) and to seize opportunities while maintaining essential skills (mean: 3.64). Their ability to recognize opportunities and threats from digitalization is slightly lower (mean: 3.38).

The degree of digital innovation among the farmers in the sample is diverse, with 36% adhering to a basic level, 24% to a moderate level, and 40% to an advanced level. However, farmers exhibit limited knowledge of advanced technologies, such as drones (2.56), sensors (2.32), and highly automated systems (2.46). In contrast, they display slightly greater familiarity with semi-automatic equipment (3.04) and weather stations (2.88).

Table 3 Cluster distribution

Cluster	Freq.	%
Cluster 1	15	30.00
Cluster 2	5	10.00
Cluster 3	11	22.00
Cluster 4	11	22.00
Cluster 5	8	16.00
Total	**50**	**100.00**

4.2 Cluster Analysis Results

The cluster analysis returns five homogeneous clusters presenting the maximum internal similarities and the highest difference among the clusters. As shown in Table 3, the first cluster contains the highest number of farms (30%), while the second cluster is the smallest (10%). The main sociodemographic characteristics of the farms in each cluster are evidenced in Table 4.

4.2.1 Cluster 1

The first cluster is characterized by a mix of young and mature farmers, most of whom hold a high school diploma. Approximately two-thirds of the farms in this cluster are managed by men, while women manage one-third.

The lack of substantial innovative activities typifies farms of the cluster: 60% of farms present a basic level of innovation, 20% of average, and 20% of high innovative activity. This partially explains a relatively stable relational architecture, which reveals a partial modification of the social capital moving from ordinary activity to innovation adoption. During ordinary agricultural activities, bonding ties, particularly family ties, serve as a critical source of information (mean: 4.20). However, their importance, despite remaining prevalent, diminishes in the context of the uptake of digital solutions (mean: 3.40). Linking ties maintain a stable influence across both contexts, while bridging social capital grows in importance, with producer organizations emerging as crucial facilitators for the uptake and implementation of digital innovations.

As shown in Table 5, farms in this cluster exhibit moderate entrepreneurial identity and dynamic capabilities, with all measured dimensions below the sample average. Risk-taking (mean: 4.00) and proactiveness (mean: 3.60) emerge as key characteristics of the entrepreneurial profile.

4.2.2 Cluster 2

Farmers in this cluster are predominantly male (60%), aged 41–64 (80%), and have low levels of education (60% have only elementary/middle school qualifications).

Table 4 Sociodemographic characteristics of the farms in each cluster

	Sample		Cluster 1		Cluster 2		Cluster 3		Cluster 4		Cluster 5		Pearson Chi-2
	Freq.	%	Freq.	%	Freq.	%	Freq.	%	Freq.	%	Freq.	%	
Gender													
Male	32	64,00	10	66,67	3	60,00	7	63,64	7	63,64	5	62,50	0,09
Female	18	36,00	5	33,33	2	40,00	4	36,36	4	36,36	3	37,50	
Age													
Under 40	30	60,00	8	53,33	1	20,00	8	72,73	9	81,82	4	50,00	13,61
41–64	17	34,00	6	40,00	4	80,00	1	9,09	2	18,18	4	50,00	
Over 65	3	6,00	1	6,67	0	0,00	2	18,18	0	0,00	0	0,00	
Educational level													
Primary/ secondary school	7	14,00	2	13,33	3	60,00	2	18,18	0	0,00	0	0,00	27,1**
High school diploma	24	48,00	11	73,33	1	20,00	2	18,18	4	36,36	6	75,00	
Bachelor	14	28,00	1	6,67	1	20,00	6	54,55	4	36,36	2	25,00	
Postgraduate diploma	5	10,00	1	6,67	0	0,00	1	9,09	3	27,27	0	0,00	
Family composition													
Single individuals	14	28,00	6	40,00	0	0,00	4	36,36	3	27,27	1	12,50	15,72
Couple without children	8	16,00	0	0,00	3	60,00	1	9,09	2	18,18	2	25,00	
Couple with children	16	32,00	6	40,00	1	20,00	2	18,18	4	36,36	3	37,50	
Single parent with children	4	8,00	1	6,67	0	0,00	1	9,09	1	9,09	1	12,50	
Family with 2 or more units	8	16,00	2	13,33	1	20,00	3	27,27	1	9,09	1	12,50	

(continued)

Table 4 (continued)

	Sample		Cluster 1		Cluster 2		Cluster 3		Cluster 4		Cluster 5		Pearson Chi-2
	Freq.	%	Freq.	%	Freq.	%	Freq.	%	Freq.	%	Freq.	%	
Utilized agricultural area													
1–5	14	28,00	5	33,33	1	20,00	3	27,27	4	36,36	1	12,50	33,4*
5–10	14	28,00	5	33,33	2	40,00	2	18,18	4	36,36	1	12,50	
10–20	12	24,00	4	26,67	0	0,00	4	36,36	3	27,27	1	12,50	
20–50	6	12,00	1	6,67	0	0,00	1	9,09	0	0,00	4	50,00	
50–100	3	6,00	0	0,00	2	40,00	0	0,00	0	0,00	1	12,50	
>100	1	2,00	0	0,00	0	0,00	1	9,09	0	0,00	0	0,00	
Standard Output (SO)													
Up to €8000	5	10,00	1	6,67	1	20,00	1	9,09	2	18,18	0	0,00	24,36
€8000–€25,000	21	42,00	9	60,00	2	40,00	2	18,18	4	36,36	4	50,00	
€25,000–€50,000	9	18,00	3	20,00	0	0,00	3	27,27	3	27,27	0	0,00	
€50,000–€100,000	9	18,00	2	13,33	0	0,00	4	36,36	2	18,18	1	12,50	
€100,000–€250,000	3	6,00	0	0,00	1	20,00	1	9,09	0	0,00	1	12,50	
€250,000–€500,000	3	6,00	0	0,00	1	20,00	0	0,00	0	0,00	2	25,00	

Note: ***$p < 0.001$; **$p < 0.01$; *$p < 0.05$

Table 5 (a) Entrepreneurial identity of farmers. (b) Dynamic capabilities

	Sample		Cluster 1	
	Mean	Std. dev.	Mean	Std. dev.
(a) *Entrepreneurial identity of farmers*				
Self-categorization	3.46	1.09	2.93	0.96
Self-efficacy	3.22	1.07	2.60	0.74
Optimism	3.48	1.09	3.00	0.65
Personal control	3.84	1.02	3.33	0.72
Risk-taking	4.26	0.85	4.00	0.93
Growth-orientation	3.98	0.94	3.67	0.82
(b) *Dynamic capabilities*				
I am able to recognize the opportunities and threats that arise from digitalization.	3.38	1.12	2.73	0.70
I am able to seize the opportunities of digitalization while maintaining and improving essential skills.	3.64	0.94	3.13	0.74
I am able to implement significant changes in my company to adapt to the digital innovations emerging around me.	3.72	0.99	3.13	0.74

Family types are dominated by couples without children (60%). Farms are medium to large, with 40% between 5 and 10 ha and 50–100 ha. Standard output is split across low (€8000–€25,000, 40%) and high (€100,000–€500,000, 40%).

Similar to the first cluster, the second one reveals a basic level of innovativeness. Moreover, it exhibits a diverse knowledge network characterized by a more balanced relational architecture, in which bridging and linking ties are more important than family ties. Producers' organizations take on the highest importance (mean increases from 4.60 to 5.00). Also, universities and training courses empower farmers' stock of knowledge (mean increases from 4.40 to 4.80). Among bridging ties, a huge increase in the importance of private advisors is registered (means increases from 3.80 to 4.80).

A strong entrepreneurial identity emerges among the cluster farms, with notably higher mean values across all dimensions, especially in terms of personal control, proactiveness, and innovativeness (Table 6a). Therefore, the farmer entrepreneur is growth-oriented and strongly believes in her/his capability to affect the farm's success through her/his entrepreneurial activity. This is also confirmed by the farms' dynamic capabilities (Table 6b). The farms demonstrate a high sensing capability (means: 4.60), effectively perceiving opportunities for innovation, while their seizing and transforming capabilities reach maximum scores (mean: 5.00). Thus, the farms can improve their essential skills and adapt their organizational structure to embrace digital solutions. Despite a sound entrepreneurial profile, adopting digital innovation seems scarce, with 80% of farms declaring a basic level of innovation, excluding digital solutions. This apparent contradiction could be attributed to the medium-to-small size of farms, which may constrain their ability to make substantial investments. The prevalence of mature families without

Table 6 (a) Entrepreneurial identity of farmers. (b) Dynamic capabilities

	Sample		Cluster 2	
	Mean	Std. dev.	Mean	Std. dev.
(a) *Entrepreneurial identity of farmers*				
Self-categorization	3.46	1.09	4.60	0.89
Self-efficacy	3.22	1.07	4.20	1.79
Optimism	3.48	1.09	4.20	1.79
Personal control	3.84	1.02	5.00	0.00
Risk-taking	4.26	0.85	5.00	0.00
Growth-orientation	3.98	0.94	5.00	0.00
(b) *Dynamic capabilities*				
I am able to recognize the opportunities and threats that arise from digitalization.	3.38	1.12	4.20	1.79
I am able to seize the opportunities of digitalization while maintaining and improving essential skills.	3.64	0.94	5.00	0.00
I am able to implement significant changes in my company to adapt to the digital innovations emerging around me.	3.72	0.99	5.00	0.00

children suggests a "cautious entrepreneurship" approach (Matacena 2020). Without generational renewal, farmers may hesitate to make long-term investments in digitalization, prioritizing stability over risk, which can hinder innovation adoption and affect farming activities' future trajectories.

4.2.3 Cluster 3

This cluster consists of males (63.64%), young (72.73% under 40), and well-educated farmers (54.55% holding a university degree). Single-parent families dominate (36.36%). Farms are medium-sized (36.36% in the 10–20-haectare range), and Standard output is mid-range (36.36% generating €50,000–€100,000).

Farms in this cluster stand out for adopting cutting-edge technologies (72.7%), reflecting the use of advanced anchoring mechanisms to integrate combinatorial knowledge and balance various dimensions of social capital. Linking ties play a pivotal role in supporting the farm's knowledge network, both in ordinary farm activity and during the adoption of digital innovations. Training courses, research centers, and financial institutions are key enablers of farm digitalization. Trust in research centers and universities is particularly relevant, with an average score of 4.27. Also, private advisors act as innovation brokers to boost digital solutions, taking high trust rates among farms (mean: 4.18). A relatively high level of dynamic capabilities characterizes this cluster, particularly in seizing opportunities (mean: 4.09). Similarly, entrepreneurial orientation is robust (Table 7), with a notable emphasis on risk-taking (mean: 4.73), personal control (mean: 4.64), and optimism and growth orientation (mean: 4.00). These attributes facilitate innovation adoption,

Table 7 (a) Entrepreneurial identity of farmers. (b) Dynamic capabilities

	Sample		Cluster 3	
	Mean	Std. dev.	Mean	Std. dev.
(a) *Entrepreneurial identity of farmers*				
Self-categorization	3.46	1.09	3.82	0.75
Self-efficacy	3.22	1.07	3.82	0.60
Optimism	3.48	1.09	4.00	0.77
Personal control	3.84	1.02	4.64	0.50
Risk-taking	4.26	0.85	4.73	0.47
Growth-orientation	3.98	0.94	4.00	0.89
(b) *Dynamic capabilities*				
I am able to recognize the opportunities and threats that arise from digitalization.	3.38	1.12	4.09	0.94
I am able to seize the opportunities of digitalization while maintaining and improving essential skills.	3.64	0.94	3.82	0.87
I am able to implement significant changes in my company to adapt to the digital innovations emerging around me.	3.72	0.99	4.00	1.18

resulting in the highest percentage of upper level technologies implemented across the clusters.

4.2.4 Cluster 4

Farmers here are predominantly male (63.64%) and young (81.82% under 40). Education levels are high, with 36.36% holding a university degree and 27.27% postgraduate qualifications. Family types include couples with children (36.36%) and single-parent families (27.27%). Farms are small, with 36.36% each in the 1–5 and 5–10-hectare ranges. Standard output is low to mid-range, with 36.36% generating €8000–€25,000.

The relational architecture of this cluster exhibits relatively lower "density," with mean values systematically lower than in other clusters. Bonding ties, however, play a fundamental role in knowledge acquisition and innovation adoption. Relatives are particularly influential, with the highest importance scores (mean: 3.18) in both ordinary activities and the uptake of digital innovations. Bonding ties, such as friends further bolster the uptake of digital innovation, and bridging ties, with private advisors notably acting as innovation brokers. Farmers and private advisors have earned significant trust within the cluster, with mean values of 3.82 and 3.73, respectively. The reliance on bonding ties does not hinder the adoption of innovative digital solutions. A 55% of farms in this cluster demonstrate a moderate approach to digital solutions, while 27.3% adopt highly innovative ones.

This propensity may be attributed to the strong entrepreneurial orientation of the farmers (Table 8a). Risk-taking (4.19), proactiveness, and growth orientation

Table 8 (a) Entrepreneurial identity of farmers. (b) Dynamic capabilities

	Sample		Cluster 4	
	Mean	Std. dev.	Mean	Std. dev.
(a) *Entrepreneurial identity of farmers*				
Self-categorization	3.46	1.09	3.82	1.08
Self-efficacy	3.22	1.07	3.55	0.93
Optimism	3.48	1.09	3.91	0.83
Personal control	3.84	1.02	3.82	0.87
Risk-taking	4.26	0.85	4.18	0.75
Growth-orientation	3.98	0.94	4.09	0.83
(b) *Dynamic capabilities*				
I am able to recognize the opportunities and threats that arise from digitalization.	3.38	1.12	3.45	0.93
I am able to seize the opportunities of digitalization while maintaining and improving essential skills.	3.64	0.94	3.82	0.75
I am able to implement significant changes in my company to adapt to the digital innovations emerging around me.	3.72	0.99	3.82	0.75

reveal the farmers' good entrepreneurial profile, being the farmers ready to accept the risk of cutting-edge digital solutions to support the farm's growth. Personal characteristics such as optimism and a strong sense of personal control further solidify the entrepreneurial mindset of farmers in Cluster 4. Combined with moderate dynamic capabilities—sensing, seizing, and adapting to change—these traits enable farms to embrace digitalization and enhance their operations effectively (Table 8b).

4.2.5 Cluster 5

This cluster is evenly divided between farmers under 40 and those aged 41–64 (50% each). Couples with children (37.50%) are the most common family structure. Half of the farms are between 20 and 50 ha and have a standard production between €8000 and €25,000. Farms in Cluster 5 evidence high levels of adoption of cutting-edge innovation, the second-highest percentage of the sample (62.5%). However, this occurs despite the lowest level of relational density in the cluster. The farmers make minimal use of available knowledge sources, both in ordinary activities and during the innovation process. Their primary sources of information include training courses (means: 2.88), relatives (mean: 2.13), and private advisors (mean: 2.13). This reliance on minimal and self-contained networks underscores that innovation here depends heavily on entrepreneurial skills and self-generated knowledge. Regarding entrepreneurial orientation (Table 9a), the cluster exhibits relatively low scores across all dimensions, with values below the sample average. Risk-taking and growth orientation seem relatively high, jointly with

Table 9 (a) Entrepreneurial identity of farmers. (b) Dynamic capabilities

	Sample		Cluster 5	
	Mean	Std. dev.	Mean	Std. dev.
(a) *Entrepreneurial identity of farmers*				
Self-categorization	3.46	1.09	2.75	1.04
Self-efficacy	3.22	1.07	2.50	0.76
Optimism	3.48	1.09	2.63	1.19
Personal control	3.84	1.02	3.00	1.20
Risk-taking	4.26	0.85	3.75	1.04
Growth-orientation	3.98	0.94	3.75	1.28
(b) *Dynamic capabilities*				
I am able to recognize the opportunities and threats that arise from digitalization.	3.38	1.12	3.00	1.07
I am able to seize the opportunities of digitalization while maintaining and improving essential skills.	3.64	0.94	3.25	1.04
I am able to implement significant changes in my company to adapt to the digital innovations emerging around me.	3.72	0.99	3.50	0.93

optimism, which addresses a decision-making dynamic oriented toward adopting highly innovative digital solutions. This is also attributable to dynamic capabilities (Table 9b), like seizing (mean: 3.25) and implementing changes (mean: 3.50). Sociodemographic variables may provide further insights on the motivation for innovating digitally: farms of the cluster show large sizes. Moreover, the significant presence of couples with children may address the decision-making process oriented toward innovation to boost generational renewal and leave future generations with solid and technological farms. Lastly, training courses and private advisors, though limited in the cluster's aggregate relational density, may still provide highly targeted and actionable knowledge, compensating for the lack of broader relational networks.

5 Discussion

The results of the empirical analysis depict a diversified scenario in which the uptake of digital solutions paves the way to a differentiated set of relational architectures boosted by "enlarged anchoring milieus" (Jeannerat and Crevoisier 2022) characterized by a relatively high number of actors and institutions acting as innovation brokers to boost innovation adoption. However, digital solutions are also boosted by different entrepreneurial mindsets and skills, jointly with heterogeneous farms' dynamic capabilities. Consequently, this chapter has provided a broader perspective of the engine of digital innovation, read through the lens of either social capital or entrepreneurial identity and capabilities of farmers.

Regarding the relational architecture, cluster analysis highlights two main groups of farms: the first, including clusters 1, 2, and 3, evidences a dense relational configuration. The second, including clusters 4 and 5, has a relatively lower relational density.

- High relational density. In the first aggregate, including clusters 1, 2, and 3, the third cluster emerges with high relational density, high entrepreneurial orientation levels, and high propensity to adopt advanced digital solutions. In this cluster, the twin transition is unfolding as a socio-technical process supported by a relational architecture where various innovation brokers boost the uptake of advanced digital solutions. These findings align with prior research identifying ambidexterity mechanisms, where a balanced relational architecture fosters cutting-edge innovations. Cofré-Bravo et al. (2019, p. 60) define ambidexterity as the "*ability to manage different types of social capital to fulfil different tasks and objectives within the innovation process, i.e., both exploration and exploitation of new technologies and practices*." Accordingly, ambidexterity is grounded on creating new combinatorial knowledge dynamics and building new territorial proximity by synthesizing geographical and organized proximities (Gallaud and Torre 2022; De Rosa et al. 2024). Therefore, a "constituent perspective" on rural embeddedness emerges (Uzzi 1996), where the rural context and rural development are reconfigured through innovative relational and entrepreneurial mechanisms. Set against this background, actors of the agricultural knowledge and innovation systems offer a relevant contribution. As a matter of fact, the role of private consultants is confirmed as essential in digital transition processes (Knierim et al. 2017). This confirms the importance of local knowledge networks in welding mobile/distant knowledge and local knowledge, and thus fostering organized proximities. As pointed out by Crevoisier and Jeannerat (2009, p. 12): "*One should not, however, misguidedly believe that since long-distance relations are becoming easier within economic dynamics, the local ones lose their pertinence*." Cluster 3 emerges as the most advanced group in knowledge of innovative technologies, with significantly higher average values for almost all variables. This inclination is correlated with relatively strong entrepreneurial traits, particularly risk-taking and personal control. Despite a good entrepreneurial orientation (in terms of both economic and individual values), farms in Cluster 2 exhibit the lowest average knowledge across most categories, suggesting a lower propensity to adopt advanced technologies. Consequently, empowering knowledge networks and anchoring is paramount to encouraging twin transition among these farms.
- Low relational density. Clusters 4 and 5 represent this group, marked by poorly developed knowledge networks but a notable inclination toward digital innovation adoption. This paradox is particularly evident in Cluster 4, where strong entrepreneurial traits like risk-taking and growth orientation drive digital technology adoption. In Cluster 5, innovation adoption relies on entrepreneurial traits like optimism and risk-taking, combined with structural factors such as farm size and family composition (e.g., couples with children). These charac-

teristics hint at a long-term perspective, focusing on generational renewal and ensuring the sustainability of farm operations. Therefore, as pointed out in other analyses (Yigit et al. 2024), entrepreneurial orientation may be characterized by a strong individualism, which does not impede the uptake of digital solutions, mostly in the presence of high levels of resilience of the family farm business. This resilience is also boosted by the will to secure solid bases for generational renewal (in a family farm with children) and, consequently, through the adoption of digital innovation raising the level of the farm's competitiveness.

In conclusion, the study reveals that the twin transition in agriculture is not a uniform process but rather a heterogeneous one, shaped by the interplay between social capital as an engine of knowledge anchoring mechanisms, entrepreneurial identity, and farms' dynamic capabilities. Supporting relational and entrepreneurial dimensions is critical for scaling digital innovations across diverse farming contexts.

6 Implications and Limits of the Study

This chapter tried to contribute to the literature on agricultural entrepreneurship by examining entrepreneurial mechanisms of knowledge anchoring and digital innovation adoption. More precisely, this chapter sheds light on the role of entrepreneurial orientation to navigate the twin transition (promoting digital and sustainable farming systems), by developing new digital sustainable business models.

This chapter shows some limits, first of all in terms of a limited sample of farms (50 farms), which need further investigation in future research. Secondly, the identification of the entrepreneurial identity of farms is based on self-categorization, which may be at risk of lack of objectivity despite it being a valid tool for identifying entrepreneurial attitudes (Vesala et al. 2007).

Despite these limits, the analysis of how farmer-entrepreneurs reconfigure their knowledge networks to anchor coherent knowledge for supporting digital solutions, jointly with their entrepreneurial competencies and dynamic capabilities, represents a contribution to the literature and evidence of how the adoption of cutting-edge digital innovations is supported:

(a) Through ambidexterity.
(b) By strong entrepreneurial traits and dynamic capabilities.
(c) By both ambidexterity and entrepreneurial orientation/dynamic capabilities.

The higher levels of both the variables (entrepreneurial traits and knowledge networks) bring about cutting-edge digital innovations.

To the best of our knowledge, our analysis makes an original contribution by emphasizing the relevance of territorial knowledge dynamics and combinatorial knowledge processes. We view these processes as the result of an entrepreneurial behavior aimed at strengthening the capability to adopt digital solutions within a new farming paradigm based on sustainable agriculture.

Future research will be necessary to provide broader evidence from other territorial contexts and agrifood systems. In particular, the role of agricultural knowledge and innovation systems in stimulating rural entrepreneurship should be further clarified, and a more dynamic perspective, one that investigates combinatorial knowledge dynamics over a longer time span, may offer a more rigorous analytical framework.

Nonetheless, the dual perspective adopted here, combining new knowledge networks with entrepreneurialism, has provided clear evidence of its role in fostering the twin transition.

This has implications regarding the activation of policy action impacting entrepreneurship and knowledge networks through innovation brokering. From a policy perspective, assuming a wider view of the innovation brings about what is identified as the "innovatization" of territorial policy, from technological to cultural, social, and societal innovations (Jeannerat and Crevoisier 2022). Recent policy trends, such as the rural development policies of the CAP 2023–27, seem oriented toward a wider perspective through encouraging either sectorial measures (for instance, SRD01 to support investment on digital solutions to improve the farm's competitiveness, or SRD07, information infrastructure and digital services) or territorial measures (such as SRG07, boosting digitalization of rural areas, and SRH06 Back-office services for knowledge transfer and innovation adoption).

Access to these measures should upgrade the farmers' capabilities of introducing digital innovation, with the support of knowledge networks which, jointly with adequate entrepreneurial orientation, could address sustainable twin transition in the next few years, in line with the Long Term Vision for rural Areas 2040 (European Commission 2021).

Acknowledgments Paper realized within the research project of national interest (Prin) funded by the Italian Ministry of University and Research: "Puzzling out smart ruralities, sound knowledge and rural (agricultural) entrepreneurial ecosystems (Smarties)," coordinated by Gianluca Brunori.

Competing Interests The author(s) has no competing interests to declare that are relevant to the content of this manuscript.

References

Aubry M, Hamadi B, Condor R, Fadil N, Fournes C (2022) Exploring digitalisation in the agri-food sector and its paradoxes: evidence from a comparative study with small French companies. Stud Agric Econ 124(2):44–58. https://doi.org/10.7896/j.2305

Autio E (2017) Digitalisation, ecosystems, entrepreneurship and policy, Policy brief n.20, Government's analysis, assessment and research activities

Baker T, Welter F (2020) Contextualizing entrepreneurship theory. Routledge Studies in Entrepreneurship, Routledge, New York

Battilana J, Leca B, Boxenbaum E (2009) How actors change institutions: towards a theory of institutional entrepreneurship. Acad Manag Ann 3(1):65–107. https://doi.org/10.1080/19416520903053598

Benedetto G, Forleo MB (2024) PDO Economy and Quality Agri-Food District in the wine sector: blockchain and digitalization as tools for the twin transition. Economia Agro-Alimentare 26(3):187–213. https://doi.org/10.3280/ecag2024oa17787

Bronson K (2019) Looking through a responsible innovation lens at uneven engagements with digital farming. NJAS Wageningen J Life Sci 90–91:100294. https://doi.org/10.1016/j.njas.2019.03.001

Brunori G (2022) Agriculture and rural areas facing the "twin transition": principles for a sustainable rural digitalization. Ital Rev Agric Econ 77(3):3–14. https://doi.org/10.36253/rea-13983

Cheriet F, Messeghem K, Lagarde V, McElwee G (2020) Agricultural entrepreneurship: challenges and perspectives. Revue de l'Entrepreneuriat 19(4):13–29. https://doi.org/10.3917/entre.194.0013

Cofré-Bravo G, Klerkx L, Englera A (2019) Combinations of bonding, bridging, and linking social capital for farm innovation: how farmers configure different support networks. J Rural Stud 69:53–64. https://doi.org/10.1016/j.jrurstud.2019.04.004

Crevoisier O (2016) The economic value of knowledge: embodied in goods or embedded in cultures? Reg Stud 50(2):189–201. https://doi.org/10.1080/00343404.2015.1070234

Crevoisier O, Jeannerat H (2009) Territorial knowledge dynamics: from the proximity paradigm to multi-location milieus. Eur Plan Stud 17:1223–1241. https://doi.org/10.1080/09654310902978231

De Rosa M, Charatsari C, Lioutas ED, Masi M, Vecchio Y, Francescone M (2024) Contextualising digitalisation through ambidexterity and new territorial proximities. J Rural Stud 109:103322. https://doi.org/10.1016/j.jrurstud.2024.103322

European Commission (2021) A long-term vision for the EU's rural areas - towards stronger, connected, resilient and prosperous rural areas by 2040, Brussels, 30.6.2021 COM(2021) 345 final

Fielke S, Taylor B, Jakku E (2020) Digitalisation of agricultural knowledge and advice networks: a state-of-the-art review. Agric Syst 180:102763. https://doi.org/10.1016/j.agsy.2019.102763

Gallaud D, Torre A (2022) Introduction: proximity relations in the 21st century. In: Gallaud D, Torre A (eds) Handbook of proximity relations. Edward Elgar Publishing, Cheltenham

Gittins P, McElwee G (2023) Constrained rural entrepreneurship: upland farmer responses to the socio-political challenges in England's beef and sheep sector. J Rural Stud 104:103141. https://doi.org/10.1016/j.jrurstud.2023.103141

Gittins P, McElwee G (2024) Farm adaptive business strategies in crisis management: COVID-19. J Rural Stud 111:103393. https://doi.org/10.1016/j.jrurstud.2024.103393

Gittins P, McElwee G, Lever J (2022) Constrained entrepreneurship in UK agriculture: a Weberian analysis. J Rural Stud 95:495–504. https://doi.org/10.1016/j.jrurstud.2022.09.021

Gremmen B, Blok V, Bovenkerk B (2019) Responsible innovation for life: five challenges agriculture offers for responsible innovation in agriculture and food, and the necessity of an ethics of innovation. J Agric Environ Ethics 32(5):673–679. https://doi.org/10.1007/s10806-019-09808-w

Halpin B (2016) Cluster analysis stopping rules in Stata. Available: https://hdl.handle.net/10344/5492

Irwin A (1995) Citizen science: a study of people, expertise and sustainable development. Routledge, London

Jeannerat H, Crevoisier O (2022) From competitiveness to territorial value: transformative territorial innovation policies and anchoring milieus. Eur Plan Stud 30(11):2157–2177. https://doi.org/10.1080/09654313.2022.2042208

Klerkx L (2020) Advisory services and transformation, plurality and disruption of agriculture and food systems: towards a new research agenda for agricultural education and extension studies. J Agric Educ Ext 26(2):131–140. https://doi.org/10.1080/1389224X.2020.1738046

Klerkx L, Rose D (2020) Dealing with the game-changing technologies of agriculture 4.0: how do we manage diversity and responsibility in food system transition pathways? Glob Food Secur 24:100347. https://doi.org/10.1016/j.gfs.2019.100347

Knierim A, Labarthe P, Laurent C, Prager K, Kania J, Madureira L, Ndah TH (2017) Pluralism of agricultural advisory service providers–facts and insights from Europe. J Rural Stud 55:45–58. https://doi.org/10.1016/j.jrurstud.2017.07.018

Korsgaard S, Muller S, Tanvig W (2015) Rural entrepreneurship or entrepreneurship in the rural—between place and space. Int J Entrep Behav Res 21(1):5–26. https://doi.org/10.1108/IJEBR-11-2013-0205

Lioutas ED, Charatsari C (2020) Smart farming and short food supply chains: are they compatible? Land Use Policy 94:104541. https://doi.org/10.1016/j.landusepol.2020.104541

Matacena R (2020) Cautious entrepreneurship: strategies and business orientation of small scale farmers in the alternative food economy. In: Vasile AJ, Subic J, Grubor A, Privitera D (eds) Handbook of research on agricultural policy, rural development, and entrepreneurship in contemporary economies. IGI Global Scientific Publishing

McElwee G, Smith R (2012) Classifying the strategic capability of farmers: a segmentation framework. Int J Entrep Ventur 4(2):111–131. https://doi.org/10.1504/IJEV.2012.046517

Metta M, Ciliberti S, Obi C, Bartolini F, Klerkx L, Brunori G (2022) An integrated socio-cyber-physical system framework to assess responsible digitalisation in agriculture: a first application with Living Labs in Europe. Agric Syst 203:103533. https://doi.org/10.1016/j.agsy.2022.103533

Pecqueur B, De Rosa M, Zumpano C (2024) The territorial approach to the crisis in the global food system. Ital Rev Agric Econ 79(1):3–4. https://doi.org/10.36253/rea-15321

Peddi B, Ludwig D, Dessein J (2023) Relating inclusive innovations to indigenous and local knowledge: a conceptual framework. Agric Hum Values 40:395–408. https://doi.org/10.1007/s10460-022-10344-z

Rijswijk K, Klerkx L, Bacco M, Bartolini F, Bulten E, Debruyne L, Dessein J, Scotti I, Brunori G (2021) Digital transformation of agriculture and rural areas: a socio-cyber-physical system framework to support responsibilisation. J Rural Stud 85:79–90. https://doi.org/10.1016/j.jrurstud.2021.05.003

Schnebelin E, Labarthe P, Touzard JM (2021) How digitalisation interacts with ecologisation? Perspectives from actors of the French Agricultural Innovation System. J Rural Stud 86:599–610. https://doi.org/10.1016/j.jrurstud.2021.07.023

Teece DJ (2007) Explicating dynamic capabilities: the nature and microfoundations of (sustainable) enterprise performance. Strateg Manag J 28(13):1319–1350. https://doi.org/10.1002/smj.640

Tisenkopfs T, Lace U, Mierina I (2008) Social capital. In: van der Ploeg JD, Marsden T (eds) Unfolding webs. van Gorcum, Assen

Toccaceli D, Pacciani A (2023) Dear old (and misunderstood) districts, let's look ahead. Ital Rev Agric Econ 78(3):3–15. https://doi.org/10.36253/rea-15062

Torre A, Wallet F, Corsi S, Steiner M, Westlund H (2021) Is there a smart development for rural areas? In: Torre A, Corsi S, Steiner M, Wallet F, Westlund H (eds) Smart development for rural areas. Routledge, London

Tsouvalis J, Seymour S, Watkins C (2000) Exploring knowledge-cultures: precision farming, yield mapping, and the expert–farmer interface. Environ Plann A Econ Space 32(5):909–924. https://doi.org/10.1068/a32138

Uzzi B (1996) The sources and consequences of embeddedness for the economic performance of organizations: the network effect. Am Sociol Rev 61(4):674–698. https://doi.org/10.2307/2096399

Verbeke W, Viaene J (1999) Beliefs, attitude and behaviour towards fresh meat consumption in Belgium: empirical evidence from a consumer survey. Food Qual Prefer 10(6):437–445

Vesala KM, Peura J, McElwee G (2007) The split entrepreneurial identity of the farmer. J Small Bus Enterp Dev 14(1):48–63

Welter F (2011) Contextualizing entrepreneurship—conceptual challenges and ways forward, Entrepreneurship. Theory and Practice 35(1):165–184. https://doi.org/10.1111/j.1540-6520.2010.00427.x

Woolcock M (1998) Social capital and economic development: toward a theoretical synthesis and policy framework. Theory Soc 27:51–208. https://doi.org/10.1023/A:1006884930135

Yigit F, Rantamäki-Lahtinen L, Sipiläinen T (2024) How does perception of success change between family and solo farmers: a perspective from strategic resources and entrepreneurial orientation. J Rural Stud 110:103359. https://doi.org/10.1016/j.jrurstud.2024.103359

Yin RK (2018) Case study research and applications: design and methods, 6th edn. Sage, London

Agriculture 4.0 Technologies for Inclusive Development: Agro-Technological Districts (ATDs) as Living Labs for Digital Transformation in Diverse Brazilian Biomes

Catarina B. Careta, Luciano Mendes, and Luiz F. Satolo

1 Introduction

In recent years, digital technologies have shaped the transformation of agrifood systems. In the context of Agriculture 4.0, technologies such as Artificial Intelligence, Internet of Things (IoT), and Big Data and Analytics stand out for promoting increased productivity, the implementation of sustainable practices, and the resilience of agricultural systems. However, the implementation of these innovations is uneven, particularly in rural areas facing structural challenges such as low connectivity, low digital literacy, and weak or nonexistent technical assistance networks.

The concept of Agriculture 4.0, while widely adopted, lacks a universally accepted definition, often being used interchangeably with terms such as smart farming, digital agriculture, and precision agriculture (Albiero et al. 2020). Broadly, it refers to the integration of cyber-physical systems and advanced digital technologies across the agrifood chain to enable data-driven decision-making, automation, and enhanced sustainability (da Silveira et al. 2021). Despite its transformative potential, the adoption of Agriculture 4.0 remains concentrated in large-scale, capital-intensive farming systems, raising concerns about inclusivity, equity, and the risk of deepening the digital divide between regions and producer profiles (Scuderi et al. 2022).

C. B. Careta (✉) · L. Mendes
University of São Paulo, Av. Pádua Dias, Piracicaba, SP, Brazil
e-mail: caretta@usp.br; mendes@usp.br

L. F. Satolo
Aeronautics Institute of Technology, Praça Marechal Eduardo Gomes, Sáo José dos Campos, SP, Brazil
e-mail: satolo@ita.br

F. da Silveira et al. (eds.), *Agriculture 4.0*,
https://doi.org/10.1007/978-3-032-24552-6_21

In response to these asymmetries, the Science Center for Development in Digital Agriculture (CCD-AD/Semear Digital) was established in Brazil in 2022, an initiative coordinated by the Brazilian Agricultural Research Corporation (Embrapa) and supported by the São Paulo Research Foundation (FAPESP). This initiative aims to design, implement, and evaluate the impact of digital solutions for small- and medium-sized producers, with a focus on economic and social development and the strengthening of local innovation ecosystems. A central element of the CCD-AD/Semear Digital strategy is the establishment of Agro-Technological Districts (ATDs), environments that serve as a breeding ground for the cocreation and implementation of digital agriculture solutions.

The ATDs were selected based on a multidimensional assessment that included agricultural production data, socioeconomic indicators, infrastructure assessments, and stakeholder engagement. The districts implement actions from the CCD-AD/Semear Digital technological axes, which involve connectivity, artificial intelligence, automation, monitoring systems, and traceability tools, always adapting the solutions to the specific biophysical and socio-institutional conditions of the location.

The CCD-AD/Semear Digital was designed to operate in five municipalities in the state of São Paulo and in five municipalities in other Brazilian regions with different biomes, cultural contexts, and production systems. In the state of São Paulo, ATDs were implemented in the cities of Alto Alegre, Caconde, Jacupiranga, Lagoinha, and São Miguel Arcanjo. The remaining Brazilian regions are represented by the cities of Breves in the state of Pará (PA), Boa Vista do Tupim in Bahia (BA), Guia Lopes da Laguna in the state of Mato Grosso do Sul (MS), Ingaí in the southern state of Minas Gerais (MG), and Vacaria in Rio Grande do Sul (RS).

This chapter discusses the actions being implemented and the opportunities for applying Agriculture 4.0 technologies in the ATDs of the states of Bahia, Mato Grosso do Sul, Minas Gerais, Pará, and Rio Grande do Sul. These regions represent a rich diversity of territorial realities, ranging from the Caatinga biome, characterized by the need for irrigation, to livestock-based systems in the Cerrado biome, and agroforestry in the Amazon biome. By analyzing these cases, this chapter aims to contribute to the global debate on inclusive digital transformation in agriculture, demonstrating how Agro-Technological Districts can function as living laboratories for context-sensitive, participatory, and equitable deployment of Agriculture 4.0 technologies.

2 Profile of Agro-Technological Districts

The Agro-Technological Districts (ATDs) established outside the state of São Paulo are located in distinct socio-environmental, demographic, and productive contexts. Their selection was coordinated by the Institute of Agricultural Economics (IEA), which was based on a combination of criteria, including economic activity profiles, agrifood systems governance, digital infrastructure, the presence of public research units, the activities of production cooperatives and rural unions, and accessibility (presence of airports and highways). Based on data published by the Brazilian Institute of Geography and Statistics (Instituto Brasileiro de Geografia e Estatística 2025), this section provides an overview of the five ATDs, which collectively represent a cross section of Brazil's biomes and rural realities.

Demographic indicators reveal significant variations in population size and density. While Breves (PA) and Vacaria (RS) are home to populations over 50,000, Ingaí (MG) has just over 2500 residents. Population density varies from less than 6 inhabitants/km^2 in Boa Vista do Tupim (BA), a municipality in the Caatinga biome, to more than 30 inhabitants/km^2 in Vacaria (RS), located in the Atlantic Forest biome. These variations reflect not only geographic and ecological diversity but also different levels of urbanization and access to infrastructure.

Economic indicators also show significant disparities. While Vacaria (RS) and Ingaí (MG) have relatively high per capita GDPs, above R$40,000, Boa Vista do Tupim (BA) and Breves (PA) remain below R$10,000. The average monthly wage of formal workers in these municipalities follows a distinct pattern, with values ranging from R$1800 in Ingaí (MG) to R$2500 in Breves (PA). These differences highlight the importance of adapting digital agriculture strategies to local economic conditions and human development levels, as reflected by the Municipal Human Development Index (MHDI), which ranges from 0.503 in Breves, Pará, to 0.721 in Vacaria, Rio Grande do Sul. Figure 1 summarizes the sociodemographic characteristics of the ATDs that are the focus of this chapter's analysis.

Regarding production profiles, each ATD specializes in different agricultural and livestock activities. Boa Vista do Tupim (BA) is located in the semiarid region of Bahia and has a strong presence of family farming, focusing on cassava, honey, and licuri. Breves (PA), on the other hand, is located in the Amazon biome and is characterized by honey production and agriculture integrated with açaí extraction. Guia Lopes da Laguna (MS), located in the Cerrado-Pantanal transition zone, focuses on livestock and grain production, including soybeans and corn. Ingaí, in southern Minas Gerais, combines coffee growing with dairy production, while Vacaria (RS) is a hub for temperate fruit production, particularly apples, grapes, and blackberries.

These diverse territorial configurations require equally diverse technological responses. While Breves (PA) faces challenges related to energy infrastructure and connectivity, Vacaria's emphasis is on data integration and traceability for commercial fruit production. Ingaí focuses on diagnostics for coffee and pasture management, while Boa Vista do Tupim requires affordable technologies compatible with low connectivity to improve production under climate stress. Guia Lopes da Laguna (MS) benefits from satellite-based pasture monitoring and decision-support systems for livestock intensification.

To synthesize and compare the environmental and productive characteristics of the five Agro-Technological Districts (ATDs) analyzed in this study, Table 1 provides a consolidated overview of dominant production systems. This comparative profile highlights the significant heterogeneity among the districts, reinforcing

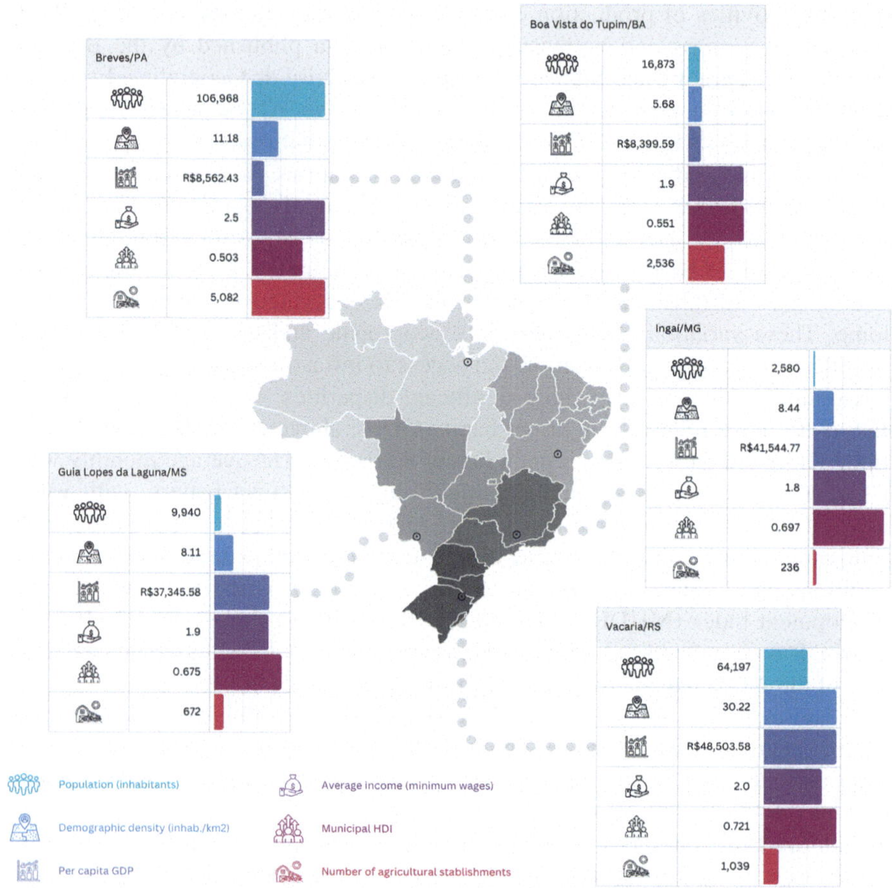

Fig. 1 Sociodemographic characterization of selected ATDs

the central argument of this chapter: The implementation of Agriculture 4.0 must be context-sensitive, adaptive, and grounded in local realities. The disparities in infrastructure, economic development, and biophysical conditions underscore the need for differentiated technological strategies, which are further explored in the following sections.

3 The Technologies Applied to Agriculture 4.0

Agriculture 4.0 is a concept derived from Industry 4.0 and refers to the incorporation of advanced digital technologies throughout agrifood supply chains, aiming to make them more efficient, integrated, sustainable, and responsive to contemporary

Table 1 Environmental and productive profile of the Agro-Technological Districts (ATDs) analyzed in this study

Municipality/state	Biome	Main product system	Key technology focus in CCD-AD/SEMEAR Digital
Boa Vista do Tupim/BA	Caatinga	Family farming, cassava, beekeeping, licuri extraction	Low-cost IoT, remote sensing for licuri mapping, AI for beekeeping (Zapbee)
Breves/PA	Amazon	Açaí extraction, beekeeping, floodplain agriculture	Satellite mapping of açaí, AI for traceability, digital extension (Infobee)
Guia Lopes da Laguna/MS	Cerrado-Pantanal transition	Beef cattle, soybeans, corn, ILPF systems	Pasture monitoring via satellite, traceability, AI for animal nutrition (Zapbee)
Ingaí/MG	Atlantic Forest	Coffee, dairy farming	Herd traceability, milk quality monitoring, pasture diagnostics
Vacaria/RS	Pampa	Apple, grape, blackberry production	Computer vision for phenology, harvest automation, precision orchard management

environmental, economic, and social challenges (Kamilaris et al. 2017; Lezoche 2020; Rose et al. 2021). While the term is widely used, it lacks a standardized definition and is often applied inconsistently across studies, sometimes conflating automation, precision farming, and digital inclusion (Albiero et al. 2020; da Silveira et al. 2021). Agriculture 4.0 technologies enable the collection, processing, and real-time analysis of large volumes of data, in addition to fostering automation and evidence-based decision-making (Maffezzoli et al. 2022).

However, their successful implementation depends not only on technological maturity but also on socioeconomic context, infrastructure availability, and institutional support, factors that are especially critical in smallholder and extractive farming systems, such as those present in the Agro-Technological Districts (ATDs) analyzed in this study.

The technologies discussed in this section are not neutral tools; their adoption and impact are shaped by power relations, access to resources, and local knowledge systems (Eastwood et al. 2019). Therefore, this section emphasizes not only the technical potential of these innovations but also their limitations and the need for context-sensitive adaptation. The main technologies discussed in this chapter are defined below.

3.1 Big Data and Analytics

Massive data analysis, or Big Data and Analytics, is emerging as one of the most promising technologies in Agriculture 4.0, enabling the transformation of large volumes of data into accessible information. This technology refers to the use of tools and methodologies for collecting, processing, storing, and analyzing data from multiple sources, such as sensors, satellite imagery, drones, weather stations, production records, and agricultural management systems, to support real-time decision-making (Wolfert et al. 2017).

The integration of Internet of Things (IoT) sensors with Big Data platforms enables continuous monitoring of agro-environmental and livestock variables, resulting in more accurate diagnoses and faster responses to production challenges (Liu et al. 2021). For example, analytical algorithms can be applied to correlate weather patterns with pest or disease outbreaks, anticipating agronomic interventions. The predictability generated by statistical and machine learning models, powered by large databases, enables robust productivity estimates, harvest forecasts, localized fertilizer recommendations, and intelligent water resource allocation (Pivoto et al. 2019).

Furthermore, the use of satellite imagery and remote sensing combined with Big Data has been increasingly employed for land use mapping, crop identification, and early detection of water or nutritional stress. The ability to integrate data of different natures, both structured and unstructured, also enables advances in traceability and certification processes, connecting production history to consumer demands for more sustainable and safe food (Kamilaris et al. 2017).

Recent studies also highlight that data-driven farming is essential for more resilient agricultural systems in the face of climate change and socioeconomic pressures. However, challenges remain, such as interoperability between platforms, data quality, connectivity in remote areas, and the technical training of end users. In tropical and peri-urban contexts, data scarcity and heterogeneity—especially for sociobiodiversity chains like açaí and licuri, can limit model accuracy and generalizability (Furuya et al. 2024). Data governance strategies, aligned with public policies and the protection of producer privacy, are necessary to consolidate the sustainable use of Big Data and Analytics in agriculture (Liu et al. 2021). Without equitable access and transparent data ownership models, there is a risk of reinforcing existing inequalities, where only large-scale producers benefit from predictive analytics, while smallholders remain data subjects rather than data owners (Scuderi et al. 2022).

3.2 Internet of Things (IoT)

The Internet of Things (IoT) refers to the network of physical devices equipped with sensors, software, and connectivity that collect and transmit data in real time, enabling remote monitoring, automation, and evidence-based decision-making. In agriculture, this technology is consolidating itself as a central component of Agriculture 4.0 by enabling precision management practices, improving traceability, and increasing production efficiency (Tzounis et al. 2017).

The application of IoT in the field integrates sensors for variables such as temperature, soil moisture, light, pH, and electrical conductivity, as well as devices installed on animals to measure physiological parameters, enabling early disease detection and improving animal welfare (Liu et al. 2021). These sensors are connected to wireless networks and cloud-based platforms, which facilitate the storage and continuous analysis of collected data. Furthermore, IoT-based systems have been used in tropical agriculture for real-time monitoring of crop environmental and operational conditions, automated irrigation, pest and disease control, and energy management in agricultural greenhouses. Systematic and automated data collection allows producers to quickly identify anomalies, optimize input use, and reduce losses, promoting greater environmental and economic sustainability on rural properties (Zhong et al. 2017).

Studies show that the use of IoT can reduce operational costs and increase productivity, while contributing to the rational use of water and fertilizers and compliance with regulatory requirements, such as those related to the traceability of agricultural products (Liu et al. 2021). This integrated approach is especially promising for small- and medium-sized producers operating in diverse tropical environments, such as those served by the Agro-Technological Districts studied in this chapter. However, the deployment of IoT in rural Brazil faces critical barriers, including limited energy access, low connectivity (especially in the Amazon and Caatinga biomes), and high maintenance costs. In many ATDs, hybrid solutions—such as solar-powered sensors and offline data synchronization—are being explored to overcome these constraints (da Silveira et al. 2021).

3.3 Remote and Proximal Sensing

Remote sensing, performed by satellites, drones, or aircraft, and proximal sensing, based on sensors positioned close to plants or on the ground, constitute key technologies for the digitalization of agriculture. In the context of Agriculture 4.0, these tools enable the continuous acquisition of spatial and temporal data, essential for precision agriculture and livestock practices (Sishodia et al. 2020).

Remote sensing images obtained by satellites (such as Landsat, Sentinel, and CBERS) have been widely used for monitoring vegetation cover, identifying cultivated areas, detecting land use changes, and estimating biomass and agricultural productivity. Drones, equipped with multispectral, hyperspectral, and RGB cameras, offer greater spatial resolution and temporal flexibility, making them particularly useful for mapping diseases, nutritional deficiencies, water stress, and localized input management (Furuya et al. 2024).

In turn, proximal sensing uses sensors installed on tractors, sprayers, or mobile platforms (such as robots) that collect data directly from the agricultural environment, including soil moisture and temperature, nutrient content, leaf chlorophyll, and ambient gas concentrations. This information is essential for automated recommendation systems, helping to reduce the use of pesticides, fertilizers, and water, while increasing production efficiency (Sishodia et al. 2020).

These technologies also have significant applications in sociobiodiversity chains, such as those for açaí, licuri, and honey, where remote sensing territorial mapping allows for estimating the extractive area, monitoring production seasonality, and identifying areas under anthropogenic pressure. Furthermore, they enable the monitoring of sustainable practices and compliance with certification of origin requirements and good environmental practices.

Finally, the integrated use of remote and proximal sensing enables the creation of intelligent systems for real-time data collection, analysis, and visualization, promoting automation and traceability throughout the production chain (Liu et al. 2021). The adoption of these tools faces technical challenges related to data heterogeneity, implementation costs, and connectivity infrastructure in rural areas but offers transformative potential, especially for small- and medium-sized producers. In regions with high cloud cover (e.g., Amazon) or fragmented land use (e.g., Caatinga), data fusion techniques and community-based validation are critical to ensure accuracy and local relevance (Silva et al. 2024).

3.4 Artificial Intelligence (AI)

Artificial intelligence is one of the key technologies in digital agriculture, enabling the transformation of large volumes of data into accessible knowledge through techniques such as machine learning, neural networks, natural language processing, and computer vision. In agriculture, its application is widespread across three major areas: prediction, automation, and intelligent recommendation (Patrício and Rieder 2018).

In the context of prediction, supervised and unsupervised learning algorithms have been used to predict yield, early detection of pests and diseases, and climate risk assessment. For example, artificial neural network models have been tested to predict the yield of crops such as soybeans and corn based on remote sensing data and meteorological variables (Patrício and Rieder 2018; Benos et al. 2021).

In automation, AI has been integrated into embedded systems in agricultural machinery and robots for pattern recognition in images, such as fruit identification, plant counting, or plant health assessment. Studies demonstrate the use of deep learning convolutional networks applied to the classification of apple leaf images for disease diagnosis, with accuracy exceeding 90%, which can support real-time interventions in the field (Patrício and Rieder 2018; Benos et al. 2021).

Another relevant area is the use of AI in recommendation systems and digital technical assistance. AI-based platforms have been developed to support farmers' decision-making, considering environmental, historical, and operational variables. Applications such as chatbots trained with technical data, practical examples, and accessible language can democratize access to agronomic knowledge, especially in regions with low in-person assistance density. Furthermore, AI can integrate information from different sources (satellite imagery, IoT sensors, climate systems, and socioeconomic data) into platforms that identify patterns and generate predictive and personalized alerts (Furuya et al. 2024; Silva et al. 2024). This enables a move beyond precision agriculture toward prescriptive and autonomous agriculture, adapted to local realities and the agroecological systems prevalent in many Brazilian territories.

Therefore, AI represents not only a technological advance but also a strategic opportunity to promote digital inclusion, sustainability, and open innovation in tropical agriculture. However, AI systems are only as good as the data they are trained on. When models are developed using data from large-scale, temperate-zone farms, they may perform poorly in tropical, smallholder, or extractive contexts, a phenomenon known as "algorithmic bias" (Rose and Chilvers 2018). To avoid this, participatory AI development is essential for equitable digital transformation, where local knowledge is integrated into model training.

3.5 Robotics and Automation

Robotics and automation are pillars of Agriculture 4.0, promoting the replacement of manual practices with mechanized and autonomous systems that operate with high precision, efficiency, and intelligence. Agricultural robotics encompasses a broad spectrum of solutions, from autonomous tractors and robotic harvesters to Unmanned Aerial Vehicles (UAVs), robotic manipulation systems for selective harvesting, and intelligent mobile robots for pest control and crop monitoring (Bechar and Vigneault 2016; Abbasi et al. 2022).

Autonomous robotic systems are one of the most widely applied technologies in the context of digital agriculture, particularly in field activities such as planting, spraying, weeding, irrigation, and harvesting. These systems operate independently or in conjunction with other digital technologies, such as sensors, IoT, and artificial intelligence algorithms, forming cyber-physical systems capable of making real-time decisions based on environmental and agronomic data (Bechar and Vigneault 2016; Abbasi et al. 2022).

The benefits associated with the adoption of robotics in agriculture include:

- Greater operational precision, which reduces losses, avoids overlapping inputs, and increases productivity
- Reduced dependence on human labor, especially in regions with a shortage of rural workers
- Reduced use of agricultural pesticides and optimization of natural resources, with localized and precise application
- Increased operational safety, by reducing human exposure to adverse conditions and chemicals
- Standardization of tasks, improving the quality of agricultural products

However, experts in the field highlight significant challenges to the widespread adoption of these systems, such as high acquisition costs, the lack of infrastructure in rural areas (e.g., connectivity and power), and the need for technical training to operate and maintain the equipment. Furthermore, many robotic applications are still at intermediate levels of technological maturity (TRL 3–6), meaning they are in the prototyping and feasibility testing phase, still far from full commercialization. Robotic systems have been primarily applied to outdoor farms, especially in the production of grains, fruits, and vegetables. However, there is growing interest in adapting these technologies to indoor production systems or controlled environments, such as greenhouses and hydroponic crops, which could further expand their disruptive potential in the agricultural sector. In the context of smallholder farming, low-cost, modular, and repairable robotic solutions are emerging as more viable alternatives to high-end commercial systems (Maffezzoli et al. 2022).

3.6 Digital Platforms and Systems Integration

The convergence of data on digital platforms represents the foundation of digital transformation in contemporary agriculture, enabling a holistic and integrated view of agricultural production. Such platforms are designed to collect, store, analyze, and visualize data from multiple sources, such as IoT sensors, geographic information systems (GIS), weather records, satellite imagery, market data, and corporate databases. This integration creates an intelligent ecosystem that enables real-time decision support, increasing producers' responsiveness to climate variations, market fluctuations, and traceability and sustainability requirements (Berthet et al. 2018; Uztürk and Büyüközkan 2024).

Digital platforms enable the automation of agricultural management processes and strengthen interconnectivity between different stakeholders in production chains. This includes everything from predictive analytics for crop planning to the traceability of production from the field to the end consumer, meeting food safety standards and environmental certifications. Systems such as Farm Management Information Systems (FMIS) and cloud-based platforms exemplify the ability of these technologies to connect field devices, drones, satellites, and mobile applications into a single operational interface (Lezoche 2020; Uztürk and Büyüközkan 2024).

Another relevant aspect is the role of digital platforms in connecting precision agriculture, robotics, and artificial intelligence. The integration of these technical modules enables the construction of cyber-physical systems that optimize the use of natural resources, reduce losses, and increase production predictability. This technological convergence is driven by open and interoperable architectures, which facilitate the connection of solutions developed by different providers and adapted to diverse contexts, from small family farms to large agro-industrial enterprises (Uztürk and Büyüközkan 2024).

Digital platforms are key tools for the strategic management of Agriculture 4.0, as they translate the complexity of collected data into actionable information aligned with economic, environmental, and social objectives. This transformation is evidenced by use cases involving pest and disease prediction, climate scenario modeling, technology package customization, and field validation of sustainability indicators. Digital platforms are not only operational management tools but also critical infrastructure for data-driven innovation and the transition from conventional models to smart and resilient agrifood systems (Uztürk and Büyüközkan 2024).

However, the adoption of integrated solutions through digital platforms faces significant challenges, such as high implementation costs, a lack of technical capacity, and uncertainties regarding interoperability between systems. Furthermore, the increasing digitalization of agriculture brings not only operational gains but also risks associated with cybersecurity. Agriculture 4.0 Systems, by integrating sensors, data networks, and management platforms, become susceptible to complex cyber threats (Bui et al. 2024; Kumar et al. 2024). Equally critical is the issue of data sovereignty: who owns, controls, and benefits from the data generated in rural territories? Without transparent governance frameworks, there is a risk that digital platforms become tools of corporate control rather than instruments of rural empowerment (Eastwood et al. 2019).

4 Agriculture 4.0 and the ATDs: Actions Under Implementation

The digitalization of agriculture is not limited to the introduction of technological devices but involves structural transformations in the production, organization, and governance of agricultural systems. The CCD-AD/Semear Digital was launched in 2022, and the Center's first 2 years were dedicated to recognizing local realities and planning research and extension initiatives. This section describes how some of the 4.0 technologies applied to agriculture are being implemented in the ATDs of Boa Vista do Tupim (BA), Breves (PA), Guia Lopes da Laguna (MS), Ingaí (MG), and Vacaria (RS).

The implementation of Agriculture 4.0 in these Agro-Technological Districts (ATDs) is not a uniform technological rollout, but a context-sensitive, participatory process of cocreation and adaptation. Each ATD functions as a living laboratory, where digital solutions are not merely deployed but codesigned with local actors—producers, cooperatives, extension agents, and community leaders—ensuring alignment with socioeconomic, ecological, and cultural realities. This approach reflects a shift from a technocentric model of innovation to one grounded in responsible research and innovation (RRI), which emphasizes inclusivity, reflexivity, and responsiveness to societal needs (Rose and Chilvers 2018). The following subsections illustrate how technologies such as Big Data, AI, and integrated systems are being operationalized in diverse biomes and production systems, from Amazonian extractivism to temperate fruit farming.

4.1 Big Data and Analytics in Boa Vista do Tupim (BA) and Breves (PA)

The adoption of Big Data and Analytics technologies in the agricultural sector has intensified with the advancement of remote sensing tools, geotechnologies, and geographic information systems (GIS). These technologies enable the massive collection, processing, and analysis of territorial, climatic, productive, and socioeconomic data, with the aim of supporting evidence-based decision-making and the formulation of territorial public policies. In the ATDs of Breves (PA) and Boa Vista do Tupim (BA), these tools will be used to improve the management of extractive chains of sociobiodiversity products, such as açaí (*Euterpe oleracea*) and licuri (*Syagrus coronata*).

In the case of Breves (PA), located in the Marajó region, açaí extraction represents an important source of income. Medium- and high-resolution satellite imagery, combined with supervised classification techniques, can be used to accurately map the areas where native and established açaí palm groves occur. These maps can be used to estimate production capacity per territorial unit, identify areas subject to degradation, and plan sustainable management actions that balance conservation and income generation.

Furthermore, cross-referencing sensor data with socioeconomic and infrastructure variables, such as the presence of riverside communities, river accessibility, and proximity to agribusinesses, allow for the construction of spatial models of vulnerability and development potential. Such approaches can be incorporated into participatory planning within the scope of the ATD, strengthening the protagonism of local communities and the effectiveness of public policies.

Similarly, in Boa Vista do Tupim (BA), in the semiarid region of Bahia, the licuri is a native palm tree of great ecological and cultural importance, whose fruit collection supports family extractive practices. Mapping the spatial distribution of licuri populations, as well as identifying priority areas for natural regeneration or intercropping, can be accomplished with the support of agro-environmental modeling based on remote sensors and historical climate data.

Integrating these data with land use and vegetation cover databases allows for the identification of land use conflicts, monitoring deforestation, and planning ecological corridors for Caatinga conservation. The data-driven approach also allows for estimating the seasonality of fruit production and planning logistical strategies for harvesting and processing, considering the context of low mechanization and water scarcity.

These applications demonstrate that Big Data and Analytics are not exclusive to large-scale, capital-intensive farming. When adapted to the logic of extractivism and family agriculture, they become powerful tools for territorial governance, environmental monitoring, and value chain enhancement. However, their success depends on the integration of local knowledge into data models. Without this integration, there is a risk of producing "digital colonialism," where external data systems override local epistemologies (Eastwood et al. 2019). The CCD-AD/Semear Digital actively involves local communities in data validation and interpretation, ensuring that analytics serve as a complement, not a substitute, for traditional knowledge.

The opportunities for the application of Big Data and Analytics in unconventional supply chains, such as açaí and licuri, are strategic and highly relevant for territories with low technological density. Adapting geospatial analysis methodologies to the context of family-based agriculture and extractivism will have positive impacts on territorial governance and the valorization of sociobiodiversity products.

4.2 Integrated Systems in Ingaí (MG) and Guia Lopes da Laguna (MS)

Agricultural traceability has evolved beyond simply recording production steps, assuming a strategic role in certifying origin, environmental sustainability, and enhancing the value of differentiated products. In this context, integrated information and monitoring systems, characteristic of 4.0 technologies, constitute the foundation for building more efficient, transparent production chains aligned with national and international standards. The cases of the ATDs in Ingaí (MG) and Guia Lopes da Laguna (MS) illustrate how the combination of sensors, databases, geotechnologies, and interoperable platforms can transform the production dynamics of dairy and beef cattle farming.

In Ingaí(MG), milk and dairy production will benefit from individual herd traceability, standardized health practices, and added value through certification of origin. In this case, the integration of sensors embedded in milking machines, livestock data collectors, management software (ERP), and cloud-based platforms can ensure continuous traceability from the animal to the final product. This structure allows data such as milk volume and quality, sanitary treatments, diet, geolocation, and welfare practices to be linked to each production unit, creating a robust and auditable digital trail.

In addition to meeting health and regulatory requirements, this integrated system allows for the certification of controlled-origin milk and the creation of territorial seals, benefiting small producers and cooperatives by accessing market niches with higher added value. The use of smart dashboards can also expand local management capacity, offering real-time reports for decision-making on nutrition, productivity, and distribution logistics, which is especially relevant in rural areas with connectivity challenges. In the Cerrado region of Mato Grosso do Sul, beef cattle farming developed in Guia Lopes da Laguna (MS) is being integrated with digital systems focused on environmental sustainability, with an emphasis on traceability and integrated crop, livestock, and forestry (ILPF). In this case, integrated systems composed of environmental sensors, satellite imagery (Landsat and Sentinel), vegetation maps, and territorial databases allow for tracking the animal's entire life cycle and measuring the environmental performance of properties.

These systems contribute to creating a digital environmental footprint for beef, enabling properties to be certified for best practices such as ILPF, pasture restoration, and maintenance of environmental preservation areas. The integrated data can be organized into interoperable platforms capable of generating automated reports for external audits required by certification agencies, institutional buyers, and international markets. Such systems are essential for integrating livestock farming into global supply chains that prioritize sustainability, traceability, and environmental governance.

The integration of digital systems in both Ingaí and Guia Lopes da Laguna illustrates a key principle of Agriculture 4.0: Technology must serve not only efficiency but also equity and sustainability. In both cases, the focus is not on replacing human judgment, but on augmenting it with data. Moreover, the emphasis on interoperability avoids vendor lock-in and promotes open innovation. This is particularly important for small producers, who are often excluded from proprietary digital ecosystems. The CCD-AD/Semear Digital promotes open standards and public digital infrastructure, positioning the state as a key enabler of digital inclusion.

The cases of Ingaí (MG) and Guia Lopes da Laguna (MS) demonstrate that systemic integration between digital technologies, geospatial data, sensors, and interoperable platforms is crucial to ensuring robust, transparent, and scalable traceability. This is essential not only for certification of origin or sustainability but also for territorial planning strategies, risk mitigation, and evidence-based public policies.

4.3 Artificial Intelligence and Robotics in Boa Vista do Tupim (BA), Breves (PA), Guia Lopes da Laguna (MS), and Vacaria (RS)

The application of artificial intelligence and robotics in Agriculture 4.0 has enabled the development of solutions tailored to the challenges of productivity, traceability, and technical assistance in diverse regions. In apple production in Vacaria, Rio Grande do Sul, the use of computer vision can support phenological monitoring, pest and disease management, and harvest automation. Cameras on mobile robots and drones can be combined with deep learning algorithms to automatically count fruit, classify them by ripeness, and detect physiological anomalies early. This approach can contribute to more accurate agronomic decisions and harvest logistics planning, reducing losses and increasing operational efficiency.

In the honey chain in Breves, Pará, AI can be applied to images obtained by camera-equipped drones to map floristic areas and monitor hive health, contributing to more efficient beekeeping management in difficult-to-access environments. Georeferenced data, processed by artificial neural networks, allow for the prediction of beekeeping productivity and harvest planning, aligning with the strategies for certifying the origin of açaí and wild honey from the Amazon. The same application of AI can be observed in Boa Vista do Tupim, Bahia, where beekeeping is also an important extractive activity. The use of drones for remote inspection of production areas, combined with AI-powered image analysis, facilitates the mapping of vegetation cover and the identification of priority areas for conservation and rational collection.

Another important advancement is the application of AI in digital technical assistance, particularly through the Infobee and Zapbee solutions, initiatives originated from the public–private partnership between Embrapa Amazônia Oriental (PA) and the company Equilibrium Web. These platforms function as intelligent customer service systems via the WhatsApp app, based on natural language processing algorithms, and are being implemented at the ATD in Breves, PA, to expand access to technical information, management protocols, and personalized health alerts. The ATDs of Boa Vista do Tupim (BA) and Guia Lopes da Laguna (MS) will also be able to receive these solutions.

Infobee, geared toward the beekeeping chain, answers questions about hive health, good collection practices, and marketability, while Zapbee focuses on guidance on animal nutrition, traceability, and land regularization. Both systems are capable of learning from interactions and generating aggregated data that assist public administrators and extension workers in decision-making.

These experiences demonstrate how AI and robotics technologies, when integrated with territorial strategies such as ATDs, can increase collective intelligence in the field, promoting digital inclusion and the targeted use of data in family and extractive farming. They also reinforce the importance of investing in digital infrastructure, interoperability between systems, and training local stakeholders, key elements for the sustainability and scalability of these solutions.

5 Discussion: Challenges of Implementing Agriculture 4.0 in ATDs

Agriculture 4.0 represents a new technological paradigm in agriculture, characterized by the intensive incorporation of digital technologies such as Artificial Intelligence, Internet of Things (IoT), and Big Data and Analytics. These innovations promote profound transformations in agricultural production, from planning to marketing, and require overcoming institutional, social, and infrastructural barriers (da Silveira et al. 2021).

Digital transformation in agriculture should be understood as a social process, whose outcomes are not predetermined by the technologies themselves, but rather shaped by the institutional choices and territorial arrangements that guide their implementation (Eastwood et al. 2019). An analysis of opportunities for implementing 4.0 technologies in ATDs in different regions of Brazil highlights the diversity of possible paths for the digital transformation of family-based agriculture and extractivism, especially when these innovations are adapted to local territorial, socioeconomic, and cultural conditions. Section 4 highlights how the use of Big Data, integrated systems, and artificial intelligence has the potential to transform production practices, expand traceability, and strengthen remote technical assistance networks, even in contexts characterized by limited digital infrastructure.

The cases presented reveal a critical insight: The success of Agriculture 4.0 in heterogeneous rural contexts depends not on technological sophistication but on the capacity for cocreation, institutional alignment, and adaptive governance. The ATDs are not merely sites of technology deployment; they function as hybrid spaces where scientific knowledge, traditional practices, and public policy converge to produce context-sensitive innovations. This model challenges the dominant narrative of Agriculture 4.0 as a top-down, capital-intensive revolution, instead positioning it as a socially embedded process of inclusive digital transformation.

Assessing the impacts of Agriculture 4.0 requires indicators that extend beyond productivity or economic efficiency. In the case of ATDs, monitoring systems could incorporate social and environmental metrics capable of capturing the multidimensional nature of digital transformation. Social indicators might include the level of digital inclusion, participation in capacity-building programs, and the distribution of benefits among producers, cooperatives, and local institutions. Environmental indicators could focus on resource-use efficiency, biodiversity conservation, and the adoption of low-emission production practices. Establishing standardized evaluation procedures for these dimensions would allow more systematic comparison across biomes and over time. Integrating such metrics into the operational design of ATDs would contribute to evidence-based policy formulation and support continuous assessment of technology adoption processes.

The application of Big Data and Analytics, as in the case of açaí mapping in Breves, Pará, and licuri in Boa Vista do Tupim, Bahia, reveals how remote sensing and geotechnologies can contribute not only to accurate spatial diagnoses but also to strengthening territorial governance and participatory public policy planning. By integrating environmental, production, and social data, these tools will enable the design of strategies more consistent with local needs and identify opportunities for adding value to sociobiodiversity products (Eastwood et al. 2019). However, these applications also expose a key tension: While satellite imagery can map palm groves, only local communities can validate phenological cycles, harvesting windows, and land use rights. Thus, the integration of traditional ecological knowledge into digital models is not an add-on, but a prerequisite for epistemic justice and long-term sustainability.

In turn, the examples of Integrated Traceability Systems, such as those to be implemented in the dairy chain in Ingaí, Minas Gerais, and the beef cattle chain in Guia Lopes da Laguna, Mato Grosso do Sul, highlight the importance of interoperability between databases and digital platforms, as well as producer adherence to certification protocols and best practices. Traceability emerges, in this context, not only as a market requirement but also as a tool to ensure transparency, sustainability, and recognition of differentiated products. Yet, the risk of "digital greenwashing" remains: Without robust auditing and community oversight, digital traceability can become a performative tool for market access, rather than a genuine mechanism for environmental accountability. The ATD model mitigates this risk by embedding traceability within participatory governance structures, ensuring that data ownership and benefits are shared equitably.

The use of AI and robotics shows promise in both precision fruit farming, such as the use of computer vision in apple production in Vacaria (RS), and extractive systems and technical assistance. The use of artificial intelligence algorithms in beekeeping, as in the case of the digital solutions Infobee and Zapbee, which will be implemented in the ATDs of the Amazon and semiarid regions, points to a new generation of personalized digital services, with great potential for democratizing access to agronomic knowledge. However, these actions also demonstrate that the challenges are not limited to technology. Building local capacity, financially supporting digital systems, interinstitutional data governance, and connectivity infrastructure remain obstacles to consolidating robust digital ecosystems in rural areas.

Moreover, the deployment of AI in low-connectivity environments demands a rethinking of digital architecture: Solutions like Infobee and Zapbee leverage WhatsApp not because it is the most advanced platform, but because it is the most accessible. This "frugal innovation" strategy is a key lesson from the ATDs. It reflects a commitment to digital inclusion that prioritizes usability, affordability, and cultural relevance over technological novelty.

At the same time, integrating emerging technologies with traditional knowledge and community leadership is essential to ensuring social inclusion and digital equity in innovation processes. Despite advances in Agriculture 4.0 research, implementation gaps remain in various production contexts, especially among smallholders. The fourth agricultural revolution demands not only the technical adoption of tools such as Artificial Intelligence and the Internet of Things but also the incorporation of responsible innovation principles, with attention to social inclusion, impact anticipation, and reflexivity in technological development and implementation processes (Rose and Chilvers 2018; Maffezzoli et al. 2022).

The CCD-AD/Semear Digital initiative exemplifies such a responsible innovation framework. By anchoring digital transformation in Agro-Technological Districts—territorially bounded, institutionally supported, and community-engaged spaces—it creates a governance model that balances technological ambition with social accountability. This approach avoids the pitfalls of "techno-solutionism" and instead fosters a coevolutionary process where technology and society shape each other.

In summary, the analysis described in this chapter indicates that the effective adoption of Agriculture 4.0 technologies in ATDs depends both on the integration of digital solutions and local territorial dynamics and on the existence of coordinated public strategies capable of integrating science, technology, and rural extension into collaborative arrangements. The CCD-AD/Semear Digital has demonstrated that this alignment is possible and desirable to build a new, more sustainable, connected, and inclusive agrifood ecosystem.

6 Final Comments

The transition to Agriculture 4.0 in the Agro-Technological Districts (ATDs) supported by the Center for Science for Development in Digital Agriculture (CCD-AD/Semear Digital) illustrates that digital transformation in rural territories must be gradual, participatory, and grounded in the realities of local contexts. Since the beginning of its activities in 2022, the CCD-AD has structured its work around a strategic, phased approach. The initial 2 years were dedicated to understanding the socio-productive specificities of each ATD, with efforts focused on mapping local production systems, identifying technological gaps, engaging stakeholders, and co-constructing agendas with community participation.

This early phase was necessary to establish trust among actors, particularly smallholder farmers, extractivist communities, and traditional populations. It allowed the development of tailored digital strategies based on participatory diagnosis and respect for sociocultural diversity. The data collected and analyzed during this period also served as the empirical foundation for designing scalable solutions aligned with local aspirations and environmental constraints.

The subsequent phase, now underway with continued support from FAPESP, emphasizes the implementation and evaluation of digital technologies in the field. Solutions such as artificial intelligence-based tools for technical assistance, platforms for traceability and certification of origin, geospatial monitoring systems, and digital advisory applications are being deployed in collaboration with Embrapa and other research and extension institutions. These technologies are not introduced in a top-down manner but integrated into territorial governance processes through partnerships with local associations, cooperatives, schools, and rural extension networks.

One of the distinguishing features of the CCD-AD's approach is its commitment to digital inclusion, ensuring that connectivity and access to information are not privileges but rights for all rural actors. Training programs, community-based laboratories, and open digital platforms play a key role in expanding local capacities and facilitating the appropriation of technological solutions by producers themselves.

By linking innovation to inclusive rural development, the CCD-AD demonstrates that the digital transformation of small-scale agriculture is not only possible but also desirable—provided that it is guided by principles of equity, social engagement, and the codesign of technologies. The ATDs emerge not only as spaces for experimentation but also as living laboratories of institutional collaboration, participatory governance, and sustainability-oriented innovation. Their consolidation will contribute to building a more resilient, fair, and connected Brazilian agrifood system.

This chapter has demonstrated that the success of Agriculture 4.0 in diverse and often marginalized territories does not depend solely on the sophistication of its technologies, but on the depth of its social embeddedness. The Agro-Technological Districts represent a paradigm shift: from a technocentric vision of digital agriculture to a socio-technical model where innovation is co-constructed, contextually adapted, and institutionally supported. By integrating advanced tools like AI, IoT, and Big Data with principles of responsible research and innovation (RRI), the CCD-AD/Semear Digital initiative challenges the dominant narrative of the Fourth Agricultural Revolution as a top-down, capital-intensive process, instead positioning it as a democratizing force for rural development.

The cases analyzed, from açaí extractivism in the Amazon to precision apple farming in the South, reveal a common thread: Digital transformation is most effective when it begins with listening, not with deploying. The emphasis on participatory diagnosis, community validation, and low-barrier digital interfaces (such as WhatsApp-based advisory tools) ensures that technology serves people, not the other way around. This "frugal innovation" approach is a critical lesson for other developing countries facing similar challenges of connectivity, equity, and scale.

However, the sustainability of these initiatives depends on long-term public investment, robust data governance frameworks, and the institutionalization of digital extension services. Without continued support, there is a risk of "pilotitis," a proliferation of short-term projects that fail to scale or endure. To avoid this, we recommend that future efforts prioritize (i) the creation of public digital infrastructure for agriculture, (ii) the integration of digital literacy into rural education, (iii) the establishment of data cooperatives to ensure producer sovereignty, and (iv) the replication of the ATD model in other socio-productive contexts.

The analysis presented in this chapter indicates that the digital transformation of agriculture in Brazil remains an uneven and context-dependent process. Future research should advance the understanding of how digital technologies affect the socioeconomic organization of smallholders and traditional communities, particularly in terms of labor relations, income diversification, and access to markets. There is also a need for comparative evaluations of technological efficiency across distinct Brazilian biomes, considering variations in infrastructure, climate conditions, and production systems. Studies adopting mixed-method approaches could contribute to identifying which combinations of technologies and governance arrangements are more compatible with each regional context. In addition, further investigation is required on institutional and policy mechanisms that enable digital inclusion and sustainability. The development of governance models that integrate research institutions, cooperatives, and local governments can support the long-term consolidation of the Agro-Technological Districts (ATDs) as territorial references for inclusive innovation. Such investigations would strengthen the empirical basis for designing public strategies aligned with social and environmental objectives.

The ATD framework offers a structured approach for the territorial implementation of digital agriculture initiatives. Its expansion to other regions depends on the capacity to adapt technological, institutional, and social components to distinct local conditions. Replication should consider differences in biome characteristics, production systems, and existing organizational networks. Scalability can be promoted through the integration of the ATD model with public programs and cooperative systems, encouraging participation of private technology developers under transparent governance arrangements. Approaches based on frugal innovation—focused on accessibility, modularity, and cost reduction—may facilitate implementation in regions with limited infrastructure or connectivity. Documenting and disseminating lessons from existing ATDs would assist in refining methodologies and establishing parameters for their replication. This process would also contribute to comparative studies on the territorialization of digital agriculture in diverse socioeconomic and environmental contexts.

In conclusion, the Agro-Technological Districts offer a replicable blueprint for inclusive digital agriculture—one that balances technological ambition with social accountability, and efficiency with equity. As global food systems face unprecedented pressures from climate change, biodiversity loss, and inequality, the Brazilian experience with the CCD-AD/Semear Digital provides a timely and compelling model for how Agriculture 4.0 can be reimagined not as a technological inevitability but as a collective, ethical, and place-based endeavor.

Competing Interests The author(s) has no competing interests to declare that are relevant to the content of this manuscript.

References

Abbasi R, Martinez P, Ahmad R (2022) The digitization of agricultural industry—a systematic literature review on agriculture 4.0. Smart Agric Technol 2:100042

Albiero D, Paulo RLD, Félix JC, Santos JDSG, Melo RP (2020). Agriculture 4.0: a terminological introduction. Revista Ciência Agronômica 51(spe):e20207737

Bechar A, Vigneault C (2016) Agricultural robots for field operations: Concepts and components. Biosyst Eng 149:94–111

Benos L, Tagarakis AC, Dolias G, Berruto R, Kateris D, Bochtis D (2021) Machine learning in agriculture: a comprehensive updated review. Sensors 21(11):3758

Berthet ETA, Hickey GM, Klerkx L (2018) Opening design and innovation processes in agriculture: insights from design and management sciences and future directions. Agric Syst 165:111–115

Bui HT, Aboutorab H, Mahboubi A, Gao Y, Sultan NH, Chauhan A, Parvez MZ, Bewong M, Islam R, Islam Z, Camtepe SA, Gauravaram P, Singh D, Babar MA, Yan S (2024) Agriculture 4.0 and beyond: evaluating cyber threat intelligence sources and techniques in smart farming ecosystems. Comput Secur 140:103754

Eastwood C, Klerkx L, Ayre M, Dela Rue B (2019) Managing socio-ethical challenges in the development of smart farming: from a fragmented to a comprehensive approach for responsible research and innovation. J Agric Environ Ethics 32:741–768

Furuya DEG, Bolfe ÉL, Parreiras TC, Barbedo JGA, Santos TT, Gebler L (2024) Combination of remote sensing and artificial intelligence in fruit growing: progress, challenges, and potential applications. Remote Sens 16(24):4805
Instituto Brasileiro de Geografia e Estatística (IBGE). *Cidades e Estados*. Disponível em: https://www.ibge.gov.br/cidades-e-estados.html. Acesso em: 28 Jul. 2025
Kamilaris A, Kartakoullis A, Prenafeta-Boldú FX (2017) A review on the practice of big data analysis in agriculture. Comput Electron Agric 143:23–37
Kumar A, Mangla SK, Kumar P (2024) Barriers for adoption of Industry 4.0 in sustainable food supply chain: a circular economy perspective. Int J Product Perform Manag 73(2):385–411
Lezoche M, Hernandez JE, Diaz MDMEA, Panetto H, Kacprzyk J (2020). Agri-food 4.0: a survey of the supply chains and technologies for the future agriculture. Comput Ind 117:103187
Liu Y, Ma X, Shu L, Hancke GP, Abu-Mahfouz AM (2021) From Industry 4.0 to Agriculture 4.0: current status, enabling technologies, and research challenges. IEEE Trans Industr Inform 17(6):4322–4334
Maffezzoli F, Ardolino M, Bacchetti A, Perona M, Renga F (2022) Agriculture 4.0: a systematic literature review on the paradigm, technologies and benefits. Futures 142:102998
Patrício DI, Rieder R (2018). Computer vision and artificial intelligence in precision agriculture for grain crops: a systematic review. Comput Electron Agric 153:69–81
Pivoto D, Barham B, Waquil PD, Foguesatto CR, Corte VFD, Zhang D, Talamini E (2019) Factors influencing the adoption of smart farming by Brazilian grain farmers. Int Food Agribus Manage Rev 22(4):571–588
Rose DC, Chilvers J (2018) Agriculture 4.0: broadening responsible innovation in an era of smart farming. Front Sustainable Food Syst 2:87
Rose DC, Wheeler R, Winter M, Lobley M, Chivers C-A (2021) Agriculture 4.0: making it work for people, production, and the planet. Land Use Policy 100:104933
Scuderi A, La Via G, Timpanaro G, Sturiale L (2022) The digital applications of "Agriculture 4.0": strategic opportunity for the development of the Italian citrus chain. Agriculture 12(3):400
Silva NRF, Chaves MED, Luciano ACS, Sanches ID, Almeida CM, Adami M (2024) Sugarcane yield estimation using satellite remote sensing data in empirical or mechanistic modeling: a systematic review. Remote Sens 16(5):863
da Silveira F, Lermen FH, Amaral FG (2021) An overview of agriculture 4.0 development: systematic review of descriptions, technologies, barriers, advantages, and disadvantages. Comput Electron Agric 189:106405
Sishodia RP, Ray RL, Singh SK (2020). Applications of remote sensing in precision agriculture: a review. Remote Sens 12(19):3136
Tzounis A, Katsoulas N, Bartzanas T, Kittas C (2017) Internet of Things in agriculture, recent advances and future challenges. Biosyst Eng 164:31–48
Uztürk D, Büyüközkan G (2024) Industry 4.0 technologies in smart agriculture: a review and a technology assessment model proposition. Technol Forecast Soc Chang 208:123640
Wolfert S, Ge L, Verdouw C, Bogaardt MJ (2017) Big data in smart farming—a review. Agric Syst 153:69–80
Zhong RY, Xu X, Klotz E, Newman ST (2017) Intelligent manufacturing in the context of industry 4.0: a review. Engineering 3(5):616–630

Implementation, Potentials, and Challenges of Agriculture 4.0 Technologies: A Case Study in Agrotechnological Districts of the State of São Paulo, Brazil

Édson Luis Bolfe, Patricia Menezes Santos, Katia de Lima Nechet, Gustavo Bayma, and Jener Fernando Leite de Moraes

1 Introduction

The Center for Science for Development of Digital Agriculture (Semear Digital Center) is a national research initiative established to enhance agricultural productivity and sustainability through the application of digital technologies and connectivity solutions. The program specifically emphasizes inclusive technological adoption, including small-to-medium producers within its innovation framework. Semear Digital Center operates under the coordination of the Brazilian Agricultural Research Corporation (Embrapa) through a collaborative consortium of multiple Research, Development, & Innovation institutions, with primary funding provided by the São Paulo Research Foundation (FAPESP) (Semear Digital Center 2024).

In this context, research priorities have been established focusing on digital technologies, including novel methodologies, advanced algorithms, and prototype systems for artificial intelligence applications, remote sensing, automation technologies, and precision agriculture solutions. Additional research domains encompass digital certification systems, blockchain-based traceability platforms, and the development of robust connectivity solutions tailored for agricultural environments. The multidisciplinary and multi-institutional research, development,

É. L. Bolfe (✉)
Embrapa Digital Agriculture, Brazilian Agricultural Research Corporation, Campinas, Brazil
e-mail: edson.bolfe@embrapa.br

P. M. Santos
Embrapa Livestock, Brazilian Agricultural Research Corporation, São Carlos, Brazil

K. de Lima Nechet · G. Bayma
Embrapa Environment, Brazilian Agricultural Research Corporation, Jaguariúna, Brazil

J. F. L. de Moraes
Instituto Agronômico de Campinas, Campinas, Brazil

F. da Silveira et al. (eds.), *Agriculture 4.0*,
https://doi.org/10.1007/978-3-032-24552-6_22

and innovation (RD&I) activities will be implemented in operational agricultural settings, specifically targeting small- and medium-scale production systems. These interventions will be conducted within designated Agrotechnological Districts (ATD) located across ten municipalities spanning multiple Brazilian states. Each ATD was delineated based on multiple indicators encompassing access to technical assistance, land tenure structure, infrastructure, economic activity, education, population, and social development (Paiva et al. 2024).

This chapter examines implemented digital technologies, explores potential applications, and analyzes key challenges associated with agricultural digital transformation within ATDs of São Paulo state: Alto Alegre, Caconde, Jacupiranga, Lagoinha, and São Miguel Arcanjo (Fig. 1). Bolfe et al. (2020) conceptualize digital agriculture as an integrated technological framework encompassing communication systems, information technologies, and analytical tools that enable comprehensive management of agricultural production systems in Brazil. This framework supports operational and strategic decision-making across all production phases: preproduction, production, and postproduction. These authors identify several core technologies currently adopted by Brazilian agricultural producers, including connectivity, mobile applications, digital platforms, geospatial databases, software, global satellite positioning systems, remote sensing, and field-based sensor networks. This pattern of technological adoption is reflected in international initiatives. A case in point is the European Union (EU), where the integration of digital technologies, including satellite positioning, field sensors, and digital platforms, has yielded positive impacts on agricultural productivity. Nonetheless, outcomes are not uniform; significant variation is observed between countries due to differing climate, soil, and crop conditions (Bocean 2024). In Asia, China's digital agriculture initiatives are expanding rapidly, employing smart farming technologies, automation, and data-driven systems to enhance productivity and sustainability (Wu et al. 2024). Conversely, adoption in the United States focuses on precision farming, sensors, digital platforms, and data analytics to achieve similar goals. However, its progression is hampered by high costs, limited rural connectivity, and cultural barriers, leading to uneven adoption where large-scale farms lead while smaller operations face significant challenges (McFadden 2023). In South America, Puntel et al. (2022) analyzed adoption patterns, revealing that while experimentation with Agriculture 4.0 technologies is increasing, significant challenges persist in infrastructure, technical support, and governance. These global and regional experiences provide useful benchmarks for contextualizing the case of Semear Digital Center in São Paulo State within the broader narrative of agriculture's digital transformation.

The 2023 agricultural production analysis in the municipalities of Alto Alegre, Caconde, Jacupiranga, Lagoinha, and São Miguel Arcanjo highlights both the diversity and specialization of agricultural activities within these regions (IBGE 2023), and all selected municipalities are situated within the Atlantic Forest biome. In Alto Alegre (SP), the main products in absolute terms are sugar cane (835,760 tons), peanuts in shell (3750 tons), and sweet potatoes (3360 tons). In terms of relative contribution to the total production of the State of São Paulo, sweet potatoes

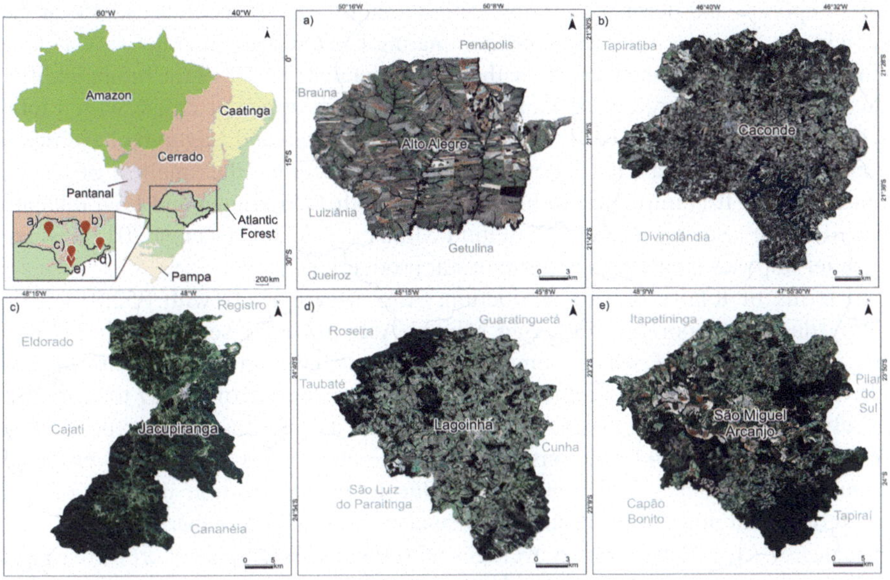

Fig. 1 Agrotechnological Districts (ATDs) map in Brazilian biomes and São Paulo state, Brazil, in the Semear Digital Center. (**a**) Alto Alegre, (**b**) Caconde, (**c**) Jacupiranga, (**d**) Lagoinha, (**e**) São Miguel Arcanjo. Sentinel−2 Satellite Image Composition Red (4)/Green (3)/Blue (2), June 2023 to September 2024

stand out, accounting for 2.12% of the State's production, followed by watermelons (0.72%) and bananas (0.57%).

In the municipality of Caconde (SP), the main products are Arabica coffee (13,500 tons), soybeans (3120 tons), and onions (300 tons). In terms of total production in the state, Arabica coffee is the highlight, contributing 4.39%, followed by onions (0.16%) and soybeans (0.06%). In Jacupiranga (SP), bananas (80,001 tons), corn (140 tons), and palm hearts (880 tons) are the most important products. Bananas stand out, accounting for 8.20% of the state's production, followed by palm hearts (2.26%) and passion fruit (0.17%) (IBGE 2023).

In Lagoinha (SP), dairy milk is the primary agricultural product, followed by corn (1040 tons), passion fruit (48 tons), and coconut (35 tons). In terms of relative contribution to the state's production, coconut accounts for 0.29%, followed by passion fruit (0.17%) and mandarins (0.02%). Finally, São Miguel Arcanjo (SP) presents a diversified agricultural production, with corn (62,500 tons), potatoes (26,320 tons), and tomatoes (21,600 tons) being the leading products. In relation to the total production of the state, persimmons stand out with 21.13%, followed by potatoes (3.47%) and tomatoes (2.49%) (IBGE 2023).

According to the Municipal Agricultural Survey 2023 (PAM), conducted by the Brazilian Institute of Geography and Statistics (IBGE), São Miguel Arcanjo represents the Agrotechnological District (ATD) with the largest cultivated area of both temporary and permanent crops, with a total of 30,130 hectares (32.4% of the municipality), with temporary crops dominating (26,680 ha). In contrast, Caconde stands out for its permanent crops, with 20,022 hectares (42.7% of the municipality). Alto Alegre shows a relative balance between the two types of crops, with a total of 12,992 hectares (40.7% of the total area). Jacupiranga and Lagoinha have much smaller cultivated areas, the latter with only 181 hectares, the smallest of all the municipalities. It is important to highlight that both Jacupiranga and Lagoinha are located in the "Serra do Mar" (Prefeitura Municipal de Jacupiranga 2017), where the relief imposes challenges to agricultural production.

In terms of total cropland production, Alto Alegre leads with 852,014 tons, mainly due to temporary crops. São Miguel Arcanjo ranks second with 271,413 tons, about a third of which comes from permanent crops. Despite its limited cultivated area, Jacupiranga achieves a substantial production of 81,427 tons, almost entirely from permanent crops. Caconde, despite its large area under permanent crops, produces only 44,784 tons. Lagoinha has the lowest agricultural cropland production with only 1809 tons, in part due to its localization in areas of low soil agricultural suitability. Productivity analysis reveals the superior overall performance of Alto Alegre, which averages 65.6 t/ha, with temporary crops reaching 68.5 t/ha. Jacupiranga shows remarkable permanent crop productivity (24.9 t/ha) despite its limited area. São Miguel Arcanjo shows moderate overall cropland productivity (9.0 t/ha), but achieves respectable permanent crop yields (26.5 t/ha). Caconde has the lowest overall cropland productivity (2.1 t/ha), reflecting poor permanent crop performance. Lagoinha maintains an average cropland productivity (10.0 t/ha), in line with its small agricultural scale.

In this chapter, we present some of the digital applications, potential, and challenges involving topics: (i) Analysis of Regional Agricultural Dynamics, (ii) Pasture Diagnosis and Monitoring, (iii) Diagnosis and Monitoring of Plant Diseases, and (iv) Hydrological Modeling and Water Resources Management.

2 Applications and Potential of Digital Technologies

2.1 Analysis of Regional Agricultural Dynamics

Satellite images and multiscale databases expand the ability to map and monitor agricultural systems. Thus, the spatial analysis of regional agricultural dynamics plays a strategic role in land use and farm planning for a more sustainable rural development. Parreiras et al. (2025) point out that changes in land use and cover have significant impacts on landscape configuration, climate change, and society. The processes of expansion, conversion, intensification, diversification, and contraction materialize these changes in the agricultural environment. This analysis examined changes in land cover and land use in the São Paulo state ATDs from 1985 to 2022. It identifies trends in forest regeneration, pasture decline, and agricultural land use shifts, along with variations in secondary vegetation, forestry plantations, and aquatic ecosystems (Fig. 2).

The characterization of the predominant land use and land cover classes was derived from the MapBiomas project data, Collection 8 (Souza Jr et al. 2020). Notably, the "mosaic of occupation" class within the MapBiomas classification system represents agricultural areas where discrimination between pasture and cropland was not feasible, and is consequently classified as an agricultural land use category.

Among the São Paulo municipalities, Jacupiranga demonstrated the most substantial agricultural expansion at 6128.7 hectares (8.7% of the municipality), while Caconde and São Miguel Arcanjo had expansions of 2702.8 hectares (15.8% of the municipality) and 5122.9 hectares (5.5% of the municipality), respectively. While comparatively smaller than other regions, these increases nevertheless reflect agricultural intensification. In Lagoinha and Alto Alegre, the expansion was more modest, with 1018.4 hectares (4% of the municipality) and 544.1 hectares (1.7% of the municipality), respectively, which may indicate a controlled or limited expansion due to local factors. In Lagoinha, for example, the predominance of steep terrains probably limits the expansion of agriculture.

Analysis of agricultural contraction revealed significant spatial variability across municipalities. Jacupiranga (SP) exhibited the most pronounced reduction (10,015.2 hectares, 14.2% of municipal area), indicating substantial land use conversion. Caconde (SP) and São Miguel Arcanjo (SP) showed comparable proportional declines, 3134.3 hectares (6.7% of the municipality) and 6219.9 hectares (6.7% of the municipality). In Lagoinha (SP) and Alto Alegre (SP), contraction was 1579.1 hectares (6.2% of the municipality) and 732.1 hectares (2.3% of the municipality), respectively, reflecting moderate decreases in agricultural areas.

2.2 *Diagnosis and Monitoring of Pastures*

Brazil is the second largest producer and the largest exporter of beef in the world, with approximately 10.6 million tons of carcass equivalent produced, representing 13.8% of world beef production in 2023 (Abiec 2024). Dairy production is also an important economic and social activity, generating employment for millions of people. It is present in nearly 98% of Brazilian municipalities and plays a crucial economic role in small- and medium-sized towns throughout the country (Brazil 2024).

Beef and dairy production in Brazil are closely linked to pastureland. The country has about 160 million hectares of pasture, with significant regional differences. Natural pastures cover about 50 million hectares and are concentrated in the Caatinga, Pampa, and Pantanal biomes. Cultivated pastures cover about 110 million hectares and are found in all regions, with forage species varying according to local climate and soil conditions (De Mori et al. 2024).

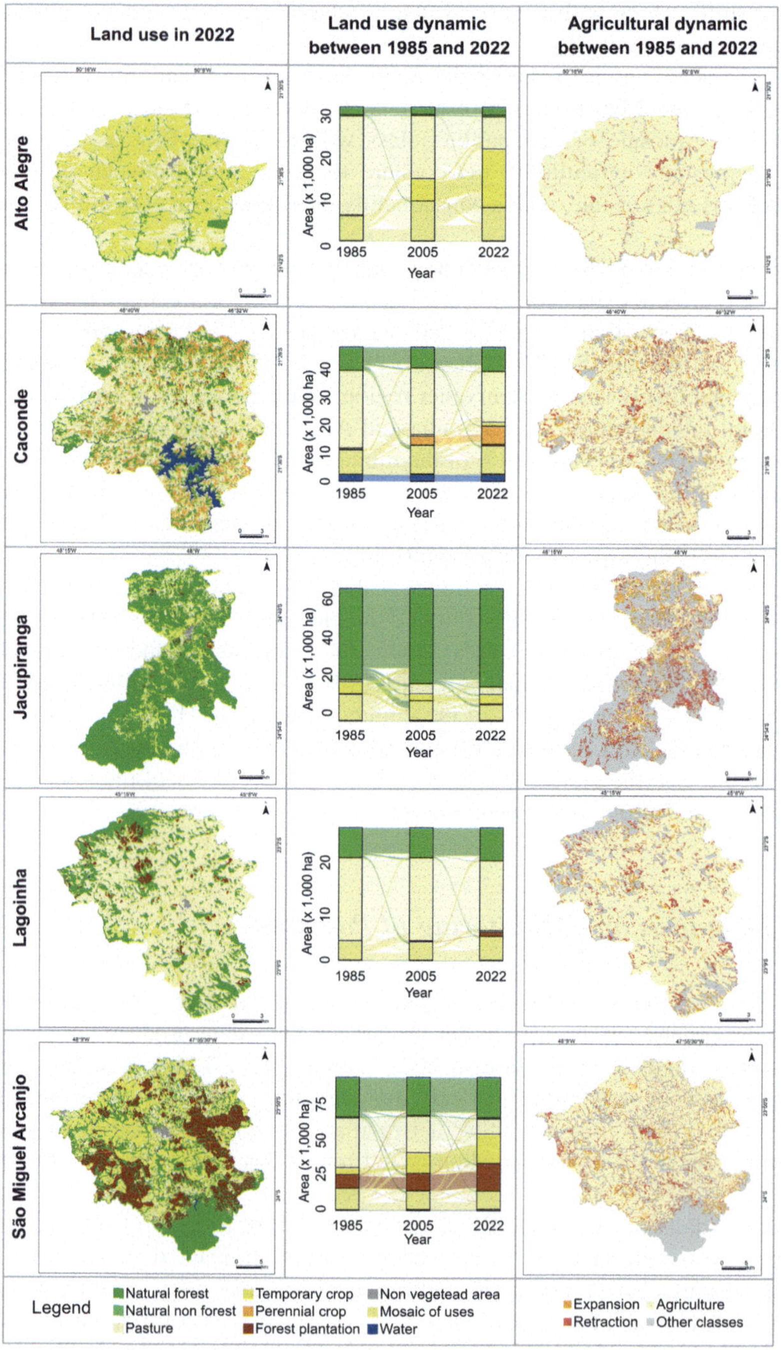

Fig. 2 Agrotechnological Districts (ATDs) map in São Paulo state, Brazil, in the Semear Digital Center. (**a**) Alto Alegre, (**b**) Caconde, (**c**) Jacupiranga, (**d**) Lagoinha, (**e**) São Miguel Arcanjo—Land Use (2022), Land Use and Agricultural Dynamic (1985/2022)

Monitoring grasslands in Brazil is a major challenge due to their vast extent, the diversity of forage species, and the unique characteristics of different biomes and production systems (Silva et al. 2024; Bayma et al. 2025; Parreiras et al. 2025). Current monitoring methods rely on indicators related to agronomic aspects and pasture productivity, such as stocking rate, carrying capacity (dos Santos et al. 2024), weed presence, bare soil exposure (Amorim et al. 2020), and pasture vigor (Santos et al. 2022). These indicators are assessed either through temporal monitoring—detecting increasing or decreasing trends—or through spatial comparisons with reference areas in good condition.

Data organization, integration, and analysis are becoming increasingly important to generate more detailed, qualified, and accurate information to support sustainable rural development planning. The use of remote sensing for pasture monitoring and productivity estimation contributes significantly to decision-making for sustainable intensification of pasture use or conversion to agricultural systems. In the central region of Brazil, yield gaps are associated with both primary productivity and pasture carrying capacity, and are mainly influenced by rainfall and temperature variability (dos Santos et al. 2024). In addition, Bolfe et al. (2024) identified approximately 28 million hectares of degraded pasture suitable for agricultural conversion. Such expansion would help meet global food demand without further deforestation and could mitigate soil erosion associated with pasture degradation. In Fig. 3, the potential for agricultural expansion in degraded pasture lands is shown for two ADTs in the state of São Paulo. Alto Alegre (SP) presents 41 ha (80.7%) classified as "severe/good," while 9.8 ha (19.3%) were classified as "intermediate/good." There are no areas classified as "severe/very good" or "intermediate/very good" (Fig. 3). In the municipality of São Miguel Arcanjo (SP), the majority of the area (87.1%) was classified as "intermediate/good," corresponding to 687 ha, while 102 hectares (12.9%) were classified as "severe/good," totaling 789 ha.

Recent evidence highlights the importance of yield gap analysis as a tool to guide sustainable intensification strategies in pasture-based systems. According to a comprehensive review by Santos et al. (2025), assessing yield gaps—defined as the difference between current and maximum achievable yields—allows the identification of regions with the highest potential for productivity gains. The study outlines several methodological approaches, such as benchmarking, climate binning, frontier methods, and production systems models, each of which is suited to different spatial scales, data availability, and research objectives. However, most existing models still lack the capacity to capture key aspects of pasture dynamics, such as grazing management, selective animal behavior, and forage quality. Despite these limitations, integrating yield gap assessments into decision-making processes can significantly improve land use efficiency, reduce pressure on native vegetation, and support food security goals through better-targeted public and private interventions.

To support these advances, the Semear Digital Center is developing a field image database aimed at improving remote sensing methods for pasture diagnosis and monitoring. This initiative aims to improve the accuracy and scalability of

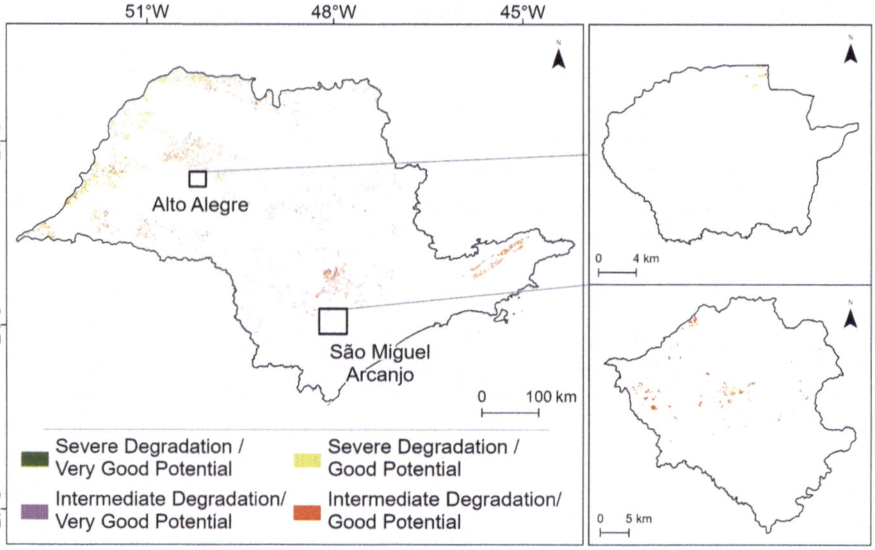

Fig. 3 Potential for Agricultural Expansion in Degraded Pasture Lands in São Paulo State, Agrotechnological Districts (ATDs)/Semear Digital Center—Alto Alegre and São Miguel Arcanjo

monitoring systems, providing robust tools for researchers, policymakers, and producers in the transition to more sustainable land use practices.

2.3 Diagnosis and Monitoring of Plant Diseases

The crop systems in the ATDs are represented by important cultures to Brazilian agriculture and comprise the main agricultural products of São Paulo State. The crops selected for agronomic studies in each ATD are: sugarcane (*Saccharum officinarum*) in Alto Alegre, coffee (*Coffea arabica*) in Caconde, grapevine (*Vitis vinifera*) in São Miguel Arcanjo—the largest grape-producing municipality in São Paulo—and banana plant (*Musa* spp.) in Jacupiranga, located in Ribeira Valley, which, along with four other municipalities, accounts for 70% of Brazil's total banana-production area. The production value of these crops is approximately $100 billion per year, with small- and medium-sized producers playing an important role in sustaining these food supply chains (IBGE 2023).

Among the current problems associated with agricultural production in each ATD, the incidence of downy mildew in grapevines in São Miguel Arcanjo and black Sigatoka in banana plants in Jacupiranga were particularly highlighted by producers as the main factors compromising productivity.

Downy mildew, caused by *Plasmopara viticola*, is one of the most significant diseases of grapevine in Brazil affecting all parts of the plant, especially the leaves,

Fig. 4 (**a**). Symptoms of downy mildew, caused by *Plasmopara viticola*, on grapevine leaf, (**b**). Symptom of black Sigatoka, caused by *Mycosphaerella fijiensis*, on a banana leaf, and (**c**) Area of banana plantation in Jacupiranga-SP. (Photos: Katia de Lima Nechet)

which develop angular or yellowish lesions with a white-cottony growth of the pathogen on the undersides (Fig. 4a). Moreover, the disease causes organoleptic defects in fruits and their derivatives (Camargo et al. 2019). The resistance of most grape cultivars is moderate or highly susceptible, and the use of fungicides is the primary method of control.

Black Sigatoka, caused by *Mycosphaerella fijiensis*, is considered the most destructive banana plant disease worldwide, causing leaf spots that progress to early death of the leaves (Fig. 4b), defoliation, and premature fruit ripening (Marin et al. 2003). Although resistant banana cultivars are available, producers in Jacupiranga typically do not adopt them. Furthermore, the presence of extensive and continuous areas of banana plantations also contributes to constant production and dispersal of fungal inoculum, resulting in uninterrupted cycles of the disease (Fig. 4c). Eventually, only big-sized producers can afford successive aerial fungicide sprays in an attempt to reduce productivity losses.

The management of plant diseases involves a series of strategies for protecting crops from losses, including prevention, diagnosis, monitoring, and adoption of control methods. Innovative methods based on digital technology, such as optical sensors, robotics, and computer vision, can provide real-time responsiveness and precision to monitor and manage plant diseases (Mahlein et al. 2024; Upadhyay et al. 2025). In ATDs, the use of digital technologies by producers must be simple and smartphone-assisted. In this sense, the studies conducted under the Semear Digital Center have focused on developing digital solutions for diagnosing and monitoring the progression of plant disease and environmental conditions.

Accurate diagnosis is the foundation of assertive decision-making control, not only for crops with well-documented diseases, such as banana plants and grapevines, but also for crops with limited information, such as peach palm (*Bactris gasipaes*). Peach palm is cultivated in Jacupiranga for the production of heart of palm as an alternative to the illegal extraction of Juçara palm heart (*Euterpe edulis*), a native species of the Atlantic Forest. Images of peach palm diseases are scarce and often restricted to scientific literature. To support these diagnoses, a database containing real-time images of disease symptoms in these crops has been recorded. This database can be used to support a mobile application that correlates symptoms, pathogen, and control methods, providing a simple and reliable decision-making tool. Additionally, the database can be used to design image processing algorithms for detecting diseases in banana, grape, and peach palm crops.

Another approach is the monitoring of black Sigatoka progress. The current method is based on tracking symptom development stages and foliar emission rate to calculate the disease's evolution state. This methodology was adapted from Fouré (1988) and modified by Moraes et al. (2011), who developed a computer application using an MS Excel spreadsheet to estimate the evolution state on a weekly basis. This technique is considered an effective tool for determining the severity and progress of black Sigatoka. However, its adoption has been limited to a restricted group in the Ribeira Valley, including big-sized farmers and research institutions. Furthermore, it required skills to identify the early disease stages using magnifying glasses (10X–20X).

In order to increase its adoption by producers, some studies have been conducted to simplify the early identification process. Among them is a software tool developed by Silva (da Silva 2008), which uses digital image processing, principal component analysis, and artificial neural networks; however, the data obtained from this software was not correlated with that from Moraes' methodology. Therefore, temporal analysis of black Sigatoka progress will be conducted in banana plantations in Jacupiranga using both monitoring methodologies to validate this digital technology and make it a reliable decision-making support tool. Additionally, monitoring data combined with environmental conditions could be used in the future to develop a forecasting system for black Sigatoka.

Other opportunities for implementing agriculture 4.0 technologies in ATDs include the use of remote sensing and unmanned aerial vehicles for detecting, monitoring, and controlling plant diseases. The unmanned aerial vehicles have gained attention for their efficiency in spraying pesticides and for their capacity to cover hilly areas, common in the Ribeira Valley, where conventional aerial fungicide spray applications often fail to reach.

2.4 *Hydrological Modeling and Water Resources Management*

The ongoing pursuit of development, coupled with rapid population growth, places significant pressure on available natural resources. In this context, land use changes—primarily driven by urban expansion and agricultural activities—have notable impacts on water resources within river basins, directly affecting both water quality and availability (Mello et al. 2020).

Another issue to be analyzed is the alteration of land cover resulting from agricultural expansion, which also directly affects the climate and water availability (Duveiller et al. 2020). Water resources undergo numerous changes over the years, either through natural processes or human-induced actions. In this context, agricultural activities and urbanization stand out for causing a series of negative environmental impacts, contributing to degradation and changes in water quality, and directly affecting the hydrological system (Mello et al. 2020).

São Paulo State is currently the leading producer of five major agricultural commodities in Brazil, particularly sugarcane, oranges, coffee, corn, and peanuts. Sugarcane is the most cultivated crop in the country and the most significant in São Paulo, which alone accounts for 420 million tons harvested over five million hectares.

One of the main challenges facing agricultural expansion in São Paulo is aligning economic development with environmental preservation, particularly in a dynamic context marked by technological change and innovation. These technologies must become increasingly accessible across the agricultural sector. In this scenario, digital technologies and precision agriculture tools are essential, as they enable greater efficiency and productivity in water use, minimizing losses and ensuring the sustainable supply of food with reduced environmental impacts (Embrapa 2014, 2023; de Araujo et al. 2020).

Technological innovation in the agricultural sector is key for Brazil to enhance its production capacity while increasing efficiency in a sustainable manner—economically, environmentally, and socially. Balancing these demands while safeguarding water availability for both agricultural and human consumption is a central objective of modern agriculture. In this context, the Semear Digital Center includes among its research lines the assessment of land use change impacts on water availability across three river basins in São Paulo State. These study areas correspond to the ATDs of Alto Alegre, Jacupiranga, and Lagoinha.

The evaluation of land use change impacts on water availability is being conducted using hydrological modeling techniques. These models simulate the behavior of the hydrological cycle within watersheds and are aimed at understanding water dynamics, including precipitation, evaporation/evapotranspiration, surface runoff, and infiltration. Hydrological models are essential tools for supporting decision-making processes and for analyzing the functioning of the hydrological cycle in diverse scenarios (Bressiani et al. 2015).

With advances in geospatial technologies, most contemporary hydrological models are integrated with Geographic Information Systems (GIS), which enhance the physical realism of simulations by spatially distributing model parameters across the landscape (Andrade et al. 2013). Among the various hydro-sedimentological models available, the Soil and Water Assessment Tool (SWAT), developed by the Agricultural Research Service of the United States Department of Agriculture (ARS-USDA), stands out for its global applicability and scientific validation (Arnold et al. 1998; Gassman et al. 2007; Neitsch et al. 2011; Bressiani et al. 2015). SWAT is widely used to simulate the hydrological impacts of land use and management practices in watershed-scale studies. It is a public-domain software designed to assess the long-term impacts of land use changes on water, sediment, nutrient, and agrochemical yields under varying scenarios (Gassman et al. 2007). Its integration with GIS platforms makes it a powerful tool for investigating watershed-scale environmental processes.

As part of the Semear Digital Center, a geospatial database is being developed for each of the selected watersheds, incorporating information on topography, land use and cover, soil types, and meteorological variables. These datasets will be input into the SWAT model via a GIS interface. Land use and cover scenarios for the years 1962, 2002, and 2022 will be analyzed to estimate water yield (streamflow, surface runoff, soil infiltration) and sediment yield across the ATDs of Alto Alegre, Jacupiranga, and Lagoinha. The resulting simulations will provide critical insights for hydrological characterization, support future scenario analysis, and guide the development of more effective and sustainable water management policies and strategies.

3 Future Research Agenda

Despite significant progress in implementing Agriculture 4.0 technologies within São Paulo State's Agrotechnological Districts, relevant research gaps persist. As emphasized in Embrapa's Vision 2030 (Embrapa 2023), emerging technologies, such as IoT, AI, machine learning, robotics, and blockchain, are driving profound digitalization across agriculture, which enhances productivity while reducing environmental impacts. These gaps underscore the need for interdisciplinary studies to guide a more inclusive, sustainable, and regionally adapted transformation in agriculture.

The impact of digitalization on rural succession remains an under-explored aspect of Agriculture 4.0. While ostensibly attractive to a new generation by integrating farming with technology and innovation, there is scant empirical evidence that digital tools actually mitigate generational issues, foster youth entrepreneurship, or secure the continuity of family farms. Future research must move beyond adoption rates to investigate the social, cultural, and psychological factors influencing young farmers' engagement with digital tools.

Brazil's agricultural digital transformation risks exacerbating existing inequalities if its benefits concentrate among large-scale producers or in well-served regions. Family farmers, a primary focus of the Semear Digital Center, are particularly vulnerable to exclusion due to limited access to connectivity, capital, and technical training. Urgent research is needed to quantify these risks, design inclusive innovation policies, and validate democratizing strategies like cooperative platforms, shared services, and public–private partnerships.

Although numerous prototypes and pilot studies show promise, the scalability, cost-effectiveness, and resilience of Agriculture 4.0 technologies remain unproven under real-world agricultural production conditions in Brazil. Future investigations should therefore track adoption over time, quantify impacts on productivity and sustainability, and assess unintended consequences, including labor market shifts and knowledge dependencies.

While this chapter's analysis is restricted to the Atlantic Forest biome in São Paulo State, Brazil's agricultural landscapes are highly heterogeneous, encompassing the Amazon, Cerrado, Caatinga, Pantanal, and Pampa. Each biome presents unique ecological, climatic, and socioeconomic conditions that will inevitably influence the implementation and outcomes of Agriculture 4.0. Therefore, expanding comparative cross-biome research is crucial for identifying context-specific opportunities, constraints, and risks, ensuring digital solutions are adapted rather than uniformly applied.

In summary, advancing this research agenda will elucidate the full transformative potential and limitations of digital agriculture in São Paulo and beyond. Such knowledge is essential to ensure that Agriculture 4.0 contributes not only to productivity but also to social inclusion, ecological resilience, and intergenerational continuity in Brazilian farming systems.

4 Final Considerations

This chapter has reviewed key methodological approaches in regional agricultural systems analysis, pasture diagnosis and monitoring, plant pathology diagnosis and surveillance, and hydrological modeling integrated with water resource management in the context of the "Center for Science for Development of Digital Agriculture" in the State of São Paulo, Brazil. Globally, regions including the European Union, the United States, Asia, and other parts of South America show parallel trends in adopting smart farming technologies aimed at increasing productivity and promoting sustainability.

We understand that digital technologies have potential, allowing us to support technicians and farmers in the process of planning, monitoring, and managing the operational and strategic activities of agricultural production systems, from preproduction, during production, and even postproduction. However, there are still a number of challenges to be overcome for greater technological advances and economic, social, and environmental sustainability. In addition to improving rural connectivity infrastructure and the availability of specific financing, it is essential to generate new training actions in Agriculture 4.0 for a multidisciplinary audience that includes agricultural researchers, policymakers, agrotechnology entrepreneurs, academics, and rural development professionals.

It is suggested that future studies related to digital technologies focus on new research questions, more specifically on the social and economic behavior of farmers in relation to the adoption of new technologies and in the context of public policies and support for digital transformation. Future research can also consider how the adoption of digital technologies will occur by the new generation, according to the level of access to connectivity in rural areas and the expectation that they will continue to manage rural properties.

Acknowledgments São Paulo Research Foundation (FAPESP) for funding projects #2022/09319-9 and to the National Council for Scientific and Technological Development (CNPq) for the research productivity fellowship grants (#302963/2025-1/Édson Luis Bolfe and #305187/2022-8/Kátia de Lima Nechet).

Competing Interests The author(s) has no competing interests to declare that are relevant to the content of this manuscript.

References

ABIEC (2024) Beef Report 2024. Available at: https://www.abiec.com.br/exportações/. Accessed 15 Apr 2025

Amorim HCS et al (2020) Soil quality indices following long-term conservation pasture management practices. Agric Ecosyst Environ 301:107060. https://doi.org/10.1016/j.agee.2020.107060

Andrade MA et al (2013) Simulação hidrológica em uma bacia hidrográfica representativa dos Latossolos na região Alto Rio Grande, MG. Rev Bras Eng Agríc Ambiental 17:69–76. https://doi.org/10.1590/S1415-43662013000100010

Arnold JG et al (1998) Large area hydrologic modeling and assessment part I: model development. J Am Water Resour Assoc 34:73–89. https://doi.org/10.1111/j.1752-1688.1998.tb05961.x

Bayma G et al (2025) Estimating forage mass in Brazilian pasture-based livestock production systems through satellite and climate data integration. Comput Electron Agric 237A:1–15. https://doi.org/10.1016/j.compag.2025.110496

Bocean CG (2024) A cross-sectional analysis of the relationship between digital technology use and agricultural productivity across EU countries. Agriculture 14(4):519. https://doi.org/10.3390/agriculture14040519

Bolfe EL et al (2020) Precision and digital agriculture: adoption of technologies and perception of Brazilian farmers. Agriculture 10:1–16. https://doi.org/10.3390/agriculture10120653

Bolfe EL et al (2024) Potential for agricultural expansion in degraded pasture lands in Brazil based on geospatial databases. Land 13:1–17. https://doi.org/10.3390/land13020200

Brasil (2024) Mapa do leite: políticas públicas e privadas para o leite. Available at: https://www.gov.br/agricultura/pt-br/assuntos/producao-animal/mapa-do-leite. Accessed 15 Apr 2025

Bressiani DA et al (2015) Review of Soil and Water Assessment Tool (SWAT) applications in Brazil: challenges and prospects. Int J Agric Biol Eng 8:9–35. https://doi.org/10.3965/j.ijabe.20150803.1765

Camargo MP et al (2019) Cryptic species and population genetic structure of Plasmopara viticola in São Paulo state, Brazil. Plant Pathol 68:719–726. https://doi.org/10.1111/ppa.12993

da Silva SHMG (2008) Processamento digital de imagens para identificação da Sigatoka negra em bananais utilizando análise de componentes principais e redes neurais. Dissertation, São Paulo State University

de Araujo MA et al (2020) Territorial analysis, development and technical assistance and rural extension for family farming. Braz J Agric 95:123–139. https://doi.org/10.37856/bja.v95i2.4236

De Mori C et al (2024) Uso das práticas de manejo de forrageiras e de patejo na bovinocultura de corte. Embrapa Pecuária Sudeste. Available at: https://www.embrapa.br/busca-de-publicacoes/-/publicacao/1164251/uso-das-praticas-de-manejo-de-forrageiras-e-de-pastejo-na-bovinocultura-de-corte. Accessed 25 Apr 2025

dos Santos ML et al (2024) Yield gap analysis framework applied to pasture-based livestock systems in Central Brazil. Field Crop Res 314:1–15. https://doi.org/10.1016/j.fcr.2024.109416

Duveiller G et al (2020) Local biophysical effects of land use and land cover change: towards an assessment tool for policy makers. Land Use Policy 9:1–11. https://doi.org/10.1016/j.landusepol.2019.104382

Embrapa (2014) Visão 2014–2034: o futuro do desenvolvimento tecnológico da agricultura brasileira. Embrapa, Brasília

Embrapa (2023) Visão 2030: AgroDigital. Centro de Ciência para o Desenvolvimento da Agricultura Digital. https://www.embrapa.br/visao-de-futuro/agrodigital

Fouré E (1988) Stratégies de lutte contre la cercosporiose noire des bananiers et des plantains provoquée par Mycosphaerella fijiensis Morelet. L'avertissement biologique au Cameroun. Evaluation des possibilités d'amélioration. Fruits 43(5):269–274

Gassman PW et al (2007) The soil and water assessment tool: historical development, applications, and future research directions. Transactions of the ASABE/ASAE 50:1211–1250. https://doi.org/10.13031/2013.23637

IBGE (2023) Produção agropecuária. Available at: https://www.ibge.gov.br/explica/producao-agropecuaria/sp. Accessed 5 Apr 2025

Mahlein AK et al (2024) From detection to protection: the role of optical sensors, robots, and artificial intelligence in modern plant disease management. Phytopathology 114:1733–1741. https://doi.org/10.1094/PHYTO-01-24-0009-PER

Marin DH et al (2003) Black Sigatoka: an increased threat to banana cultivation. Plant Dis 87:208–222. https://apsjournals.apsnet.org/doi/pdf/10.1094/pdis.2003.87.3.208

McFadden J (2023) Precision agriculture in the digital era: recent adoption on U.S. farms. Economic Research Service, U.S. Department of Agriculture. https://ers.usda.gov/sites/default/files/_laserfiche/publications/105894/EIB-248.pdf

Mello KD et al (2020) Multiscale land use impacts on water quality: assessment, planning, and future perspectives in Brazil. J Environ Manag 270:1–16. https://doi.org/10.1016/j.jenvman.2020.110879

Moraes WS et al (2011) Técnica de monitoramento da Sigatoka-negra na cultura da banana. Pesquisa & Tecnologia 8:1–9

Neitsch SL et al (2011) Soil and Water Assessment Tool: theoretical documentation: version 2009. Texas Water Resources Institute Report No. 406. Texas Water Resources Institute, Texas

Paiva CMN et al (2024) Pathways to rural sustainability: opportunities and challenges in the creation of an agrotechnological district in Ingaí City, Brazil. Agriculture 14:1–21. https://doi.org/10.3390/agriculture14122185

Parreiras TC et al (2025) Applications, challenges and perspectives for monitoring agricultural dynamics in the Brazilian savanna with multispectral remote sensing. Remote Sens Appl Soc Environ 37:1–18. https://doi.org/10.1016/j.rsase.2025.101448

Prefeitura Municipal de Jacupiranga (2017) Plano municipal de turismo: Jacupiranga/SP: 2017–2027. Available at: https://www.jacupiranga.sp.gov.br/public/admin/globalarq/uploads/files/20170519170427.pdf. Accessed 5 Apr 2025

Puntel LA et al (2022) How digital is agriculture in a subset of countries from South America? Adoption and limitations. Crop Pasture Sci 73(9):949–963. https://doi.org/10.1071/CP21516

Santos F et al (2022) Assessing the wall-to-wall spatial and qualitative dynamics of the Brazilian pasturelands 2010-2018, based on the analysis of the Landsat data archive. Remote Sens 14:1024. https://doi.org/10.3390/rs14041024

Santos PM et al (2025) Assessment and analysis of yield gaps in pasture-based livestock systems: a review of methods. Agric Syst 226:1–13. https://doi.org/10.1016/j.agsy.2025.104323

Semear Digital (2024) Sobre o Projeto Semear Digital: Centro de Ciência para o Desenvolvimento em Agricultura Digital. Available at: https://www.semear-digital.cnptia.embrapa.br/sobre/. Accessed 5 Apr 2025

Silva Y et al (2024) Intense pasture management in Brazil in an integrated crop-livestock system simulated by the DayCent model. Sustainability 14:3517. https://doi.org/10.3390/su14063517

Souza CM Jr et al (2020) Reconstructing three decades of land use and land cover changes in Brazilian biomes with Landsat archive and Earth Engine. Remote Sens 12:1–27. https://doi.org/10.3390/rs12172735

Upadhyay A et al (2025) Deep learning and computer vision in plant disease detection: a comprehensive review of techniques, models, and trends in precision agriculture. Artif Intell Rev 58:1–64. https://doi.org/10.1007/s10462-024-11100-x

Wu W et al (2024) The rise of smart agriculture in China: current situation and suggestions for further development. Exp Agric 60(2):1–15. https://doi.org/10.1017/S001447972400022X

Regional Differences in Digital Agriculture: The Role of the Digital Agri Hub Dashboard in Providing Insights on the Sector and Its Sustainability

Inder Kumar, Sander Janssen, Tomaso Ceccarelli, Giacomo Rambaldi, Kudzanai Chimhanda, Eliya Buyukkaya, Rob Lokers, and Marlies de Keizer

1 Introduction

Digital technologies have a strong potential to support the transformation of agriculture, building a resilient, sustainable, and inclusive agrifood system (MacPherson et al. 2022; Hamidon and Ahamed 2024). In the era of agriculture 4.0 and Digitization for Agriculture (D4Ag) (Fragomeli et al. 2024), digital innovations can play a catalytic role in improving the capacity of Small-Scale Producers (SSPs) to sustain their livelihood, increase climate resilience, farm productivity, and income (The World Bank 2019). D4Ag actors[1] (Agyekumhene 2021) working in Low- and Middle-Income Countries (LMICs) deal with great vulnerabilities to climate change, such as droughts, floods, and an unprecedented change in temperature. Therefore, they require data and insights on early warning signs, precision farming, industry trends, and uptake of digital technologies in agriculture for capacity building and creating an enabling environment in LMICs, leading to the sustainability and climate resilience of the sector (Tsan et al. 2019). To ensure a strong and sustainable growth of digital innovations in the D4Ag sector, a trusted and reliable knowledge broker that tracks and monitors the D4Ag landscape is

[1] Small Scale Producers (SSPs), AgriTech startups/digital solution providers, agribusinesses/value chain actors, government and public sector, development agencies and donors, investors and financial institutions, Mobile Network Operators (MNOs), research and academia, NGOs and civil society organizations, multilateral organizations (e.g., FAO, World Bank), data providers, data intermediaries, extension officers, farmer organizations and cooperatives.

I. Kumar (✉) · S. Janssen · T. Ceccarelli · G. Rambaldi · K. Chimhanda · E. Buyukkaya · R. Lokers · M. de Keizer
Team Earth Informatics (EI), Environmental Science Group, Wageningen University and Research, Wageningen, The Netherlands
e-mail: inder.kumar@wur.nl

F. da Silveira et al. (eds.), *Agriculture 4.0*,
https://doi.org/10.1007/978-3-032-24552-6_23

essential to support the data-driven decision-making process of D4Ag actors across the agrifood system (Abbasi et al. 2022; Abiri et al. 2023).

The Digital Agri Hub tracks the evolution of the D4Ag ecosystem in LMICs. In collaboration with sectoral stakeholders, it publishes curated information on digital innovations to generate new insights and updates on developments in the domain of digital agriculture.

1.1 Challenges in D4Ag and the Role of Digital Platforms in LMICs

Despite a rapid expansion of the D4Ag sector across LMICs, the reach, adoption, and retention of D4Ag solutions[2] remain fairly low, especially among SSPs. Inadequate access to the internet and digital services, coupled with issues of affordability, digital literacy, and digital skills, are among the factors that have widened the disconnect (Tsan et al. 2019). One of the main barriers holding back investment in D4Ag solutions by SSPs is a lack of cost-effective ways to assess technology readiness, maturity of business model, ability to scale, etc. (Beanstalk Agtech 2023). Due to this, actors such as donor organizations, governments, and impact investors still struggle with informed decision-making on investment readiness and impact assessment of digital solutions in the D4Ag sector.

To bridge the gap between the outreach (one-time access), adoption, and retention (prolonged adoption) of digital technologies by SSPs globally, and especially in developing countries, it is necessary to better implement user-centric design and adoption drivers in the development of digital solutions and services. This is an underlying necessity to scale investments (grant-based projects, blended finance, PPPs, etc.) and enhance ecosystem federation by creating synergies between data platforms and D4Ag stakeholders utilizing data and insights from such platforms.

To stimulate large-scale investments in D4Ag solution development, a better understanding of the regional differences in the D4Ag sector is required. The scope of this chapter is to assess the importance of monitoring and tracking the D4Ag landscape and adopting an ecosystem approach to enhance the sustainability of digital agriculture solutions. Furthermore, this chapter will expand on the following:

1. What regional differences exist in the adoption and deployment of digital solutions across Low- and Middle-Income Countries (LMICs), and what underlying factors contribute to these differences?

[2] Digital solutions refer to the digitally enabled business models and technologies to address farmers and food system actors' constraints (e.g., around market access, inputs, financing and climate).

2. What role does a platform such as the Digital Agri Hub play in fostering D4Ag ecosystem coordination and inclusive growth of digital innovations through data-driven insights?

2 The Digital Agri Hub—Description and Definitions

Definition The Digital Agri Hub (https://digitalagrihub.org/), hosted and maintained by Wageningen University and Research (WUR), is a public good and acts as a trusted knowledge broker, performing due diligence for technology assessment, bringing partners together, and connecting D4Ag actors. It supports them with access to reliable, up-to-date, and locally relevant data and insights on digital technologies, associated use cases[3] for outreach to the farmers.

Currently, the Digital Agri Hub is the only global platform/repository of digital solutions that exclusively focuses on the D4Ag sector in LMICs. It provides data and insights that are required for informed decision-making at the farm and farmer level, scaling investments in digital products and services, and provides objective knowledge and information for the D4Ag audience. The hub strengthens the D4Ag ecosystem by engaging with its stakeholders and reinforcing their networking. The DigitalAgriHub actively seeks to enhance interoperability between data platforms by aligning to common data standards and open Application Programming Interface (APIs) for access. The Digital Agri Hub platform and its home page are shown in Fig. 1.

3 Methodology

This chapter implements a quantitative, data-driven approach utilizing the Digital Agri Hub's dashboard and its repository of 1400+ digital solutions deployed in LMICs as the primary data source, with a time stamp of September 2025. The LMICs are identified and aligned with the United Nations (UN) Food and Agriculture Organisation's State of Food Agriculture (SOFA) report published in 2022 (FAO 2022). All data on digital solutions (e.g., digital solutions deployed per country, launch year, use case, and technology) were obtained directly from the Digital Agri Hub dashboard over the past 5 years. These data are combined from several sources, such as self-reported by solution providers, curated datasets

[3] Farm management and advisory, market linkages, supply chain management, finance and ecosystem support.

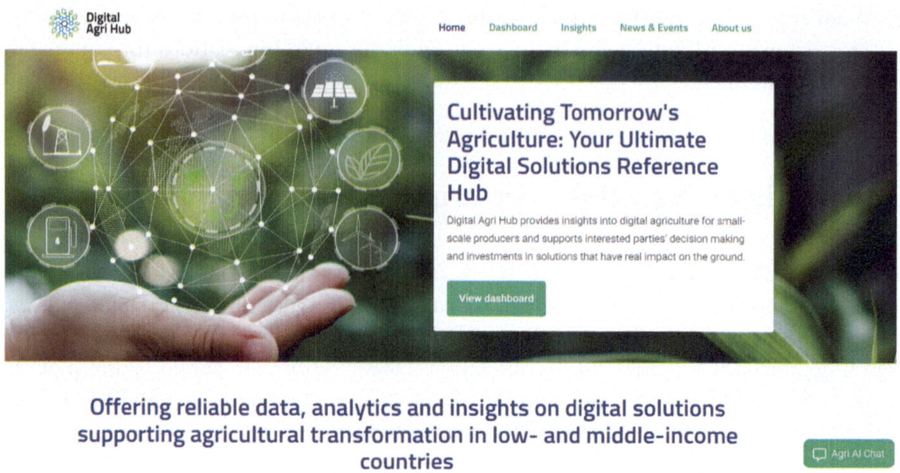

Fig. 1 The Digital Agri Hub platform's home page (Source: Digital Agri Hub, 2025)

by the Hub, and harmonizing databases from various data platforms,[4] ensuring a comprehensive and up-to-date repository of D4Ag initiatives.

The dashboard's metrics included for this study are the number of digital solutions, their regional deployments,[5] primary use cases, digital technologies, year of launch, and whether they have been removed or ceased operation. Therefore, the study is essentially a secondary analysis of the Digital Agri Hub's own data. This approach means that all findings are gathered from the Hub's curated data and expert interpretations, aligning with the chapter's aim to reflect on the role of the Digital Agri Hub as both a data platform and knowledge broker. The data reflects the state of digital solutions as reported by the Hub up to 2024; any biases or gaps are inherent in those of the Digital Agri Hub dataset itself.

The analysis was executed around three key themes derived from the dashboard data: (1) regional disparities in digital solution deployments, (2) growth patterns in the main use cases and technologies utilized by these solutions, (3) trends in the launch of new solutions deployed, and trends in the retirements/withdrawal of digital solutions that are no longer operational. Together, these insights highlight the evolutionary trends of the D4Ag sector in the LMICs.

[4] Technical Centre for Agricultural and Rural Cooperation (CTA), Grow Asia, SkyQuest, AgriTerra, MercyCorps, IICA, Enabel and CCARDESA.

[5] The sum of total deployments in a country or region is independent of the sum of digital solutions, as one solution can be deployed in multiple countries.

4 Results

4.1 Regional Deployment Differences in LMICs

Noticeable regional fragmentation in the deployment of digital solutions exists (Fig. 2). Sub-Saharan Africa (SSA) reported the highest number of deployments in the 2019–2023 period, with 414 deployments; the largest share of any region. In contrast, other regions, for example, Oceania, had only 22 deployments over the same 5 year span, the lowest among the regions examined. Regions such as Northern Africa and Western and Central Asia also registered very limited deployment numbers in the double digits, further emphasizing a visible gap between SSA and less active regions. Deployments in the regions such as Southern Asia and Latin America and the Caribbean "combined" underline lower deployments as compared to the SSA. This wide variation indicates that digital agriculture solutions are unevenly distributed geographically, with certain regions (notably SSA) being more active and visible, while others receive relatively less attention. The disparity raises concerns about primary factors: SSA's visible position may be linked to concerted donor support and prior initiatives (the Hub's dataset itself builds on the Technical Centre for Agricultural and Rural Cooperation (CTA) 2018–2019 inventory), whereas regions like Oceania may suffer from limited market size, less investor attention, or infrastructural barriers. However, also more geographical or structural factors may play a role here, as the number of small-scale producers per region is different (Lowder et al. 2016). SSA and Asia have a relatively large number of SSPs compared to other continents, and especially SSA has a large number of low-income countries with their own languages and legal frameworks, leading to new D4Ag solutions starting in each of these countries. Despite high density of SSPs

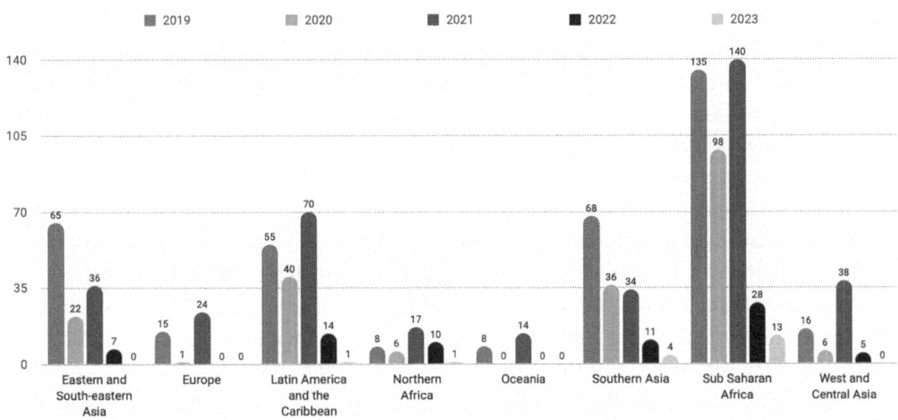

Fig. 2 The sum of digital solutions deployments launched per region in the last 5 years (Source: Digital Agri Hub, 2025). The sum of total deployments is independent of the sum of digital solutions, as one solution can be deployed in multiple countries

and new digital innovations in the SSA region, it differs significantly to develop and scale innovations, for example, for the Digital Climate Advisory Services (DCAS) it depends on three dimensions: (1) the institutional environment (the policy/regulatory framework, e.g., for data sharing); (2) the technical/ infrastructure environment (e.g., penetration of mobile phones); and (3) the business environment (e.g., possible business carriers to partner with) (Ceccarelli 2021).

4.2 Growth Patterns in the Main Use Cases and Technologies Utilized

Since the launch of the Digital Agri Hub in 2021, corresponding with the overall increase in digital solutions, there has been substantial growth in the use cases addressed and technologies used by the digital products and service providers. When it comes to the use case categories (Fig. 3), the solutions focusing on Farm Management and Advisory services form the largest segment and have seen remarkable growth. In 2022, the Hub recorded 489 solutions offering farm management or advisory services; by 2024, this number had risen 67% (819 solutions), reflecting the high demand for services that help farmers plan, monitor, and improve their farm operations. This makes Farm Management and Advisory the dominant use case across the D4Ag landscape. Market Linkage solutions—which connect producers to buyers or markets, etc., increased by 96% between 2022 and 2024 (growing from 241 to 471). Solutions addressing Supply Chain Management—including logistics and traceability, etc., grew by 56% from 214 in 2022 to 334 in 2024. Digital Finance for agriculture—e.g., credit, insurance, payments, etc., more than doubled, increasing by 116% from 135 in 2022 to 291 in 2024. Moreover, Smart Farming—precision agriculture or IoT-based on-farm solutions, etc., experienced rapid early growth and then plateaued: from 94 to 262 solutions between 2022 and 2023, then only marginally to 263 in 2024. Overall, the consistent increase in most use case categories indicates expansion of D4Ag across different domains with a particularly strong emphasis on farm management and market linkage-based digital products and services, which directly benefit producers.

In parallel, the underlying technologies enabling digital agriculture solutions have also seen an upward trend (Fig. 3). The Hub's data highlights eight leading technologies utilized by the majority of D4Ag solutions, all of which saw an upward adoption trend from 2022 to 2024. The most widely utilized are data analytics and business intelligence tools, which were used by 155 solutions in 2022 and grew to 289 solutions by 2024. This trend reflects the importance of data-driven decision support in agriculture, as solutions increasingly leverage data analytics for insights on crop performance, market trends, and farmer behavior. Artificial Intelligence (AI) is another leading technology: AI-enabled solutions increased from 141 to 203, indicating greater incorporation of machine learning algorithms, predictive models, and AI-driven automation in Agri-Tech services. Cloud-based services saw

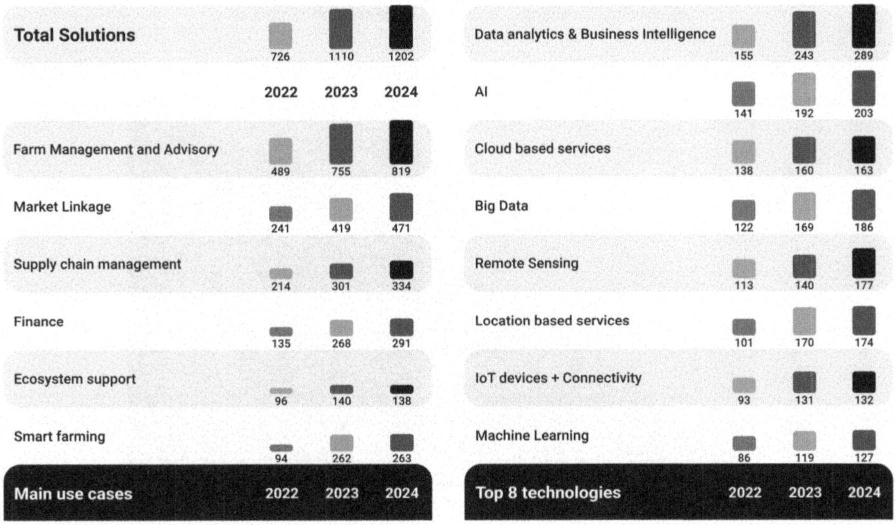

Fig. 3 Growth of digital solutions by main use case (left) and technology employed (right), from 2022 to 2024. (Source: Digital Agri Hub 2025)

moderate growth (138–163 solutions), which is expected as cloud infrastructure reinforces many digital platforms for scalability and remote access. Big Data grew from 122 to 186 solutions, reflecting the growth of large datasets in agriculture and the tools to process them. Notably, a technology that showed signs of saturation is IoT devices and connectivity: solutions utilizing IoT grew from 93 to 131 between 2022 and 2023, but only marginally grew to 132 in 2024.

4.3 *Trends in Solution Launches and Retirements*

The rate at which new digital solutions are being launched (Fig. 4) has decreased noticeably in recent years (Digital Agri Hub 2025). The dashboard records show that 146 new solutions were introduced in 2019, followed by 105 in 2020, 76 in 2021, 60 in 2022, and only 15 in 2023. This trend, illustrated in Fig. 4, exhibits a steady year-by-year decline in the number of new D4Ag solutions launched. The year 2019 was spotlighted as the peak of innovation in terms of new competitors, after which each subsequent year saw fewer launches (notably, the 2023 figure is partial-year data up to the point of analysis; moreover, it indicates a sharp drop). This downward trajectory suggests a maturation of the sector: the flurry of new startups and digital initiatives that characterized the late 2010s has given way to a period of consolidation. Additionally, some solutions in development might not yet be captured in deployment counts due to the time needed to reach the operational stage or gain traction and visibility.

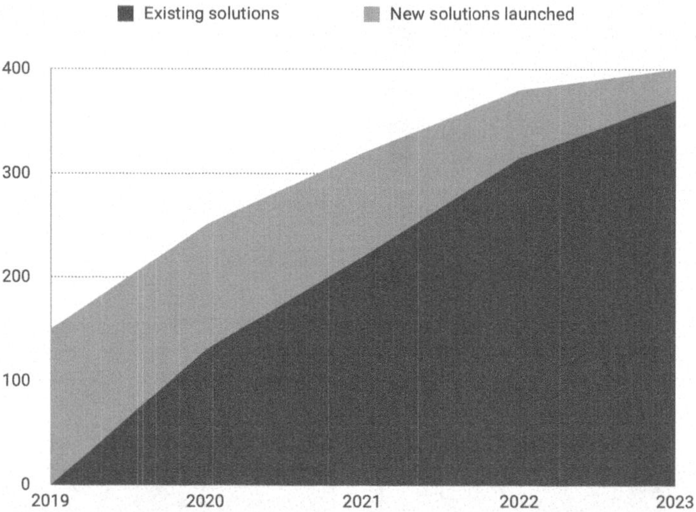

Fig. 4 Number of new digital solutions launched each year (2019–2023). The grey segment represents new solutions added that year, while the black area represents existing solutions. (Source: Digital Agri Hub 2025)

In parallel with the slowdown in new launches, the sector has seen a noteworthy number of existing solutions phasing out or ceasing operations (Fig. 5). According to the Digital Agri Hub dashboard data, a total of 113 solutions that were at one point listed as active have since been removed or withdrawn from its dashboard. Figure 5 illustrates the timing of these withdrawals concerning the year in which those solutions were originally launched. The pattern reveals that the bulk of the solution ceasing to exist corresponds to solutions launched in the mid-2010s. Specifically, solutions that were launched around 2015–2018 have the highest termination counts—for instance, those launched in 2016 account for the largest number of retirements (over 20 solutions reported ceased operations), and similarly high failure numbers are observed for 2017, 2018, and 2019 launches. In contrast, very few of the solutions launched in 2020 (as reported up until 2024) have been retired—for example, almost no 2021 launches have been reported as phased out yet. This indicates a time lag in the lifecycle. Digital services rarely succeed on their own in reaching SSPs unless their strategies are incorporating Community-Based Agents (CBAs), e.g., farmer champions, rural promoters, and agro dealers. CBAs play an instrumental role in the adoption and retention of digital services by SSPs as they help translate DAS information into local dialects and contextualize technical recommendations that can be trusted (Gray and Iyer 2025).

The launch and retirement data together illustrate that the D4Ag sector is past an "explosive growth" phase and is entering a phase where ensuring the longevity of digital innovations is significantly influenced by the roles played by co-creators (e.g., CBAs for DAS), working with SSPs, and is fundamental to the sustainability of digital innovations in agriculture.

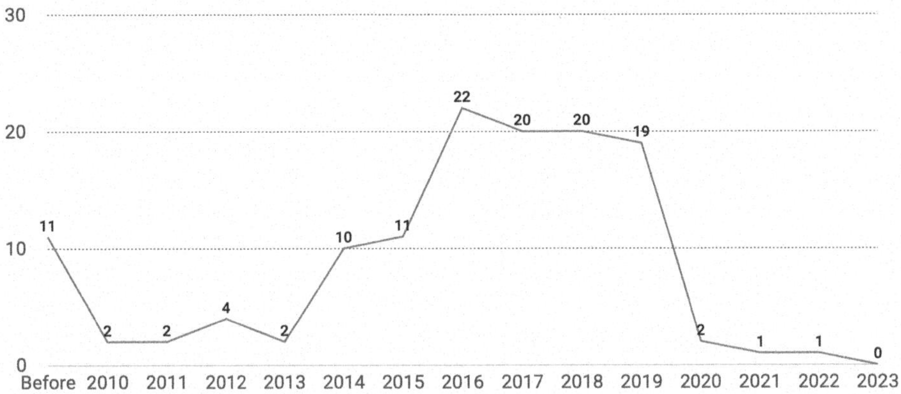

Fig. 5 Number of digital solutions that have been removed (retired), grouped by their launch year (Before 2010 through 2023). (Source: Digital Agri Hub, 2025)

5 Discussion

Our results show a vibrant yet irregularly developing D4Ag ecosystem, highlighting both the progress and continuing challenges. A key observation is the *regional imbalance* in digital solution deployments. SSA's most visible position (414 deployments in 2019–2023) suggests that strong drivers, such as targeted investments by donor organizations and impact investors, and a large SSPs base in need of services, have catalyzed D4Ag growth in the region. The CTA Dalberg report on Digital Agriculture in Africa 2018–2019 (Tsan et al. 2019) may have given SSA a head start in setting the foundation of the Digital Agri Hub database. In contrast, regions like Northern Africa and Central Asia, which show very low deployment numbers, could face *structural barriers* that hinder the launch and adoption of digital solutions. These can include insignificant and a lack of dynamic policy infrastructure for public, private, and PPP-based innovations to flourish, limited digital infrastructure (poor internet connectivity, low smartphone penetration), and reduced attainable market size (i.e., the number of digital solutions deployed to the percentage of agriculture to regional GDP).

Alternatively, it could be an attention gap: the high visibility of some regions suggests a dynamic, attractive market, while less attention from international investors or development programs suggests more difficult market conditions. This might not necessarily be true in practice, and a rigorous analysis of market maturity and attractiveness on objective parameters is missing. The wide gap between the "high potential markets," for example, SSA, and the "low potential markets," for example, island countries in Oceania, raises concerns about *equitable access* to digital innovation. It implies that without mediation, certain regions risk being left behind. Notably, the regional differences also prompt reflection on inclusive growth: is the high number of solutions in SSA sustainable and translating into tangible impact, or could it be revealing many pilot programs and overlapping projects? On the contrary, are the low numbers in other regions a sign of untapped opportunity or unsuitability of generic innovations in those contexts? These questions highlight that *fundamental factors* (geopolitical, socioeconomic, and environmental) critically influence where and how digital solutions prosper.

Another relevant theme is the *chronological cycle* of digital innovation in the D4Ag sector. The decline in new digital solution launches each year since 2019 points to a sector that is no longer in a phase of rapid expansion. This could be interpreted as market saturation or higher entry barriers—a natural progression as the sector matures. Early interest and funding may have led to a surge of start-ups and digital projects around 2018–2020, with many undertakings similar to these initiatives. By 2021–2022, the duplication of efforts and competitiveness for user retention likely made it harder for newcomers to scale their digital innovations, and thus fewer new solutions emerged. The data on solution retirements supports this narrative: a considerable number of the solutions that were launched in the mid-2010s did not survive much beyond a few years. The peak in retirements for solutions launched around 2016–2017 shows that a wave of initiatives from that period eventually *failed to scale or sustain*. The insights reflect classic start-up challenges that are sector agnostic (Ritter and Pedersen 2022). Many digital solutions struggle with *long-term viability* in the agriculture sector, especially for those whose end users are SSPs, as it is a difficult market with low margins. Many start-ups that are investor-driven struggle to achieve a sustainable long-term cash flow to run their operations beyond their initial series of investment funding (Series A, B, C). As an alternative strategy to investor funding, acquiring *project-based donor funding*, which often means that when grant funding cycles end, the products and services often retire if no self-sustaining model is established. Ultimately, the challenges for D4Ag solution providers are creating a sustainable cash flow beyond donor or investor funding at low enough operational costs. *Economic Sustainability may be more of a challenge* in the D4Ag sector.

The combination of various new digital solutions being founded in the past decade and many retiring or stagnating leads to an important insight: the digital agriculture sector has been *highly fragmented*. Over 1400 digital solutions are deployed across LMICs, often with overlapping use cases in farm advisory, market linkage, supply-chain management, and financial services. The current landscape features numerous digital initiatives "*operating in silos*," often competing for similar audiences, funding, and impact goals, which ultimately limits the collective potential for inclusive sectoral growth. In practical terms, this means many digital products and services are reinventing the wheel—for example, multiple separate apps might all provide weather alerts to farmers in the same region, each developed by different organizations not collaborating. Such parallel efforts can lead to inefficient use of resources and *user fatigue*, as end users receive nonexclusive offerings. The *lack of coordination and interoperability* among these digital solutions has been identified as a critical issue that needs to be addressed to avoid redundancies and inefficiencies. These insights spotlight an ongoing issue in the D4Ag ecosystem: while the sector is often described as highly fragmented, it overlooks how these innovations connect to the people who actually use them. Community-Based Agents

(CBAs) should be involved in the early stage of the innovation development, and not as end-users. CBAs act as co-designers and cultural navigators to close the feedback loop, highlighting challenges faced by SSPs in their daily routine. In parallel to this, the data trends we observed (numerous launches, many short-lived) emphasize this insight: fragmentation can lead to a cycle where new solutions emerge rapidly but then disappear if they fail to bring innovative products and services, depending on the following factors:

- A lack of robust business strategies to increase sustainable adoption among small-scale producers while dealing with the fundamental challenges on the ground, such as weak digital infrastructure in remote areas, digital literacy, and trust in technical insights from such innovations.
- Underestimating the role of learning through user feedback and end-user validations, where farmers validate the effective use of technologies such as AI, IoT, big data, and remote sensing.
- Weak data governance and standards supporting interoperability among different digital solutions and data platforms to minimize user fatigue, redundancies in the D4Ag ecosystem, and regional differences for the deployment of digital innovations.

The role of *Digital Agri Hub* is valuable *as a knowledge broker and data platform, based on the analysis shared in this chapter*. The Hub serves as a *repository and observatory* for the D4Ag sector, synthesizing relevant insights for its stakeholders. By providing a curated dashboard of digital solutions, the Hub makes the fragmented landscape visible and easy to navigate for D4Ag actors. This transparency is the first step toward ecosystem coordination: when donors, governments, and implementers can see existing digital solutions in a country or region of choice, they are less likely to reinvent the wheel and more likely to identify gaps that truly need new interventions.

Building on both dashboard insights and practitioner dialogues, a number of questions remain unresolved and deserve closer attention as an avenue for future research. One concerns the long-term balance between SSPs and digital technology: at what stage can CBAs step back, and which types of digital advisory tools allow farmers to operate more independently? Another concern is business frameworks and strategies for the use of such technologies by SSPs. Who pays for the work of CBAs—donors, cooperatives, governments, or private firms—and how sustainable are such business models? Gender and generational gap is a third area that needs deeper inquiry. Recruiting and retaining female CBAs continues to be difficult, yet where it succeeds, it changes both household dynamics and community trust in digital services. Finally, there is the question of farmer voice. Several contributors in the dialogue reminded us that "no entrepreneur likes to lose." Ensuring that farmers can validate advice, challenge it, or request clarification is essential if adoption is to last beyond pilot phases.

6 Conclusion

The Digital Agri Hub's dashboard provides data and insights on emerging trends in the D4Ag sector, including a unique perspective into the state of digital agriculture in LMICs, revealing both inclusive growth (digital solutions, use cases, and technology focus) and critical challenges (fragmentation, uneven regional progress, and questions of longevity) to effectively work with SSPs. The results show slowing launch rates of digital innovations in the sector and significant figures on the retirement of digital solutions. The sector is at a crossroads where scaling up and sustaining proven digital solutions may take priority over sheer experimentation. Whereas regional differences in the D4Ag deployments point to the need for enhancing market access potential and investment-ready corridors in lagging regions. There is a need for objective knowledge and information dissemination for ecosystem coordination and federating the synergies between Digital platforms, focusing on mobilizing their resources in LMICs.

Competing Interests The author(s) has no competing interests to declare that are relevant to the content of this manuscript.

References

Abbasi R, Martinez P, Ahmad R (2022) The digitisation of the agricultural industry – a systematic literature review on agriculture 4.0. Smart Agric Technol 2:100042

Abiri R, Rizan N, Balasundram SK, Shahbazi AB, Abdul-Hamid H (2023) Application of digital technologies for ensuring agricultural productivity. Heliyon

Agyekumhene C (2021) Digitalisation of smallholder value chain lending partnerships: an interplay of trust and inclusion. Doctoral dissertation, Wageningen University

Beanstalk Agtech (2023) State of the Digital Agriculture Sector. Harnessing the potential of digital for impact across agricultural value chains in low- and middle-income countries. Retrieved on March 31, 2025, from https://farmfitinsightshub.org/resources/state-of-the-digital-agriculture-sector

Ceccarelli T.(2021) Digital Climate Advisory Services in Sub-Saharan Africa: insights for developing and scaling DCAS for small-scale producers. Retrieved on September 19, 2025, from https://digitalagrihub.org/documents/d/guest/dcas-in-sub-saharan-africa

FAO (2022) The State of Food and Agriculture 2022, leveraging automation in agriculture for transforming agrifood systems. FAO, Rome. https://doi.org/10.4060/cb9479en

Fragomeli R, Annunziata A, Punzo G (2024) Promoting the transition towards Agriculture 4.0: a systematic literature review on drivers and barriers. Sustainability 16(6):2425

Gray B, Iyer L (2025) Summary of the 9th eConversation 2nd series: Is there a world where digital agriculture advisory services can scale without agents? Retrieved on September 19, 2025, from https://digitalagrihub.org/documents/d/guest/summary-of-the-9th-e-coversation-series-2

Hamidon MH, Ahamed T (2024) Digital innovations in agrifood systems to achieve the SDGs and Society 5.0. In: IoT and AI in agriculture: smart automation systems for increasing agricultural productivity to achieve SDGs and Society 5.0. Springer Nature Singapore, Singapore, pp 1–23. https://doi.org/10.1016/j.worlddev.2015.10.041

Lowder SK, Skoet J, Raney T (2016) The number, size, and distribution of farms, smallholder farms, and family farms worldwide. World Dev 87:16–29

MacPherson J, Voglhuber-Slavinsky A, Olbrisch M, Schöbel P, Dönitz E, Mouratiadou I, Helming K (2022) Future agricultural systems and the role of digitalisation for achieving sustainability goals. A review. Agron Sustain Dev 42(4):70

Ritter T, Pedersen CL (2022) An entrepreneur's guide to surviving the "death valley curve". Harv Bus Rev. Retrieved on June 18, 2025 from https://hbr.org/2022/04/an-entrepreneurs-guide-to-surviving-the-death-valley-curve

Tsan M, Totapally S, Hailu M, Addom BK (2019) The digitalisation of african agriculture report 2018–2019. Wageningen, The Netherlands: CTA/Dalberg Advisers

The World Bank (2019) Retrieved on October 3, 2024, from https://www.worldbank.org/en/topic/agriculture/publication/future-of-food-harnessing-digital-technologies-to-improve-food-system-outcomes

Integration of Robotics, IoT, 5G, and Edge Computing for Vineyard Monitoring: A Case Study

David Orbea, Luz K Atencia, Hipatia Delgado, Belén Diezma, Antonio Barrientos, María Gómez del Campo, and Constantino Valero

1 Introduction

The use of sensors and information and telecommunication technologies in the agricultural sector allowed advanced vineyard monitoring, mainly due to the high end-value of the product and the use of technology for this crop (Barriguinha et al. 2021). The number of recent research publications related to precision viticulture corresponds to 5% of publications related to precision agriculture (Ferro and Catania 2023). A key objective of using these technologies is to obtain and supply data to grape growers and wine producers as a basis to improve land and vine management through a more-informed decision-making process, recently with the aid of artificial intelligence modeling (Tardaguila et al. 2021). Among the variety of technologies, proximal sensing has been widely developed (Moreno and Andújar 2023), as well as robotics for crop scouting and data acquisition (Roure et al. 2020; Botta et al. 2022). The vast amount of data generated by these sensors could benefit from last generation communication technologies such as 5G mobile networks and edge computing. Only recent projects have explored this path (Edemetti et al. 2022). Previous research lacks an integrated approach that combines data-intensive sensors, wireless plant monitoring, fast communications, and edge computing for real-

D. Orbea · A. Barrientos
Universidad Politécnica de Madrid, Centre for Automation and Robotics (UPM-CSIC), Madrid, Spain
e-mail: david.orbea@upm.es; antonio.barrientos@upm.es

L. K. Atencia · H. Delgado · M. Gómez del Campo
Universidad Politécnica de Madrid, CEIGRAM, ETSIAAB, Madrid, Spain
e-mail: maria.gomezdelcampo@upm.es

B. Diezma · C. Valero (✉)
Universidad Politécnica de Madrid, LPF TAGRALIA. ETSIAAB., Madrid, Spain
e-mail: belen.diezma@upm.es; constantino.valero@upm.es

F. da Silveira et al. (eds.), *Agriculture 4.0*,
https://doi.org/10.1007/978-3-032-24552-6_24

time data collection and modeling. This aim of this work is to develop a case study through the implementation of a robotic prototype for vineyard phenotyping with onboard sensors combined with field sensors, working in collaboration with a specifically deployed 5G network for real-time database sending and cloud computing. The working hypothesis is that the combination of data will allow estimation and forecasting of crop evolution and final harvest.

2 Methodology

2.1 Vineyard Characteristics

The chosen plot was Cabernet Sauvignon, located at Latitude 39°56'35", longitude 3°42'55", altitude 564 m, with 20-year-old plants grafted onto SO4 (Oppenheim 4 Selection), without oriented rows, with a planting frame of 2.60 × 1.10 m. From the selected plot, a 190 m long line with 50 plants is chosen, based on high NDVI vigor index values (greater than 0.8), to ensure work with healthy vines. Five treatment zones are defined to increase the natural variability of the vines by applying different intensities of pruning and leaf removal.

The treatments are described in Table 1.

The case study is located at Bodegas y Viñedos Casa del Valle, Yepes, Spain. The plot selected for the study is indicated in Fig. 1, based on the following requirements:

- 5G antenna visibility
- Various single varieties
- Homogeneity of the plants and adequate development
- Proximity to the winery

2.2 Data Acquisition

Data acquisition is performed through the robotic platform shown in Fig. 1, composed of a mobile robot Summit XL, a robotic arm Unitree Z1 Pro, navigation

Table 1 Description of applied treatments

Name	Number of plants	Treatment
Treatment 1	10	50 % additional leaf shedding
Treatment 2	10	25 % additional leaf shedding
Treatment 3	10	No modifications (reference condition)
Treatment 4	10	50 % additional pruning (stumping)
Treatment 5	10	25 % additional pruning (stumping)

Fig. 1 Robotic system in Casa del Valle vineyard. Yepes, Spain

Fig. 2 Sencrop weather station and Plantae humidity sensors. Source: Authors

sensors (3D Lidar Velodyne VLP-16 and GPS RTK geolocation system), 5G communication modem Netgear, multispectral camera Micasense Altum, and 2D LiDAR Hokuyo URG-04LX. This system is integrated through the Robot Operating System (ROS) framework, on an onboard computer with Ubuntu 20.04 and ROS1 Noetic.

In addition, a Sencrop weather station was integrated into the system, equipped with sensors for relative humidity, temperature, leaf wetness, wind speed, and solar radiation. Plantae soil moisture sensors are also integrated for each treatment block, Fig. 2.

The robotic system is in charge of the 3D reconstruction of the vines, the acquisition of multispectral images, and the uploading of data to a remote server using a 5G modem.

For the 3D reconstruction of the vine, vertical scanning and horizontal scanning movements are performed. The positions of the manipulator arm are shown in Fig. 3.

The point clouds reconstructed in the ROS framework are shown in Fig. 4.

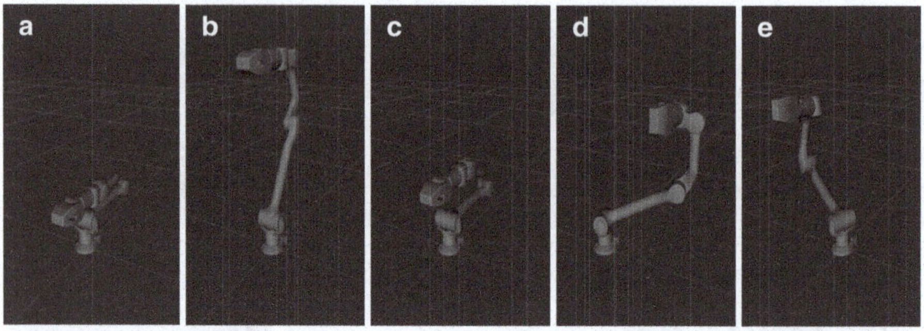

Fig. 3 Robotic arm positions for 3D scanning of the vine. (**a**) Sleep, (**b**) start of vertical scan, (**c**) end of vertical scan, (**d**) start of horizontal scan, and (**e**) end of horizontal scan. Source: Authors

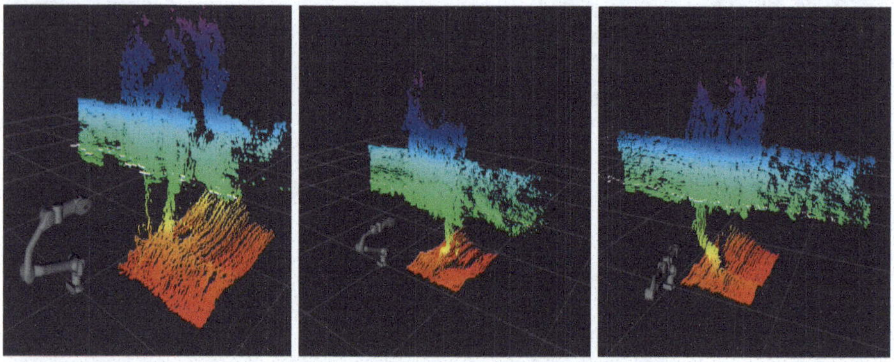

Fig. 4 Point clouds of reconstructed vines. Source: Authors

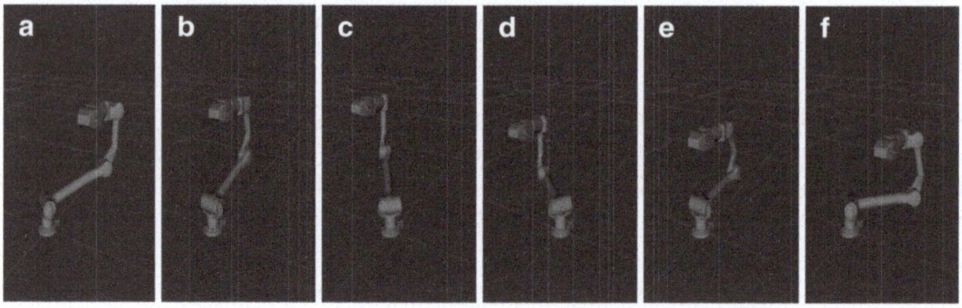

Fig. 5 Robotic arm positions for multispectral image acquisition. (**a**) Position 1, (**b**) position 2, (**c**) position 3, (**d**) position 4, and (**e**) position 5. Source: Authors

Multispectral image acquisition is structured with six manipulator positions to cover areas of interest in detail. The arrangement of the robotic arm is shown in Fig. 5 and the areas of interest of the vine in Fig. 6.

The acquired images are shown in Fig. 7.

Fig. 6 Zones of interest of the plant according to robotic arm positions. Source: Authors

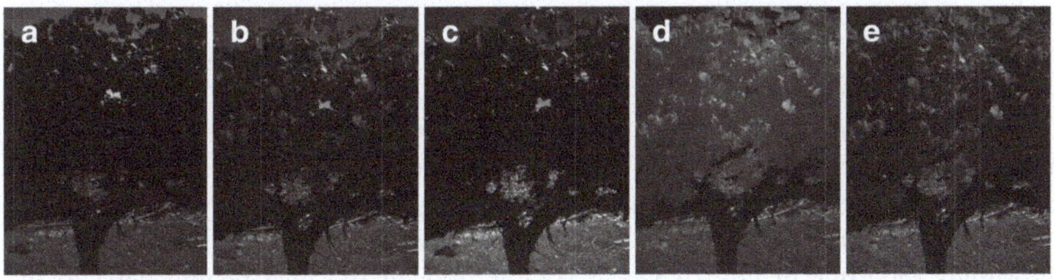

Fig. 7 Individual channels of acquired multispectral image. (**a**) blue, (**b**) green, (**c**) red, (**d**) red edge, and (**e**) near-infrared (NIR). Source: Authors

2.3 Image Processing for Grape Segmentation

In multispectral cameras, the misalignment of the planes caused by the sensor arrangement is magnified when the distance to the object is small, as in this study. To ensure consistent and comparable spectral signatures for each pixel across bands, the different planes of the multispectral image must be spatially aligned. This is achieved through a registration process that corrects shifts between channels by applying geometric transformations. This ensures that each pixel represents the same point on the object or surface in all bands. Image registration of the RGB channels was carried out using the *imreg* function (a Python package specializing in image registration), and similarity transformations were used to ensure correct alignment of the red, green, and blue channels. These channels were generated from captured sequences from two perspectives (front and back for each side of the vine

Fig. 8 Grayscale images of individual channels with reconstructed RGB images. Source: Authors

row). The result of the recording process is a synthetic color image for each row and side. For example, images recorded from several vine rows are shown in Fig. 8. Areas where the three channels do not overlap are visible, resulting in color artifacts. These areas are eliminated in subsequent image analysis steps. The different effects of illumination in each image can also be seen and must be considered in subsequent segmentation processes.

The pixels containing grape clusters were identified in color images by scanning different color spaces and transforming the images from RGB format to CIE L*a*b*, xyz, and HSV formats (Connolly and Fleiss 1997; Jack 2008; Luszczkiewicz-Pikatek 2014). A Matlab script was used to identify segmentation thresholds in different color spaces; HSV produced the best results. Due to the variability in image brightness, segmentation thresholds were optimized for groups of similar brightness images producing a preliminary mask or a binary image, which was then improved by implementing various morphological and filtering operations showing the results of segmenting the images from front and back perspectives (Fig. 9).

2.4 Image Processing for Reflectance Estimation

For the estimation of reflectance, an image is used in the position that reflects the highest leaf concentration, so the image in position 2 is used in the different channels: blue, green, red, red edge, and near-infrared (NIR). In these images, an area of interest is cropped, and masks are applied through the OpenCV library to omit the incidence of direct light and shadows in overly dark areas; then, this mask is applied to the original image, and the mean, median, and mode values of nonzero pixel values are calculated to estimate the reflectance, exported in a file

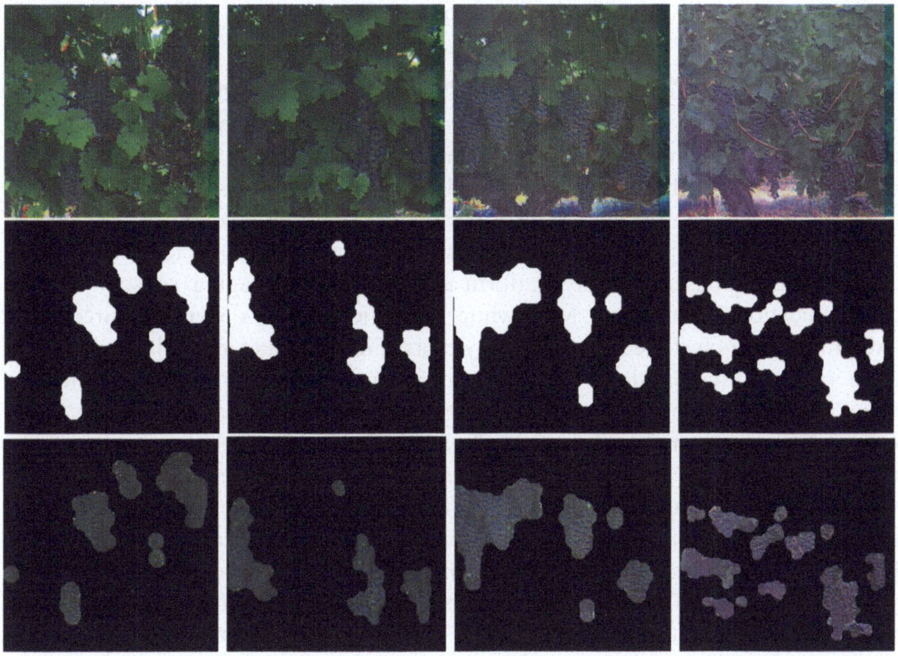

Fig. 9 Process of grape segmentation. **(top)** RGB images, **(middle)** binary mask, and **(low)** segmented images. Source: Authors

Fig. 10 Image processing of region of interest for reflectance estimation in individual channels of multispectral images. Source: Authors

along with the generation of a concatenated image of the different channels to verify that the processing has been performed correctly. This process is shown graphically in Fig. 10.

2.5 Stem Water Potential prediction

Multispectral data and environmental variables were integrated in real time to estimate Stem Water Potential (SWP) in a vineyard. The state of the vine water was assessed using the pressure chamber technique (Scholander et al. 1965). SWP was measured in leaves that had been covered with an opaque plastic bag 1 hour prior to measurement. These leaves were collected after the robot captured multispectral images of the corresponding vine. Multispectral data were collected on the move using a ground-based robotic platform equipped with imaging sensors and 5G communication capabilities. Meanwhile, climatic variables were acquired from fixed IoT sensors installed near the study plots (Fig. 2). Data were collected in close temporal coordination along a test row of 50 vines over eight measurement days along the growing season. A total of 400 observations were taken. Measurements were taken between 10:00 and 12:00 local time. A correlation analysis was conducted to identify the most influential variables. Measurement time and date were expressed as numerical variables (Hour_num and Date_num, respectively) to evaluate their monotonic association with SWP using Spearman's correlation. Following, a Random Forest machine learning model was developed using a random split of the dataset: 70% of the observations were used for training, and the rest 30% were used as the independent testing set. The RF model was then evaluated using the variables that showed the strongest associations with SWP in the Spearman correlation analysis.

2.6 Yield Prediction

Grape yield prediction was performed using data collected on September 2, 2024, when the study vines were harvested. Measurements were taken from a total of 50 vines. After data filtering, 12 records were discarded as outliers, leaving 38 records for model development. The percentage of pixels that the grape clusters occupied within the total image was quantified once they had been segmented. The Support Vector Regression (SVR) model has been implemented in Matlab for grape harvest prediction. SVR is an extension of the SVM model that is used to predict continuous numerical values instead of classifying data. This makes it useful for applications such as agricultural production prediction. The SVR model attempts to identify a function that best fits the training data, thus minimizing errors within an acceptable tolerance while maintaining a simple model (Teja et al. 2022; Umoh et al. 2022). The model was trained using the percentage of area occupied by pixels corresponding to grape clusters and leaf volume as independent variables, while actual yield was used as the response variable. Once training of the Support Vector Regression (SVR) model in MATLAB was complete, the model was stored in a structure called "trainedModel." This structure contains the training variables (percentage of cluster area and leaf volume), as well as the model coefficients, support vectors, kernel type, regularization parameter C, and other hyperparameters associated with the SVR model.

3 Results

3.1 Stem Water Potential Prediction Model

The relationship between all variables assessed in this study, quantified with Spearman's ρ, is shown in Fig. 11. The Stem Water Potential (SWP) shows a significant correlation with several climatic variables, particularly mean temperature, relative humidity, measurement time (Hour_num), and NIR spectral band. Among these, the

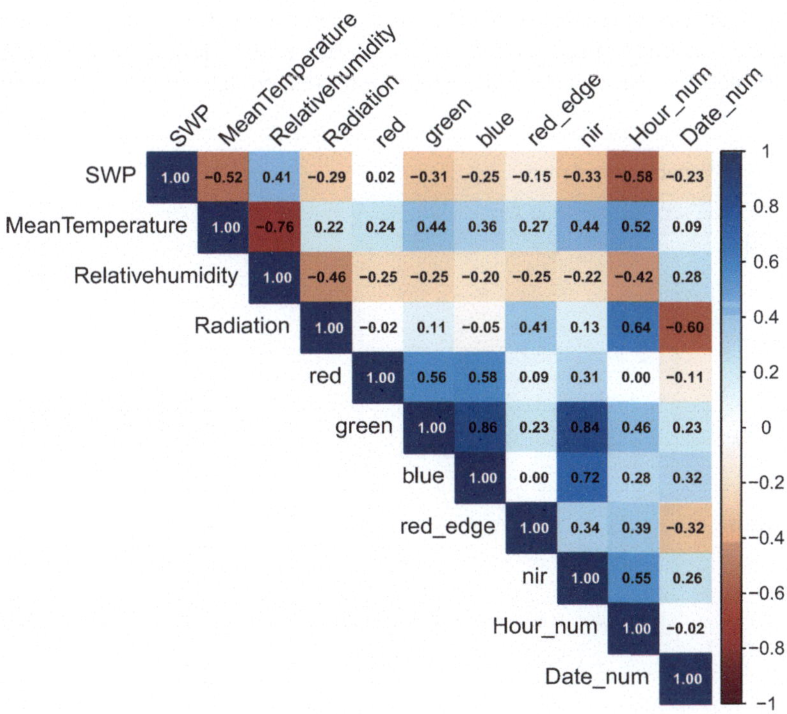

Fig. 11 Spearman correlation matrix among all predictor variables and Stem Water Potential (SWP). The strength and direction of correlations are represented by the color and intensity of the cells, with blue indicating positive and red indicating negative correlations. Source: Authors

measurement time exhibits the strongest relationship with SWP. This relationship could reflect, at least in part, the temporal evolution of vine water potential during the morning, as measurements were taken sequentially. Previous studies have shown that vine water potential changes throughout the morning and reach a stable minimum around midday (Choné et al. 2001). However, because measurements were performed sequentially, Hour_num may also partially capture the sampling order and concurrent changes in environmental conditions.

The Spearman correlation observed between the green and NIR bands ($\rho = 0.84$) indicates that both spectral variables showed a strong positive monotonic association and therefore contained partially redundant information. To reduce multicollinearity among spectral predictors and improve model interpretability, only the NIR band was retained, as it showed the strongest individual relationship with SWP.NIR reflectance has been reported to be sensitive to vegetation properties associated with plant water status (Romero et al. 2018). The Random Forest model output (Fig. 12) indicated that meteorological variables had higher predictive importance than the spectral predictor. Mean temperature and measurement time (Hour_num) were the most important variables in the model, consistent with the expected temporal dynamics of SWP during the morning. This pattern is consistent with grapevine physiology, as SWP typically decreases during the morning and tends to stabilize around midday. Hour_num helped account for the temporal

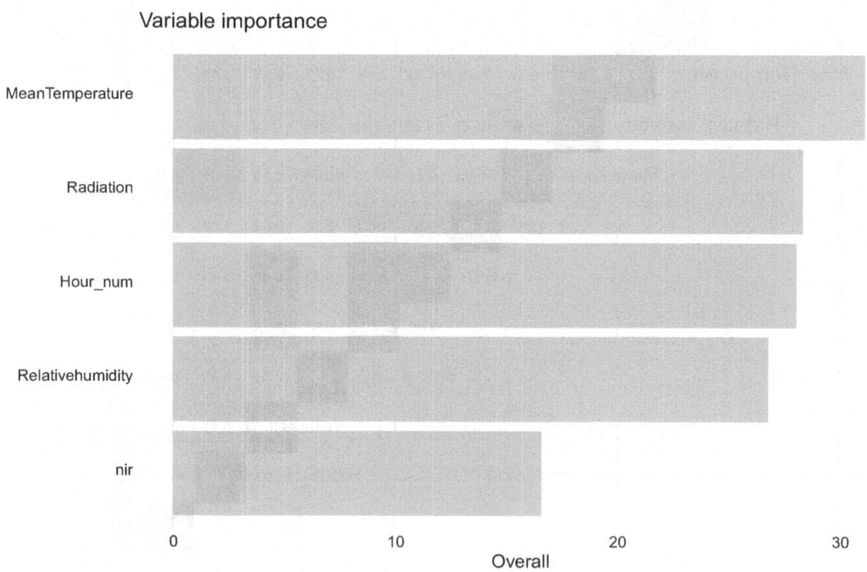

Fig. 12 Variable importance in the Random Forest model for predicting Stem Water Potential (SWP). Source: Authors

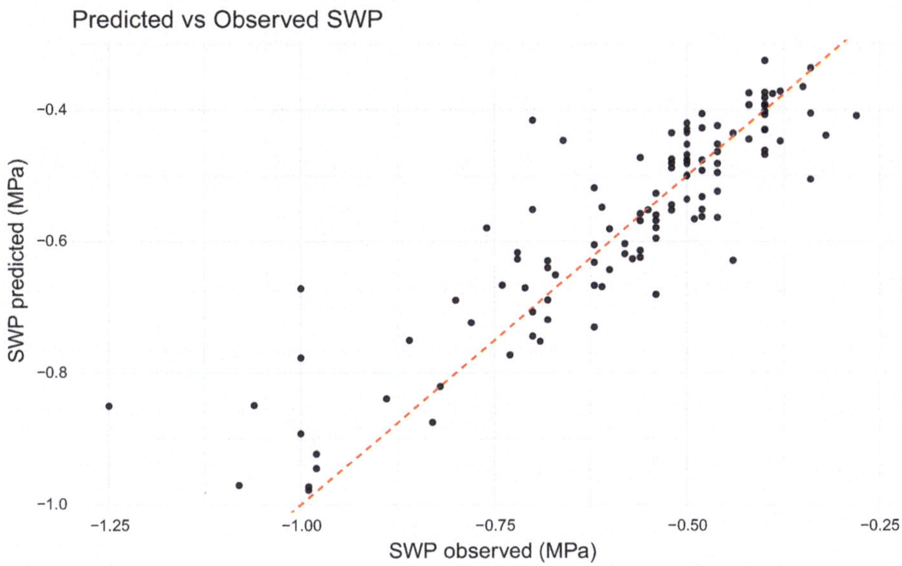

Fig. 13 Relationship between observed and predicted Stem Water Potential (SWP) values using the Random Forest model. Each point represents an individual sample, and the red dashed line indicates the 1:1 reference line. Source: Authors

variation of SWP within the morning measurement window. Relative humidity and solar radiation also showed considerable predictive importance, suggesting that atmospheric demand contributed to variations in vine water status. Although the NIR spectral band was less influential than the climatic variables, it contributed meaningfully to the model and was retained as the most informative variable among the original set of spectral bands.

Figure 13 shows the relationship between the observed SWP values and those predicted by the Random Forest model. Each point represents an individual measurement. The clustering of points along this line indicates good predictive performance of the model. Slight deviations are observed for extreme values (both high and low), suggesting a slight underestimation of the most negative SWP values. Nevertheless, the model achieves a coefficient of determination (R^2) of 0.78 and a root mean square error (RMSE) of 0.087 MPa.

The residual histogram (Fig. 14) shows a symmetric distribution centered on zero, indicating that there is no strong bias in the model. Most errors are small, between −0.25 and 0.25. The shape suggests approximate normality, which supports the model reliability.

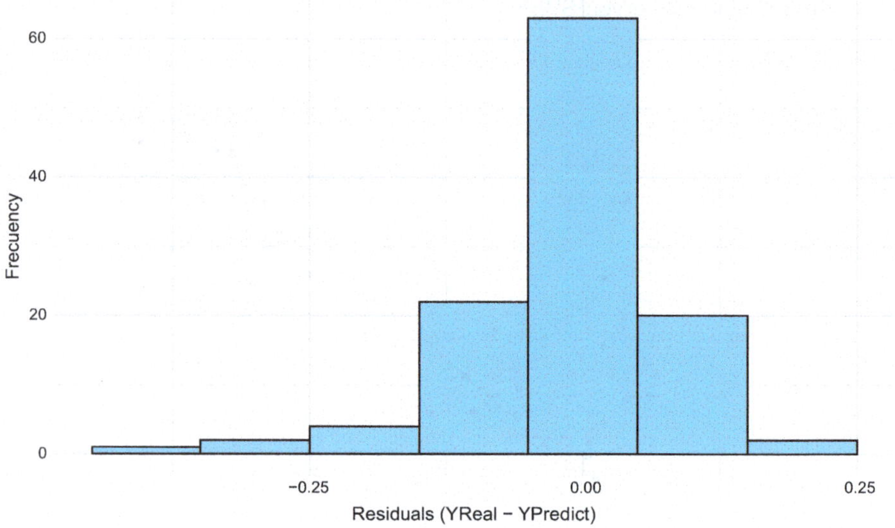

Fig. 14 Histogram of errors (residuals). Source: Authors

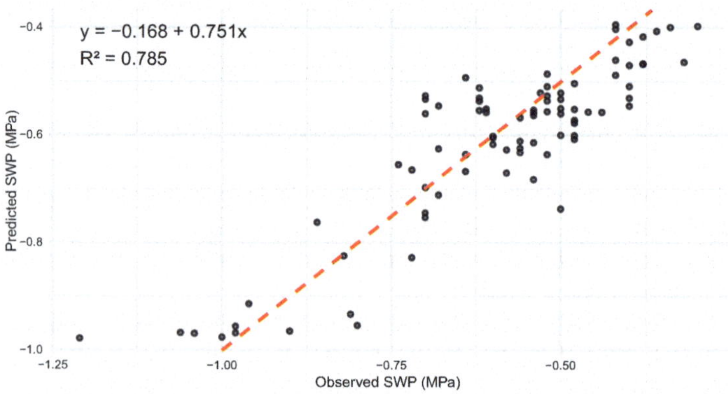

Fig. 15 Yield prediction. Source: Authors

3.2 *Yield Prediction Model*

The performance of the yield prediction model was evaluated using the coefficient of determination (R^2) (Fig. 15) showing the predictive capacity of the model when applied to grape yield prediction. The model achieves acceptable accuracy, explaining 70% of the variability observed in the production data.

In addition, a residuals histogram (Fig. 16) was constructed to evaluate the effectiveness of the crop production prediction model. The residuals were calculated

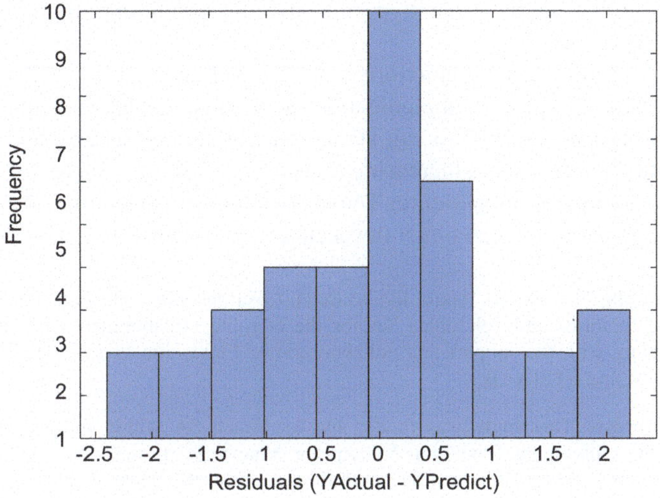

Fig. 16 Histogram of errors (residuals). Source: Authors

by subtracting the actual production values (YReal) from the production values estimated by the model (YPredict).

Most of the residuals are concentrated around zero, suggesting that the predictions were generally accurate and close to the true values. However, some outliers were also observed, both positive and negative, indicating that the model sometimes exhibits significant errors. These errors may be due to outliers or limitations in the model's ability to capture certain variations in the data. Additionally, the root mean square error (RMSE) is 1.08 units, providing a quantitative measure of the mean prediction error, as well as the mean absolute error (MAE) of 0.8378. These values indicate that, on average, the model's predictions deviate from the true value by approximately 0.84 units. Furthermore, large errors have a moderate impact, as reflected by the RMSE. Overall, these metrics and the residual analysis suggest that the model performs relatively accurately and satisfactorily, although adjustments or improvements could be explored to reduce errors in more extreme cases.

4 Conclusions

The yield prediction model performs satisfactorily and has a relatively high level of accuracy. Its performance is acceptable and functional for practical applications. However, to improve its accuracy, it is recommended that data from previous harvests be collected and that the number of samples and variables be increased. This would enable the model to be trained using a larger and more representative dataset. This would enable the model to capture variations in production more effectively, optimizing its predictive capacity and providing more reliable and realistic estimates.

The random forest-based prediction model yielded satisfactory results in estimating the SWP. However, future implementations should consider reducing the robot's data acquisition time, as SWP is not static and can change rapidly during the day due to climatic factors such as increasing transpiration rates driven by solar radiation and increasing ambient temperature. These physiological dynamics may introduce temporal mismatches that can affect the accuracy of the model.

Acknowledgments The authors would like to thank Casa del Valle vineyards (Bodegas Olarra Group) for the facilities and Telefonica SA for the ongoing collaboration with its Innovation Department, along with the support and collaboration of Fundación Premio Arce, CEIGRAM, CAR CSIC-UPM and ETSIAAB.

Competing Interests This study was funded by Telefonica SA within the framework of the "Smart Agro 5G: Intelligent Prediction System for Vineyards" project. The authors have no conflicts of interest to declare that are relevant to the content of this chapter.

References

Barriguinha A, Castro Neto M, Gil A (2021) Vineyard yield estimation, prediction, and forecasting: a systematic literature review. Agronomy 11. https://www.mdpi.com/2073-4395/11/9/1789

Botta A, Cavallone P, Baglieri L, Colucci G, Tagliavini L, Quaglia G (2022) A review of robots, perception, and tasks in precision agriculture. Appl Mech 3:830–854. https://www.mdpi.com/2673-3161/3/3/49

Choné X, van Leeuwen C, Dubourdieu D, Gaudillère JP (2001) Stem water potential is a sensitive indicator of grapevine water status. Ann Bot 87(4):477–483

Connolly C, Fleiss T (1997) A study of efficiency and accuracy in the transformation from RGB to CIELAB color space. IEEE Trans Image Process 6:1046–1048

Edemetti F, Maiale A, Carlini C, D'Auria O, Llorca J, Maria Tulino A (2022) Vineyard Digital Twin: construction and characterization via UAV images—DIWINE Proof of Concept. In: 2022 IEEE 23rd International Symposium on a World of Wireless, Mobile And Multimedia Networks (WoWMoM), pp 601–606

Ferro M, Catania P (2023) Technologies and innovative methods for precision viticulture: a comprehensive review. Horticulturae 9. https://www.mdpi.com/2311-7524/9/3/399

Jack K (2008) Chapter 2—Color Spaces. In: Digital Video And DSP, pp 15–29. https://www.sciencedirect.com/science/article/pii/B9780750689755000029

Luszczkiewicz-Pikatek M (2014) Which color space should be chosen for robust color image retrieval based on mixture modeling. Image Process Commun Challenges 5:55–64

Moreno H, Andújar D (2023) Proximal sensing for geometric characterization of vines: A review of the latest advances. Comput Electron Agric 210:107901

Romero M, Luo Y, Su B, Fuentes S (2018) Vineyard water status estimation using multispectral imagery from an UAV platform and machine learning algorithms for irrigation scheduling management. Comput Electron Agric 147:109–117. https://doi.org/10.1016/j.compag.2018.02.013

Roure F, Bascetta L, Soler M, Matteucci M, Faconti D, Gonzalez J, Serrano D (2020) Lessons learned in vineyard monitoring and protection from a ground autonomous vehicle. In: Advances in Robotics Research: from lab to market: ECHORD++: Robotic Science Supporting Innovation, pp 81–105 . https://doi.org/10.1007/978-3-030-22327-4_5

Scholander PF, Hammel HT, Hemmingsen EA, Bradstreet ED (1965) Sap pressure in vascular plants. Science 148(3668):339–346

Tardaguila J, Stoll M, Gutiérrez S, Proffitt T, Diago M (2021) Smart applications and digital technologies in viticulture: a review. Smart Agric Technol 1:100005. https://www.sciencedirect.com/science/article/pii/S2772375521000058

Teja M, Preetham T, Sujihelen L, Christy, Jancy S, Selvan M (2022) Crop recommendation and yield production using SVM algorithm. In: 2022 6th International Conference on Intelligent Computing and Control Systems (ICICCS), pp. 1768–1771

Umoh U, Asuquo D, Eyoh I, Abayomi A, Nyoho E, Vincent H (2022) A fuzzy-based support vector regression framework for crop yield prediction. In: Soft computing: theories and applications, pp 173–185

A Review of IoT in Agriculture Through Bibliometric Analysis

Dimitris Folinas, Konstantinos Rotsios, and Maria-Theodora Folina

1 Introduction

The global agricultural sector is under pressure to sustainably increase productivity and efficiency, with estimates that food production must rise by roughly 60% by 2050 to feed an expected 9.7 billion people (Shahab et al. 2024). In this context, the Internet of Things (IoT) has emerged as a transformative force in "smart agriculture" (or smart farms), leveraging networks of sensors and devices to collect real-time data and automate farm management. IoT-based smart farming systems allow continuous real-time monitoring of critical farm parameters, from soil moisture and nutrient levels to microclimate conditions and pest activity, enabling farmers to make data-driven decisions remotely (Miller et al. 2023). Key applications of IoT in agriculture include automated irrigation systems that use sensor feedback to optimize water use, precision agriculture techniques that apply water, fertilizers, and pesticides only where and when needed, and cloud-connected decision support platforms that analyze farm data to provide actionable recommendations (Miller et al. 2023; Shahab et al. 2024). For example, IoT soil-moisture sensors combined with adaptive irrigation controllers have been shown to reduce on-farm water usage by about 30% without compromising crop yields in farm trials (Dong et al. 2024). Likewise, integrating machine vision and AI analytics with IoT sensor networks enables early

D. Folinas (✉)
Department of Supply Chain Management, International Hellenic University, Thessaloniki, Greece
e-mail: folinasd@ihu.gr

K. Rotsios
School of Business, Anatolia American University, Thessaloniki, Greece

M.-T. Folina
Department of Applied Informatics, University of Macedonia, Thessaloniki, Greece

F. da Silveira et al. (eds.), *Agriculture 4.0*,
https://doi.org/10.1007/978-3-032-24552-6_25

detection of crop diseases and stress, so that robotic sprayers or drones can precisely target interventions, a capability already demonstrated to improve yields and reduce environmental impacts in IoT-enabled farms (Miller et al. 2023; Shahab et al. 2024). Such innovations underscore IoT's potential to enhance agricultural productivity, resource efficiency, and sustainability across diverse global contexts, from high-tech greenhouses in the Netherlands to smallholder farms in India and Brazil (Abdullahi et al. 2023).

Despite the rapid growth of IoT applications in agriculture worldwide, there remains a notable research gap in the literature. Many peer-reviewed studies in recent years have explored specific IoT-based solutions, for instance, designing IoT-enabled environmental monitoring systems for crop management or deploying AI-driven IoT platforms for automated irrigation (Miller et al. 2023; Dong et al. 2024). While these case-specific studies (often involving experimental farm trials or prototype deployments) provide valuable insights, they do not offer a consolidated view of the broader academic discourse. In particular, no recent work has comprehensively mapped the global research trends, thematic clusters, and evolution of knowledge on IoT in smart agriculture. Existing reviews tend to focus on technological developments or regional outcomes, and only a handful of bibliometric or scientometric analyses exist. For example, Abdullahi et al. (2023) examined IoT-in-agriculture publications from 2016 to 2023 and confirmed a surge of research interest, particularly from India, the United States, and China. However, that analysis was limited to highly cited papers and a single database, leaving a need for an updated, more inclusive mapping of global publication patterns. In short, the literature lacks a recent consolidated bibliometric review that globally charts the academic publication trends, keyword themes, and thematic evolution in IoT-driven smart farming research.

In response to this gap, this chapter aims to provide a global bibliometric analysis of IoT applications in smart farms. The study's objectives are to: (1) examine the growth of scholarly publications and geographical distribution of research on IoT in agriculture, (2) identify dominant research themes and keywords in this domain, and (3) elucidate the evolving thematic structure of the farm using advanced text clustering techniques. By analyzing keyword frequencies and co-occurrence patterns in the literature and by applying machine learning clustering (K-means) coupled with Principal Component Analysis (PCA) for visualization, we seek to uncover major topic groupings and trend dynamics that have characterized IoT-in-agriculture research in recent years.

To achieve these objectives, a bibliometric study was conducted using publications retrieved from the Scopus database (covering global peer-reviewed literature). Bibliographic data (including titles, abstracts, and author keywords) were analyzed to generate a network of co-occurring keywords and to quantify research output over time. We then employed unsupervised machine learning techniques, specifically K-means clustering on the high-frequency keywords, and used PCA to project these clusters for visualization of thematic groupings. This approach allows us to cluster related research topics and map how subfarms (e.g., IoT-based irrigation management, sensor networks for crop monitoring, AI integration in smart farming) interrelate within the broader knowledge domain. Notably, whereas most recent IoT-agriculture studies employ experimental designs or develop specific IoT systems (such as IoT-enabled monitoring systems for greenhouses or AI-based decision support platforms for precision farming), our study takes a meta-analytical approach (Miller et al. 2023; Dong et al. 2024). By aggregating and analyzing publication metadata, we provide a higher level "big picture" view of global research trends that complements the on-the-ground technological studies in the literature.

The remainder of this chapter is organized as follows. First, we present a focused literature review of IoT applications in smart agriculture, highlighting major developments and case studies across the world. Next, the methodology section details the data retrieval and bibliometric analysis procedures, including the keyword co-occurrence and clustering techniques. We then report the results of the bibliometric analysis in terms of publication trends, keyword networks, and identified research clusters. This is followed by a discussion section that interprets the findings, compares them with prior work, and addresses implications for researchers and practitioners. Finally, the chapter concludes by summarizing the key insights and suggesting future research directions in the IoT-powered smart agriculture domain.

2 IoT in Agriculture Industry

The Internet of Things (IoT) technology is rapidly developing and today represents a huge opportunity for overall development and enhanced efficiency in business operations (Tzounis et al. 2017; Ayaz et al. 2019). Practically all objects that surround us in everyday life can be connected to the network, enhancing the efficiency of the food value chain from seed planting to distribution and sales to the end consumer (Thilakarathne et al. 2021; Acharya et al. 2022). IoT can be viewed as a network of connected devices that enable information sharing across platforms (Zhang et al. 2018). Although its impact across sectors presents certain similarities, it is important to distinguish among them and examine each sector separately (Chataut et al. 2023). This technology can truly enhance the effectiveness and efficiency of the agricultural sector and provide solutions to the challenges the sector is facing (Quy et al. 2022).

The agricultural sector is currently facing several challenges and undergoing dramatic changes. The world population is growing and is expected to reach about nine billion by 2050 (Ayaz et al. 2019; Passiou et al. 2023). This trend is putting pressure on agriculture to increase food production and improve the efficiency of food systems to meet rising demand (Shahab et al. 2024). Food security thus emerges as a key challenge to be addressed (Acharya et al. 2022). Today, the food value chain is under pressure to maintain ecological balance, use fewer resources, reduce waste, deliver products promptly, and comply with strict health and environmental regulations (Shahab et al. 2024). As traditional agricultural practices cannot achieve all these goals, the food chain needs to drastically transform to address these challenges (Thilakarathne et al. 2021).

Through technology, the sector can gain significantly in effectiveness and efficiency (Subeesh and Mehta 2021). Future developments are expected to further advance IoT applications, though significant challenges remain; however, given the rapid changes in IoT technology, a deeper understanding of how IoT has evolved and how it will develop in the future is required. The outlook indicates continued growth, albeit with persisting barriers to adoption, since various stakeholders in the food value chain now recognize the role IoT can play in these efforts. The increased demand for automation has led to the rapid development and advancement of IoT (Kim et al. 2020). At the same time, available technological tools continue to improve in order to ensure the security of IoT devices and networks (Demestichas et al. 2020). As a result, the agricultural sector today is smarter and more technology-oriented than it has ever been (Ayaz et al. 2019).

2.1 Critical Areas Within the Chain Where IoT Can Add Value

The agricultural sector can benefit from technology at several stages of the production process and the postharvest chain. For example, it can improve soil management, chemical control, crop monitoring, disease prevention, irrigation control, supply chain traceability, and the control of vehicles and machinery (Navarro et al. 2020). Practically every component of the agricultural production chain can be digitalized. Some of the key areas in the chain that can benefit from IoT include:

- Provide data and support data-based decision-making: As data refers to facts and numbers collected for analysis, its importance in modern agriculture is critical. Based on accurate data, growers can make informed decisions on many aspects of production. They can identify weaknesses and work on addressing them efficiently. The collection and analysis of agricultural data over time enable the comparison of different resources and production parameters. Machine learning tools can analyze large amounts of such data to identify patterns and correlations among the various parameters. Data-driven decision-making can optimize farm operations while reducing their environmental impact (Dewi and Chen 2019; Acharya et al. 2022).
- Develop and implement automated tasks: IoT can assist agriculture by enabling automated tasks, particularly routine and repeatable ones. IoT allows for continuous monitoring and performance of required tasks. This is especially valuable in large-scale operations where precision is key to cost efficiency and productivity. In addition to detecting problems and inefficiencies in agricultural production, IoT can automate processes and take responsive actions to resolve them. Among the most common automated tasks are farm management decision-making, communication between equipment, irrigation control, weed control, and storage

management. The rate of automation in agriculture has been increasing, and this trend is expected to continue in the future (Kim et al. 2020; Subeesh and Mehta 2021).

- Provide solutions for day-to-day operations: The application of IoT has revolutionized agriculture and impacted day-to-day operations. It has enhanced data-based decision-making processes and enabled growers to make more effective decisions. It is imperative to increase production to satisfy global food needs, but equally critical to do so in an environmentally responsible way. IoT enables farmers to remotely monitor their operations (both plants and animals) and to gather timely, accurate information, giving them a detailed and up-to-date view of their farmland. Farmers can access weather information and calculate precise production levels. The connection of tools such as sensors with cloud platforms informs decisions about crop needs (e.g., water and nutrients) and soil conditions, enabling growers to address those needs more efficiently. Resource use has been optimized—IoT has led to better utilization of land, water, chemicals, labor, fertilizers, and pesticides. All of the above can be achieved without the growers' physical presence on the farm. Data collection, processing, and the corresponding responsive actions can all be conducted remotely. Furthermore, yields have increased drastically as a result (Dhanaraju et al. 2022; Xu et al. 2022).

2.2 *Objectives of IoT Applications*

IoT applications in agriculture can be grouped into five main categories based on their objective: monitoring, tracking and traceability, smart precision farming, and greenhouse production.

- Monitoring: All parameters that affect the agricultural production process can be monitored and recorded in real time. For example, growers can obtain real-time information on air temperature, precipitation, humidity, soil moisture, salinity, solar radiation, pest status, soil nutrient levels, and more. This information provides a holistic picture of farm conditions. In addition, IoT systems can suggest timely recommendations to farmers on how to deal with any issues that arise, and they can issue warning signals in cases where recorded data exceed certain thresholds. IoT systems can even apply recommended solutions automatically through interconnected devices and infrastructure. Several IoT solutions are available to facilitate large farm operations and to increase productivity and profitability (Kim et al. 2020; Navarro et al. 2020).
- Tracking and traceability: Food traceability is becoming a major consumer demand. Consumers increasingly seek food products with production processes and supply chain flows that are easily accessible and traceable. Improved traceability increases consumer trust, so food processors are striving to enhance the traceability of their products. Furthermore, traceability enables actors across

the food supply chain to increase efficiency, detect and resolve issues along the chain, and optimize their operations. Several IoT solutions are now available to build consumer trust by providing full background information on the products people consume. It is anticipated that IoT use will continue to grow in order to facilitate food tracking, traceability, and safety (Yu et al. 2022).

- Smart precision farming: IoT can facilitate smart precision farming. The use of GPS and global navigation satellite systems has led to the creation of detailed and accurate farm maps. In addition, sensors can report soil characteristics, moisture, temperature, and all other factors affecting plant growth. These data have enabled the automation of farm machinery. Today, farm machinery can operate autonomously, reducing operational expenses, increasing efficiency, and improving the accuracy of tasks. This, in turn, can lead to higher efficiency and returns. In practice, smart precision farming is used mostly in large-scale farming operations (Zhang et al. 2018; Xu et al. 2022).
- Greenhouse production: In greenhouse farming (as opposed to outdoor farming), plants are cultivated in a highly controlled environment, enabling year-round production of virtually any crop. Several monitoring systems are used to record the factors that affect plant growth and productivity, providing effective interventions and solutions when needed. Another key area where IoT can add value in greenhouses is the management of labor and energy requirements. In both cases, IoT solutions can significantly decrease labor and energy costs (Kim et al. 2020; Navarro et al. 2020). Notably, these advanced solutions currently apply largely to large farming operations. IoT adoption remains at rather low levels among smaller farms, due to the high cost of acquiring the systems and a knowledge deficit among farmers regarding how to manage and operate them (Thilakarathne et al. 2021; Xu et al. 2022). Efforts should continue to enable the use of IoT solutions in smaller farms so that IoT can contribute to the overall sustainable development of the agricultural sector.
- Increase environmental sustainability and decrease food waste: Currently, on a global level, it is estimated that roughly 50% of food production does not reach final consumers and goes to waste at the farm level due to inefficiencies in the food system (Navarro et al. 2020; Rotsios et al. 2025). IoT helps decrease waste and optimizes efficiency by remotely monitoring and controlling the various stages of food systems (Navarro et al. 2020; Shahab et al. 2024). By monitoring the farm environment, an integrated IoT system can provide specific recommendations in terms of which crops to cultivate, irrigation scheduling, fertilization and nutrition needs, optimal harvest time, distribution planning, and so on. With IoT, agricultural tasks can be performed with great precision. This leads not only to economic gains but also to a more sustainable agricultural sector, one that develops with respect for the environment (Acharya et al. 2022; Shahab et al. 2024). (This point is especially important in today's increasingly competitive global agricultural landscape.)

2.3 Challenges in Implementing IoT in the Agriculture Sector

There is consensus that the use of IoT in agriculture has expanded and will continue to expand in the future (Chataut et al. 2023). Despite the many benefits outlined in this section, several factors are hindering IoT adoption by farmers. A key limiting factor is the high cost of acquiring, implementing, and maintaining IoT systems in agricultural operations (Navarro et al. 2020; Thilakarathne et al. 2021). For example, the number of wireless sensors needed increases as farm size increases, which drives up the cost of the required devices and infrastructure.

Another major issue is connectivity. In many parts of the world, farms are in remote locations with weak signals and limited network capacity, leading to connectivity issues for IoT devices (Thilakarathne et al. 2021; Xu et al. 2022).

A further limiting factor is the knowledge and skills required to operate these systems. The farming population in most countries is aging, and many farmers either lack the technical expertise to operate IoT systems or are not aware of the available technology and its capabilities. Developing IoT awareness and providing training for farmers to operate these systems is therefore necessary (Thilakarathne et al. 2021; Quy et al. 2022).

Concerns have also been expressed about the reliability and accuracy of information provided by IoT, since false information can have devastating results for farmers. Reliability issues may stem from environmental conditions. Sensors placed outdoors are exposed to harsh weather (e.g., extreme temperatures, humidity) and thus need to function properly under those conditions. Large fluctuations in weather can affect sensor operation and the accuracy of the data provided. Sensors need to be checked regularly and replaced if needed, and new, more durable materials may be required to improve sensor longevity (Demestichas et al. 2020; Quy et al. 2022).

The issue of security is another hurdle to wider IoT use in agriculture. The agricultural sector appears more sensitive to security issues than other sectors (Demestichas et al. 2020). Furthermore, ownership of data (whether it belongs to IoT service providers or to farms) remains under debate, which raises security and privacy concerns. With the use of IoT, a huge amount of information is gathered and transferred online, and the overall expansion of IoT raises concerns about data protection (Demestichas et al. 2020). For example, if data on production levels and other sensitive information about farm productivity become available to government authorities, it may lead to increased taxes or other fees for farmers. Likewise, if such data become accessible to third parties, they may be used unethically. For instance, if production projections are obtained by third parties, they could manipulate market prices to the disadvantage of farmers (Acharya et al. 2022). Additionally, cyberattacks can disrupt interconnected agricultural enterprises. Farmers could suffer serious financial and personal damage from data theft. Data could potentially be stolen by other stakeholders in the food supply chain and used unethically to minimize farmers' profits for their own gain. Overall, despite progress made in IoT security, many issues remain unresolved. This means that all stakeholders in the food supply chain should exercise caution in their use of IoT. It is also important to allocate resources toward enhanced security measures and tools to ensure the security of IoT, so that its use in the agricultural sector can continue to grow (Demestichas et al. 2020).

3 Bibliometric Analysis

This section presents findings of a bibliometric analysis focused on the intersection of IoT, agriculture, and "smart farms." The Scopus database was queried using the keywords: KEY(agriculture) AND KEY("smart farms") AND KEY(IoT), filtered to include only English-language journal articles and conference proceedings. This search returned a total of 83 relevant publications. Bibliographic data (titles, abstracts, keywords) were exported from Scopus and analyzed using VOSviewer for keyword co-occurrence and network visualization. Additionally, K-means clustering with PCA was employed to validate and deepen the thematic interpretation of the data.

3.1 Distribution of Articles by Year

The search revealed that early research on IoT-based smart agriculture was sparse before 2015, but publications began rising significantly around 2020. This trend corresponds to the broader adoption of IoT in agricultural applications and a global push for "smart farming" solutions during this period (Abdullahi et al. 2023; Xu et al. 2024; Abdelmoneim et al. 2025). In fact, a recent systematic review observed a rapid surge in IoT-agriculture publications from 2020 onward, attributing it to advancements in low-power sensors, widespread connectivity (e.g., cloud platforms) and AI-based analytics that together enabled more sophisticated farming applications (Yousaf et al. 2023). Consistent with this, bibliometric studies confirm that 2020–2023 marked a peak in IoT-related agricultural research output, reflecting heightened scholarly and practical interest in digital farming (Maulana et al. 2022; Abdelmoneim et al. 2025). These years saw not only more papers being published but also greater citation impact, indicating that IoT in agriculture had gained recognition as a critical research frontier by the early 2020s (Fig. 1).

3.2 Distribution of Articles by Type

Conference proceedings made up 83.1% of the dataset, while journal articles accounted for 16.9%. This imbalance reflects the farm's technological orientation and the fast-paced nature of IoT development, where new solutions are often

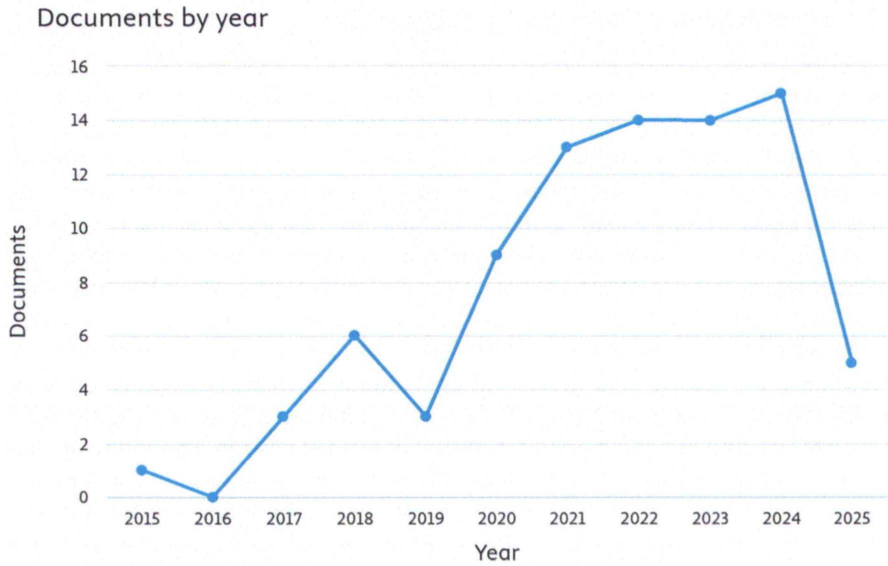

Fig. 1 Distribution of scientific articles by year. (Source: Scopus)

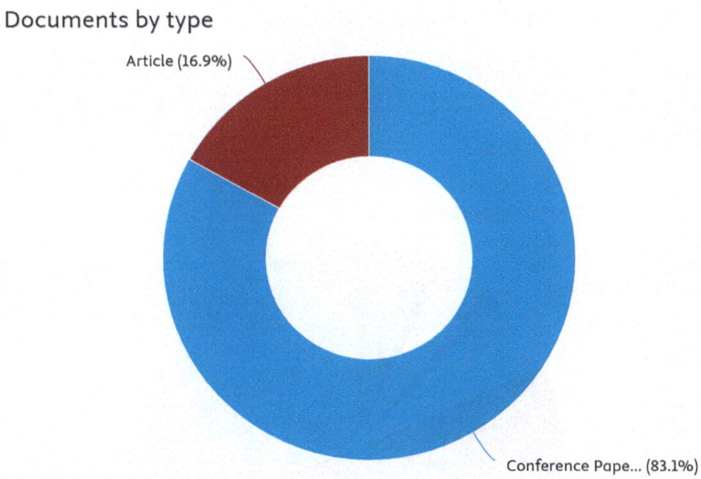

Fig. 2 Percentage of articles by type (conference papers vs. journal papers)

disseminated at conferences before being formalized into journal studies (Abdullahi et al. 2023; Kushartadi et al. 2023; Bhuyan 2024). Such a trend is typical in emerging tech farms where applied innovations lead theoretical advancements (Fig. 2).

3.3 Distribution of Articles by Subject Area

The dominant subject areas of these publications (according to Scopus classifications) were Computer Science (30.2% of documents) and Engineering (25.9%), followed by smaller contributions from areas like Decision Sciences, Physics and Astronomy, and Mathematics. This distribution aligns with the core technologies underpinning smart agriculture, such as IoT systems, data analytics, cloud computing, and robotics, which are fundamentally rooted in computing and engineering disciplines (Suebsombut et al. 2017; Wolfert et al. 2017; Pathmudi et al. 2023; Abdelmoneim et al. 2025). In other words, much of the research in "smart farms" is driven by technological innovation (sensors, networks, AI algorithms, etc.) rather than traditional agricultural science. The relative absence of publications in agronomy or plant science journals underscores the need for more interdisciplinary collaboration between domain experts in agriculture and the developers of IoT technologies. Bridging this gap is critical to ensure that high-tech solutions address real agronomic needs and are usable by farmers (Abdelmoneim et al. 2025). Encouragingly, some recent initiatives and studies emphasize cross-disciplinary approaches, for example, integrating agronomists, engineers, and data scientists in project teams, to harness IoT's full potential in farming environments (Abdelmoneim et al. 2025). This suggests that while the current knowledge base is tech-heavy, there is growing recognition that agricultural and computer sciences must work hand-in-hand to achieve sustainable "smart agriculture" outcomes (Fig. 3).

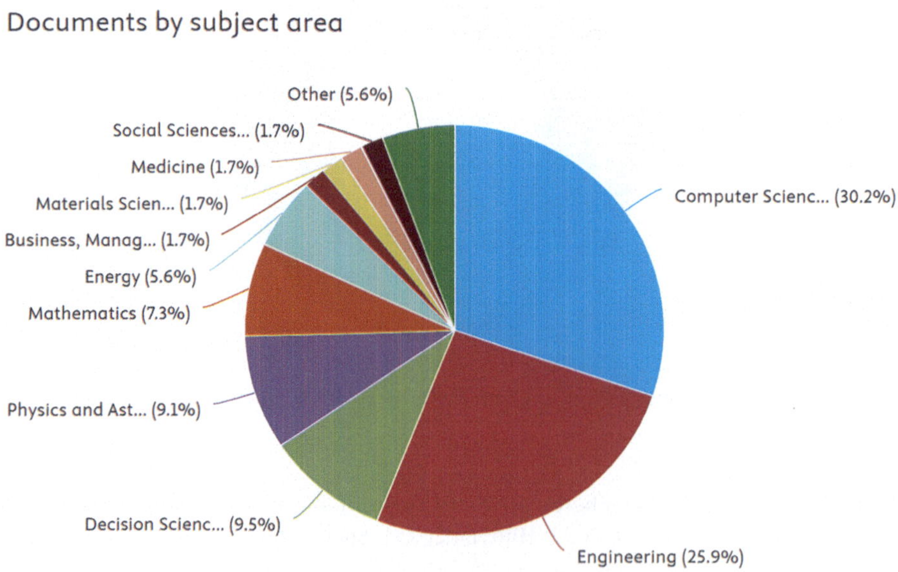

Fig. 3 Percentage of articles by subject area

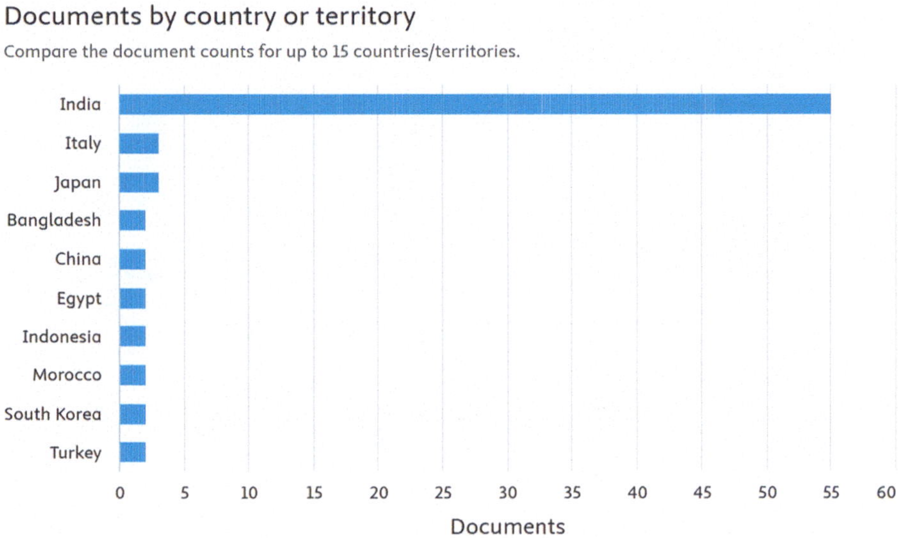

Fig. 4 Productivity of articles by country (number of publications per country)

Figure 4 illustrates the ten most productive countries in terms of scientific publications related to the application of IoT in the agricultural sector. These countries represent the global centers of research activity driving innovation and technological advancement in smart agriculture.

The country-level distribution of publications indicates that India is by far the leading contributor to research on IoT in agriculture, accounting for the vast majority of documents in the dataset, nearly 60 of the 83 articles. This finding is consistent with previous large-scale reviews that identify India as a global leader in IoT-agriculture research, followed distantly by China, the United States, and European countries (Singh et al. 2020; Abdullahi et al. 2023; Safeer et al. 2024). This dominant output underscores India's active engagement in smart farming research, likely driven by its large and diverse agricultural sector, strong emphasis on digital transformation, and national initiatives supporting agri-tech innovation. Other countries such as Italy, Japan, Bangladesh, China, and Egypt appear with significantly fewer publications, highlighting a regional imbalance in research activity. This suggests that while IoT in agriculture is a global concern, scholarly contributions are currently concentrated in specific countries with strong technical or agricultural research infrastructures.

3.4 The Most Frequent Keywords of the 83 Articles

To understand the thematic focus of the literature, a bibliometric map was generated to analyze keyword co-occurrence among the 83 articles. We used VOSviewer to build a network of all keywords from these publications, with co-occurrence

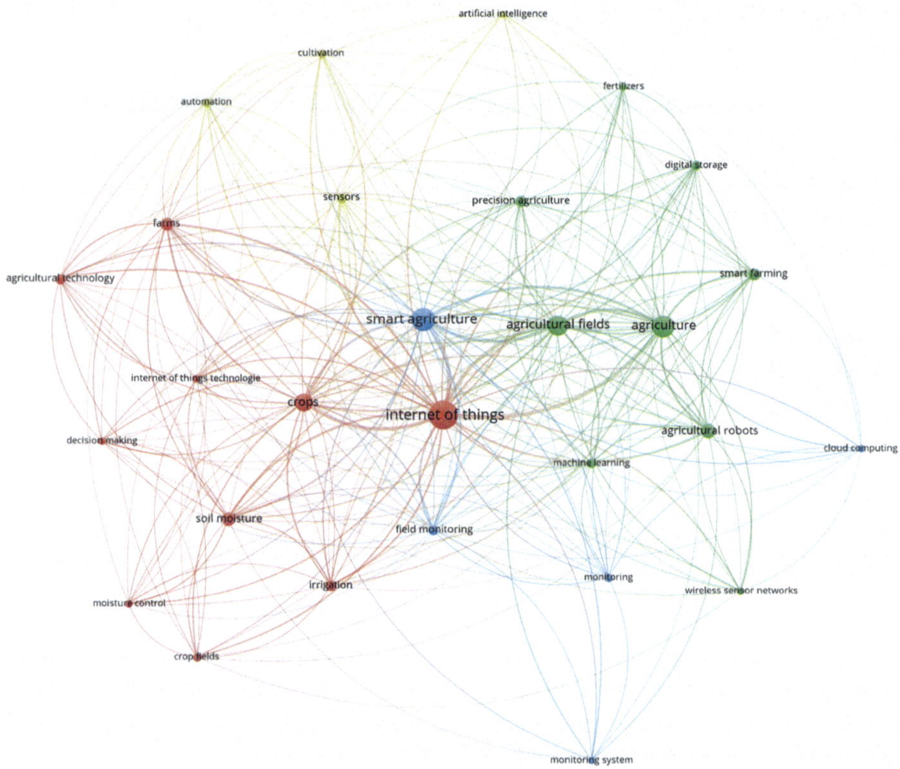

Fig. 5 Keyword density visualization of the 83-article dataset (high-frequency keywords and their co-occurrence clusters)

serving as the measure of relatedness. In the VOSviewer mapping, each keyword is represented as a node, and the proximity between nodes reflects their frequency of appearing together in documents. For inclusion in the visualization, a minimum threshold of 10 occurrences per keyword was set (given the moderate size of the dataset) to highlight only the most significant terms. Figure 5 presents the keyword density visualization, where color intensity denotes the concentration of frequently co-occurring keywords. Areas in yellow indicate higher keyword density (i.e., many keywords in close semantic proximity), whereas areas in darker blue indicate lower density. By inspecting this map, one can discern clusters of related keywords and infer broader research themes. Closely positioned keywords represent higher thematic similarity, for example, keywords related to IoT architecture and wireless sensors might cluster together, whereas words plotted farther apart are less directly related within the literature. The initial bibliometric mapping delineated several distinct clusters of keywords within the IoT-agriculture research domain, offering a visual snapshot of how topics are grouped and their relative prominence.

Table 1 All keywords for clusters

Color	Cluster	Keywords (ascending alphabetical order)
Red	Cluster 1	Agricultural farms, agricultural robots, agriculture, digital storage, fertilizers, internet of things, machine learning, precision agriculture, smart farming, wireless sensor networks
Green	Cluster 2	Agricultural technology, artificial intelligence, automation, crops, cultivation, farms, internet of things, sensors, smart agricultures
Blue	Cluster 3	Cloud computing, farm monitoring, monitoring, monitoring system
Yellow	Cluster 4	Crop farms, irrigation, soil moisture

The keywords extracted from the analyzed articles were organized into four thematic clusters based on their semantic relationships. Table 1 presents these clusters in a structured format, including the designated color for each cluster, the corresponding thematic focus with the number of related publications, and a list of representative keywords. This categorization enables a clearer understanding of how the research landscape is structured around distinct areas of focus.

Upon examining these clusters, we can conceptually group the research themes as follows: Cluster 1 relates to core IoT technologies and general smart farming concepts; Cluster 2 centers on advanced farming techniques (often AI-enhanced) and automation; Cluster 3 emphasizes supporting infrastructure like cloud and monitoring platforms; and Cluster 4 is focused on specific farm applications (especially irrigation) which intersects with the other clusters.

These results align well with patterns identified in the previous literature. For instance, in a comprehensive bibliometric study, Abdullahi et al. (2023) found that common keyword themes in smart agriculture research include IoT and sensor networks, precision agriculture and machine learning, and smart farming management systems, closely mirroring the clusters identified here. Similarly, other reviews have noted that topics such as precision agriculture, "smart agriculture" (broadly speaking), remote sensing technologies, and climate-smart farming tend to form distinct clusters in the literature (Yousaf et al. 2023). Furthermore, Chamara et al. 2022 review of Ag-IoT for crop monitoring supports the prominence of IoT sensors, data analytics, and precision techniques as key themes. Our clustering outcomes reinforce those observations, indicating that the knowledge structure of IoT in agriculture consistently revolves around a few key thematic pillars: technological infrastructure (IoT/WSN, cloud), data-driven farming practices (precision/AI), and domain-specific applications (crop management, irrigation, etc.).

After delineating the clusters with VOSviewer, we generated a visual map (Fig. 6) to depict the distribution of keywords and their interrelations across the dataset. In this network visualization, each keyword is a node and links between nodes represent co-occurrences in the same publications. The nodes are colored according to the cluster membership (red, green, blue, and yellow as described earlier), and the size of each node reflects the frequency of that keyword's appearance in the dataset. Thus, larger nodes (keywords like "Internet of Things," "smart farming," etc.) indicate topics that appear most frequently across the 83 articles.

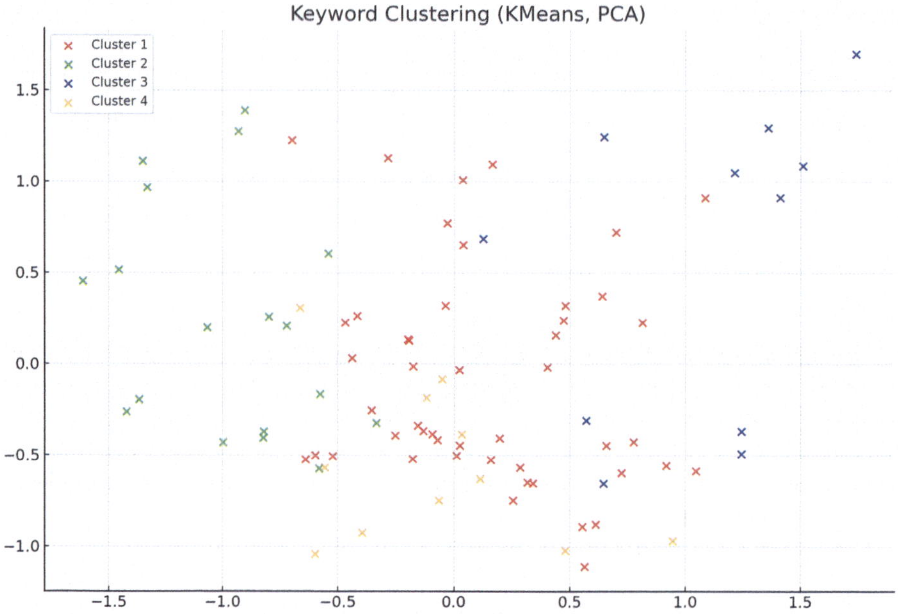

Fig. 6 Keyword clustering using K-Means and PCA dimensionality reduction

The mapped network provides an intuitive overview of the thematic landscape: one can see clusters of tightly connected keywords, demonstrating how certain concepts commonly appear together (for example, "wireless sensor networks" might be closely linked with "precision agriculture" and "machine learning" in many papers, indicating a cluster on data-driven precision farming). The distance and linkage among clusters also illustrate the degree of overlap; in our map, the clusters are relatively well-defined but not isolated, suggesting an interdisciplinary interplay (e.g., the "irrigation" cluster overlaps with the IoT-tech cluster, reflecting that irrigation studies often involve IoT devices and AI algorithms).

To validate the thematic structure of the bibliometric data, we compared the clustering results from VOSviewer with those generated through an independent machine learning approach (K-means clustering combined with PCA for dimensionality reduction). The goal was to see if an algorithmic classification of the high-frequency keywords would converge on a similar grouping as the manual VOSviewer approach. The K-means clustering was applied to the same keyword co-occurrence data, specifying four clusters for direct comparability. The results were highly consistent with the VOSviewer findings, which strengthens confidence in our interpretation. In both cases, four conceptually coherent groups emerged:

- Cluster 1 (IoT Applications in Agriculture), K-means grouped keywords related to IoT deployments on farms, smart farming practices, and general agricultural IoT use-cases. This closely mirrors VOSviewer's Cluster 1 (IoT and core smart farming concepts).

- Cluster 2 (Precision Agriculture and AI), Terms involving data-driven decision support, AI, machine learning, and automation in farming clustered together, similar to VOSviewer's precision agriculture/AI-enhanced farming cluster (Cluster 2 in our description).
- Cluster 3 (Monitoring Infrastructure), Keywords such as cloud computing, farm monitoring, and sensor networks were clustered by K-means, aligning with the technological infrastructure theme (VOSviewer's Cluster 3).
- Cluster 4 (General Enabling Technologies), A somewhat broader cluster including terms like sensors, automation, and possibly miscellaneous smart-agri tools emerged, comparable to the remaining general tools/enablers category (which in VOSviewer was partly combined with others but conceptually present as the foundational tech and techniques enabling smart agriculture).

The high degree of thematic overlap between the manual (VOSviewer) and algorithmic (K-means) approaches confirms the robustness of the identified research trends. In essence, both methods independently recognized the same underlying structure in the data: one cluster focusing on IoT systems in agriculture, one on precision farming and AI methods, one on supporting IT infrastructure, and one on miscellaneous or general topics that bridge the others. This convergence suggests that these four areas truly represent the core contours of IoT-in-agriculture research circa 2015–2023. It also lends credibility to the subsequent analyses, as the coherence of the farm's knowledge structure is now verified by two different techniques.

Figure 7 provides a network visualization of the keyword co-occurrence map, with an overlay that illustrates the clusters in different colors (as previously described) and highlights the prominence of each keyword by node size. In this figure, the color of each node corresponds to its cluster membership (red, green, blue, or yellow) and the links between nodes show co-occurrence relationships (thicker/stronger links meaning more frequent co-occurrence). The size of each node is proportional to the number of publications in which the keyword appears; effectively, a larger node means that the keyword is a major topic across many of the 83 publications. This allows us to see not only the cluster groupings but also which topics are most central in the IoT-agriculture research landscape. For example, terms like "Internet of Things" and "smart farming" appear as large nodes (indicating they are widely discussed across the literature), whereas a term that might be more niche (like "soil moisture," even though it forms a cluster, could be slightly smaller if it appears in fewer papers overall). The network visualization helps illustrate how the farm is structured: a few big themes dominate (IoT, smart/precision agriculture, sensors), and these are interconnected, whereas more specialized topics orbit around these main hubs.

In addition to the network visualization, an overlay visualization was generated (not shown here, but conceptually described), which scores each keyword based on a common attribute, in this case, the average year of publication of the articles in which the keyword appears. This overlay adds a temporal dimension to the bibliometric map. The background of the overlay map is deep blue, and keywords

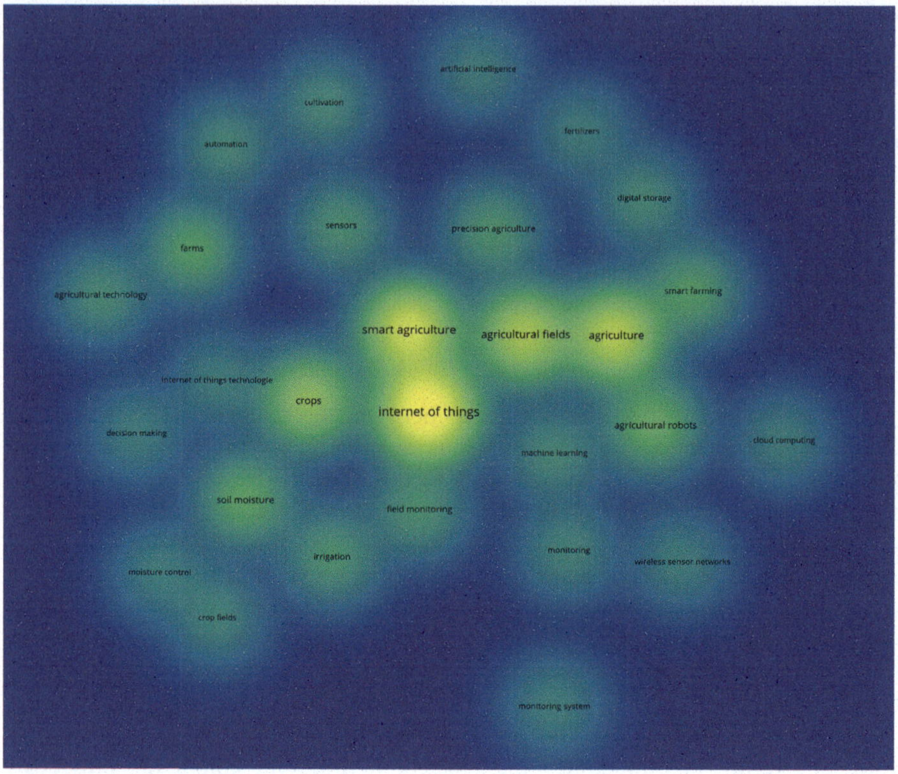

Fig. 7 Network visualization of keyword co-occurrence, with node color indicating cluster membership and node size indicating keyword frequency across the dataset

(nodes) are colored from blue to yellow according to their "average publication year" score. If a keyword appears mostly in older publications (e.g., around 2015–2018), it will tend toward blue; if it appears more in recent publications (e.g., 2021–2023), it will be shaded yellow. In the density view, areas that are brightly colored (yellow) indicate regions of the map where many keywords from recent years co-occur, whereas darker areas indicate older clusters. This kind of overlay helps identify emerging versus established topics. For instance, one might observe that terms related to AI and machine learning in agriculture glow yellow, suggesting they have become prominent in very recent years, whereas a term like "wireless sensor networks" might be more evenly blue-green, indicating it has been a steady topic over the whole period. Such temporal insights are valuable for understanding how the focus of IoT in agriculture has shifted over time, for example, from early emphasis on basic sensor networks to more recent emphasis on AI-driven analytics. In our analysis, the overlay confirmed that many of the precision agriculture and AI-related keywords have become more frequent in the latest years (2020–2023), whereas foundational IoT terms were present throughout the decade. This dynamic view underscores how the farm has evolved and where it is potentially heading (e.g., greater integration of AI, edge computing, and advanced analytics in the coming years). Areas of the map that are bright yellow highlight current "hot spots" in research, whereas any that remain blue might indicate topics that peaked earlier or have not seen as much recent activity.

4 Conclusions

This study offers an extensive bibliometric analysis of global research activity on the application of the Internet of Things (IoT) in agriculture, with an emphasis on "smart farms." The findings show a significant and rapidly increasing research output from around 2020 onwards, reflecting the broader digital transformation of the agricultural sector in the past few years. In particular, the number of IoT-in-agriculture publications and their impact grew exponentially after 2019, indicating that the scientific community has swiftly recognized IoT as a catalyst for innovation in farming (Yousaf et al. 2023). Geographically, India emerged as the most prolific producer of research work in this domain, followed by China, the United States, and selected European countries. India's leadership in publication volume corroborates other analyses and is likely driven by a combination of its pressing agricultural needs and strong ICT expertise (Singh et al. 2020; Abdullahi et al. 2023).

Through keyword co-occurrence analysis and clustering, we identified four main research thematic areas within the IoT-agriculture literature: (1) IoT applications and systems for agriculture (e.g., smart farming frameworks, IoT architectures, sensor deployments on farms); (2) Precision agriculture and AI-enhanced farming practices (data-driven approaches, machine learning for crop monitoring, decision support systems); (3) Technological infrastructure for monitoring and automation (cloud computing, IoT communication networks, farm monitoring systems); and (4) Specific applications in resource management (with a focus on irrigation management, soil moisture monitoring and related smart farm practices). The strong convergence between the results of our machine-learning clustering and the manual VOSviewer clustering confirms the reliability of these thematic conclusions, multiple methods point to the same set of core topics. These areas are also consistent with the priorities highlighted in contemporary smart agriculture research: improving efficiency and precision in farm operations, leveraging big data and AI for better decision-making and developing robust IoT infrastructures to support these goals (Abdullahi et al. 2023; Yousaf et al. 2023).

Despite the considerable potential of IoT to transform agricultural production, our analysis and the broader literature, indicate that on-the-ground adoption of these technologies remains limited, especially among smallholder and resource-constrained farmers. Key barriers frequently cited include high upfront costs of IoT devices and systems, lack of rural connectivity (reliable Internet and communication networks in farming areas), concerns about data privacy and security and limited technical expertise or digital literacy among farmers who are the end-users (Yousaf et al. 2023; Abdelmoneim et al. 2025; Nxumalo and Chauke 2025). For example, small farms may find it prohibitively expensive to deploy sensor networks and may lack access to technical support, leading to a slow uptake of IoT solutions in practice. Overcoming these barriers is critical to realizing the benefits of smart agriculture on a global scale (Kapari et al. 2024; Abdelmoneim et al. 2025). Policy interventions and support programs can play a role, for instance, subsidies or financial incentives for adopting precision irrigation systems, investment in rural Internet infrastructure and training programs to build farmers' capacity to use digital tools. Additionally, technology design can be made more accessible (both in terms of cost and user-friendliness) to suit the needs of farmers with varying levels of expertise. Removing these barriers through supportive policies, education, and inclusive design is essential if IoT innovations are to move beyond pilot projects and affluent farms into mainstream agriculture.

In summary, IoT holds transformative potential for enabling more sustainable and efficient agricultural practices, as evidenced by the surge in research and development efforts in recent years. The global research landscape for IoT in agriculture is rapidly evolving, with certain countries and topics leading the way. Going forward, further interdisciplinary studies are needed, bringing together agronomists, computer engineers, data scientists and social scientists, especially in areas that are currently underrepresented (such as socioeconomic impacts, farmer adoption strategies, and policy frameworks for smart farming). Such research will help ensure the effective, safe, and socially equitable implementation of "smart" agricultural practices on a global scale. By addressing both technical challenges and human-centric factors, the next wave of IoT-agriculture innovations can be more widely adopted and truly contribute to global food security and sustainable farming in the coming decades.

Future research could place greater emphasis on the development of cost-effective and scalable IoT solutions that can be realistically adopted by smallholder and resource-constrained farmers. Such technologies should be designed with affordability, ease of use, and adaptability to diverse farming contexts in mind, ensuring that the benefits of digital agriculture are not restricted to large-scale or technologically advanced farms. At the same time, investments in rural connectivity infrastructure—such as reliable Internet access, low-power communication networks, and cloud or edge computing support—are essential for enabling the consistent and efficient operation of IoT-based systems in remote agricultural areas. Another critical direction involves the establishment of robust governance frameworks that address issues of data security, privacy, and ownership, safeguarding farmers' rights while promoting trust in IoT adoption. Without clear regulatory mechanisms and ethical guidelines, concerns about misuse of agricultural data may continue to hinder widespread implementation.

In addition, there is an urgent need for interdisciplinary research efforts that bridge the current gap between agricultural science, computer engineering, and data analytics. While technological advancements have dominated the literature to date, more integrated approaches that involve agronomists, social scientists, and policymakers will be crucial for designing solutions that are both technologically advanced and socially sustainable. Such collaborations could, for example, explore how IoT technologies influence farmers' decision-making processes, economic viability, and long-term sustainability outcomes. By aligning digital innovation with agronomic expertise and socioeconomic realities, future research can ensure that IoT not only advances technical efficiency but also contributes to broader goals of food security, environmental sustainability, and equitable agricultural development.

Competing Interests The author(s) has no competing interests to declare that are relevant to the content of this manuscript.

References

Abdelmoneim AA, Kimaita HN, Al Kalaany CM (2025) IoT sensing for advanced irrigation management: a systematic review of trends, challenges and future prospects. Sensors 25(7):2291. https://doi.org/10.3390/s25072291

Abdullahi HO, Mahmud M, Hassan AA, Ali AF (2023) A bibliometric analysis of the evolution of IoT applications in smart agriculture. Ingénierie des Systèmes d'Inform 28(6):1495–1504

Acharya B, Garikapati K, Yarlagadda A, Dash S (2022) Internet of things (IoT) and data analytics in smart agriculture: benefits and challenges. In: AI, edge and IoT-based smart agriculture. Academic Press, pp 3–16

Ayaz M, Ammad-Uddin M, Sharif Z, Mansour A, Aggoune EHM (2019) Internet-of-Things (IoT)-based smart agriculture: toward making the farms talk, vol 7. IEEE Access, pp 129551–129583

Bhuyan AS (2024) Smart farming with IoT: a bibliometric analysis of trends and developments. In: Proceedings of the 2024 International conference on progressive innovations in intelligent systems and data science (ICPIDS), Pattaya, Thailand, 27–28 December 2024, pp 1–6

Chamara N, Islam MD, Bai G, Shi Y, Ge Y (2022) Ag-IoT for crop and environment monitoring: past, present and future. Agric Syst 203:103497

Chataut R, Phoummalayvane A, Akl R (2023) Unleashing the power of IoT: a comprehensive review of IoT applications and future prospects in healthcare, agriculture, smart homes, smart cities, and industry 4.0. Sensors 23(16):7194

Demestichas K, Peppes N, Alexakis T (2020) Survey on security threats in agricultural IoT and smart farming. Sensors 20(22):6458

Dewi C, Chen RC (2019) Decision-making based on IoT data collection for precision agriculture. In: Asian conference on intelligent information and database systems. Springer International Publishing, Cham, pp 31–42

Dhanaraju M, Chenniappan P, Ramalingam K, Pazhanivelan S, Kaliaperumal R (2022) Smart farming: Internet of Things (IoT)-based sustainable agriculture. Agriculture 12(10):1745

Dong Y, Werling B, Cao Z, Li G (2024) Implementation of an in-farm IoT system for precision irrigation management. Front Water 6:Article 1353597

Kapari M, Mapiye C, Gumbi RV (2024) Barriers to digital technology adoption among smallholder farmers in South Africa. Afr J Sci Technol Innov Dev 16(1):123–135

Kim WS, Lee WS, Kim YJ (2020) A review of the applications of the internet of things (IoT) for agricultural automation. J Biosyst Eng 45(4):385–400

Kushartadi T, Mulyono AE, Al Hamdi AH et al (2023) Theme mapping and bibliometric analysis of two decades of smart farming. Information 14(7):396

Maulana FI, Purnomo A, Pratama FC (2022) Mapping and visualizing research trends of smart farming with IoT in the last ten years using bibliometric analysis. In: Proceedings of the 7th NA International conference on industrial engineering and operations management (IEOM 2022), pp 2593–2602

Miller T, Mikiciuk G, Durlik I, Mikiciuk M, Łobodzińska A, Śnieg M (2023) The IoT and AI in agriculture: the time is now – a systematic review of smart sensing technologies. Sensors 25(12):3583

Navarro E, Costa N, Pereira A (2020) A systematic review of IoT solutions for smart farming. Sensors 20(15):4231

Nxumalo GS, Chauke H (2025) Challenges and opportunities in smallholder agriculture digitization in South Africa. Front Sustainable Food Syst 9:1583224. https://doi.org/10.3389/fsufs.2025.1583224

Passiou KE, Agapitou C, Rotsios K, Folinas D (2023) Plant-based protein food products: perceptions from the Greek food industry. Econ Balkan East Eur Countries J 2023:119–137

Pathmudi VR, Khatri N, Kumar S, Abdul-Qawy AS, Vyas AK (2023) A systematic review of IoT technologies and their constituents for smart and sustainable agriculture applications. Sci Afr 19:e01577

Quy VK, Hau NV, Anh DV, Quy NM, Ban NT, Lanza S, Muzirafuti A (2022) IoT-enabled smart agriculture: architecture, applications, and challenges. Appl Sci 12(7):3396

Rotsios K, Folinas D, Mouchtari C, Andreou A, Fotiadis T, Folina MT, Gasteratos A (2025) Exploring the factors influencing the acceptance of IoT applications in food packaging. Foods 14(4):575

Safeer S, De Mastro G, Pulvento C (2024) IoT-based climate-smart agriculture succeeded by blockchain database – a bibliometric analysis. Front Sustain Food Syst 8:1406871

Shahab H, Iqbal M, Sohaib A, Khan FU, Waqas M (2024) IoT-based agriculture management techniques for sustainable farming: a comprehensive review. Comput Electron Agric 220:108851

Singh S, Haneef F, Kumar S, Ongsakul V (2020) Internet of Things and agriculture relationship: a bibliometric analysis. J Global Bus Adv 13(4):643–664

Subeesh A, Mehta CR (2021) Automation and digitization of agriculture using artificial intelligence and internet of things. Artif Intell Agric 5:278–291

Suebsombut P, Sekhari A, Ueasangkomsate P, Bouras A (2017) Using bibliometric analysis to classify trends and future directions on "smart farm". In: 2017 International conference on digital arts, media and technology (ICDAMT) Chiang Mai, Thailand, 1–4 March 2017, pp 205–210

Thilakarathne NN, Bakar MS, Abas PE, Yassin H (2021) Towards making the fields talks: a real-time cloud enabled IoT crop management platform for smart agriculture. Front Plant Sci 13:1030168

Tzounis A, Katsoulas N, Bartzanas T, Kittas C (2017) Internet of Things in agriculture, recent advances and future challenges. Biosyst Eng 164:31–48

Wolfert S, Ge L, Verdouw C, Bogaardt MJ (2017) Big data in smart farming-a review. Agricultural systems, 153:69–80. https://doi.org/10.1016/j.agsy.2017.01.023

Xu J, Gu B, Tian G (2022) Review of agricultural IoT technology. Artif Intell Agric 6:10–22

Xu J, Li Y, Zhang M, Zhang S (2024) Sustainable agriculture in the digital era: past, present and future trends by bibliometric analysis. Heliyon 10(7):e034612

Yousaf A, Kayvanfar V, Mazzoni A, Elomri A (2023) Artificial intelligence-based decision support systems in smart agriculture: bibliometric analysis for operational insights and future directions. Front Sustain Food Syst 6:1053921. https://doi.org/10.3389/fsufs.2022.1053921

Yu Z, Jung D, Park S, Hu Y, Huang K, Rasco BA, Chen J (2022) Smart traceability for food safety. Crit Rev Food Sci Nutr 62(4):905–916

Zhang L, Dabipi IK, Brown WL Jr (2018) Internet of Things applications for agriculture. In: Internet of Things A to Z: technologies and applications. IEEE Press, pp 507–528

Part III
Consolidation and Systemic Impacts

Digital Technologies and Sustainable Development in Agriculture

Ariel Zajdband, Fabiana Garcia, Janaina Mitsue Kimpara, Michelle Pinheiro Vetorelli, and Wagner C. Valenti

Cooperation for mutual benefit, a survival strategy very common in natural systems, is one that humanity needs to emulate.

Eugene Odum

1 Introduction

The transformation of agricultural systems through digital technologies has become a central topic in the discussions on sustainability and global food security. As agriculture faces challenges from climate change and resource scarcity to demographic pressures and geopolitical instability, technological innovation emerges as a strategic pathway toward sustainability. Digital agriculture encompasses a broad set of tools that have the potential to contribute to environmentally sound, socially inclusive, and economically viable food systems. In this complex context, this chapter explores the relationship between digital technologies and sustainable agriculture, grounded in the framework of the Sustainable Development Goals (SDGs). It analyzes recent empirical evidence and international case studies, investigates the conceptual benefits and challenges of this digital transformation, and evaluates its overall impact on sustainable development.

A. Zajdband
Planet PBC, San Francisco, CA, USA

F. Garcia
Fisheries Institute/APTA/SAA, Votuporanga, Brazil

J. M. Kimpara (✉)
Aquaculture Systems, Embrapa Digital Agriculture, Campinas, Brazil
e-mail: janaina.kimpara@embrapa.br

M. P. Vetorelli
Federal University of Grande Dourados, UFGD, Dourados, Brazil

W. C. Valenti
Aquaculture Center, São Paulo State University, Jaboticabal, Brazil

F. da Silveira et al. (eds.), *Agriculture 4.0*,
https://doi.org/10.1007/978-3-032-24552-6_26

The chapter is structured as follows: First, it maps the key digital technologies, from precision farming and IoT to big data analytics, directly to their potential contributions to specific Sustainable Development Goals, particularly SDG 2 (Zero Hunger) and SDG 12 (Responsible Consumption and Production). Second, it critically examines the barriers to adoption, such as the digital divide, data privacy concerns, and the high cost of implementation, using illustrative case studies from diverse geopolitical contexts. Finally, the chapter concludes by synthesizing these findings to propose a framework for policymakers and practitioners aiming to harness digital agriculture's potential while mitigating its risks, ensuring a truly inclusive and sustainable food future.

2 A Brief Conceptualization of Sustainability

Current civilization depends on a continuous relationship between the environment, society, and its production systems. The environment provides natural resources to production systems, which generate goods and services for society while returning residues and waste to the environment. Society, in turn, consumes these products and provides the workforce for production systems. Therefore, production systems, including agriculture, utilize human and natural resources to generate economic outputs as well as byproducts and discards, which can be collectively referred to as pollution (Fig. 1). The continuity of civilization depends on the dynamic equilibrium between these processes. This long-term resilience and balance are at the heart of sustainability.

The concept of sustainability has evolved over the past 40 years, largely through conferences promoted by the United Nations. The fundamental driver for this concept is the recognition that the global economy relies on finite natural resources. Based on international documents like Agenda 21 and Agenda 2030, sustainability can be defined as the management of natural, human, financial, technological, and institutional resources to ensure the continued satisfaction of human needs for present and future generations. Consequently, sustainable agriculture can be defined as perennial, resilient, and profitable farming that fairly remunerates all factors of production, generates and distributes income to local populations with

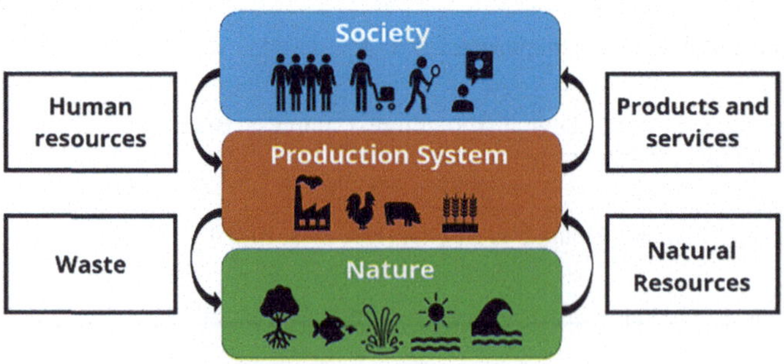

Fig. 1 The current base of human civilization

equity, and conserves or regenerates natural resources. Furthermore, sustainable agricultural systems must be adapted to a continuously changing world, maintaining productivity and profitability in different scenarios.

In this chapter, we use the term agriculture in a broad sense, including crops, terrestrial animal husbandry, and aquaculture. Although agriculture involves living organisms, it is not a bioeconomy per se. The concept of bioeconomy encompasses the utilization of renewable biological resources in innovative technologies and sustainable processes that conserve or regenerate natural resources. Therefore, traditional agribusinesses are not bioeconomic activities. However, agriculture can innovate by exploiting new candidate species from biodiversity to produce organisms or molecules in a sustainable manner that recycles residues and wastes while revitalizing the environment or a specific natural resource. In this way, new agricultural systems can be categorized as bioeconomic businesses. If aquatic organisms are produced, the system can be referred to as a Blue Bioeconomy.

The traditional economic model is a Linear Economy, based on the principle of "extract-use-discard" (Fig. 2). This model involves extracting materials and energy from nature, using them to produce goods and services, and discarding by-products and waste. While this approach has long existed, the Industrial Revolution marked a turning point, leading to exponential growth in resource use and pollution. The linear economy remained dominant because a relatively low global population and an abundance of natural resources masked its long-term unsustainability. By the twenty-first century, however, the human population reached approximately 8.5 billion people, leading to the clear overexploitation of nature.

Consequently, a Circular Economy model was developed. This model is based on the principle of extracting the minimum, then using, reusing, redesigning, utilizing byproducts, and recycling waste (Fig. 2). The circular economy aims to adapt civilization to the new reality of resource scarcity by closing loops.

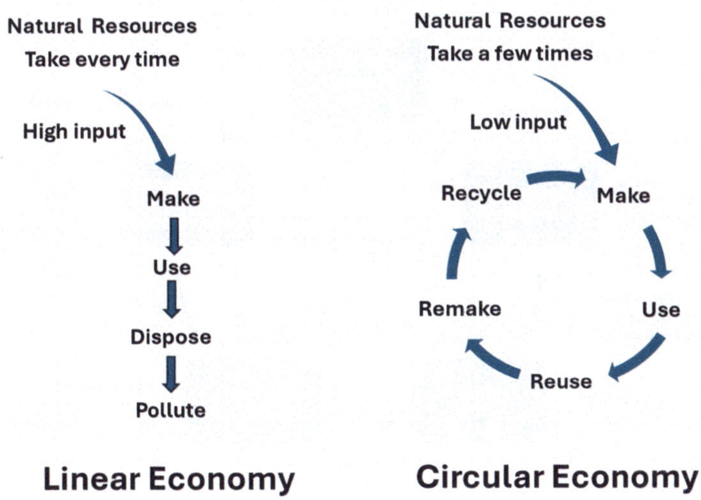

Fig. 2 Linear and circular economy models

Just as the Industrial Revolution transformed society and production, the current digital revolution—often called Industry 4.0 or Society 5.0—is causing changes that are just as radical. This new era is defined by the Internet of Things, artificial intelligence, and cyber-physical systems. In agriculture, this movement is called Agriculture 4.0 (or smart farming/digital agriculture) and will be explored in the next section.

3 Understanding Digital Technologies and how they Relate to Sustainability

Digital transformation in agriculture, often conceptualized as Agriculture 4.0, involves the synergistic application of advanced digital tools to enhance farming practices. While observation has traditionally guided agricultural decision-making, the current shift is based on the ability to collect, store, and process high-resolution data in near-real time. This enables more precise, timely, and informed decisions, positioning these technologies as key enablers of sustainability by improving resource use efficiency, reducing environmental impacts, and enhancing socioeconomic outcomes.

To understand how these technologies function in an integrated way, it is helpful to conceptualize their application in four main components: (1) data collection, (2) data analysis, (3) a trust protocol, and (4) a simulation layer. Figure 3 provides a conceptual diagram of this process, showing how data is collected, transmitted, and processed to enable automated decisions. The specific tools involved in this cycle are detailed in Table 1, which provides a summary of key technologies and their relation

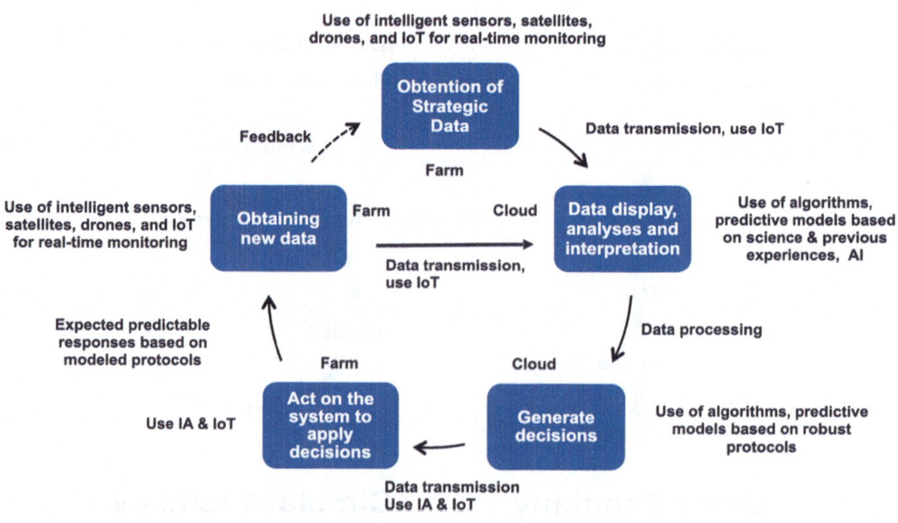

Fig. 3 Conceptual diagram of Agriculture 4.0

Table 1 Brief description of Agriculture 4.0 technologies and their relation to sustainability

Technology	Description	Relation to sustainability	Reference
Big data	Large datasets that can be analyzed to reveal patterns, trends, and associations	Enables the measurement, monitoring, and optimization of resource use and carbon footprints	Belaud et al. (2019)
Smart sensors	Devices that collect, process, and transmit data about their environment, often in real time	Help industries reduce energy waste, monitor resource usage, and improve efficiency	Brewster et al. (2017)
Internet of Things (IoT)	Network of interconnected devices and systems that exchange data to automate and optimize processes	Enables efficient resource management, reduces energy consumption, and improves environmental monitoring	Farooq et al. (2020)
Remote sensing	The science of acquiring information about the Earth's environment from a distance, typically using satellites or aircraft	Provides comprehensive, repetitive, and cost-effective monitoring of environmental changes	Jafarbiglu and Pourreza (2022)
Cloud computing	Delivers computing services—servers, storage, databases, networking, software—over the internet, reducing the need for infrastructure	Promotes energy efficiency and resource optimization, with providers increasingly sourcing renewable energy	Goldstein et al. (2022)
Artificial Intelligence (AI) and Natural Language Processing (NLP)	Systems that perform tasks typically such as learning, reasoning, and problem-solving Branch of AI that enables computers to understand, interpret, and generate human language	Allows access to information and services to a broader public. Provides quick responses using vast data	Holzinger et al. (2024)
Robotics	Design, construction, and operation of robots, which are programmable machines capable of carrying out complex tasks	Enables greater production efficiency through precision in input application and labor productivity, reducing input losses and losses during the production process	Oliveira et al. (2021)
Extended reality (XR)	Virtual, augmented, and mixed reality technologies that blend digital and physical environments	Costs, time, and carbon emissions reductions by enabling immersive experiences; accelerated solution development; animal welfare through noninvasive virtual simulations	Selva et al. (2023)

(continued)

Table 1 (continued)

Technology	Description	Relation to sustainability	Reference
Blockchain	Decentralized digital ledger that records transactions across multiple computers securely and transparently	Enhancing traceability, preventing corruption, and ensuring information security, thus promoting transparent, ethical, and responsible supply chains	Sajja et al. (2023)
6G networks	Next generation of wireless communication technology, promising ultra-fast speeds, low latency, and massive connectivity	Enables the delivery of essential services. Supply chain automation optimizes resource use and minimizes waste	Selva et al. (2023)
Digitalization as a whole	Integration of digital technologies into all areas of business and society, transforming how organizations operate and deliver value	Improved efficiency and governance, with enhanced capacity for data-driven decision-making	Abbasi et al. (2022)

to sustainability. Finally, Fig. 4 illustrates the broad potential of these technologies to help achieve the Sustainable Development Goals (SDGs).

The data collection component gathers information about the physical environment and turns it into a continuous stream of data. IoT sensors in the field, as well as remote sensing devices from UAVs to satellites have the ability to collect data in real-time on a diverse set of variables, including soil moisture, pH, water quality, and weather, among others. A network of sensors, machinery, and other devices can be connected to enable the data transmission (Shafi et al. 2019; Li et al. 2021).

The processing of the data collected can leverage some of the technologies discussed above, including AI and ML. The processing component can involve different functions such as forecasting future events (e.g., pest outbreaks, nutrient requirements), using computer vision to analyze images for different purposes (e.g., weed mapping, plant disease diagnosis), and drive resource use by creating variable-rate prescriptions and powering autonomous systems to perform operations (e.g., weeding, harvest) (Kempelis et al. 2024). The capacity to anticipate and act in near-real time is key to increase resource use efficiency by matching crop and animal needs with resource supply such as water, nutrients, and feed. The third component is a group of technologies that provides a digital ledger, which builds trust throughout the value chain. Blockchain can be used to verify the origin and production practices of agricultural products (Kamble et al. 2020). In addition, smart contracts can be used to automate transactions such as trigger payments upon delivery or the verification of ecosystem services, including carbon sequestration (Yang et al. 2025). The final component is a digital twin, which has the ability to simulate different scenarios by replicating virtually the physical farming system. This virtual twin is continuously updated with the data collected from the first component. The results from the simulations are feeding the data processing,

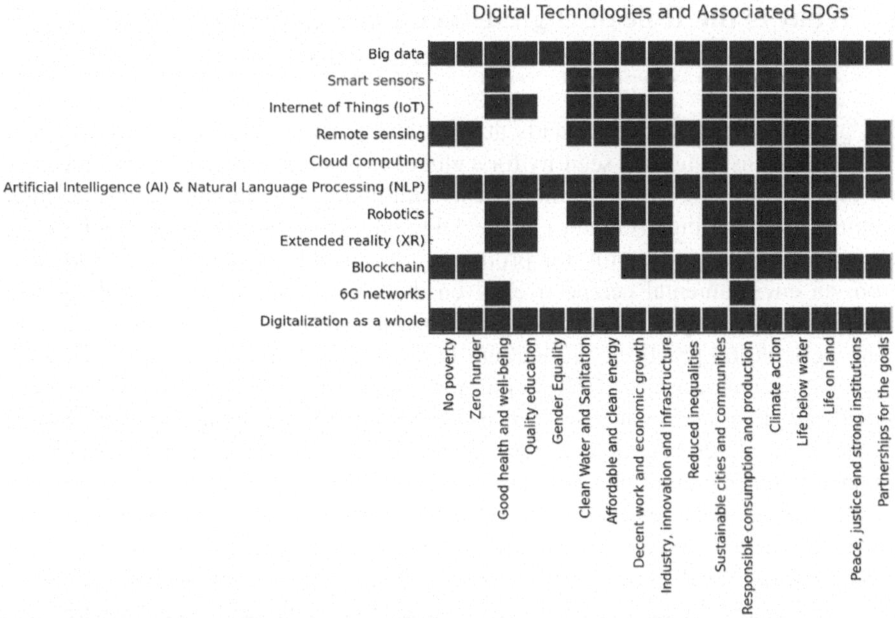

Fig. 4 Technologies and related SDGs. (Source: Authors)

generating a feedback loop where operational decisions can be simulated and, once executed, evaluated in real time (Escribà-Gelonch et al. 2024; Zhang et al. 2025). In this context, the combination of technologies with innovative agricultural management practices enables the continuous testing and refinement in the farm environment.

Digital technologies contribute to the achievement of several Sustainable Development Goals (SDGs), particularly those related to resource efficiency, namely SDG 12 (Responsible Consumption and Production), SDG 13 (Climate Action), and SDG 15 (Life on Land). In contrast, the social dimensions of sustainability receive less attention (Fig. 4). While digital technologies can help address social issues by improving financial access (SDG 1: No Poverty), enabling more inclusive education (SDG 4: Quality Education), and empowering women through access to information in remote areas (SDG 5: Gender Equality), inequitable access to these technologies remain a major barrier. This digital divide limits the broader social impact of digitalization (Ariansyah et al. 2024). However, the true impact of technologies on sustainability requires not only technical performance but also inclusive, equitable deployment that benefits people, the planet, and profit alike.

4 What Is the Role of Digital Technologies in Promoting Sustainable Development in Agriculture?

The global agrifood system stands at a critical juncture, confronted by the dual challenge of ensuring food security for a global population projected to reach nearly ten billion by 2050 while simultaneously mitigating its profound environmental footprint and adapting to climate change. Digital technologies are increasingly being recognized as essential tools for promoting sustainable development worldwide. From an environmental perspective, given the sector's significant contribution to climate change, digital solutions are particularly relevant for enabling climate-smart agricultural practices. Agriculture is responsible for approximately 13–21% of global greenhouse gas (GHG) emissions, and when the entire food system is considered, this share can rise to nearly one-third of the total global emissions (Tubiello et al. 2021). In addition, deforestation, often driven by agricultural expansion for commodities such as beef, soybeans, and palm oil, contributes significantly to these emissions by releasing carbon stored in biomass and soils while simultaneously eliminating vital carbon sinks (Li et al. 2025).

The conventional agricultural paradigm, largely defined by intensification, is facing increasing threats to its long-term sustainability and resilience. Regenerative and circular agriculture are two concepts that surge as an alternative to the mainstream paradigm. While these concepts are interrelated, regenerative agriculture focuses on promoting practices that improve soil health and biodiversity (Giller et al. 2021). Circular agriculture has its origins in the concept of the circular economy, and focuses on closing nutrient and energy feedback loops in agricultural systems to minimize the reliance on external inputs (Newton et al. 2020). Parallel to these ecological agricultural practices, a technological revolution is reshaping the agricultural landscape. There is an opportunity to transform agricultural systems by combining these ecological principles with digital technologies, especially given the need to understand each environment to adapt the practices promoted by the regenerative and circular agriculture frameworks.

Various agricultural practices have offered promising mitigation potential for GHG. Precision agriculture, using GPS, IoT, and sensor-based technologies, enhances the spatial and temporal efficiency of input applications such as fertilizers and water, thereby reducing emissions (Balafoutis et al. 2017). The combination of these technologies is a key enabler for transitioning agriculture from a linear to a circular model. IoT sensors, for example, can be deployed to manage and monitor closed-loop systems by providing essential real-time data. These sensors are crucial for tracking nutrient recycling, such as by analyzing soil conditions or the composition of organic fertilizers derived from farm waste to ensure nutrients are returned to the system effectively. They also allow for the precise monitoring of waste streams, quantifying the volume and quality of byproducts like manure or crop residues so they can be efficiently repurposed as energy, feed, or other inputs (Nguyen et al. 2023; Parra-López et al. 2025). Data analytics then processes this continuous stream of information, verifying inputs for circular models and providing the data-driven decisions needed to optimize the reuse, redesigning, and recycling of resources throughout the production system (Ahmed and Shakoor 2025).

Other examples are conservation practices, including cover cropping and no-till systems, that contribute to improved soil structure, increased organic matter, and enhanced carbon sequestration (Hobbs et al. 2008; Wagena and Easton 2018). Also, in flooded rice systems, the implementation of Alternate Wetting and Drying (AWD) reduces the duration of anaerobic soil conditions, significantly lowering methane emissions (Dahlgreen and Parr 2024). In addition, in livestock systems, rotational grazing improves forage quality and pasture productivity. So, the higher digestibility of the forage resource reduces enteric methane emissions by improving rumen fermentation efficiency (Sanguinetti et al. 2025).

Digital technologies play a pivotal role in both enabling the adoption of these kinds of practices and verifying their implementation. Remote sensing, in particular, serves a dual function: it supports site-specific management through high-resolution spatial and temporal monitoring, and it enables the large-scale verification needed for carbon credit generation, thereby creating economic incentives that reinforce sustainable practices. For example, variable-rate input application, guided by remote data, reduces resource use and minimizes runoff (Radočaj et al. 2022). These tools also support soil management by tracking indicators related to conservation tillage, cover cropping, and land-use changes, while enabling early detection of crop stress from pests, disease, or drought, helping to reduce agrochemical use (Abdullah et al. 2023).

At a broader scale, remote sensing is critical for monitoring deforestation and enforcing land-use regulations. Given that land-use change remains a major driver of agricultural emissions, integrating remote sensing with other digital tools is essential for climate mitigation and evidence-based policymaking (Correa Gonzaga et al. 2024). Together, these technologies offer a scalable, data-driven foundation for advancing sustainable and climate-resilient agriculture (Getahun et al. 2024).

In animal confined systems, such as those used in pork and poultry production, advancements in computer vision and sensor technologies have enabled real-time monitoring of animal behavior, stress levels, and health indicators (Balasubramaniam et al. 2025). These systems can automatically assess weight gain, detect early signs of illness or discomfort, and optimize feeding and environmental conditions, ultimately contributing to both productivity and animal welfare. In addition, the salmon industry uses AI tools to enhance farming management, including exoparasite control, monitoring growth performance, and estimating biomass for harvest decisions. Digital twins have also been used to simulate fish behavior and environmental conditions to develop effective management strategies (Budaev et al. 2024).

Digital transformation also holds significant potential for enhancing public trust and fostering greater cooperation with civil society by improving transparency and access to reliable information. By enabling more accurate, timely, and disaggregated data, it can contribute to better design of policies for poverty reduction, mitigation of rural inequalities, education quality, while facilitating greater integration into the global economy. These advancements suggest that digital transformation can serve as a powerful enabler for several Sustainable Development Goals (SDGs), particularly SDGs 1, 4, 8, 9, 10, 11, 16, and 17 (ElMassaha and Mohieldin 2020).

A notable example of digital technologies advancing sustainable agriculture is Colombia's decision-support tool for rice farmers, developed by the International Centre for Tropical Agriculture (CIAT) and cited by ElMassaha and Mohieldin (2020). Leveraging big data and machine learning, the tool predicts optimal planting periods based on weather and harvest data, aiding farmers in adapting to climate change and contributing to SDGs 1, 2, and 13. Similarly, Brazil's ZARC system, developed by Embrapa and other state institutions, uses agrometeorological models to identify climate risk and optimal planting windows. Accessible via web and mobile, ZARC supports inclusive rural planning, informs rural credit policies, and enhances resilience, supporting SDGs 1, 12, 13, and 16. In India, Ferguson et al. (2024) examined the role of digital agricultural technologies in supporting food security among smallholder farmers in Odisha during the COVID-19 pandemic. The eKutir platform, which provides farmers with timely access to information, was instrumental in maintaining income levels and market access despite the crisis. The study underscores that when digital tools are designed around farmers' needs, they can significantly enhance food security and system resilience, even amid ongoing debates surrounding data-driven agriculture, in compliance with SDGs 1, 2, 10, and 12.

Digital technologies are not merely tools to boost productivity but foundational to transforming agriculture into a more sustainable, resilient, and equitable sector. As illustrated by experiences in Colombia, Brazil, and India, these technologies enhance decision-making, strengthen climate risk management, support evidence-based public policies, and empower smallholder farmers. By bridging data gaps, increasing transparency, and fostering innovation, they occupy a central role in advancing and localizing the Sustainable Development Goals.

The implementation of digital technologies can shorten the challenging and costly transition to sustainable agricultural practices. This period often involves a temporary drop in profitability and crop yields, creating a significant financial barrier for farmers. Precision agriculture helps mitigate these challenges by optimizing the efficiency of inputs (Walter et al. 2017). Furthermore, technologies combining the Internet of Things, AI, and Blockchain provide robust Measurement, Reporting, and Verification (MRV). This trustworthy MRV enables farmers to access new revenue streams by validating ecosystem services, such as selling carbon credits. This additional income can offset initial losses, making the transition economically viable (Kamilaris et al. 2019; Klerkx et al. 2019).

Digital agriculture can also promote a transition beyond traditional yield metrics toward a holistic, data-driven evaluation of the entire agroecosystem. The combination of digital technologies discussed can provide high-resolution, real-time data to measure the system's performance and quantify its environmental footprint (Aijaz et al. 2025). In addition, the adoption of digital twins can enable the measurement of the agricultural system's resilience by assessing its capacity to anticipate, absorb, and recover from shocks like extreme weather or supply chain disruptions through the future's scenario methodology (Walker et al. 2002; Carpenter et al. 2006). The ultimate framework would synthesize these technology-derived data streams into a dynamic dashboard of interconnected indicators, reflecting not just productivity but the long-term sustainability and adaptive capacity of the agricultural system.

5 Challenges for the Implementation of Digital Technologies to Achieve Sustainability

Although digital technologies hold significant promise for making agriculture sustainable, their effective implementation is constrained by a series of deep, interconnected challenges. The fundamental challenge is related to the access and affordability of digital technologies, especially in rural and low-income areas. Although over two-thirds of the global population now uses the Internet, digital exclusion persists because of geographic disparities in infrastructure and connectivity (UN 2024). Many farming communities, particularly in rural and low-income areas, lack reliable or affordable Internet, which is a prerequisite for accessing digital technologies. This lack of foundational infrastructure restricts the transformative potential of digital tools before they can even be deployed.

This access barrier is compounded by the cost of technology. For instance, precision agriculture depends on sensors, drones, and software systems that are often unaffordable for smallholder farmers and remain cost-prohibitive even for medium-sized operations. The additional financial burden of specialized labor for installation, calibration, and maintenance means the digital transformation of agriculture remains uneven and exclusionary (Dibbern et al. 2024). Without public policies that ensure connectivity, including subsidies for Internet access and digital equipment, and targeted investment in rural digital infrastructure, these foundational barriers will continue to constrain progress toward sustainability goals.

Even where access is available, structural and social inequalities strongly influence who can appropriate and benefit from digital technologies. Factors including gender, age, educational level, land tenure, and financial capacity create further divides. As Hackfort (2021) identified, inequality manifests in five key patterns within digital agriculture: (1) technology development, (2) benefit distribution, (3) sovereignty over data and infrastructure, (4) skills and knowledge, and (5) problem-solving capacities. From this perspective, digitalization can reproduce existing power relations rather than acting as a revolutionary force.

Gender inequality presents a particularly relevant example. Codesign workshops often see low participation from women, whose household responsibilities conflict with extensive meeting formats. Furthermore, women who do participate may hesitate to express opinions in predominantly male groups. This reflects deeply rooted structural inequalities (i.e., unequal divisions of labor, restricted access to technology, and persistent social norms) that are replicated within innovation systems. Addressing these challenges related to the human dimensions including social inequalities and digital literacy requires a deliberate focus on building human capacity. Yuan et al. (2025) demonstrated that digital literacy strengthens farmers' adoption of low-carbon technologies and contributes to ESG-aligned outcomes. Therefore, strengthening social inclusion with diverse representation, supported by inclusive digital literacy programs and peer-learning networks, is fundamental. Building autonomy in knowledge and skills is crucial to ensure users can adapt technological solutions to their local realities, empowering farmers rather than deepening existing disparities.

The design of the technologies themselves constitutes another major challenge. Many digital solutions are not created to be inclusive, accessible, or responsive to the diverse realities of farmers. As Steinke et al. (2022) highlight, designing for an *average user* can inadvertently reinforce exclusion. For instance, the Ushauri platform in Kenya, which assumed access to mobile phones and digital familiarity, tended to benefit more privileged groups while marginalizing remote or resource-poor farmers. Adopting inclusive design methodologies that explicitly prioritize marginalized users, such as women or illiterate farmers, during the ideation and prototyping phases is key to consider different needs from farmers and other users.

This design flaw is linked to significant data availability and quality hurdles. Sustainable practices depend on accurate, timely, and disaggregated data, yet many countries lack the capacity to generate or use it effectively. Critically, the focus of existing digital tools has largely been on monitoring large-scale industrial crop systems like extensive grain cropping systems. Smallholder farmers, who produce a significant portion of the world's food, often find these solutions irrelevant or inaccessible. Participatory codesign, where farmers, researchers, and policymakers collaboratively develop tools, is essential to bridge this gap, ensuring that technologies are aligned with users' real needs and operational contexts.

Other relevant issues for achieving sustainable agriculture in a digital context include governance grounded in ethics, security and responsible data sharing, and inclusivity. Technological innovation must align with data sovereignty and social equity to ensure that digital transformation benefits all stakeholders.

Data governance models should be guided by ethical frameworks that emphasize autonomy, justice, and accountability. This involves engaging diverse stakeholders, including farmers, policymakers, and researchers, to foster knowledge exchange and address emerging ethical challenges in an AI-driven agricultural system. To ensure data governance models that embed ethics, security, transparency, and responsible sharing, several dedicated data frameworks have been developed. One example is the Common European Agricultural Data Space (CEADS), which establishes technical standards and common rules for interoperating public and private data under a federated governance structure (Marquardt 2024). Additionally, the FAIR (Findable, Accessible, Interoperable, Reusable) and CARE (Collective Benefit, Authority to Control, Responsibility, Ethics) principles were developed to guide data governance for both scientific and Indigenous or traditional knowledge systems, ensuring justice and self-determination (GO FAIR 2025).

Another important mechanism in this context is the data stewardship model, often implemented through data cooperatives, which ensures collective control, democratic decision-making, and fair value distribution. Such models are particularly relevant for small-scale producers, enabling them to negotiate data-use conditions, define anonymization protocols, and secure economic and social returns (Mendonça et al. 2025). Furthermore, the Data Governance Act (DGA) establishes neutral data intermediaries, frameworks for data altruism and exchange, and compliance labelling, while the Data Act reinforces rights to access and portability of data from devices and machines (e.g., tractors, sensors). Together, these instruments anchor the creation of sectoral data spaces and strengthen trust-based digital ecosystems. Ultimately, these mechanisms align digital agriculture with sustainability goals, as the integration of advanced digital platforms becomes essential for enhancing efficiency, transparency, and long-term environmental resilience (Gavai et al. 2025).

In addition, farmer-centric data rights and transparency mechanisms are essential to mitigate risks associated with data opacity and technological dependency (Sangwa and Mutabazi 2025; Zangana et al. 2025). Regulations and public–private partnerships that provide incentives for collaboration between governments and enterprises can facilitate responsible data sharing and resource management, aligning digital innovation with the principles of sustainable development (Chen 2024). Regional adaptations are equally important to tailor regulations and incentives to specific socio-environmental contexts, ensuring that local needs and conditions are met effectively (Sangwa and Mutabazi 2025). This approach aligns with Sustainable Development Goal 16 (UN 2025), which aims to promote peaceful and inclusive societies, ensure access to justice for all, and build effective, accountable, and inclusive institutions at all levels.

Finally, realizing the potential of digital agriculture requires confronting broader political and economic forces. As Béné (2022) argues, global food systems remain locked in unsustainable patterns due to the dominance of large corporations, conflicting policy agendas, and cultural inertia. Innovation alone cannot lead to sustainability if it remains profit-driven and unguided by public policy that actively steers it toward sustainable outcomes. Furthermore, technologies designed without considering social impacts can lead to the loss of traditional knowledge, labor displacement, and unequal access to benefits (Rose et al. 2021).

The core of this systemic challenge is that most digital tools are designed to optimize linear, monoculture-based systems. However, true sustainability demands a transition toward integrated, multifunctional systems based on the principles of the circular economy and regenerative agriculture. Many farmers lack the tools and protocols to make this shift. Ultimately, these regenerative systems are inherently complex and face a critical economic barrier. Current business models do not value or compensate for the ecosystem services they provide, such as microclimatic regulation and biodiversity maintenance. Without mechanisms like payments for ecosystem services that account for both positive and negative externalities, regenerative models will not be financially sustainable when compared with conventional business models. It is therefore imperative to broaden our economic perspective, ensuring that digital innovation contributes not just to productivity, but to an agricultural system that is environmentally, socially, and economically sustainable over the long term, by supporting this transition through monitoring and quantifying indicators for a broad systemic efficiency, as long as they are designed for this purpose.

6 Conclusion: Toward an Inclusive and Regenerative Digital Agriculture

Digital transformation may play a key role in achieving more sustainable outcomes in agriculture. To realize this transformative potential, significant challenges must be addressed. The main one is related to ensuring equitable access to these technologies. Although digital innovations have advanced rapidly, their adoption remains uneven, particularly in rural and low-income areas. The high cost of infrastructure, limited digital literacy, and unequal data-governance frameworks risk marginalizing vulnerable populations, concentrating benefits among a few actors, and exacerbating global inequalities. This broader perspective calls into question the dominant growth paradigm.

Scholars such as Saito (2020) and Odum and Odum (2001) have emphasized the necessity of transitioning toward degrowth, a deliberate scaling down of natural resource inputs to achieve ecological integrity and social equity. In this vision, digital transformation should serve as a strategic enabler of structural changes, rather than a vehicle to perpetuate the current models of agriculture based on natural resources extraction.

A sustainability-oriented digital transformation requires inclusive governance and participatory design. In this context, SDGs such as no poverty, reduced inequalities, and gender equality may be achieved, in addition to the environmental efficiency metrics, as the majority of technologies are currently aiming. Technologies must be open, interoperable, and adapted to local contexts. As discussed, without the robust, farmer-centric data governance frameworks previously outlined (such as data cooperatives and the FAIR/CARE principles), these digital tools risk reinforcing exclusion rather than empowerment.

Competing Interests The author(s) has no competing interests to declare that are relevant to the content of this manuscript.

References

Abbasi R, Martinez P, Ahmad R (2022) The digitization of agricultural industry – a systematic literature review on agriculture 4.0. Smart Agric Technol 2:100042. https://doi.org/10.1016/j.atech.2022.100042

Abdullah HM, Mohana NT, Khan BM et al (2023) Present and future scopes and challenges of plant pest and disease (P&D) monitoring: remote sensing, image processing, and artificial intelligence perspectives. Remote Sens Appl Soc Environ 32:100996. https://doi.org/10.1016/j.rsase.2023.100996

Ahmed N, Shakoor N (2025) Advancing agriculture through IoT, Big Data, and AI: a review of smart technologies enabling sustainability. Smart Agric Technol 10:100848. https://doi.org/10.1016/j.atech.2025.100848

Aijaz N, Lan H, Raza T, Yaqub M, Iqbal R, Pathan MS (2025) Artificial intelligence in agriculture: advancing crop productivity and sustainability. J Agric Food Res 20:101762. https://doi.org/10.1016/j.jafr.2025.101762

Ariansyah A, Prayogi S, Kurnia N, Bilad M, Sutarto S (2024) Digital technology to support Sustainable Development Goals (SDGs): literature review. Lensa Jurnal Kependidikan Fisika 12(2):315–358. https://doi.org/10.33394/j-lkf.v12i2.13557

Balafoutis A, Beck B, Fountas S, Vangeyte J, Wal TV, Soto I, Gómez-Barbero M, Barnes A, Eory V (2017) Precision agriculture technologies positively contributing to GHG emissions mitigation. Farm Prod Econ Sustain 9(8):1339. https://doi.org/10.3390/su9081339

Balasubramaniam S, Vijesh Joe C, Prasanth A, Kumar KS (2025) Computer vision systems in livestock farming, poultry farming, and fish farming. In: Dhanaraj RK, Balusamy B, Samuel P, Sathyamoorthy M, Bashir AK (eds) Computer vision in smart agriculture and crop management, pp 221–258. https://doi.org/10.1002/9781394186686.ch10

Belaud J-P, Prioux N, Vialle C, Sablayrolles C (2019) Big data for agri-food 4.0: application to sustainability management for by-products supply chain. Comput Ind 111:41–50. https://doi.org/10.1016/j.compind.2019.06.006

Béné C (2022) Why the great food transformation may not happen – a deep-dive into our food systems' political economy, controversies and politics of evidence. World Dev 154:105881. https://doi.org/10.1016/j.worlddev.2022.105881

Brewster C, Roussaki I, Kalatzis N, Doolin K, Ellis K (2017) IoT in agriculture: designing a Europe-wide large-scale pilot. IEEE Commun Mag 55(9):26–33. https://doi.org/10.1109/MCOM.2017.1600528

Budaev S, Dumitru ML, Enberg K et al (2024) Premises for a digital twin of the Atlantic salmon in its world: agency, robustness, subjectivity and prediction. Aquac Fish Fish 4:e153. https://doi.org/10.1002/aff2.153

Carpenter SR, Bennett EM, Peterson GD (2006) Scenarios for ecosystem services: an overview. Ecol Soc 11(1):29. http://www.jstor.org/stable/26267787

Chen X (2024) AI ethics and sustainable agricultural transformation in emerging economies. J Agric Policy Innov 12(3):207–216

Correa Gonzaga CA, Roquette JG, Angeoletto F et al (2024) The public prosecutor's office's experience using Global Forest Watch to monitor and deter deforestation in the Cerrado. Environ Conserv 51(3):236–241. https://doi.org/10.1017/S0376892924000110

Dahlgreen J, Parr A (2024) Exploring the impact of alternate wetting and drying and the system of rice intensification on greenhouse gas emissions: a review of rice cultivation practices. Agronomy 14(2):378. https://doi.org/10.3390/agronomy14020378

Dibbern T, Romani LAS, Silveira Massruhá SMF (2024) Main drivers and barriers to the adoption of digital agriculture technologies. Smart Agric Technol 8:100459. https://doi.org/10.1016/j.atech.2024.100459

ElMassaha S, Mohieldin M (2020) Digital transformation and localizing the Sustainable Development Goals (SDGs). Ecol Econ 169(1):10. https://doi.org/10.1016/j.ecolecon.2019.106490

Escribà-Gelonch M, Liang S, van Schalkwyk P, Fisk I, Long NVD, Hessel V (2024) Digital twins in agriculture: orchestration and applications. J Agric Food Chem 72(19):10737–10752. https://doi.org/10.1021/acs.jafc.4c01934

Farooq MS, Riaz S, Abid A, Umer T, Zikria YB (2020) Role of IoT technology in agriculture: a systematic literature review. Electronics 9(2):319. https://doi.org/10.3390/electronics9020319

Ferguson R, Mishra S, Qureshi S, Injac L, Talukder B, Orbinski J (2024) Digital technologies and food security during crisis: COVID-19 experiences from smallholder farmers in Odisha, India. Front Sustainable Food Syst 7:1150197. https://doi.org/10.3389/fsufs.2023.1150197

Gavai AK, Bouzembrak Y, Xhani D, Sedrakyan G, Meuwissen MPM, Souza RGS, Marvin HJP, van Hillegersberg J (2025) Agricultural data privacy: emerging platforms & strategies. Food Hum 4:100542. https://doi.org/10.1016/j.foohum.2025.100542

Getahun S, Kefale H, Gelaye Y (2024) Application of precision agriculture technologies for sustainable crop production and environmental sustainability: a systematic review. Sci World J 2024:2126734. https://doi.org/10.1155/2024/2126734
Giller KE, Hijbeek R, Andersson JA, Sumberg J (2021) Regenerative agriculture: an agronomic perspective. Outlook Agric 50(1):13–25. https://doi.org/10.1177/0030727021998063
GO FAIR (2025) FAIR principles. Retrieved from https://www.go-fair.org/fair-principles/
Goldstein A, Fink L, Ravid G (2022) A cloud-based framework for agricultural data integration: a top-down-bottom-up approach. IEEE Access 10:88527–88537. https://doi.org/10.1109/ACCESS.2022.3198099
Hackfort S (2021) Patterns of inequalities in digital agriculture: a systematic literature review. Sustainability 13(22):12345. https://doi.org/10.3390/su132212345
Hobbs PR, Ken S, Raj G (2008) The role of conservation agriculture in sustainable agriculture. Philos Trans R Soc B 363:543–555. https://doi.org/10.1098/rstb.2007.2169
Holzinger A, Fister I, Fister I, Kaul HP, Asseng S (2024) Human-centered AI in smart farming: towards agriculture 5.0. IEEE Access 12:62199–62212. https://doi.org/10.1109/ACCESS.2024.3395532
Jafarbiglu H, Pourreza A (2022) A comprehensive review of remote sensing platforms, sensors, and applications in nut crops. Comput Electron Agric 196:106844. https://doi.org/10.1016/j.compag.2022.106844
Kamble SS, Gunasekaran A, Sharma R (2020) Modeling the blockchain enabled traceability in agriculture supply chain. Int J Inf Manag 52:101967. https://doi.org/10.1016/j.ijinfomgt.2019.05.023
Kamilaris A, Fonts A, Prenafeta-Boldύ FX (2019) The rise of blockchain technology in agriculture and food supply chains. Trends Food Sci Technol 91:640–652. https://doi.org/10.1016/j.tifs.2019.07.034
Kempelis A, Polaka I, Romanovs A, Patlins A (2024) Computer vision and machine learning-based predictive analysis for urban agricultural systems. Futur Internet 16(2):44. https://doi.org/10.3390/fi16020044
Klerkx L, Jakku E, Labarthe P (2019) A review of social science on digital agriculture, smart farming and agriculture 4.0: new contributions and a future research agenda. NJAS Wageningen J Life Sci 90-91:100315. https://doi.org/10.1016/j.njas.2019.100315
Li L, Zhang S, Wang B (2021) Plant disease detection and classification by deep learning—a review. IEEE Access 9:56683–56698. https://doi.org/10.1109/ACCESS.2021.3069646
Li L, Awada T, Shi Y, Jin VL, Kaiser M (2025) Global greenhouse gas emissions from agriculture: pathways to sustainable reductions. Glob Change Biol 31:e70015. https://doi.org/10.1111/gcb.70015
Marquardt D (2024) Blueprint proposal for the Common European Agricultural Data Space. Shaping Europe's Digital Future. Available in: https://digital-strategy.ec.europa.eu/en/library/blueprint-proposal-common-european-agricultural-data-space. Accessed on 2 May 2025
Mendonça F, Di Marzo G, Abdennadher N (2025) Data cooperatives: democratic models for ethical data stewardship. ArXiv. University of Geneva; University of Applied Sciences of Western Switzerland. Preprint. https://arxiv.org/abs/2504.10058
Newton P, Civita N, Frankel-Goldwater L, Bartel K, Johns C (2020) What is regenerative agriculture? A review of scholar and practitioner definitions based on processes and outcomes. Front Sustain Food Syst 4:577723. https://www.frontiersin.org/articles/10.3389/fsufs.2020.577723
Nguyen VD, Pham TC, Pham TT, Phan AN (2023) Digital and circular technologies for climate-smart and sustainable agriculture: the case of Vietnamese coffee. IOP Conf Ser Earth Environ Sci 1278:012003
Odum HT, Odum EC (2001) A prosperous way down. University Press of Colorado, Boulder
Oliveira LFP, Moreira AP, Silva MF (2021) Advances in agriculture robotics: a state-of-the-art review and challenges ahead. Robotics 10(2):52. https://doi.org/10.3390/robotics10020052
Parra-López C, Ben Abdallah S, Garcia-Garcia G, Hassoun A, Trollman H, Jagtap S, Gupta S, Aït-Kaddour A, Makmuang S, Carmona-Torres C (2025) Digital technologies for water use and management in agriculture: recent applications and future outlook. Agric Water Manag 309:109347. https://doi.org/10.1016/j.agwat.2025.109347

Radočaj D, Jurišić M, Gašparović M (2022) The role of remote sensing data and methods in a modern approach to fertilization in precision agriculture. Remote Sens 14:778. https://doi.org/10.3390/rs14030778

Rose DC, Wheeler R, Winter M, Lobley M, Chivers C-A (2021) Agriculture 4.0: making it work for people, production, and the planet. Land Use Policy 100:104933. https://doi.org/10.1016/j.landusepol.2020.104933

Saito K (2020) Capital in the Anthropocene [Hitoshinsei no "Shihoron"]. Shueisha, Tokyo

Sajja GS, Rane KP, Phasinam K, Kassanuk T, Okoronkwo E, Prabhu P (2023) Towards applicability of blockchain in agriculture sector. Mater Today Proc 80(3):3705–3708. https://doi.org/10.1016/j.matpr.2021.07.366

Sanguinetti L, Sanguinetti VM, Remnant J, Orsel K, Ganshorn H, Tang M, Lhermie G (2025) A scoping review on the impact of rotational grazing in beef cattle systems on greenhouse gas emissions, soil health, plant diversity, and plant productivity parameters. Can J Anim Sci 105:1–17. https://doi.org/10.1139/cjas-2025-0010

Sangwa N, Mutabazi K (2025) Inclusive governance and digital agriculture in Sub-Saharan Africa. Agric Syst 228:104947

Selva E, Gati A, Hamon M-H et al (2023) Towards a 6G embedding sustainability. In: 2023 IEEE international conference on communications workshops (ICC Workshops), pp 1588–1593. https://doi.org/10.1109/ICCWorkshops57953.2023.10283655

Shafi U, Mumtaz R, García-Nieto J, Hassan SA, Zaidi SAR, Iqbal N (2019) Precision agriculture techniques and practices: from considerations to applications. Sensors 19(17):3796. https://doi.org/10.3390/s19173796

Steinke J, van Etten J, Müller A, Ortiz-Crespo B, van de Gevel J, Silvestri S, Priebe J, Macours K (2022) Participatory design of digital innovation in agriculture: insights from ICT4Ag research-for-development projects. Agric Syst 195:103298. https://doi.org/10.1016/j.agsy.2021.103298

Tubiello FN, Rosenzweig C, Conchedda G et al (2021) Greenhouse gas emissions from food systems: building the evidence base. Environ Res Lett 16:065007. https://doi.org/10.1088/1748-9326/ac018e

United Nations (2024) Global Internet use continues to rise but disparities remain. Department of Economic and Social Affairs—Social Development Network. Retrieved October 27, 2025, from https://social.desa.un.org/sdn/global-internet-use-continues-to-rise-but-disparities-remain

United Nations (2025) Sustainable development goals report 2025. United Nations Publications, New York

Wagena MB, Easton ZM (2018) Agricultural conservation practices can help mitigate the impact of climate change. Sci Total Environ 635:132–143. https://doi.org/10.1016/j.scitotenv.2018.04.110

Walker B, Carpenter S, Anderies J, Abel N, Cumming G, Janssen M, Lebel L, Norberg J, Peterson GD, Pritchard R (2002) Resilience management in social-ecological systems: a working hypothesis for a participatory approach. Conserv Ecol 6(1):14. http://www.jstor.org/stable/26271859

Walter A, Finger R, Huber R, Buchmann N (2017) Smart farming is key to developing sustainable agriculture. Proc Natl Acad Sci 114(24):6148–6150. https://doi.org/10.1073/pnas.1707462114

Yang Y, Lin M, Lin Y, Zhang C, Wu C (2025) A survey of Blockchain applications for management in agriculture and livestock internet of things. Futur Internet 17(1):40. https://doi.org/10.3390/fi17010040

Yuan Y, Jiang H, Zhao L, Chen C (2025) The influence of digital literacy on low-carbon agricultural technology adoption: evidence from Chinese farmers. Sustainability 17(3):2187. https://doi.org/10.3390/su17032187

Zangana HM, Amelia P, Luckyardi S, Mustafa FM, Li S (2025) Governing AI-driven agriculture. In: Handbook of research on AI applications in the agricultural industry. IGI Global, pp 285–316. https://doi.org/10.4018/979-8-3373-4862-9.ch011

Zhang R, Zhu H, Chang Q, Mao Q (2025) A comprehensive review of digital twins technology in agriculture. Agriculture 15(9):903. https://doi.org/10.3390/agriculture15090903

Policies for a Sustainable and Digital Agri-Food System: Accelerating the Adoption of Agriculture 4.0

Carlos Parra-López and Carmen Carmona-Torres

1 Introduction

The global agri-food system is facing an unprecedented convergence of crises, including climate change, pandemics, and geopolitical conflicts (Galanakis 2023) and structural issues, such as policy incoherence and governance failures (Béné 2022), that threaten food security, safety, and sustainability. A rapidly growing world population, projected to reach 9.7–9.8 billion by 2050, will require significant increases in food production, even as climate change intensifies, natural resources become increasingly scarce, and environmental degradation threatens the very basis of agricultural productivity (Chawla et al. 2020; Dong et al. 2020; Singh 2021). These intertwined pressures require a paradigm shift that moves beyond conventional agricultural practices to innovative and sustainable solutions (Klerkx and Begemann 2020; Mondejar et al. 2021; Jacquet et al. 2022). Agriculture 4.0, signifying the deep integration of digital technologies throughout the agri-food value chain, is emerging as a potential solution to these complex and interrelated challenges (Hassoun et al. 2023; Parra-López et al. 2024).

This technological revolution, characterized by the synergistic convergence of Artificial Intelligence (AI), the Internet of Things (IoT), big data analytics, remote sensing, robotics, blockchain, nanotechnology, and other digital technolo-

C. Parra-López (✉)
Department of Agrifood System Economics, Institute of Agricultural and Fisheries Research and Training (IFAPA), Granada, Spain
e-mail: carlos.parra@juntadeandalucia.es

C. Carmona-Torres
Institute of Regional Development, University of Granada, Granada, Spain

Department of Agrifood System Economics, Institute of Agricultural and Fisheries Research and Training (IFAPA), Granada, Spain

F. da Silveira et al. (eds.), *Agriculture 4.0*,
https://doi.org/10.1007/978-3-032-24552-6_27

gies, promises to revolutionize all aspects of agriculture, from precision farming techniques in crop production to smart supply chain management and enhanced food safety protocols (Yadav et al. 2022; Hassoun et al. 2023; Sharma et al. 2023). Agriculture 4.0 has the potential to optimize resource use, increase productivity, minimize environmental impacts, improve food quality and safety, and build resilience to climate change and other external shocks (Hassoun et al. 2022; Finger 2023; Parra-López et al. 2024; Parra-López et al. 2025).

However, realizing this transformative potential is not an automatic process. The mere existence of these advanced technologies does not guarantee their widespread adoption, equitable access, or responsible and sustainable implementation in the diverse and complex global agrifood system (Klerkx and Rose 2020; Lioutas et al. 2021; Shang et al. 2021). For example, adoption has been slower and more uneven than expected, particularly among smallholder farmers who lack the resources, infrastructure and knowledge to effectively implement these solutions (Fragomeli et al. 2024). This raises concerns about the digital divide and the potential exclusion of marginalized farming communities from the benefits of digital agriculture.

This chapter focuses on the critical and often underappreciated role of well-designed and effectively implemented policies in accelerating the adoption of Agriculture 4.0 and steering its trajectory toward a truly sustainable agri-food system. While technological innovation is the engine of Agriculture 4.0, policies serve as the compass and roadmap, guiding its development and deployment in a way that is consistent with broader societal goals. This chapter embarks on a comprehensive exploration of the policy landscape, meticulously examining key policy instruments, strategic interventions, and essential considerations for advancing the digital transformation of agriculture, consistently emphasizing the multifaceted dimensions of sustainability, inclusiveness, and ethical responsibility. By analyzing the rationale for policy interventions, dissecting the array of available policy instruments, and addressing the challenges and barriers to adoption, this chapter aims to provide actionable insights and practical recommendations for policymakers, stakeholders, and researchers seeking to harness the transformative power of Agriculture 4.0 for a more sustainable and food-secure future.

2 The Rationale for Policy Interventions in Agriculture 4.0: Addressing Systemic Needs and Guiding Transformation

While the digital transformation of agriculture holds great promise, it is not a self-executing or inherently equitable process. The Agriculture 4.0 utopia of optimized food production, improved sustainability, and widespread prosperity will not spontaneously emerge from technological advances alone. The agri-food sector, a complex and multifaceted system deeply intertwined with societal well-being, environmental health, and economic stability, requires proactive and strategic policy interventions to guide and accelerate the adoption of digital technologies in a direction that serves the broader public good. Without carefully crafted and effectively implemented policies, the digital revolution in agriculture risks exacerbating existing inequalities, overlooking critical sustainability imperatives, and ultimately failing to realize its full potential to address pressing global challenges facing the agri-food system.

2.1 Addressing Market Failures: Correcting Inefficiencies and Promoting Public Goods

Several inherent characteristics of the agri-food sector create market failures that impede the natural evolution and efficient adoption of Agriculture 4.0 technologies. Information asymmetries, a pervasive feature of agricultural markets, are a significant barrier. Farmers, especially smallholder farmers who make up the majority of the world's agricultural population, often lack the information, technical expertise, and resources necessary to adequately assess the potential benefits and navigate the complexities of digital solutions (Shang et al. 2021). This information deficit creates a knowledge gap that hinders informed decision-making and prevents many farmers from adopting potentially transformative technologies.

In addition to knowledge gaps, financial constraints represent a critical market failure in the adoption of Agriculture 4.0. Many digital farming technologies require high initial investments in equipment, software, and infrastructure, with uncertain returns on investment. For example, Barnes et al. (2019) found that without clear financial incentives or immediate cost savings, farmers, especially small and medium-sized enterprises (SMEs), are reluctant to adopt precision agriculture tools. In addition, financial institutions (banks, credit providers) may be reluctant to lend to farmers for Agriculture 4.0 technologies due to the perceived high risk of new technologies. This highlights the need for more targeted financial support mechanisms, such as subsidies, low-interest loans or cost-sharing initiatives, to reduce the economic barriers to adoption.

Network externalities, another key market failure, further complicate the adoption landscape. The value and effectiveness of many digital technologies in agriculture are inextricably linked to the extent of their adoption across the value chain (Rijswijk et al. 2021). For example, while traceability systems based on blockchain technology offer increased transparency and food safety across the supply chain, they require coordinated investment and participation from multiple actors, including producers, processors, distributors, and retailers (Kamble et al. 2020). Individual actors may not have sufficient incentives to make such coordinated investments on their own, leading to suboptimal levels of adoption and a failure to fully realize the systemic benefits of digital agriculture. Policies that act as catalysts for collective action can effectively address these coordination challenges and unlock the systemic value of digital technologies.

In addition, many sustainable agricultural practices that will be significantly enhanced and enabled by Agriculture 4.0 generate critical public goods, such as improved environmental quality, enhanced biodiversity, and climate change mitigation (Maffezzoli et al. 2022). While these public goods are essential to societal well-being and long-term sustainability, they are often undervalued or not rewarded by traditional market mechanisms. As a result, private sector actors driven by profit motives may underinvest in these practices, leading to suboptimal environmental and societal outcomes. Strategic policy interventions, including incentives, regulations, and public awareness campaigns, are therefore essential to correct these market failures, align private incentives with broader public interests, and ensure that Agriculture 4.0 becomes a powerful force for environmental protection and sustainable resource management. For example, studies have shown that aligning private incentives with broader sustainability goals through regulatory frameworks and financial mechanisms can significantly increase the adoption of environmentally friendly precision farming techniques (Finger 2023).

2.2 Guiding Responsible Innovation: Navigating Ethical Dilemmas and Societal Impacts

Beyond addressing market inefficiencies, policy has a critical role to play in guiding responsible innovation in Agriculture 4.0 and proactively addressing and mitigating potential ethical, social, and environmental risks associated with rapid technological advances. The increasing reliance on data-driven technologies and artificial intelligence in agriculture raises profound ethical and societal issues that require careful consideration and robust governance frameworks (Galaz et al. 2021).

For example, algorithmic bias is a key concern that policymakers need to address. AI-driven decision-making in agriculture, such as automated credit scoring for agricultural loans, predictive pricing models, and yield forecasting, can reinforce existing inequalities if these systems are trained on biased data or developed without transparency (van der Burg et al. 2021). This could lead to discriminatory outcomes in access to credit, insurance, or market opportunities for certain groups of farmers, exacerbating existing social inequalities. Unequal access to digital technologies and infrastructure, the digital divide, exacerbates these concerns, creating a scenario where the benefits of Agriculture 4.0 disproportionately accrue to privileged segments of society, while marginalized farming communities are left behind, further widening the digital haves and have-nots gap (Shang et al. 2021).

Policies need to promote algorithmic transparency to ensure that AI-driven decision-making processes are explainable and accountable, thereby fostering trust in the use of these powerful technologies in the agri-food sector. Lioutas et al. (2021) aptly capture this challenge as the "trolley dilemma" of digital agriculture, highlighting the need for policymakers to navigate complex trade-offs and ensure that the pursuit of efficiency and productivity gains does not come at the expense of social justice, ethical principles, or environmental sustainability. Policy therefore serves as an ethical compass and guiding hand to steer the digital transformation of agriculture toward a future where technology enables sustainable, inclusive and responsible agricultural development that benefits all stakeholders and contributes to a more just and equitable food system.

Concerns about data ownership, privacy, and security are paramount in the data-rich ecosystem of digital agriculture. Farmer data generated by sensors, drones, and digital platforms is a valuable asset that requires robust protection and transparent governance (Rijswijk et al. 2021; Parra-López et al. 2024). Policies must establish clear frameworks that protect farmers' data rights and ensure that farmers retain control over their data and benefit fairly from its use. Therefore, robust data privacy regulations, aligned with international standards such as the General Data Protection Regulation (GDPR), are essential to prevent unauthorized access, misuse, or exploitation of sensitive farmer data. In this context, emerging regulatory frameworks such as the European Commission's AI Act emphasize the need for "high-risk" AI applications to adhere to stricter compliance and transparency measures (European_Commission 2021). Applying similar AI governance standards in agriculture can help mitigate ethical risks.

3 Challenges and Strategies for Adopting Agriculture 4.0

The digital transformation of agriculture, or Agriculture 4.0, promises to revolutionize the sector by increasing productivity, sustainability and efficiency through advanced digital technologies. However, the path to achieving this transformation is fraught with challenges that need to be addressed to ensure its success and equity. This section explores five key challenges and strategies to address them.

3.1 Bridging the Digital Divide: A Persistent Challenge in Rural Areas and Developing Countries

The persistent digital divide, characterized by large disparities in access to digital technologies, robust infrastructure, and basic digital literacy skills, remains a significant and pervasive barrier to the widespread adoption of Agriculture 4.0, particularly in rural areas and developing countries (Rotz et al. 2019; Mehrabi et al. 2021). Overcoming this digital divide is not simply a matter of providing access to technology; it requires a multifaceted and sustained effort to address the underlying causes of digital exclusion, including limited internet connectivity, the high cost of digital devices and services, and insufficient digital literacy and skills training within farming communities. Policies must prioritize targeted investments in rural digital infrastructure, strategically extending broadband internet access and mobile coverage to even the most remote agricultural areas. In addition, comprehensive digital literacy programs tailored to the specific needs and learning styles of diverse farmer groups are essential to empower farmers to effectively use digital technologies, navigate the digital landscape, and fully participate in the digital economy. Public–private partnerships, community-based initiatives, and innovative financing models are critical to bridging this persistent digital divide and ensuring equitable access to digital opportunities for all farming communities.

3.2 Overcoming Affordability Constraints: Addressing Economic Barriers to Technology Adoption

The significant upfront costs associated with the adoption of many Agriculture 4.0 technologies, including sensors, drones, data analytics platforms, and precision machinery, pose a significant economic barrier, especially for smallholder farmers and agri-food SMEs operating with limited financial resources. In this sense, a systematic review by El Bakali et al. (2023) assesses the impact of financial incentives on the adoption of conservation agriculture, identifies their benefits and limitations, and provides evidence-based recommendations to inform policy development. Policies need to proactively address these affordability constraints through the strategic design and implementation of targeted financial incentive programs, innovative sharing models for technology access, and the promotion of low-cost digital solutions tailored to the economic realities of different farming contexts. Financial incentive programs can play a critical role in reducing the initial investment burden for farmers, making digital technologies more economically accessible. Leasing programs, equipment sharing cooperatives, and pay-per-use models offer alternative ways to access technology, allowing farmers to benefit from advanced digital tools without incurring prohibitive upfront costs. In addition, encouraging the development and dissemination of low-cost digital solutions that leverage open source technologies and affordable hardware is essential to ensuring that Agriculture 4.0 is not the domain of large-scale, capital-intensive operations, but rather a transformative force that empowers farmers of all sizes and resource levels.

3.3 Navigating the Complexities of Data Management and Interoperability: Promoting Seamless Data Flow and Collaboration

While the data-rich environment of Agriculture 4.0 holds great promise, it also presents significant technical and organizational complexities related to data management and interoperability (Wolfert et al. 2017). Lack of standardized data formats, fragmented data systems, and privacy and security concerns can create barriers to effective data sharing and collaboration across the agri-food value chain, limiting the potential for data-driven innovation and optimized decision-making

(Jouanjean et al. 2020). Policies must actively address these challenges by promoting the development and adoption of open data standards and protocols, fostering interoperability between different digital platforms and systems, and establishing secure data sharing frameworks that protect farmers' data rights while facilitating data sharing and collaboration. Investing in data infrastructure, such as cloud-based data platforms and data marketplaces, can further improve data accessibility, foster data-driven innovation, and unlock the full potential of agricultural data for improved decision-making and sustainability.

3.4 Addressing Technical Expertise and Capacity Gaps: Building a Digitally Literate Agricultural Workforce

The successful adoption and effective use of Agriculture 4.0 technologies depend on the availability of a digitally literate workforce across the agri-food sector. Limited digital skills and technical expertise, particularly among farmers and agricultural support services, can limit the adoption and effective use of digital tools, thereby limiting access to critical resources such as credit, modern technology, and market information (Cheng et al. 2024). Policies must strategically address this skills gap through targeted investments in education and training programs, fostering collaboration between technology providers and agricultural stakeholders, and actively attracting and retaining digital talent in agriculture. Education and training initiatives should cover a wide range of digital skills, from basic digital literacy to advanced data analytics and AI applications, to meet the diverse needs of farmers, extension agents, agribusinesses, and policymakers. In addition, promoting vocational training programs, apprenticeships, and on-the-job training opportunities in digital agriculture is essential to building a digitally savvy agricultural workforce capable of driving innovation and leading the digital transformation of the sector.

3.5 Tailoring Policies to Contextual Needs: Embracing Diversity and Local Realities

Agriculture is characterized by its inherent context specificity, encompassing a wide variety of farming systems, agro-ecological conditions, socio-economic realities and cultural landscapes across regions, countries and communities. Policies for Agriculture 4.0 must recognize and embrace this diversity, adopting a nuanced and context-sensitive approach that avoids one-size-fits-all solutions that may be ineffective or even counterproductive in specific local contexts (Diogo et al. 2022). Participatory policy development processes that actively involve local stakeholders, including farmers, extension agents, community leaders, and local organizations, are critical to ensuring that policies are relevant, effective, and responsive to the unique needs, priorities, and challenges of diverse farming communities. Policies should be designed with flexibility and adaptability in mind, allowing for local customization and innovation, and ensuring that they can evolve and adapt to changing technological landscapes, evolving agricultural practices, and the specific needs of different agricultural regions and sectors. A deep understanding of local contexts, coupled with participatory policy design processes, is essential to ensure that Agriculture 4.0 policies are not only technologically sound but also socially equitable, culturally appropriate, and environmentally sustainable in diverse agricultural landscapes around the world.

4 Policy Imperatives for Accelerating Agriculture 4.0

This section explores how policy can accelerate the adoption of digital technologies in agriculture, known as Agriculture 4.0. It is divided into four main areas, each with specific strategies to help farmers and agribusinesses embrace these changes. Figure 1 presents a comprehensive model illustrating how four distinct yet interconnected policy domains converge, providing a visualization of this ecosystem. These policies act as catalysts, accelerating the adoption of digital technologies in the agri-food system. As the model shows, this acceleration is intended to generate positive, synergistic economic, social and environmental impacts, which will be explored in the following sections.

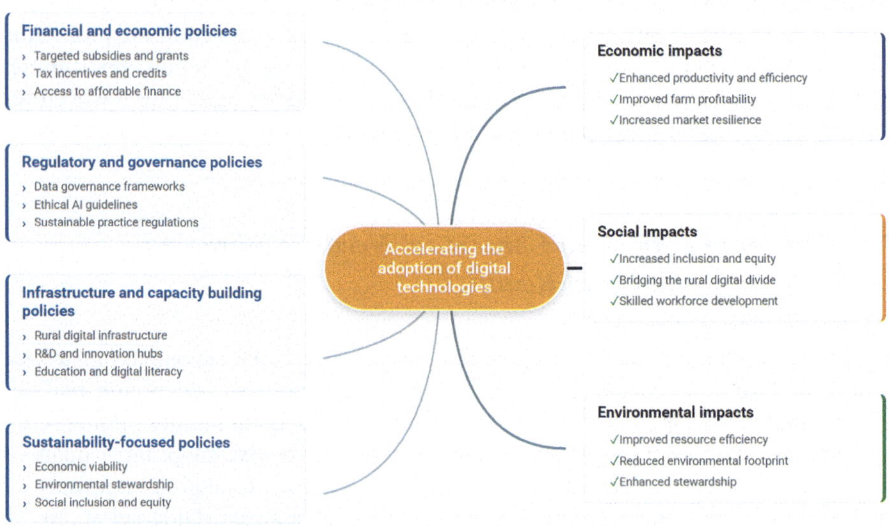

Fig. 1 The policy ecosystem for accelerating the adoption of Agriculture 4.0

4.1 Financial and Economic Policies: Catalyzing Adoption Through Economic Incentives and Investment

Financial and economic policies are a cornerstone of any effective strategy to accelerate the adoption of Agriculture 4.0, as they directly address the economic barriers that often hinder the adoption of new technologies, especially in the inherently risk-averse and resource-constrained agricultural sector. The strategic use of financial incentives, targeted subsidies, and robust investment support can significantly lower the economic barriers for farmers and agribusinesses, making digital solutions more accessible, economically viable, and ultimately more attractive.

4.1.1 Targeted Subsidies and Grants: Direct Financial Support for Technology Adoption

Direct financial subsidies and grants are a powerful policy instrument for reducing the immediate financial burden associated with the adoption of Agriculture 4.0 technologies. These instruments can be strategically designed and targeted to incentivize the adoption of specific technologies and sustainable practices that align with broader policy objectives. For example, governments can offer subsidies to farmers who invest in precision irrigation systems, effectively lowering the upfront costs of adopting water-saving technologies and promoting responsible water management in agriculture (Finger 2023; Parra-López et al. 2025). In support of this approach, Barnes et al. (2019) found that financial incentives, including subsidies, play an important role in motivating European arable farmers to adopt precision agriculture technologies, supporting the argument for targeted financial support. Similarly, grant programs can be established to support the adoption of smart fertilizer application technologies, directly incentivizing farmers to minimize nutrient runoff, reduce fertilizer consumption, and mitigate environmental pollution from excessive fertilizer use (Javed et al. 2022).

Financial incentives have also been effectively used to develop and implement digital platforms that improve traceability and transparency across food supply chains (Ali et al. 2021; Chandan et al. 2023). Subsidies could be provided to support the adoption of blockchain-based traceability systems, enabling food producers and processors to enhance food safety, build consumer trust, and access premium markets that demand verifiable sustainability credentials. By directly addressing the economic disincentives and financial barriers to adoption, targeted subsidies and grants serve as powerful catalysts, accelerating the transition to a more digital, efficient, and sustainable agri-food system.

4.1.2 Tax Incentives and Credits: Stimulating Private Sector Investment and Innovation

Beyond direct financial support to farmers, to foster a vibrant and innovative Agriculture 4.0 landscape, policies should proactively focus on stimulating broader private sector investment. Governments can strategically use tax incentives and credits to make investments in digital agriculture more attractive to agribusinesses, technology companies, and financial institutions. For example, tax breaks for companies that invest in rural broadband infrastructure can be a key strategy for bridging the digital divide in agriculture. This can, for example, encourage telecommunications providers to expand connectivity to underserved rural areas, thereby facilitating more equitable adoption of digital agricultural technologies and ensuring that the benefits of Agriculture 4.0 are more widely available (Shepherd et al. 2020; Rijswijk et al. 2021). Similarly, tax credits can be offered to technology companies that devote resources to research, development, and deployment of sustainable Agriculture 4.0 solutions, spurring innovation and accelerating the availability of green digital technologies specifically tailored to agricultural needs.

4.1.3 Improving Access to Affordable Finance and Credit: Empowering Farmers and Agribusinesses

Financial constraints, including limited access to affordable credit and lack of government support, are significant barriers to the adoption of Industry 4.0 practices in the agricultural supply chain. For example, Mhlanga and Ndhlovu (2023) highlight that financial constraints are a significant barrier for smallholder farmers in Africa when considering the adoption of digital agricultural technologies. Addressing these financial challenges through targeted policy interventions could facilitate the transition to more digital and sustainable agricultural systems (Kumar et al. 2021; Yadav et al. 2022). This could include establishing specialized agricultural credit lines within existing financial institutions, promoting the growth of microfinance institutions that serve smallholder farmers, and exploring innovative financing models such as crowdfunding platforms and peer-to-peer lending schemes.

In addition, risk-sharing mechanisms and loan guarantees offered by governments or development agencies can play a critical role in mitigating the perceived risks for financial institutions lending to the agricultural sector. By sharing the risks associated with agricultural lending, these mechanisms can encourage financial institutions to lend to farmers and agribusinesses, facilitate technology adoption, and enable them to invest in digital solutions that improve their productivity, profitability, and sustainability.

4.2 *Regulatory and Governance Policies: Ensuring Responsible and Ethical Digital Transformation*

Regulatory and governance policies are essential tools for shaping a responsible, ethical, and trustworthy digital agriculture ecosystem. They serve to set the rules of the game and foster innovation while preserving critical societal values, protecting privacy, ensuring data security, and promoting environmental sustainability. Clear, predictable and enabling regulatory frameworks are essential to build trust among all stakeholders and ensure that the digital transformation of agriculture benefits society as a whole.

4.2.1 Establishing Robust Data Governance Frameworks and Standards: Promoting Trust and Interoperability

As Agriculture 4.0 becomes increasingly data-driven, the establishment of robust data governance frameworks is critical to ensure responsible and ethical data handling. In this sense, Jouanjean et al. (2020) emphasize that clear and robust data governance frameworks are crucial for building trust among farmers and other stakeholders in the digital transformation of agriculture. They also highlight that fragmented and unclear data governance arrangements can hinder farmers' willingness to adopt digital solutions, underscoring the need for policies that ensure data privacy, security, and ownership. Therefore, these frameworks should prioritize clear regulations that address privacy and security concerns, which have been identified as key barriers to the adoption of digital agriculture (Parra-López et al. 2024). In addition, ensuring equitable access to digital benefits and mitigating systemic risks associated with artificial intelligence in agriculture remain important considerations (Galaz et al. 2021). While data governance design plays a role in facilitating the responsible use of digital technologies (Shepherd et al. 2020), further efforts are needed to develop comprehensive regulatory policies governing data ownership, security, and interoperability to foster trust and promote sustainable digital agriculture.

For farmers to trust and actively participate in data-driven Agriculture 4.0, data ownership rights must be clearly defined and legally protected. Policies should prioritize frameworks that enable farmers to maintain control over the data generated on their farms, ensuring that they benefit fairly from its potential value and are not disadvantaged by the digital transformation of agriculture (Klerkx and Begemann 2020; Shepherd et al. 2020; Rijswijk et al. 2021). Data privacy regulations aligned with international best practices, such as the GDPR, are essential to protect farmers' sensitive data from unauthorized access, misuse, or commercial exploitation.

In addition, establishing common standards and ensuring that data can be shared seamlessly between different systems and actors is essential to building a truly integrated and efficient Agriculture 4.0 ecosystem (Rijswijk et al. 2021; Latino et al. 2022; Parra-López et al. 2024b). Policies should encourage the development and adoption of common data standards, protocols, and APIs (application programming interfaces) that facilitate seamless data exchange and integration between different digital platforms and systems. This interoperability will enable farmers, researchers, and agribusinesses to share data securely and efficiently, fostering collaboration, promoting innovation, and maximizing the collective benefits of digital agriculture. Open source data platforms and data sharing initiatives are critical infrastructure for unlocking the full potential of data-driven innovation in agriculture. Supported by transparent and robust governance frameworks, these initiatives can democratize data access, foster collaboration, and create a more inclusive and thriving digital ecosystem for all stakeholders (Chandan et al. 2023; Zheng et al. 2023; Parra-López et al. 2024).

4.2.2 Implement Ethical Guidelines and Promote Algorithmic Transparency: Ensuring Fairness and Accountability in AI-Driven Agriculture

Beyond data governance, policies need to proactively address the ethical implications of AI and automation to ensure the responsible and ethical development of Agriculture 4.0. In this sense, Dara et al. (2022) identify key challenges such as fairness, transparency, accountability, sustainability, privacy, and robustness. They emphasize the need for clear guidelines and transparency in AI algorithms to ensure that these technologies are used responsibly and ethically in the agricultural sector. This includes establishing frameworks that ensure fairness, accountability, and transparency in algorithmic decision-making processes and mitigate potential risks of bias and unfair outcomes (Galaz et al. 2021; Lioutas et al. 2021; Rijswijk et al. 2021; Dara et al. 2022).

Ethical guidelines and standards should be developed and implemented for the design, development, and deployment of AI-enabled agricultural technologies that address potential biases that may be embedded in algorithms, ensure transparency in how AI systems arrive at their recommendations and decisions, and mitigate the risk of unintended consequences. In this regard, Mark (2019) explores the ethics of using AI and big data in agriculture through a case study of a large multinational company, providing insights into how ethical issues are being addressed in practice. Dara et al. (2022) provide recommendations for the ethical and responsible use of AI in digital agriculture, emphasizing the need for robust data governance. In addition, van der Burg et al. (2019) discuss current ethical issues in smart agriculture and provide directions for responsible innovation, highlighting the importance of integrating ethical considerations from the beginning of technology development.

Algorithmic transparency is critical to building trust and fostering user confidence in AI-driven agricultural systems. Farmers and other stakeholders need to understand how AI algorithms work, how they process data, and how they arrive at their recommendations in order to make informed decisions and promote accountability. Policies should mandate transparency in the design and use of algorithms, and require developers and users of AI-driven agricultural technologies to adhere to ethical guidelines and transparency standards.

Public policy should actively promote public dialogue and stakeholder engagement on the ethical implications of digital agriculture, fostering a shared understanding of the values and principles that should guide responsible innovation and ensure that technology serves the broader public good. Ethical standards and certification schemes for digital agriculture technologies and practices can be developed to promote responsible innovation, build consumer trust in digitally-enabled food systems, and foster a culture of ethical data use and algorithmic accountability across the agri-food sector. In addition, policies should support ongoing research on the ethical and societal impacts of digital agriculture, continuously assess and adapt regulatory frameworks to address emerging ethical challenges, and ensure that innovation is guided by ethical principles and contributes to a socially just and equitable agri-food future.

4.2.3 Enacting Regulations for Sustainable Practices: Driving Environmental Stewardship Through Policy

Strategic regulatory policies that incentivize and support digital innovation are critical to advancing sustainable agricultural practices. Digital technologies can enhance these policies by optimizing resource efficiency and improving environmental stewardship. Well-designed regulatory frameworks, therefore, amplify the impact of sustainability efforts in the agricultural sector (Kumar et al. 2021; Shang et al. 2021; Parra-López et al. 2024a). For example, environmental performance standards for agricultural inputs such as fertilizers and pesticides can be set to encourage the adoption of precision application technologies that minimize pollution and reduce reliance on harmful chemicals (Jacquet et al. 2022). Regulations can also promote water conservation by setting water-use efficiency standards for irrigation practices, encouraging the use of smart irrigation systems that optimize water use based on real-time data, and rewarding farmers who adopt water-saving technologies (Parra-López et al. 2025). Elsewhere, Finger (2023) explores the role of digital innovation in fostering sustainable and resilient agricultural systems, emphasizing that policy measures are critical to creating an enabling environment for the adoption of these technologies, and suggests that policies should focus on strengthening digital infrastructure, legal frameworks, and education to amplify the impact of digital agriculture on sustainability.

In addition, regulatory frameworks have the potential to support carbon sequestration initiatives in agriculture by promoting sustainable soil management practices. Effective soil management plays a critical role in enhancing soil carbon storage and mitigating climate change (Lal et al. 2021). While policies that promote climate-smart agricultural practices can help reduce GHG emissions, more research is needed to determine the extent to which direct regulatory incentives drive widespread adoption. Blockchain-based traceability systems mandated or incentivized by regulation can also increase transparency and accountability in agricultural supply chains, promoting sustainable sourcing, responsible production practices, and consumer confidence in the environmental integrity of food (Saurabh and Dey 2021; Zheng et al. 2023). Through the strategic use of regulatory policy, governments can create a powerful framework for promoting environmental sustainability in Agriculture 4.0 and fostering a greener and more responsible agri-food sector.

4.3 Infrastructure and Capacity Building Policies: Creating an Enabling Ecosystem for Digital Agriculture

Effective policy frameworks for Agriculture 4.0 go beyond financial and regulatory instruments to include fundamental infrastructure and capacity-building initiatives that are critical for widespread and impactful digital transformation. Strategic investments in digital infrastructure, targeted R&D support, and comprehensive education and training programs are essential to creating an enabling ecosystem that fosters innovation, facilitates adoption, and empowers stakeholders to fully participate in the digital agricultural revolution.

4.3.1 Strategic Investment in Rural Digital Infrastructure and Connectivity: Bridging the Digital Divide

Bridging the digital divide requires sustained and substantial public and private investment in rural digital infrastructure and connectivity. Recognizing digital connectivity as a fundamental enabler of Agriculture 4.0 and a critical public utility for modern agricultural practices, governments must prioritize the expansion of broadband internet access and reliable mobile network coverage in rural areas (Parra-López et al. 2021). A multi-pronged approach is essential, including direct public investment in infrastructure development, strategic policies that incentivize private sector participation, and the exploration of innovative infrastructure solutions tailored to the unique challenges of rural areas (Kumar et al. 2021). A study by Cao et al. (2022) highlights that digital villages can effectively bridge the urban-rural digital divide through comprehensive empowerment of rural subjects, agro-industrial structure, public services, and rural governance. This underscores the importance of strategic investment in rural digital infrastructure to ensure that digital technologies are accessible and beneficial to all segments of the agricultural community.

Encouraging private sector investment can be achieved through a variety of policy instruments, including spectrum allocation policies that prioritize rural coverage, infrastructure sharing agreements that reduce deployment costs for telecommunications companies, and targeted tax incentives that reward private sector investment in rural digital infrastructure. Innovative infrastructure solutions, such as community-owned networks that leverage local resources and expertise, and satellite-based Internet access that provides connectivity to remote and underserved areas, can further extend digital reach and bridge the connectivity gap. Addressing the digital divide is not just about providing access to technology; it is about ensuring digital inclusion, empowering rural communities to fully participate in the digital economy, and creating a level playing field for all farmers to benefit from the transformative potential of Agriculture 4.0.

4.3.2 Targeted R&D Investments and Collaborative Innovation Ecosystems: Fostering Innovation and Tailoring Solutions to Agricultural Needs

Sustained and strategically targeted public investment in research and development (R&D) is essential to drive innovation in Agriculture 4.0 and address the unique challenges and opportunities of building a truly sustainable agriculture sector (Shepherd et al. 2020). In this sense, Wolfert et al. (2023) emphasize the importance of designing digital innovation ecosystems in the agri-food sector and provide principles and an organizational framework that can guide the development of such ecosystems. In addition, Finger (2023) highlights the role of digital innovations in promoting sustainable and resilient agricultural systems, underscoring the need for targeted R&D investments to effectively develop and implement these technologies.

Governments should prioritize funding for interdisciplinary research collaborations that intentionally bring together agricultural scientists, engineers, data scientists, social scientists, and ethicists to foster a holistic and multifaceted approach to innovation. These collaborative initiatives should focus on co-creating and co-innovating digital solutions that are specifically tailored to diverse agricultural contexts, addressing the unique needs of different farming systems, agro-ecological zones, and socio-economic realities.

Research priorities should cover a wide range of areas, including the development of affordable and accessible digital technologies for smallholder farmers, the application of AI and advanced data analytics for sustainable resource management, the exploration of novel sensing technologies for precision agriculture, and the rigorous assessment of the ethical and social implications of digital agriculture. Fostering collaborative innovation ecosystems is also critical to accelerate the translation of research into real-world applications and to ensure that innovations are effectively adopted and scaled up across the agri-food sector. Innovation hubs, living labs and agri-tech parks can serve as dynamic platforms to bring together researchers, technology companies, farmers, policymakers and other stakeholders, creating vibrant spaces for knowledge sharing, experimentation, technology transfer and the co-creation of innovative solutions deeply rooted in real-world agricultural needs and challenges.

4.3.3 Comprehensive Education, Training and Digital Literacy Programs: Building Human Capital for a Digital Future

A digitally literate workforce is the cornerstone of a thriving and effective Agriculture 4.0 ecosystem. Policymakers must prioritize the development and implementation of comprehensive education, training, and digital literacy programs that are strategically designed to equip farmers, extension agents, agri-food professionals, and policymakers with the essential knowledge, practical skills, and digital competencies needed to effectively engage with and benefit from digital technologies (Mmbando 2025). Digital literacy programs should cover a wide range of skills, from basic computer literacy and internet navigation to advanced data management, data analytics, AI applications, and the responsible use of digital tools for sustainable agriculture. For example, a study by Magesa et al. (2023) assessed the digital literacy of smallholder farmers in Tanzania and found that although they had access to the internet and mobile phones, their digital literacy levels were low in terms of managing, integrating, evaluating, and creating information. Education and training initiatives should therefore be carefully tailored to the specific needs, skill levels and learning styles of different target groups, using a variety of delivery methods, including online courses, interactive hands-on workshops, on-farm demonstrations and peer-to-peer learning networks.

In addition, curricula at agricultural universities, vocational training institutions and extension services should be comprehensively modernized to seamlessly integrate digital agriculture skills into core agricultural education, preparing the next generation of agricultural professionals for a digitally-driven future and ensuring that the existing agricultural workforce is equipped to navigate and thrive in this rapidly evolving technological landscape. Promoting digital literacy and fostering a culture of digital fluency is not just about technology adoption; it is fundamentally about empowering stakeholders to become active participants, informed decision-makers, and agents of innovation within the digital transformation, ensuring that the benefits of Agriculture 4.0 are widely shared and contribute to a more inclusive and prosperous agricultural sector.

4.4 Sustainability-Focused Policies: Integrating Environmental, Social and Economic Dimensions into Digital Transformation

Policies to accelerate the adoption of Agriculture 4.0 must be firmly and explicitly grounded in a holistic understanding of sustainability that encompasses the interrelated environmental, social, and economic dimensions. A siloed policy approach that focuses narrowly on technology adoption without considering the broader sustainability implications risks unintended consequences, and may ultimately fail to achieve the long-term goals of a truly sustainable and resilient agri-food system. Sustainability must be woven into the very fabric of Agriculture 4.0 policies, guiding digital transformation toward a future where technology not only enables increased productivity and economic growth but also environmental stewardship, social equity, and ethical responsibility.

4.4.1 Prioritizing Economic Viability and Resilience: Building a Prosperous and Adaptive Agri-Food Sector in the Digital Age

Economic viability and resilience are essential pillars of a truly sustainable agri-food system. Policies for Agriculture 4.0 must be strategically designed to enhance the economic competitiveness and long-term profitability of farms and agribusinesses, while strengthening the sector's resilience to economic shocks, climate change impacts, and other external pressures (Finger 2023). Digital technologies offer a powerful set of tools to enhance economic viability by increasing farm productivity, reducing input costs through precision resource management, improving market access through e-commerce platforms and digital marketplaces, and enabling the development of new value-added services such as direct-to-consumer sales, agritourism initiatives, and data-driven agricultural extension services. For example, a comprehensive review by Papadopoulos et al. (2024) highlights that digital agricultural technologies can lead to significant reductions in fertilizer and pesticide use, thereby reducing input costs and improving economic viability for farmers. In addition, Geng et al. (2024) conducted a mixed-methods study in China and found that farmers' adoption of digital technologies positively affects their economic benefits, underscoring the importance of policies that encourage such adoption.

Policies should actively support the development of digital business models that are economically viable for farmers of all sizes and promote the adoption of digital marketing platforms, e-commerce solutions and precision farming technologies that increase farm efficiency, profitability and market competitiveness. In addition, policies must prioritize building resilience within agri-food systems to enable them to withstand economic volatility, adapt to climate change, and respond effectively to other unforeseen risks and disruptions. Digital technologies play a critical role in enhancing resilience through improved risk management tools, early warning systems for climate-related risks and pest and disease outbreaks, and diversified production strategies that use data-driven insights to optimize crop selection, resource allocation, and market responsiveness. Policies can incentivize the adoption of digital tools for climate risk assessment, drought monitoring, pest and disease forecasting, and supply chain optimization, enabling farmers and agribusinesses to make more informed decisions, adapt to changing environmental and economic conditions, and build more resilient and adaptive agribusinesses.

4.4.2 Improving Environmental Sustainability and Resource Efficiency: Harnessing Digital Tools for a Greener Agriculture

Improving environmental sustainability and resource efficiency must be a central and overarching imperative for all policies related to Agriculture 4.0 (Hassoun et al. 2022). Digital technologies offer an unprecedented array of powerful tools to minimize the environmental footprint of agricultural practices, significantly reduce reliance on chemical inputs, optimize resource use, and promote ecological balance. In this sense, Zhang et al. (2023) emphasize the central role of the digital economy in driving agricultural technology innovation and improving agricultural green total factor productivity, highlighting the transformative potential of digitalization to advance environmental sustainability across the agricultural sector. In addition, Basso and Antle (2020) highlight that digital agriculture, through the use of digital and geospatial technologies, can significantly contribute to the design of sustainable agricultural systems by effectively monitoring, assessing and managing soil, climate and genetic resources. Also, Shamshiri et al. (2024) provide a use-case review of digitalization in agriculture for sustainable crop production, highlighting specific examples where digital tools have been successfully applied to improve environmental sustainability.

Policies should strategically prioritize the adoption of precision agriculture technologies, such as variable-rate fertilizer and pesticide application systems, sensor-based irrigation systems that optimize water delivery based on real-time crop needs, and drone-based crop monitoring platforms that enable early detection of pests and diseases, allowing for targeted and minimized interventions. For example, remote sensing, advanced geospatial analytics, and AI-powered platforms can be used to monitor soil health at scale, assess water quality in agricultural landscapes, optimize the timing and dosage of pesticide applications to minimize off-target effects, and promote biodiversity conservation by identifying and protecting critical habitats within agricultural areas (Jung et al. 2021; Blickensdörfer et al. 2022). These digital tools provide invaluable data and insights for informed decision-making, enabling farmers, land managers and policymakers to adopt more sustainable land management practices, optimize resource allocation and minimize the environmental impact of agricultural production.

Policies can actively promote the adoption of climate-smart agricultural practices such as no-till farming, cover cropping, agroforestry and precision livestock management, all of which are significantly enhanced by digital technologies and contribute to carbon sequestration, improved soil health, enhanced climate resilience and reduced greenhouse gas emissions from agriculture. In addition, robust policies should promote the use of digital technologies for comprehensive environmental monitoring and transparent reporting, ensuring accountability in agricultural sustainability efforts and fostering consumer trust in the environmental integrity of food products.

4.4.3 Ensuring Social Inclusion and Equity: Bridging the Digital Divide and Empowering Marginalized Communities

Social inclusion and equity must be at the heart of all policy initiatives to promote Agriculture 4.0. Digital transformation should not inadvertently exacerbate existing inequalities within the agri-food system, creating a two-tier agriculture where only large, technologically advanced farms benefit, while smallholders and marginalized communities are left behind. Instead, policies should actively seek to ensure that the benefits of digital agriculture are broadly and equitably shared across all segments of the farming population, contributing to a more equitable and inclusive agricultural landscape. In this sense, Yu et al. (2022) highlight the importance of recognizing the unique characteristics of smallholder farming systems, emphasizing that policies must be sensitive to the diverse realities of agricultural landscapes and avoid one-size-fits-all approaches that may inadvertently disadvantage smallholder producers. In addition, Dittmer et al. (2022) highlight the need for socially inclusive digital tools in agriculture and provide recommendations on how such tools can be designed to support diverse smallholder farmers. A comprehensive review by Klerkx et al. (2019) provides an overview of social science research on digital agriculture, identifying key themes such as social inclusion and equity. In addition, Prause (2021) discusses the challenges related to labor and social sustainability in digital agriculture, highlighting the importance of considering these aspects in policy development.

Policies must proactively address the digital divide by implementing targeted programs and initiatives to bridge the gap in access to digital technologies, infrastructure and digital literacy to ensure that all farmers, regardless of farm size, geographic location or socio-economic status, have the opportunity to participate in and benefit from the digital revolution. Digital literacy programs should be specifically tailored to the needs of smallholder farmers, women farmers and other marginalized groups, taking into account their specific challenges, cultural contexts and learning preferences. In addition, policies should consider the potential social impacts of digital agriculture on labor markets and rural livelihoods, and implement proactive measures to support labor force transitions, create new economic opportunities in rural areas, and mitigate the potential negative impacts of automation and technological change on agricultural employment. Promoting digital entrepreneurship and innovation in rural communities can foster local economic development, empower marginalized groups and ensure that the digital transformation of agriculture contributes to rural prosperity and social well-being, rather than exacerbating existing inequalities.

5 Actionable Policy Recommendations and Future Directions: Setting the Course for a Sustainable and Thriving Agriculture 4.0 Ecosystem

To effectively navigate the complexities and challenges of accelerating the adoption of Sustainable Agriculture 4.0 and fully realize its transformative potential for the global agri-food sector, a set of strategic, forward-looking, and actionable policy recommendations is essential. These recommendations, based on analyzed policies and challenges, aim to chart a course for a sustainable and thriving Agriculture 4.0 ecosystem.

A fundamental step toward realizing this vision is the development and implementation of integrated digital agriculture strategies. Governments should take a proactive and leading role in creating comprehensive national and supranational strategies that articulate a clear vision for the digital transformation of the agriculture and food sector. These strategies need to set ambitious yet achievable goals, along with measurable targets and concrete action plans for policy implementation in key areas. These include infrastructure development, targeted research and development investments, comprehensive education and training programs, enabling regulatory frameworks, and the promotion of robust international cooperation. Crucially, these digital agriculture strategies should be strategically aligned with broader national and supranational sustainability goals that encompass environmental stewardship, social equity, and economic performance. This alignment will ensure that the digital transformation of agriculture truly contributes to a more sustainable and prosperous future for the entire agri-food system.

Building on strategic frameworks, it is critical to prioritize strategic investments in rural digital infrastructure and connectivity. Bridging the persistent digital divide must be treated as a national priority. Governments must commit to sustained and substantial public and private investment to expand broadband Internet access, improve mobile coverage, and deploy robust digital infrastructure, specifically in rural areas. Digital connectivity should be recognized as a fundamental enabler of Agriculture 4.0 and a critical public utility for modern agricultural practices. Policies should proactively encourage private sector participation through targeted financial incentives, promote infrastructure sharing agreements to optimize resource use, and actively explore innovative infrastructure solutions. This includes considering community-owned networks and satellite-based Internet access to ensure digital inclusion and equitable access to digital opportunities for all farming communities, regardless of their location or remoteness.

In addition to infrastructure development, the design and implementation of targeted and results-based financial incentive programs are critical to addressing affordability barriers and promoting sustainable technology adoption. These programs should be strategically designed and carefully targeted to effectively address the financial barriers farmers face in adopting Agriculture 4.0 technologies and practices. Targeted subsidies, grants, and tax breaks should be thoughtfully designed to reduce the upfront costs of technology adoption, making digital solutions more accessible and economically viable, especially for smallholder farmers and SMEs. It is critical that these financial incentives are strategically coupled with rigorous monitoring and evaluation mechanisms. This will ensure that they are demonstrably effective in achieving desired outcomes, such as increased technology adoption rates, tangible improvements in environmental sustainability, improved farm profitability, and greater social equity within the agricultural sector.

In addition, the creation of enabling and adaptive regulatory frameworks for data governance and ethical use is critical in the digital age. Governments must proactively create clear, enabling, and adaptive frameworks for data governance and ethical data use in agriculture that recognize data as a critical asset. These frameworks should prioritize data privacy, data security, data interoperability, and algorithmic transparency. This approach is essential to build trust and promote responsible data sharing and innovation, while remaining flexible and adaptable to the rapidly evolving technological landscape and emerging ethical considerations. Robust data governance policies should be developed through participatory processes that actively engage stakeholders from across the agri-food sector. This collaborative approach will ensure that policies are both effective and equitable, effectively protecting farmers' data rights while fostering data-driven innovation for the public good.

Fostering robust R&D ecosystems and public-private innovation partnerships is essential to truly foster innovation and drive practical solutions. Sustained and strategically targeted public funding for R&D is essential to drive continued innovation in sustainable Agriculture 4.0 technologies and practices. Governments should actively promote collaborative research initiatives, public-private partnerships and vibrant innovation ecosystems. These ecosystems should bring together researchers from different disciplines, technology companies, farmers, extension agents and other relevant stakeholders. These collaborative platforms should be specifically designed to accelerate the development, rigorous testing and efficient scaling of digital solutions. These solutions need to be tailored to the diverse needs and unique challenges of sustainable agriculture in different contexts and farming systems.

Addressing the digital skills gap is also critical to ensuring that the agricultural workforce is equipped to thrive in this digital age. Implementing comprehensive digital literacy and skills development initiatives is therefore essential to creating a digitally literate agricultural workforce. Governments, educational institutions, industry stakeholders, and civil society organizations must work together to develop and implement comprehensive programs. These programs should strategically target farmers, extension agents, agri-food professionals and policy makers. These initiatives should be carefully tailored to different needs and learning styles, use a variety of delivery methods, and focus on practical, hands-on training. This approach will enable stakeholders to confidently and effectively use digital technologies for sustainable agricultural practices.

Finally, addressing the global challenges facing the agriculture and food sector requires a collaborative and globally coordinated approach. Encouraging and facilitating international cooperation and knowledge sharing is essential to accelerate the adoption of sustainable Agriculture 4.0 practices and policies worldwide. The establishment of robust platforms for knowledge sharing, technology transfer and joint research initiatives is paramount. These platforms will facilitate the exchange of best practices, lessons learned from different contexts, and technological innovations across countries and regions, thus fostering a collective global effort toward a more sustainable, resilient and food-secure future for agriculture.

6 Conclusions

Policies are not just passive enablers of the digital transformation of agriculture; they are the active catalysts, the guiding frameworks, and the essential cornerstones upon which a truly sustainable and resilient Agriculture 4.0 ecosystem can be built. Strategic, well-designed and effectively implemented policies are essential to unlock the transformative potential of digital technologies in the agri-food sector and to ensure that this technological revolution serves the broader public good, contributing to a more prosperous, equitable and environmentally sound future for food production worldwide. By proactively addressing market failures, mitigating potential risks, fostering a vibrant culture of innovation, and consistently prioritizing sustainability and inclusivity, policymakers can chart a course for a digital agriculture revolution that truly benefits all stakeholders and ensures a sustainable and food-secure future for generations to come. The path to this envisioned future will require continued commitment, adaptive policy frameworks that can respond to rapid technological advances, and a collaborative spirit among all stakeholders working together to build an Agriculture 4.0 ecosystem that is not only technologically advanced, but also deeply rooted in ethical principles, social justice, and environmental stewardship. Future research should focus on rigorously evaluating the long-term impacts of different policy instruments, continuously refining policy design based on empirical evidence and the evolving technological landscape, and exploring innovative policy approaches that can effectively navigate the complex and rapidly changing terrain of digital agriculture to ensure that its transformative potential is fully realized for the benefit of humanity and the planet.

Competing Interests The author(s) have no competing interests to declare that are relevant to the content of this manuscript.

References

Ali MH, Chung LN, Kumar A, Zailani S, Tan KH (2021) A sustainable blockchain framework for the halal food supply chain: lessons from Malaysia. Technol Forecast Soc Chang 170:120870

Barnes AP, Soto I, Eory V, Beck B, Balafoutis AT, Sanchez B, Vangeyte J, Fountas S, van der Wal T, Gómez-Barbero M (2019) Influencing incentives for precision agricultural technologies within European arable farming systems. Environ Sci Pol 93:66–74

Basso B, Antle J (2020) Digital agriculture to design sustainable agricultural systems. Nat Sustainability 3:254–256

Béné C (2022) Why the great food transformation may not happen – a deep-dive into our food systems' political economy, controversies and politics of evidence. World Dev 154:105881

Blickensdörfer L, Schwieder M, Pflugmacher D, Nendel C, Erasmi S, Hostert P (2022) Mapping of crop types and crop sequences with combined time series of Sentinel-1, Sentinel-2 and Landsat 8 data for Germany. Remote Sens Environ 269:112831

Cao L, Niu H, Wang Y (2022) Utility analysis of digital villages to empower balanced urban-rural development based on the three-stage DEA-Malmquist model. PLoS One 17:e0270952

Chandan A, John M, Potdar V (2023) Achieving UN SDGs in food supply chain using blockchain technology. Sustainability 15:2109

Chawla I, Karthikeyan L, Mishra AK (2020) A review of remote sensing applications for water security: quantity, quality, and extremes. J Hydrol 585:124826

Cheng C, Gao Q, Ju K, Ma Y (2024) How digital skills affect farmers' agricultural entrepreneurship? An explanation from factor availability. J Innov Knowl 9:100477

Dara R, Hazrati Fard SM, Kaur J (2022) Recommendations for ethical and responsible use of artificial intelligence in digital agriculture. Front Artif Intell 5:884192

Diogo V, Helfenstein J, Mohr F, Varghese V, Debonne N, Levers C, Swart R, Sonderegger G, Nemecek T, Schader C, Walter A, Ziv G, Herzog F, Verburg PH, Bürgi M (2022) Developing context-specific frameworks for integrated sustainability assessment of agricultural intensity change: an application for Europe. Environ Sci Pol 137:128–142

Dittmer KM, Wollenberg E, Burns S (2022) Socially inclusive digital tools for agriculture: a way forward. TRANSITIONS: inclusive digital tools to enable climate-informed Agroecological Transitions (ATDT). Alliance of Bioversity & CIAT, Cali, p 7

Dong SK, Shang ZH, Gao JX, Boone RB (2020) Enhancing sustainability of grassland ecosystems through ecological restoration and grazing management in an era of climate change on Qinghai-Tibetan Plateau. Agric Ecosyst Environ 287:106684

El Bakali I, Ait El Mekki A, Maatala N, Harbouze R (2023) A systematic review on the impact of incentives on the adoption of conservation agriculture: new guidelines for policymakers and researchers. Int J Agric Sustain 21:2290415

European Commission (2021) Proposal for a Regulation on Artificial Intelligence. COM/2021/206 final. Retrieved from https://eur-lex.europa.eu/legal-content/EN/TXT/?uri=CELEX%3A52021PC0206

Finger R (2023) Digital innovations for sustainable and resilient agricultural systems. Eur Rev Agric Econ 50:1277–1309

Fragomeli R, Annunziata A, Punzo G (2024) Promoting the transition towards Agriculture 4.0: a systematic literature review on drivers and barriers. Sustainability 16:2425

Galanakis CM (2023) The "Vertigo" of the food sector within the triangle of climate change, the post-pandemic world, and the Russian-Ukrainian War. Foods 12:721

Galaz V, Centeno MA, Callahan PW, Causevic A, Patterson T, Brass I, Baum S, Farber D, Fischer J, Garcia D, McPhearson T, Jimenez D, King B, Larcey P, Levy K (2021) Artificial intelligence, systemic risks, and sustainability. Technol Soc 67:101741

Geng W, Liu L, Zhao J, Kang X, Wang W (2024) Digital technologies adoption and economic benefits in agriculture: a mixed-methods approach. Sustainability 16:4431

Hassoun A, Boukid F, Pasqualone A, Bryant CJ, Garcia GG, Parra-Lopez C, Jagtap S, Trollman H, Cropotova J, Barba FJ (2022) Emerging trends in the agri-food sector: digitalisation and shift to plant-based diets. Curr Res Food Sci 5:2261–2269

Hassoun A, Ait-Kaddour A, Abu-Mahfouz AM, Rathod NB, Bader F, Barba FJ, Biancolillo A, Cropotova J, Galanakis CM, Jambrak AR, Lorenzo JM, Mage I, Ozogul F, Regenstein J (2023) The fourth industrial revolution in the food industry-part I: Industry 4.0 technologies. Crit Rev Food Sci Nutr 63:6547–6563

Jacquet F, Jeuffroy MH, Jouan J, Le Cadre E, Litrico I, Malausa T, Reboud X, Huyghe C (2022) Pesticide-free agriculture as a new paradigm for research. Agron Sustain Dev 42:8

Javed T, I I, Singhal RK, Shabbir R, Shah AN, Kumar P, Jinger D, Dharmappa PM, Shad MA, Saha D, Anuragi H, Adamski R, Siuta D (2022) Recent advances in agronomic and physio-molecular approaches for improving nitrogen use efficiency in crop plants. Front Plant Sci 13:877544

Jouanjean M-A, Casalini F, Wiseman L, Gray E (2020) Issues around data governance in the digital transformation of agriculture: the farmers' perspective, OECD Food, Agriculture and Fisheries Papers, No. 146. OECD Publishing, Paris. https://doi.org/10.1787/53ecf2ab-en

Jung J, Maeda M, Chang A, Bhandari M, Ashapure A, Landivar-Bowles J (2021) The potential of remote sensing and artificial intelligence as tools to improve the resilience of agriculture production systems. Curr Opin Biotechnol 70:15–22

Kamble SS, Gunasekaran A, Sharma R (2020) Modeling the blockchain enabled traceability in agriculture supply chain. Int J Inf Manag 52:101967

Klerkx L, Begemann S (2020) Supporting food systems transformation: the what, why, who, where and how of mission-oriented agricultural innovation systems. Agric Syst 184:102901

Klerkx L, Rose D (2020) Dealing with the game-changing technologies of Agriculture 4.0: how do we manage diversity and responsibility in food system transition pathways? Global Food Secur 24:100347

Klerkx L, Jakku E, Labarthe P (2019) A review of social science on digital agriculture, smart farming and agriculture 4.0: new contributions and a future research agenda. NJAS Wageningen J Life Sci 90–91:100315

Kumar S, Raut RD, Nayal K, Kraus S, Yadav VS, Narkhede BE (2021) To identify industry 4.0 and circular economy adoption barriers in the agriculture supply chain by using ISM-ANP. J Clean Prod 293:126023

Lal R, Bouma J, Brevik E, Dawson L, Field DJ, Glaser B, Hatano R, Hartemink AE, Kosaki T, Lascelles B, Monger C, Muggler C, Ndzana GM, Norra S, Pan XC, Paradelo R, Reyes-Sánchez LB, Sandén T, Singh BR, Spiegel H, Yanai J, Zhang JB (2021) Soils and sustainable development goals of the United Nations: an International Union of Soil Sciences perspective. Geoderma Reg 25:e00398

Latino ME, Menegoli M, Lazoi M, Corallo A (2022) Voluntary traceability in food supply chain: a framework leading its implementation in Agriculture 4.0. Technol Forecast Soc Chang 178:121564

Lioutas ED, Charatsari C, De Rosa M (2021) Digitalization of agriculture: a way to solve the food problem or a trolley dilemma? Technol Soc 67:101744

Maffezzoli F, Ardolino M, Bacchetti A, Perona M, Renga F (2022) Agriculture 4.0: a systematic literature review on the paradigm, technologies and benefits. Futures 142:102998

Magesa M, Jonathan J, Urassa J (2023) Digital literacy of smallholder farmers in Tanzania. Sustainability 15:13149

Mark R (2019) Ethics of using AI and Big Data in agriculture: the case of a large agriculture multinational. The ORBIT J 2:1–27

Mehrabi Z, McDowell MJ, Ricciardi V, Levers C, Martinez JD, Mehrabi N, Wittman H, Ramankutty N, Jarvis A (2021) The global divide in data-driven farming. Nat Sustainability 4:154–160

Mhlanga D, Ndhlovu E (2023) Digital technology adoption in the agriculture sector: challenges and complexities in Africa. Hum Behav Emerging Technol 2023:1–10

Mmbando GS (2025) Harnessing artificial intelligence and remote sensing in climate-smart agriculture: the current strategies needed for enhancing global food security. Cogent Food Agric 11:2454354

Mondejar ME, Avtar R, Diaz HLB, Dubey RK, Esteban J, Gomez-Morales A, Hallam B, Mbungu NT, Okolo CC, Prasad KA, She Q, Garcia-Segura S (2021) Digitalization to achieve sustainable development goals: steps towards a smart green planet. Sci Total Environ 794:148539

Papadopoulos G, Arduini S, Uyar H, Psiroukis V, Kasimati A, Fountas S (2024) Economic and environmental benefits of digital agricultural technologies in crop production: a review. Smart Agric Technol 8:100441

Parra-López C, Reina-Usuga L, Carmona-Torres C, Sayadi S, Klerkx L (2021) Digital transformation of the agrifood system: quantifying the conditioning factors to inform policy planning in the olive sector. Land Use Policy 108:105537–105537

Parra-López C, Ben Abdallah S, Garcia-Garcia G, Hassoun A, Sánchez-Zamora P, Trollman H, Jagtap S, Carmona-Torres C (2024) Integrating digital technologies in agriculture for climate change adaptation and mitigation: state of the art and future perspectives. Comput Electron Agric 226:109412

Parra-López C, Reina-Usuga L, Garcia-Garcia G, Carmona-Torres C (2024a) Designing policies to promote the adoption of digital phytosanitation towards sustainability: the case of the olive sector in Andalusia. Agric Syst 221:104147

Parra-López C, Reina-Usuga L, Garcia-Garcia G, Carmona-Torres C (2024b) Functional analysis of technological innovation systems enabling digital transformation: a semi-quantitative multi-criteria framework applied in the olive sector. Agric Syst 214:103848

Parra-López C, Ben Abdallah S, Garcia-Garcia G, Hassoun A, Trollman H, Jagtap S, Gupta S, Aït-Kaddour A, Makmuang S, Carmona-Torres C (2025) Digital technologies for water use and management in agriculture: recent applications and future outlook. Agric Water Manag 309:109347

Prause L (2021) Digital agriculture and labor: a few challenges for social sustainability. Sustainability 13:5980

Rijswijk K, Klerkx L, Bacco M, Bartolini F, Bulten E, Debruyne L, Dessein J, Scotti I, Brunori G (2021) Digital transformation of agriculture and rural areas: a socio-cyber-physical system framework to support responsibilisation. J Rural Stud 85:79–90

Rotz S, Gravely E, Mosby I, Duncan E, Finnis E, Horgan M, LeBlanc J, Martin R, Neufeld HT, Nixon A, Pant L, Shalla V, Fraser E (2019) Automated pastures and the digital divide: how agricultural technologies are shaping labour and rural communities. J Rural Stud 68:112–122

Saurabh S, Dey K (2021) Blockchain technology adoption, architecture, and sustainable agri-food supply chains. J Clean Prod 284:124731

Shamshiri RR, Sturm B, Weltzien C, Fulton J, Khosla R, Schirrmann M, Raut S, Basavegowda DH, Yamin M, Hameed IA (2024) Digitalization of agriculture for sustainable crop production: a use-case review. Front Environ Sci 12

Shang LM, Heckelei T, Gerullis MK, Börner J, Rasch S (2021) Adoption and diffusion of digital farming technologies-integrating farm-level evidence and system interaction. Agric Syst 190:103074

Sharma B, Tiwari S, Kumawat KC, Cardinale M (2023) Nano-biofertilizers as bio-emerging strategies for sustainable agriculture development: potentiality and their limitations. Sci Total Environ 860:160476

Shepherd M, Turner JA, Small B, Wheeler D (2020) Priorities for science to overcome hurdles thwarting the full promise of the 'digital agriculture' revolution. J Sci Food Agric 100:5083–5092

Singh A (2021) Soil salinity: a global threat to sustainable development. Soil Use Manag 38:39–67

van der Burg S, Bogaardt M-J, Wolfert S (2019) Ethics of smart farming: current questions and directions for responsible innovation towards the future. NJAS Wageningen J Life Sci 90-91:1–10

van der Burg S, Wiseman L, Krkeljas J (2021) Trust in farm data sharing: reflections on the EU code of conduct for agricultural data sharing. Ethics Inf Technol 23:185–198

Wolfert S, Ge L, Verdouw C, Bogaardt M-J (2017) Big data in smart farming – a review. Agric Syst 153:69–80

Wolfert S, Verdouw C, van Wassenaer L, Dolfsma W, Klerkx L (2023) Digital innovation ecosystems in agri-food: design principles and organizational framework. Agric Syst 204:103558

Yadav VS, Singh AR, Gunasekaran A, Raut RD, Narkhede BE (2022) A systematic literature review of the agro-food supply chain: challenges, network design, and performance measurement perspectives. Sustainable Prod Consumption 29:685–704

Yu PH, Fennell S, Chen YY, Liu H, Xu L, Pan JW, Bai SY, Gu SX (2022) Positive impacts of farmland fragmentation on agricultural production efficiency in Qilu Lake watershed: implications for appropriate scale management. Land Use Policy 117:106108

Zhang YF, Ji MX, Zheng XZ (2023) Digital economy, agricultural technology innovation, and agricultural green total factor productivity. SAGE Open 13:1961

Zheng Y, Xu YQ, Qiu ZG (2023) Blockchain traceability adoption in agricultural supply chain coordination: an evolutionary game analysis. Agriculture-Basel 13:184

Guidelines for Promoting the Adoption and Effective Use of Digital Agricultural Technologies by Farmers

Karen McGrath, Áine Regan, and Tomás Russell

1 A New Era of Farming

Imagining a future farm without digital technologies or smart farming systems is becoming increasingly difficult, owing to the rapid integration of digitalisation into the sector. Digital advances are ushering in a new, once-unimaginable era of farming, profoundly changing what farming is, how farming is done, and what it means to be a farmer. As these transformations unfold, we face some uncertainty about what the future farm will truly look like and what the ripple effects of digitalisation will bring. However, the growing body of social science research in this area offers vital insights into the transformative power of digital innovations, aiding our understanding and anticipation of their impacts.

It is accepted that these effects could be drastic, potentially making some hesitant, fearful, or resistant to a new 'digital farming' future. Yet, it is important to note that such radical shifts in agriculture are not new. This current digital transformation represents the latest era in agriculture's long history of profound change, which has evolved through three major revolutions (Fig. 1). Each of these revolutions had different drivers of change, and at their core was the advancement and adoption of new tools and techniques that replaced, enhanced, and reshaped existing systems.

K. McGrath (✉) · T. Russell
School of Agriculture & Food Science, University College Dublin, Dublin, Ireland
e-mail: karen.mcgrath@ucd.ie

Á. Regan
Department of Agri-food Business & Spatial Analysis, Teagasc Athenry, Galway, Ireland

F. da Silveira et al. (eds.), *Agriculture 4.0*,
https://doi.org/10.1007/978-3-032-24552-6_28

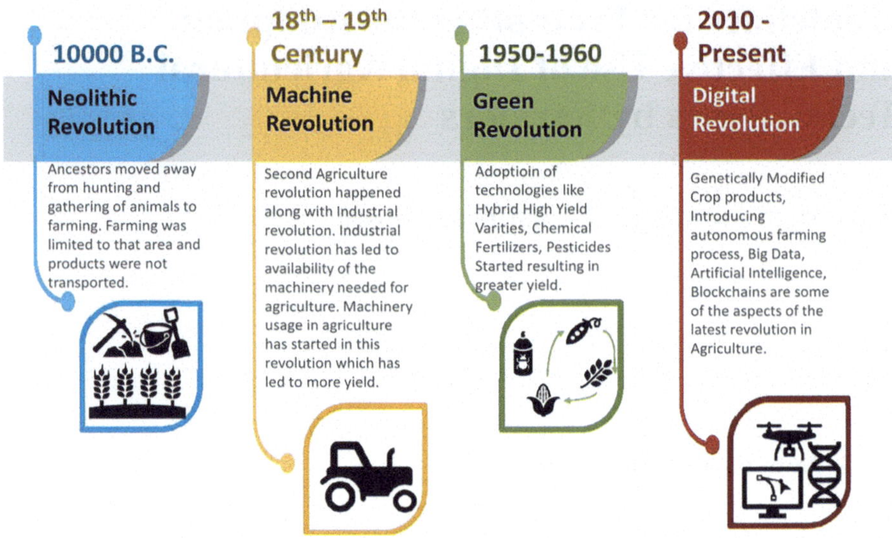

Fig. 1 Agricultural revolutions. (Figure taken from Bapatla et al. 2022)

2 Evolution of Agricultural Tools and Technologies

Casting back to the beginning, the first agricultural revolution (circa 10,000 BC) saw the gradual transition of humanity from a society of hunter-gatherers to that of a settled agricultural community. Heavy manual labour and the development of basic indigenous tools such as hoes, shovels, and rakes underpinned early agricultural practices and supported the emergence of farming as a stable livelihood. Over time, as humans settled and became accustomed to growing their own food, they sought ways to maximise food production. This gave rise to the second agricultural revolution (eighteenth to nineteenth century), which was characterised by a significant rise in productivity driven by the development of innovations aimed at maximising yields. Within this revolution, new techniques such as crop rotation were adopted, and modifications to early tools led to the development of the plough, seeders, and harrows, all of which were drawn by traction animals. The development of these tools greatly reduced labour demand and increased farm efficiency and productivity. Next came the third agricultural revolution (1950s—1960s) which was associated with newer methods of mechanisation and cultivation, as well as technology transfer to developing countries (Rose and Chilvers 2018). Advancements to tools continued, and further modernisation and mechanisation led to the development of the internal combustion engine. This paved the way for machines such as tractors and combine harvesters, which replaced animal-drawn equipment and significantly boosted farm productivity and labour efficiency.

Building on its history, agriculture is now moving into a new, fourth revolution, brought about by the introduction and proliferation of digital technology into the sector. This transition from mechanisation to digitalisation is marked by the emergence of robotics, automation, and data-driven systems. It is well accepted that the integration of these digital tools will unlock unprecedented levels of efficiency and insight, enabling farmers to operate in previously unimaginable ways. For example, we are now seeing self-driving tractors in our fields and robots milking our cows, practices and concepts which would have been thought of as science fiction five or six decades ago. These types of advancements are greatly reducing the labour burden of farming, with robotics and automation capable of taking over and replacing most of the hands-on, labour-intensive tasks, transforming the very nature of 'traditional' farm work. The labour-saving and production potential of these tools are paving the way for more sustainable, productive, and resilient farming systems, with some positioning digital innovations as the 'silver bullet' solution to many of the sector's challenges.

3 Questions of Adoption

Given the revolutionary potential and high expectations for these digital tools, one might naturally assume their swift and widespread adoption. However, we know that this is not the case. Real-world adoption rates often lag expectations, and issues with digital technology uptake persist in the sector. The underutilisation of these tools represents a considerable opportunity and value lost, with adoption now being seen as a critical barrier to the transition to and success of the digital revolution. Adding to this challenge, given the escalating urgency of environmental, food security, and labour availability issues, low uptake is fast becoming a major concern requiring prompt attention.

Consequently, the issue of technology adoption has dominated conversations within the agriculture space, with considerable focus on understanding and promoting the uptake and use of these tools amongst the farming population. This has naturally instigated a significant academic inquiry, with a large body of social science research dedicated to understanding farmer engagement with digital innovations. Early research in this area (2014–2020) has extensively honed in on identifying determinants of adoption, uncovering a wide range of influencing factors (both positive and negative) associated with six key themes (McGrath et al. 2023) (Fig. 2).

A plethora of examples can be listed under each theme, albeit arguably at risk of saturation in its reporting on some (farmer age, level of education, and farm size to name a few). Scholarship continues to research and report on this adoption area, producing consistent findings from increasingly different countries and contexts, providing an overall comprehensive view of digital technology adoption. Whilst some repetition is seen in this literature, the identification of these barriers and facilitators has been crucial to furthering our understanding of farmer engagement

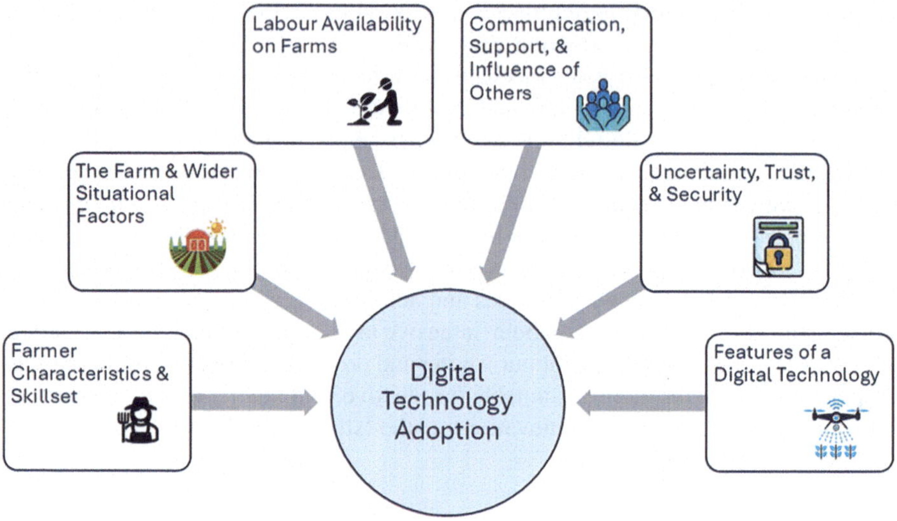

Fig. 2 Factors influencing digital technology adoption

with digital technology, and of the digital revolution overall, marking this body of work as hugely significant.

3.1 Challenging Adoption Narratives

However, as we have progressed a couple of years further into this digital transition, our knowledge has evolved, things have become more contextualised, and new insights have emerged. These new discoveries can challenge or add nuance to previous lines of thought. Among these is the potential over-repetition of certain narratives associated with digital innovation adoption. These over-emphases risk oversimplifying the broader context of technology uptake, leading us to think about and approach the issue of adoption unilaterally, and potentially exaggerating the extent and nature of the problem itself. For instance, some scholarship and narratives continually place such intense focus on adoption that it is, perhaps unintentionally, suggested and entrenched as the sole hurdle to the success of this digital revolution. There is also an assumption that an increase in uptake will automatically yield the benefits of digital tools for farmers. However, adoption is multi-faceted and complex, and we cannot approach this topic with a simplistic or narrow perspective. There are critical nuances associated with digital innovation adoption that are worth exploring and explaining. Considering these factors will help us to move beyond simplistic explanations and provide new perspectives that will help us to re-frame, critically assess, and more holistically understand digital technology adoption by farmers.

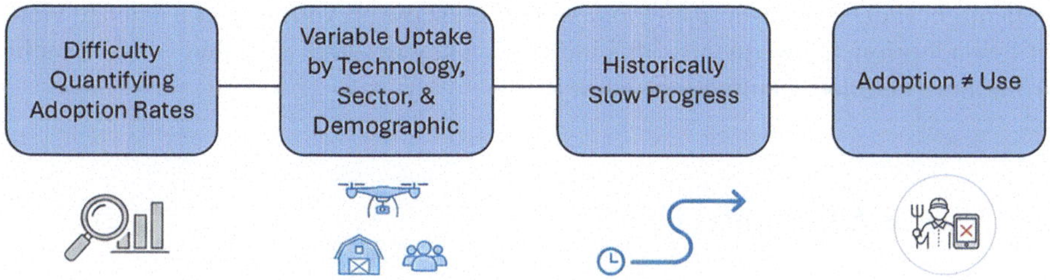

Fig. 3 Considerations for a better understanding of adoption issues

4 Contextualising Digital Technology Adoption Issues

The remainder of this chapter will outline some of these critical nuances (Fig. 3). These include recognising the challenges to quantifying actual adoption rates; understanding that low uptake is not applicable to all technologies, sectors or demographics; and realising that the embracing of technology has been historically slow. More critically, it will discuss that whilst we may pour energy into promoting the uptake of digital tools, adoption of a digital innovation does not necessarily equate effective use of a digital innovation.

4.1 Challenges to Quantify Adoption

One of the key challenges when considering technology adoption by farmers is in quantifying actual uptake rates. As agricultural digitalisation is a relatively new phenomenon, robust data on the adoption of digital technologies remains sparse, as no country or internal organisation yet systematically collects data on their use (FAO 2022). Despite the absence of this hard data, social science research has attempted to establish trends and patterns of technology uptake across farmer and other user segments through survey methods. This cumulative volume of research consistently supports the conclusion that overall adoption remains low, despite positive promotion and concerted efforts to encourage its use. For example, Denmark is a country with a strong extension system, educated farmers and rigorous environmental laws, which, in theory, should encourage the use of digital innovations such as sensor-based nitrogen fertiliser systems. However, in 2021, it was reported that only 2% of Danish farms used these types of tools (FAO 2022). On a more global scale, despite its wide availability, only 20% of farmers have adopted Variable Rate Technology (VRT); the rate of adoption of remote sensing and yield soil mapping technologies has been somewhat slow (Shepherd et al. 2018; Lowenberg-DeBoer and Erickson 2019); and although artificial intelligence is a much-discussed topic in agriculture, there are few examples of its practical use in the industry (FAO 2022). Despite these findings, the lack of comprehensive measurable data might fuel scepticism from those who contest the prevailing social science view of low adoption. Consequently, this indicates a key research area that would benefit from more detailed quantitative enquiry.

4.2 Not All Technology

That being said, it is crucial to recognise that low adoption levels are not applicable to all technologies, sectors, or demographics. Some technologies, such as Global Navigation Satellite Systems (GNSS), have gained traction quite quickly and have been widely embraced (Lowenberg-DeBoer and Erickson 2019). Further findings from the social science literature indicate that technology adoption is relatively strong in specific sectors such as dairy and tillage, as well as amongst younger demographics.

Moreover, the future outlook in some countries suggests promising trends. A survey report from Ireland published in 2019 shows that 86% of responding farmers ($n = 710$) indicated that they already use technology on their farm or are planning to embrace it in the future (IFA 2019). Unsurprisingly, given adoption patterns, most of these were from a dairy or tillage background. These findings suggest that whilst overall adoption rates might be a concern, there are pockets of engagement and optimism that challenge the blanket narrative of universal under-adoption—these caveats and considerations further highlight agri-tech adoption as a nuanced and complex concept.

4.3 Adoption Takes Time

To add to this, an important nuance that Barrett and Rose (2020) rightly point out is that agricultural changes are generally small and incremental, and uptake of new technology can take time. Reflecting on the pace of adoption of new innovations and techniques in past revolutions, consistent trends of slow uptake can be seen. For example, although the first successful internal combustion engine tractor was commercialised in 1902, fully motorised farms did not become common in the US until after the introduction of pneumatic tractor tyres in 1932. Similarly, hybrid maize took approximately 20 years from its first commercialisation in the 1920s to become widely adopted in the US (Lowenberg-DeBoer and Erickson 2019), and it has taken 22 years for 77% of Australian grain growers to adopt Autosteer (GPS guidance) in tractors.

Therefore, despite the prevailing expectation that farmers must rapidly adopt new technologies, perhaps the current rate of adoption is consistent with historical trends. This would suggest that current 'low' adoption is not a new or necessarily negative phenomenon, but a normal progression rather than an exceptional issue. As such, the continually repeated argument of 'poor adoption' might be oversimplified and suggests a need for a more patient and measured stance regarding uptake rates.

4.4 *Adoption Does Not Equate to Use*

The final, and perhaps most critical, nuance requiring consideration is what happens after farmers adopt these technologies. There is a concerted research effort dedicated to understanding drivers and barriers to uptake, with the purpose of using this information to leverage facilitators and mitigate barriers, ultimately aiming to increase adoption levels (i.e., getting farmers to purchase or engage with technology). With this goal in mind, two crucial questions arise:

1. Does the success of this digital transition hinge solely on high adoption rates?
2. Are we disproportionately emphasising initial adoption and overlooking or under-prioritising what happens post-adoption, that is, whether these technologies are being fully utilised or used at all?

To place these questions into context, let us consider some of the post-adoption challenges experienced by farmers which have been identified in the social science literature (see McGrath et al. 2023). Some findings from this scholarship indicate that once adopted, farmers can face a number of difficulties primarily associated with three areas: adapting to new digital systems, successfully managing data, and obtaining sufficient implementation support (Table 1). Each of these post-adoption challenges can lead to farmers dismissing, only partly adopting, or discontinuing the use of digital tools, thereby not realising or reaching the full benefit or potential of agri-tech tools and data.

Table 1 Post-adoption challenges faced by farmers

Adapting to a new digital system	Struggle with basic level usability
	Heavy learning load
	Stressful transition period as the farmer and farm adjust to new ways of 'digitally' doing things
Managing digital data	More data leads to more complex decision-making
	Lack digital literacy and skills to manage, utilise, and interpret data effectively
	Concern that farmers might inadvertently misuse or misinterpret information
Absence of post-adoption supports	Gap between industry support and user ability to operate tools
	Lack of corresponding support on various aspects of implementation available to farmers

A finding from McGrath et al. (2024b) provides a prime example in depicting this adoption—utilisation issue. As part of this research, a focus group was held with a group of farm animal veterinarians, with the purpose of discussing the potential of digital tools to manage animal health issues on-farm. During this discussion, the topic of farmers' use (or non-use) of digital technologies for animal health arose, with one veterinarian giving his thoughts on post-adoption use of innovations by recalling this encounter with a farmer client to demonstrate his point.

> Obviously, the equipment is only as good as the user. I saw two cows for a fella on Sunday morning, and they were definitely sick for several days and they had [monitoring] collars on them. And I said [to the farmer], 'did you not get an alert?' and he says, 'you see I don't look at that often enough'. So, the technology is only as good as the user

This example, coupled with the scholarly findings in Table 1, demonstrates a critical nuance that is often overlooked in the push for increasing adoption of technology: adoption is not always synonymous with utilisation. Whilst a farmer might adopt a technology (i.e., engage in the act of purchasing a technology or a tool), it does not guarantee its practical use. Thus, whilst the objective may be to increase adoption levels, more consideration must be given to post-adoption realities. Focusing solely on adoption rates is insufficient for measuring the success of agriculture's digital transformation. Indeed, if these post-adoption realities are not addressed, our efforts to increase adoption may prove futile, as the full potential and value of digital tools will not be realised on-farm.

5 Promoting Adoption and Effective Use

Therefore, when it comes to the topic of digital technology adoption in agriculture, a dual focus is now needed: promoting both innovation adoption *and* effective utilisation. Achieving this requires identifying and addressing the underlying reasons for challenges in successful on-farm technology use, some of which are outlined in Table 1. This will necessitate a holistic and multi-faceted approach, considering farmer-centred, technology-centred, and broader social context factors (Fig. 4). Some examples include enhancing farmer digital literacy, optimising innovation design for improved usability, and improving the provision of post-adoption supports for farmers.

5.1 Farmer-Centred Factors: Enhancing Digital Literacy

It is well reported that farmers often struggle to interpret, manage, understand, and utilise data (Drewry et al. 2019) which significantly impacts both their adoption and effective use of tools. Farmers who do not possess strong digital literacy may be deterred from adopting technologies altogether, primarily due to a lack of

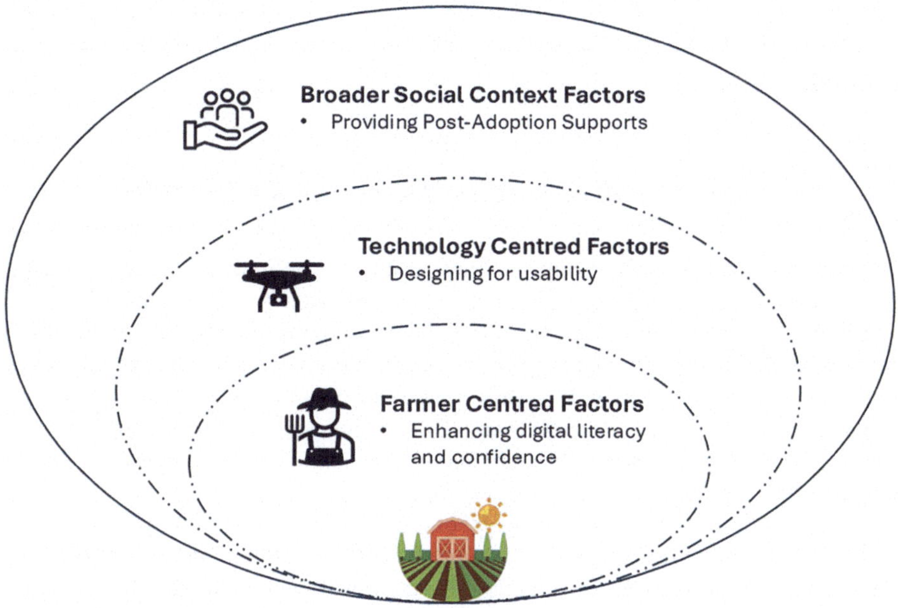

Fig. 4 Holistic approach to practically enhancing adoption and effective use of digital innovations

confidence in their ability to navigate them, which can induce technological anxiety. Furthermore, even farmers with relatively strong digital literacy may still struggle with practical use. This may be due to factors such as technological complexity, a lack of standardisation or interface functionality across different apps and digital platforms, and the rapid, increasingly complex advancements of innovations over time. Consequently, this lack of digital skillset can directly hinder the application of innovations on farms. Users may struggle not only with data input, but also with understanding and interpreting the specialised language of digital systems, limiting their ability to translate that information into informed farm management decisions.

Provision of training, education, and support in learning how to use and implement digital technologies is therefore an essential requirement for this digital transition. However, these provisions are often underdeveloped or poorly delivered in many contexts. This presents a clear gap and an opportunity for developing and providing specialised training and extension services aimed at technologically up-skilling farmers. Given that the average age of farmers in most countries is relatively high (50–60+ years of age), many did not grow up surrounded by digital technology and may have limited prior experience with their use. Consequently, some of the training that is delivered should be designed and developed cognisant of this demographic of farmers in mind and begin at a fundamental level (e.g., navigating a laptop screen, using email, etc.). Empowering farmers with this basic skillset and knowledge would instil technological confidence, providing a foundation for further learning. Ultimately, this ensures farmers are better positioned to embrace and effectively utilise and extract full value from innovations, thereby enhancing the effects of adoption and contributing to bridging the digital divide apparent in some regions.

5.2 *Technology-Centred Factors: Designing for Usability*

Whilst farmers have room to improve their ability to use innovations, a significant opportunity lies in developing technologies in a more user-friendly manner from the outset. This, however, demands a mindset shift among those involved in innovation development, including technology developers and researchers. It requires these actors to become aware of and move beyond their own biases, instead adopting an empathetic understanding of their users. This means actively considering farmers' wants, needs, limitations, and broader social contexts when designing technology functionality.

This paradigm shift moves away from 'top down' to 'bottom-up' innovation design, where future users become actively involved in shaping innovations (Fig. 5). This approach ensures that technologies are developed with and for farmers, preventing developers from merely assuming their wants and needs and designing technologies in isolation. Such user-centred design can be effectively facilitated by participatory and co-design methodologies. These approaches effectively merge the scientific knowledge of technology developers and researchers, with the experiential knowledge of farmers (Moretti et al. 2021), ensuring innovations are both techno-

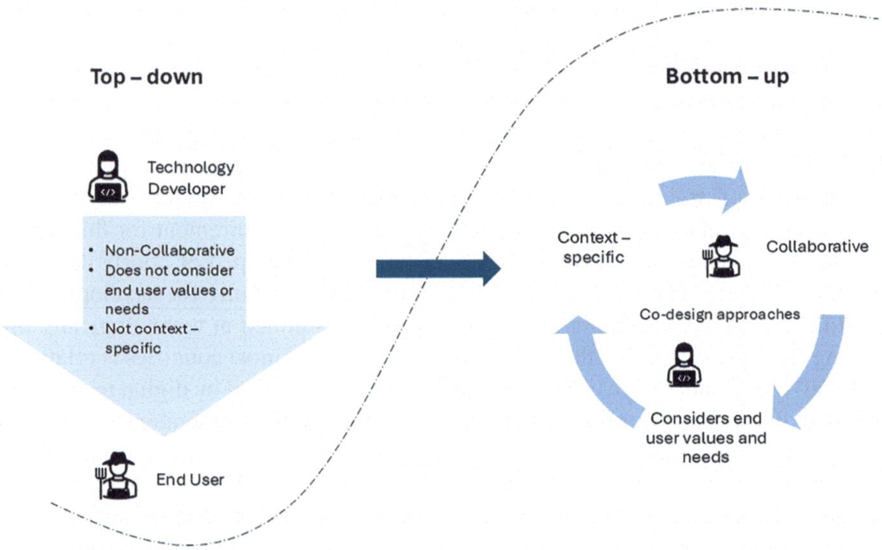

Fig. 5 Shift from top down to bottom-up approaches for technology design

logically sound and practically viable. Scholarly research on co-design approaches for agricultural innovation consistently highlight farmers' emphasis on basic, user-friendly, and context specific tools. An illustration of this comes from McGrath et al. (2024a) who, when co-designing a digital diagnostic tool for animal health, found that farmers wanted simplified, easy to read results (such as colour coded results with a reference sheet to eliminate the risk of misinterpretation). They also emphasised physical features suitable for on-farm use, such as being waterproof and durable. Using these bottom-up and co-design approaches provides invaluable user information that ag-tech developers can leverage to guide and shape the design of agricultural innovations.

As more examples of successful co-design emerge, these methods ensure innovations are purposefully developed with farmers in mind, leading to more usable innovations. By intentionally designing technologies to address how farmers will use them, innovations become better aligned with their values and needs (promoting adoption). Crucially, these approaches can also ensure that the information and data generated by technology are designed with farmers in mind, communicating useful and understandable information to the user (promoting effective utilisation). Overall, employing approaches to create more user-centred and user-friendly innovations will increase the likelihood of successful and impactful implementation in the agricultural sector.

5.3 *Broader Social Context Factors: Provision of Post-adoption Supports*

Finally, an important consideration for the wider agricultural sector is farmers' need for post-adoption support. Research has indicated a significant lack of corresponding support for various aspects of technology implementation, as well as a lack of training in how to use digital systems (Contillo and Tiongco 2019; Drewry et al. 2019). The absence of these supports can lead to an underutilisation of the tool, foster frustrations among farmers potentially resulting in technology abandonment, and generate negative word-of-mouth that may deter other farmers from adopting.

Therefore, there is scope for the development of post-adoption support from technology providers. For instance, these services could include training in how to use the innovation, technical troubleshooting, ongoing education, and readily available helplines. Reflecting this, McGrath et al. (2024a), in the same study mentioned previously, found a strong emphasis from farmers on support and education in using the device. Farmers in this study specifically requested detailed technical training (including step-by-step training videos and graphical instructions), alongside having post-adoption support features. These features included online AI (e.g., ChatGPT) support, an accompanying FAQ or 'Standard of Practice' (SOP) document, confident technical assistance, and a human-staffed customer care line. To ensure successful implementation at farm level, well beyond initial adoption, consideration of post-adoption supports like these must be a core focus and not viewed as an optional add-on.

6 Conclusion

Agriculture's digital transition presents an opportunity for positive change in farming. Whilst promoting the uptake of digital innovations receives significant attention, their success and the realisation of their full potential hinge on more than just initial adoption. We must understand that adoption is not a linear process, but rather a multi-faceted and complex phenomenon which necessitates a holistic approach to effectively integrate these tools into farm systems.

Therefore, as we progress into the digital transition, a broader perspective on digital innovation adoption is essential. This means challenging current adoption narratives and extending our focus beyond initial uptake of innovations to what happens post-adoption. Critical consideration needs to be given to facets such as farmer readiness and ability to use technology, the design of technologies themselves, and the presence of important environmental and situational provisions. Ultimately, integrating these social science elements into our understanding of technology uptake will be imperative for both adoption and effective use of innovations on farms, ensuring a successful, responsible, and sustainable transition into this new digital farming era.

Competing Interests The authors have no conflicts of interest that are relevant to the content of this chapter.

References

Bapatla AK, Mohanty SP, Kougianos E (2022) sFarm: a distributed ledger based remote crop monitoring system for smart farming. In: Camarinha-Matos LM, Heijenk G, Katkoori S, Strous L (eds) Internet of Things technology and applications. IFIPIoT 2021. IFIP advances in information and communication technology, vol 641. Springer, Cham. https://doi.org/10.1007/978-3-030-96466-5_2

Barrett H, Rose DC (2020) Perceptions of the fourth agricultural revolution: what's in, what's out, and what consequences are anticipated? Sociologia Ruralis 62:162–189

Contillo G, Tiongco M (2019) Determinants of adoption of the rice crop manager system among farmers in Pangasinan, Philippines. In: 2019 IEEE 11th international conference on humanoid, nanotechnology, information technology, communication and control, environment, and management, HNICEM. IEEE

Drewry JL, Shutske JM, Trechter D, Luck BD, Pitman L (2019) Assessment of digital technology adoption and access barriers among crop, dairy and livestock producers in Wisconsin. Comput Electron Agric 165:104960

FAO (2022) Economics of adoption for digital automated technologies in agriculture. Retrieved from FAO. Available at: https://www.fao.org/3/cc2624en/cc2624en.pdf

IFA (2019) Digital agriculture technology: adoption and attitudes study. Retrieved from https://www.akillnetireland.ie/wp-content/uploads/2019/11/IFA-Skillnet-Digital AgricultureTechnology.pdf

Lowenberg-DeBoer J, Erickson B (2019) Setting the record straight on precision agriculture adoption. Agron J 111(4):1552–1569. https://doi.org/10.2134/agronj2018.12.077

McGrath K, Brown C, Regan Á, Russell T (2023) Investigating narratives and trends in digital agriculture: a scoping study of social and behavioural science studies. Agric Syst 207:103616

McGrath K, Regan Á, Russell T (2024a) A user-centred future for agricultural digital innovation: Demonstrating the value of design thinking in an animal health context. J Agric Educ Ext, 31(3), 417–435. https://doi.org/10.1080/1389224X.2024.2397968

McGrath K, Regan Á, Russell T (2024b) Development of an animal health testing tool to reduce antimicrobial use on farms: perceptions, implications, and needs of irish dairy farmers and farm veterinarians. Ir Vet J 77(1):12. https://doi.org/10.1186/s13620-024-00268-x

Moretti D, Baum C, Wustmans M, Bröring S (2021) Application of journey maps to the development of emergent sustainability-oriented technologies: lessons for user involvement in agriculture. Bus Strategy Dev 5(3):209–221. https://doi.org/10.1002/bsd2.192

Rose DC, Chilvers J (2018) Agriculture 4.0: broadening responsible innovation in an era of smart farming. Front Sustain Food Syst 2:87

Shepherd M, Turner JA, Small B, Wheeler D (2018) Priorities for science to overcome hurdles thwarting the full promise of the 'digital agriculture' revolution. J Sci Food Agric 100:5083–5092

Teachers' Contributions to the Development of the Agriculture 4.0 Workforce

Amy Cosby, Patricia Menchon, Nicole McDonald, Patricia Colusso, and Jaime Manning

1 Introduction

The future workforce in the agricultural sector depends on the next generation seeing the industry as an attractive career option Hendrix and Morrison (2018). Furthermore, as the agricultural industry has become more technologically driven and complex, attracting and retaining skilled talent has become crucial to having a suitable workforce (Lowenberg-DeBoer 2022; Pratley et al. 2022; Baierle et al. 2023). The fourth agricultural revolution (Agriculture 4.0) demands a range of specific skills and knowledge including the use of artificial intelligence, Internet of Things, gene editing, drones, robotics and others (Javaid et al. 2022). Rapid technological advances have created a disparity between employees' current skills and the changing demands of their jobs. There are several strategies available to address the development of new skills within the in-demand agricultural workforce. Formal education at all levels, industry-led workforce training, apprenticeship and partnership programmes, and collaborations between businesses, educational institutions, and governments are some examples (O'Dea et al. 2022; Pratley et al. 2022; Manning et al. 2022b, 2024). One essential solution to this skills shortage is to strategically attract school students interested in STEM (Science, Technology, Engineering, and Mathematics) careers to consider applying their talents to work in agriculture. This requires students to have greater knowledge and experience in applying their STEM skills to a farming context. However, despite career advisers and teachers playing a key role in raising agricultural career aspirations, agricultural

A. Cosby (✉) · P. Menchon · N. McDonald · P. Colusso · J. Manning
Institute for Future Farming Systems, School of Health, Medical and Applied Sciences, CQUniversity Australia, Rockhampton, QLD, Australia
e-mail: a.cosby@cqu.edu.au; p.menchon@cqu.edu.au; n.mcdonald3@cqu.edu.au; p.colusso@cqu.edu.au; j.k.manning@cqu.edu.au

F. da Silveira et al. (eds.), *Agriculture 4.0*,
https://doi.org/10.1007/978-3-032-24552-6_29

education programmes in schools are limited across the globe (Weeks et al. 2020; Manning et al. 2022b).

The dearth in agricultural education programmes, and ensuring the skills of Agriculture 4.0 are being developed with students, highlights the need for more agricultural technology (AgTech) specific professional development for educators (Manning et al. 2022b). Integrating technology into agriculture/STEM classrooms could increase student awareness of existing agricultural career paths and the continuously evolving nature of the world of digital agriculture work (Cosby et al. 2024). The role of teachers is fundamental, as they can (a) promote agricultural concepts among school-aged children, (b) provide linkages between STEM skills and the diverse career paths available, whilst also (c) dispelling negative perceptions and outdated stereotypes of agricultural work. Teachers who deliver quality agricultural education experiences provide students with greater self-efficacy to build their AgTech skills, promote understanding of the rewarding nature of agricultural work and inspire interest in exploring agriculture STEM careers (McDonald et al. 2022). All of which encourages their future participation in the labour market (Cosby et al. 2022b).

The objective of this chapter is to explore the role of teachers in the transfer of AgTech knowledge and the development of student interest and awareness of agricultural careers. It is argued that teachers and the educational experiences they provide students with are the mechanism through which the next generation becomes the workforce that will continue to drive the digital revolution in agriculture. The following sections of this chapter will detail the activities and insights obtained by the Agri-tech Education and Extension team within CQUniversity's Institute for Future Farming Systems in Australia. This team, composed of specialists in various academic disciplines including education, organizational psychology and career development, agricultural science and AgTech, have carried out a series of research activities over the last 8 years with the overall objective of developing innovative programmes to increase the skills and knowledge of the current and next generation agricultural workforce in AgTech tools and systems. This chapter will explore, firstly, the factors that influence teachers' use of technology in classrooms to increase awareness about careers and encourage the next workforce to consider a career in agriculture. Secondly, the chapter describes different pathways for professional development in agriculture in Australia for teachers. Case studies will be presented across the chapter to explore the effectiveness of different ways to develop the Agriculture 4.0 workforce through agricultural education.

2 Factors that Influence Teachers' Delivery of Agricultural Education

Perceptions about a specific career significantly influence an individual's assessment of whether that work is a viable option (Lent et al. 1994). Unfortunately, outdated ideas about agriculture persist across the community members of many developed nations, including educators (McIlveen and McDonald 2019; Voconiq 2022). According to Fouad and Guillen (2006), past performance experiences, indirect experiences, verbal persuasion, and affective states are the four types of learning that shape people's beliefs about the outcomes that can be achieved in the pursuit of vocational careers. Therefore, it is crucial to better understand the presence or absence of these sources of outcome expectations that could influence teachers' understanding of agricultural careers and discussions of the world of agricultural work with students. This is important because of the essential role teachers play in providing the personal and environmental resources that in turn foster students' self-efficacy, outcome expectations, and career interests in agriculture (Cosby et al. 2022a). Inaccurate perceptions of the work conditions and agricultural job outcomes (i.e. unsatisfactory pay, only dirty manual work) have direct impacts on the quality of agricultural education delivered to students and in turn their career aspirations. Recruiting and developing teachers who are interested and knowledgeable in agriculture is a key factor in the knowledge transfer process.

To understand teachers' perceptions regarding agriculture in high schools in Australia, Cosby et al. (2022a) conducted a study to answer the following research questions: "Does having a formal qualification in agriculture impact a teacher's perception of, or connection to agriculture, including their perception of job opportunities in the field?" (p. 94) and "Does a teacher's perception of agriculture impact the likelihood of encouraging students to pursue a career in the field?" (p. 94). The participants of this study were Technology Mandatory teachers from New South Wales (NSW, Australia) ($n = 185$), who are a compulsory subject for students (12–14 years old) in NSW schools. The teacher participants completed a workshop for the GPS Cows programme (see "case study" in the chapter) co-developed by CQUniversity researchers and NSW Department of Education teachers. To answer the research questions, a post-workshop survey was completed and the relationships between the responses were analysed through statistical analysis. Interactions between the following factors were studied: current school location; agriculture qualification; childhood location, perceptions on job opportunities; encouragement to pursue a career in agriculture; and connection to agriculture.

Several relationships were found that explain how factors influence teachers' knowledge of agricultural job opportunities and their subsequent likelihood of encouraging technology students to consider a career in the sector. The results of this study suggest that direct experiences with agriculture are crucial to deepen teachers' knowledge and perceptions of the agricultural sector and to promote agriculture as a possible career for students (Cosby et al. 2022a). For instance, teachers' connection to agriculture was influenced by their childhood and the current location of their schools. Furthermore, those who had a rural upbringing or whose current school was in a rural area showed a stronger connection to agriculture. A notable relationship was observed between teachers' formal agricultural academic qualifications and their perception of the numerous career opportunities offered by the agricultural sector. This highlighted the importance of valuing and encouraging professional development that exposes teachers to the agricultural industry. These initiatives could, in turn, encourage teachers to support their students in exploring careers in the agricultural sector. More recently, McDonald et al. (2025) demonstrate the importance of educators being aware of the variety of jobs and possible careers in agriculture so that agricultural career exploration is considered relevant to all subject disciplines.

3 Interventions to Enhance Teacher's Knowledge of the World of Agricultural Work

As mentioned earlier in this chapter, teacher professional development has a significant impact on encouraging students to pursue agricultural-related careers. However, some challenges arise, such as high participation in the workshop but low implementation in the classroom, because teachers may not be engaged with professional readings (Easterly Iii and Myers (2019). After a comprehensive global review, Manning et al. (2022b) reported that the main constraints to teaching AgTech in the classroom were associated with teachers' time and skill set, as well as access to funding and equipment resources. Different training and capacity building strategies for teachers were developed, delivered and analysed by the Agri-tech Education and Extension team at CQUniversity in Australia (Fig. 1):

As previously mentioned, integrating technology into agriculture/STEM classes can boost teachers and students' awareness of current agricultural careers and the fast-changing technology in agriculture. Practical examples of these interventions are listed in Table 1.

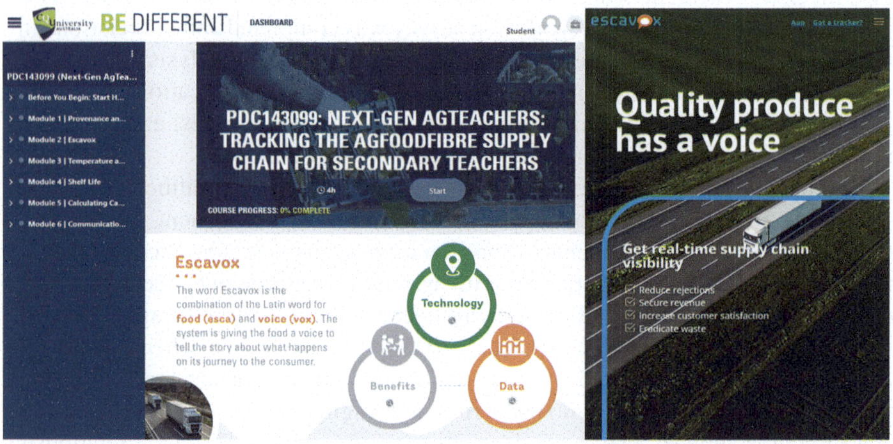

Fig. 1 Tracking the AgFoodFibre supply chain for secondary teachers micro-credential, showcasing Escavox, a real-world agricultural platform addressing food waste

Table 1 Examples of interventions and case studies that integrate Science, Technology, Engineering, and Mathematics knowledge (STEM) and AgTech

Intervention	Case study	STEM example	AgTech example
(a) Observation/participation in student programmes	EXCITED 4 careers in agriculture	Testing of agronomic concepts based on organic chemistry, such as soil fertility	Digital soil sensors and satellite imaging apply chemistry principles of soil fertility to guide precise land and crop management decisions
(b) Targeted professional development for teachers	GPS Cows	Use of geographic information with ArcGIS whilst tracking livestock	Livestock tracking technologies provide real-time location data, helping to address animal health, welfare, and management concerns
	RACE Gippsland	Traceability to the supply chain, including monitoring temperature and condition of food	Temperature and humidity sensors monitor the condition of food throughout the supply chain, reducing spoilage and food waste.
	Women in Agri-tech	Drone technology and mapping for plant science and environmental monitoring	Distinguishing between crops and weeds of local or national significance
(c) Micro-credentials	Next-Gen AgTeachers	Developing coding skills, including sequencing, logic, and problem-solving	Using Edison robots to model autonomous tractors and demonstrate their applications in modern agriculture

3.1 Observation/Participation in Student Programmes

It is possible to support the transfer of agricultural knowledge in the classroom and raise awareness about possible paths to pursue in agricultural-related careers by involving teachers in student programmes. The assumed role of teachers when an external provider engages directly with students is that of supervision to support student behaviour and engagement. However, in research with teachers, it appears that a majority do view these programmes as an opportunity for professional development. Findings in Table 2 highlight how teachers were more likely to

Table 2 Teacher responses to role in student programme, learning, and potential to integrate learning into their classrooms in the Excited 4 Careers in Agriculture programme

Item	Strongly disagree (1)	Disagree (2)	Neither agree or disagree (3)	Agree (4)	Strongly Agree (5)
Today was about more than supervising my students than my own professional development	14.9%	28.4%	18.9%	24.3%	13.5%
I learnt something new today about the range of skills that are required for different careers in agriculture	1.4%	2.7%	12.2%	45.9%	37.8%
Today has shown me that technology is a key component of modern agricultural careers	4.1%	4.1%	10.8%	54.1%	27.0%
What I learnt I can incorporate into the classroom	4.1%	4.1%	10.8%	54.1%	27.0%

Note. Excited 4 Careers in Agriculture ($n = 74$) is a combination of an incursion and excursion

report that they were experiencing the student activities as professional development when these were a classroom incursion activity, compared to fulfilling more of a supervision role when supporting students on excursions. Regardless, the majority of teachers did report learning more about the role of technology in Australian agriculture, the diverse skill sets required in the agricultural industry and that these learnings could be incorporated into their classrooms in the future. This demonstrates the value of learning through vicarious experiences as teachers observing lessons being delivered and gain exposure to agricultural workplaces through student centred incursion and excursion programmes. A case study EXCITED 4 Careers in Agriculture programme is presented below.

Case Study: EXCITED 4 Careers in Agriculture

The EXCITED 4 Careers in Agriculture programme was funded through a National Careers Institute Partnership grant to support agricultural career exploration by primary and secondary school students across Australia. The programme was delivered between 2022 and 2023 and was a joint initiative between CQUniversity and seven industry partners, targeting students from kindergarten to grade 10 in regional and rural communities. The industry partners were Sugar Research Australia, AgForce Queensland, Cotton Australia, NT Farmers, Australian Prawn Farmer Association, GippsDairy and Wine Australia. The EXCITED 4 Careers in Agriculture programme developed short learning activities aligned to Holland's RIASEC model (Prediger and Vansickle 1992; Armstrong and Anthoney 2009), a leading theory in career development which allowed them to explore their own career interests and how these align with various agricultural careers. Students engaged in practical classroom activities that focused on realistic (R), investigative (I), artistic (A), social (S), entrepreneurial (E), and conventional (C) tasks relevant to a wide range of rewarding jobs in agriculture. For example, to highlight careers in mango production (NTFarmers industry partner) students participated in activities associated with the following roles in the industry, such as "Packer or Quality Assurance Officer" (R), "Biosecurity Officer or Research Agronomist" (I), "Marketing Manager" (A), "Extension Officer" (S), "Farmer" (E) and "Truck Driver or Logistic Officer" (C). Additionally, students participated in field trips outside the classroom, where they were encouraged to learn about the world of work in different agricultural industries. During this programme, 300 teachers and industry members and 2118 students from 61 schools across Australia participated. The results showcased that 100% of teachers reported after the programme, that they have the required knowledge to use the RIASEC framework when discussing agriculture careers with their students. Additionally, 96% of teachers saw the value of integrating career education into the classroom. The EXCITED 4 Careers programme with the RIASEC model has promoted the integration of career development into the school curriculum so that subjects and skills development are translated into the world of agricultural work.

3.2 Targeted Professional Development for Teachers

The professional development (PD) of teachers is a fundamental requirement to achieve the levels of awareness about possible careers in agriculture that they could communicate to students in the classroom. This can be achieved in five broad ways: (a) immersion, when teachers perform the requisite work under the guidance of a trained professional; (b) curriculum development, refers to teachers developing new resources for use in the classroom; (c) curriculum implementation, applying and refining existing resources; (d) examining practice, refers to the analysis of real classroom instruction and (e) collaborative work, refers to peer coaching and mentoring (Loucks-Horsley et al. 1998). In a comprehensive review to explore the utilization of AgTech in secondary schools across the globe, Manning et al. (2022b) found that the main theme was the need for additional AgTech-specific PD for teachers. The delivery methods used to achieve the different professional development paths may vary depending on the participants, geographic scope and context of the schools. For instance, in-person activities, webinars, conferences, guided or self-learning workshops are some variations of delivery methods. According to Easterly Iii and Myers (2019), agricultural education workshops should remain a key component of the professional development for agriculture teachers. These PD workshops are crucial to increase the quality of agricultural literacy, especially on subjects where many teachers are teaching out of scope and do not have the necessary training (Manning et al. 2022a). Additionally, Cosby et al. (2019) reported that teachers not only need professional development, but also networking opportunities, to effectively implement learnings.

The following case studies showcase a variety of targeted professional development activities for teachers including: (a) face-to-face PD workshops focused on supporting the implementation of curriculum-aligned lessons developed by the project team (GPS Cows); (b) full day teacher PD programmes that include industry tours and that develop greater connections between educators and the current agriculture industry workforce (RACE Gippsland); (c) hybrid teacher PD experiences where the project team coached teachers to deliver an agricultural education lesson to students (RACE Gippsland); and (d) a comprehensive PD programme that supported teachers to develop their own curriculum-aligned AgTech lessons and share these with their peers (Women in Agri-tech).

Case Study: GPS Cows
An example of PD through curriculum implementation and collaborative work is GPS Cows, a module co-designed by CQUniversity researchers and the NSW Department of Education staff. This module was aligned with several of the learning outcomes in the New South Wales (NSW) Stage 4 (students from 12 to 14 years old) Technology Mandatory syllabus (NSW Education Standards Authority 2017). The GPS Cows Module involves 10 weeks (25 h) of content and was developed to provide teachers with a complete module to apply in their Technology Mandatory teaching programme, focusing on the relevant outcomes of the agriculture, food technologies and digital contexts (Manning et al. 2022a).

With the aim of training teachers in the content of the module before implementing it in their classroom, one-day teacher professional development workshops were delivered across NSW. A total of 185 teachers participated in the workshops. Firstly, participants received an overview of GPS Cows and insights into technological advancements in livestock production. This provided a foundational understanding of the agricultural sector's challenges and how technology can help address them. Secondly, a video showcased a case study farm and highlighted current issues that could be tackled with livestock tracking technology. Next, participants engaged with a real-life GPS dataset to perform various activities utilizing tools like Microsoft Excel and Esri ArcGIS Online (a Geographic Information System software, Esri (2021)) (Manning et al. 2022a). A post-workshop survey was conducted to evaluate the GPS Cows module and the suitability of content related to Technology Mandatory subject. As a result of the workshops, more than 90% of teachers expressed agreement that the content addressed some Agriculture and Food Technologies and Digital Technologies outcomes of the curriculum (Manning et al. 2023). One of the unique features of this programme is that it uses authentic data related to animal tracking using GPS technology in Australia. If schools did not have access to livestock, a sample data set from a case study farm was provided. However, if access to livestock was possible, students were encouraged to collect their own data and develop their own research questions. This is crucial since using authentic data in the classroom ensures a real-world context and enables students to understand the accurate application of knowledge (Kjelvik and Schultheis 2019). For the workshop, the main positive aspects identified by teachers were the presenter's knowledge, activities and engagement with data and access to a resource which provided links to the real world and value of technology in agriculture (Manning et al. 2022a). However, the most common limitation identified by teachers was that more time was required to acquire and understand the new knowledge. The results of this programme indicate that professional development, which includes immersion activities, curriculum implementation, and collaborative work, provides teachers with the knowledge and confidence to use resources that enhance student learning outcomes, even when the content is outside their discipline of instruction.

Case Study: RACE Gippsland

An example of a programme that includes professional development for teachers was RACE Gippsland. This programme aimed to deliver a series of activities to raise rural and regional students' aspirations, understanding and confidence in pursuing education and training options and enhance their life skills in Australia. The programme was developed by CQUniversity, funded by the Victorian State Government and also included partner organizations (Gippsland Local Learning and Employment Networks, GippsDairy, VegeNet Gippsland and TAFE Gippsland). Part of the RACE Gippsland programme included professional development workshops and webinars for teachers. Using agriculture to encourage STEM in the classroom, the free two-day teacher professional development workshops expanded teachers' knowledge of educational and career pathways in the food and fibre sector. The workshops also included a visit to a food and fibre business in the Gippsland region of Victoria for educators to gain hands-on experiences on farm or in a manufacturing/processing facility. Webinars were designed to empower educators with innovative agricultural concepts and were recorded so they could be accessed outside of the initial viewing.

During the National Science Week in Australia in 2021, webinars were practical and showcased curriculum-aligned classroom activities. During the COVID-19 pandemic, when many schools were unable to host incursions or go on excursions, teachers who registered for a webinar were sent a resource pack in advance, so they could complete the activity being demonstrated during the webinar. Other webinar opportunities exposed educators to evidence-based strategies for integrating vocational training into the classroom, expanding their understanding of agricultural careers and inspiring their students. They also had the opportunity to connect with other educators, fostering an ongoing support network and exchanging ideas. The feedback from participants in this programme was positive, and the impact on educators' career knowledge, confidence to incorporate food and fibre into their classroom, and support for their students to explore agricultural careers ensures strong legacy impacts from the activities delivered.

Case Study: Women in Agri-tech

According to Manning et al. (2024), for teachers to serve as role models for future agricultural professionals and provide up-to-date information on career opportunities in the sector, they need to be familiar with authentic industry practices that can be developed through collaboration with industry. An example of this is the Women in Agri-tech programme aimed to create and foster a strong network of teachers who will become leaders in digital literacy, STEM, and entrepreneurship in regional, rural, and remote areas. In this programme, researchers, practitioners, and entrepreneurs guided teachers in developing engaging learning resources that, in turn, inspired girls in the classroom. The programme was delivered by CQUniversity and funded by the Federal Department of Industry, Science, Energy and Resources, NSW Department of Education, Queensland Department of Agriculture and Fisheries and Cotton Australia. This PD programme was based on curriculum development and collaborative work, according to the concepts developed by Loucks-Horsley et al. (1998). During the programme, 15 teachers were selected from 68 applications. Successful applicants focused on how their participation would benefit female students ranked highly, as did those who demonstrated how they would use their participation to develop their leadership and teaching skills. The key project activities delivered relevant to teachers included:

- Women in Agri-tech Teacher Programme: Teachers attended a 2-day symposium to increase their skills, knowledge and awareness of new and emerging AgTech, as well as opportunities to network with researchers, tech developers, professionals and fellow teachers (participants: 15 teachers).
- Best practice guidelines: Teacher participants and industry representatives were interviewed to identify common factors that culminate effective engagement between teachers and industry when co-developing learning material (participants: 9 teachers and 6 industry stakeholders).
- Innovations in Agriculture Learning Module: A series of six interactive lessons were developed to provide teachers and students with knowledge on issues within the agricultural industry, AgTech innovations designed to overcome these issues and the process of the innovation cycle to devise and prepare a pitch (participants: 125 teachers and students).
- Agri-tech Learning Modules: Female teachers were supported by researchers, professionals and entrepreneurs to develop a learning module showcasing the use of an AgTech tool or system to collect and analyse data to implement in their own school (participants: 14 teachers) (Table 3).

As a general result of the programme Women in Agri-tech teachers involved reported various personal and professional benefits. These individual achievements are likely to be shared with peers and colleagues, potentially leading participants to become role models and leaders in the agricultural industry due to their enhanced knowledge and confidence.

During the programme, Manning et al. (2024) explored the characteristics of effective partnerships between educators and industry to increase knowledge and understanding of agricultural technology in schools and enhance the visibility of agricultural careers. From this research, a Best Practice Guide was developed which included advice from the participants in building an industry-school partnership to integrate AgTech into the classroom (Table 4). Successfully navigating the

Table 3 Examples of Agri-tech Learning Modules incorporating STEM and agricultural concepts developed by teachers during women in Agri-tech

Agri-tech Learning Modules developed by teacher participants in Women in Agri-tech	
Geo-locating bees using tablets/mobile phones Students will learn how to build a native bee hotel and using a geo-location app identify and map insects	*Using Soil Moisture Probes* Students will understand how farmers can use soil moisture data to support decision making in their business. They will learn more about soil, how to test and analyse soil moisture, including using soil moisture probes
Developments in Agricultural Technologies—Farmbrite Students will use a farm management software, Farmbrite to complete a series of tasks and evaluate how technology can be used to improve management decisions on farm	*Mapping with DataFarming* Students will understand how satellite imagery can be used to map farms and how DataFarming can be used to determine health of crops. They will then analyse the relationship between crop health and soil quality
Using 'Ask Bill' to make on-farm decisions Students will learn how to collect data, input information and utilize AskBill, a web-based software to look at sheep wellbeing and productivity to make better informed management decisions	*Drone Technology in Agriculture* Students get a taste of how UAV's can be used in a real-life, local agricultural setting. A weed of local significance (Hudson pear (*Cylindropuntia rosea*)) was used, with schools able to identify a weed of national or local significance relevant to their area
Radio Frequency Identification (RFID) for Livestock Management Students will gain an insight into RFID and other associated technologies that are enhancing the sustainability and productivity of livestock production in Australia and the world	*Using weather data in agriculture* Weather data is now readily available through a variety of apps and websites. This unit explores the different ways weather data is accessed and used in different businesses related to agriculture
Using NDVI data to develop zone transections for soil testing Students will use DataFarming to learn about NDVI data and how it can be utilized to develop zone transections for soil testing, thereby reducing the high costs associated with traditional grid soil testing	*Managing mangoes* Students are introduced to the biology and physiology of mangoes to understand the management of the school's mango orchard to optimize production. Students will use technologies like CyberTracker and Microsoft forms to assess trees and measure the impact of trimming

development of a partnership is crucial because effective partnerships between teachers and industry can establish knowledge platforms that foster engaged, real-world learning and offer a well-informed pathway into careers within the sector (Manning et al. 2024). Additionally, school-industry partnerships are crucial because they promote the industries and their associated career opportunities among young people and, additionally, generate goodwill in the community (O'Dea et al. 2022).

Table 4 Advice for teachers and industry on developing a partnership

Top 10 tips for teachers	Top 10 tips for industry
1. Seek out industry contacts by approaching companies at local field days and agricultural shows, or ask for recommendations from rural supplies stores or stock and station agents	1. Initiate contact with schools by approaching them directly, contacting members of the school council or getting in touch with the agricultural teacher association in your state or territory
2. Get up to speed on new technologies by reading articles from media outlets, for example, print, social media, to see what's happening in industry. Many industry groups publish relevant information regularly	2. Be open to working with teachers from a range of disciplines. Food and fibre encompasses a range of subjects (e.g. geography, science and business), not just agriculture
3. Take advantage of free trials or ask about discounted education packages before committing to purchasing the technology	3. Not all teachers have a background in AgTech or agriculture in general. Find out what they do and don't know and help build their knowledge and confidence
4. Before you purchase an AgTech tool or system, ask to speak to a farmer or other end user who has experienced the technology themselves and get their feedback on how it is working on their property	4. High school teachers will want to focus on one subject area for an individual resource, but this doesn't mean technology doesn't have application to other disciplines within a school
5. Just because you've contacted an industry rep it doesn't mean you have to commit to purchasing or collaborating further if it doesn't suit you or your school	5. Consider how your software/tech can be applied in a school setting. Modifications such as individual accounts for each student and downsizing for small ag plots will be needed
6. Don't expect the industry rep to respond to your questions overnight, these things take time. Often several levels of approval are required before they can reply	6. Teachers often want the option to view students' work and ensure they can make changes that don't change the master data or software file
7. Keep in mind new companies and start-ups often can't afford to give schools everything for free	7. School farms are not commercial operations and their ag plots are generally small (if they are lucky to have one at all)
8. AgTech tools and systems aren't made for education so don't always expect a perfect fit straightaway. Commercial products are often unable to accommodate multi-user access and may take some work to modify for the classroom	8. Create real-life context by connecting commercial farms that are using your product with schools so they can see how the technology is being used to make improvements
9. Collaborate with schools in your area to share resources and costs	9. Understand that schools rarely have large budgets for their agricultural programmes to purchase AgTech tools and systems
10. Your industry guest speaker should not be expected to run your class on their own; facilitation is required, and you will need to be there to keep the class on track	10. Don't assume all schools have reliable internet coverage or access to devices, such as iPads and laptops

3.3 Micro-credentials

The use of short formal training courses is a growing option to support the needs of the workforce (Rosenbaum et al. 2021). These courses, called micro-credentials, are usually developed from the learning content of larger educational programmes. According to Oliver (2022), micro-credentials are a focused learning achievement record that verifies the learner's knowledge, understanding, or skills. They also include assessments based on clear standards and are awarded by a credible provider. Other benefits of micro-credential training are that they can complement other credentials. Therefore, it is crucial to identify micro-credential skill training for agricultural careers and develop workforce skills requirements. With that objective, Cosby et al. (2023) explored potential skills training pathways in Australia through stakeholder surveys and found 17 technical and compliance skills as priorities. For instance, digital agriculture technologies demand skills training in data management, programming of automation and digital literacy. This is particularly crucial due to Agriculture 4.0, which is founded on the key pillars of innovative tool-based production, data management, sustainability, and professionalization (Javaid et al. 2022). As mentioned earlier in this chapter, training the future workforce could be addressed through the role of teachers in the classroom. For instance, integrating technology into agriculture/STEM classes could encourage agricultural career paths for students, which is critical given the severe skills shortage in this sector (Cosby et al. 2024).

Case Study: Next-Gen AgTeachers

A successful example of teacher training through micro-credentials is through the project launched in 2023, Next-Gen AgTeachers: Upskilling Teachers in AgFoodFibre for the twenty-first century. Developed and delivered by the Agri-tech Education team from CQUniversity and Queensland Department of Employment, Small Business and Training, five micro-credentials were created to improve primary and/or secondary teacher knowledge of modern agriculture and the digital capability required for work in agriculture:

- Next-Gen AgTeachers: Establishing and Running a School Livestock Farm—teachers developed their knowledge and an understanding of the operational and legislative requirements involved in running a school livestock farm.
- Next-Gen Ag Teachers: Agriculture in the twenty-first century for Secondary Teachers—teachers developed their knowledge and skills necessary to teach AgTech concepts and twenty-first-century agriculture skills and practices into a secondary school teaching programme.
- Next-Gen Ag Teachers: Agriculture in the twenty-first century for Primary Teachers—teachers develop their knowledge and skills necessary to teach agri-tech concepts and twenty-first-century agriculture skills and practices into a secondary school teaching programme.
- Next-Gen Ag Teachers: Tracking the AgFoodFibre Supply Chain for Secondary Teachers—teachers developed their knowledge and skills to integrate technology

in the agricultural supply chain operations to demonstrate innovative approaches to improve agricultural productivity and efficiencies.

- Next-Gen Ag Teachers: Remote Farm Management for Secondary Teachers—teachers developed their knowledge and skills using cloud-based farm management platforms that facilitate evidence-based decision-making within farming enterprises and provide real-world learning opportunities for critical data analysis and evaluation by students.

There were 331 people enrolled and 450 micro-credential completions. However, only 166 people completed at least one micro-credential and received their digital badge. Teacher motivations and priority for the micro-credentials were to upskill and develop ideas for lessons (60.4%) followed by the content being relevant to their teaching (24.2%). In-person workshops were developed alongside the micro-credentials, with 97.8% of participants agreeing that the workshops and micro-credentials were informative and relevant to their teaching. As a result, a significant number of participants (63.3%) plan to incorporate at least one of the micro-credential learnings into their teaching programmes, highlighting an additional direct employment benefit. Furthermore, 77.8% of participants felt confident in delivering the micro-credentials learnings without further assistance, underscoring the practical relevance of the programme.

Participant quotes and feedback:

> *One of the most valuable in services [professional development] I have attended in quite some time. Well presented*

> *Great job - keep more micros coming. Can CQU run any hands on practical micro to help upskill science teachers/make the teacher switch for Ag*

> *Presentations were excellent, engaging and presenters were helpful and very knowledgeable on content. Hands on and practical ideas ready to take away.*

> *We are implementing a new course next year, so the advice we received today was valuable . . .*

4 Limitations and Further Research

The programmes developed and presented as case studies in this chapter have faced challenges and were not without limitations. The main challenges are the geographical dispersion of schools and the development and coordination of partnerships between schools, teachers, and industry. Lack of time by teachers was a recurring limitation across programmes presented in this chapter, especially those outside of school hours including professional development. Additionally, teachers' perceptions of the agricultural industry, their networks, and their knowledge of the diversity of jobs within it impact how interventions are incorporated into the classroom. Therefore, the voluntary participation of teachers involved in these programmes could constitute a bias and interfere with the programme's effectiveness. If teachers' perceptions of agriculture differ based on their exposure to the agricultural industry, further research should explore whether programme outcomes differ based on the geographic location of the schools (remote areas versus metropolitan areas).

Future research should include studies of the impact of teacher's knowledge and perception of the agricultural industry on student academic outcomes and career intentions. With agriculture teacher shortages common across the globe, it is not uncommon for those educating our students not to be formally trained in agriculture. However, there is limited research which provides insight into the impact of this on the agricultural knowledge of students and their aspirations to enter the workforce. Longitudinal studies are also required to understand the long-term impact of various methods of professional development on both teacher knowledge and student outcomes. These longitudinal studies could be extended to students to determine the agricultural educational activities, both at school and extra-curricular, that influence their decision to undertake further study and a career in the agricultural industry.

5 Conclusion

This chapter has focused on the potential that primary and secondary school teachers hold in giving students the opportunity to explore a range of diverse careers involved in Agriculture 4.0. A variety of interventions that increase teachers' knowledge, confidence and motivation to integrate agriculture concepts and AgTech into their classrooms have been presented. In considering the diverse range of professional development delivery pathways identified, we argue there is no 'one-size fits all' approach to these interventions. Indeed, teacher's different past experiences, connections to agriculture, and resources or barriers that enable or prevent their engagement in teacher professional development experiences strengthen the need for different learning experiences to be offered. Teacher professional development and consequent improvements in agricultural education are the mechanisms through which the agriculture industry can progress the next generation's awareness and aspirations to become part of the dynamic and ever-evolving technology driven world of agricultural work.

Competing Interests The author(s) has no competing interests to declare that are relevant to the content of this manuscript.

References

Armstrong PI, Anthoney SF (2009) Personality facets and RIASEC interests: an integrated model. J Vocat Behav 75(3):346–359. https://doi.org/10.1016/j.jvb.2009.05.004

Cosby A, Manning J, Trotter M (2019) TeacherFX – building the capacity of STEM, agriculture and digital technologies teachers in Western Australia. Int J Innov Sci Math Educ 27(4):76–87

Cosby A, Manning J, Fogarty E, McDonald N, Harreveld B (2022a) High school technology teacher's perceptions of agriculture and careers: an Australian perspective. J Agric Educ Ext 30(1):1–22. https://doi.org/10.1080/1389224x.2022.2153887

Cosby A, Manning J, Power D, Harreveld B (2022b) New decade, same concerns: a systematic review of agricultural literacy of school students. Educ Sci 12(4):235. https://doi.org/10.3390/educsci12040235

Cosby A, McDonald N, Lovric K (2023) Designing a skills training pathway for the agricultural workforce from the employer perspective: skills micro-credentials from seasonal worker to supervisor. J Agric Educ Ext 30(4):1–18. https://doi.org/10.1080/1389224x.2023.2249435

Cosby A, Sullivan M, Manning J, Harreveld B (2024) Tech-ready teachers for agriculture 4.0: a teacher-industry partnership case study. Educ Train 66(6):668–691. https://doi.org/10.1108/ET-05-2023-0166

Easterly Iii RG, Myers BE (2019) Professional development engagement and career satisfaction of Agriscience teachers. J Agric Educ 60(2):69–84. https://doi.org/10.5032/jae.2019.02069

Esri (2021) ArcGis online. In https://www.esri.com/en-us/arcgis/products/arcgis-online/overview

Fouad NA, Guillen A (2006) Outcome expectations: looking to the past and potential future. J Career Assess 14(1):130–142. https://doi.org/10.1177/1069072705281370

Hendrix R, Morrison CC (2018) Student perceptions of workforce readiness in agriculture. J Agric Educ 59(3):213–228. https://doi.org/10.5032/jae.2018.03213

Javaid M, Haleem A, Singh RP, Suman R (2022) Enhancing smart farming through the applications of agriculture 4.0 technologies. Int J Intell Networks 3:150–164. https://doi.org/10.1016/j.ijin.2022.09.004

Kjelvik MK, Schultheis EH (2019) Getting messy with authentic data: exploring the potential of using data from scientific research to support student data literacy. CBE Life Sci Educ 18(2):es2. https://doi.org/10.1187/cbe.18-02-0023

Lent RW, Brown SD, Hackett G (1994) Toward a unifying social cognitive theory of career and academic interest, choice, and performance. J Vocat Behav 45(1):79–122. https://doi.org/10.1006/jvbe.1994.1027

Loucks-Horsley S, Hewson P, Love N, Stiles K (1998) Designing professional development for teachers of science and mathematics. Corwin Press. https://doi.org/10.4135/9781452219103

Lowenberg-DeBoer J (2022) Economics of adoption for digital automated technologies in agriculture. Background paper for the state of food and agriculture. FAO, Rome

Manning J, Cosby A, Fogarty E, Harreveld B (2022a) Do teacher workshops as a professional development activity provide the adequate skills, knowledge and confidence to deliver the GPS cows NSW stage 4 technology mandatory module? Aust Int J Rural Educ 32(3):22–37. https://doi.org/10.47381/aijre.v32i3.338

Manning J, Cosby A, Power D, Fogarty E, Harreveld B (2022b) A systematic review of the emergence and utilisation of agricultural technologies into the classroom. Agriculture 12(6):818. https://doi.org/10.3390/agriculture12060818

Manning J, Cosby A, Fogarty E, Harreveld B (2023) Teachers' perspectives about the content, context, and educational suitability of the GPS cows module for Australian teachers in New South Wales. Int J Technol Des Educ 34(1):229–247. https://doi.org/10.1007/s10798-023-09817-x

Manning JK, Sullivan M, Harreveld B, Cosby A (2024) A dynamic capabilities framework for agriculture education partnerships. J Agric Educ Ext 31(5):1–21. https://doi.org/10.1080/1389224x.2024.2429502

McDonald N, Lovric K, Black R, McIlveen P, Cosby A (2022) Career development and agriculture: We don't need a marketing campaign. CQUniversity. Journal contribution. https://hdl.handle.net/10779/cqu.26932609.v1

McDonald N, Cosby A, Luke J, Manning JK (2025) Enhancing agricultural career advice in schools: New South Wales secondary educators' perceptions. Aust J Career Dev 34(2):174–183. https://doi.org/10.1177/10384162251348287

McIlveen P, McDonald N (2019) The vocational psychology of agriculture: Fiat Panis. In: Athanasou JA, Perera HN (eds) International handbook of career guidance. Springer International Publishing, pp 459–474. https://doi.org/10.1007/978-3-030-25153-6_21

O'Dea M, Cosby A, Manning J, McDonald N, Harreveld B (2022) Industry perspectives of industry school partnerships: what can agriculture learn? Aust Int J Rural Educ 32:1–21. https://doi.org/10.47381/aijre.v32i3.334
Oliver B (2022) Towards a common definition of micro-credentials. UNESCO, Paris, p 36
Pratley J, Graham S, Manser H, Gilbert J (2022) The employer of choice or a sector without workforce. Farm Policy J Winter 2022:32–42. https://www.farminstitute.org.au/product/fpj1902-pratley-j-et-al-2022-the-employer-of-choice-or-a-sector-without-workforce/
Rosenbaum MS, Russell-Bennett R, Contreras-Ramírez G (2021) Editorial: business education in profound disruption. J Serv Mark 35(5):553–558. https://doi.org/10.1108/jsm-08-2021-528
da Silva FT, Baierle IC, Correa RG d F, Sellitto MA, Peres FAP, Kipper LM (2023) Open innovation in agribusiness: barriers and challenges in the transition to agriculture 4.0. Sustainability 15(11):8562. https://doi.org/10.3390/su15118562
Prediger DJ, Vansickle TR (1992) Locating occupations on Holland's hexagon: beyond RIASEC. J Vocat Behav 40:111–128
Voconiq (2022) Community perceptions and worker experiences program: quantitative research results. AgriFutures Australia. https://agrifutures.com.au/product/community-perceptions-and-worker-experiences-program-quantitative-research-results/
Weeks KJ, Lawver RG, Sorensen TJ, Warnick BK (2020) Do teachers have the skills: 21st century skills in the agricultural education classroom? J Agric Educ 61(4):127–142. https://doi.org/10.5032/jae.2020.04127

Women in Agriculture 4.0: Inequalities, Competencies, and Sociotechnical Innovation

Carla Geovana do Nascimento Macário, Márcia Izabel Fugisawa Souza, Lucimar Santiago de Abreu, Laura Martins de Carvalho, and Tércia Zavaglia Torres

1 Introduction

Women play a fundamental role in rural development and community sustenance, making decisive contributions to food security and local economies. According to the FAO (2011), women were protagonists in the earliest stages of agriculture, initiating seed cultivation. They currently constitute, on average, 43% of the agricultural labor force in developing countries—approximately 20% in Latin America and up to 50% in regions of Africa and Asia.

The digitalization of agriculture, often referred to as Agriculture 4.0, is an expanding phenomenon that is profoundly reshaping the production, circulation, and application of knowledge, technologies, and decision-making in rural areas. Rooted in emerging and interconnected technologies—such as the Internet of Things (IoT), Artificial Intelligence (AI), and digital data platforms—this transformation extends beyond technical and productive domains. It affects sociotechnical ecosystems, labor dynamics, and structures of power and knowledge in the countryside (Eastwood et al. 2019; Klerkx et al. 2019).

Sociotechnical ecosystems comprise the interplay between technical elements—such as digital infrastructures, devices, and algorithms—and social elements, including people, institutions, knowledge, norms, and practices. Their configuration is continuously shaped and reshaped by the manner in which technologies are

C. G. do Nascimento Macário (✉) · M. I. F. Souza · T. Z. Torres
Embrapa Digital Agriculture, Campinas, SP, Brazil
e-mail: carla.macario@embrapa.br

L. S. de Abreu
Embrapa Environment, Jaguariúna, SP, Brazil

L. M. de Carvalho
CEUCI/University of Campinas, Campinas, Brazil

F. da Silveira et al. (eds.), *Agriculture 4.0*,
https://doi.org/10.1007/978-3-032-24552-6_30

appropriated, deployed, and governed within specific local contexts (Klerkx et al. 2019).

These transformations are not value-neutral: they intersect with preexisting social inequalities, specially those related to gender, and, in many settings, actively contribute to their reproduction (Burg et al. 2019). Although women act as producers, innovators, and custodians of traditional knowledge, rural women remain underrepresented in decision-making processes and in the development and governance of digital agricultural systems (FAO 2023). Building on Agarwal's analysis (Agarwal 1994; Mohiuddin 1994), it is contended that the exclusion of women from access to productive resources and decision-making spaces constitutes a persistent structural inequality, which is now renewed through the technological-modernization dynamics associated with Agriculture 4.0.

This exclusion manifests in multifaceted ways, constituting what is widely recognized as the Gender Digital Divide (GDD)—the systematic disparities between women and men in access to infrastructure and devices, affordable connectivity, digital skills and capabilities, and meaningful participation in data governance and decision-making. In line with international frameworks, the divide encompasses not only material access but also the power to shape how technologies are designed and used (Hilbert 2020).

In the context of family farming, however, women's growing protagonism is evident across various productive fronts. As Siliprandi (2011) highlights, women farmers operate within multiple systems—ranging from household gardens and the cultivation of medicinal plants to orchards, animal husbandry, and agro-industrial production. Furthermore, they play a strategic role in marketing both fresh and processed products, particularly within local markets.

Women constitute the primary labor force in food systems, exerting a direct influence on household food security and sovereignty. Their involvement in food production and commercialization unfolds within an ethical-relational framework that prioritizes product quality and their destination.

Their engagement also encompasses environmental stewardship and the formation of sociopolitical networks of resistance. Abreu and Watanabe (2020) observe that women have spearheaded initiatives focused on biodiversity conservation and the sustainable provision of food and water, particularly in contexts of socio-environmental vulnerability. These practices manifest through multi-layered networks of care for life, deeply anchored in female leadership.

Within this scenario, digital technologies have increasingly emerged as strategic instruments for women engaged in family farming. Their uptake has intensified in recent years—particularly for commercialization and communication—facilitating cost reduction, customer retention, and value-addition to products (Carvalho et al. 2020). Platforms such as Facebook, Instagram, WhatsApp, and other social networks have thereby become pivotal allies in these endeavors.

A convergence is therefore evident between digital media and short food supply chains. As Darolt et al. (2016) observe, these circuits are characterized by a reduction of intermediaries, enhanced traceability, and greater transparency in consumption relationships. Digital technologies are pivotal in enabling these attributes, bringing producers and consumers into closer contact and reinforcing territorial linkages.

Belletti and Marescotti (2020) further argue that digital platforms, websites, and social networks constitute essential channels for local and regional marketing within proximity markets. For ecologically oriented family farmers, digital media thus provide an effective means of distributing production, expanding commercial reach, and adding value to sustainable practices.

Against this backdrop, this chapter undertakes a critical analysis of the role of women in Agriculture 4.0 through an intersectional lens. This perspective recognizes that gender, race, class, territorial location, and other social markers intersect to produce inequalities. Guided by Tavenner et al. (2022), the approach elucidates the multiple forms of exclusion influencing women's participation in agricultural digitalization processes and frames the discussion around principles of inclusion, recognition, and empowerment.

The chapter seeks to identify the main structural barriers to women's participation in Agriculture 4.0, while also highlighting the innovative practices and strategic actions they led across diverse rural contexts (Arora-Jonsson 2011; FAO 2023). Anchored in the scientific landscape of digital agricultural ecosystems (Klerkx et al. 2019) and aligning a gender perspective with ongoing technological transformations, it proposes an analytical framework for fostering digital transition processes that are more just, inclusive, and responsive to structural inequalities in rural areas.

The chapter is organized into four sections: (i) theoretical framework, (ii) methodology, (iii) results and thematic discussion, and (iv) final considerations.

2 Theoretical Framework

2.1 Gender, Rurality, and Power Asymmetries

Gender relations in rural contexts are shaped by historical processes that have rendered women materially and symbolically invisible, relegating them to subaltern positions. For decades, feminist scholarship—understood here as the corpus of critical inquiry exposing structural and epistemological gender inequalities—has consistently denounced women's exclusion from decision-making processes, control over productive resources, and the formal recognition of their knowledge and practices.

Agarwal (1994) analysis of unequal land access in South Asia illustrates that patriarchal structures and agrarian policies systematically marginalize rural women, revealing that such inequalities are upheld by deeply entrenched institutional and cultural arrangements.

Haraway (1988) introduced the concept of situated knowledge, positing that all epistemic claims stem from a specific social, historical, and geographical standpoint. The author critiques "technoscientific truth regimes"—hegemonic systems of knowledge validation that frequently delegitimize local epistemologies traditionally associated with women and Indigenous peoples. This tension between traditional knowledges and technological innovation is also evident in rural Brazilian women's adoption of Information and Communication Technologies (ICTs), which, despite their emancipatory potential, continue to operate within entrenched structures of inequality and exclusion (Escosteguy et al. 2017).

Halpern (2019) extends this analytical lens to scientific communication, contending that active listening to marginalized groups is indispensable for more inclusive processes of knowledge-production and dissemination. Complementarly, Federici (2019) argues that women's labor in rural areas—particularly work related to social reproduction and the management of common goods—is systematically devalued and appropriated by capitalist dynamics, thereby deepening inequalities and intensifying exploitation in rural contexts.

A confluence of structural factors—restricted access to land, credit, technical assistance, education, and technological infrastructure—further exacerbates these asymmetries (FAO 2023). Gender inequalities in rural contexts are thus not isolated; rather, they intersect with other forms of exclusion linked to social class, race, territorial location, and technological access. Tavenner et al. (2022) contend that an intersectional approach is critical to understanding how these social markers converge in rural women's experiences with agricultural digitalization and to crafting policies that are sensitive to local specificities and the multiple dimensions of inequality.

2.2 Agriculture 4.0 and Sociotechnical Inclusion

Agriculture 4.0 is frequently promoted as a technical remedy for the agri-food sector's challenges of productivity, sustainability, and efficiency. A critical perspective, however, reveals that digital systems are not value-neutral; they embed social, political, and economic choices that affect different social groups unequally (Burg et al. 2019; Eastwood et al. 2019). As Klerkx et al. (2019) emphasize, agricultural digitalization reconfigures sociotechnical ecosystems, introducing new forms of technological dependence and power concentration, particularly concerning data control and the logic of digital platforms.

The notion of sociotechnical inclusion recognizes that mere access to the Internet or digital devices does not guarantee equity. For digitalization to advance social justice, it must address digital literacy, content relevance, technological adaptation to local contexts, and the active participation of stakeholder groups in the system design and governance (Hernandez et al. 2024). In the absence of these elements, digitalization processes tend to reproduce—and even deepen—preexisting inequalities, with disproportionate impacts on women, Indigenous peoples, and family farmers.

From an analytical standpoint, it is essential to distinguish—while recognizing their interrelation—between sociotechnical inclusion and sociotechnical innovation. Sociotechnical inclusion concerns dismantling the structural barriers that limit access, meaningful use, and active engagement of particular social groups (e. g., rural women) in processes mediated by digital technologies. Sociotechnical innovation, by contrast, denotes the co-production of novel practices, institutional arrangements, and technological solutions that emerge from social, cultural, and technical interactions, often catalyzed by historically marginalized actors. In this sense, inclusion is a necessary precondition for innovation to advance social justice, whereas sociotechnical innovation, in turn, broadens the meanings and application of technologies within local contexts.

2.3 Capabilities and Agency

The capability approach, originally advanced by Sen (1999) and later expanded by Nussbaum (2000), offers a theoretical lens through which competencies can be understood as manifestations of *substantive freedoms*—the real capacity to choose and to act. Within the context of Agriculture 4.0, these competencies extend beyond the technical operation tools to encompass strategic, social, and cognitive dimensions that enable women to access, reinterpret, and transform technological systems for the benefit of their communities.

Recent empirical evidence supports this perspective. Kosec et al. (2023) demonstrate that women's adoption of agricultural technologies translates into measurable gains in productivity and income only when accompanied by complementary resources and services, including secure land tenure, credit, capacity building, adequate infrastructure, and market access. In the absence of these elements, expected benefits are curtailed and existing structural inequalities can deepen.

Agency—women's capacity to act autonomously, deliberately, and creatively within structural constraints—manifests, for example, in the formation of cooperative networks, the mediating between traditional knowledge and digital technologies, and the development of localized solutions to everyday challenges (Arora-Jonsson 2011). These forms of action move beyond viewing women as passive *end users*, recognizing them instead as co-producers of sociotechnical innovation.

Strengthening these technical, strategic, and interpersonal competencies is therefore crucial for constructing Agriculture 4.0 models that are more just, inclusive, and resilient.

3 Methodology

3.1 Type of Study

This study employed a targeted thematic review grounded in the principles of integrative reviews and narrative mappings (Souza et al. 2010; Snyder 2019). This approach ensured conceptual coherence, analytical depth, aligning with the manuscript's scope, without requiring strict protocols such as PRISMA—whose suitability for qualitative, context-driven reviews has been critically questioned (Greenhalgh et al. 2018). The strategy facilitated the organization of the bibliographic corpus into predefined analytical categories, combining rigorous source systematization with a flexible interpretative approach that better reflects the complexity and emergent nature of the field under analysis.

3.2 Information Sources

Source selection followed a targeted search strategy spanning academic databases, institutional repositories, and gray literature. Priority was given to sources focusing on the following categories:

- *Indexed bibliographic databases*: Platforms such as Scopus and Web of Science were required for peer-reviewed publications most pertinent to the topic.
- *International organizations repositories*: Reference materials produced by entities such as the Food and Agriculture Organization of the United Nations (FAO), the Consultative Group on International Agricultural Research (CGIAR), and the United Nations Entity for Gender Equality and the Empowerment of Women (*UN Women*) informed the study's theoretical framing and contextualization. It should be noted, however, that not all holdings within these repositories were systematically reviewed.
- *Complementary sources*: To capture diverse perspectives, additional materials were selected for their thematic relevance and alignment with the study's objectives. Owing to the limited scholarly literature at the intersection of Agriculture 4.0 and women's roles, non-conventional sources and practice-based evidence documented in technical and institutional reports were also incorporated.

The review spans 2014-2024, a decade witnessing a marked surge in scientific output on digital agriculture and rapid advances in smart sensors, the Internet of Things (IoT), machine learning, and big data analytics for the agri-food sector (Barman et al. 2024, 2025; Bhardwaj et al. 2025; Paudel et al. 2025). The corpus was restricted to English- and Portuguese—language publications that interrogate the intersections of women, Agriculture 4.0, and sociotechnical inclusion in rural contexts.

3.3 Inclusion and Exclusion Criteria

Inclusion criteria:

- Empirical or theoretical studies that explicitly examine women's participation in agricultural digitalization processes;
- Research analyzing the barriers rural women encounter in accessing digital technologies;
- Documents discussing on digital literacy, skills development, or innovation initiatives through a gender-informed perspective.

Exclusion criteria:

- Publications on Agriculture 4.0 lacking any reference to gender considerations;
- Duplicate records or documents with incomplete bibliographic information;
- Studies purely technical in nature that omit social, cultural, or gender equity dimensions.

3.4 Organization and Analysis Strategy

The analysis employed a structured thematic spreadsheet that captured the following variables: source, year, type of document, region, identified barriers, women's competencies highlighted, initiatives and policies examined, and other analytical observations.

The analytical workflow comprised four sequential steps:

- *Definition of Thematic Axes*: central analytical dimensions were established—(i) structural barriers, (ii) digital competencies, (iii) female innovation and entrepreneurship, and (iv) gaps in public policies and available data. These axes were derived from an initial exploratory reading of the corpus combined with recurring categories identified in the specialized literature.
- *Selection and reading of sources*: up to 30 documents considered representative of the field's thematic, methodological, and institutional diversity were purposively chosen for their analytical relevance, without pursuing exhaustiveness.
- *Data extraction*: data was systematically extracted from each selected text into the spreadsheet.
- *Interpretative synthesis*: patterns, gaps, and critical interrelationships were identified and organized around the predefined thematic axes.

The analysis was supported by SciSape, an Artificial Intelligence (AI) platform that was used to generate automatic summaries of each document, extracting key elements such as objectives, methodologies, results, conclusions, and contributions. This functionality enabled rapid identification of each source's central ideas, thereby streamlining the initial categorization phase.

AI assistance was exploratory and complementary, intended to enhance—rather than replace—the critical reading of the most relevant texts. Aware of the risks of oversimplification or loss of nuance, the automated outputs were consistently cross-checked against a rigorous analytical assessment.

In addition to SciSpace, ChatGPT served as a support tool for analytical systematization, aiding in idea structuring, preliminary drafting of explanatory passages, and linguistic refinement. Its use was carefully supervised to ensure all interpretations remained grounded in critical reading and aligned with the study's objectives.

3.5 Methodological Limitations

Although structured, this review was not exhaustive; instead, it offers a representative analytical perspective on women's roles within Agriculture 4.0. The scarcity of empirical studies directly addressing this topic justified the use of tailored inclusion and exclusion criteria, as well as bespoke procedures for organizing and analyzing the literature. This limitation also motivated the incorporation of illustrative case studies to broaden both the scope and the relevance of the evidence examined.

The absence of automated bibliometric tools (e.g., VOSviewer) was mitigated through a manual categorization process, supported by critical interpretation grounded in clearly defined theoretical and analytical criteria. As Snyder (2019) and Souza et al. (2010) observe, targeted reviews are particularly suited to contextualized, exploratory analyses. Likewise, Okoli (2015) recognizes the applicability of such approaches to emerging research domains, while Greenhalgh et al. (2018) contend that methodological coherence and analytical transparency can preserve systematic rigor in qualitative reviews even without extensive automation.

4 Results and Discussion

The analysis of the selected documents—focusing on peer-reviewed scientific publications and representative case studies—revealed consistent interpretative patterns and recurring analytical categories related to the role of women in Agriculture 4.0. The findings are organized into four thematic axes: (i) structural barriers; (ii) digital competencies; (iii) innovation and female leadership; and (iv) gaps in public and data policies. The discussion is further substantiated with illustrative case studies that showcase concrete instances of women-led sociotechnical inclusion across diverse rural digital contexts.

4.1 Structural Barriers to the Digital Inclusion of Rural Women

The transformations ushered in by Agriculture 4.0 are not distributed equitably across rural territories; on the contrary, they frequently expose—and, in some cases, exacerbate—preexisting inequalities. These inequalities constitute structural barriers of a systemic and interconnected nature, limiting the access and full participation of historically marginalized groups, such as rural women, in digital agricultural ecosystems. Such barriers extend beyond the technological sphere to encompass social, economic, and cultural dimensions, manifested in gender norms, the unequal division of labor, and restricted access to resources—factors that become even more critical in the current context of Agriculture 4.0.

The principal barriers encountered by women include:

- Limited access to digital connectivity, particularly in remote, low-population-density areas;
- Restricted access to land, credit, and productive inputs, compounded by unequal social and legal norms;
- The overload of domestic and reproductive labor, which constrains participation in capacity building, innovation, and digital-governance processes;
- The absence of public policies incorporating a gender perspective, thereby undermining the effectiveness of digital-inclusion strategies.

In addition, the scarcity of sex-disaggregated data on technology access, use, and appropriation in rural areas hampers evidence-based policy planning and contributes to the invisibilization of women's contributions to digital rural development (Hilbert 2020; FAO 2023).

These obstacles constitute what is commonly referred to as the gender digital divide—a term that denotes the systematic inequalities between men and women with respect to their access to, meaningful use of, and participation in the creation and governance of digital technologies. This divide is particularly pronounced in rural contexts, where factors such as poor infrastructure, restrictive sociocultural norms, and the burden of reproductive labor further constrain women's digital inclusion.

Overcoming this situation requires coordinated actions on multiple fronts, encompassing public policies, technical capacity building, equitable access to resources, and the transformation of sociocultural norms (Agarwal 1994).

These barriers reinforce structural inequalities and hinder women's full integration into Agriculture 4.0 ecosystems, undermining not only their access to digital technologies but also their active participation in digital production, commercialization, and innovation circuits.

4.2 Digital Competencies in Rural Contexts

In contrast to perspectives that underestimate their capabilities, rural women demonstrate a diverse repertoire of digital competencies, including:

- Creative use of digital tools and social media for commercialization, traceability, and direct communication with consumers;
- Integration of traditional knowledge with digital tools, such as using climate applications in agroecological management;
- Community organization for the collective use of technologies, strengthening networks of cooperation and learning.

These competencies extend beyond technical proficiency, encompassing strategic, interpersonal, and organizational capacities. As argued by Nussbaum (2000) and Sen (1999), such capacities can be understood through the lens of the capability approach—conceived as substantive freedoms to act, choose, and transform one's reality. Table 1 synthesizes the different expressions of these competencies—sometimes contrasting between women and men—highlighting technical, strategic, interpersonal, and infrastructure-related dimensions.

However, these forms of competence are frequently rendered invisible by conventional metrics, hindering their valuation in formal capacity-building programs. Recognizing them is essential for developing digital-literacy models sensitive to rural contexts and to the specific realities experienced by women. In this regard, Brito and Dias-Baptista (2022) demonstrate how the integration of ICTs, combined with female leadership, has become a distinguishing factor in Brazilian family farming, strengthening women's autonomy and expanding their role within value chains.

Considering the multiple dimensions of sociotechnical inclusion, it is possible to identify concrete practices whereby women not only access and use digital technologies but also reinterpret and transform them into context-specific solutions. Such practices attest to women's creative agency and active participation in sociotechnical innovation processes, which will be elaborated on in detail in the following section.

4.3 Innovation, Entrepreneurship, and Women's Leadership

The leadership of women in rural digital innovation is evidenced by several practical experiences:

- women-led cooperatives that employ digital technologies and participatory certification schemes to add value and reach new markets;
- startups founded by women farmers marketing local products via mobile applications;

Table 1 Digital competencies in rural contexts: a gender perspective

Dimension	Women	Men	Remarks
Technical	Adaptive use of mobile technologies, sensors, and market applications	Greater exposure and prior familiarity with hardware, software, and infrastructure	The technology access gap remains critical for rural women
Strategic	Integration of local knowledge with digital solutions; engagement in community network	Decision-making at commercial scales; negotiation with large platforms	Women tend to innovate in collaborative environments and at local and regional scales
Cognitive-interpersonal	Resilience, continuous learning, conflict mediation	Technical specialization and focus on individual performance	Women act as mediators between traditional knowledge and digital technologies
Symbiotic infrastructure	Limited access to connectivity, credit, and land ownership	Broader access to inputs, land, connectivity, and equipment	Infrastructure remains a critical driver of gender inequality in rural areas

- active participation in incubation, acceleration, and digital literacy programs, particularly within networks dedicated to family farming, agroecology, and food security.

  These initiatives challenge the technocratic, market-oriented logic of conventional agricultural systems, proposing pathways grounded in cooperation, territorial identity, and sustainability. An illustrative example is the digital coffee-traceability project led by women in Rwanda, which employs blockchain technology to ensure transparency across the international market (Ronja and Kahnert 2020). As Eastwood et al. (2019) and Klerkx et al. (2019) observe, innovation processes must be understood as sociotechnical and situated phenomena, shaped by power dynamics and by the leadership of historically marginalized actors.

4.4 Gaps in Public Policies and Gender-Sensitive Data

Despite isolated advances, the analysis revealed the absence—or weakness—of public policies designed to promote Agriculture 4.0 through a gender focus. The principal shortcomings include:

- Lack of sex-disaggregated indicators on connectivity, access to credit, capacity building, and availability of technological equipment
- Under-representation of women in decision-making arenas related to technological innovation in rural areas
- Insufficient coordination among gender policies, digital-inclusion, and sustainable rural development strategies

Although in many countries the absence of consistent public policies for the digital inclusion of rural women still predominates, international experiences provide relevant references that can inspire strategies in Brazil and Latin America. The FAO has repeatedly emphasized the centrality of gender equity for the digital transformation of rural areas, highlighting in its reports *The Status of Women in Agri-food Systems* (FAO 2023) and *Towards Digital Inclusion in Rural Transformation* (Hernandez et al. 2024) the necessity of integrating connectivity, capacity building, and the recognition of women's competencies into rural development programs. Complementarily, Molinier (2024), in *Placing gender equality at the heart of the Global Digital Compact,* advocates a human rights-based approach to ensure that digitalization does not deepen historical inequalities but instead expands opportunities for women and girls. The European Union, through its *Gender Equality Strategy 2020–2025* and *Digital Education Action Plan 2021–2027* (European Commission 2020a, b), has also established an articulated set of measures linking gender equality, digital education, and the formal recognition of competencies, providing a benchmark for the design of inclusive public policies.

In the absence of such information, it becomes impossible to monitor the impact of digitalization or accurately gauge the reach of implemented strategies. Heeks and Renken (2016) and Taylor (2017) contend that effective innovation policies must integrate principles of data justice and foster participatory governance processes. This conception involves not only the equitable distribution of benefits derived from data, but also transparent procedures for data collection and use, along with the recognition of gender specificities that often remain invisible in agricultural digitalization processes.

Given these gaps, a broader understanding of innovation in digital agriculture is required. This chapter conceives Agriculture 4.0 as a sociotechnical process, recognizing that digital technologies do not operate neutrally or in isolation; rather, they are co-produced by networks of actors, knowledge systems, institutions, and cultural values (Geels 2004; Smith et al. 2005). Within this perspective, sociotechnical innovations transcend the mere deployment of tools or devices, embodying profound shifts in labor organization, everyday practices, and power relations.

Initiatives led by rural women—such as the creative use of digital platforms, the articulation of cooperative networks, and the mediation between local knowledge and digital technologies—constitute concrete manifestations of this mode of innovation, rooted in situated practices, collective values, and strategies that challenge structural asymmetries. Udisha and Philomina (2024) demonstrated that the use of mobile technologies by women farmers in Kerala strengthened their market participation, reduced dependence on intermediaries, and contributed to their socioeconomic empowerment.

5 Illustrative Cases

The following section presents five cases that exemplify concrete practices of technological appropriation by rural women. Each case demonstrates how digital solutions were incorporated into specific contexts, fostering sociotechnical innovation rooted in collaborative practices, local values, and resistance to structural inequalities.

Case 1: Women's Cooperative with a Digital Traceability System
Facing challenges in accessing more demanding markets, a women's cooperative strategically adopted digital tools to enhance both the quality and perceived value. Key measures encompassed the implementation of participatory certification protocols and digital traceability systems, combined with the integration of platforms for production, logistics, and distribution management. As a result, the women farmers increased the added value of their food products, improved transparency concerning product origin and production methods, and strengthened their participation in more equitable commercialization circuits. This experience highlights the pivotal role of digital technologies in promoting productive autonomy and reinforcing trust between producers and consumers (Charlebois et al. 2024; Sharmila 2024).

Case 2: Rural Startup Founded by a Woman Farmer
Motivated by the challenge of ensuring the fair distribution of agroecological products, a woman farmer conceived and developed a mobile application for direct marketing. The tool connects women producers with end consumers, thereby eliminating intermediaries. The technological solution contributed to increased income, enhanced customer loyalty, and promoted economic autonomy. The case exemplifies how female leadership can drive innovations centered on sustainability and social justice in rural areas (Pooja 2024).

Case 3: Digital Literacy Program in Rural Areas
To mitigate the digital exclusion affecting many women farmers, universities partnered with rural women's associations to implement a digital-literacy program tailored to local realities. The initiative provided capacity building in the use of social media, farm-management applications, and fundamental principles of digital security. Mastery of these tools enhanced participants' organizational, communicative, and commercial capacities while also strengthening their engagement in collective forums. This case demonstrates that digital capacity building can act as a catalyst for sociotechnical inclusion and the empowerment of women's agency (Baseer 2024; Dayanand et al. 2025).

Case 4: MST Women's Brigades and Digital Platforms for Solidarity-Based Marketing
In agrarian reform territories, networks of peasant women from the Landless Workers' Movement (Movimento dos Trabalhadores Rurais Sem Terra—MST)

frequently encounter obstacles when placing their products in conventional market channels. To surmount these barriers, they organized brigades that employ digital platforms—such as WhatsApp and Instagram—alongside graphic design and geolocation tools to promote direct food sales. The initiative strengthened agroecological practices, increased product visibility, and consolidated short supply chains. Beyond generating income, it advanced an alternative development model grounded in the solidarity economy and food sovereignty (Melo 2024).

Case 5: Women Coffee Producers' Collective and Digital Marketing in Paraná (Brazil)
Women producers from the Norte Pioneiro region of Paraná faced persistent challenges in gaining recognition for their work and accessing differentiated markets. To confront these obstacles, they formed collectives and adopted digital platforms as a marketing strategy. Digitalization enhanced product visibility, improved communication with consumers, and strengthened the group's collective identity. As a result, the women consolidated their productive integration and achieved greater autonomy. This experience demonstrates how the articulation between social organization and digital tools can drive significant transformations in gender and labor relations in rural areas (Moriconi et al. 2023).

These examples reveal that women's engagement in digitalization processes is not merely instrumental but transformative, incorporating the values of solidarity, social justice, and food sovereignty.

Case 6: Social Agriculture and Women's Digital Inclusion in Ghana and Nigeria
In West Africa, women farmers have increasingly turned to social platforms as a means to strengthen their economic participation in agricultural value chains. A recent initiative documented by Schoemaker (2024) shows how rural women in Ghana and Nigeria employ WhatsApp, Facebook, and Instagram to market agricultural products, exchange information on production techniques, and coordinate logistics with local cooperatives. These digital platforms have enhanced women's visibility in the market, expanded consumer networks, and contributed to greater autonomy in decision-making processes.

Despite their positive impacts, these initiatives face recurrent challenges. Poor connectivity in rural zones, dependency on donor funding to sustain capacity building and platform development, and the need for continuous capacity-building remain major barriers. Furthermore, cultural resistance to women's leadership in digital markets continues to constrain the full potential of these technologies. The case illustrates how digital social agriculture can generate new opportunities for rural women while also exposing the structural constraints that must be addressed to achieve sustainable inclusion (Schoemaker 2024).

Case 7: Digital Applications for Women Farmers in Vietnam
Recent initiatives in Vietnam highlight how digital tools can both support women's agricultural production and reduce structural inequalities. The *Reaping Equality*

project, led by McCormick and documented by the International Finance Corporation (2023), worked with women pepper producers to promote equal access to capacity building, digital resources, and sustainable agricultural practices. By integrating women farmers into global value chains through digital traceability systems and mobile-based capacity building, the initiative not only improved production standards but also enhanced women's agency within traditionally male-dominated supply chains.

Complementing this, the Uy et al. (2023) examined the use of mobile phones for farm-related activities among ethnic minority farmers in Quang Tri Province, Central Vietnam, with a particular attention to gender differences. The study found that women farmers increasingly rely on mobile devices and applications to access market information, obtain weather forecasts, communicate with extension agents, and coordinate agricultural tasks. These practices illustrate how mobile technologies, even in contexts of limited infrastructure, can expand women's technical knowledge and strengthen their participation in decision-making and commercial processes. At the same time, the study underscores structural barriers, including deficits in digital literacy, gendered cultural expectations that limit women's leadership roles, and unequal access both devices and reliable connectivity (International Finance Corporation 2023; Uy et al. 2023).

Taken together, the seven illustrative cases highlight the diverse ways in which rural women are appropriating digital tools to strengthen market participation, build cooperative networks, and advance social and environmental sustainability. From women-led cooperatives in Brazil and MST brigades promoting solidarity-based marketing, to blockchain-enabled coffee traceability in Rwanda and mobile-based literacy programs in Kerala, India, these experiences showcase the transformative potential of digital innovation when embedded in inclusive and context-sensitive frameworks. The coffee producers' collective in Paraná demonstrates how regional identity and social organization can be reinforced through digital platforms, while programs in Ghana and Nigeria illustrate the role of e-agriculture in improving market transparency and access to extension services. Similarly, initiatives in Vietnam and the Philippines reveal the promise of mobile applications in supporting pest management, financial literacy, and cooperative participation.

At the same time, these initiatives reveal that digitalization in rural contexts is neither linear nor unproblematic. Structural constraints—such as limited digital literacy, restrictive cultural norms that hinder women's leardship, and persistent inequalities in access to devices, connectivity, and financial resources—continue to shape women's experiences. In several cases, the sustainability of projects also depends on external funding and donor support, raising concerns about long-term viability once international cooperation is reduced (International Finance Corporation 2023; Uy et al. 2023; Schoemaker 2024).

6 Conclusions

The digitalization of agriculture should be understood not merely as a technical-operational advancement, but as a broad transformation that reconfigures power relations, modes of knowledge production, and organizational dynamics in rural territories. As Klerkx et al. (2019) emphasize, this transformation is not socially neutral: it can reproduce or even deepen historical inequalities, particularly in terms of access to digital infrastructure, connectivity, capability, and participation in decision-making processes.

The promise of Agriculture 4.0 for efficiency, sustainability, and inclusion will remain unfulfiled if the structural gender asymmetries embedded in both agricultural practices and technological innovation systems are not addressed. Ignoring these asymmetries means disregarding the tangible constraints on access, agency, and recognition experienced by rural women across diverse socioeconomic contexts.

By articulating the scientific mapping of digital ecosystems with an intersectional, gender sensitive approach, it becomes evident that women are not passive technology users. On the contrary, they act as central agents in the co-construction of a more inclusive Agriculture 4.0, mobilizing local knowledge, agroecological practices, solidarity networks, and digital tools to develop sustainable and transformative solutions. Even when confronted with multiple barriers—such as limited access to resources, reliable connectivity, technical capacity building, and decision-making spaces—they have built trajectories of social and technological innovation that challenge the dominant logics of rural development (Nussbaum 2000; Sachs 2019).

The illustrative cases analyzed in this chapter confirm that women's digital inclusion is advancing through multiple pathways. From women-led cooperatives in Brazil and MST brigades in solidarity-based markets, to blockchain-enabled coffee traceability in Rwanda and digital literacy programs in Kerala, India, these experiences demonstrate the emancipatory potential of digital technologies. Additional cases from Ghana and Nigeria illustrate the contribution of e-agriculture platforms to market transparency and access to extension services, while initiatives in Vietnam and the Philippines highlight the role of mobile applications in pest management, financial literacy, and cooperative participation. At the same time, these initiatives reveal persistent obstacles, such as poor connectivity, dependency on external funding, and cultural norms that restrict women's leadership in technological spaces (International Finance Corporation 2023; Uy et al. 2023; Schoemaker 2024).

The analysis underscores an urgent need to reconfigure technological infrastructures, capacity-building initiatives, and public policies through the lens of data justice, gender equity, and sociotechnical inclusion (Taylor 2017; European Commission 2020a, b; FAO 2023; Molinier 2024). Beyond simply expanding access to technology, it is essential to recognize and value the diverse forms of digital appropriation already led by women, thereby ensuring the conditions for these initiatives to consolidate, multiply, and shape the trajectory of digital transformation in rural areas.

The convergence of women's capabilities and digital systems—particularly those developed under conditions of scarcity, adversity, and resistance—constitutes a strategic asset of considerable value. These capabilities combine creativity, empirical knowledge, collaborative practices, and a strong sense of territorial belonging, all of which are fundamental to building more resilient, equitable, and sustainable agricultural futures.

It is therefore imperative that research agendas, public policies, institutional actions, and regulatory frameworks converge to promote just digital transitions that place women at the center of the sociotechnical project of Agriculture 4.0. This centrality is not merely a question of equity or symbolic representation; it is an epistemological, ethical, and political imperative. Without the active recognition of women's experiences, knowledge, and leadership, any digitalization strategy risks reproducing the exclusionary patterns that have historically marginalized rural women. The construction of a just and sustainable Agriculture 4.0 thus depends on acknowledging gender inequalities, valuing local knowledge, and embracing principles of social equity.

Acknowledgments The authors acknowledge that language editing services were utilized to improve the clarity and consistency of this chapter. The scientific content, conceptual framing, and resultant interpretations are the sole responsibility of the authors.

The authors also acknowledge the support of the São Paulo Research Foundation (FAPESP), grant no. 2023/04126-0.

Competing Interests The author(s) has no competing interests to declare that are relevant to the content of this manuscript.

References

Abreu LS, Watanabe MA (2020) Multi-connected network involving women in agriculture in South Amazon for food sovereignty. In: Arzabe C, Costa VC (eds) Gender equality: contributions of Embrapa. Embrapa, pp 45–49. www.alice.cnptia.embrapa.br/alice/bitstream/doc/1120923/1/GenderEquality.pdf

Agarwal B (1994) A field of one's own: gender and land rights in South Asia. Cambridge University Press, p 572

Arora-Jonsson D (2011) Virtue and vulnerability: discourses on women, gender and climate change. Glob Environ Chang 21:744–751. https://doi.org/10.1016/j.gloenvcha.2011.01.005

Barman B, Singh R, Quader SW, Nath RK (2024) The rise of agriculture 4.0: a bibliometric analysis of research trends and future directions. Res Square. https://doi.org/10.21203/rs.3.rs-4948484/v1

Barman B, Singh R, Padaria RN, Quader SS, Das A, Nath RK, Ghosh B, Munshi SA, Sultana SNM (2025) Mapping the evolution of agriculture 4.0: a bibliometric analysis of research trends. Res Square. https://doi.org/10.21203/rs.3.rs-4948484/v2

Baseer A (2024) Empowering rural women through digital literacy. The Agricultural Economist. https://www.agrieconomist.com/empowering-rural-women-through-digital-literacy. Accessed 7 June 2026

Belletti G, Marescotti A (2020) Short food supply chains for promoting local food on local markets. United Nations Industrial Development Organization. https://hub.unido.org/node/2879

Bhardwaj M, Kumar P, Singh A (2025) Bibliometric review of digital transformation in agriculture: innovations, trend and sustainable futures. J Agribus Dev Emerg Econ. https://doi.org/10.1108/JADEE-02-2025-0083

Brito AVR, Dias-Baptista RD (2022) Mulheres rurais e a inclusão da Tecnologia da Informação e Comunicação como um diferencial para agricultura familiar. In: Moreira FM (ed) Transformação digital no contexto dos pequenos e médios produtores rurais: Os dados como diferencial para os desafios do Século XXI. Faculdade de Ciências e Engenharia UNESP, pp 167–189

Burg S, Bogaardt M-J, Wolfert S (2019) Ethical smart farming: current questions and directions for responsible innovation towards the future. NJAS Wageningen J Life Sci 90-91(1):1–10. https://doi.org/10.1016/j.njas.2019.01.001

Carvalho JRC, Mangini ER, Rosa JCC (2020) Aplicação e uso do marketing digital: A criação do plano de mídias sociais em cooperativa rural. In: Even3. Proceedings of the 10th Colóquio REIAd–Redes, Empreendedorismo e Inovação em Administração. https://www.even3.com.br/anais/reiad/321898-aplicacao-e-uso-do-marketing-digital%2D%2Da-criacao-de-plano-de-midias-sociais-em-cooperativa-rural/

Charlebois S, Latif N, Ilahi I, Sarker B, Music J, Vezeau J (2024) Digital traceability in agri-food supply chains: a comparative analysis of OECD member countries. Foods 13(7):1–29. https://doi.org/10.3390/foods13071075

Darolt MR, Lamine C, Brandenburg A, de Alencar MCF, Abreu LS (2016) Alternative food networks and new producer-consumer relations in France and Brazil. Ambiente Sociedade 19(2):1–22. https://doi.org/10.1590/1809-4422ASOC121132V1922016

Dayanand R, Khan Z, Kalwar M, Kumar A (2025) Digital literacy for rural women's empowerment and socioeconomic participation: a comprehensive study. SSRN Agric Econ J. Paper ID: 5093591. https://doi.org/10.2139/ssrn.5093591

Eastwood C, Klerkx L, Ayre M, Della Rue B (2019) Managing socio-ethical challenges in the development of smart farming: from a fragmented to a comprehensive approach for responsible research and innovation. J Agric Environ Ethics 32:741–768. https://doi.org/10.1007/s10806-017-9704-5

Escosteguy ACD, Sifuentes L, Bianchini A (2017) Rural women and their mediated uses of ICTs: tensions and continuities in gender inequalities. Intercom – Revista Brasileira de Ciências da Comunicação 40(1):195–210. https://doi.org/10.1590/1809-58442017111

European Commission (2020a) Digital Education Action Plan 2021–2027: resetting education and training for the digital age. https://eur-lex.europa.eu/legal-content/EN/TXT/PDF/?uri=CELEX:52020DC0624

European Commission (2020b) A union of equality: gender equality strategy 2020–2025. https://eur-lex.europa.eu/legal-content/EN/TXT/PDF/?uri=CELEX:52020DC0152

FAO (2011) The state of food and agriculture. Women in agriculture: closing the gender gap for development – 2010–2011. FAO. https://www.fao.org/4/i2050e/i2050e.pdf

FAO (2023) The status of women in agrifood systems. FAO. https://doi.org/10.4060/cc5343en

Federici S (2019) Revolution at point zero: housework, reproduction, and feminist struggle (2nd). PM Press

Geels FW (2004) From sectoral systems of innovation to socio-technical systems: insights about dynamics and change from sociology and institutional theory. Res Policy 33(6-7):897–920. https://doi.org/10.1016/j.respol.2004.01.015

Greenhalgh T, Thorne S, Malterud K (2018) Time to challenge the spurious hierarchy of systematic over narrative reviews? Eur J Clin Investig 48(6):1–6. https://doi.org/10.1111/eci.12931

Halpern M (2019) Feminist standpoint theory and science communication. J Sci Commun 18(04):1–6. https://doi.org/10.22323/2.18040302. https://jcom.sissa.it/article/pubid/JCOM_1804_2019_C02/

Haraway D (1988) Situated knowledge: the science question in feminism and the privilege of partial perspective. Fem Stud 14(3):575–599. https://doi.org/10.2307/3178066

Heeks R, Renken J (2016) Data justice for development: what would it mean? Centre for Development Informatics. https://doi.org/10.1177/0266666916678282

Hernandez K, Flynn J, He J, Alsashi H (2024) Towards digital inclusion in rural transformation. FAO. https://doi.org/10.4060/cc9816en

Hilbert M (2020) Digital gender divide or technologically empowered women in developing countries? A typical case of lies, damned lies, and statistics. Women's Stud Int Forum 34(6):479–489

International Finance Corporation (2023) Reaping equality: creating sustainable pepper supply chains by strengthening the role of female farmers in Viet Nam. https://www.ifc.org/content/dam/ifc/doc/2024/Reaping-Equality-McCormick-Case-Study-web-optimized.pdf

Klerkx L, Jakku E, Labarthe P (2019) A review of social science on digital agriculture, smart farming and agriculture 4.0: new contributions and a future research agenda. NJAS Wageningen J Life Sci 90-91:1–16. https://doi.org/10.1016/j.njas.2019.100315

Kosec K, Hidrobo M, Gartaula H, Van Campenhout B, Carrillo L (2023) Making complementary agricultural resources, technologies and services more gender responsive. Glob Food Sec 42:100778. CGIAR Gender Impact Platform. https://hdl.handle.net/10568/129706

Melo JG (2024) Framing the intersectional representation of the landless workers' movement on Instagram. Iberoam Nordic J Lat Am Caribb Stud 53(1):120–133. https://doi.org/10.16993/iberoamericana.622

Mohiuddin Y (1994) Book reviews. The University of South Sewanee. https://file.pide.org.pk/pdfpdr/1996/191-196.pdf

Molinier H (2024) Placing gender equality at the heart of the global digital Compact: taking forward the recommendations of the sixty-seventh session of the commission on the status of women. UN Women. https://www.unwomen.org/sites/default/files/2024-03/placing-gender-equality-at-the-heart-of-the-global-digital-compact-en.pdf

Moriconi A, Abreu LS, Montebello AES (2023) Estratégia de resistência: A comercialização digital de produtoras de café do Norte Pioneiro do Paraná, Brasil. In: Trabalhos do 10° Encontro da Rede de Estudos Rurais. Rede de Estudos Rurais. https://www.alice.cnptia.embrapa.br/alice/bitstream/doc/1160566/1/Abreu-Estrategia-resistencia-2023.pdf

Nussbaum MC (2000) Women and human development: the capabilities approach. Cambridge University Press. https://genderbudgeting.wordpress.com/wp-content/uploads/2012/12/nussbaum_women_capabilityapproach2000.pdf

Okoli C (2015) A guide to conducting a standalone systematic literature review. Commun Assoc Inf Syst (CAIS) 37:879–910. https://doi.org/10.17705/1CAIS.03743

Paudel B, Riaz S, Teng SW, Kolluri RR, Sandhu H (2025) The digital future of farming: a bibliometric analysis of big data in smart farming research. Cleaner Circ Bioecon 10:1–17. https://doi.org/10.1016/j.clcb.2024.100132

Pooja D (2024) Women-led agri-business innovation through startups: an opinion survey. AMC Indian J Entrep 7(2):18–30. https://doi.org/10.17010/amcije/2024/v7i2/174822

Ronja P, Kahnert L (2020) Digital traceability for women's coffee from Rwanda: making supply chains more transparent using blockchain. SNRD Africa. https://www.snrd-africa.net/digital-traceability-for-womens-coffee-from-rwanda/

Sachs CE (ed) (2019) Gender, agriculture and agrarian transformation: changing relation in Africa. Routledge, Latin America and Asia

Schoemaker E (2024) Social agriculture in Ghana and Nigeria: key takeaways. Caribou Digital Publishing, Farnham. (36 pp.) https://caribou.global/wp-content/uploads/2025/04/Takeaways_SocialAg_Nigeria_Ghana_April2024_web.pdf

Sen A (1999) Development as freedom. Alfred A. Knopf. https://kuangaliablog.wordpress.com/wp-content/uploads/2017/07/amartya_kumar_sen_development_as_freedombookfi.pdf

Sharmila V (2024) A study on digital literacy of rural women based on their educational qualification. GRT J Educ Sci Technol 2(2):52–56. https://doi.org/10.26452/grtjest.v2i2.44

Siliprandi E (2011) Mulheres agricultoras no Brasil: sujeitos políticos na luta por soberania e segurança alimentar. Pensamiento Iberoamericano 9:169–183. https://dialnet.unirioja.es/servlet/articulo?codigo=3710909

Smith A, Stirling A, Berkhout F (2005) The governance of sustainable socio-technical transitions. Res Policy 34(10):1491–1520. https://doi.org/10.1016/j.respol.2005.07.005

Snyder H (2019) Literature review as a research methodology: an overview and guidelines. J Bus Res 104:333–339. https://doi.org/10.1016/j.jbusres.2019.07.039
Souza MT et al (2010) Integrative review: what is it? How to do it? Einstein 8(1):102–106. https://doi.org/10.1590/S1679-45082010RW1134
Tavenner K, Crane TA, Bullock R, Galié A (2022) Intersectionality in gender and agriculture: toward an applied research design. Gend Technol Dev 26(13):385–403. https://doi.org/10.1080/09718524.2022.2140383
Taylor L (2017) What is data justice? The case for connecting digital rights and freedoms globally. Big Data Soc 4(2):205395171773633. https://doi.org/10.1177/2053951717736335
Udisha O, Philomina IGA (2024) Bridging the digital divide: empowering rural women's farmers through mobile technology in Kerala. Sustainability 16(21):1–14. https://doi.org/10.3390/su16219188
Uy TC, Ha HD, Truyen NN, Chung NV, Phuoc DN, Nam LV, Phuong LTH, Thuyet V (2023) Mobile phone use for farm-related activities by ethnic minority farmers during the Covid-19 pandemic in Quanq Tri Province, Central Vietnam. Int Serv Use 43(1):27–37. https://doi.org/10.3233/ISU-230179

The Role of Green Technologies in Sustainable Food Processing for Agriculture 4.0

Abul Hossain, Anubhav Pratap-Singh, and Rickey Y. Yada

1 Introduction

Growing global demand for sustainable food production, fueled by an increasing population, is pushing the food industry to adopt innovative and environmentally friendly food processing practices. Stricter regulations aimed at reducing carbon footprint and food waste are being implemented, alongside a rising consumer demand for natural and minimally processed foods. This can be accomplished by implementing green food technologies, which focus on reducing environmental impact and preserving food quality without destroying nutritional attributes. Green food processing refers to energy-efficient, sustainable, and eco-friendly methods and works based on green chemistry and green engineering approaches, minimizing environmental impact while maintaining food safety and quality. In contrast, conventional food processing often leads to high energy consumption, high processing time, nutritional loss, excessive water usage, and substantial waste generation (Lisboa et al. 2024). For example, thermal sterilization, a conventional technique prominently used in microbial inactivation, demands elevated temperatures. However, this process can compromise heat-sensitive compounds (e.g., vitamins and antioxidants such as anthocyanins and curcumin), leading to a decline in the product's nutritional value and potentially altering its sensory characteristics such as flavor, color, and texture (White et al. 2025). Adopting green technologies can extend shelf life, enhance food safety and quality, reduce reliance on synthetic additives, and recycle by-products, which typically include microwave

A. Hossain · A. Pratap-Singh
Faculty of Land and Food System, University of British Columbia, Vancouver, Canada

R. Y. Yada (✉)
Faculty of Agricultural, Life and Environmental Sciences, University of Alberta, Edmonton, Canada
e-mail: ryada@ualberta.ca

F. da Silveira et al. (eds.), *Agriculture 4.0*,
https://doi.org/10.1007/978-3-032-24552-6_31

processing, Supercritical CO_2 Processing (SCP), High-Pressure Processing (HPP), Pulsed Electric Field (PEF), Cold Plasma (CP) technology, membrane filtration and minimal processing, irradiation, ohmic heating, ultrasonication, nanotechnology, and encapsulation, among others (Onyeaka and Nwabor 2022; Lisboa et al. 2024). Among them, nonthermal processing methods, including HPP, PEF, CP, and ultrasound, have become increasingly significant as alternatives or supplements to conventional methods. They effectively prevent the denaturation of nutrients, avoid side toxic reactions, and maintain the physical and mechanical properties of the products (Safwa et al. 2024; White et al. 2025).

In the context of Agriculture 4.0, these green food processing technologies can be further enhanced through the integration of smart and digital tools. Technologies such as the Internet of Things (IoT) enable real-time monitoring of processing parameters (e.g., temperature, pressure, and flow rates), while Artificial Intelligence (AI) and Machine Learning (ML) algorithms can be used to optimize processes, reduce waste, and predict maintenance needs. By combining green processing with these Agriculture 4.0 technologies, it becomes possible to build intelligent, adaptive, and sustainable food systems that respond to environmental, economic, and consumer-driven challenges (Fig. 1) (Ercan et al. 2023; Hassoun et al. 2024). Beyond IoT and AI, digital twins allow virtual simulation of food processes, blockchain ensures full traceability and certification of sustainable practices, and adaptive automation systems can dynamically adjust processing conditions based

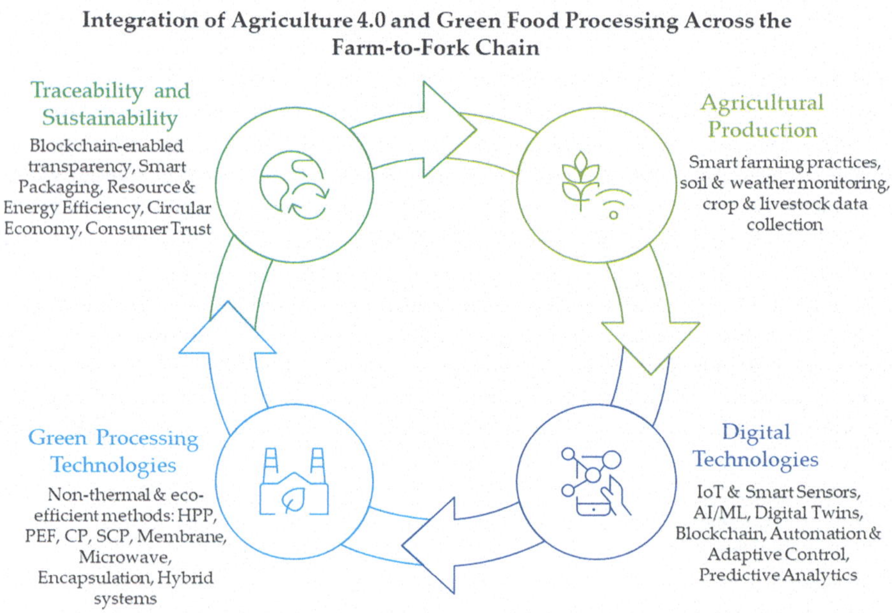

Fig. 1 Conceptual framework showing how Agriculture 4.0 integrates the food system from farm to fork

on raw material variability (Cordeiro and Ferreira 2025; Hassoun 2025). Together, these digital tools reinforce the connection between Agriculture 4.0 and green processing. It should be noted, however, that research on the integration of these digital technologies into green food processing is still in its early stages, and direct applications remain limited, though they hold immense potential for future development. Therefore, this chapter aims to provide a comprehensive understanding of green food processing technologies by exploring both foundational principles and cutting-edge innovations. It seeks to highlight the benefits, limitations, and industrial applications of sustainable approaches, while also addressing how digital innovations may complement these technologies in the broader transition toward a modern agri-food ecosystem.

2 Emerging Green Technologies in Food Processing

2.1 High-Pressure Processing (HPP)

HPP is an advanced nonthermal food preservation technology that inactivates harmful microorganisms while preserving the nutritional and sensory qualities of food. Unlike traditional heat-based pasteurization, which can degrade vitamins, antioxidants, flavor, and texture, HPP relies on hydrostatic pressure to achieve microbial inactivation and extend shelf life. HPP is an energy-efficient technique compared to heat-based pasteurization and is commonly used in juice, dairy, meat, and seafood processing, retaining nutrients and sensory properties (Hossain et al. 2022a). In the context of Agriculture 4.0, the integration of digital tools such as real-time sensors, AI, and IoT devices is enhancing the performance and traceability of HPP systems. Smart sensors can continuously monitor pressure, temperature, and microbial load, allowing for dynamic control and optimization of processing conditions. In addition, adaptive control systems are now being developed that automatically adjust pressure and holding times based on raw material variability, ensuring consistent quality and reducing energy use (Hassoun 2025). For example, ML model can be implemented in HPP production lines to enhance predictive maintenance accuracy, thereby improving cost efficiency, reducing industrial maintenance expenses up to 10%, and positively influencing overall product costs (Srisuwan and Innet 2024). Furthermore, AI and ML algorithms can analyze historical and real-time data to predict processing outcomes, minimize energy consumption, and ensure product safety. Moreover, blockchain technology can be employed to document and verify each stage of the HPP process, improving transparency and consumer trust (Srisuwan and Innet 2024; Dhal and Kar 2025).

2.1.1 Principles of HPP

HPP works by subjecting food products, typically sealed in flexible, water-resistant packaging, to extremely high pressures ranging from 100 to 800 MPa for a specific duration (from a few seconds to several minutes) at a low temperature (5–10 °C). The principle of adiabatic compression heating states that increasing the pressure by 100 MPa causes the water temperature to rise by about 3 °C. Therefore, HPP ensures that the temperature does not surpass 30 °C when the pressure is increased to 800 MPa (Onyeaka and Nwabor 2022; Safwa et al. 2024). HPP uses two key principles, the isostatic and Pascal's principles, ensuring even pressure distribution on food products in a sealed vessel. Pascal's principle explains that any pressure change applied to a fluid at rest in a closed container is uniformly transmitted throughout the fluid and to the container's walls without any loss. The isostatic principle states that when pressure is applied to a liquid in a closed environment, it is evenly distributed to all objects within, regardless of their shape or size, maintaining products' quality (Huang et al. 2020; Onyeaka and Nwabor 2022). Therefore, under high pressure, the pressure is transmitted instantly and uniformly throughout the entire product in a short period, indicating that pressure plays a more critical role than holding time (Hossain et al. 2022a). Moreover, when high pressure is applied through water, the system (in this case, the food and its microbial contaminants) responds by reducing its volume. This affects the molecular interactions within the microorganisms, leading to the disruption of noncovalent bonds (e.g., ionic bonds, hydrogen bonds, and hydrophobic interactions) in proteins and other cellular components. This goes to Le Chatelier's principle, where the system (microorganisms) responds to the pressure by minimizing volume changes, leading to the disruption of cellular structures and inactivation of pathogens. Le Chatelier's principle states that a system at equilibrium will adjust to counteract any applied change (López-Pedrouso et al. 2019; Waghmare 2024). As a result, HPP is effective regardless of the shape and size of food products, allowing for the processing of various volumes in a single batch.

2.1.2 Common Applications of HPP

HPP is widely used across various food categories to enhance food safety and extend shelf life while maintaining product integrity. HPP preserves the natural color, aroma, texture, and flavor of food, making it ideal for clean-label products that do not require added preservatives. Moreover, HHP is also used to modify functional properties in food, including protein structural properties and functionality, proteolysis reaction to produce peptides, and starch properties. Apart from that, it can also extract various bioactive compounds, mainly thermosensitive compounds, and improve the bioaccessibility and bioavailability of these components (Trych et al. 2020; Waghmare 2024).

HPP can directly eliminate harmful and spoilage-causing microorganisms, slow down chemical and enzymatic reactions that lead to food deterioration, and help maintain the essential physical and chemical properties of food. Thus, the application of HPP in the food industry includes various areas, such as the production of probiotic yogurt, apple puree, and wine, the pretreatment of meat, poultry, and seafood, and the preservation of fruits and vegetables, juices, beverages, sausages, dairy products (e.g., soft cheeses), salsa, guacamole, smoothies, baby food, pet food, and ready-to-eat foods, among others. For example, HPP-treated guacamole maintains its vibrant green color by preventing enzymatic browning. Moreover, HPP helps in shucking shellfish (e.g., oysters, mussels, and lobsters) by breaking the connective tissues, making meat extraction easier. HHP is used in seafood to eliminate/control pathogenic *Vibrio* and viruses, while it is applied in fresh meat products to control risks associated with *Listeria monocytogenes*, *E. coli*, and *Salmonella* spp. (López-Pedrouso et al. 2019). To ensure microbiological safety, indicator pathogens (e.g., *E. coli*) should be reduced to at least 5-log for any time/pressure combination (Onyeaka and Nwabor 2022). Although the exact mechanism that is responsible for microbial inactivation has not been fully understood, a combination of a series of functions, including disrupting their cell membranes, unfolding proteins and enzymes, altering their shape, affecting their metabolism, and interfering with their genetic processes, plays the main role (Hossain et al. 2022a; Waghmare 2024).

Recently, HPP has been studied for extracting valuable bioactive compounds, such as phenolics and pigments, offering higher yields in less time while preserving their chemical structure and enhancing bioaccessibility by avoiding heat damage. HPP can be used as a pretreatment before extraction or as a direct extraction method using solvents/enzymes inside the HPP vessels. For instance, HPP pretreatment was applied to extract phenolic compounds from sea cucumbers, resulting in an over 30% improvement in extraction yield (Hossain et al. 2022a, b). HPP disrupts cell membranes, thereby enhancing solvent penetration, promoting mass transfer, increasing cellular permeability, and significantly improving extraction efficiency. Additionally, HPP can inactivate enzymes like peroxidase (POD) and polyphenol oxidase (PPO), which are responsible for anthocyanin degradation and browning in foods (Hossain et al. 2022b). Another application in the food industry involves using HPP before freezing to minimize weight loss during thawing while preserving food properties and extending shelf life. In addition, applying HPP to other products, such as edible flowers, emulsions and nano-emulsions, edible films, and biopolymers, can also be found (Lamela 2023).

Despite its numerous advantages, HPP has some limitations that need to be addressed. For example, HPP equipment is expensive and requires substantial capital investment, which may be a barrier for small-scale food processors. Moreover, products must be packaged in flexible, water-resistant materials (e.g., vacuum-sealed pouches, plastic bottles) since glass or rigid containers may break under high pressure. HPP is primarily effective for moist or liquid-based products and is unsuitable for dry foods like powders or cereals. The lack of sufficient water in dry or low-moisture products hinders pressure transmission and uniform compression, thereby reducing the effectiveness of microbial inactivation and structural disruption essential for efficient processing. Additionally, HPP alone cannot effectively eliminate microbial spores (e.g., *Clostridium* and *Bacillus* spores), as they are more resistant to pressure than regular bacterial cells. Because of this limitation, HPP foods are not available at room temperature and must be transported through cold chain systems to ensure safety (Onyeaka and Nwabor 2022; Lamela 2023).

2.2 *Pulsed Electric Field (PEF) Treatment*

PEF, also known as electropermeabilization or electroporation, is an innovative nonthermal food processing technology that uses short, high-voltage electrical pulses to disrupt microbial cell membranes and enhance the permeability of plant and animal tissues. Unlike traditional thermal processing methods, PEF operates at low temperatures, preserving the nutritional, sensory, and functional properties of food while ensuring microbial safety. With applications in microbial inactivation, juice extraction, and plant-based ingredient extraction, PEF is a promising green technology that aligns with the food industry's sustainability goals by reducing energy consumption and improving product quality (White et al. 2025). When integrated with smart sensors, AI-driven controls, and real-time data analytics, PEF systems can further optimize energy efficiency, enhance precision, and support traceable, adaptive processing, the core goals of Agriculture 4.0. In addition to detection and monitoring, AI has the potential to improve the effectiveness of PEF, playing a crucial role in enhancing food safety and prolonging shelf life (Dhal and Kar 2025). Recently, predictive analytics platforms have been applied in PEF systems to forecast equipment failures or nutrient degradation, enabling early intervention and minimizing process losses (Dhal and Kar 2025; Semercioz-Oduncuoglu and Luning 2025).

2.2.1 Principles of PEF

PEF technology applies short-duration (microseconds to milliseconds) high-voltage electrical pulses to food materials placed between two electrodes. This induces a phenomenon known as electroporation, where the cell membranes of microbes and plant cells become permeabilized or ruptured due to the high voltage (Yan et al. 2024). The actual mechanism of action of membrane permeabilization remains unclear, but electroporation is generally understood to occur in four phases: formation of small hydrophilic pores once a threshold potential (0.2–1.0 V) is reached, an increase in transmembrane potential due to external electric field application, expansion in the number and/or size of these pores during PEF process, and a post-treatment phase where intracellular compounds may leak out, extracellular substances may enter (irreversible electroporation), or the membrane may reseal and recover (reversible electroporation) (White et al. 2025). Therefore, the effectiveness of PEF depends on electric field strength (typically 1–50 kV/cm), pulse duration (microseconds to milliseconds), pulse frequency (number of pulses per second), pulse shape, pulse width, and total treatment time and energy input (Bocker and Silva 2022).

2.2.2 Common Applications of PEF

PEF has diverse applications across the food, beverage, and nutraceutical industries. For example, PEF treatment for fresh juices is a common application because traditional thermal pasteurization can degrade vitamin C and polyphenols. PEF treatment at 15–35 kV/cm effectively inactivates *E. coli, Listeria,* and *Salmonella* while retaining natural color, flavor, and nutrition. Moreover, PEF can inactivate pathogens in milk, dairy beverages, and plant-based dairy substitutes without affecting protein structures (Cavalcanti et al. 2023). On the other hand, PEF improves juice (e.g., grapes and apples) extraction (up to 10–20%) from fruits and vegetables by enhancing cell membrane permeability (Naliyadhara et al. 2022). This treatment can also be used for the extraction of various bioactive compounds, including polyphenols from berries, lycopene from tomatoes, and anthocyanins from grapes, as it assists in the critical acceleration of the solid-liquid extraction (Yan et al. 2024). PEF is also useful for texture modification and preservation of solid foods, such as reducing frying time and oil absorption in French fries and chips production due to softened potatoes. As a result, it leads to a healthier end product with lower fat content. For instance, PEF-treated Lamoka potato chips had lower acrylamide levels (28.9%), lower oil content (8%), less reducing sugars (19.22%), less frying time (10%), and better textural attributes compared to continuous frying cooking (Santiago-Mora et al. 2024). Furthermore, it improves marination efficiency in meat and seafood, enhancing flavor and tenderness as well as improving juiciness in processed meats. This is because PEF enhances the permeability of food tissues, facilitating better brining and marination of meats and seafood as well as faster drying in food dehydration processes (Naliyadhara et al. 2022).

2.3 Cold Plasma (CP)Technology

Cold Plasma (CP) technology is an emerging nonthermal food processing technique that utilizes ionized gases to inactivate bacteria, mold, and viruses on food surfaces and packaging materials. This innovative approach provides an effective, chemical-free method for ensuring food safety, extending shelf life, and reducing the reliance on conventional preservatives. As part of green food processing technologies, CP offers low-energy consumption, minimal impact on sensory attributes, and environmental sustainability (Bayati et al. 2024). When combined with Agriculture 4.0 tools such as real-time sensors, ML algorithms, and automated control systems, CP can deliver more precise, consistent, and adaptive treatments tailored to specific food types and microbial loads (Ercan et al. 2023). For instance, ML techniques could be used to predict the antimicrobial activity of CP-treated water (Özdemir et al. 2023). Similar to HPP and PEF, there is potential for adaptive control platforms to be integrated with CP systems such that plasma intensity and treatment time could be dynamically adjusted in response to microbial load or product surface variability, helping avoid over-processing while ensuring safety. While Hassoun (2025) discuss the promise of such automation in green processing, concrete case studies in CP remain limited.

2.3.1 Principles of CP

Plasma is often referred to as the fourth state of matter, alongside solids, liquids, and gases. When a gas, such as air, oxygen, nitrogen, or argon, is exposed to an electric field, microwave, or radiofrequency energy, it becomes ionized, forming cold plasma. CP consists of reactive oxygen and nitrogen species (RONS), free radicals, electrons, and UV photons, which interact with microbial cell walls, DNA, and proteins, leading to microbial inactivation. In particular, it produces over 75 distinct species, including short-lived (e.g., atomic oxygen (O), singlet oxygen ($^1\Delta_g O_2$), hydroxyl radicals (OH•), hydrogen radicals (H•), hydroperoxyl radicals (OOH•), superoxide (O_2^-•), nitric oxide radicals (NO•), nitrate radicals (NO_3•), and peroxynitrous acid (ONOOH)) and long-lived species (e.g., ozone (O_3), hydrogen peroxide (H_2O_2), nitrogen dioxide (NO_2), nitrous oxide (N_2O), dinitrogen pentoxide (N_2O_5), nitrite (NO_2^-) and nitrate (NO_3^-), oxoazanide (NO^-), nitrous oxide anions (N_2O^-), and peroxynitrite ($ONOO^-$)) (Gupta and Routray 2025; Mohammadi et al. 2025). The production of RONS occurs in two steps: electron collisions cause ionization, excitation, dissociation, and electron attachment, and then interactions between heavy particles further generate RONS as electron energy rises. Unlike traditional thermal plasma, CP uses high-energy electrons rather than heat and operates at near-room temperatures, making it safe for heat-sensitive foods. Plasma production is acquired by numerous methods, including plasma jet, dielectric barrier discharge, corona discharge, microwave-induced plasma, radio frequency plasma, and multi-pin atmospheric cold plasma (Mandal et al. 2018; Oguzlu et al. 2024; Gupta and Routray 2025).

Dielectric barrier discharge alternates current to dielectric materials placed between two electrodes: a high-voltage electrode and a ground electrode. The samples are positioned between them with a small gap (ranging from 0.1 mm to several cm) for surface treatment, while direct treatment is applied directly to the sample, with the sample itself acting as part of the dielectric barrier. The effectiveness of the cold plasma treatment depends on the distance between the plasma source and the target surface, as some reactive species have very short lifetimes, suggesting a higher effectiveness upon direct treatment (Gupta and Routray 2025). On the other hand, the plasma jet system generates cold plasma by applying high-voltage electrical discharge to a gas (e.g., helium, argon, or air) as it flows through a nozzle. The energized gas exits as a plasma plume containing reactive species, which interact with the target surface without significant heat transfer. This allows for precise surface modifications, sterilization, and material processing without thermal damage. Unlike dielectric barrier discharge, plasma jets release reactive species into an open environment, enabling direct treatment of objects of various sizes without spatial limitations and producing more uniform results (Oguzlu et al. 2024).

2.3.2 Common Applications of CP

CP is gaining traction in various food industry sectors due to its versatility and safety, and it has been considered for a wide range of applications, including fresh produce decontamination, meat and seafood safety, dairy and egg processing, and food packaging and storage, among others. For example, CP is an effective treatment for decontaminating fresh fruits and vegetables as it removes pesticide residues, bacteria, and mold spores from apples, lettuce, berries, and spinach (Onyeaka and Nwabor 2022; Yawut et al. 2024; Yasir et al. 2025). Moreover, it can extend shelf life by inhibiting fungal growth (e.g., *Botrytis* in strawberries), preventing mold growth and mycotoxin production, and reducing aflatoxin contamination in peanuts, maize grains, nuts, and cereals. It reduces *Salmonella* and *Listeria* on raw chicken and beef without affecting freshness, while it controls microbial load and delays the oxidation of fish fillets, shrimp, and shellfish. Furthermore, CP has been used in dairy and egg processing as an alternative to thermal pasteurization, preserving probiotics and heat-sensitive enzymes in dairy (Gupta and Routray 2025; MRI et al. 2025). For instance, CP deactivates 92.20% of *S. enterica* and 98.74% of *E. coli* on egg surfaces (Movasaghi et al. 2025).

CP inactivates microorganisms through multiple mechanisms, including oxidative damage, UV radiation and DNA disruption, and electrostatic disruption of cell membranes, among others. For example, the reactive oxygen species (ROS), mainly NO_2 and O_3, in CP disrupt microbial cell walls, causing permeability loss and oxidative stress, leading to cell death. Lipids, proteins, and nucleic acids in bacteria and mold are oxidized due to the activity of NO_2 and NO, preventing replication (Zhao et al. 2024). In particular, the double bond of unsaturated fatty acids in the cell membrane could be damaged through lipid peroxidation. ROS generated by plasma are predicted to break key bonds (C–O, C–N, and C–C) in the peptidoglycan of Gram-positive bacteria, while in Gram-negative bacteria, plasma exposure chemically alters the lipopolysaccharides. Moreover, plasma-generated charged particles create electrostatic stress on microbial membranes, causing leakage and cell death. On the other hand, plasma emits UV light, which damages DNA and RNA in bacteria, viruses, and molds, preventing their reproduction. In particular, it is effective against foodborne pathogens, such as *Salmonella*, *Listeria*, *Aspergillus flavus*, and *E. coli* (Gupta and Routray 2025; MRI et al. 2025).

CP enhances the extraction of certain bioactive compounds (e.g., anthocyanins, lutein, quercetin, hesperidin, naringin, curcumin, resveratrol, carvacrol, and thymol) and improves the solubility and emulsifying properties of food ingredients. RONS produced by CP interact with cell walls, causing oxidative stress and damage that enhance the release of intracellular compounds. This process boosts extraction efficiency, shortens processing time, and preserves the structural integrity of target compounds (Gupta and Routray 2025). Apart from this, cold plasma can be used for active packaging, mainly to prepare plastic films, paperboard, and biodegradable packaging to create antimicrobial surfaces. For example, CP generates ozonated or reactive water, which can be used for sanitizing equipment and surfaces in food processing facilities. In particular, CP has gained interest in modifying packaging materials to improve barrier properties due to its ability to alter the surface chemistry of packaging materials by introducing polar functional groups (e.g., hydroxyl, carbonyl, or carboxyl). This enhances surface energy and leads to better film uniformity and coating adhesion, which can reduce gas and moisture permeability (Ranjha et al. 2023).

2.4 *Supercritical CO_2 Processing (SCP)*

Supercritical CO_2 processing is an advanced green food processing and extraction technologies that utilize supercritical CO_2 instead of organic solvents to extract bioactive compounds, volatiles, essential oils, and flavors from various food and natural products. This is because carbon dioxide is widely available, inexpensive, and has relatively mild critical conditions. It is nonflammable, nontoxic, chemically stable in many environments, and exists as a gas under atmospheric pressure. SCP is commonly used for microbial inactivation/pasteurization, drying, particle formation/encapsulation, impregnation/ fortification, decontamination & detoxification, and food structuring/foam formation, while near-critical CO_2 is used for enzyme inactivation, delicate bioactive stabilization, and texture modification, among others (Onyeaka and Nwabor 2022; da Veiga et al. 2024). These methods have gained significant attention due to their high efficiency, environmental sustainability, and ability to preserve heat-sensitive compounds while producing solvent-free extracts. Moreover, integration with smart sensors, AI-driven predictive modeling, and IoT-enabled control systems can enhance the precision and adaptability of SCP processes (Roach et al. 2023). For example, real-time monitoring of pressure and temperature parameters allows dynamic optimization during bioactive extraction, reducing energy use and maximizing yield (Li et al. 2024). In addition, digital twins have been shown in thermal food processing to enable virtual simulation, rapid feedback, and improved control of process conditions (Kannapinn et al. 2022). While their application in SCP is still emerging, such virtual models hold promise for energy efficiency and parameter optimization as well.

2.4.1 Principles of SCP

Supercritical fluids are substances that exist at temperatures and pressures above their critical point (31.1 °C and 73.8 bar), where they exhibit properties of both liquids and gases. In contrast, CO_2 is in a near-critical state when it is close to, but not fully at, the supercritical point, typically just below the critical temperature and pressure (da Veiga et al. 2024). Supercritical fluids have unique, tunable physical properties that can be adjusted by changing pressure and/or temperature above their critical point. In the supercritical state, there is no distinct boundary between liquid and gas phases, eliminating surface tension. Supercritical fluids combine the low viscosity of gases, the high density of liquids, and intermediate diffusivity, making them ideal for extracting compounds from within cells. Additionally, even small changes in solvent density can significantly affect properties like solubility, dielectric constant, and partitioning behavior, greatly enhancing the mass transfer of solutes during extraction. SCP can be performed in batch or continuous systems, depending on the application and scale. Batch systems are ideal for small-scale, high-value products and allow for precise control over-processing conditions, while continuous systems offer higher throughput and are better suited for industrial-scale operations requiring consistent and efficient production. A continuous system is used for extraction and drying operations (da Veiga et al. 2024; Ren et al. 2025).

2.4.2 Common Applications of SCP

SCP is the most used fluid in supercritical fluid extraction due to it has a low critical temperature and pressure, making it suitable for extracting heat-sensitive compounds, behaves like a gas, allowing it to penetrate porous materials efficiently, and acts like a liquid, dissolving and carrying bioactive compounds (e.g., essential oils, polyphenols, and lipids) out of the matrix (Onyeaka and Nwabor 2022; da Veiga et al. 2024). Extraction conditions are controlled using a pressure release valve and temperature regulators on the vessel, ensuring a uniform flow of supercritical solvent through the solid material. The solvent remains in contact with the matrix for a set time to achieve equilibrium. After extraction, the compounds are collected as CO_2 returns to gas form, which is then either recycled or safely released. Recycling in closed-loop systems helps reduce emissions and costs. Moreover, CO_2 is nontoxic, nonflammable, and leaves no harmful residues in the final product, which makes SCP particularly valuable for functional food, nutraceutical, and pharmaceutical applications, where purity is essential. Additionally, CO_2 is generally sourced as a by-product from fermentation, natural gas processing, or industrial processes, making it a recycled and sustainable solvent. In particular, it is used to decaffeinate coffee while maintaining its aroma and flavor. Unlike chemical methods using methylene chloride or ethyl acetate, SCP yields a cleaner and safer final product. Apart from this, it is commonly used for the extraction of essential oils from citrus peels, rosemary, basil, and oregano, resveratrol and proanthocyanidins from grapes, curcumin from turmeric, omega-3 fatty acids from fish oils and microalgae, and removing pesticide residues, heavy metals, and mycotoxins, without altering the natural composition (López-Pedrouso et al. 2019; Onyeaka and Nwabor 2022; Ren et al. 2025). For example, Nam et al. (2025) identified 47 volatile compounds in edible insect oils, with significantly higher concentrations by SCP than those obtained through ultrasound-assisted extraction.

In addition to extraction, supercritical fluids are also used for food transformation and preservation. SCP preserves food by inactivating microorganisms and enzymes due to its low viscosity, lack of surface tension, and moderate diffusivity. These properties enable CO_2 to quickly penetrate cells and food particles when pressurized in a reactor chamber under controlled temperature and pressure conditions. The exact mechanism of enzyme inactivation in fruit and vegetable tissues remains unclear, but it is likely that processing conditions cause conformational changes in the enzymes, especially under pH levels different from those in the raw materials (Hart et al. 2022). Apart from this, drying, precipitation or crystallization of food compounds, textural modifications, and fractionation of aromas using SCP have attracted growing interest for its increasing applications in green food processing (da Veiga et al. 2024).

2.5 Membrane Filtration

Membrane filtration is a nonthermal separation technology that removes unwanted compounds, microorganisms, and contaminants while preserving essential nutrients and bioactive compounds (e.g., phenolic compounds, anthocyanins, antioxidants, and carotenoids). This technique is widely used in dairy processing, production of nonalcoholic beverages, recovery of aromas, clarification and concentration of natural extracts and juices, water purification, and wastewater recycling in the food industry. As part of minimal processing technologies, membrane filtration helps enhance food safety, extend shelf life, and reduce the need for preservatives without compromising flavor, texture, or nutritional quality. Membrane technologies offer several advantages over traditional separation methods. They operate under mild pressure and temperature conditions, which help preserve bioactivities of bioactives in natural and processed products. These processes do not require chemicals or biological agents, minimizing contamination risks. Additionally, membrane techniques are highly selective, easy to implement and scale up, require fewer operational steps, and offer significant energy savings (Park et al. 2024). In addition, real-time monitoring of fouling, flow rates, and pressure differentials allows predictive maintenance and dynamic process adjustments, ensuring consistent product quality, resource efficiency, and system longevity (Wang et al. 2024). Similar to the above techniques, there is potential for adaptive control platforms to be integrated with membrane filtration systems, where real-time monitoring of fouling, flow rates, and pressure differentials could dynamically adjust operation parameters. AI-driven predictive maintenance has already been applied in membrane bioreactors, and extending these tools to food applications could improve consistency, efficiency, and product quality (Frontistis et al. 2023; Wang et al. 2024).

2.5.1 Principles of Membrane Filtration

Membrane filtration works on the principle of size-based separation, where a semipermeable membrane allows selective passage of molecules based on pore size and molecular weight cut-off. The process is driven by pressure differentials that push fluids through membranes, retaining larger particles while allowing smaller molecules to pass.

There are four primary types of membrane filtration used in food processing: Microfiltration (MF), Ultrafiltration (UF), Nanofiltration (NF), and reverse osmosis (RO) (White et al. 2025). MF removes suspended solids, bacteria, and fat globules (pore size: 0.1–10 μm), which is mainly used in milk standardization and clarification of fruit juices and beer. In contrast, UF retains proteins, enzymes, and large molecules, such as polymers, colloidal particles, emulsions, and micelles, having a molecular weight range of 0.3–1000 kDa, allowing water, minerals, and smaller molecules to pass (pore size: 1–100 nm), which is applied in cheese-making, whey protein concentration, and plant-based milk production. NF, on the other hand, selectively removes dissolved salts, sugars, and organic compounds in the molecular weight range of 0.2–2 kDa while retaining some essential nutrients (pore size: 0.1–5 nm), which is used for reducing lactose in dairy products and softening water. Additionally, RO removes water molecules from a solution, leaving behind salts, sugars, and other solutes (pore size: <1 nm), which is commonly applied in water purification, juice concentration, and desalination. These systems function mainly based on pressure-driven separation, where a feed solution (e.g., milk, juice, or wastewater) is forced through a membrane under different pressures, such as 1–2 bar (MF), 2–10 bar (UF), 4–30 bar (NF), and 30–100 bar (RO) (Park et al. 2024).

The filtration process separates the feed into: retentate, the portion retained by the membrane (e.g., proteins, bacteria, and large molecules), and permeate, the filtered liquid that passes through the membrane (e.g., purified water and lactose-free milk). The effectiveness of a membrane process is primarily assessed using two key parameters: separation efficiency, which reflects how well components are retained during concentration, and productivity (Gao et al. 2025). Apart from pressure-driven membrane operations, electrically driven and vapor pressure gradient membranes are also used in food processing. Electrically driven membranes use ion exchange membranes and an electric field to separate ionic from nonionic species. This method is commonly used for demineralizing liquid streams, such as in the classic example of whey demineralization, or for recovering and concentrating ionized compounds, like organic salts from fermentation processes. In contrast, vapor pressure gradient membranes utilize differences in vapor pressure across the membrane to drive separation, making them effective for processes like pervaporation and membrane distillation, particularly in the separation of volatile compounds or water from mixtures (Park et al. 2024; White et al. 2025).

2.5.2 Common Applications of Membrane Filtration

The most common applications of membrane filtration are purification, concentration/extraction, and separation/fractionation in food processing. For example, the dairy industry generates large volumes of whey waste, which contains valuable proteins and lactose. Therefore, UF concentrates casein and whey proteins, allowing manufacturers to adjust protein content in cheese and yogurt production, while NF removes lactose from milk, making it suitable for lactose-intolerant consumers. Furthermore, UF separates whey proteins from lactose and minerals, producing whey protein isolates and concentrates for sports nutrition and functional foods (Mishra et al. 2025). For instance, filtration at 50 °C using membranes with pore sizes of 0.10, 0.14, and 0.20 μm showed that the 0.20 μm membrane allowed casein to pass through, resulting in a permeate with a roughly 1:1 ratio of casein to serum proteins. This suggests the potential of the 0.20 μm membrane for creating novel dairy ingredients (Song et al. 2022). On the other hand, MF and UF remove pulp, pectin, and microbial contaminants, creating clear, stable juices without chemical additives, while RO is used to concentrate fruit juices without thermal evaporation, preserving vitamin C, antioxidants, and flavor compounds (Gao et al. 2025). In addition, RO removes dissolved salts, heavy metals, and microbial contaminants, ensuring safe and high-quality drinking water. Food processing plants use membrane filtration (MF, UF, and RO) to treat wastewater, allowing for water recycling and reduced environmental impact. For example, Qi et al. (2021) found about 90% water recovery from swine wastewater using an UF membrane bioreactor and RO process. On the other hand, the application of ML algorithms in membrane bioreactor systems can improve the prediction of membrane fouling, enhance system control and optimization, and enable early fault detection, ultimately supporting the development of innovative cleaning strategies (Frontistis et al. 2023).

2.6 Microwave Technology

Microwave technology has emerged as a promising green processing method in the food industry due to its ability to preserve nutritional and functional quality. As the demand for sustainable and eco-friendly food processing technologies grows, microwave-assisted techniques offer significant advantages over conventional methods, including reduced processing time, lower energy consumption, and minimal solvent usage (Pratap-Singh and Ahmed 2024). These characteristics make microwave processing an ideal candidate for integration into green food processing systems, aligning with principles of environmental stewardship and food quality enhancement. Furthermore, with the integration of Agriculture 4.0 tools, microwave systems can achieve more uniform energy distribution, prevent overheating, and optimize drying or extraction parameters (Hassoun et al. 2024). In addition, microwave processing systems could also benefit from predictive analytics and digital twin applications, where real-time dielectric property measurements and AI-based control models dynamically adjust power distribution and heating uniformity. Such integration would reduce energy use, prevent overheating, and improve nutrient preservation across diverse food matrices (Núñez et al. 2024).

2.6.1 Principles of Microwave Technology

Microwave heating involves the application of electromagnetic waves, typically at frequencies of 915 MHz or 2.45 GHz, to heat polar molecules in food materials. The primary mechanism behind microwave heating is dielectric heating, wherein polar molecules such as water align with the oscillating electric field, generating heat through molecular friction and dipole rotation (Salakhi et al. 2025). Microwaves penetrate the food material and cause internal volumetric heating, which leads to a rapid rise in temperature. In particular, water's dipolar nature causes it to absorb microwave energy through dielectric heating. As the electric field rapidly changes direction, water molecules continuously realign, creating internal friction that generates heat throughout the food's volume (Pratap-Singh and Ahmed 2024). This contrasts with conventional heating methods that rely on surface heat transfer and subsequent conduction. The depth of microwave penetration and the rate of heating depend on the dielectric properties of the food matrix, moisture content, and frequency of the microwave energy. In particular, the efficiency and uniformity of microwave heating are affected by several parameters, such as dielectric constant (ε'), which indicates the ability of a material to store electrical energy, dielectric loss factor (ε''), which represents the ability to convert electromagnetic energy into heat, and penetration depth, which refers to the depth at which microwave power decreases to 1/e of its surface value (Pratap-Singh and Ahmed 2024; Salakhi et al. 2025).

2.6.2 Common Applications of Microwave Technology

Microwave drying is widely used to dehydrate fruits, vegetables, and herbs while preserving color, flavor, and bioactive compounds. Unlike hot air drying, which may degrade heat-sensitive nutrients, microwave-assisted drying significantly shortens drying time and improves product quality, such as instant foods, spices, seasonings, beef jerky, snacks, and ready-to-eat foods, among others. Microwave vacuum dehydration uses microwave energy under vacuum conditions to dry food quickly and at low temperatures, reducing heat damage through rapid, uniform heating and shorter drying time (Pratap-Singh et al. 2024). For example, in a study by Zhou et al. (2021), microwave-assisted vacuum drying (MWVD) was used to dehydrate cranberries. The process retained higher levels of phenolic compounds and antioxidants compared to conventional hot air drying. The drying time was reduced by 54%, demonstrating the energy efficiency of MWVD. On the other hand, microwave heating can be used for rapid pasteurization and sterilization of liquid and semi-solid foods, including milk, soups, and ready-to-eat meals. The quick heating ensures microbial safety while minimizing thermal degradation (Salakhi et al. 2025).

Microwave-assisted extraction (MAE) is a green alternative to conventional solvent extraction methods. It enhances mass transfer, reduces solvent use, and preserves sensitive phytochemicals. MAE has been successfully applied to extract polyphenols from plant matrices such as grape pomace, green tea, and olive leaves. These processes typically use water or ethanol as solvents and operate at lower temperatures, aligning with green chemistry principles. Moreover, enzymes such as PPO and POD, which cause browning and spoilage in fruits and vegetables, can be effectively inactivated using microwave heating. This improves the shelf life and sensory quality of minimally processed foods. In addition, microwave ovens are extensively used for cooking and reheating food products at the consumer level. In industrial settings, microwave-assisted baking and cooking help maintain moisture, improve uniformity, and reduce processing time. Apart from this, microwave energy is used in roasting nuts and grains and puffing snacks, leading to better control over final moisture content and expanded structure (Vicente et al. 2024; Salakhi et al. 2025).

2.7 *Encapsulation*

Encapsulation involves entrapping active compounds, such as vitamins, flavors, probiotics, essential oils, and antioxidants, within a protective matrix to improve their stability, bioavailability, and controlled release. In green food processing, encapsulation aligns with sustainability principles by reducing processing severity, protecting sensitive compounds from degradation/oxidation, and enabling targeted delivery of nutrients or bioactives, thus minimizing losses and enhancing efficiency (Xu et al. 2024). Besides, the integration of digital tools like AI-assisted formulation design, ML for predicting release profiles, and smart encapsulation systems equipped with IoT-enabled microfluidics allows for precise control over particle size, release timing, and encapsulant-material interactions (Serrano et al. 2024). For example, IoT-enabled microfluidic encapsulation platforms have the potential to self-adjust droplet size, release timing, or coating properties based on continuous sensor feedback, ensuring greater reproducibility and efficiency in producing functional food ingredients (Serrano et al. 2024).

2.7.1 Principles of Encapsulation

Encapsulation protects active or sensitive ingredients within a carrier material to form micro- or nano-sized particles, resulting in improved delivery of bioactive compounds. During the encapsulation process, a range of particle sizes can be produced, including microparticles (1–1000 μm), submicron particles (less than 1 μm), and nanoparticles (typically between 1 and a few hundred nanometers), depending on the specific micro- or nanoencapsulation method used (Xu et al. 2024). The core material (the ingredient to be protected or delivered) is surrounded by a shell or matrix known as the wall material, which may be composed of natural biopolymers like proteins, polysaccharides, or lipids. This protective layer shields the core from environmental stresses such as heat, light, oxygen, and moisture. Microencapsulation is employed for several key purposes, including protecting sensitive compounds from adverse environmental conditions (e.g., heat, light, oxygen, and moisture), enabling controlled release, preserving antioxidant properties until the point of consumption, and masking undesirable tastes/odors such as astringency (Onyeaka and Nwabor 2022). Nanoencapsulation is widely used to enhance flavor retention and controlled release while also maintaining nutritional value in food products. For example, anthocyanins, plant pigments known for their vibrant color and diverse biological activities, are highly unstable and reactive, which can be protected using nanoencapsulation. Encapsulation improved the thermal stability and photostability of cyanidin-3-*O*-glucoside (anthocyanin) (Xue et al. 2024). Various encapsulation techniques are employed in food processing, including spray drying, freeze drying, coacervation, extrusion, emulsion-based methods, electrospinning, and the formation of liposomes or nanoparticles (Dhiman and Anshul 2025). These methods are chosen based on the nature of the core material, the desired release behavior, and the application in the final food product.

2.7.2 Common Applications of Encapsulation

Polyphenols and antioxidants are prone to degradation during processing and storage. Encapsulation helps preserve their activity and ensures their controlled release in the digestive system. For instance, Tyagi et al. (2021) encapsulated green tea catechins using chitosan and sodium tripolyphosphate, and the encapsulated product showed significantly improved stability, bioavailability, and bioactivity. Moreover, probiotics are sensitive to environmental stresses such as heat, oxygen, and gastric acidity. Microencapsulation enhances their viability during processing, storage, and gastrointestinal transit (Paul et al. 2025). Encapsulation of volatile aroma compounds and flavors helps retain sensory quality during processing and storage and allows controlled release during consumption. For example, essential oil encapsulation effectively masked volatility and bitterness and improved sensory attributes (Onyeaka and Nwabor 2022). In addition, omega-3 fatty acids, especially eicosapentaenoic acid (EPA) and docosahexaenoic acid (DHA), are highly susceptible to oxidation. Encapsulation can improve the shelf life and sensory attributes of these fatty acids. Apart from this, encapsulation enables the targeted release of bioactives in response to specific conditions such as pH or enzymatic activity in the gastrointestinal tract. For example, microencapsulation of curcumin improved its bioaccessibility and stability (Xu et al. 2024, b).

3 Challenges and Future Perspectives

Despite the growing promise and adoption of green food processing technologies, several challenges hinder their full-scale implementation. These include technological limitations, high capital costs, regulatory gaps, and consumer acceptance. For example, one of the most significant challenges is the high initial investment cost associated with equipment such as HPP, PEF, or SCP systems, among others. For small- and medium-scale enterprises (SMEs), the financial burden remains a barrier to adoption. For instance, while HPP offers superior microbial inactivation and nutrient preservation, the cost of installation can range from USD 0.5 million to over four million, depending on the capacity (Janahar and Balasubramaniam 2022). Similarly, CP and microwave processing systems, though effective at preserving quality and extending shelf life, often require specialized design to accommodate different food matrices. Issues like inconsistent energy distribution and scalability constraints limit their broader industrial use (Gupta and Routray 2025). On the other hand, most emerging green technologies lack harmonized international regulations and food safety standards. For instance, CP and encapsulation techniques, despite their potential, face hurdles in regulatory approval for certain applications in food preservation or functional ingredient delivery (Gupta and Routray 2025). Regulatory bodies, such as the Food and Drug Administration (FDA) and European Food Safety Authority (EFSA), evaluate any chemical residues or by-products generated during CP treatment to confirm they are safe for human health and do not present environmental hazards. Furthermore, the absence of clear labeling guidelines for nonthermal technologies may confuse consumers and limit market growth. Aside from this, consumer attitude toward minimally processed or nonthermally treated food can also pose a barrier. Studies show that while consumers are increasingly concerned with sustainability, a lack of awareness about technologies like PEF or SCP can lead to hesitation in purchasing such products (Giacalone and Jaeger 2023). Transparency in processing methods and clear communication of benefits, such as "clean-label," nutrient retention, and safety, are essential for market success.

Research is also expanding into hybrid technologies, such as combining PEF with cold plasma or SCP with encapsulation, to enhance microbial inactivation and bioactive preservation while maintaining process efficiency. In addition, integration with other preservation technologies can enhance their effectiveness. For example, combining HPP with ultraviolet (UV) treatment, natural antimicrobials, or mild heat treatments to enhance microbial inactivation (López-Pedrouso et al. 2019; Yang and Liao 2024). In parallel, sustainable packaging is emerging as a critical component of green food processing. One of the major contributors to food waste is the misinterpretation of food labeling, such as best-before dates, which leads to the premature disposal of edible products (Hossain 2023). Recent advancements in smart and biodegradable packaging systems are helping to address this issue by extending shelf life and ensuring product safety. These innovative packaging solutions integrate sensors, indicators, or natural preservatives that enable real-time monitoring of product condition, providing consumers and supply chain actors with clear, data-driven information on freshness and spoilage traceability (Sobhan et al. 2025). Such smart packaging systems represent a confluence of green and Agriculture 4.0 technologies, combining biodegradable materials with embedded electronics for intelligent monitoring and communication. Moreover, waste reduction and by-product utilization are increasingly prioritized in green food technologies. The food and seafood industries generate significant volumes of waste, much of which holds untapped potential for conversion into value-added products such as functional foods, nutraceuticals, bioactive ingredients, and biodegradable materials (Senadheera et al. 2023a, b). As these innovations become more cost-effective and scalable, they offer promising pathways to reduce food waste, improve sustainability, and support circular economy goals in the food sector. In the future, the convergence of green food processing and Agriculture 4.0 offers a transformative opportunity to build more resilient, efficient, and sustainable food systems. With appropriate investment, cross-sector collaboration, and supportive regulatory frameworks, this integration can catalyze a new era of smart, eco-innovative agri-food solutions.

In the evolving landscape of Agriculture 4.0, the synergy between green food processing technologies and AI offers transformative potential for building intelligent, efficient, and sustainable food systems. While green processing methods focus on reducing environmental impacts and preserving food quality, the application of digital technologies enhances control, optimization, and adaptability across the value chain. The integration of AI, ML, and real-time sensors into food processing can optimize energy use, process control, and quality assurance, as well as ensure safety and traceability (Sobhan et al. 2025). For example, predictive modeling in SCP could fine-tune temperature and pressure for targeted bioactive extraction, reducing energy inputs (Roach et al. 2023). Additionally, portable, modular systems for HPP and CP are being developed to serve decentralized and mobile food units, making the technologies more accessible to smaller producers. These advancements reflect the broader movement toward Agriculture 4.0, where digital innovation is reshaping the agri-food ecosystem. Smart sensors and IoT devices can be embedded in processing lines to continuously monitor temperature, pressure, and microbial activity, ensuring optimal conditions are maintained in real time. AI algorithms can not only control process parameters but also adapt them dynamically based on food characteristics and environmental inputs, improving both product quality and energy efficiency. Furthermore, blockchain technologies can be employed to ensure transparency and traceability across the supply chain, documenting the use of green technologies and sustainable practices from farm to fork (Li et al. 2024; Srisuwan and Innet 2024; Dhal and Kar 2025). Future directions also involve tighter integration between smart farming and green processing, where cultivation data (e.g., soil, weather, and crop maturity) can directly inform processing parameters to optimize extraction, preservation, and nutrient retention. This integration supports true farm-to-fork traceability through blockchain and IoT-enabled certification (Cordeiro and Ferreira 2025; Hassoun 2025). Table 1 highlights the areas of the application of AI and digital technologies already seeing progress in green food processing.

Table 1 Applications of AI and digital technologies in green food processing technologies

AI/ digital tool	Application in green food processing technologies	Associated technologies	Benefits	References
ML (support vector machine, random forest, artificial neural network, etc.)	Predict product quality (e.g., nutrient retention, microbial stability); optimize parameters (e.g., temperature, pressure); and detect outliers in real time	HPP, PEF, CP, and ultrasonication	Enhanced safety, minimal nutrient loss, reduced trial-and-error, and improved reproducibility	Safwa et al. (2024) and White et al. (2025)
Digital twins	Simulate and optimize internal product responses; predict heat/mass transfer behavior; and design safer sterilization protocols	Modeling, analysis, testing, and simulation (MATS), ohmic heating, and ultrasonication	Increased energy efficiency, uniform microbial inactivation, and reduced thermal gradients	White et al. (2025) and Lisboa et al. (2024)
Blockchain technology	Ensure secure, immutable tracking of raw material origin and processing steps; verify sustainability claims and green certifications	Upcycled ingredients, minimally processed foods, and decentralized co-manufacturing	Enhances traceability, builds consumer trust, prevents food fraud, and supports regulatory compliance	Hassoun et al. (2024) and Ercan et al. (2023)
Smart sensors and IoT	Real-time monitoring of pressure, pH, conductivity, and flow; sensor fusion for automation and multivariate control	CP (ozone), SCP, membrane filtration, and ultrasonication	Continuous monitoring, automation, traceability, and reduced manual sampling	Ercan et al. (2023) and Hassoun et al. (2024)
Computer vision (AI-enhanced)	Visual inspection for color, defects, texture; inline evaluation of biodegradable films and post-processed food	Edible coatings, PEF-treated produce, and HPP-treated vegetables	Reduced manual inspection, increased throughput, and improved product consistency	Onyeaka and Nwabor (2022) and Lisboa et al. (2024)
AI for sustainability assessment	Perform life cycle assessment (LCA) using AI models; optimize supply chains and by-product valorization	Encapsulation systems, by-product reuse, and fermentation-based alternatives	Reduces carbon footprint, boosts circularity, and supports green certifications	Lisboa et al. (2024) and Hassoun et al. (2024)
Predictive maintenance (AI-based)	Detect anomalies in mechanical/ electrical systems; forecast failures in pumps and energy delivery hardware	Ultrasonication, ohmic heating, MATS, and CP units	Minimizes downtime, extends equipment lifespan, improves reliability	Ercan et al. (2023)
Edge AI/edge computing	Real-time, on-device AI processing for monitoring and control in remote or mobile setups	Mobile CP units, portable ultrasonication, and low-power IoT nodes	Low-latency response, reduced bandwidth needs, and ensures uptime in decentralized operations	Still exploratory
Natural Language Processing (NLP) and chatbots	Natural language interaction with operators, supply chain actors, or consumers; automate sustainability reports or consumer transparency tools	Blockchain traceability apps, smart certification labels, and training portals	Enhances education, supports compliance communication, and improves digital transparency for end-users	Still exploratory

4 Conclusion

Green food processing technologies represent a transformative shift toward more sustainable, efficient, and health-conscious food production systems. The food industry can significantly reduce its environmental footprint by adopting energy efficiency, water conservation, waste minimization, and sustainable packaging. Emerging nonthermal and minimal processing techniques, such as HPP, PEF, SCP, membrane filtration, and encapsulation, demonstrate promising capabilities to retain nutritional quality, extend shelf life, and valorize by-products without compromising safety or functionality. While these technologies offer immense environmental and economic potential, challenges such as high capital costs, regulatory ambiguity, technological scalability, and consumer acceptance must be addressed through interdisciplinary collaboration and innovation. Importantly, the future of green food processing will not rely solely on eco-efficient physical methods but will be increasingly defined by its integration with digital innovations. The convergence of smart sensors, real-time data analytics, AI, and blockchain technologies is enabling smarter, more adaptive, and resilient food systems, core components of the Fourth Agricultural Revolution. These tools support dynamic process optimization, traceability, predictive maintenance, and intelligent quality control, further reinforcing the sustainability and safety of food production. By embedding green technologies within a digital ecosystem, the food industry can move toward closed-loop systems that not only minimize waste and energy but also respond to environmental changes and market demands with agility. Ultimately, Agriculture 4.0 provides the digital backbone linking agricultural production, eco-efficient processing, smart packaging, and blockchain-enabled traceability into a cohesive and sustainable food system. With continued investment in research, policy support, and consumer education, these technologies can play a pivotal role in shaping a sustainable future for global food production.

Competing Interests The author(s) has no competing interests to declare that are relevant to the content of this manuscript.

References

Bayati M, Lund MN, Tiwari BK, Poojary MM (2024) Chemical and physical changes induced by cold plasma treatment of foods: a critical review. Compr Rev Food Sci Food Saf 23:e13376. https://doi.org/10.1111/1541-4337.13376

Bocker R, Silva EK (2022) Pulsed electric field assisted extraction of natural food pigments and colorings from plant matrices. Food Chem X 15:100398. https://doi.org/10.1016/J.FOCHX.2022.100398

Cavalcanti RN, Balthazar CF, Margalho LP et al (2023) Pulsed electric field-based technology for microbial inactivation in milk and dairy products. Curr Opin Food Sci 54:101087. https://doi.org/10.1016/J.COFS.2023.101087

Cordeiro M, Ferreira JC (2025) Beyond traceability: decentralised identity and digital twins for verifiable product identity in agri-food supply chains. Appl Sci 15(11):6062. https://doi.org/10.3390/app15116062

da Veiga GC, Mafaldo ´iM, Barão CE et al (2024) Supercritical carbon dioxide technology in food processing: insightful comprehension of the mechanisms of microbial inactivation and impacts on quality and safety aspects. Compr Rev Food Sci Food Saf 23:e13345. https://doi.org/10.1111/1541-4337.13345

Dhal SB, Kar D (2025) Leveraging artificial intelligence and advanced food processing techniques for enhanced food safety, quality, and security: a comprehensive review. Discov Appl Sci 7:75. https://doi.org/10.1007/s42452-025-06472-w

Dhiman D, Anshul A (2025) Encapsulation techniques of sludge generated from wastewater treatment. J Environ Manag 387:125788. https://doi.org/10.1016/J.JENVMAN.2025.125788

Ercan UK, Özdemir GD, Özdemir MA, Güren O (2023) Plasma medicine: the era of artificial intelligence. Plasma Process Polym 20:e2300066. https://doi.org/10.1002/ppap.202300066

Frontistis Z, Lykogiannis G, Sarmpanis A (2023) Machine learning implementation in membrane bioreactor systems: progress, challenges, and future perspectives: a review. Environments 10(7):127. https://doi.org/10.3390/environments10070127

Gao SS, Zhang XH, Geng MY, Tian JY (2025) Effect of membrane material and pore size on membrane fouling during filtration of algae-laden water. Water Sci Eng 18(3):335–344. https://doi.org/10.1016/J.WSE.2025.04.001

Giacalone D, Jaeger SR (2023) Consumer acceptance of novel sustainable food technologies: a multi-country survey. J Clean Prod 408:137119. https://doi.org/10.1016/J.JCLEPRO.2023.137119

Gupta KK, Routray W (2025) Cold plasma: a nonthermal pretreatment, extraction, and solvent activation technique for obtaining bioactive compounds from agro-food industrial biomass. Food Chem 472:142960. https://doi.org/10.1016/J.FOODCHEM.2025.142960

Hart A, Anumudu C, Onyeaka H, Miri T (2022) Application of supercritical fluid carbon dioxide in improving food shelf-life and safety by inactivating spores: a review. J Food Sci Technol 59:417–428. https://doi.org/10.1007/s13197-021-05022-7

Hassoun A (2025) Food sustainability 4.0: harnessing fourth industrial revolution technologies for sustainable food systems. Discov Food 5:171. https://doi.org/10.1007/s44187-025-00461-z

Hassoun A, Aït-Kaddour A, Dankar I et al (2024) The significance of industry 4.0 technologies in enhancing various unit operations applied in the food sector: focus on food drying. Food Bioprocess Technol 18:109–128. https://doi.org/10.1007/s11947-024-03465-2

Hossain A (2023) Do date codes cause food waste? Smart packaging might tackle the problem. J Food Bioactives 24:6–12. https://doi.org/10.31665/jfb.2023.18359

Hossain A, Dave D, Shahidi F (2022a) Effect of high-pressure processing (HPP) on phenolics of North Atlantic Sea cucumber (*Cucumaria frondosa*). J Agric Food Chem 70:3489–3501. https://doi.org/10.1021/acs.jafc.2c00140

Hossain A, Yeo JD, Dave D, Shahidi F (2022b) Phenolic compounds and antioxidant capacity of sea cucumber (*Cucumaria frondosa*) processing discards as affected by high-pressure processing (HPP). Antioxidants 11:337. https://doi.org/10.3390/antiox11020337

Huang HW, Hsu CP, Wang CY (2020) Healthy expectations of high hydrostatic pressure treatment in food processing industry. J Food Drug Anal 28:1–13. https://doi.org/10.1016/J.JFDA.2019.10.002

Janahar JJ, Balasubramaniam VM (2022) Application of high-pressure-based technologies in the food industry. In: https://ohioline.osu.edu/factsheet/fst-fabe-1001

Kannapinn M, Pham MK, Schäfer M (2022) Physics-based digital twins for autonomous thermal food processing: efficient, non-intrusive reduced-order modeling. Innov Food Sci Emerg Technol 81:103143. https://arxiv.org/abs/2209.03062

Lamela CP (2023) Green technologies for sustainable food production and preservation: high-pressure processing. In: Sustainable food science - a comprehensive approach, vol 1–4. Elsevier, pp V4-158–V4-183

Li M, Jiang W, Zhao S et al (2024) Employment of artificial intelligence approach for optimizing the solubility of drug in the supercritical CO_2 system. Case Stud Therm Eng 57:104326. https://doi.org/10.1016/J.CSITE.2024.104326

Lisboa HM, Pasquali MB, dos Anjos AI et al (2024) Innovative and sustainable food preservation techniques: enhancing food quality, safety, and environmental sustainability. Sustainability 16(18):8223. https://doi.org/10.3390/su16188223

López-Pedrouso M, Díaz-Reinoso B, Lorenzo JM et al (2019) Green technologies for food processing: principal considerations. In: Innovative thermal and non-thermal processing, bioaccessibility and bioavailability of nutrients and bioactive compounds. Elsevier, pp 55–103

Mandal R, Singh A, Pratap Singh A (2018) Recent developments in cold plasma decontamination technology in the food industry. Trends Food Sci Technol 80:93–103. https://doi.org/10.1016/J.TIFS.2018.07.014

Mishra A, Salunke P, Sunkesula V, Metzger LE (2025) Manufacture of high-solids milk protein concentrate retentate using plate-and-frame membrane filtration system at low and high temperatures. J Dairy Sci 108:4785–4795. https://doi.org/10.3168/JDS.2024-25472

Mohammadi X, Hossain A, Kontogiorgos V, Pratap-Singh A (2025) Cold oxygen plasma-induced modifications in chitosan films: effects of solution and film treatment on structure and functionality. Food Hydrocoll 169:111572. https://doi.org/10.1016/j.foodhyd.2025.111572

Movasaghi M, Heydari MM, Schwean-Lardner K et al (2025) Investigating cold plasma jet effectiveness for eggshell surface decontamination. Food Cont 168:110928. https://doi.org/10.1016/J.FOODCONT.2024.110928

MRI S, Kamal MM, Karim N et al (2025) Cold plasma and integrated approaches for fresh produce preservation: mechanisms, quality attributes, challenges, and future directions. Trends Food Sci Technol 159:104988. https://doi.org/10.1016/J.TIFS.2025.104988

Naliyadhara N, Kumar A, Girisa S et al (2022) Pulsed electric field (PEF): Avant-garde extraction escalation technology in food industry. Trends Food Sci Technol 122:238–255. https://doi.org/10.1016/J.TIFS.2022.02.019

Nam JK, Kwak BM, Jang HW (2025) Optimization and comparative analysis of quality characteristics and volatile profiles in edible insect oils extracted using supercritical fluid extraction and ultrasound-assisted extraction methods. Food Chem 474:143237. https://doi.org/10.1016/J.FOODCHEM.2025.143237

Núñez RV, Hernández A, Rodríguez I et al (2024) Artificial intelligence applied to microwave heating systems: prediction of temperature profile through convolutional neural networks. Thermo 4:346–372. https://doi.org/10.3390/thermo4030018

Oguzlu H, Baldelli A, Mohammadi X et al (2024) Cold plasma for the modification of the surface roughness of microparticles. ACS Omega 9:35634–35644. https://doi.org/10.1021/acsomega.4c03787

Onyeaka HN, Nwabor OF (2022) Green technology in food processing and preservation. In: Food preservation and safety of natural products. Elsevier, pp 87–118

Özdemir MA, Özdemir GD, Gül M et al (2023) Machine learning to predict the antimicrobial activity of cold atmospheric plasma-activated liquids. Mach Learn Sci Technol 4:015030. https://doi.org/10.1088/2632-2153/acc1c0

Park HW, Baek J, Kim WJ (2024) Forward osmosis and direct contact membrane distillation: emerging membrane technologies in food and beverage processing. Innov Food Sci Emerg Tech 93:103626. https://doi.org/10.1016/J.IFSET.2024.103626

Paul B, Xie L, Yahia ZO, Chen W (2025) Recent review on the stability of bioactive substances through encapsulation and their application in dairy products. Food Rev Int 42(2):605–631. https://doi.org/10.1080/87559129.2025.2492338

Pratap-Singh A, Ahmed W (2024) Effect of microwaves on food quality parameters. In: Pratap Singh A, Erdogdu F, Wang S, Ramaswamy HS (eds) Microwave processing of foods: challenges, advances and prospects. Springer, Cham, pp 133–150

Pratap-Singh A, Karmakar P, Mandal R (2024) Microwave-vacuum dehydration in food processing. In: Pratap-Singh A, Erdogdu F, Wang S, Ramaswamy HS (eds) Microwave processing of foods: challenges, advances and prospects. Springer, Cham, pp 387–410

Qi B, Jiang X, Wang H et al (2021) Resource recovery from liquid digestate of swine wastewater by an ultrafiltration membrane bioreactor (UF-MBR) and reverse osmosis (RO) process. Environ Technol Innov 24:101830. https://doi.org/10.1016/J.ETI.2021.101830

Ranjha MMAN, Shafique B, Aadil RM et al (2023) Modification in cellulose films through ascent cold plasma treatment and polymerization for food products packaging. Trends Food Sci Technol 134:162–176. https://doi.org/10.1016/J.TIFS.2023.03.011

Ren S, He Y, Liu C et al (2025) A novel Visbreaking-Supercritical Fluid extraction (SFE) strategy for efficient upgrading of vacuum residue: experimental optimization and molecular dynamics insights. Sep Purif Technol 373:133554. https://doi.org/10.1016/J.SEPPUR.2025.133554

Roach L, Rignanese GM, Erriguible A, Aymonier C (2023) Applications of machine learning in supercritical fluids research. J Supercrit Fluids 202:106051. https://doi.org/10.1016/J.SUPFLU.2023.106051

Safwa SM, Ahmed T, Talukder S et al (2024) Applications of non-thermal technologies in food processing industries-a review. J Agric Food Res 18:100917. https://doi.org/10.1016/J.JAFR.2023.100917

Salakhi M, Di Liddo L, Thomson MJ (2025) Modeling microwave heating in fluidized bed reactors: revealing the interaction between microwave absorption and fluidization hydrodynamics. Chem Eng Sci 315:121884. https://doi.org/10.1016/J.CES.2025.121884

Santiago-Mora P, Skinner M, Hendricks A et al (2024) Pulsed electric field effect on acrylamide reduction and quality attributes of continuous-style Lamoka potato chips. Heliyon 10:e31790. https://doi.org/10.1016/J.HELIYON.2024.E31790

Semercioz-Oduncuoglu SA, Luning PA (2025) Industry 4.0 technologies in quality and safety control in food industry: a systematic techno-managerial analysis on benefits and barriers. Trends Food Sci Technol 163:105144. https://doi.org/10.1016/j.tifs.2025.105144

Senadheera TRL, Hossain A, Dave D, Shahidi F (2023a) Antioxidant and ACE-inhibitory activity of protein hydrolysates produced from Atlantic Sea cucumber (*Cucumaria frondosa*). Molecules 28(13):5263. https://doi.org/10.3390/molecules28135263

Senadheera TRL, Hossain A, Shahidi F (2023b) Marine bioactives and their application in the food industry: a review. Appl Sci 13:12088. https://doi.org/10.3390/app132112088

Serrano DR, Luciano FC, Anaya BJ et al (2024) Artificial intelligence (AI) applications in drug discovery and drug delivery: revolutionizing personalized medicine. Pharmaceutics 16:1328. https://doi.org/10.3390/pharmaceutics16101328

Sobhan A, Hossain A, Wei L et al (2025) IoT-enabled biosensors in food packaging: a breakthrough in food safety for monitoring risks in real time. Foods 14:1403. https://doi.org/10.3390/foods14081403

Song B, Zhang S, Zhang Y et al (2022) Ceramic membrane filtration of skim milk for the production of a casein-enriched permeate. Int Dairy J 135:105478. https://doi.org/10.1016/J.IDAIRYJ.2022.105478

Srisuwan S, Innet S (2024) An investigation of machine learning algorithms for predictive maintenance in high pressure processing systems. In: Proceedings - 21st International Joint Conference on Computer Science and Software Engineering, JCSSE 2024. Institute of Electrical and Electronics Engineers Inc., pp 94–98. https://doi.org/10.1109/JCSSE61278.2024.10613634

Trych U, Buniowska M, Skąpska S et al (2020) Impact of HPP on the bioaccessibility/bioavailability of nutrients and bioactive compounds as a key factor in the development of food processing. In: Francisco JB, Carole TS, Eduardo P, María L (eds) Present and future of high pressure processing: a tool for developing innovative, sustainable, safe and healthy foods. Elsevier, pp 87–109. https://doi.org/10.1016/B978-0-12-816405-1.00004-2

Tyagi T, Garlapati PK, Yadav P et al (2021) Development of nano-encapsulated green tea catechins: studies on optimization, characterization, release dynamics, and in-vitro toxicity. J Food Biochem 45:e13951. https://doi.org/10.1111/jfbc.13951

Vicente A, Villanueva M, Caballero PA et al (2024) Microwave treatment enhances the physical and sensory quality of quinoa-enriched gluten-free bread. Food Hydrocoll 155:110244. https://doi.org/10.1016/J.FOODHYD.2024.110244

Waghmare R (2024) High pressure processing of fruit beverages: a recent trend. Food Humanit 2:100232. https://doi.org/10.1016/J.FOOHUM.2024.100232
Wang L, Li Z, Fan J, Han Z (2024) The intelligent prediction of membrane fouling during membrane filtration by mathematical models and artificial intelligence models. Chemosphere 349:141031. https://doi.org/10.1016/J.CHEMOSPHERE.2023.141031
White S, Jackson-Davis A, Gordon K et al (2025) A review of non-thermal interventions in food processing technologies. J Food Prot 88:100508. https://doi.org/10.1016/J.JFP.2025.100508
Xu Y, Yan X, Zheng H et al (2024) The application of encapsulation technology in the food industry: classifications, recent advances, and perspectives. Food Chem X 21:101240. https://doi.org/10.1016/J.FOCHX.2024.101240
Xue H, Zhao J, Wang Y et al (2024) Factors affecting the stability of anthocyanins and strategies for improving their stability: a review. Food Chem X 24:101883. https://doi.org/10.1016/J.FOCHX.2024.101883
Yan B, Li J, Liang QC et al (2024) From laboratory to industry: the evolution and impact of pulsed electric field technology in food processing. Food Rev Int 41:373–398. https://doi.org/10.1080/87559129.2024.2401433
Yang P, Liao X (2024) High pressure processing plus technologies: enhancing the inactivation of vegetative microorganisms. Adv Food Nutr Res 110:145–195. https://doi.org/10.1016/BS.AFNR.2024.02.003
Yasir M, Hossain A, Pratap-Singh A (2025) Pesticide degradation: impacts on soil fertility and nutrient cycling. Environments 12(8):272. https://doi.org/10.3390/environments12080272
Yawut N, Mekwilai T, Vichiansan N et al (2024) Cold plasma technology: transforming food processing for safety and sustainability. J Agric Food Res 18:101383. https://doi.org/10.1016/J.JAFR.2024.101383
Zhao X, Wang N, Lu Y et al (2024) Effects of cold plasma on the growth and aflatoxin production of aspergillus flavus. Food Biosci 61:104552. https://doi.org/10.1016/J.FBIO.2024.104552
Zhou YH, Staniszewska I, Liu ZL et al (2021) Microwave-vacuum-assisted drying of pretreated cranberries: drying kinetics, bioactive compounds and antioxidant activity. LWT 146:111464. https://doi.org/10.1016/J.LWT.2021.111464

The manufacturer's authorised representative in the EU is Springer Nature Customer Service Centre GmbH, Europaplatz 3, 69115 Heidelberg, Germany. If you have any concerns regarding our products, please contact ProductSafety@springernature.com

Printed and bound by CPI Group (UK) Ltd, Croydon, CR0 4YY
21/09/2026
02219991-0019